CONTEMPORARY MATHEMATICS

382

Israel Mathematical Conference Proceedings

Complex Analysis and Dynamical Systems II

A conference in Honor of
Professor Lawrence Zalcman's Sixtieth Birthday
June 9–12, 2003
Nahariya, Israel

Mark Agranovsky
Lavi Karp
David Shoikhet
Editors

American Mathematical Society
Providence, Rhode Island

Bar-Ilan University
Ramat Gan, Israel

Editorial Board of Contemporary Mathematics

Dennis DeTurck, managing editor

George Andrews   Carlos Berenstein   Andreas Blass   Abel Klein

Editorial Board of Israel Mathematical Conference Proceedings

Louis Rowen, *Bar-Ilan University*, managing editor

J. Bernstein, *Tel-Aviv University*   M. Katz, *Bar-Ilan University*
H. Furstenberg, *Hebrew University*   S. Shnider, *Bar-Ilan University*
S. Gelbart, *Weizmann Institute*   L. Small, *University of California at San Diego*
V. Gol′dshtein, *Ben-Gurion University* L. Zalcman, *Bar-Ilan University*

Miriam Beller, *Technical Editor*

Founding editors include: Z. Arad and B. Pinchuk, *Bar-Ilan University*

2000 *Mathematics Subject Classification.* Primary 30–XX, 32–XX, 37–XX; Secondary 34–XX, 35–XX, 46–XX, 47–XX.

Library of Congress Cataloging-in-Publication Data

International Conference on Complex Analysis and Dynamical Systems (2nd : 2003 : Nahariya, Israel)

  Complex analysis and dynamical systems II : a conference in honor of Professor Lawrence Zalcman's sixtieth birthday, June 9–12, 2003, Nahariya, Israel / Mark Agranovsky, Lavi Karp, David Shoikhet, editors.
   p. cm. — (Israel mathematical conference proceedings) (Contemporary mathematics, ISSN 0271-4132 ; 382)
  ISBN 0-8218-3709-5 (alk. paper)
  1. Functions of complex variables–Congresses. 2. Differentiable dynamical systems-Congresses. I. Zalcman, Lawrence Allen. II. Agranovsky, M. L. (Mark L'vovich) III. Karp, Lavi, 1955– IV. Shoikhet, David, 1953– V. Title. VI. Series. VII. Contemporary mathematics (American Mathematical Society); v. 382.
QA331.7.I58 2003
515′.9—dc22   2005041245

Copying and reprinting. Material in this book may be reproduced by any means for educational and scientific purposes without fee or permission with the exception of reproduction by services that collect fees for delivery of documents and provided that the customary acknowledgment of the source is given. This consent does not extend to other kinds of copying for general distribution, for advertising or promotional purposes, or for resale. Requests for permission for commercial use of material should be addressed to the Managing Editor, IMCP, Department of Mathematics, Bar-Ilan University, Ramat-Gan, 52900 Israel. Requests can also be made by email to rowen@macs.biu.ac.il.

  Excluded from these provisions is material in articles for which the author holds copyright. In such cases, requests for permission to use or reprint should be addressed directly to the author(s). (Copyright ownership is indicated in the notice in the lower right-hand corner of the first page of each article.)

© 2005 by Bar-Ilan University. Printed in the United States of America.

∞ The paper used in this book is acid-free and falls within the guidelines established to ensure permanence and durability.
Visit the AMS home page at http://www.ams.org/

10 9 8 7 6 5 4 3 2 1   10 09 08 07 06 05

Lawrence Zalcman

Contents

Preface	ix
List of Participants	xi
Conference Program	xv
Lawrence Zalcman at Sixty MARK AGRANOVSKY and DAVID SHOIKHET	1
Bibliography of Lawrence Zalcman	7
A Local Two Radii Theorem for the Twisted Spherical Means on $\mathbb{C}^n$ M. L. AGRANOVSKY and E. K. NARAYANAN	13
A Multiplicator Problem and Characteristics of Growth of Entire Functions VLADIMIR AZARIN	29
Are They Limit Periodic? J. BELLISSARD, J. GERONIMO, A. VOLBERG, and P. YUDITSKII	43
Quasinormal Families and Periodic Points WALTER BERGWEILER	55
Local Center Conditions for the Abel Equation and Cyclicity of its Zero Solution M. BLINOV, M. BRISKIN, and Y. YOMDIN	65
Univalent Functions Starlike with Respect to a Boundary Point D. BSHOUTY and A. LYZZAIK	83
On the Geometry Induced by a Grusin Operator O. CALIN, D.-C. CHANG, P. GREINER, and Y. KANNAI	89
The Kœnigs Embedding Problem for Operator Affine Mappings MARK ELIN and VICTOR KHATSKEVICH	113
On an Arithmetical Function II HERSHEL M. FARKAS	121
A Glance at Wiman-Valiron Theory P. C. FENTON	131
Billiards in an Ellipse LEOPOLD FLATTO	141

(p, q, r)–Kleinian Groups and the Margulis Constant F. W. GEHRING and G. J. MARTIN	149
Holomorphic Extendibility and the Argument Principle JOSIP GLOBEVNIK	171
Homeomorphisms with Finite Mean Dilatations ANATOLY GOLBERG	177
On a Connection Between the Number of Poles of a Meromorphic Function and the Number of Zeros of its Derivatives ANATOLII GOL'DBERG	187
The General Solution of the Loewner Differential Equation on the Unit Ball in $\mathbb{C}^n$ IAN GRAHAM, GABRIELA KOHR, and JOHN A. PFALTZGRAFF	191
On the Zeros of a q-Bessel Function W. K. HAYMAN	205
Entire Functions with No Unbounded Fatou Components A. HINKKANEN	217
A Note on a Theorem of J. Globevnik DMITRY KHAVINSON	227
Behaviour of a Dynamical System Far from its Equilibrium F. C. KLEBANER	229
A Tauberian Theorem for Laplace Transforms with Pseudofunction Boundary Behavior JAAP KOREVAAR	233
The Schwarzian Derivative and Complex Finsler Metrics SAMUEL L. KRUSHKAL	243
On Evaluation of the Cauchy Principal Value of the Singular Cauchy-Szegö Integral in a Ball of $\mathbb{C}^n$ A. M. KYTMANOV and S. G. MYSLIVETS	263
Boundary Properties of Convex Functions ADAM LECKO	273
Regularization of a Solution to the Cauchy Problem for the System of Thermoelasticity O. I. MAKHMUDOV and I. E. NIYOZOV	285
Modules of Vector Measures on the Heisenberg Group IRINA MARKINA	291
An Analogue of the Fuglede Formula in Integral Geometry on Matrix Spaces ELENA OURNYCHEVA and BORIS RUBIN	305
Characteristic Problems for the Spherical Mean Transform V. P. PALAMODOV	321

On the Essential Spectrum of Electromagnetic Schrödinger Operators
  V. S. Rabinovich   331

A Critical Example for the Necessary and Sufficient Condition for Unique
Quasiconformal Extremality
  Edgar Reich   343

Generic Convergence of Iterates for a Class of Nonlinear Mappings in
Hyperbolic Spaces
  Simeon Reich and Alexander J. Zaslavski   349

The Beltrami Equation and FMO Functions
  V. Ryazanov, U. Srebro, and E. Yakubov   357

Pseudodifferential Operators with Operator-Valued Symbols
  Bert-Wolfgang Schulze and Nikolai Tarkhanov   365

Pluripolar Sets and Pseudocontinuation
  Józef Siciak   385

Convolution Inverses
  Herb Silverman and Evelyn M. Silvia   395

Composition Operators on Sobolev Spaces
  S. K. Vodopyanov   401

New Results in Integral Geometry
  V. V. Volchkov and Vit. V. Volchkov   417

Preface

The Second International Conference on Complex Analysis and Dynamical Systems (*CA&DS II*), sponsored by ORT Braude College (Karmiel, Israel), Bar-Ilan University, and the University of Potsdam, took place at the Carlton Hotel in Nahariya, Israel, during June 9–12, 2003. This was the fourth in a series of mathematics conferences organized by ORT Braude over the past several years. Altogether, 78 participants from over a dozen countries attended the Conference, which was held in honor of Professor Lawrence Zalcman's sixtieth birthday.

This volume is the tangible record of *CA&DS II*. Most of the papers collected here have been contributed by participants in the Conference. In some cases, they have chosen to submit manuscripts which depart from the texts of their lectures. There are also contributions from participants who did not speak at the Conference, as well as several papers by friends and admirers of Larry Zalcman who were unable to attend. All submissions have been carefully refereed. Taken together, the papers presented here cover an extraordinarily broad range of subjects within complex analysis and bordering areas and testify to the continuing vitality of the interplay between classical and modern analysis.

We acknowledge with thanks the support provided for the Conference by the Gelbart Research Institute for Mathematical Sciences and the Emmy Noether Research Institute for Mathematics of Bar-Ilan University and the Sacta-Rashi Foundation and for the preparation of this volume by the Gelbart and Noether Institutes. Special thanks to the technical editor, Miriam Beller, for her heroic efforts to produce this volume in a timely fashion and to Jeremy Schiff for superb technical assistance when it mattered most.

The Editors

List of Participants

M. Agranovsky
Bar-Ilan University, Israel

D. Aharonov
Technion, Israel

L. Aizenberg
Bar-Ilan University, Israel

J. M. Anderson
University College London, UK

V. Azarin
Bar-Ilan University, Israel

W. Bergweiler
Christian-Albrechts-Universität zu Kiel, Germany

D. Bshouty
Technion, Israel

D.-C. Chang
Georgetown University, USA

I. Chavel
The City College of CUNY, USA

M. Chuaqui
Pontificia Universidad Católica de Chile, Chile

M. Cwikel
Technion, Israel

H. Dym
Weizmann Institute of Science, Israel

M. Elin
ORT Braude College, Israel

G. Enden
ORT Braude College, Israel

H. M. Farkas
Hebrew University, Israel

J. Fiadin
ORT Braude College, Israel

C. H. FitzGerald
University of California, San Diego, USA

L. Flatto
Bell Labs, USA

J. Gevirtz
Pontificia Universidad Católica de Chile, Chile

J. Globevnik
Univerza v Ljubljani, Slovenia

A. Golberg
Bar-Ilan University, Israel

A. Gol'dberg
Bar-Ilan University, Israel

A. Goldvard
ORT Braude College, Israel

V. Goryainov
Volgograd State University, Russia

E. Grinberg
University of New Hampshire, USA

V. Gromak
Belarusian State University, Belarus

W. K. Hayman
Imperial College, UK

A. Hinkkanen
University of Illinois, USA

PARTICIPANTS

T. Ishankulov
Samarkand State University,
Uzbekistan

F. Jacobson
ORT Braude College, Israel

L. Karp
ORT Braude College, Israel

V. Katsnelson
Weizmann Institute of Science, Israel

R. Kerdman
ORT Braude College, Israel

V. Khatskevich
ORT Braude College, Israel

F. Klebaner
Monash University, Australia

M. Klein
Universität Potsdam, Germany

S. Krushkal
Bar-Ilan University, Israel

A. Kytmanov
Krasnoyarsk State University, Russia

A. Lecko
Politechnika Rzeszowska, Poland

G. Levin
Hebrew University, Israel

E. Liflyand
Bar-Ilan University, Israel

A. Losev
Volgograd State University, Russia

Y. Lutsky
ORT Braude College, Israel

L.S. Maergoiz
Krasnoyarsk Civil Engineering
Institute, Russia

O. I. Makhmudov
Samarkand State University,
Uzbekistan

I. Markina
Universidad Técnica Federico Santa
María, Chile

R. Miniowitz
Technion, Israel

S. Myslivets
Krasnoyarsk State University, Russia

S. Nevo
Bar-Ilan University, Israel

A. Olevskii
Tel-Aviv University, Israel

E. Ournycheva
Hebrew University, Israel

V. Palamodov
Tel-Aviv University, Israel

X. C. Pang
East China Normal University, China

B. Pinchuk
Bar-Ilan University, Israel

A. Pinkus
Technion, Israel

V. Rabinovich
Instituto Politécnico Nacional, Mexico

S. Reich
Technion, Israel

S. Ruscheweyh
Bayerische Julius-Maximilians
Universität Würzburg, Germany

E. Sattorov
Samarkand State University,
Uzbekistan

J. Schiff
Bar-Ilan University, Israel

B. Schreiber
Wayne State University, USA

B. W. Schulze
Universität Potsdam, Germany

D. Shoikhet
ORT Braude College, Israel

L. Shvartsman
ORT Braude College, Israel

J. Siciak
Uniwersytet Jagielloński, Poland

H. Silverman
College of Charleston, USA

M. Sodin
Tel-Aviv University, Israel

U. Srebro
Technion, Israel

N. Tarkhanov
Universität Potsdam, Germany

A. Vasil'ev
Universidad Técnica Federico Santa
María, Chile

S. Vodopyanov
Siberian Branch of the Russian
Academy of Sciences, Russia

Z. Volkovich
ORT Braude College, Israel

Y. Weit
University of Haifa, Israel

E. Yakubov
Holon Academic Institute of
Technology, Israel

Y. Yomdin
Weizmann Institute of Science, Israel

P. Yuditskii
Johannes Kepler Universität Linz,
Austria

L. Zalcman
Bar-Ilan University, Israel

J. Zemánek
Polska Akademia Nauk, Poland

Conference Program

Monday, June 9

    09:00 - 09:30   Refreshments and Registration

    09:30 - 10:00   Opening Ceremony

Plenary Morning Session, Gallery Hall

    10:10 - 11:00   W. K. Hayman
    Zeros of solutions to the functional equations
$$\sum_{j=1}^{m} a_j(z) f(c^j z) = Q(z)$$

    11:10 - 12:00   W. Bergweiler
    Normal families and exceptional values of derivatives

    12:20 - 12:50   A. Hinkkanen
    Growth estimates for certain analytic functions

    12:55 - 13:25   C. H. FitzGerald
    The image of the sum of slit mappings

Afternoon Session, Gallery Hall

    14:45 - 15:15   J. M. Anderson
    The dyadic parametrization of curves

    15:20 - 15:50   A. Gol'dberg
    On a connection between the number of poles of a meromorphic function and the number of zeros of its derivatives

    15:55 - 16:25   X. C. Pang
    Quasinormal families of meromorphic functions omitting a holomorphic function

    16:40 - 17:10   J. Globevnik
    Analyticity on circles

    17:15 - 17:45   M. Sodin
    Harmonic functions and sign changes

Tuesday, June 10

Plenary Morning Session, Gallery Hall

    09:00 - 09:50   V. Palamodov
                               Darboux equation and reconstruction from spherical means

    10:00 - 10:50   L. Zalcman
                                 A glance backward, a long look ahead

Morning Session 1, Gallery Hall

    11:10 - 11:40   L. Flatto
                                 Billiards in an ellipse

    11:45 - 12:15   D.-C. Chang
                                 Sub-Riemannian geometry on a step 2k sub-Riemannian manifold

    12:20 - 12:50   E. Grinberg
                                 Mean value theorems for PDE and fitting convex bodies

Morning Session 2, La Caprice Hall

    11:10 - 11:40   B.-W. Schulze
                                 Ellipticity on spaces with corners

    11:45 - 12:15   J. Schiff
                                 A modification of the modified equations method

    12:20 - 12:50   F. Klebaner
                                 The most visited region of the Lotka-Volterra system

Afternoon Session 1, Gallery Hall

    14:30 - 15:00   J. Gevirtz
                                 Variational methods for first-order univalence criteria

    15:05 - 15:35   V. Goryainov
                                 Evolution families of analytic functions

    15:40 - 16:10   A. Lecko
                                 On univalent functions starlike with respect to a boundary point

    16:30 - 17:00   S. Krushkal
                                 The Schwarzian derivative and complex Finsler metrics

    17:05 - 17:35   H. Silverman
                                 Inverses under convolution

Afternoon Session 2, La Caprice Hall

    14:30 - 15:00   V. Katsnelson
                            Rational solutions of the Schlesinger system

    15:05 - 15:35   H. Dym
                            Riccati equations and bitangential interpolation problems

    15:40 - 16:10   Y. Yomdin
                            Center-focus problem, moments, and compositions: some new developments

    16:30 - 17:00   E. Liflyand
                            Hausdorff operators on the real Hardy space

    17:05 - 17:35   V. Rabinovich
                            Essential spectrum of Schrödinger operators

Wednesday, June 11

Plenary Morning Session, Gallery Hall

    09:00 - 09:50   J. Siciak
                            Pluripolar hulls and pseudocontinuation

    09:55 - 10:45   A. Olevskii
                            Representation of functions by power series

Morning Session 1, Gallery Hall

    11:00 - 11:30   D. Aharonov
                            Minimal area problems and quadrature domains

    11:35 - 12:05   A. Vasil'ev
                            Flows on homogeneous spaces and a parametric method for conformal maps with quasiconformal extension

    12:20 - 12:50   V. Azarin
                            Some characteristics of asymptotic behavior of subharmonic and entire functions and their independence

    12:55 - 13:25   A. Golberg
                            Homeomorphisms with mean integral dilatations

Morning Session 2, La Caprice Hall

    11:00 - 11:30   S. Vodopyanov
        Geometry of Carnot-Carathéodory spaces and differentiability of mappings

    11:35 - 12:05   M. Cwikel
        Moduli of continuity and K-divisibility of K-functionals

    12:20 - 12:50   P. Yuditskii
        Szegö, Killip-Simon and others

    12:55 - 13:25   M. Chuaqui
        Simple curves in $\mathbb{R}^n$ and Ahlfors' Schwarzian derivative

Thursday, June 12

Plenary Morning Session, Gallery Hall

    09:00 - 09:50   N. Tarkhanov
        Holomorphic Lefschetz formula

Morning Session 1, Gallery Hall

    10:10 - 10:40   H. M. Farkas
        The solutions of some Diophantine equations

    11:10 - 11:40   B. Schreiber
        Representation of isotropic harmonizable covariances

    11:45 - 12:15   G. Levin
        A straightening theorem for a class of meromorphic functions

    12:20 - 12:50   E. Yakubov
        The Beltrami equation and finite mean oscillation

Morning Session 2, La Caprice Hall

    10:10 - 10:40   L. Maergoiz
        Optimal estimate for extrapolation from a finite set in the Wiener class

    11:10 - 11:40   S. Myslivets
        Holomorphic Lefschetz formula for manifolds with boundary

    11:45 - 12:15   D. Bshouty
        Planar harmonic mappings

    12:20 - 12:50   I. Markina
        Extremal widths on the Heisenberg group

Afternoon Session 1, Gallery Hall

    14:30 - 15:00   A. Kytmanov
Removable singularities of CR functions on singular hypersurfaces

    15:05 - 15:35   T. Ishankulov
On the possibility of continuation of functions from a part of the boundary of a domain to the whole domain as solutions of the Moisil-Theodoresco system of equations

Afternoon Session 2, La Caprice Hall

    14:30 - 15:00   E. Ournycheva
Radon transforms on matrix spaces

    15:05 - 15:35   J. Zemánek
Cesàro means and resolvents of linear operators

Concluding Session, Gallery Hall

    16:00 - 16:30   L. Aizenberg
The classical theorem of Rogosinski on partial sums of power series, its generalizations and applications

Lawrence Zalcman at Sixty

Mark Agranovsky and David Shoikhet

The Second International Conference on Complex Analysis and Dynamical Systems held in Nahariya, Israel, on June 9–12, 2003, in honor of Professor Lawrence Zalcman's sixtieth birthday assembled close to eighty of Larry's friends, colleagues, and admirers from over a dozen countries on five continents in a tribute to one of Israel's leading analysts.

Lawrence Zalcman was born on June 9, 1943, in Kansas City, Missouri (USA), and received his primary and secondary education in the public schools of that city. He attended Dartmouth College, where he learned complex variables from A.S. Besicovitch and functional analysis from Mischa Cotlar, graduating in three years, first in his class, in 1964. His last year in college, he was one of the five highest ranking individuals in the William Lowell Putnam Mathematical Competition. It was at Dartmouth also that he wrote his first mathematical paper [1][1] (with Michael Voichick), based on his undergraduate thesis. Even today, almost forty years later, this article continues to be cited in the literature.

From Dartmouth, Zalcman went to MIT, to study with Kenneth Hoffman. These were the heady days of function algebras, and Larry fell in easily with the circle of talented young mathematicians that had grouped around Hoffman. When he left MIT in 1968, he had already published six papers, as well as the monograph *Analytic Capacity and Rational Approximation*, which immediately attracted international attention. Reviewing this volume in *Mathematical Reviews*, G.G. Lorentz wrote [Lo], "An interested reader will consult this learned and witty book with pleasure and profit."

Zalcman's first academic position was an assistant professorship at Stanford University, where he enjoyed day-to-day contact with Stefan Bergman, George Pólya, H.L. Royden, and M.M. Schiffer. In 1972, he accepted a position as associate professor at the University of Maryland, where his senior colleagues included J.L. Walsh, Edoardo Vesentini, and M.H. Heins. In 1974, just two years after coming to Maryland and barely six years out of graduate school, Larry was promoted to the rank of professor.

2000 *Mathematics Subject Classification*. Primary 01A70.

[1]Numbered references refer to the list of Zalcman's mathematical papers included in the comprehensive bibliography of his works in this volume. All other references are to the bibliography at the end of this article.

Larry's doctoral dissertation [**7**] had opened the study of Banach algebras of bounded analytic functions on domains of infinite connectivity, with an emphasis on the new phenomena which arise in this case. While he was still at Stanford, his work took a sharp turn. Inspired by a question from a student, he proved that if $f \in C(\mathbb{R}^2)$ satisfies

$$\int_\Gamma f(z)dz = 0$$

for every circle of radius r_1 and r_2, then f must be holomorphic in the whole complex plane (entire) so long as r_1/r_2 is not a quotient of zeros of the Bessel function J_1; if r_1/r_2 *is* such a quotient, f need not be holomorphic anywhere. This is the original Zalcman two circle theorem; and its effect, even at Stanford, was electrifying. This and related results appeared in [**11**]. This was followed by the extraordinarily rich paper [**12**] where, among many other things, it was shown that the weak solutions of *any* differential equation of the form $P(D)f = 0$, where P is a homogeneous polynomial in n variables and $D = \left(\frac{\partial}{\partial x_1}, \frac{\partial}{\partial x_2}, \ldots, \frac{\partial}{\partial x_n}\right)$, may be characterized by appropriate two radius theorems; in the version cited above, $P(x_1, x_2) = \frac{1}{2}(x_1 + ix_2)$ and $P(D) = \frac{\partial}{\partial \bar{z}}$. (Note that even the "trivial" case $P(x) = 1$, $P(D) = I$ is highly nontrivial!) In [**21**], written with his Maryland colleague Carlos Berenstein, these results are extended to rank one symmetric spaces. The beautiful mix of function theory, PDE, and harmonic analysis and integral geometry on euclidean and noneuclidean spaces contained in these papers was summarized in the prize-winning survey [**20**]. The field remains active even today, and Larry continues to contribute to it [**41**], [**53**].

While still actively engaged in the research described above, Larry turned his attention to a completely different problem, the explication of Bloch's Principle. Attributed to André Bloch, this heuristic principle asserts that "a family of holomorphic (meromorphic) functions which have a property P in common in a domain D is (apt to be) a normal family in D if P cannot be possessed by non-constant entire (meromorphic) functions in the finite plane" [**Hi**, p.250]. In his retiring presidential address to the Association for Symbolic Logic [**R**], Abraham Robinson listed the explication of this principle as one of twelve problem worthy of the attention of logicians and, by extension, mathematicians in general. In [**15**], Larry offered such an explication. Specifically, he proved a theorem which contains many of the most important instances of Bloch's Principle and also explains its failure in those cases where it fails to hold. The key here is a renormalization result which has become known as Zalcman's Lemma [**Ch**], [**Sch**]. (This idea is sometimes attributed [**Gro**, pp.344-5] to Robert Brody, who discovered a very similar result a few years later and used it to prove that a compact complex manifold is hyperbolic if and only if it contains no complex lines [**Br**]. As noted by H.H. Wu [**W**, p.95], Brody's Theorem follows at once from the argument used to prove Zalcman's Lemma.) Over the years, Zalcman's Lemma has been generalized in various directions and the explication of Bloch's Principle correspondingly extended; for details, see [**40**]. Much of Larry's current research finds its roots, one way or another, in [**15**].

During all this time, Larry's service to the profession was no less distinguished than his mathematics. From 1976 to 1982, he was a member of the Editorial Board of *Proceedings of the American Mathematical Society*. He also sat on the Council of the AMS (1979-1982) and chaired a number of important committees of the Mathematical Association of America. During the period 1982-1984, he served as Chairman of the Joint AMS-ASL-IMS Committee on Translations from Russian and Other Foreign Languages, from which he waged a tenacious and, ultimately, completely successful struggle to secure translation and publication rights for articles and books by mathematicians from the Soviet Union who were in official disfavor, whether because of their Jewish origins, dissident politics, or the simple fact of their having emigrated to the West. To this day, this remains one of his proudest accomplishments.

The Zalcmans spent almost a third of the period 1970-1981 in Israel. In 1970-71, Larry was a NATO Postdoctoral Fellow at the Hebrew University and the Weizmann Institute; and in 1975, he was a visiting professor at the Technion. He was at the Hebrew University again in 1979-80 as a Lady Davis Visiting Professor and continued on in 1980-81 as a visiting professor at both the Hebrew University and Bar-Ilan University. Larry's frequent long-term stays in Israel inevitably prompted informal offers of employment. Finally, in 1985, he accepted the Lady Davis Chair in Mathematics at Bar-Ilan University, and the Zalcmans moved to Israel for good.

During the next ten years, Larry brought literally dozens of distinguished analysts from all over the world to Israel and to Bar-Ilan as month-long visitors. The Analysis Seminar which he founded at Bar-Ilan became the speaking venue of choice for analysts visiting Israel. Throughout his career, Larry has always made a point of fostering mathematical talent whenever he has encountered it; and more than a few well-known mathematicians have benefited from his encouragement, advice, and personal support at an early stage in their careers. In Israel, Larry's efforts on behalf of immigrant mathematicians from the former Soviet Union were unstinting and largely successful; one result of these was the creation of the exceptionally strong group in analysis at Bar-Ilan. Another expression of the same urge for excellence is seen in Larry's mathematical contacts with China, which he has visited three times since coming to Israel. In fact, in recent years, he has hosted some of the most talented of the younger generation of Chinese complex analysts at Bar-Ilan as long-term visitors.

Larry's participation in mathematical life in Israel has not been restricted to his activities at Bar-Ilan. Since 1987, he has been Editor of *Journal d'Analyse Mathématique*, in which role he has enhanced yet further the prestige of the world's premier journal of classical analysis. And, from 1997 to 1999, he served as President of the Israel Mathematical Union.

This is not the place for a detailed catalogue of Larry's contributions to mathematics. We have already mentioned his two radius theorems and the whole area of research they opened up. Zalcman's Lemma has become an indispensable tool in the study of normal families; its usefulness in function theory has been seriously compared to that of Schwarz' Lemma. (For an up-to-date statement and an elementary proof, see [**44**]. Numerous applications are discussed in [**40**]; cf. [**Bg**].) There are also Zalcman domains [**HKN**], [**J**] (which play a role in the classification theory of Riemann surfaces) and Zalcman functions [**St**] (which arise in complex dynamics); and there is the Pizzetti-Zalcman formula in PDE [**CS**, p.96]. The list could be extended.

Looking over Larry's published work, one notices some recurrent themes. Perhaps the most prominent is the constant search for the proper point of view.

Another is "that even the basic theorems of complex analysis, the classical corpus covered in a first-year course in function theory, can afford an ample arena for interesting and worthwhile research" [23, p.76]. And throughout, there is a persistent attempt to avoid the purely technical and to focus on mathematical issues of general interest. His success in this endeavor may be judged by the fact that his results appear not only in basic texts on complex variables [**BG**], [**Bu**], [**Ga**], but in treatises on logic [**Str**], partial differential equations [**CS**], [**F**], and operator theory [**BS**], [**N**], as well. Another hallmark of Zalcman's work is the emphasis he places on scholarship, something quite rare in our profession. A glance at the bibliographies of [**11**] or [**12**], for instance, will be enough to persuade the reader of the enormous effort expended to trace down earlier work in the area and to assign proper credit, an effort that persists to this day [**35**], [**45**]. But what is perhaps most impressive of all is the continued vitality, the sheer staying power of Larry's ideas. Papers of his written 25, 30, or even 35 years ago are cited regularly even today. At the same time, Larry's research activity continues unabated; indeed, the past few years have been among the most productive of his career. His recent work has focused mainly (though not exclusively) on the theory of normal and quasinormal families of meromorphic functions on plane domains, specifically, on criteria for (quasi)normality formulated in terms of the values shared or omitted by functions and their derivatives [**43**], [**44**], [**47**]–[**52**], [**54**]–[**59**].

Larry's skills as an expositor, both on paper and in person, are legendary. His paper [**14**] received both a Lester R. Ford Award and the prestigious Chauvenet Prize from the Mathematical Association of America. The article [**20**], based on an invited address to the Annual Meeting of the MAA, won another Ford Award; many people consider this paper a masterpiece of mathematical exposition, and it is unusual for it to be cited in the literature without some accompanying superlative. Of [**23**], based on an invited address to the American Mathematical Society, Peter Lappan writes in *Mathematical Reviews* [**La**], "This very interesting and amusingly written article ... should make interesting reading for all mathematicians ... [and] should be required reading for all students in function theory courses." The more recent survey [**40**] is cut of very much the same cloth.

Finally, we must say something about two special qualities of Larry's: his humor and his erudition. Humor has always played an important part in Larry's approach to life in general and mathematics in particular. Nothing illustrates this better than his paper (with Dov Aharonov and Max Schiffer) "Potato Kugel" [**22**]. That this paper is serious mathematics (it contains the converse to a famous theorem of Isaac Newton, together with generalizations, and was cited in [**Gru**] as one of "seven pearls of convexity") has not prevented the authors from having fun with the title and the bibliography. (Larry's coauthors are on record as saying that all the jokes in the paper are due to him.) In fact, the title of the paper is a pun which, on the one hand, encapsulates the main result of the paper, viz., that a "potato" (i.e., a homogeneous solid) which exerts a gravitational attraction on the rest of the universe identical to that exerted by a point mass must be a "Kugel" (i.e., a ball) centered at the point in question and, on the other hand, names one of Larry's favorite dishes, a savory pudding made of potatoes and eggs which is the perfect complement to roast chicken. Larry likes to tell the story of how, when speaking on this topic at the Mathematics Colloquium of the Hebrew University, he arranged to have the usual cookies and cakes of the pre-colloquium tea replaced by generous servings of potato kugel and how, having seen to the necessary advance publicity, the tea (if not the colloquium) was the best attended event of its kind in history. (Since someone is sure to ask, yes, there is also a paper entitled "Matzoh Ball Soup" – in fact, there are *two* [**A**], [**MS**] – and, yes, Larry is responsible for that title, too [**31**].)

Although he wears his learning lightly, Larry's extra-mathematical intellectual horizons are unusually broad, and his expertise in subjects which interest him unusually deep. He has published articles in leading academic journals in such areas as philosophy, history of science, literature, and rabbinics. Special mention should be made of Larry's interest in the textual criticism of the Hebrew Bible. He has written over a dozen articles in this last area (including two invited encyclopedia entries), and his suggestions can be found cited with approval in standard academic biblical commentaries. We hasten to add that not all of Larry's interests are so "heavy." He also maintains a long-term and lively interest in the cinema; in fact, as a graduate student, he almost became the film critic for *Boston After Dark*.

No account of Larry's accomplishments would be complete without mention of his charming wife Adrienne. They met on his twenty-first birthday (he calls her "the best birthday present I ever got"), married the following year, and have been together ever since. And, for someone who has written over a dozen and a half papers on normal families, it seems especially appropriate that his own very normal, very happy family (wife, two married children, their spouses, and four – now six – extremely cute grandchildren) should have played so prominent a role in the Conference festivities.

In the lecture he delivered at the Conference, entitled "A Glance Backward, A Long Look Ahead," Larry concluded by saying that he considered himself "incredibly lucky" in both his personal and his professional life and thanked the members of the audience for their friendship as well as their attention. We know we speak for many others when we say that it is we who are the lucky ones to have benefited over the years from Larry Zalcman's friendship, intelligence, energy, and personal and professional support.

Happy Birthday, Larry!

ACKNOWLEDGMENT. Obviously, this piece could not have been written without Larry's cooperation. We thank him for having shared some of his memories with us.

References

[A] Giovanni Alessandrini *Matzoh ball soup: a symmetry result for the heat equation*, J. Analyse Math. **54** (1990), 229–236.

[BG] Carlos A. Berenstein and Roger Gay, *Complex Variables. An Introduction*, Springer-Verlag, 1991.

[Bg] Walter Bergweiler, *A new proof of the Ahlfors five islands theorem*, J. Analyse Math. **76** (1998), 337–347.

[BS] Albrecht Böttcher and Bernd Silbermann, *Analysis of Toeplitz Operators*, Springer-Verlag, 1990.

[Br] Robert Brody, *Compact manifolds in hyperbolicity*, Trans. Amer. Math. Soc. **235** (1978), 213–219.

[Bu] Robert B. Burckel, *An Introduction to Classical Complex Analysis*, Vol. 1, Academic Press, 1979.

[CS] R.W. Carroll and R.E. Showalter, *Singular and Degenerate Cauchy Problems*, Academic Press, 1976.

[Ch] Chi-Tai Chuang, *Normal Families and Meromorphic Functions*, World Scientific, 1993.

[F]   L.E. Fraenkel, *Introduction to Maximum Principles and Symmetry in Elliptic Problems*, Cambridge University Press, 2000.
[Ga] Theodore W. Gamelin, *Complex Analysis*, Springer-Verlag, 2001.
[Gro] Misha Gromov, *Metric Structures for Riemannian and Non-Riemannian Spaces*, Birkhäuser, 1999.
[Gru] Peter M. Gruber, *Seven small pearls from convexity*, Math. Intelligencer **5** (1983), 16–19.
[HKN] Mikihiro Hayashi, Yasuyuki Kobayashi, and Mitsuru Nakai, *A uniqueness theorem and the Myrberg phenomenon for a Zalcman domain*, J. Analyse Math. **82** (2000), 267–283.
[Hi] Einar Hille, *Analytic Function Theory*, Vol. 2, Ginn, 1962.
[J]   Piotr Jucha, *Bergman completeness of Zalcman type domains*, Studia Math. **163** (2004), 71–83.
[La] P. Lappan, Review of [23], MR 83d:30003.
[Lo] G.G. Lorentz, Review of *Analytic Capacity and Rational Approximation*, MR 37 #3018.
[MS] Rolando Magnanini and Shigeru Sakaguchi, *Matzoh ball soup: heat conductors with a stationary isothermic surface*, Ann. of Math. (2) **156** (2002), 931–946.
[N]   N.K. Nikol'skiĭ, *Treatise on the Shift Operator*, Springer-Verlag, 1986.
[R]   Abraham Robinson, *Metamathematical problems*, J. Symbolic Logic **38** (1973), 500–516.
[Sch] Joel L. Schiff, *Normal Families*, Springer-Verlag, 1993.
[St] Norbert Steinmetz, *Zalcman functions and rational dynamics*, New Zealand J. Math. **32** (2003), 91–104.
[Str] K.D. Stroyan, *Introduction to the Theory of Infinitesimals*, Academic Press, 1976.
[W]   H. Wu, *Some theorems on projective hyperbolicity*, J. Math. Soc. Japan **33** (1981), 79–104.

DEPARTMENT OF MATHEMATICS, BAR-ILAN UNIVERSITY, 52900 RAMAT-GAN, ISRAEL
E-mail address: `agranovs@macs.biu.ac.il`

DEPARTMENT OF MATHEMATICS, ORT BRAUDE COLLEGE, 21982 KARMIEL, ISRAEL
E-mail address: `davs27@netvision.net.il`

Bibliography of Lawrence Zalcman

Mathematical Writings

Books

1. *Analytic Capacity and Rational Approximation*, Springer Lecture Notes in Mathematics, vol. 50, 1968.
2. *Advances in Complex Function Theory*, Springer Lecture Notes in Mathematics, vol. 505, 1976 (editor with W.E. Kirwan).
3. *Proceedings of the Ashkelon Workshop on Complex Function Theory (May 1996)*, IMCP, vol. 11, 1997 (editor).
4. *Entire Functions in Modern Analysis. Boris Levin Memorial Conference*, IMCP, vol. 15, 2001 (editor with Yuri Lyubich, Vitali Milman, Iossif Ostrovskii, Mikhail Sodin and Vadim Tkachenko).
5. *Complex Analysis and Dynamical Systems*, Contemp. Math. **364** (2004) (editor with Mark Agranovsky, Lavi Karp and David Shoikhet).

Papers

1. *Inner and outer functions on Riemann surfaces*, Proc. Amer. Math. Soc. **16** (1965), 1200-1204. (with Michael Voichick)
2. *Null sets for a class of analytic functions*, Amer. Math. Monthly **75** (1968), 462-470.
3. *Analytic functions and Jordan arcs*, Proc. Amer. Math. Soc. **19** (1968), 508; **21** (1969), 507.
4. *A note on invariant subspaces*, Illinois J. Math. **12** (1968), 303-306.
5. *Hadamard products of schlicht functions*, Proc. Amer. Math. Soc. **19** (1968), 544-548.
6. *A note on absolute continuity*, J. Math. Anal. Appl. **24** (1968), 527-529.
7. *Bounded analytic functions on domains of infinite connectivity*, Trans. Amer. Math. Soc. **144** (1969), 241-269.
8. *Derivation pairs on algebras of analytic functions*, J. Functional Analysis **5** (1970), 329-333.
9. $H^\infty + C$, Indiana Univ. Math. J. **20** (1971), 971-974.
10. *The essential range of a function of class $H^\infty + C$*, Mich. Math. J. **19** (1972), 385-386.
11. *Analyticity and the Pompeiu problem*, Arch. Rational Mech. Anal. **47** (1972), 237-254.
12. *Mean values and differential equations*, Israel J. Math. **14** (1973), 339-352.

13. *Analytic functions and harmonic analysis*, Proceedings of the Symposium on Complex Analysis (Canterbury, 1973), Cambridge University Press, 1974, pp. 139-142.
14. *Real proofs of complex theorems (and vice versa)*, Amer. Math. Monthly **81** (1974), 115-137; reprinted in *The Chauvenet Papers* (J.C. Abbott, ed.), Mathematical Association of America, 1978, vol. 2, pp. 573-595
15. *A heuristic principle in complex function theory*, Amer. Math. Monthly **82** (1975), 813-817.
16. *A mathematician looks at mathematics*, FSC Journal of Mathematics Education **10** (1975), 1-12.
17. *Pompeiu's problem on spaces of constant curvature*, J. Analyse Math. **30** (1976), 113-130. (with Carlos A. Berenstein)
18. *A coefficient problem for bounded nonvanishing functions*, J. Analyse Math. **31** (1977), 169-190. (with J. A. Hummel and Stephen Scheinberg)
19. *Picard's theorem without tears*, Amer. Math. Monthly **85** (1978), 265-268.
20. *Offbeat integral geometry*, Amer. Math. Monthly **87** (1980), 161-175.
21. *Pompeiu's problem on symmetric spaces*, Comment. Math. Helv. **55** (1980), 593-621. (with Carlos A. Berenstein)
22. *Potato kugel*, Israel J. Math. **40** (1981), 331-339. (with Dov Aharonov and M. M. Schiffer)
23. *Modern perspectives on classical function theory*, Rocky Mountain J. Math. **12** (1982), 75-92.
24. *Uniqueness and nonuniqueness for the Radon transform*, Bull. London Math. Soc. **14** (1982), 241-245.
25. *Polynomial approximation with bounds*, J. Approx. Theory **34** (1982), 379-383.
26. *Netradiční integrální geometrie*, Pokroky Mat. Fyz. Astronom. **27** (1982), 9-23 (translation of [**20**]).
27. *Moderní perspektivy klasické teorie funkcí*, Pokroky Mat. Fyz. Astronom. **29** (1984), 257-273 (translation of [**23**]).
28. *S"vremenni perspektivi na klasiceskata teoriya na funktsiite*, Fiz.-Mat. Spis. B"lgar. Akad. Nauk **27** (**60**) (1985), 48-64 (translation of [**23**]).
29. *Determining sets for functions and measures*, Real Anal. Exchange **11** (1985-86), 40-55.
30. *In memoriam: Elisha Netanyahu 1912-1986*, J. Analyse Math. **46** (1986), 1-10.
31. *Some inverse problems of potential theory*, Integral Geometry, Contemp. Math. **63** (1987), 337-350.
32. *Positive derivatives and increasing functions*, Elem. Math. **43** (1988), 120-121.
33. *Mathematicians sweep 1988 Wolf Prizes*, Math. Intelligencer **11**/2 (1989), 39-48.
34. *Matematika a Wolfovy ceny za rok 1988*, Pokroky Mat. Fyz. Astronom. **36** (1991), 129-140 (translation of [**33**]).
35. *A bibliographic survey of the Pompeiu problem*, Approximation by Solutions of Partial Differential Equations (B. Fuglede et al., eds.), Kluwer Acad. Publ., 1992, pp. 185-194.

36. *Variations on the theorem of Morera*, The Madison Symposium on Complex Analysis, Contemp. Math. **137** (1992), 63-78. (with Carlos Berenstein, Der-Chen Chang and Daniel Pascuas)
37. *Normal families revisited*, Complex Analysis and Related Topics (J. Wiegerinck, ed.), University of Amsterdam, 1993, pp. 149-164.
38. *Instability phenomena for the moment problem*, Ann. Scuola Norm. Pisa Cl. Sci. (4) **22** (1995), 95-107. (with Lev Aizenberg)
39. *New light on normal families*, Proceedings of the Ashkelon Workshop on Complex Function Theory (May 1996), (Lawrence Zalcman, ed.), IMCP, vol. 11, 1997, pp. 237-245.
40. *Normal families: new perspectives*, Bull. Amer. Math. Soc. (N.S.) **35** (1998), 215-230.
41. *Conical uniqueness sets for the spherical Radon transform*, Bull. London Math. Soc. **31** (1999), 231-236. (with M. L. Agranovsky and V. V. Volchkov)
42. *On theorems of Hayman and Clunie*, New Zealand J. Math. **28** (1999), 71-75. (with Xuecheng Pang)
43. *Normality and shared values*, Ark. Mat. **38** (2000), 171-182. (with Xuecheng Pang)
44. *Normal families and shared values*, Bull. London Math. Soc. **32** (2000), 325-331. (with Xuecheng Pang)
45. *Supplementary bibliography to "A bibliographic survey of the Pompeiu problem,"* Radon Transforms and Tomography, Contemp. Math. **278** (2001), 69-74.
46. *Mean-value characterization*, in Encyclopaedia of Mathematics, Supplement III (Michiel Hazewinkel, ed.), Kluwer Acad. Publ., 2001, pp. 258-261. (with L. Aizenberg)
47. *Normal families and shared values of meromorphic functions*, Ann. Polon. Math. **80** (2003), 133-141. (with Mingliang Fang)
48. *Normal families and shared values of meromorphic functions II*, Comput. Methods Funct. Theory **1** (2001), 289-299. (with Mingliang Fang)
49. *Normal families and shared values of meromorphic functions III*, Comput. Methods Funct. Theory **2** (2002), 385-395. (with Mingliang Fang)
50. *Normal families of meromorphic functions whose derivatives omit a function*, Comput. Methods Funct. Theory **2** (2002), 257-265. (with Xuecheng Pang and Degui Yang)
51. *Normal families and uniqueness theorems for entire functions*, J. Math. Anal. Appl. **280** (2003), 273-283. (with Mingliang Fang)
52. *Normal families of meromorphic functions with multiple zeros and poles*, Israel J. Math. **136** (2003), 1-9. (with Xuecheng Pang)
53. *Moment versions of the Morera problem in $\mathbf{C}^n$ and $\mathbf{H}^n$*, Adv. in Appl. Math. **31** (2003), 273-300. (with Carlos Berenstein, Der-Chen Chang and Wayne Eby)
54. *A note on normality and shared values*, J. Aust. Math. Soc. **76** (2004), 141-150. (with Mingliang Fang)
55. *Normal families of holomorphic functions*, Illinois J. Math. **48** (2004), 319-337. (with Jianming Chang and Mingliang Fang)

56. *Quasinormal families of meromophic functions*, Rev. Mat. Iberoamericana **21** (2005), 249-262. (with Xuecheng Pang and Shahar Nevo)
57. *Quasinormal families of meromorphic functions II*, Selected Topics in Complex Analysis (V. Ya. Eiderman and M. V. Samokhin, eds.), Birkhäuser, 2005, pp. 177-189. (with Xuecheng Pang and Shahar Nevo)
58. *Normal families and omitted functions*, Indiana Univ. Math. J. **54** (2005), 223-236. (with Xuecheng Pang and Degui Yang)
59. *Normality and fixed points of meromorphic functions*, Ark. Mat. **43** (2005). (with Jianming Chang and Mingliang Fang)

Unpublished Manuscripts

1. *Selected Topics in Complex Analysis*, Technion, mimeographed lecture notes, 1976, 89pp. (Hebrew).
2. *On some questions of Hayman*, 1994, 5pp.

Reviews

1. Review of *Fourier Series and Integrals* by H. Dym and H. P. McKean, Bull. Amer. Math. Soc. **79** (1973), 641-645.
2. Review of *The Radon Transform* by Sigurdur Helgason, SIAM Review **25** (1983), 275-278.
3. Review of *Mishnah Tractate Rosh Hashanah* with commentary by Rabbi Ovadiah MiBartinura (Jeffrey R. Cohen, trans.), Archaeoastronomy **7** (1984), 176-177.
4. Review of *The Mysterious Numbers of the Hebrew Kings* by Edwin R. Thiele, Archaeoastronomy **8** (1985), 176-177.
5. Review of *Littlewood's Miscellany* (Bela Bollobás, ed.), Math. Intelligencer **11**/1 (1989), 63-65.
6. Review of *Darstellung und Begründung einiger neuerer Ergebnisse der Funktionentheorie* (dritte, erweiterte Auflage) by Edmund Landau and Dieter Gaier, Math. Intelligencer **11**/4 (1989), 61-63.
7. Review of *Discrete Thoughts: Essays in Mathematics, Science and Philosophy* by Mark Kac, Gian-Carlo Rota and Jacob T. Schwartz, Math. Intelligencer **12**/3 (1990), 81-83.
8. Review of *Complex Analysis* by Joseph Bak and Donald J. Newman and *Invitation to Complex Analysis* by Ralph Philip Boas, Amer. Math. Monthly **97** (1990), 262-266.
9. Review of *The Apprenticeship of a Mathematician* by André Weil, Math. Intelligencer **15**/4 (1993), 64-68.
10. Review of *Indiscrete Thoughts* by Gian-Carlo Rota, Math. Intelligencer **21**/2 (1999), 72-74; **22**/2 (2000), 1.

Nonmathematical Writings

History of Science
1. *Counting: It's child's play*, J. Irr. Res. **23** (1978), 5-7.
2. *The great schism and the supernova of 1054*, Physis **21** (1979), 55-59.

Philosophy
1. *I'm glad you asked me that question*, Analysis **48** (1988), 160.
2. *The prisoner's other dilemma*, Erkenntnis **34** (1991), 83-85.

Bible
1. *Piercing the darkness at* bôqēr, Vetus Testamentum **30** (1980), 252-255.
2. *Di sera, desert, dessert*, Expository Times **91** (1980), 311.
3. *Astronomical illusions in Amos*, Journal of Biblical Literature **100** (1981), 53-58.
4. *Ambiguity and assonance at Zephaniah II 4*, Vetus Testamentum **36** (1986), 365-371.
5. *"Orion,"* in Dictionary of Deities and Demons in the Bible, (Karel van der Toorn, Bob Becking, Pieter W. van der Horst, eds.), 2nd ed., Brill, 1999, pp. 648-649.
6. *"Pleiades,"* ibid., pp. 657-659.
7. *Drums on the water in Fair Puzai*, Zeitschrift für die alttestamentliche Wissenschaft **111** (1999), 616-618. (with Aron Pinker)
8. *Laying* dmšq 'rś *to rest (Amos III 12)*, Vetus Testamentum **52** (2002), 557-559.
9. *Proverbs 5,19c*, Zeitschrift für die alttestamentliche Wissenschaft **115** (2003), 433-434.
10. *The righteous sage: pleonasm or oxymoron? (Kohelet 7,16-18)*, Zeitschrift für die alttestamentliche Wissenschaft **115** (2003), 435-439. (with Steven Shnider)
11. *Philistines on the threshold at Amos 9:1?*, Revue Biblique **110** (2003), 481-486.
12. *Intertextuality at Nahum 1,7*, Zeitschrift für die alttestamentliche Wissenschaft **116** (2004), 614-615.
13. *Shield of Abraham, Fear of Isaac, Dread of Esau*, Zeitschrift für die alttestamentliche Wissenschaft **117** (2005).

Rabbinics
1. *Christians,* Noṣerim, *and Nebuchadnezzar's daughter*, Jewish Quarterly Review **81** (1991), 411-426.
2. *The Eternal City: Rome or Jerusalem?*, Journal of Jewish Studies **48** (1997), 312-313.

Literature
1. *Death and the calendar*, Hebrew University Studies in Literature and the Arts **16** (1988), 97-112.
2. *La muerte y el calendario*, Hispamerica **45** (1986), 17-29 (translation of [1]).

A Local Two Radii Theorem for the Twisted Spherical Means on $\mathbb{C}^n$

M. L. Agranovsky and E. K. Narayanan

To Larry Zalcman, with our friendship and admiration

ABSTRACT. We prove a local two radii theorem for the twisted spherical means on $\mathbb{C}^n$. More precisely, if the twisted convolutions $f \times \mu_{r_1} = f \times \mu_{r_2} = 0$ in a ball B_R and $r_1 + r_2 < R$, then f is the zero function provided the confluent hypergeometric function ${}_1F_1(a, n, \frac{r^2}{2})$ does not vanish simultaneously at the points r_1 and r_2 for any $a \in \mathbb{R}$.

1. Introduction

1.1. Let ν_r be the normalized surface measure on the sphere $S_r = \{x \in \mathbb{R}^n : |x| = r\}$. For a continuous function f on $\mathbb{R}^n$, the spherical means are defined by the convolution

$$(f * \nu_r)(x) = \int_{|y|=r} f(x-y) d\nu_r(y).$$

This is nothing but the average of f over the sphere of radius r centered at x. A natural question associated to this operator is whether these averages determine the function uniquely. In other words, is the operator $f \to f * \nu_r$ injective on the space of continuous functions on $\mathbb{R}^n$? In general, the answer is negative. A counterexample is produced by considering the Bessel functions. Let

$$\phi_\lambda(x) = c_n \frac{J_{\frac{n}{2}-1}(\lambda|x|)}{(\lambda|x|)^{\frac{n}{2}-1}}.$$

Then it is well-known that

(1.1) $$(\varphi_\lambda * \nu_r)(x) = \varphi_\lambda(r)\varphi_\lambda(x);$$

so if $\varphi_\lambda(r) = 0$, the convolution vanishes identically.

Nevertheless, a well-known result due to Zalcman (see [21]) says that two well-chosen radii determine the function uniquely. In [3], Berenstein and Gay proved a

---

2000 *Mathematics Subject Classification.* Primary 53C65, 44A35; Secondary 35L05, 43A85.

Key words and phrases. Twisted convolution, special Hermite operator, wave equation, harmonic oscillator, two radii theorem.

The first author was supported by Israel Scientific Foundation, grant No. 279/02-01.

local version of Zalcman's result. More precisely, the conditions $(f * \nu_{r_k})(x) = 0$ for $|x| \leq R - r_k$, $k = 1, 2$, imply that f is the zero function, provided $r_1 + r_2 < R$ and $\frac{r_1}{r_2}$ is not a ratio of zeros of a particular Bessel function. In a series of papers by V.V. Volchkov and Vit.V. Volchkov [12]-[19], local versions of Zalcman's two-radii theorem were obtained for non-Euclidean homogeneous spaces, even in a refined form, and the sharpness of the conditions was shown. For these results, as well as for an extended bibliography on the problem, see also the recent book [20] by V.V. Volchkov. Note that the condition on the radii means precisely that there exist no radial eigenfunctions of the Laplacian which vanish simultaneously on the spheres of radius r_1 and r_2.

The aim of this paper is to prove a local two-radii theorem for the twisted spherical means on $\mathbb{C}^n$. Recall that the twisted spherical means of a function f on $\mathbb{C}^n$ are defined as

$$(f \times \mu_r)(z) = \int_{|w|=r} f(z-w) e^{\frac{i}{2} Im\, z \cdot \bar{w}}\, d\mu_r(w),$$

where μ_r is the normalized surface measure on the sphere $S_r = \{|w| = r\} \subset \mathbb{C}^n$. More generally, the twisted convolution of two functions f and g on $\mathbb{C}^n$ is defined as

$$(f \times g)(z) = \int_{\mathbb{C}^n} f(z-w)\, g(w)\, e^{\frac{i}{2} Im\, z \cdot \bar{w}}\, dw.$$

For more about this convolution and its relation to the group convolution on the Heisenberg group, we refer the reader to [9] and [11].

Let ${}_1F_1(a, n, z)$ be the confluent hypergeometric function (see [4]). Our main result is the following.

THEOREM 1.1. *Let r_1, r_2 and R be positive real numbers such that $r_1 + r_2 < R$. Let $f \in L_{loc}(B_R)$ and suppose that*

(1.2) $$(f \times \mu_{r_k})(z) = 0 \quad \text{for} \quad |z| < R - r_k, \quad k = 1, 2.$$

If the equation

(1.3) $${}_1F_1\left(a, n, \frac{r_1^2}{2}\right) = {}_1F_1\left(a, n, \frac{r_2^2}{2}\right) = 0$$

has no solution $a \in \mathbb{R}$, then f is the zero function. On the other hand, any solution $a \in \mathbb{R}$ of (1.3) gives rise to a nontrivial solution of (1.2).

REMARK. We shall see that the above condition on the radii holds if and only if no radial eigenfunction of the special Hermite operator (see definition below) vanishes simultaneously on the spheres of radius r_1 and r_2.

Our proof mainly follows the scheme elaborated in [12]-[19]. Namely, the first step is to obtain the spectral decomposition of functions f satisfying the single convolution equation $f \times \mu_{r_1} = 0$ in the ball of radius R. To this end, we decompose f in the ball $|z| < r_1$ into a series of eigenfunctions of the sub-Laplacian associated to the Heisenberg group, with Dirichlet boundary conditions on the sphere $|z| = r_1$. The next step is extending the decomposition in the larger ball $|z| < R$. First this is done for radial functions, and then the general case is reduced to the radial one by decreasing homogeneity by means of differential operators from the Lie algebra. Proving the spectral decomposition for radial functions is a crucial step. In the above mentioned works [12]-[19], this is done by reducing the problem to specific convolution equations in one variable. These equations vary for different homogeneous spaces, and proving uniqueness requires specific methods for each case. Our approach differs at this point and is based on general properties of the wave equation. We hope that it may lead to a uniform approach to local theorems of the above type for more or less general Riemannian spaces (at least, those associated to Gelfand pairs).

1.2. We use some preliminaries. Define $2n$ vector fields on $\mathbb{C}^n$ by

$$Z_j = \frac{\partial}{\partial z_j} + \frac{1}{2}\bar{z}_j, \quad \bar{Z}_j = \frac{\partial}{\partial \bar{z}_j} - \frac{1}{2}z_j, \quad j = 1, 2, \cdots n.$$

These vector fields generate a Lie algebra, which is isomorphic to the Heisenberg Lie algebra. We define a second order elliptic operator $\mathcal{L}$, the sub-Laplacian (special Hermite operator), on $\mathbb{C}^n$ by

$$\mathcal{L} = -\frac{1}{2}\sum_{j=1}^{n}(Z_j\bar{Z}_j + \bar{Z}_jZ_j).$$

An explicit computation shows that $\mathcal{L}$ can be written in the form

$$\mathcal{L} = -\Delta_z + \frac{1}{4}|z|^2 - iD = L - iD,$$

where L is the radial part of $\mathcal{L}$ and D is the rotation operator

$$D = \sum_{j=1}^{n}\left(x_j\frac{\partial}{\partial x_j} - y_j\frac{\partial}{\partial x_j}\right).$$

We are interested in radial eigenfunctions of $\mathcal{L}$. For this purpose, let

$$\mathcal{L}\psi = \beta\psi.$$

If ψ is radial, then $D\psi = 0$ and consequently the above equation reduces to the eigenfunction equation for the harmonic oscillator

$$L\psi = -\Delta_z\psi + \frac{1}{4}|z|^2\psi = \beta\psi.$$

Using the polar coordinates $z = r\omega, |\omega| = 1$, we have

(1.4) $$\psi''(r) + \frac{2n-1}{r}\psi'(r) - \frac{1}{4}r^2\psi(r) + \beta\psi(r) = 0.$$

Put $\psi(r) = u(\frac{1}{2}r^2)$. Then the above equation reduces to

$$ru''(r) + nu'(r) - \frac{r - 2\beta}{4}u(r) = 0.$$

A final substitution $v(r) = e^{\frac{r}{2}}u(r)$ reduces the above to the following confluent hypergeometric equation:

$$rv''(r) + (n-r)v'(r) - \frac{n-\beta}{2}v(r) = 0.$$

Let $a = \frac{n-\beta}{2}$. Then a solution of the above equation is given by

$$v(r) = v_a(r) = M(a, n, r),$$

where $M(a, n, r)$ denotes the confluent hypergeometric function $_1F_1(a, n, r)$, defined by

$$M(a, c, z) = {}_1F_1(a, c, z) = \sum_{s=0}^{\infty} \frac{(a)_s}{(c)_s} \frac{z^s}{s!}, \quad z \in \mathbb{C}, \ a \in \mathbb{C}, \ c \in \mathbb{C} \quad (c \neq 0, -1, -2, \cdots);$$

here we use the notation

$$(a)_0 = 1$$

and

$$(a)_s = a(a+1)(a+2)\cdots(a+s-1), \ s = 1, 2, \cdots.$$

Retracing the various substitutions made, we see that a radial eigenfunction of $\mathcal{L}$-corresponding to the eigenvalue β is given by

$$(1.5) \qquad \phi_\beta(z) = M\left(\frac{n-\beta}{2}, n, \frac{1}{2}|z|^2\right) e^{-\frac{1}{4}|z|^2}.$$

We use the following asymptotics of the above hypergeometric functions (see [**4**, pp. 279-80]). Assume β is positive; then for r bounded and bounded away from zero, we have

$$(1.6) \qquad M(a, n, r) = 2^{\frac{1}{2}} \Gamma(n) e^{\frac{1}{2}r} \left(\frac{\beta}{2}\right)^{\frac{3}{4}-\frac{n}{2}} r^{\frac{3}{4}-\frac{n}{2}} \{c_1 e^{i\sqrt{2\beta r}} + c_2 e^{-i\sqrt{2\beta r}}\}$$

$$+ \left(\frac{\beta r}{2}\right)^{-\frac{1}{2}} O(1),$$

where c_1 and c_2 are constants of modulus one depending on a.

For $|2r||\beta| \leq 1$, we also have

$$(1.7) \qquad M(a, n, r) = \Gamma(n) \left(\frac{\beta r}{2}\right)^{\frac{1}{2}-\frac{n}{2}} e^{\frac{1}{2}r} J_{\frac{1}{n}}(\sqrt{(2\beta r)}) + O(|\beta|^{-1}).$$

Before we go further, let us notice that if φ is a radial eigenfunction of the operator $\mathcal{L}$, then so is $\varphi \times \mu_s$ for any $s > 0$ with the same eigenvalue. Hence we have

$$(\varphi \times \mu_s)(z) = C_s \ \varphi(z),$$

as both are smooth at the origin and the origin is a regular singular point for the differential equation (1.2). Evaluating the above at $z = 0$ and using the fact that φ is radial, we obtain

$$(1.8) \qquad (\varphi \times \mu_s)(z) = \frac{\varphi(s)}{\varphi(0)} \ \varphi(z).$$

In particular, $(\varphi \times \mu_s)(z) = 0$ if $\varphi(s) = 0$, where by $\varphi(s)$ we mean the value of φ on a sphere of radius s.

In the course of the proof, we show that any smooth radial function f in the kernel of the operator $f \to f \times \mu_s$ has an expansion in terms of the radial eigenfunctions of $\mathcal{L}$ satisfying the same convolution equation. Using this expansion, we show that a local two radii theorem holds for radial functions. In the subsequent section, we deduce the general result from the special case. In the last section, we use these results to obtain a local two radii theorem on the Heisenberg group.

We conclude this section by introducing some notation. The open ball of radius $s > 0$ centered at 0 will be denoted by B_s. Let N denote the positive zeros of the radial eigenfunctions of the differential operator $\mathcal{L}$. According to the above description of the radial eigenfunctions, N consists of positive solutions r to the equation

$$\phi_\beta(r) = M\left(\frac{n-\beta}{2}, n, \frac{1}{2}r^2\right) = 0$$

for some $\beta > 0$. Note that 0 does not belong to the set N.

REMARK. After this article was written, the authors became aware of a manuscript by V.V. Volchkov and Vit.V. Volchkov where they announce, without proof, a local version of the two radii theorem for the reduced Heisenberg group, in an even sharper form, including analysis of the limiting case $r_1 + r_2 = R$. To our understanding, their approach differs at the main point (Theorem 2.1), where we use the wave equation method.

2. Two radii theorem for radial functions

THEOREM 2.1. *Let $R > r > 0$ and suppose that $f \in C^\infty(B_R)$ is radial. Suppose that $(f \times \mu_r)(z) = 0$ for $|z| \leq R - r$. If f is zero in B_r, then f vanishes in B_R.*

The proof of the above theorem rests on an analysis of the solution to the wave equation associated to the special Hermite operator $\mathcal{L}$. In the sequel, $u(z,t)$ is always a function of $2n+1$ variables with $z \in \mathbb{C}^n$ and $t \in \mathbb{R}$; and u_t, u_{tt}, etc., denote partial derivatives with respect to the variable t. The statement *u is radial* means that u is radial as a function of z for every t. At times we also write $z = x + iy$, where $x, y \in \mathbb{R}^n$.

LEMMA 2.2. *Let $u(z,t) \in C^2$ be a radial solution of*

(2.1) $$u_{tt}(z,t) = -\mathcal{L}u(z,t).$$

Suppose that $u = u_t = 0$ on the ball $B = \{(z,0) : |z| \leq r_0\}$ in the hyperplane $t = 0$. Then $u(z,t) = 0$ in the conical region

$$\Omega = \{(z,t) : 0 \leq t \leq r_0, \ |z| + t \leq r_0\}.$$

PROOF. This is a well-known result on the domain of dependence for the wave equation in Euclidean space. The same proof works here as well (see [**5**]); we include it here for the sake of completeness.

As u is radial, we have $Du = 0$, where D is the rotation operator. Hence u satisfies the wave equation for the radial part of $\mathcal{L}$:

(2.2) $$u_{tt}(z,t) = \left(\Delta - \frac{1}{4}|z|^2\right)u(z,t).$$

By taking real and imaginary parts, we may assume that u is real-valued. Let $B_{r_0-t} = \{z : |z| \leq r_0 - t\}$. Now consider the energy integral in the ball B_{r_0-t} at time t:

$$E(t) = \frac{1}{2}\int_{B_{r_0-t}} G(z,t)\, dV(z),$$

where

$$G = u_t^2 + |\nabla_z u|^2 + \frac{1}{4}|z|^2 u^2$$

and the gradient ∇_z is understood with respect to the real variables x_k, y_k such that $z_k = x_k + iy_k$, $k = 1, \cdots, n$.

Differentiating with respect to t yields

$$E'(t) = \int_{B_{r_0-t}} \left\{ u_t u_{tt} + \langle \nabla_z u, \nabla_z u_t \rangle + \frac{1}{4}|z|^2 u u_t \right\} dV(z) - \frac{1}{2} \int_{\partial B_{r_0-t}} G(z,t) d\sigma(z),$$

where $d\sigma(z)$ is the surface area measure on the sphere ∂B_{r_0-t}.

If we write
$$\frac{\partial u}{\partial x_j} \frac{\partial^2 u}{\partial x_j \partial t} = \frac{\partial}{\partial x_j}\left[\frac{\partial u}{\partial x_j}\frac{\partial u}{\partial t}\right] - \frac{\partial^2 u}{\partial x_j^2}\frac{\partial u}{\partial t},$$

and similarly for the y-variables and simplify, the first integral in $E'(t)$ reduces to

$$\int_{B_{r_0-t}} \left\{ \sum_{j=1}^n \frac{\partial}{\partial x_j}\left[\frac{\partial u}{\partial x_j}\frac{\partial u}{\partial t}\right] + \sum_{j=1}^n \frac{\partial}{\partial y_j}\left[\frac{\partial u}{\partial y_j}\frac{\partial u}{\partial t}\right] + u_t\left(u_{tt} - \Delta u + \frac{1}{4}|z|^2 u\right) \right\} dV(z).$$

Since u satisfies the wave equation (2.2), the last expression transforms, after applying the divergence theorem, to

$$\int_{\partial B_{r_0-t}} u_t \langle \nabla u, \nu_z \rangle d\sigma(z),$$

where ν_z is the outward unit normal vector to ∂B_{r_0-t} in $\mathbb{C}^n$. The above is bounded in modulus by

$$\int_{\partial B_{r_0-t}} \left|\frac{\partial u}{\partial t}\right| \left(\sum_{j=1}^n \left|\frac{\partial u}{\partial x_j}\right|^2 + \left|\frac{\partial u}{\partial y_j}\right|^2\right)^{\frac{1}{2}} d\sigma(z)$$

which in turn is bounded by

$$\frac{1}{2}\int_{\partial B_{r_0-t}} \left\{ \left|\frac{\partial u}{\partial t}\right|^2 + \sum_{j=1}^n \left(\left|\frac{\partial u}{\partial x_j}\right|^2 + \left|\frac{\partial u}{\partial y_j}\right|^2\right) \right\} dz = \frac{1}{2}\int_{\partial B_{r_0-t}} G(z,t)d\sigma(z).$$

Hence it follows from the above expression for $E'(t)$ that $E'(t) \leq 0$. But clearly, $E \geq 0$ and $E(0) = 0$ because the Cauchy data vanish. Hence $E(t) = 0$, and so $u = 0$ on Ω. $\square$

LEMMA 2.3. *Let $u(z,t)$ be a radial solution to the wave equation (2.1). If $u(z,t) = 0$ for $|z| \leq r \leq R$ and $0 \leq t \leq R - r$, then u vanishes in the region $\{(z,t) : |z| \leq r, |z| + t < R\}$.*

PROOF. Pick $s \leq R - r$. Then the function $v(z,t) = u(z, t+s)$ solves the wave equation with the Cauchy data $v(z,0) = 0$, $v_t(z,0) = 0$ for $|z| \leq r$. Hence the proof follows from Lemma 2.2. $\square$

LEMMA 2.4. *Let $u(z,t)$ be a radial solution to the wave equation (2.1). Suppose also that $u = u_t = 0$ for $|z| \leq r$, $t = 0$, and $u(z,t) = 0$ for $|z| = r$, $0 \leq t \leq R-r$. Then u vanishes in the solid cylinder $\Omega = \{(z,t) : |z| \leq r, 0 \leq t < R - r\}$.*

PROOF. As earlier, we may assume that u is real valued. Let $s < R - r$ and
$$\Omega_s = \{(z,t) : |z| \leq r, 0 \leq t \leq s\}.$$

We have
$$\left(u_t^2 + |\nabla u|^2 + \frac{1}{4}|z|^2 u^2\right)_t - 2\nabla.(u_t \nabla u) = 2u_t\left(u_{tt} - \Delta u + \frac{1}{4}|z|^2\right) = 0 \text{ in } \Omega_s.$$

Hence, by Green's theorem,
$$\int_{\partial \Omega_s} \left[\left(u_t^2 + |\nabla u|^2 + \frac{1}{4}|z|^2 u^2\right) \nu_t - 2u_t \langle \nabla u, \nu_z \rangle \right] dA(z,t) = 0,$$
where $\nu = (\nu_z, \nu_t)$ is the unit outward normal vector and $dA(z,t)$ is surface area measure on $\partial \Omega_s$.

Now $\partial \Omega_s = C_1 \cup C_2 \cup C_3$, where $C_1 = \{(z,0) : |z| \leq r\}$, $C_2 = \{(z,s) : |z| \leq r\}$ and $C_3 = \{(z,t) : |z| = r, 0 \leq t \leq s\}$. We calculate each integral separately. The integral over C_1 vanishes, as $u(z,0) = 0$, $u_t(z,0) = 0$ and $\nu_z = 0$ since $\nu = (0,-1)$. The integral over C_3 also vanishes, as $u_t = 0$ there (since $u(z,t) = 0$ for $|z| = r$ and $0 \leq t \leq R-r$) and $\nu_t = 0$ since $\nu = (\nu_z, 0)$. What remains is the integral over C_2. On C_2 the normal vector is $\nu = (0,1)$; hence
$$\int_{C_2} \left(u_t^2 + |\nabla u|^2 + \frac{1}{4}|z|^2 u^2\right) dA(z,t) = 0,$$
which implies that $u(z,s) = 0$ for $|z| \leq r$ as the integrand is nonnegative. This finishes the proof. $\square$

We also need to use the following result due to Baouendi and Zachmanoglou [**2**].

THEOREM 2.5. *Let $D_R = \{(x,t) \in \mathbb{R}^{n+1}; |x| + |t| < R\}$. Let $u \in C^2(D_R)$ be a radial function (in x) such that*
$$|\Delta_x u - \partial_t^2 u| \leq C \left(|x|^{-1} |\nabla u| + |x|^{-2} |u|\right) \quad \text{in} \quad D_R.$$
If u and ∇u vanish to infinite order on the line segment $\{x = 0, |t| < R\}$, then u vanishes identically in D_R.

REMARK. The above result from [**2**] is true for nonradial functions u as well under an additional condition. We refer the reader to [**2**, p. 7] for further details.

PROOF OF THEOREM 2.1. We are now in a position to prove Theorem 2.1. Let u be the solution to the Cauchy problem,
$$(2.3) \qquad u_{tt} = \left(\Delta - \frac{1}{4}|z|^2\right) u, \quad u(z,0) = 0, \quad u_t(z,0) = f(z).$$

Since f is radial and the operator $\Delta - \frac{1}{4}|z|^2$ commutes with rotations, by uniqueness of the solution, u is radial too. Now that $(f \times \mu_r)(z) = 0$ for $|z| \leq R-r$. We claim that $u(z,t) = 0$ for $|z| = r$ and $0 \leq t \leq R-r$.

Consider the equation
$$(2.4) \qquad v_{tt} = \left(\Delta - \frac{1}{4}|z|^2\right) v, \quad v(z,0) = 0, \quad v_t(z,0) = (f \times \mu_r)(z).$$
Then
$$v(z,t) = (u(.,t) \times \mu_r)(z)$$
is easily seen to be a radial solution to (2.4). Since the Cauchy data for (2.4) vanish when $|z| \leq R-r$, it follows from Lemma 2.2 that $v(z,t) = 0$ for $|z| + t \leq R-r$. But
$$v(0,t) = (u(.,t) \times \mu_r)(0) = \int_{|y|=r} u(-y,t) d\mu_r(y) = u(z,t), \quad |z| = r,$$
which finishes the proof of the claim.

Combining the above with the fact that $u_t(z,0) = f(z) = 0$ for $|z| \leq r$ and Lemma 2.4 we see that $u(z,t)$ is zero in the solid cylinder

$$\{(z,t) : |z| \leq r, \quad 0 \leq t \leq R - r\}.$$

Applying Lemma 2.3 shows that $u(z,t) = 0$ for $|z| \leq r$ and $|z| + t \leq R$. Extending the solution for negative time t as an odd function, we see that for any $s < R$ there is a neighborhood of origin in $\mathbb{C}^n$, N_s, such that $u(z,t)$ is zero for $(z,t) \in N_s \times (-s, s)$. By Theorem 2.5, it follows that $u(z,t) = 0$ in the domain $D_R = \{|z| + t \leq R\}$, which in turn implies that $f(z) = 0$ for $z \in B_R$ since $f(z) = u_t(z,0)$. This completes the proof of Theorem 2.1 □

REMARK. Alternatively, we may use Holmgren's uniqueness theorem [6] to conclude that $u(z,t)$ is zero in a neighborhood of the cylinder and deduce that the function f vanishes in a slightly larger ball than B_r. Then we may apply a general result due to Quinto [7] to conclude that f is zero everywhere. Quinto's result holds true also for integrals over geodesic spheres with a real analytic weight depending analytically on the center.

Before we state the two radii theorem for radial functions, a few remarks are in order. Restriction of the special Hermite operator $\mathcal{L}$ to the space $L^2(B_r)$ is self adjoint with Dirichlet boundary conditions. Hence there exists an orthonormal basis $\{\phi_l^r\}_{l=0}^\infty$ of eigenfunctions satisfying $\mathcal{L}\phi_l^r = \lambda_l\, \phi_l^r$ and $\phi_l^r(x) = 0$ for $|x| = r$. As $\mathcal{L}$ is a positive unbounded operator, the eigenvalues satisfy

(2.5) $$0 < \lambda_1 \leq \lambda_2 \leq \cdots\cdots \to \infty.$$

By Weyl's asymptotic formula for the distribution of eigenvalues (see [8]),

(2.6) $$\lambda_l \geq C l^b \quad \text{for some positive } b.$$

Let $\{\phi_{k_s}^r\}_{s=0}^\infty$ be the radial eigenfunctions in the above orthonormal basis with eigenvalues $\{\lambda_{k_s}\}_{s=0}^\infty$. Note that $k_s \geq s$. Hence from (2.6) we have

(2.7) $$\lambda_{k_s} \geq C\, k_s^b \geq C\, s^b$$

for some positive b.

The functions $\phi_{k_s}^{r_1}$ are of the form

(2.8) $$w_s \,{}_1F_1\left(a_{k_s}, n, \frac{1}{2}|z|^2\right) e^{-\frac{1}{4}|z|^2}$$

(see (1.5)), where $a_{k_s} = \frac{n - \lambda_{k_s}}{2}$ and w_s is the reciprocal of the $L^2(B_r)$-norm of the functions ${}_1F_1(a_{k_s}, n, \frac{1}{2}|z|^2)\, e^{-\frac{1}{4}|z|^2}$.

We are now ready to state the main result of this section.

THEOREM 2.6. *Let r_1, r_2 and R be positive reals such that $r_1 + r_2 < R$. Let $f \in L_{loc}(B_R)$ be radial. Suppose that*

(2.9) $$(f \times \mu_{r_k})(z) = 0 \quad \text{for} \quad |z| < R - r_k, \quad k = 1, 2.$$

If the equation

(2.10) $${}_1F_1\left(a, n, \frac{r_1^2}{2}\right) = {}_1F_1\left(a, n, \frac{r_2^2}{2}\right) = 0$$

has no solution $a \in \mathbb{R}$, then f is the zero function. On the other hand, any solution of (2.10) gives rise to a nontrivial solution of (2.9).

PROOF. Since $r_1 + r_2 < R$, convolving with smooth radial functions supported in small neighborhoods of the origin, we may assume that f itself is smooth. Note that $f(z) = 0$ for $|z| = r_1$. Hence f may be expanded in terms of the radial eigenfunctions of $\mathcal{L}$ which vanish on the sphere $|z| = r_1$,

$$(2.11) \quad f(z) = \sum_{s=0}^{\infty} c_s\, \phi_{k_s}^{r_1}(z) = \sum_{s=0}^{\infty} c_s \omega_s\, {}_1F_1\left(a_{k_s}, n, \frac{1}{2}|z|^2\right) e^{-\frac{1}{4}|z|^2}, \quad |z| \leq r_1.$$

Applying the operator $\mathcal{L}$ repeatedly to (2.11), we get

$$(2.12) \quad \sum_{s=0}^{\infty} |c_s|^2 \lambda_{k_s}^{2N} < \infty \quad \text{for all } N.$$

Using the asymptotics (1.7) to estimate the L^2-norm of ${}_1F_1(a_{k_s}, n, \frac{1}{2}|z|^2)\, e^{-\frac{1}{4}|z|^2}$ from below, we obtain

$$(2.13) \quad w_s \leq C \lambda_{k_s}^b$$

for some real b.

Now define

$$(2.14) \quad \tilde{f}(z) = \sum_{s=0}^{\infty} c_s \phi_{k_s}^{r_1}(z)$$

for all $z \in B_R$. From (1.6) and (2.12), we see that $|\phi_{k_s}^{r_1}(z)| \leq \lambda_{k_s}^{\beta}$ for some $\beta \in \mathbb{R}$ for $r_1 \leq |z| \leq R$. Hence the series (2.14) converges absolutely in B_R and equals $f(z)$ for $|z| \leq r_1$. Repeating the same arguments for the series

$$\sum_{s=0}^{\infty} c_s\, \lambda_{k_s}^N\, \phi_{k_s}^{r_1}(z),$$

we see that $\mathcal{L}^N \tilde{f}$, initially defined in the weak sense, belongs to the space $L^2(B_R)$ for all $N > 0$. Since $\mathcal{L}$ is an elliptic operator, it follows from the elliptic regularity theorem (see [**5**, Theorem 6.33]) that $\tilde{f}$ is a C^∞-function.

Define $g(z) = f(z) - \tilde{f}(z)$ for $z \in B_R$. Then g is a smooth radial function in B_R satisfying

$$(g \times \mu_{r_1})(z) = 0$$

for $|z| \leq R - r_1$. Since g vanishes in the ball B_{r_1}, by Theorem 2.1 g is identically equal to zero in B_R. This proves that

$$f(z) = \sum_{s=0}^{\infty} c_s \phi_{k_s}^{r_1}(z)$$

for all $z \in B_R$.

Now we make use of the second convolution equation. The condition $f \times \mu_{r_2} = 0$ implies

$$(2.15) \quad \sum_{s=0}^{\infty} c_s\, \frac{\phi_{k_s}^{r_1}(r_2)}{\phi_{k_s}^{r_1}(0)}\, \phi_{k_s}^{r_1}(z) = 0$$

for $|z| \leq R - r_2$, and in particular for $z \in B_{r_1}$, as $r_1 + r_2 < R$. Hence, by the uniqueness of the expansion in terms of $\phi_{k_s}^{r_1}$, we have

$$c_s \phi_{k_s}^{r_1}(r_2) = 0 \quad \text{for all } s,$$

which in turn implies that $c_s = 0$ for all s.

Conversely, if $_1F_1(a, n, \frac{r_1^2}{2}) = {}_1F_1(a, n, \frac{r_2^2}{2}) = 0$ for some $a \in \mathbb{R}$, then the function $f(z) = {}_1F_1(a, n, \frac{|z|^2}{2}) \, e^{-\frac{1}{4}|z|^2}$ satisfies

$$(f \times \mu_{r_1})(z) = (f \times \mu_{r_2})(z) = 0 \quad \text{for all} \quad z.$$

This completes the proof of Theorem 2.6. $\square$

3. A general two radii theorem

In this section, we use the result established in the previous section to deduce a general two radii theorem.

THEOREM 3.1. *Let r_1, r_2 and R be positive reals such that $r_1 + r_2 < R$. Let $f \in L_{loc}(B_R)$ and suppose that*

(3.1) $$(f \times \mu_{r_k})(z) = 0 \quad for \quad |z| < R - r_k, \quad k = 1, 2.$$

If the equation

(3.2) $$_1F_1\left(a, n, \frac{r_1^2}{2}\right) = {}_1F_1\left(a, n, \frac{r_2^2}{2}\right) = 0$$

has no solution $a \in \mathbb{R}$, then f is the zero function. On the other hand, any solution $a \in \mathbb{R}$ of (3.2) gives rise to a nontrivial solution of (3.1).

Before we embark on the proof, observe that

$$(f \times \mu_r)(z) = \overline{(\mu_r \times \bar{f})(z)}.$$

Hence, replacing $\bar{f}$ by f, we may (and do) assume that $(\mu_{r_k} \times f)(z) = 0$ for $k = 1, 2$. Notice also that for radial functions f, we have $f \times \mu_r = \mu_r \times f$.

As in [12], we prove the above theorem by reducing to the radial case and using the results of the previous section. For this, we need an expansion of the function which is well-adapted to the representation of the unitary group $U(n)$. We briefly recall the necessary details.

For each pair of non-negative integers p and q, let $\mathcal{P}_{pq}$ be the space of all polynomials P in z and $\bar{z}$ of the form

$$P(z) = \sum_{|\alpha|=p} \sum_{|\beta|=q} c_{\alpha\beta} z^\alpha \bar{z}^\beta.$$

Let $\mathcal{H}_{pq} = \{P \in \mathcal{P}_{pq} : \Delta P = 0\}$, where Δ is the standard Laplacian on $\mathbb{C}^n$. Elements of $\mathcal{H}_{pq}$ are called bigraded solid harmonics on $\mathbb{C}^n$. It is known that $L^2(S^{2n-1})$ is the orthogonal sum of $\mathcal{H}_{pq}$'s.

For non-negative integers p and q, define

$$\alpha(p, q) = \int_{S^{2n-1}} |z_1|^{2p} |z_2|^{2q} \, d\sigma(z).$$

Let $\{S_{pq}^j\}_{j=1}^{d(p,q)}$ be an orthonormal basis for $\mathcal{H}_{pq}$ with

$$S_{pq}^1(z) = \frac{1}{\sqrt{\alpha(p,q)}} \frac{z_1^p \bar{z}_2^q}{|z|^{p+q}}.$$

Here $d(p, q)$ is the dimension of $\mathcal{H}_{pq}$.

It is well-known that the natural action of the unitary group $U(n)$ on the spaces $\mathcal{H}_{pq}$ is irreducible. Hence, if $\tau \in U(n)$,

$$(3.3) \qquad S_{pq}^i(\tau^{-1}z) = \sum_{j=0}^{d(p,q)} t_{ij}^{p,q}(\tau) \, S_{pq}^j(z),$$

where $\tau \to t_{ij}^{p,q}(\tau)$ are the matrix entries of the above representation. We associate with each function $f \in L_{loc}(B_R)$ the Fourier series

$$(3.4) \qquad f(z) = \sum_{p,q} \sum_{j=1}^{d(p,q)} f_{pq}^j(|z|) \, S_{pq}^j(z'),$$

where $z' = \frac{z}{|z|}$. Using (3.3) and the orthogonality of the matrix entries, we have

$$(3.5) \qquad f_{pq}^j(|z|) \, S_{pq}^i(z') = a_{pq} \int_{U(n)} f(\tau^{-1}z) t_{ij}^{p,q}(\tau) \, d\tau$$

for $1 \leq i,\, j \leq d(p,q)$.

As in the previous section, we may assume that f is smooth without loss of generality. Define the twisted translation of a function f by

$$(3.6) \qquad \tau_w f(z) = f(z - w) \, e^{\frac{i}{2} Im\, z \cdot \bar{w}}.$$

Then it is easy to see that

$$\tau_w(\mu_r \times f) = \mu_r \times (\tau_w f).$$

For each $|w| \leq \frac{R - r_1 - r_2}{2}$, define the radial function

$$f_w(z) = \int_{U(n)} (\tau_w f)(\tau z) d\tau.$$

Since the measure μ_r and the symplectic form $z \cdot \bar{w}$ are invariant under $U(n)$, we have

$$(\mu_{r_k} \times f_w)(z) = 0 \quad \text{for} \quad k = 1, 2,$$

in B_{R_1}, where $R > R_1 > r_1 + r_2$. Hence, by Theorem 2.6, $f_w(z) = 0$ for all z. Evaluating at $z = 0$, we obtain $f(z) = 0$ for z in a neighborhood of the origin, which implies that the spherical harmonic coefficients of f, namely the f_{pq}^j, also vanish in the same neighborhood.

Our aim is to show that each of these coefficients as functions of one variable satisfy an ordinary differential equation with smooth coefficients. Since they all vanish in an open interval, they have to vanish everywhere by the uniqueness for solutions to ODE, which finishes the proof.

For $s > 0$, let us denote by $V_s(B_R)$ the class of smooth functions satisfying $(\mu_s \times f)(z) = 0$ for $|z| \leq R - s$.

LEMMA 3.2. *Suppose that* $f \in V_s(B_R)$. *Then*

$$f_{pq}^j(|z|) \, Y_{pq}(z') \in V_s(B_R)$$

for any Y_{pq} *in* $\mathcal{H}_{pq}$.

PROOF. From (3.5), we have

$$f_{pq}^j(|z|) \, S_{pq}^i(z') \in V_s(B_R).$$

Since i is arbitrary, the lemma follows. □

If Z_j and $\bar{Z}_j$ are the operators defined in the introduction, we have (see [9])
$$Z_j(\mu_s \times f) = \mu_s \times (Z_j f) \quad \bar{Z}_j(\mu_s \times f) = \mu_s \times (\bar{Z}_j f).$$
So $Z_j f$ and $\bar{Z}_j f$ belong to the class $V_s(B_R)$ provided f belongs to the same class.

LEMMA 3.3. *a) Let $f(|z|) Y(z') \in V_s(B_R)$ for some nonzero $Y \in \mathcal{H}_{p0}$. Then*
$$\left[\frac{\alpha(p,0)}{\sqrt{\alpha(p-1,0)}} \left(f'(r) - \frac{pf(r)}{r} + \frac{r^2 f(r)}{2} \right) + \frac{pf(r)}{r} \right] S^1_{p-1\,0}(z') \in V_s(B_R).$$

b) Let $f(|z|) Y(z') \in V_s(B_R)$ for some nonzero $Y \in \mathcal{H}_{pq}$, $p \geq 0$ and $q > 0$. Then
$$\left[\frac{\alpha(p,q)}{\sqrt{\alpha(p,q-1)}} \left(f'(r) - \frac{(p+q)}{r} f(r) - \frac{r^2 f(r)}{2} \right) + \frac{qf(r)}{r} \right] S^1_{p\,q-1}(z') \in V_s(B_R).$$

PROOF. From Lemma 3.2, we have
$$f(|z|) \frac{z_1^p}{|z|^p} \in V_s(B_R).$$
Applying the differential operator Z_1, we get
$$(3.7) \quad \left(f'(r) - \frac{pf(r)}{r} + \frac{f(r)}{2r^2} \right) \frac{z_1^{p-1} |z_1|^2}{|z|^{p-1}} + \frac{pf(r)}{r} \frac{z_1^{p-1}}{|z|^{p-1}} \in V_s(B_R).$$
Now the function $\dfrac{z_1^{p-1} |z_1|^2}{|z|^p}$ can be expanded in terms of bi-graded spherical harmonics on the sphere. Observe that the projection of the above function to the span of $S^1_{p-1\,0}(z) = \dfrac{1}{\sqrt{\alpha(p,0)}} \dfrac{z_1^{p-1}}{|z|^{p-1}}$ is nonzero since
$$\int_{S^{2n-1}} z_1^{p-1} |z_1|^2 \bar{z}_1^{p-1} \, d\sigma(z) = \alpha(p,0) > 0.$$
So the expression in (3.5) may be written as
$$\left[\frac{\alpha(p,0)}{\sqrt{\alpha(p-1,0)}} \left(f'(r) - \frac{pf(r)}{r} + \frac{r^2 f(r)}{2} \right) + \frac{pf(r)}{r} \right] S^1_{p-1\,0}(z')$$
$$+ \quad \text{orthogonal terms}.$$
Hence (a) is proved by Lemma 3.2.

To prove assertion (b), we proceed as follows. Again from Lemma 3.2, we have
$$\frac{f(|z|)}{|z|^{p+q}} z_1^p \bar{z}_2^q \in V_s(B_R).$$
Applying the operator $\bar{Z}_2$ to the above expression, we obtain
$$(3.8) \quad \left(f'(r) - \frac{(p+q)f(r)}{r} - r^2 f(r) \right) \frac{z_1^p \bar{z}_2^{q-1} |z_2|^2}{|z|^{p+q+1}} + \frac{qf(r)}{r} \frac{z_1^p \bar{z}_2^{q-1}}{|z|^{p+q-1}} \in V_s(B_R).$$
Now, the projection of $\dfrac{z_1^p \bar{z}_2^{q-1} |z_2|^2}{|z|^{p+q+1}}$ to the span of $\dfrac{1}{\sqrt{\alpha(p,q-1)}} \dfrac{z_1^p \bar{z}_2^{q-1}}{|z|^{p+q-1}}$ is given by $\dfrac{\alpha(p,q)}{\sqrt{\alpha(p,q-1)}} \dfrac{z_1^p \bar{z}_2^{q-1}}{|z|^{p+q-1}}$. Proceeding as in the previous case (a), we finish the proof. $\square$

We are now in a position to complete the proof of Theorem 3.1. By Lemma 3.2, we have that

(3.9) $$f^j_{pq}(|z|)\, Y_{pq}(z') \in V_{r_k}(B_r) \quad k=1,2$$

for any $Y_{pq} \in \mathcal{H}_{pq}$ and for all p, q and $1 \leq i$, $j \leq d(p,q)$. To show that f^j_{pq} vanishes we proceed by induction. The result for $(p,q) = (0,0)$ follows from Theorem 2.6. When $(p,q) = (1,0)$ we apply Lemma 3.3(a) and obtain that the radial function

(3.10) $$\alpha(1,0) \left[(f^j_{1\,0})'(r) - \frac{f^j_{1\,0}(r)}{r} + \frac{r^2\, f^j_{1\,0}(r)}{2} \right] + \frac{f^j_{1\,0}(r)}{r} \in V_{r_k}(B_R)\ k=1,2,$$

so the above expression vanishes for $0 \leq r \leq R$ by Theorem 2.6. As $f^j_{1\,0}(r) = 0$ for $r \in (0,\epsilon)$ for some fixed $\epsilon > 0$, it follows from the uniqueness for solutions of ordinary differential equations that $f^j_{1\,0}(r)$ vanishes for all $r \in [0,R]$. By induction and repeated application of Lemma 3.3(a), we can prove that $f^j_{p\,0}(r) = 0$ for $r \in [0, R]$ and for all p and j. Now fix $p > 0$. Using Lemma 3.3(b) and induction in q along with the result for the case $(p, 0)$, we can show that f^j_{pq} vanish too, which finishes the proof. □

4. Two radii theorem on the Heisenberg group

In this section, we deduce a two radii theorem on the Heisenberg group from the results in the previous section. Recall that $H^n = \mathbb{C}^n \times \mathbb{R}$ with the group law

$$(z,t)(w,s) = \left(z + w, t + s + \frac{1}{2} Im\, z \cdot \bar{w} \right).$$

If μ_r is the normalized surface measure on the sphere $\{(z,0): |z| = r\} \subset \mathbb{C}^n$, the spherical means of a function f on H^n are defined by

(4.1) $$(f * \mu_r)(z,t) = \int_{|w|=r} f\left(z - w, t - \frac{1}{2} Im\, z \cdot \bar{w} \right) d\mu_r(w).$$

A two radii theorem for bounded continuous functions on H^n was established in [1] (see [10] for a one radius theorem for L^p-functions). The following is a local version of this result.

THEOREM 4.1. *Let f be a continuous function defined in the cylinder $B_R \times \mathbb{R} \subset H^n$ which is of tempered growth in the t variable. Let r_1 and r_2 be positive numbers such that $r_1 + r_2 < R$. Suppose that*

(4.2) $$(f * \mu_{r_k})(z,t) = 0 \quad for \quad k = 1, 2.$$

Then f is the zero function if for no $a \in \mathbb{R}$ is the ratio $\frac{r_1}{r_2}$ a ratio of two solutions (in r) to the equation $_1F_1(a, n, \frac{r^2}{2}) = 0$. If $\frac{r_1}{r_2}$ is such a ratio, there is a nontrivial solution to (4.2).

We prove the above theorem by taking Fourier transform in the t variable and reducing the problem to a system of twisted convolutions (see [10]). For a nonzero real λ, define the λ-twisted convolution of two functions g and h defined on $\mathbb{C}^n$ by

$$(g *_\lambda h)(z) = \int_{\mathbb{C}^n} g(z-w)\, h(w)\, e^{\frac{i\lambda}{2} Im\, z \cdot \bar{w}}\, dw.$$

Observe that when $\lambda = 1$, the above is the twisted convolution considered in the previous sections. Of course, the above convolution can be extended to measures as well.

LEMMA 4.2. *Let R, r_1 and r_2 be positive reals such that $r_1 + r_2 < R$. Let λ be a nonzero real number and g a continuous function defined in B_R. Suppose that*

(4.3) $$(g *_\lambda \mu_{r_k})(z) = 0 \quad for \quad k = 1, 2.$$

Then g vanishes identically provided that there does not exist $a \in \mathbb{R}$ such that $_1F_1(a, n, \frac{|\lambda|r_1^2}{2}) = {}_1F_1(a, n, \frac{|\lambda|r_2^2}{2}) = 0$.

PROOF. Assume that λ is positive and define $g_\lambda(z) = g(z/\sqrt{\lambda})$. Then from (4.3),

(4.4) $$(g_\lambda \times \mu_{\sqrt{\lambda}r_k})(\sqrt{\lambda}z) = g *_\lambda \mu_r(z) = 0$$

for $k = 1, 2$ and for $\sqrt{\lambda}|z| \le \sqrt{\lambda}(R - r_k)$. Since $\sqrt{\lambda}r_1 + \sqrt{\lambda}r_2 < \sqrt{\lambda}R$, it follows from Theorem 3.1 that g_λ is zero. The case $\lambda < 0$ can be handled similarly. $\square$

PROOF OF THEOREM 4.1. Taking Fourier transform in the t-variable (see [**10**]) reduces the problem in hand to the equations

(4.5) $$(f^\lambda *_\lambda \mu_{r_k})(z) = 0$$

for $|z| < R$, $\lambda \ne 0$ and $k = 1, 2$, where f^λ is defined by

$$f^\lambda(z) = \int_\mathbb{R} f(z, t)\, e^{i\lambda t}\, dt.$$

By Lemma 4.2, f^λ is zero for all λ, thus f itself vanishes, which completes the "if" part of the proof.

In the opposite direction, if $\frac{r_1}{r_2} = \frac{\sqrt{\lambda}s_1}{\sqrt{\lambda}s_2}$ where

$$_1F_1\left(a, n, \frac{\lambda s_1^2}{2}\right) = {}_1F_1\left(a, n, \frac{\lambda s_2^2}{2}\right) = 0$$

for some $a \in \mathbb{R}$ and $\lambda > 0$, then the function

$$f(z, t) = e^{i\lambda t}\, _1F_1\left(a, n, \frac{\lambda|z|^2}{2}\right) e^{-\frac{1}{4}\lambda|z|^2}$$

satisfies

$$(f * \mu_{r_1})(z, t) = (f * \mu_{r_2})(z, t) = 0$$

for all $(z, t) \in H^n$. This completes the proof. $\square$

ACKNOWLEDGMENTS. The authors thank E.T. Quinto for very useful discussions on applications of his results in [**7**]. This work was done when the second author was visiting the Department of Mathematics and Statistics of Bar-Ilan University. He thanks the Department for financial support and hospitality during the stay.

References

[1] M. L. Agranovsky, C. Berenstein, D-C. Chang and D. Pascuas, *Injectivity of the Pompeiu transform in the Heisenberg group*, J. Anal. Math. **63** (1994), 131-173.
[2] M. S. Baouendi and E. C. Zachmanoglou, *Unique continuation of solutions of partial differential equations and inequalities from manifolds of any dimension*, Duke. Math. J. **45** (1978), 1-13.
[3] C. A. Berenstein and R. Gay, *A local version of the two circles theorem*, Israel J. Math. **55** (1986), 267-288.
[4] A. Erdelyi, W. Magnus, F. Oberhettinger and F. G. Tricomi, *Higher Transcendental Functions*, Vol. 1, McGraw-Hill, New York, 1953.

[5] G. B. Folland, *Introduction to Partial Differential Equations*, Princeton University Press, Princeton, NJ, 1995.

[6] L. Hörmander, *The Analysis of Linear Partial Differential Operators I*, Springer-Verlag, 1983.

[7] E. T Quinto, *Pompeiu transform on geodesic spheres in real analytic manifolds*, Israel J. Math. **84** (1993), 353-363.

[8] M. Reed and B. Simon *Methods of Modern Mathematical Physics, Vol. IV. Analysis of Operators*, Academic Press, 1978.

[9] S. Thangavelu, *Lectures on Hermite and Laguerre Expansions*, Princeton Univ. Press, Princeton, 1993.

[10] S. Thangavelu, *Spherical means and CR functions on the Heisenberg group*, J. Anal. Math. **63** (1994), 255-286.

[11] S. Thangavelu, *Harmonic Analysis on the Heisenberg Group*, Birkhäuser Boston, Boston, 1998.

[12] V. V. Volchkov, *A definitive version of the local two- radii theorem*, Sb. Math. **186** (1995), 783-802.

[13] V. V. Volchkov, *Two radii theorems on spaces of constant curvature*, Dokl. Akad. Nauk. **347** (1996), 300-302.

[14] V. V. Volchkov, *Injectivity sets for the Radon transform on spheres*, Izv. Ross. Akad. Nauk Ser. Mat. **63** (1999), No. 3, 63-76; translation in Izv. Math. 63(1999), 481-493.

[15] V. V. Volchkov, *A definitive version of the local two radius theorem on hyperbolic spaces*, Izv. Ross. Akad. Nauk Ser. Mat. **65** (2001) No. 2, 3-26; translation in Izv. Math. **65** (2001), 207-229.

[16] V. V. Volchkov, *A local two radius theorem on symmetric spaces*, Dokl. Akad. Nauk **381** (2001), 727-731.

[17] Vit. V. Volchkov, *Theorems on spherical means on complex hyperbolic spaces*, Dopov. Nats. Akad. Nauk Ukr. Mat. Prirodozn. Tehk. Nauki. 2000, No. 4, 7-10.

[18] Vit. V. Volchkov, *On functions with zero spherical means on complex hyperbolic spaces*, Mat. Zametki 68 (2000), 504-512; translation in Math. Notes **68** (2000), 436-443.

[19] Vit. V. Volchkov, *A definitive version of local two radius theorem on a quaternion hyperbolic space*, Dokl. Akad. Nauk. **384** (2002) 449-451.

[20] V. V. Volchkov, *Integral Geometry and Convolution Equations*, Kluwer Acad. Publ., Dordrecht-Boston-London, 2003.

[21] L. Zalcman, *Offbeat integral geometry*, Amer. Math. Monthly **87** (1980), 161-175.

DEPARTMENT OF MATHEMATICS, BAR-ILAN UNIVERSITY, RAMAT-GAN 52900, ISRAEL
E-mail address: `agranovs@macs.biu.ac.il`

DEPARTMENT OF MATHEMATICS, INDIAN INSTITUTE OF SCIENCE, BANGALORE - 560 012, INDIA
E-mail address: `naru@math.iisc.ernet.in`

A Multiplicator Problem and Characteristics of Growth of Entire Functions

Vladimir Azarin

Dedicated to Professor Lawrence Zalcman on his sixtieth birthday

ABSTRACT. It is known that using any *total* and *independent* family of asymptotic characteristics such as the indicator and lower indicator or asymptotics of the Fourier coefficients, one can determine whether an entire function of finite order is a function of completely regular growth. However, such characteristics do not determine whether a function has an H-multiplicator.

1. Introduction and main result

1.1. Let Φ be an entire function of order ρ and normal type; we write $\Phi \in A(\rho)$. The *Phragmén-Lindelöf indicator* is defined by

$$h(\Phi, \phi) := \limsup_{r \to \infty} r^{-\rho} \log |\Phi(re^{i\phi})|.$$

The function $h(\Phi, \phi)$ is 2π-periodic and ρ-trigonometrically convex. This means that on the unit circle $\mathbb{T}$ it satisfies the differential inequality

(1.1.1) $$h'' + \rho^2 h := \nu \geq 0,$$

in the sense of distributions.

Let $H(\phi)$ be a fixed ρ-trigonometrically convex function. The function $g \in A(\rho)$ is called an *H-multiplicator* of Φ if the Phragmén-Lindelöf indicator $h(g\Phi, \phi)$ of the product $g\Phi$ satisfies the condition

$$h(g\Phi, \phi) \leq H(\phi).$$

The property of having an H-multiplicator depends only on the asymptotic behavior of Φ. In certain problems in analysis, it is important to know whether the function Φ has an H-multiplicator (for details, see [5],[6]).

2000 *Mathematics Subject Classification*. Primary 30D20, 30D16.

1.2. A function $f \in A(\rho)$ is called a *function of completely regular growth* in the sense of Levin-Pfluger (CRG-function) if

$$|\lambda|^{-\rho} \log |f(\lambda)| - h(f, \arg \lambda) \to 0,$$

when $\lambda \to \infty$ outside some small set (a C_0–set) (for details, see [**9**], Ch.2,3; [**10**]).

To be a CRG-function is an asymptotic property of an entire function. This class of entire functions has been studied thoroughly and plays an important role in applications [**9**], Ch.2; [**10**].

1.3. There exist numerous asymptotic characteristics of the growth of entire functions. We recall some of them and their connections to the CRG-functions.

The lower indicator in the sense of A. Gol'dberg is defined by

$$(1.3.1) \qquad \underline{h}(f, \phi) := \sup_{E \in \mathcal{E}_0} \left\{ \liminf_{r \to \infty, r \notin E} \frac{\log |f(re^{i\phi}|}{r^\rho} \right\},$$

where $\mathcal{E}_0$ is a class of "small" sets $E \subset [0, \infty)$ satisfying the condition

$$\mathrm{mes}\{E \cap [0, R]\} = o(R)$$

as $R \to \infty$.

The sets E serve to exclude the influence of zeros of the function f. If the function has no zeros in some open angle containing the ray $\{\arg \lambda = \phi\}$, then the right side of (1.3.1) can be replaced by the usual lower limit.

The family $\mathcal{H} := \{h(f, \phi), \underline{h}(f, \phi) : \phi \in [0, 2\pi)\}$ is *total*, i.e., using it, one can determine whether a function $f \in A(\rho)$ is a CRG-function.

THEOREM A. (A. Gol'dberg) ([**8**], cf.[**7**]) *A function $f \in A(\rho)$ is a CRG-function if and only if*

$$(1.3.2) \qquad h(f, \phi) = \underline{h}(f, \phi), \ \forall \phi \in [0, 2\pi).$$

The family $\mathcal{H}$ is "minimal" in some sense. Namely, if (1.3.2) holds for $\phi \in [0, 2\pi) \setminus U$, where U is an open set, the function f need not be a CRG-function. Moreover, this family is also *independent*. Namely, for any open set U, there exists a function $f \in A(\rho)$ such that (1.3.2) fails exactly for $\phi \in U$ ([**2**], [**3**]).

Consider another family of asymptotic characteristics. Set

$$c_k(f, r) := \begin{cases} \frac{1}{2\pi} \int_0^{2\pi} \log |f(re^{i\phi})| \cos k\phi \, d\phi, & \text{for } k = 0, 1, 2, ..; \\ -\frac{1}{2\pi} \int_0^{2\pi} \log |f(re^{i\phi})| \sin k\phi \, d\phi, & \text{for } k = -1, -2, ..., \end{cases}$$

the Fourier coefficients of $\log |f|$.

The corresponding asymptotic characteristics are defined by

$$\bar{c}_k(f) := \limsup_{r \to \infty} \frac{c_k(f, r)}{r^\rho}, \quad \underline{c}_k(f) := \liminf_{r \to \infty} \frac{c_k(f, r)}{r^\rho}, \quad k \in \mathbb{Z}.$$

The family $\mathcal{F} := \{\underline{c}_k(\cdot), \bar{c}_k(\cdot) : k \in \mathbb{Z}\}$ is also total:

THEOREM B. (V. Azarin, A. Gol'dberg) ([1], cf. [9], p.54) *A function $f \in A(\rho)$ is a CRG-function if and only if*

$$\overline{c}_k(f) = \underline{c}_k(f) \tag{1.3.3}$$

for $k \in \mathbb{Z}$.

The family $\mathcal{F}$ is also "minimal" and independent. If (1.3.3) fails to hold for all $k \in \mathbb{Z}$, the function f need not be a CRG-function. Moreover, for any decomposition of $\mathbb{Z}$ into two subsets A and $\mathbb{Z} \setminus A$, there exists $f \in A(\rho)$ such that (1.3.3) holds for $k \in A$ and does not hold for $k \in \mathbb{Z} \setminus A$ (see [1]).

Let us consider one more family of characteristics which were exploited by A. Gol'dberg in the proof of Theorem B. Let $G \subset [0, 2\pi)$ be an open set. Set

$$I_G(r, f) := \int_G \log |f(re^{i\phi})| d\phi,$$

$$\overline{I}_G(f) := \limsup_{r \to \infty} \frac{I_G(f, r)}{r^\rho}, \qquad \underline{I}_G(f) := \liminf_{r \to \infty} \frac{I_G(f, r)}{r^\rho}, \qquad G \subset [0, 2\pi).$$

The family of characteristics $\mathcal{I} := \{\overline{I}_G, \underline{I}_G\}$ is total even if all these G's are only open intervals. It is also independent in the sense that for any G, there exists a function $f \in A(\rho)$ such that $\underline{I}_G(f) \neq \overline{I}_G(f)$ and $\underline{I}_{G'}(f) = \overline{I}_{G'}(f)$ for all $G' \subset [0, 2\pi) \setminus G$.

1.4. It is natural to study the connection between the property of a function Φ having a multiplicator and the asymptotic characteristics described above.

Perhaps suprisingly, it turns out that there is *no* such connection.

THEOREM 1.4.1. *For any smooth, strictly[1] ρ-trigonometrically convex function $H(\phi)$, there exist two functions $\Phi_1, \Phi_2 \in A(\rho)$, $\rho > 1/2$, such that*

$$\tau(\Phi_1) = \tau(\Phi_2), \ \tau \in \mathcal{H} \cup \mathcal{F} \cup \mathcal{I}, \tag{1.4.1}$$

and Φ_1 has an H-multiplicator while Φ_2 does not.

From the proof of Theorem 1.4.1, one can see (see Remark 2 in §3.1) that the set of characteristics for which (1.4.1) holds can be considerably enlarged without changing the construction. For example, the characteristics

$$\sigma(f) := \limsup_{r \to \infty} \frac{\log M(r, f)}{r^\rho}, \qquad \underline{\sigma}(f) := \liminf_{r \to \infty} \frac{\log M(r, f)}{r^\rho}, \tag{1.4.2}$$

where $M(r, f) := \max_\phi |f(re^{i\phi})|$ and

$$\overline{T}(f) := \limsup_{r \to \infty} \frac{T(r, f)}{r^\rho}, \qquad \underline{T}(f) := \liminf_{r \to \infty} \frac{T(r, f)}{r^\rho} \tag{1.4.3}$$

where $T(r, f) := \int_0^{2\pi} \log^+ |f(re^{i\phi}| d\phi$ (i.e., the Nevanlinna function) can be also included.

1.5. I thank Prof. A. Gol'dberg for very valuable discussions and suggestions. I also thank Prof. L. Zalcman for many important critical remarks.

---

[1]This means that ν from (1.1.1) is strictly positive as a function in L^1.

2. Auxiliary information

2.1. Recall the following facts on limit sets of entire functions, which will be required in the sequel (see [9], Ch.3).

Let $f \in A(\rho)$. Then the function $u(\lambda) := \ln|f(\lambda)|$ is subharmonic in $\mathbb{C}$. Consider the family of subharmonic functions

$$(2.1.1) \qquad u_t(\lambda) := u(t\lambda)t^{-\rho}, \ t > 0.$$

This family is precompact in the $\mathcal{D}'$-topology of L. Schwartz distributions in $\mathbb{C}$ when $t \to \infty$. That is to say, for any sequence $\{u_{t_j}, t_j \to \infty\}$ there exists a subsequence $\{u_{t_{j'}}, j' \to \infty\}$ and a function v subharmonic in $\mathbb{C}$ such that $u_{t_{j'}} \to v$ in the $\mathcal{D}'$-topology.

The set of all such v is called the *limit set* of f and is denoted by $\mathbf{Fr}f$. This is a connected compact set contained, for $\sigma \geq \sigma_f$, in

$$(2.1.2) \qquad U[\rho,\sigma] := \{v \text{ is subharmonic} : v(0) = 0; \ v(\lambda) \leq \sigma|\lambda|^\rho, \ \lambda \in \mathbb{C}\}.$$

$\mathbf{Fr}f$ is invariant with respect to transformations of the form (2.1.1), i.e., $v \in \mathbf{Fr}f$ if and only if $v_t \in \mathbf{Fr}f$.

The indicator and lower indicator can be expressed in terms of $\mathbf{Fr}f$:

$$(2.1.3) \qquad h(f,\varphi) = \sup\{v(e^{i\varphi}) : v \in \mathbf{Fr}f\};$$

$$\underline{h}(f,\varphi) = \inf\{v(e^{i\varphi}) : v \in \mathbf{Fr}f\}.$$

If $\underline{h}(f,\varphi) = h(f,\varphi)$ for all $e^{i\varphi} \in \mathbb{T}$, then $\mathbf{Fr}f$ consists of the single function $v_0 = h(f,\varphi)r^\rho$ and f is a CRG-function.

The characteristics $\overline{c}_k(\cdot)$ and $\underline{c}_k(\cdot)$ can be expressed in terms of $\mathbf{Fr}f$ in the following way.

Let us define

$$(2.1.4) \qquad c_k(v) := \begin{cases} \frac{1}{2\pi}\int_0^{2\pi} v(e^{i\phi})\cos k\phi \, d\phi, & \text{for } k = 0,1,2,\ldots; \\ -\frac{1}{2\pi}\int_0^{2\pi} v(e^{i\phi})\sin k\phi \, d\phi, & \text{for } k = -1,-2,\ldots. \end{cases}$$

Then ([9], Ch.3, Theorem 17)

$$(2.1.5) \qquad \overline{c}_k(f) = \sup\{c_k(v) : v \in \mathbf{Fr}f\}; \ \underline{c}_k(f) = \inf\{c_k(v) : v \in \mathbf{Fr}f\}, \ k \in \mathbb{Z}.$$

If $\underline{c}_k(f) = \overline{c}_k(f)$ for all $k \in \mathbb{Z}$, then $\mathbf{Fr}f$ consists of a single function v_0 and f is a CRG-function.

The characteristics of the family $\mathcal{I}$ can be expressed in the form

$$(2.1.6) \qquad \left.\begin{matrix}\overline{I}_G(f) \\ \underline{I}_G(f)\end{matrix}\right\} = \begin{cases} \sup \\ \inf \end{cases}\{I_G(v) : v \in \mathbf{Fr}f\}, \ I_G(v) := \int_G v(e^{i\phi})d\phi.$$

If $\overline{I}_G(f) = \underline{I}_G(f)$ for all $G \subset [0,2\pi)$, then $\mathbf{Fr}f$ consists of a single function v_0 and f is a CRG-function.

We express the existence of an H-multiplicator in terms of the limit set. For $\lambda = re^{i\phi}$, set

$$\hat{H}(\lambda) := H(\phi)r^\rho.$$

This is a subharmonic function belonging to $U[\rho,\sigma]$ (see (2.1.2)).

THEOREM C. ([5], Th.1) *A function $f \in A(\rho)$ has a multiplicator if and only if for any $v \in \mathbf{Fr}f$, the function*

(2.1.7) $$\hat{m}(\lambda, v) := \hat{H}(\lambda) - v(\lambda)$$

has a minorant $w(\cdot, v) \in U[\rho, \sigma]$.

As matter of fact, if the subharmonic minorant of m exists and vanishes at zero, then it belongs to $U[\rho, \sigma]$.

2.2. We now consider a special class of limit sets – the *periodic* limit sets (see [9], p.54).

Let $v \in U[\rho, \sigma]$ satisfy the condition

(2.2.1) $$v(\lambda e^P) = e^{\rho P} v(\lambda)$$

for some positive P and for all $\lambda \in \mathbb{C}$.

Then the operation
$$T_t v(\cdot) := e^{-\rho t} v(\cdot e^t)$$
satisfies the condition
$$T_{t+P} v = T_t v$$
for all $t \in (-\infty, \infty)$.

The set

(2.2.2) $$S[v] := \{T_t v : t \in (0, P)\} = \{v_t : t \in (1, e^P)\}$$

is a limit set for some $f \in A(\rho)$. It is a periodic limit set.

Let us note that $T_t \hat{H}(\lambda) = \hat{H}(\lambda)$ for all t and the operation T_t preserves subharmonicity like $(\cdot)_t$ from (2.1.1). Hence the function $\hat{m}(\cdot, T_t v) = \hat{H}(\cdot) - T_t v(\cdot)$ has a subharmonic minorant in $\mathbb{C}$ for all $t \in (-\infty, \infty)$ if and only if the same holds for $\hat{m}(\cdot, v) = \hat{H}(\cdot) - v(\cdot)$. Thus, we have the following corollary of Theorem C:

PROPOSITION 2.2.1. *Let $f \in A(\rho)$ have a periodic limit set $S[v]$. Then it has a multiplicator if and only if the function $\hat{m}(\cdot, v)$ has a subharmonic minorant.*

2.3. Here we give an exposition of some results from [4] and their corollaries. Let v be a subharmonic function that satisfies (2.2.1). Set

(2.3.1) $$q(z) := e^{-\rho x} v(e^z), \quad z = x + iy.$$

This function is 2π–periodic in y and P–periodic in x. Thus it can be considered as a function on a torus $\mathbb{T}_P^2$ obtained by identifying the opposite sides of the rectangle $\Pi := (0, P) \times (0, 2\pi)$.

Set
$$L_\rho := \frac{\partial^2}{\partial x^2} + \frac{\partial^2}{\partial y^2} + 2\rho \frac{\partial}{\partial x} + \rho^2 = \Delta + 2\rho \frac{\partial}{\partial x} + \rho^2,$$
where Δ is the Laplace operator.

The function q is a *subfunction with respect to L_ρ*. That is to say, it is upper semicontinuous and satisfies the inequality $L_\rho q := \nu > 0$ as a distribution over $\mathbb{T}_P^2$. For sufficiently smooth v, this follows from the equality

$$\Delta[e^{\rho x} q(z)] = e^{\rho x} L_\rho q(z)$$

and the subharmonicity of $v(e^z)$.

An L_ρ-subfunction $g(z), z \in \mathbb{T}_P^2$, which satisfies the inequality $g(z) \leq m(z)$ for $z \in \mathbb{T}_P^2$ is called an L_ρ-subminorant of m.

Proposition 2.2.1 can be reformulated in the following form.

PROPOSITION 2.3.1. *Let $f \in A(\rho)$ have a periodic limit set $S[v]$. Then it has a multiplicator if and only if the function*

(2.3.2) $$m(z, v) := H(y) - q(z)$$

has an L_ρ-subminorant, where q is defined by (2.3.1).

Let $M \subset \{z : 0 < \Im z < 2\pi\}$ be a P-periodic simply-connected curvilinear strip (i.e., $M \cap \{\Re z = x_0\}$ is an interval for all $x_0 \in \mathbb{R}$) with sufficiently smooth (for example, Lipschitz) boundary. We can consider it as a domain on the torus $\mathbb{T}_P^2$.

Let us consider the *spectral problem* for the *operator pencil* (family of operators):

(2.3.3) $$L_r q := (\Delta + 2r\frac{\partial}{\partial x} + r^2)q(z) = 0, \ z \in M \subset \mathbb{T}_P^2;$$

(2.3.4) $$q\,|_{\partial M} = 0.$$

The spectrum Spec M of this pencil is nonempty and consists of complex numbers r_k having no finite accumulation point. If $r_k \in \text{Spec}\,M$, then $r_k + \frac{2\pi i}{P}m \in \text{Spec}\,M$ for all $m \in \mathbb{Z}$, i.e., points of Spec M form a lattice on vertical lines. The imaginary axis $\{\Re z = 0\}$ is not one of these lines. Take the vertical line from the half-plane $\{\Re z > 0\}$ containing points of Spec M nearest to the imaginary axis. It is known (see [**4**], Prop.1.37) that the intersection of this vertical line with the real axis belongs to Spec M. We denote this point by $\rho(D)$; it is the *positive point of the spectrum that is nearest to zero*.

Note that if $|r| < \rho(M)$, the Dirichlet problem (2.3.3)-(2.3.4) has only the trivial solution.

Let us note that the solution of the problem (2.3.3)-(2.3.4) that corresponds to $r = \rho(M)$ preserves its sign in M and does not vanish in M. We also exploit the following two properties of $\rho(M)$: the *strict monotonicity* in M and *continuity from the left*.

THEOREM E. ([**4**], Theorem 0.17) *Let $M_1, M_2 \subset \mathbb{T}_P^2$ be domains such that $M_1 \subset M_2$ and $\text{cap}(M_2 \setminus M_1) \neq 0$ where cap means logarithmic capacity. Then $\rho(M_1) > \rho(M_2)$.*

THEOREM F. ([**4**], Prop.6.6) *If $M_n \uparrow M$, then $\rho(M_n) \downarrow \rho(M)$.*

A corollary of these properties is

COROLLARY G. ([**4**], Lemma 9.17) *If $\rho(M) < \rho$, there exists $M_1 \subset\subset M$ with arbitrarily smooth boundary such that $\rho(M_1) = \rho$*

Let $m(z), z \in \mathbb{T}_P^2$, be a continuous function and suppose that the open set

$$E^+(m) := \{z \in \mathbb{T}_P^2 : m(z) > 0\}$$

has a finite number of connected components C_j which are images in the torus $\mathbb{T}_P^2$ of the P-periodic, simply-connected curvilinear strips with sufficiently smooth boundary mentioned above.

Let
$$\rho(E^+(m)) := \min_j \rho(C_j),$$
where the minimum is taken over all the components of $E^+(m)$ and is positive.

THEOREM H. ([4], Theorem 0.19)[2] *In order that $m(z)$ have a nonzero L_ρ-subminorant, it is necessary that*
$$\rho(E^+(m)) \leq \rho.$$

Thus, if this condition fails and $m(z_0) < 0$ at some point $z_0 \in \mathbb{T}_P^2$, then $m(z)$ does not have any L_ρ-subminorant.

Recall that every quadrilateral $ABCD$ can be conformally mapped to a rectangle with vertices $0, 1, 1+il, il$ such that $A \mapsto 0$, $B \mapsto 1$, $C \mapsto 1+il$, $D \mapsto il$. Then l is called the *conformal modulus* of $ABCD$ with *marked sides* AB and CD ($l := \mathrm{mod}(ABCD)(AB, CD)$).

We are going to use the following assertion, which connects $\rho(M)$ and the conformal modulus of the quadrilateral $M(I_0, I_P) := M \cap \{z : 0 < \Re z < P\}$ with marked sides $I_0 = M \cap \{\Re z = 0\}$ and $I_P = M \cap \{\Re z = P\}$.

THEOREM I. ([4], Corollary 4.9) *If M is symmetric with respect to I_0, then*
$$\rho(M) = (\pi/P)\,\mathrm{mod}\, M(I_0, I_P).$$

For an entire function f with a periodic limit set (2.2.2) and $q(z)$ defined by (2.3.1), the equalities (2.1.3), (2.1.5) and (2.1.6) have the following form:

(2.3.5) $\quad \overline{h}(f, \phi) = \sup\limits_{0 \leq x \leq P} q(x + i\phi); \ \underline{h}(f, \phi) = \inf\limits_{0 \leq x \leq P} q(x + i\phi), \ e^{i\phi} \in \mathbb{T};$

(2.3.6) $\quad \overline{c}_k(f) = \sup\limits_{0 \leq x \leq P} c_k(q(x + i\cdot)); \ \underline{c}_k(f) = \inf\limits_{0 \leq x \leq P} c_k(q(x + i\cdot)); \ k \in \mathbb{Z};$

(2.3.7) $\quad \left.\begin{array}{r}\overline{I}_G(f)\\ \underline{I}_G(f)\end{array}\right\} = \left\{\begin{array}{l}\sup\limits_{0 \leq x \leq P}\\ \inf\limits_{0 \leq x \leq P}\end{array}\right. I_G(q(x + i\cdot)),$

where "·" marks the variable on which the corresponding functional is applied.

3. Proof of Theorem 1.4.1

3.1. To begin with, we consider the case $1/2 < \rho \leq 1$. We consider domains in the torus obtained from periodic strips as described in §2.

LEMMA 3.1.1. *For any domain $D \subset \mathbb{T}_P^2$ with Lipschitz boundary such that $\rho(D) < \rho$ ($1/2 < \rho \leq 1$), there exists an L_ρ-subfunction $Y(z)$ such that*
$$\{z : Y(z) < 0\} \supset \mathbb{T}_P^2 \setminus \overline{D}.$$

This lemma is only needed for proving the following lemma.

Denote by $\tau[\cdot] \in C^*(\mathbb{T})$ any linear functional on the space $C(\mathbb{T})$ of continuous functions on the unit circle.

---

[2] In [4], this theorem is proved in more general form.

LEMMA 3.1.2. *There exists an open set $D \subset \mathbb{T}_P^2$ and twice continuously differentiable functions $r_1(z)$, $r_2(z)$, $z \in \mathbb{T}_P^2$, such that*
  (1) *for some L_ρ-subfunction Y, which is positive in D and negative outside $\overline{D}$, $r_1(z) > Y(z)$ for $z \in \mathbb{T}_P^2$ and $\rho(E^+(r_2)) > \rho$, while $r_2(z_0) < 0$ for some $z_0 \in \mathbb{T}_P^2 \setminus \overline{D}$;*
  (2) *for all $\tau \in C^*(\mathbb{T})$,*
$$\sup_{0 \leq x \leq P} \tau[r_1(x + i\cdot)] = \sup_{0 \leq x \leq P} \tau[r_2(x + i\cdot)]$$

where "·" indicates the variable on which the functional acts.

For example, if $\tau[f] = f(y_0)$, then $\tau[r(x + i\cdot)] = r(x + iy_0)$. If

$$\tau[f] = \int_\alpha^\beta f(y) dy,$$

then

$$\tau[r(x + i\cdot)] = \int_\alpha^\beta r(x + iy) dy.$$

REMARK 1. From the proof of Lemma 3.1.2 for $1/2 < \rho \leq 1$, we will see that $E^+(r_2)$ is the image on the $\mathbb{T}_P^2$ of a curvilinear periodic strip, contained in the strip $S_{2\pi} = \{z : 0 < y < 2\pi\}$.

Now we prove Theorem 1.4.1 for $1/2 < \rho \leq 1$ using these Lemmas, which will be proved in the sequel.

Let $H(y) \in C^2(\mathbb{T})$ be a strictly ρ-trigonometrically convex function. Set

(3.1.1) $$q_1(z) = H(y) - \epsilon r_1(z),$$

where ϵ is taken so small that $L_\rho q_1(z) > 0$ for $z \in \mathbb{T}_P^2$. This is possible because $L_\rho H = H'' + \rho^2 H > 0$ and $\mathbb{T}_P^2$ is compact.

Then $v_1(\lambda) := q_1(\log \lambda)|\lambda|^\rho \in U(\rho, \sigma)$ for some $\sigma > 0$ and satisfies the condition (2.2.1). Hence the set $S[v_1]$ from (2.2.2) is a limit set for some entire function Φ_1. Since $m(z, v_1) = H(y) - q_1(z) = \epsilon r_1(z) > \epsilon Y(z)$ by Lemma 3.1.2, Proposition 2.3.1 shows that Φ_1 has a multiplicator.

Using $r_2(z)$, we define in the same way $q_2(z)$ and an entire function Φ_2 with periodic limit set $S[v_2]$ with $v_2(\lambda) := q_2(\log \lambda)|\lambda|^\rho$. Let us prove that Φ_2 has no H-multiplicators. We have $m(z, v_2) = \epsilon r_2(z)$; hence $E^+(m(\cdot, v_2)) = E^+(r_2)$. By Lemma 3.1.2, $\rho(E^+(m(\cdot, v_2))) > \rho$ and $m(z_0, v_2) = \epsilon r_2(z_0, v_2) < 0$ for some $z_0 \in \mathbb{T}_P^2$. Thus $m(z, v_2)$ has no L_ρ - subminorant by Theorem H, and hence Φ_2 has no H-multiplicator by Proposition 2.3.1.

Let us verify that the conditions (1.4.1) hold, for example, for $\tau \in \mathcal{H}$. Apply property 2 of Lemma 3.1.2 for proving

$$\underline{h}(\Phi_1, \phi) = \underline{h}(\Phi_2, \phi), \ h(\Phi_1, \phi) = h(\Phi_2, \phi), \ \phi \in [0, 2\pi).$$

Take $\tau[f] = f(\phi)$ in Lemma 3.1.2 (2). Then

$$\sup_{0 \leq x \leq P} r_1(x + i\phi) = \sup_{0 \leq x \leq P} \tau[r_1(x + i\cdot)] = \sup_{0 \leq x \leq P} \tau[r_2(x + i\cdot)] = \sup_{0 \leq x \leq P} r_2(x + i\phi).$$

Applying $\tau[f] := -f(\phi)$, we obtain
$$\inf_{0 \le x \le P} r_1(x + i\phi) = \inf_{0 \le x \le P} r_2(x + i\phi).$$

Using (2.3.5), we obtain
$$h(\Phi_1, \phi) = H(\phi) - \inf_{0 \le x \le P} r_1(x + i\phi) = H(\phi) - \inf_{0 \le x \le P} r_2(x + i\phi) = h(\Phi_2, \phi);$$

$$\underline{h}(\Phi_1, \phi) = H(\phi) - \sup_{0 \le x \le P} r_1(x + i\phi) = H(\phi) - \sup_{0 \le x \le P} r_2(x + i\phi) = \underline{h}(\Phi_2, \phi).$$

The proofs for other characteristics are analogous. So Theorem 1.4.1 for the case $1/2 < \rho \le 1$ is proved up to Lemmas 3.1.1 and 3.1.2.

REMARK 2. For an entire function with periodic limit set determined by an L_ρ-subfunction $q(z)$, the characteristics $\overline{\sigma}(f), \underline{\sigma}(f), \overline{T}(f), \underline{T}(f)$ (see (1.4.2), (1.4.3)) can be made explicit as in (2.3.5)-(2.3.7):

$$\left.\begin{array}{c}\overline{\sigma}(f) \\ \underline{\sigma}(f)\end{array}\right\} = \left\{\begin{array}{c}\max_{0 \le x \le P} \\ \min_{0 \le x \le P}\end{array} \max_{0 \le y \le 2\pi} q(x + iy)\right. ; \quad \left.\begin{array}{c}\overline{T}(f) \\ \underline{T}(f)\end{array}\right\} = \left\{\begin{array}{c}\max_{0 \le x \le P} \\ \min_{0 \le x \le P}\end{array} \int_0^{2\pi} q^+(x + iy)dy\right..$$

Thus they are connected with the *convex* functionals $\max_y f(y), \int_0^{2\pi} f^+(y)dy$. However, every convex functional is determined by some family of linear functionals and condition 2 of Lemma 3.1.2 can be applied to them.

3.2. For proving Lemma 3.1.2, we require the assertion below. Denote by $S_l := \{z : 0 < \Im z < l\}$, $0 < l < 2\pi$, a strip that can be considered as a domain on the torus $\mathbb{T}_P^2$. Let us make cuts of form $Ct_m := \{z : \Re z = mP; l - h \le \Im z \le l\}$. The strip
$$D_{l,h} := S_l \setminus \bigcup_{m \in \mathbb{Z}} Ct_m$$
is P-periodic and can also be considered as a domain in $\mathbb{T}_P^2$. By Theorem I (§2.3), we have
$$\rho(S_l) = \pi/l,$$
since $\mod S_l(I_0, I_P) = P/l$. Let us compute $\rho(D_{l,h})$.

Set
$$sn^{-1}(w, k) := \int_0^w [(1 - t^2)(1 - k^2 t^2)]^{-1/2} dt, \quad \Im w \ge 0.$$

As is well-known (see, e.g., [**11**], Ch.4, §19), this function maps the upper half-plane to the rectangle with vertices K, $K + iK'$, $-K + iK'$, $-K$, where $K = sn^{-1}(1, k)$, $K' = sn^{-1}(1, (1 - k^2)^{1/2})$ such that the points $1, 1/k, -1/k, -1$ map to the corresponding vertices.

The function $sn(\cdot, k)$ is the inverse function that maps the rectangle to the half-plane.

Let us map the quadrilateral $D_{l,h}(I_0, I_P)$, where
$$I_0 = \{z = iy : 0 < y < l - h\}, \quad I_P = \{z = P + iy : 0 < y < l - h\},$$
to a standard rectangle $R(l_1) = \{\zeta = \xi + i\eta : 0 < \xi < P, 0 < \eta < l_1\}$ such that I_0 and I_P are mapped to the vertical sides of the rectangle.

PROPOSITION 3.2.1. *We have*

(3.2.1) $$\rho(D_{l,h}) = \pi/l_1 = \frac{\pi}{P}\frac{sn^{-1}(1,k_1)}{sn^{-1}(1,\sqrt{1-k_1^2})},$$

while

(3.2.2) $$1/k_1 = sn(zP^{-1}sn^{-1}(1,k),k)|_{z=P+i(l-h)},$$

and for $h \to 0$,

(3.2.3) $$\rho(D_{l,h}) = \rho(S_l) + Bh^2 + o(h^2), B > 0.$$

PROOF. Let $k = k(l, P)$ be determined by the equation

(3.2.4) $$\frac{l}{P} = \frac{sn^{-1}(1,\sqrt{1-k^2})}{sn^{-1}(1,k)}.$$

The function
$$z = P\frac{sn^{-1}(w,k)}{sn^{-1}(1,k)}$$
maps the upper half-plane to the rectangle $R(l)$ in such a way that $1 \mapsto P$, $1/k \mapsto P + il$. Denote by $1 < 1/k_1 < 1/k$ the point defined by (3.2.2), i.e., such that $P + i(l-h) \mapsto 1/k_1$. At the point $P + il$, the derivative of the function has a second order zero because the angle of the rectangle is doubled by this map. Thus

(3.2.5) $$1/k_1 = (1/k) + Ch^2 + o(h^2).$$

Now the function
$$\zeta = P\frac{sn^{-1}(w,k_1)}{sn^{-1}(1,k_1)},$$
where
$$w = sn(zP^{-1}sn^{-1}(1,k),k),$$
maps $D_{l,h}(I_0, I_P)$ to $R(l_1)$ with l_1 obtained from (3.2.4), for $k = k_1$. Thus (3.2.1) holds. From (3.2.1) we obtain (3.2.3), since by (3.2.4), $l'(k) \neq 0, \infty$ for all $k < 1$. □

Let us also record the following simple fact, which we have already exploited.

PROPOSITION 3.2.2. *We have*

(3.2.6) $$\rho(S_{l-h}) = \rho(S_l) + Ah + o(h), \quad A > 0, \quad h \to 0.$$

This is because $\rho(S_l) = \pi/l$.

3.3.

PROOF OF LEMMA 3.1.1. Let $D_1 \subset\subset D$ be a domain with smooth boundary such that $\rho(D_1) = \rho$; such exist by Corollary H, §2.3. Let $Y_1(z)$, $z \in D_1$, be the eigenfunction of the problem (2.3.3), (2.3.4) in the domain D_1. It is positive in D_1 and has positive inner normal derivative.

Now we use the following

THEOREM J. ([**5**], Theorem 3.7)[3] *If $\rho(D) \leq 1$, then*

(3.3.1) $$\rho(\mathbb{T}_P^2 \setminus D) > 1$$

if $D \neq \{z \in \mathbb{T}_P^2 : 0 < \Im z < \pi\}$.

Thus the Green function $G(z,\zeta)$ (< 0) of $\mathbb{T}_P^2 \setminus D_1$ for the differential operator L_ρ exists, and we may define the Green potential (see [**4, 8.7**])

$$\Pi(z) := \int_{\mathbb{T}_P^2 \setminus D_1} G(z,\zeta) g(\zeta) d\xi d\eta,$$

where $\zeta = \xi + i\eta$ and $d\xi d\eta$ is the element of area. Here the function g must be chosen positive and sufficiently smooth for the potential $\Pi(z)$ to be smooth and negative in $\mathbb{T}_P^2 \setminus \overline{D}_1$ and vanish on the boundary. Its normal derivative is smooth on the boundary.

Thus we can find a sufficiently large constant $c > 0$ such that the function

$$Y(z) := \begin{cases} cY_1(z), & z \in D_1 \\ \Pi(z), & z \in \mathbb{T}_P^2 \setminus D_1 \end{cases}$$

is an L_ρ-subfunction. This is possible because the jump of the inner normal derivative (with respect to $\mathbb{T}_P^2 \setminus D_1$) of this function on the boundary is negative. □

Observe that Y is not smooth on $\mathbb{T}_P^2$.

3.4. We represent domains in $\mathbb{T}_P^2$ by their images on the basic rectangle $\Pi = [0,P) \times [0,2\pi)$. Consider the following two domains in $\mathbb{T}_P^2$:

$$S_{l,h,a} := S_l \setminus \{z = x + iy : a \leq x \leq P, l - h \leq y \leq l\};$$

$$D_{l,2h,a} := S_l \setminus \{z = x + iy : 0 \leq x \leq a, l - 2h \leq y \leq l\},$$

where $S_l = [0,P) \times [0,l)$, $0 < l < 2\pi$.

LEMMA 3.4.1. *There exist l, h, a, ϵ such that $\rho(S_{l+\epsilon, h+\epsilon, a}) > \rho$ and $\rho(D_{l,2h,a}) < \rho$.*

PROOF. Take l_0 such that $\rho(S_{l_0}) = \rho$. By (3.2.6) and (3.2.3),

$$\rho(S_{l_0+h^2}) = \rho(S_{l_0}) - Ah^2 + o(h^2),$$

$$\rho(D_{l,h}) = \rho(S_l) + Bh^2 + o(h^2), h \to 0$$

Choose $0 < k < (1/2)\sqrt{A/B}$. Then

(3.4.1) $$\rho(D_{l_0+h^2, 2kh}) = \rho(S_{l_0}) - (A - 4k^2 B)h^2 + o(h^2) < \rho$$

for small h.

At the same time,

(3.4.2) $$\rho(S_{l_0 - kh}) = \rho(S_{l_0}) + Akh + o(h) > \rho$$

for small h.

Using Theorems E and F from §2.3, we can choose a and ϵ such that the inequalities (3.4.1) and (3.4.2) are preserved. Setting $l := l_0 + h^2$, $h := kh$, we obtain the assertion of the lemma. □

---

[3]The formulation of this theorem in [5] contains a misprint.

3.5.

PROOF OF LEMMA 3.1.2. Changing the boundary of $D_{l,2h,a}$ in a small neighborhood of points $A_1 = (0,l)$, $A_2 = (0,l-2h)$, $A_3 = (a,l-2h)$, $A_4 = (a,l)$, one can make this boundary smooth and preserve the property of Lemma 3.4.1.

So we obtain two domains $D, S := S_{l+\epsilon, h+\epsilon, a} \subset \mathbb{T}_P^2$ such that

(3.5.1) $$\rho(D) < \rho,\ \rho(S) > \rho.$$

Let us emphasize that the boundary of $D_{l,2h,a}$ was not changed on the intervals $I_1 = (d_1, d_2) \subset (0,a)$ and $I_2 = (d_3, d_4) \subset (a, P)$.

Let $Y(z)$ be an L_ρ-subfunction which satisfies the condition of Lemma 3.1.1. For $y \in [0, 2\pi)$, let $g_1(y)$ be a smooth function satisfying the conditions

(3.5.1a) $\quad g_1(y) > \max\limits_{x \in [0,P)} Y(x,y),\ y \in [0, 2\pi);\ g_1(y) < 0,\ y \in [l+\epsilon, P);$

$$g_1(0) = g_1(2\pi).$$

This is possible because $Y(z) < 0, z \notin D$.

By $g_2(y)$ we denote a function on $[0, 2\pi)$ satisfying the following conditions:

(3.5.1b) $\quad g_2(y) < 0,\ l - h - \epsilon < y < 2\pi;\ g_2(y) \geq \max\limits_{x \in I_1} Y(x,y);$

$$g_2(0) = g_2(2\pi) = 0.$$

This is possible because $Y(z) < 0, z \notin D$.

One can find a smooth function $b_1(x)$ on $(0, P)$ satisfying the conditions $b_1|_{I_1} = 0$, $1 - b_1|_{I_2} = 0$, $0 \leq b_1 \leq 1$, such that

$$p_1(x,y) := g_1(y)(1 - b_1(x)) + g_2(y)b_1(x)$$

satisfies the inequality

$$p_1(x,y) > Y(x,y), (x,y) \in \mathbb{T}_P^2$$

and

$$p_2(x,y) := g_1(y)b_1(x) + g_2(y)(1 - b_1(x))$$

satisfies

(3.5.1c) $$E^+(p_2) \subset S.$$

This is possible because of (3.5.1a) and (3.5.1b).

Now let us take two positive numbers $X_{\min} < 1 < X_{\max}$ such that the functions $X_{\min} p_1(x,y)$ and $X_{\max} p_1(x,y)$ also majorize $Y(x,y)$.

Take two points $x_{1,\min}$ and $x_{1,\max}$ in the interval I_1 and two points $x_{2,\min}$ and $x_{2,\max}$ in the interval I_2. Consider a smooth function $X(x)$ that has minimum equal to $X_{\min}$ at the points $x_{1,\min}$ and $x_{2,\min}$ and maximum equal to $X_{\max}$ at the points $x_{1,\max}$ and $x_{2,\max}$.

Set
$$r_1(x,y) := X(x)p_1(x,y).$$

Then

(3.5.2) $$r_1(z) > Y(z),\ z \in \mathbb{T}_P^2$$

and

$$\tau[r_1(x,\cdot)] = X(x)[(1 - b_1(x))\tau[g_1] + b_1(x)\tau[g_2]].$$

Hence, for any x,

(3.5.3) $$\tau[r_1(x,\cdot)] \le X(x)\max\{\tau[g_1],\tau[g_2]\}.$$

If at least one of the numbers $\tau[g_1], \tau[g_2]$ is positive, the maximum is attained at least at one of the points $x_{1,\max}, x_{2,\max}$. If both numbers are negative, the maximum is attained at least at one of the points $x_{1,\min}, x_{2,\min}$.

In other words, $\max\limits_x \tau[r_1(x,\cdot)]$ is attained at least at one of these four points and

(3.5.4) $$\max_{x\in[0,P)} \tau[r_1(x,\cdot)] = \begin{cases} X_{\max}\max\{\tau[g_1],\tau[g_2]\}, & \text{if } \max\{\tau[g_1],\tau[g_2]\}\ge 0 \\ X_{\min}\max\{\tau[g_1],\tau[g_2]\}, & \text{if } \max\{\tau[g_1],\tau[g_2]\}\le 0. \end{cases}$$

Now we construct the function $r_2(x,y)$. Set

$$r_2(x,y) := X(x)p_2(x,y).$$

The function $r_2(x,y)$ satisfies the conditions of Lemma 3.1.2, (1) because of (3.5.1c) and the monotonicity of $\rho(D)$. Since $\max\limits_{x\in[0,P)} \tau[r_2(x,\cdot)]$ can be also computed by (3.5.4), the assertions of Lemma 3.1.2 ,(2) are also fulfilled. $\square$

3.6. Now we prove Theorem 1.4.1 for arbitrary $\rho > 1/2$. It is sufficient to prove Lemma 3.1.2 for arbitrary $\rho > 1/2$.

Let p be an integer such that $p/2 < \rho \le (p+1)/2$. Then $\rho' := \rho/p$ satisfies the inequalities $1/2 < \rho' \le (1+1/p)/2 < 1$ if we exclude the case $p=1$, which has been already considered.

Set $P' = Pp$. Let $D' \subset \mathbb{T}^2_{P'}, Y'(z'), r'_1(z'), r'_2(z')$ be from Lemma 3.1.2. Let us check that the functions

$$Y(z) := Y'(pz), r_1(z) := r'_1(pz), r_2(z) := r'_2(pz)$$

also satisfy the assertions of Lemma 3.1.2.

Actually, by changing variables in the operator

$$L_{\rho'} := \frac{\partial^2}{\partial x'^2} + \frac{\partial^2}{\partial y'^2} + 2\rho'\frac{\partial}{\partial x'} + \rho'^2 = \Delta_{z'} + 2\rho'\frac{\partial}{\partial x'} + \rho'^2,$$

where $\Delta_{z'}$ is the Laplace operator on z', we obtain $p^2 L_{\rho'} = L_\rho$.

Hence $Y(z)$ is an L_ρ-subfunction, which is P-periodic in x and $(2\pi/p)$-periodic and hence 2π-periodic in y.

The functions r_1 and r_2 are also P-periodic in x and $(2\pi/p)$-periodic in y. Hence, every linear functional on these functions can be represented as a linear functional on 2π-periodic functions $r'_1(z'), r'_2(z')$. Thus assertion (2) of Lemma 3.1.2 holds.

The set

$$E^+(r_2) = \bigcup_0^{p-1} E_j,$$

where each E_j is an image of the curvilinear strip, included in the strip $S_{2\pi/p,j} := \{z : (2\pi/p)j < y < (2\pi/p)(j+1)\}$, and hence they do not intersect. For each $j = 0,...,p-1$, we have $\rho(E_j) > \rho$. Thus $\rho(E^+(r_2)) > \rho$ by definition. So all the assertions of Lemma 3.1.2 hold. This completes the proof of Lemma 3.1.2 for arbitrary $\rho > 1/2$ and the proof of Theorem 1.4.1. $\square$

References

1. V. Azarin, *On the regularity of the growth of the Fourier coefficients of the logarithm of the modulus of an entire function*, Selecta Math. Soviet. **2** (1982), 51-63.
2. V. Azarin, *On rays of completely regular growth of an entire function*, Math. USSR-Sb. **8** (1969), 437-450.
3. V. Azarin, *An example of an entire function with a given indicator and lower indicator*, Math. USSR-Sb. **18** (1974), 541-558.
4. V. Azarin, D. Drasin and P. Poggi-Corradini, *A generalization of ρ-trigonometric convexity and its relation to positive harmonic functions in homogeneous domains*, J. Anal. Math., in press.
5. V. Azarin and V. Giner, *Limit sets and multiplicators of entire functions*, Entire and Subharmonic Functions, Adv. Soviet Math. **11**, Amer. Math. Soc., Providence, RI, 1992, pp. 251-275.
6. V. Azarin and V. Giner, *Limit sets of entire functions and completeness of exponential systems*, Mat. Fiz. Anal. Geom. **1** (1994), 1-30.
7. V. Azarin and L. Podoshev, *Limit sets and indicators of entire functions*, Siberian Math. J. **25** (1984), 833-844.
8. A.A. Gol'dberg, *Indicators of entire functions and an integral with respect to a non-additive measure*, Contemporary Problems in the Theory of Analytic Functions, Nauka, Moscow, 1966, pp. 88-99.
9. A.A. Gol'dberg, B.Ya. Levin and I.V. Ostrovskii, *Entire and Meromorphic Functions*, Encyclopaedia Math. Sci. **85**, Springer, 1997, pp. 4-172.
10. B.Ya. Levin, *Distribution of Zeros of Entire Functions*, Amer. Math. Soc., Providence, RI, 1980.
11. A.I. Markushevich, *Theory of Functions of a Complex Variable,* Vol. II, Prentice-Hall, 1965.

DEPARTMENT OF MATHEMATICS, BAR-ILAN UNIVERSITY, RAMAT-GAN 52900, ISRAEL
E-mail address: `azarin@macs.biu.ac.il`

Are They Limit Periodic?

J. Bellissard[1], J. Geronimo[2], A. Volberg[3], and P. Yuditskii[4]

Dedicated to Professor Lawrence Zalcman on the occasion of his sixtieth birthday

ABSTRACT. We prove a partial result concerning the long-standing problem on limit periodicity of the Jacobi matrix associated with the balanced measure on the Julia set of an expanding polynomial. Besides this, connections of the problem with the Faybusovich-Gekhtman flow and many other objects (the Hilbert transform, the Schwarz derivative, the Ruelle and Laplace operators) of independent interest are discussed.

1. Introduction

In the 1980's it was discovered that the spectral measure of an almost periodic Jacobi matrix can be singular continuous (supported on a Cantor type set of zero Lebesgue measure). The effect was studied from both sides: from coefficient sequences to spectral data [1], [4] and from spectral data to Jacobi matrices.

The second, usually more elegant, approach produced the following example [3], [2]. Let $T(z) = z^2 - C$. For $C > 2$, the Julia set E of T is a real Cantor type set, $|E| = 0$. Denote by μ the balanced measure on E, $\mu(T^{-1}(F)) = \mu(F)$ for all $F \subset E$. Let

$$(1) \qquad J = \begin{bmatrix} q_0 & p_1 & & \\ p_1 & q_1 & p_2 & \\ & \ddots & \ddots & \ddots \end{bmatrix}$$

be the Jacobi matrix associated to the given measure. Note that to construct $J : l^2(\mathbb{Z}_+) \to l^2(\mathbb{Z}_+)$, one uses the three term recurrence relation for polynomials orthonormal in $L^2_{d\mu}$:

$$(2) \qquad \lambda P_k(\lambda) = p_k P_{k-1}(\lambda) + q_k P_k(\lambda) + p_{k+1} P_{k+1}(\lambda)$$

and the unitary map $L^2_{d\mu} \to l^2(\mathbb{Z}_+)$, defined by $P_k \mapsto |k\rangle$, where $\{|k\rangle\}$ is the standard basis in $l^2(\mathbb{Z}_+)$.

2000 *Mathematics Subject Classification.* Primary 47B36; Secondary 37F15, 37J35.

This work supported by [1]NSF Grant 0300398; [2]NSF Grant DMS-0200219; [3]NSF Grant DMS-0200713; [4]Austrian Founds FWF, Project Number: P16390–N04.

Then the given matrix satisfies the renormalization equation:
$$V^*T(J)V = J,$$
where $V|k\rangle = |2k\rangle$. In fact, this is a system of nonlinear equations for the p_n's ($q_n = 0$ in this case) from which, at least for $C > 3$, one gets inductively that
$$|p_{2^n l + m} - p_m| \leq \epsilon_n, \quad \text{for all } l, m; \quad \epsilon_n \to 0 \quad (n \to \infty).$$
That is, the sequence $\{p_n\}$ and (by definition) the matrix itself are limit periodic. It seems very natural to conjecture that if T is an arbitrary expanding polynomial in the sense of complex dynamics [7], then its balanced measure produces a limit periodic Jacobi matrix. Several research groups attacked this problem (in full generality) but failed. Even the case of a quadratic polynomial with $C > 2$ is still open.

Recall some properties of Jacobi matrices. Let J be a Jacobi matrix, $J^* = J$, acting on $\mathbb{C}^d$ or $l^2(\mathbb{Z}_+)$. Under the assumption $p_k \neq 0$ the vector $|0\rangle$ of the standard basis is cyclic for J. The resolvent function is a function of the form

(3) $$r(z) = \langle 0 | (J - z)^{-1} | 0 \rangle.$$

It has positive imaginary part in the upper half plane and hence possesses the representation

(4) $$r(z) = \int \frac{d\sigma}{\lambda - z} = \langle \mathbf{1} | (\lambda - z)^{-1} | \mathbf{1} \rangle_{L^2_{d\sigma}},$$

where $\lambda\cdot$ is the operator multiplication by the independent variable in $L^2_{d\sigma}$ and $\mathbf{1}$ is the function that equals one identically. Formulas (3) and (4) give a one-to-one correspondence between triples $\{L^2_{d\sigma}, \lambda\cdot, \mathbf{1}\}$ and $\{l^2(\mathbb{Z}_+), J, |0\rangle\}$ or $\{\mathbb{C}^d, J, |0\rangle\}$, in the finite dimensional case. To recapture J starting from the nonnegative measure σ, one uses (2).

Our first object is the following

CONJECTURE 1.1. *Let $T(z)$ be an expanding polynomial of degree d with a real Julia set E, $E \subset [-\xi, \xi]$, $T^{-1} : [-\xi, \xi] \to [-\xi, \xi]$. Define $J = J(x)$ by*

(5) $$\langle 0 | (z - J(x))^{-1} | 0 \rangle = \frac{T'(z)/d}{T(z) - x}, \quad x \in [-\xi, \xi].$$

In a similar way, let $J_n(x)$ be associated with the iteration $T_n = T^{\circ n}$, $\deg T_n = d_n$. Then for every ϵ, there exists n such that

(6) $$\|J_n(x) - J_n(0)\| \leq \epsilon.$$

Note that eigenvalues of $J_n(x)$ and $J_n(0)$ are close, so the nontrivial part deals with eigenvectors.

Let us explain how this conjecture is related to the general one. If μ is the balanced measure on E, then the resolvent of $J = J(\mu)$ satisfies the Renormalization Equation

(7) $$V^*(z - J)^{-1}V = (T(z) - J)^{-1}T'(z)/d,$$

where $V|k\rangle = |kd\rangle$. Let us embed J in a chain $\{J_n(t)\}_{t \in [0,1]}$ defined by
$$V_n^*(z - J_n(t))^{-1}V_n = (T_n(z) - tJ)^{-1}T_n'(z)/d_n$$
(compare the last equation with (5)). Then the main goal is to show that
$$\|J_n(1) - J_n(0)\| \leq \epsilon \quad \text{for } n > n_0,$$

since this would imply immediately that $J(\mu)$ is limit periodic. Thus, trying to prove Conjecture 1.1 is a good model problem on the way to proving limit periodicity of $J(\mu)$.

The following approach looks very natural: to get (6), we have to estimate $J'(x)$. The given derivative has the special representation

$$\tag{8} \frac{dJ(x)}{dx} = F(J) + [G, J]$$

with $F(J) = \{T'(J)\}^{-1}$. It is a certain flow on Jacobi matrices which is, in a sense, dual to the well-known Toda flow. We call it the FG flow [**6**] (see Section 2). The first term at the right hand side in (8) is small by the characteristic property of expanding polynomials: $|T'_n(x)| \geq Ac^n$, $x \in E$, with $A > 0$, $c > 1$. It appears that the estimate we get for G is not enough to conclude that the commutator $[G, J]$ is sufficiently small (Proposition 2.6). However, on the way, we found some remarkable formulas and connections with a number of objects (the Hilbert transform, the Schwarz derivative, the Ruelle and Laplace operators), which we feel are of independent interest.

In the framework of this approach, initiated in [**9**], we have proved the following theorem, which partially confirms the main hypothesis.

THEOREM 1.2. *Let J be the Jacobi matrix associated with iterations of an expanding polynomial T. Then for every ϵ there exists n such that*

$$\tag{9} |p_{k+sd_n^2} - p_k| \leq \epsilon, \quad |q_{k+sd_n^2} - q_k| \leq \epsilon,$$

for all $s \geq 0$ and $k = 1, 2, \ldots, d_n$.

Note that our goal is actually to prove (9) when $k = 1, 2, \ldots, d_n^2$. A proof of the theorem is given in Section 3.

Acknowledgment. We thank Misha Shapiro for calling our attention to the results of [**6**].

2. FG flow

2.1. Definition. Let $J : \mathbb{C}^d \to \mathbb{C}^d$. Consider the resolvent function

$$\tag{10} \langle 0 | (z - J)^{-1} | 0 \rangle = \sum_{k=1}^{d} \frac{\sigma_k}{z - \lambda_k}.$$

Under the Toda flow, the spectrum is stable, $\lambda_k = \text{Const}$, but the masses vary with time $\sigma_k = \sigma_k(t)$. In the FG flow case, $\lambda_k = \lambda_k(t)$ but $\sigma_k = \text{Const}$. Moreover, in our case (5), time is x, $\sigma_k = 1/d$ and $T(\lambda_k(x)) = x$. Recall that $T(z)$ is an expanding polynomial of degree d with real Julia set E, $E \subset [-\xi, \xi]$, $T^{-1} : [-\xi, \xi] \to [-\xi, \xi]$.

We want to get a differential equation for J. Let $\mathfrak{B}$ be a unitary matrix such that

$$J\mathfrak{B} = \mathfrak{B}\Lambda,$$

where $\Lambda = \text{diag}\{\lambda_k\}$. Since we can choose

$$\lambda_1(x) < \lambda_2(x) < \cdots < \lambda_d(x)$$

to hold for all x, $\mathfrak{B}$ essentially is well-defined. We put

$$\mathfrak{B} = \frac{1}{\sqrt{d}} \begin{bmatrix} P_0(\lambda_1) & \ldots & P_0(\lambda_d) \\ \vdots & & \vdots \\ P_{d-1}(\lambda_1) & \ldots & P_{d-1}(\lambda_d) \end{bmatrix},$$

where $P_k(z)$ is the orthonormal polynomial.

Differentiating J with respect to x, we obtain

$$\dot{J} = \mathfrak{B}\dot{\Lambda}\mathfrak{B}^{-1} + \dot{\mathfrak{B}}\Lambda\mathfrak{B}^{-1} - \mathfrak{B}\Lambda\mathfrak{B}^{-1}\dot{\mathfrak{B}}\mathfrak{B}^{-1} = F + GJ - JG,$$

where $F := \mathfrak{B}\dot{\Lambda}\mathfrak{B}^{-1}$, $G := \dot{\mathfrak{B}}\mathfrak{B}^{-1}$. By definition, $F = f(J)$ with $f(\lambda_k) = \dot{\lambda}_k$. Thus $F = T'(J)^{-1}$. The next step is to determine G.

Note some evident facts. G is skew-symmetric and $\langle 0|G = 0$, so $G|0\rangle = 0$. Also, it is easy to show, say by induction, that

$$\frac{d}{dx}J^n = nJ^{n-1}F + GJ^n - J^nG.$$

Finally, since $P_k(J)|0\rangle = |k\rangle$ and

$$\frac{d}{dx}P_k(J) - \frac{\partial}{\partial x}P_k(J) = FP'_k(J) + GP_k(J) - P_k(J)G,$$

we get

(11) $$-\frac{\partial}{\partial x}P_k(J)|0\rangle = FP'_k(J)|0\rangle + GP_k(J)|0\rangle.$$

Let G_+ be the lower triangular matrix with zeros on the main diagonal such that $G = G_+ - G_+^*$. Then (11) implies that

(12) $$G_+|k\rangle = G_+P_k(J)|0\rangle = -(FP'_k(J)|0\rangle)^{(k)}_+.$$

Here $h^{(k)}_+$ means that in the vector $h = \{h_j\}_{j=0}^{d-1}$ we have to replace all coordinates h_j, $0 \leq j \leq k$, by zeros.

Let us rewrite (12) in other terms. Define an operator D by

$$D|k\rangle = DP_k(J)|0\rangle := P'_k(J)|0\rangle.$$

Then

$$G_+ = -(FD)_+.$$

It is easy to check, using the functional representation in $L^2_{d\sigma}$, that

(13) $$DJ - JD = I - |(p_dP_d)'\rangle\langle P_{d-1}|,$$

where $p_dP_d(\lambda)$ is defined by (2). Note that $p_dP_d(z) = 0$ in $L^2_{d\sigma}$, that is, it has the same roots $\{\lambda_k(x)\}$ as $T(z) - x$. Thus $p_dP_d(z) = C(T(z) - x)$ and $(p_dP_d)'(z) = CT'(z)$.

DEFINITION 2.1. *The FG flow is given by a differential equation of the form*

(14) $$\dot{J} = F + GJ - JG,$$

with $F = f(J)$ and $G = G_+ - G_+^*$, where $G_+ = -(FD)_+$ and D is an (upper triangular) matrix such that commutant $[D, J]$ equals the identity matrix up to a one-dimensional perturbation [6].

2.2. (FD) as a Hilbert transform.

LEMMA 2.2. *The matrix of the operator (FD) with respect to the basis of eigenvectors of J has the form*

(15)
$$\begin{bmatrix} \frac{1}{2}\frac{T''(\lambda_1)}{T'(\lambda_1)} & \cdots & \frac{1}{\lambda_1-\lambda_d} \\ \vdots & & \vdots \\ \frac{1}{\lambda_d-\lambda_1} & \cdots & \frac{1}{2}\frac{T''(\lambda_d)}{T'(\lambda_d)} \end{bmatrix} \begin{bmatrix} \frac{1}{T'(\lambda_1)} & & \\ & \ddots & \\ & & \frac{1}{T'(\lambda_d)} \end{bmatrix}.$$

PROOF. Let us evaluate D in the basis of eigenvectors of J. In this basis,

$$|P(x)\rangle \to \frac{1}{\sqrt{d}}\begin{bmatrix} P(\lambda_1) \\ \vdots \\ P(\lambda_d) \end{bmatrix}.$$

As we know, $(p_d P_d)'(\lambda_k) = CT'(\lambda_k)$. Taking into account that now J is diagonal, we conclude that the diagonal entries of $DJ - JD$ are zeros. Therefore, $P_{d-1}(\lambda_j) = d/\{CT'(\lambda_j)\}$. Thus the right hand side of (13) is of the form

$$I - \begin{bmatrix} T'(\lambda_1) \\ \vdots \\ T'(\lambda_d) \end{bmatrix} \begin{bmatrix} \frac{1}{T'(\lambda_1)}, & \cdots, & \frac{1}{T'(\lambda_d)} \end{bmatrix}.$$

Referring again to a diagonal form of J, we solve (13) and get

(16) $$D_{ij} = \frac{1}{\lambda_i - \lambda_j}\frac{T'(\lambda_i)}{T'(\lambda_j)}, \quad i \neq j.$$

To find the diagonal entries D_{ii}, we have to use $D|0\rangle = 0$. Since

$$|0\rangle \to \mathbf{1} \to \frac{1}{\sqrt{d}}\begin{bmatrix} 1 \\ \vdots \\ 1 \end{bmatrix},$$

we get

$$D_{ii} = T'(\lambda_i)\sum_{k \neq i}\frac{1}{T'(\lambda_k)}\frac{1}{\lambda_k - \lambda_i}.$$

Note that

$$\sum_{k=1}^{d}\frac{1}{T'(\lambda_k)}\frac{1}{\lambda_k - z} = -\frac{1}{T(z) - x}.$$

Therefore,

(17)
$$\frac{D_{ii}}{T'(\lambda_i)} = \lim_{z \to \lambda_i}\left\{-\frac{1}{T'(\lambda_i)}\frac{1}{\lambda_i - z} - \frac{1}{T(z) - x}\right\}$$
$$= \lim_{z \to \lambda_i}\frac{\frac{T(z)-x}{z-\lambda_i}\frac{1}{T'(\lambda_i)} - 1}{T(z) - x} = \frac{\frac{1}{2}T''(\lambda_i)}{(T'(\lambda_i))^2}.$$

Thus (16) and (17) finish the proof. □

2.3. Trace of $(FD)^*(FD)$.

LEMMA 2.3. *Let L_2 be a Ruelle operator of the form*

$$L_2 g(x) = \frac{1}{d} \sum_{Ty=x} \left(\frac{g}{T'^2}\right)(y) \tag{18}$$

and let $S(T)$ be the Schwarz derivative of T, $S(T) = \frac{T'''}{T'} - \frac{3}{2}\left(\frac{T''}{T'}\right)^2$. Then

$$\frac{1}{d} \operatorname{tr}\{(FD)^*(FD)\} = -\frac{1}{3} L_2\{S(T)\}.$$

PROOF. First we simplify

$$u_i = \sum_{k \neq i} \frac{1}{(\lambda_i - \lambda_k)^2} = \lim_{z \to \lambda_i} \left\{ \sum_{T(\lambda)=x} \frac{1}{(z-\lambda)^2} - \frac{1}{(z-\lambda_i)^2} \right\}. \tag{19}$$

Note that

$$\sum_{T(\lambda)=x} \frac{1}{(z-\lambda)} = \frac{T'(z)}{T(z)-x}.$$

That is,

$$\sum_{T(\lambda)=x} \frac{1}{(z-\lambda)^2} = \frac{T'^2(z) - T''(z)(T(z)-x)}{(T(z)-x)^2}.$$

So, passing in a usual way to the limit in (19), we get

$$u_i = \frac{\frac{1}{2}(T'')^2 - \frac{2}{3}T'T'''}{2(T')^2}(\lambda_i). \tag{20}$$

This means that a diagonal entry of the operator $(FD)^*(FD)$ with respect to the basis of eigenvectors of J has the form

$$\frac{1}{T'^2}\left\{\left(\frac{1}{2}\frac{T''}{T'}\right)^2(\lambda_i) + u_i\right\} = \frac{1}{T'^2}\frac{(T'')^2 - \frac{2}{3}T'T'''}{2(T')^2}(\lambda_i) = -\frac{1}{3}\left(\frac{S(T)}{T'^2}\right)(\lambda_i).$$

□

Naturally, in the same way, we can find off-diagonal entries of the matrix of the operator $(FD)^*(FD)$.

LEMMA 2.4. *For $i \neq j$,*

$$\{(FD)^*(FD)\}_{ij} = \frac{1}{T'(\lambda_i)} \frac{2}{(\lambda_i - \lambda_j)^2} \frac{1}{T'(\lambda_i)}.$$

We consider $\Delta := (FDF^{-1})^*(FDF^{-1})$ a counterpart of the Laplacian in light of the following proposition.

COROLLARY 2.5. *Δ is a positive operator which satisfies*

$$[J, [J, \Delta]] = 2d|0\rangle\langle 0| - 2.$$

PROOF. See Lemma 2.4. □

Our plan to estimate $[G, J]$ in (8) was based on the conjecture $||(FD)_n|| \sim \kappa^n$ with $\kappa < 1$ (typically everything that goes to zero in the subject converges geometrically). Since $(G_n)_+ = -(FD)_{n+}$, that would give the estimate

$$||(G_n)_+|| \sim \kappa^n n \log d,$$

and we are done. However, the following proposition shows that $||(FD)_n|| \not\to 0$.

PROPOSITION 2.6. *We have*

(21) $$\lim_{n\to\infty} \frac{1}{d^n} \operatorname{tr}\{(FD)_n^*(FD)_n\} = \frac{1}{3}(I - L_2)^{-1} L_2 S(T).$$

PROOF. Let us use the chain rule for the Schwarzian derivative:

$$S(T_{n+1}) = S(T_n) \circ TT'^2 + S(T).$$

Since $L_2\{g \circ TT'^2\} = g$ for every function g, we have $L_2^{n+1}\{S(T_n) \circ TT'^2\} = L_2^n S(T_n)$ and therefore

(22) $$L_2^{n+1} S(T_{n+1}) = L_2^n S(T_n) + L_2^{n+1} S(T) = L_2 S(T) + \cdots + L_2^{n+1} S(T).$$

Since the spectral radius of L_2 less than $1/d^2$ (see Lemma 3.2), (22) completes the proof. □

REMARK. We still believe that $J(\mu)$ is limit periodic. Recall that we have to estimate not $(FD)_n$ itself but the commutator $[G_n, J_n]$. It is worth mentioning that the right hand side of the commutant identity for $(FD)_n$

$$(FD)_n J_n - J_n (FD)_n = F_n - d^n |0\rangle\langle 0| F_n$$

goes to zero in norm (again Lemma 3.2). That is, asymptotically $(FD)_n$ and J_n commute.

3. Partial result in the right direction

3.1. Renormalization equation. Let

$$Lg(x) = \frac{1}{d} \sum_{Ty=x} g(y)$$

be the Ruelle operator associated with the expanding polynomial $T(z)$. If $\tilde{J}$ is the Jacobi matrix associated with a measure $\tilde{\sigma}$ supported on E, $\tilde{J} := \tilde{J}(\tilde{\sigma})$, then the Renormalization Equation

(23) $$V^*(z - J)^{-1} V = (T(z) - \tilde{J})^{-1} T'(z)/d, \quad V|k\rangle = |kd\rangle,$$

has a unique solution $J := J(\sigma)$, where $\sigma := L^*(\tilde{\sigma})$ [2], [8]. This follows basically from the identity

$$\left(L \frac{1}{z-y}(g \circ T)(y)\right)(x) = \frac{T'(z)/d}{T(z) - x} g(x)$$

and the functional representations of both operators in $L^2_{d\sigma}$ and $L^2_{d\tilde{\sigma}}$, respectively. Note that (23) becomes (7) if $\tilde{\sigma} = \mu$, since for the balanced measure we have $\mu = L^*(\mu)$.

LEMMA 3.1. *Let $J^{(s)}$ be the s-th $d \times d$ block of the matrix J, that is,*

$$
(24) \qquad J^{(s)} = \begin{bmatrix} q_{sd} & p_{sd+1} & & & \\ p_{sd+1} & q_{sd+1} & p_{sd+2} & & \\ & \ddots & \ddots & \ddots & \\ & & p_{sd+d-2} & q_{sd+d-2} & p_{sd+d-1} \\ & & & p_{sd+d-1} & q_{sd+d-1} \end{bmatrix}.
$$

Then its resolvent function is of the form

$$
(25) \qquad \left\langle 0 \left| (z - J^{(s)})^{-1} \right| 0 \right\rangle = \frac{T'(z)/d}{T^{(s)}(z)}.
$$

Moreover, at the critical points $\{c : T'(c) = 0\}$, we have the continued fraction decomposition

$$
(26) \qquad T^{(s)}(c) = T(c) - \tilde{q}_s - \cfrac{\tilde{p}_s^2}{T(c) - \tilde{q}_{s-1} - \cdots}.
$$

PROOF. We write J as a $d \times d$ block matrix (each block is of infinite size):

$$
(27) \qquad J = \begin{bmatrix} \mathcal{Q}_0 & \mathcal{P}_1 & & & & S_+\mathcal{P}_d \\ \mathcal{P}_1 & \mathcal{Q}_1 & \mathcal{P}_2 & & & \\ & \ddots & \ddots & \ddots & & \\ & & \mathcal{P}_{d-2} & \mathcal{Q}_{d-2} & \mathcal{P}_{d-1} & \\ \mathcal{P}_d S_+^* & & & \mathcal{P}_{d-1} & \mathcal{Q}_{d-1} \end{bmatrix}.
$$

Here $\mathcal{P}_k$ (respectively, $\mathcal{Q}_k$) is a diagonal matrix $\mathcal{P}_k = \mathrm{diag}\{p_{k+sd}\}_{s \geq 0}$ and S_+ is the one-sided shift. In this case, V^* is the projection on the first block-component.

Using this representation and the well-known identity for block matrices

$$
\begin{bmatrix} A & B \\ C & D \end{bmatrix}^{-1} = \begin{bmatrix} (A - BD^{-1}C)^{-1} & * \\ * & * \end{bmatrix},
$$

we get

$$
(28) \qquad \frac{T(z) - \tilde{J}}{T'(z)/d} = z - \mathcal{Q}_0 - [\mathcal{P}_1, \ldots, S_+\mathcal{P}_d]\{z - J_1\}^{-1}\begin{bmatrix} \mathcal{P}_1 \\ \vdots \\ \mathcal{P}_d S_+^* \end{bmatrix},
$$

where J_1 is the matrix that we obtain from J by deleting the first block-row and the first block-column in (27). Note that in $(z - J_1)$, each block is a diagonal matrix; thus we can easily get an inverse matrix in terms of orthogonal polynomials.

Let us introduce the following notation: everything related to $J^{(s)}$ has superscript s. For instance: $p_k^{(s)} = p_{sd+k}$, $1 \leq k \leq d$; and $P_d^{(s)}$ and $Q_d^{(s)}$ denote, respectively, orthonormal polynomials of the first and second kind. In these terms, equation (28) is equivalent to the two series of scalar relations corresponding to the diagonal and off-diagonal entries

$$
(29) \qquad \frac{T(z) - \tilde{q}_{s+1}}{T'(z)/d} = \frac{P_d^{(s+1)}(z)}{Q_d^{(s+1)}(z)} - p_{ds}^2 \frac{Q_{d-1}^{(s)}(z)/p_{ds}}{Q_d^{(s)}(z)}
$$

and

$$
(30) \qquad \frac{\tilde{p}_{s+1}}{T'(z)/d} = \frac{p_1^{(s)} \cdots p_d^{(s)}}{z^{d-1} + \cdots} = \frac{1}{Q_d^{(s)}(z)}.
$$

Recall (see (25) and (30)) that
$$\frac{Q_d^{(s)}(z)}{P_d^{(s)}(z)} = \frac{z^{d-1}+\ldots}{z^d+\ldots} = \frac{T'(z)/d}{T^{(s)}(z)}.$$

Now by the Wronskian identity, if $T'(c) = 0$, then
$$(31) \qquad -p_{ds} Q_{d-1}^{(s)}(c) = \frac{1}{P_d^{(s)}(c)}.$$

So, combining (29), (30) and (31), we get the recurrence relation
$$(32) \qquad T(c) - \tilde{q}_{s+1} = T^{(s+1)}(c) + \frac{\tilde{p}_{s+1}^2}{T^{(s)}(c)}$$

with initial data
$$T^{(0)}(c) = T(c) - \tilde{q}_0.$$

□

3.2. p_{sd_n} are exponentially small.

LEMMA 3.2. Let J be the Jacobi matrix associated with the iterations $\{T_n\}_{n\geq 1}$ of an expanding polynomial T, that is, $J = J(\mu)$ where $L^*\mu = \mu$. Then
$$(33) \qquad C_-(\rho d)^n p_s \leq p_{sd_n} \leq C_+(\rho d)^n p_s$$
with $C_\pm > 0$ and $0 < \rho < 1/d$.

PROOF. Recall that $P_{sd} = P_s \circ T$ and $Q_{sd} = (T'/d) Q_s \circ T$ [2], [8]. We use the interpolation formula
$$(34) \qquad \int R \, d\mu = \sum_{y: P_{sd}(y)=0} R(y) \frac{Q_{sd}}{P'_{sd}}(y), \quad \deg R < sd,$$

and the Wronskian identity
$$(35) \qquad p_{sd}\{P_{sd-1}Q_{sd} - Q_{sd-1}P_{sd}\} = 1.$$

Substituting (35) in (34), we obtain
$$p_{sd}^2 = \int \{p_{sd}P_{sd-1}\}^2 d\mu = \sum_{y:P_{sd}(y)=0} \{p_{sd}P_{sd-1}(y)\}^2 \frac{Q_{sd}}{P'_{sd}}(y) = \sum_{y:P_{sd}(y)=0} \frac{1}{(Q_{sd}P'_{sd})(y)}.$$

Therefore,
$$p_{sd}^2 = \sum_{x:P_s(x)=0} \sum_{y:T(y)=x} \frac{1}{((T'^2/d)(Q_s P'_s) \circ T)(y)}$$
$$= \sum_{x:P_s(x)=0} \frac{1}{(Q_s P'_s)(x)} \left\{ \frac{1}{d} \sum_{y:T(y)=x} \frac{d^2}{T'^2(y)} \right\}$$
$$= \sum_{x:P_s(x)=0} \{p_s P_{s-1}(x)\}^2 \frac{Q_s}{P'_s}(x) \left\{ \frac{1}{d} \sum_{y:T(y)=x} \frac{d^2}{T'^2(y)} \right\}.$$

Now we use the Ruelle version of the Perron-Frobenius theorem [7], [5] with respect to L_2 (18). According to this theorem,
$$\frac{1}{\rho^{2n}} L_2^n g \to h(x) \int g \, d\nu,$$

uniformly in x for a certain continuous function $h > 0$ and positive measure ν; ρ^2 is the spectral radius of L_2. Combining this with the interpolation formula, we get the two-sided estimate (33).

We only have to show that $(\rho d) < 1$. Let $b(z)$ be the complex Green's function of the domain $\overline{\mathbb{C}} \setminus E$ with respect to infinity. Consider the sequence of functions $\{f_n\}_{n \geq 1}$, where $f_n(z) := (b^{d_n - 1} P_{d_n - 1})(z)$. Each f_n is a multiple-valued function in the domain $\overline{\mathbb{C}} \setminus E$ with single-valued modulus whose square has a harmonic majorant $u_n(z) \geq |f_n(z)|^2$. Moreover, $u_n(\infty) = \|P_{d_n-1}\|^2_{L^2_{d\mu}} = 1$. We claim that f_n tends to zero pointwise. If not, we can find a subsequence $\{f_{n_k}\}$ that converges to a nontrivial function f. However, in this case, $(bf)(z)$ is a nontrivial single-valued function in $\overline{\mathbb{C}} \setminus E$, $|(bf)(z)|^2$ has a harmonic majorant and $(bf)(\infty) = 0$. This contradicts the well-known fact that the Lebesgue measure of E is zero. Indeed, $|bf|^2$ has a harmonic majorant in the upper/lower half-plane and therefore belongs to H^2 in the upper/lower half-plane (with respect to harmonic measure). However, since $|E| = 0$ we have also $(bf)(x + i0) = (bf)(x - i0)$ for a.e. x on $\mathbb{R}$ (with respect to the Lebesgue measure). Thus $bf = $ const, so by the normalization, $bf = 0$ identically.

Therefore, the sequence converges to zero. In particular,

$$(b^{d_n - 1} P_{d_n - 1})(\infty) = \frac{1}{p_1 \ldots p_{d_n - 1}} = \frac{p_{d_n}}{p_1} \to 0, \quad n \to \infty.$$

But $\frac{p_{d_n}}{p_1} \sim (\rho d)^n$; thus $(\rho d) < 1$. □

REMARK 3.3. Let us mention here that $q_{sd} = q_0$, since

$$q_{sd} = \int y P_{sd}^2 \, d\mu = \int y P_s^2 \circ T \, dL^* \mu = \int (Ly) P_s^2 \, d\mu$$

and $Ly = q_0$.

3.3. The result. First we prove (the undoubtedly well-known and simple)

LEMMA 3.4. *Assume the two measures σ and $\tilde{\sigma}$ are mutually absolutely continuous and $d\tilde{\sigma} = f \, d\sigma$, where $1 - \epsilon \leq f \leq (1 - \epsilon)^{-1}$. Let us associate with these measures Jacobi matrices $J = J(\sigma)$, $\tilde{J} = J(\tilde{\sigma})$. Then for their coefficients we have*

$$|\tilde{p}_s - p_s| \leq \frac{\epsilon}{1 - \epsilon} \|J\|.$$

PROOF. Let us use the extreme property of orthogonal polynomials,

$$\tilde{p}_1^2 \ldots \tilde{p}_s^2 = \int \tilde{p}_1^2 \ldots \tilde{p}_s^2 \tilde{P}_s^2 \, d\tilde{\sigma} \geq (1 - \epsilon) \int \{z^s + \ldots\}^2 \, d\sigma$$

$$\geq (1 - \epsilon) \inf_{\{P = z^s + \ldots\}} \int P^2 \, d\sigma = (1 - \epsilon) p_1^2 \ldots p_s^2.$$

Similarly,

$$p_1^2 \ldots p_{s-1}^2 \geq (1 - \epsilon) \tilde{p}_1^2 \ldots \tilde{p}_{s-1}^2.$$

Therefore,

$$\frac{1}{(1 - \epsilon)^2} p_s^2 \geq \tilde{p}_s^2 \geq (1 - \epsilon)^2 p_s^2;$$

and hence

$$-\epsilon p_s \leq \tilde{p}_s - p_s \leq \frac{\epsilon}{1 - \epsilon} p_s.$$

□

Now we are in position to prove Theorem 1.2.

PROOF. As follows from Lemma 3.1,
$$T^{(s)}(c) = T(c) - q_s - p_s^2 \int \frac{d\nu^{(s)}(x)}{T(c) - x}.$$

Here $\nu^{(s)}$ is a discrete measure such that $\operatorname{supp}\{\nu^{(s)}\} \subset [-\xi, \xi]$, $\nu^{(s)}([-\xi, \xi]) = 1$; recall that $[-\xi, \xi]$ is the smallest interval containing the Julia set E. In particular,
$$T^{(sd_n)}(c) = T(c) - q_0 - p_{sd_n}^2 \int \frac{d\nu^{(sd_n)}(x)}{T(c) - x}.$$

Now, since
$$\operatorname{dist}_{\{c:T'(c)=0\}}\{T(c), [-\xi, \xi]\} = \delta > 0,$$
for every $\epsilon > 0$, there exists n such that
$$(1-\epsilon) \leq \frac{T^{(sd_n)}(c)}{T(c) - q_0} \leq (1-\epsilon)^{-1}$$

(here we have used Lemma 3.2). Recalling (25), we see that Lemma 3.4 completes the proof. □

References

1. J. Avron and B. Simon, *Singular continuous spectrum for a class of almost periodic Jacobi matrices* Bull. Amer. Math. Soc. (N.S.) **6** (1982), 81–85.
2. M. F. Barnsley, J. S. Geronimo and A. N. Harrington, *Almost periodic Jacobi matrices associated with Julia sets for polynomials*, Comm. Math. Phys. **99** (1985), 303–317.
3. J. Bellissard, D. Bessis and P. Moussa, *Chaotic states of almost periodic Schrödinger operators*, Phys. Rev. Lett. **49** (1982), 701–704.
4. J. Bellissard and B. Simon, *Cantor spectrum for the almost Mathieu equation* J. Funct. Anal. **48** (1982), 408–419.
5. R. Bowen, *Equilibrium States and the Ergodic Theory of Anosov Diffeomorphisms*, Springer, Berlin, Heidelberg, New York, 1975.
6. L. Faybusovich and M. Gekhtman, *Poisson brackets on rational functions and multi-Hamiltonian structure for integrable lattices*, Phys. Lett. A **272** (2000), 236–244.
7. A. Eremenko and M. Lyubich, *The dynamics of analytic transformations*, Algebra i Analiz 1 (1989), no. 3, 1–70; translation in Leningrad Math. J. **1** (1990), 563–634.
8. G. Levin, M. Sodin and P. Yuditskii, *A Ruelle operator for a real Julia set*, Comm. Math. Phys. **141** (1991), 119–132.
9. M. Sodin and P. Yuditskii, *The limit-periodic finite-difference operator on $l^2(\mathbf{Z})$ associated with iterations of quadratic polynomials*, J. Statist. Phys. **60** (1990), 863–873.

SCHOOL OF MATHEMATICS, GEORGIA INSTITUTE OF TECHNOLOGY, ATLANTA, GA 30332-0160, U.S.A.
E-mail address: `jeanbel@math.gatech.edu`

SCHOOL OF MATHEMATICS, GEORGIA INSTITUTE OF TECHNOLOGY, ATLANTA, GA 30332-0160, U.S.A.
E-mail address: `geronimo@math.gatech.edu`

DEPARTMENT OF MATHEMATICS, MICHIGAN STATE UNIVERSITY, EAST LANSING, MI 48824, U.S.A.
E-mail address: `volberg@math.msu.edu`

INSTITUTE FOR ANALYSIS, JOHANNES KEPLER UNIVERSITY OF LINZ, A-4040 LINZ, AUSTRIA
E-mail address: `Petro.Yudytskiy@jku.at`

Quasinormal Families and Periodic Points

Walter Bergweiler

Dedicated to Larry Zalcman on his 60th birthday

ABSTRACT. Let $n \geq 2$ be an integer and $K > 1$. By f^n we denote the n-th iterate of a function f. Let $\mathcal{F}$ be the family of all functions f holomorphic in some domain such that $|(f^n)'(\xi)| \leq K$ whenever $f^n(\xi) = \xi$. We show that $\mathcal{F}$ is quasinormal of order 1. If K is sufficiently small, then $\mathcal{F}$ is normal. We also show that if f is a transcendental entire function, then f has a sequence (ξ_k) of periodic points of period n such that $(f^n)'(\xi_k) \to \infty$ as $k \to \infty$.

1. Introduction and main results

Let $D \subset \mathbb{C}$ be a domain and let $f : D \to \mathbb{C}$ be a holomorphic function. The iterates $f^n : D_n \to \mathbb{C}$ of f are defined by $D_1 := D$, $f^1 := f$ and $D_n := f^{-1}(D_{n-1})$, $f^n := f^{n-1} \circ f$ for $n \in \mathbb{N}$, $n \geq 2$. Note that $D_2 = f^{-1}(D_1) \subset D = D_1$ and thus $D_{n+1} \subset D_n \subset D$ for all $n \in \mathbb{N}$.

A point $\xi \in D$ is called a *periodic point of period n* of f if $\xi \in D_n$ and $f^n(\xi) = \xi$, but $f^m(\xi) \neq \xi$ for $1 \leq m \leq n-1$. A periodic point of period 1 is called a *fixed point*. The periodic points of period n are thus the fixed points of f^n which are not fixed points of f^m for any m less than n. Let ξ be a periodic point of period n of f. We say that ξ is *repelling* if $|(f^n)'(\xi)| > 1$.

The periodic points play an important role in complex dynamics. For example, the Julia set of a rational or entire function, which is defined as the set where the iterates fail to be normal, is the closure of the set of repelling periodic points.

The following result is due to M. Essén and S. Wu [11, Theorem 1].

THEOREM A. *Let $D \subset \mathbb{C}$ be a domain and let $\mathcal{F}$ be the family of all holomorphic functions $f : D \to \mathbb{C}$ for which there exists $n = n(f) > 1$ such that f^n has no repelling fixed point. Then $\mathcal{F}$ is normal.*

Without the word "repelling" the same authors had proved this result earlier in [10]. We mention that the results by Essén and Wu [10, 11] answered a question of L. Yang [21, Problem 8].

2000 *Mathematics Subject Classification.* Primary 30D45; Secondary 30D05, 37F10.

This research was supported by the German-Israeli Foundation for Scientific Research and Development (G.I.F.), grant no. G-643-117.6/1999.

We note that in Theorem A the condition that f^n has no repelling fixed point cannot be replaced by the condition that f has no periodic point of period n. In fact, the family $\mathcal{F} = \{nz\}_{n\in\mathbb{N}}$ is not normal at 0, and the functions in $\mathcal{F}$ do not have periodic points of period greater than one. Also, $\mathcal{F} = \{-z + az^2\}_{a\in\mathbb{C}\setminus\{0\}}$ is not normal at 0, and the functions in $\mathcal{F}$ do not have periodic points of period 2. It was shown in [**3**, Theorem 2] that – in a suitable sense – non-normal sequences of holomorphic functions which fail to have periodic points of some period greater than one always arise from these examples.

Moreover, the following result was proved in [**3**, Theorem 3].

THEOREM B. *Let $D \subset \mathbb{C}$ be a domain and let $\mathcal{F}$ be the family of all holomorphic functions $f : D \to \mathbb{C}$ for which there exists $n = n(f) > 1$ such that f has no repelling periodic point of period n. Then $\mathcal{F}$ is quasinormal.*

Recall here that a family $\mathcal{F}$ of functions holomorphic in a domain D is called *quasinormal* (cf. [**8, 14, 19**]) if for each sequence (f_k) in $\mathcal{F}$, there exists a subsequence (f_{k_j}) and a finite set $E \subset D$ such that (f_{k_j}) converges locally uniformly in $D\setminus E$. If the cardinality of the exceptional set E can be bounded independently of the sequence (f_k), and if q is the smallest such bound, then we say that $\mathcal{F}$ is quasinormal of *order q*.

Our first result is an improvement of Theorem B.

THEOREM 1.1. *Let $K > 1$ and let $D \subset \mathbb{C}$ be a domain. Let $\mathcal{F}$ be the family of all holomorphic functions $f : D \to \mathbb{C}$ for which there exists $n = n(f) > 1$ such that $|(f^n)'(\xi)| \leq K$ for every periodic point ξ of period n of f. Then $\mathcal{F}$ is quasinormal of order 1.*

The following results are closely connected to Theorems A and B.

THEOREM C. *Let f be a transcendental entire function and let $n \in \mathbb{N}$, $n \geq 2$. Then f^n has infinitely many fixed points.*

THEOREM D. *Let f be a transcendental entire function and let $n \in \mathbb{N}$, $n \geq 2$. Then f has infinitely many repelling periodic points of period n.*

Theorem C is due to P. C. Rosenbloom [**18**], while Theorem D can be found in [**4**, Theorem 1]. We mention that already P. Fatou [**12**, p. 345] had proved that the second iterate of a transcendental entire function has a fixed point; he had used this result to show [**12**, p. 348–350] that the Julia set of an entire transcendental function is always non-empty.

The connection between Theorems A and B on the one hand and Theorems C and D on the other hand is given by a heuristic principle attributed to A. Bloch which relates normal families and entire functions. This principle says that the family of all holomorphic functions with a certain property is likely to be normal if all entire functions with this property are constant. More generally, one may expect normality or at least quasinormality, if there are only "few" entire functions with this property. For an excellent discussion of Bloch's Principle, we refer to the book by J. Schiff [**19**] and two papers by L. Zalcman [**22, 23**].

The following sharpening of Theorem D can be considered as an analogue of Theorem 1.1 for entire functions according to (the converse of) Bloch's Principle.

THEOREM 1.2. *Let f be a transcendental entire function and let $n \in \mathbb{N}$, $n \geq 2$. Then f has a sequence (ξ_k) of periodic points of period n such that $(f^n)'(\xi_k) \to \infty$ as $k \to \infty$.*

Theorem 1.1 shows in particular that if $K > 1$, then the family $\mathcal{F}_K$ of all holomorphic functions f for which there exists $n \geq 2$ such that $|(f^n)'(\xi)| \leq K$ for all fixed points ξ of f^n is quasinormal. Considering the family $\mathcal{G} = \{az^2\}_{a \in \mathbb{C} \setminus \{0\}}$, we see that $\mathcal{F}_K$ fails to be normal if $K \geq 4$. Indeed, for $f(z) = az^2 \in \mathcal{G}$, the fixed points of f^2 are given by $\xi = 0$ and $\xi = \omega/a$, where ω is a third root of unity; and we have $(f^2)'(0) = 0$ and $(f^2)'(\omega/a) = 4$, so that $\mathcal{G} \subset \mathcal{F}_4$. Clearly $\mathcal{G}$ is not normal at 0, and thus $\mathcal{F}_K$ is not normal if $K \geq 4$.

It does not seem unlikely that $\mathcal{F}_K$ is normal if $K < 4$. In this direction we have the following result.

THEOREM 1.3. *For each integer n greater than 1, there exists a constant $K_n > 1$ with the following property: if $D \subset \mathbb{C}$ is a domain and $\mathcal{F}$ is a family of holomorphic functions $f : D \to \mathbb{C}$ such that $|(f^n)'(\xi)| \leq K_n$ for all fixed points ξ of f^n, then $\mathcal{F}$ is normal.*

The example mentioned above shows that the conclusion of Theorem 1.3 does not hold for $K_n = 2^n$. Possibly it holds for each $K_n < 2^n$. This would follow if one could show that for any polynomial p of degree at least 2, p^n has a fixed point ξ such that $|(p^n)'(\xi)| \geq 2^n$ (cf. Lemma 5.2 below).

ACKNOWLEDGMENTS. I thank Jürgen Grahl for useful comments on an early version of this paper.

2. Preliminary Lemmas

We shall build on the ideas developed in [**3, 10, 11**]. As in these papers, one of the central tools comes from the Ahlfors theory of covering surfaces; see [**1**], [**13**, Chapter 5] or [**15**, Chapter XIII] for an account of this theory. The idea to use the Ahlfors theory to prove the existence of repelling periodic points is due to I. N. Baker [**2**], who had used it to show that such points are dense in the Julia set of an entire transcendental function. For a survey of further applications of the Ahlfors theory in complex dynamics, we refer to [**6**].

Given a holomorphic function $f : D \to \mathbb{C}$ and a Jordan domain $V \subset \mathbb{C}$, we say that f has a *simple island* over V if $f^{-1}(V)$ has a component U with $\overline{U} \subset D$ such that $f|_U : U \to V$ is bijective. Such a component U is then called a *simple island* over V. We often use the trivial observation that if f has a simple island U over V and if V' is a Jordan domain contained in V, then f has a simple island $U' \subset U$ over V'.

The result from the Ahlfors theory that we need is the following Lemma 2.1. Besides the references already mentioned, we refer to [**5**] for a proof.

LEMMA 2.1. *Let $D \subset \mathbb{C}$ be a domain and let D_1, D_2 and D_3 be Jordan domains with pairwise disjoint closures. Let $\mathcal{F}$ be a family of functions holomorphic in D which is not normal. Then there exists a function $f \in \mathcal{F}$ which has a simple island over D_1, D_2 or D_3.*

This result has a counterpart for entire functions according to the principle by Bloch already mentioned: if f is a non-constant entire function and if D_1, D_2, D_3 are Jordan domains with pairwise disjoint closures, then f has a simple island over at least one of these domains. This result (and similarly Lemma 2.1) does not hold if we take only two domains D_1 and D_2, as shown by the example $f(z) = \cos z$, $D_1 = D(-1, \frac{1}{2})$, $D_2 = D(1, \frac{1}{2})$, with $D(a, r) := \{z \in \mathbb{C} : |z - a| < r\}$. A rather simple but, for the purposes of this paper, quite important observation is that if f is a polynomial, then we need only two domains D_1 and D_2 in the above statement. More generally, this holds for polynomial-like mappings f.

By definition, if $U, V \subset \mathbb{C}$ are Jordan domains with $\overline{U} \subset V$, and if $f : U \to V$ is a proper holomorphic map (of degree d), then the triple (f, U, V) is called a *polynomial-like map* (of *degree* d). The basic result about polynomial-like maps (see [7, Theorem VI.1.1] or [9, Theorem 1]) says that polynomial-like maps are quasiconformally conjugate to polynomials (of the same degree), but we do not need this result. The concept of a polynomial-like mapping, introduced by A. Douady and J. H. Hubbard [9], was also used in [3, 4, 10, 11] to prove the existence of (repelling) fixed points and periodic points.

LEMMA 2.2. *Let (f, U, V) be a polynomial-like map of degree d and let D_1 and D_2 be Jordan domains with pairwise disjoint closures contained in V. Then there exist two domains $U_1, U_2 \subset U$ which are simple islands over D_1 or D_2.*

Note that U_1 and U_2 need not be islands over the same domain. We allow the possibility that U_1 is an island over D_1 and U_2 is an island over D_2, or vice versa.

PROOF OF LEMMA 2.2. We denote by $V_1, \ldots, V_m$ the components of $f^{-1}(D_1)$ and by $W_1, \ldots, W_n$ the components of $f^{-1}(D_2)$. Let μ_j be the degree of the proper map $f|_{V_j} : V_j \to D_1$ and let ν_j be the degree of the proper map $f|_{W_j} : W_j \to D_2$. Then $d = \sum_{j=1}^m \mu_j = \sum_{j=1}^n \nu_j$. By the Riemann-Hurwitz formula ([20, p.7]; observe that since D_1 and D_2 are simply connected, so are the V_j and W_j), f has $d - 1$ critical points in U; and of these critical points, there are $\mu_j - 1$ critical points in V_j and $\nu_j - 1$ critical points in W_j. Thus

$$d - 1 \geq \sum_{j=1}^m (\mu_j - 1) + \sum_{j=1}^n (\nu_j - 1) = \sum_{j=1}^m \mu_j + \sum_{j=1}^n \nu_j - m - n = 2d - m - n.$$

This yields $m + n \geq d + 1$. Since f has $d - 1$ critical points, at least two of the domains V_j, W_j do not contain a critical point and thus are simple islands. $\square$

The following lemma is simple and well-known.

LEMMA 2.3. *Let $0 < \delta < \frac{\varepsilon}{2}$ and let $U \subset D(a, \delta)$ be a simply-connected domain. Let $f : U \to D(a, \varepsilon)$ be holomorphic and bijective. Then f has a fixed point ξ in U which satisfies $|f'(\xi)| \geq \varepsilon/4\delta$.*

PROOF. We consider the inverse function $g : D(a, \varepsilon) \to U$ of f. It follows easily from Rouché's Theorem that g has a fixed point $\xi \in U \subset D(a, \delta)$. The function

$$h(z) = \frac{1}{2\delta} \left(g\left(\frac{\varepsilon}{2} z + \xi\right) - \xi \right)$$

then maps the unit disk $D(0, 1)$ into itself and satisfies $h(0) = 0$. From Schwarz's Lemma, we have $1 \geq |h'(0)| = \varepsilon |g'(\xi)|/4\delta$. The conclusion follows since $f'(\xi) = 1/g'(\xi)$. $\square$

We recall some elementary graph theoretic notions used in [3] (and implicitly in [10, 11]). For a set V and a set $E \subset V \times V$, we call the pair $G = (V, E)$ a *digraph*. The elements of V are called *vertices* and those of E are called *edges*. In contrast to the usual terminology, we allow loops; that is, we do not exclude edges e of the form $e = (v, v)$ with $v \in V$.

We call $w = (v_0, v_1, \ldots, v_n) \in V^{n+1}$ a *closed walk of length n* if $v_0 = v_n$ and $(v_{k-1}, v_k) \in E$ for $k = 1, \ldots, n$. Note that we do not assume that $v_j \neq v_k$ for $0 \leq j < k \leq n-1$. We call a closed walk $w = (v_0, v_1, \ldots, v_n)$ *primitive* if there does not exist $p \in \mathbb{N}$, $1 \leq p < n$, such that $p|n$ and $v_j = v_k$ for all j, k satisfying $p|(j - k)$. A primitive closed walk is thus a closed walk which is not obtained by running through a closed walk of smaller length several times. Finally recall that the outdegree of a vertex v is defined to be the cardinality of $\{u \in V : (v, u) \in E\}$.

As in [**3**, §5], we use the following lemma.

LEMMA 2.4. *Let $q, n \in \mathbb{N}$, $q \geq 6$, $n \geq 2$. Let G be a digraph with q vertices such that the outdegree of each vertex is a least $q - 2$. Then G contains a primitive closed walk of length n.*

We also use the following well-known result (see [**8**, p. 131] or [**19**, Proposition A.2]) about quasinormal families, whose simple proof we include for completeness.

LEMMA 2.5. *Let $\mathcal{F}$ be a family of functions holomorphic in a domain D. Suppose that $\mathcal{F}$ is quasinormal of order q. Let (f_k) be a sequence in $\mathcal{F}$ and let $a_1, \ldots, a_q \in D$. If no subsequence of (f_k) is normal at any of the points a_j, then $f_k \to \infty$ in $D \setminus \{a_1, \ldots, a_q\}$.*

PROOF. Since $\mathcal{F}$ is quasinormal of order q and no subsequence of (f_k) is normal at the points a_j, the sequence (f_k) is normal in $D \setminus \{a_1, \ldots, a_q\}$. If a subsequence of (f_k) would tend to a finite limit function in $D \setminus \{a_1, \ldots, a_q\}$, then, by the maximum principle, this subsequence would be locally bounded and thus normal in D. Thus $f_k \to \infty$ in $D \setminus \{a_1, \ldots, a_q\}$. □

The following result [**19**, Theorem A.6] is a simple consequence of Lemma 2.5.

LEMMA 2.6. *Let $\mathcal{F}$ be a family of functions holomorphic in a domain D. Suppose that $\mathcal{F}$ is quasinormal of order q. If the functions in $\mathcal{F}$ are bounded at $q + 1$ points of D, then $\mathcal{F}$ is normal.*

3. Proof of Theorem 1.1

First we prove that $\mathcal{F}$ is quasinormal of order at most 5. The argument in this part of the proof is essentially the same as in [**3**]. We refer to this paper for further details.

Assuming that $\mathcal{F}$ is not quasinormal of order at most 5, we find six points $a_1, \ldots, a_6 \in D$ and a sequence (f_k) in $\mathcal{F}$ such that no subsequence of (f_k) is normal in a neighborhood of any a_j. We choose $\varepsilon < \min_{i \neq j} |a_i - a_j|$ and $\delta < \varepsilon/4K$. For fixed k we consider the digraph $G = (V, E)$ whose vertices are the a_j and whose edges are all pairs (a_i, a_j) for which f_k has a simple island over $D(a_j, \varepsilon)$ which is contained in $D(a_i, \delta)$. It follows from Lemma 2.1 that if k is large enough, then the outdegree of each vertex is at least 4. Lemma 2.4 now shows that if $n \geq 2$, then G contains a primitive closed walk $(a_{i_0}, a_{i_1}, \ldots, a_{i_n})$ of length n. Thus $D(a_{i_{n-1}}, \delta)$ contains a simple island U_{n-1} over $D(a_{i_n}, \varepsilon)$. Next, $D(a_{i_{n-2}}, \delta)$ contains a simple island over $D(a_{i_{n-1}}, \varepsilon)$ and thus, in particular, a simple island U_{n-2} over U_{n-1}. Inductively we find simple islands $U_j \subset D(a_{i_j}, \delta)$ over U_{j+1} for $j = 0, 1, \ldots, n - 2$. We deduce that f_k^n maps U_0 bijectively onto $D(a_{i_n}, \varepsilon)$. Since $U_0 \subset D(a_{i_0}, \delta) = D(a_{i_n}, \delta)$, it follows from Lemma 2.3 that f_k^n has a fixed point $\xi \in U_0 \subset D(a_{i_0}, \delta)$ with

$|(f_k^n)'(\xi)| \geq \varepsilon/4\delta > K$. Since the walk $(a_{i_0}, a_{i_1}, \ldots, a_{i_n})$ is primitive, this fixed point ξ of f_k^n is in fact a periodic point of f_k of period n. This contradicts the assumption that $f_k \in \mathcal{F}$.

Hence $\mathcal{F}$ is quasinormal of order q for some $q \leq 5$. We have to show that $q = 1$ and thus assume that $q \geq 2$. As before, there exist q points $a_1, \ldots, a_q \in D$ and a sequence (f_k) in $\mathcal{F}$ such that no subsequence of (f_k) is normal in a neighborhood of any a_j. By Lemma 2.5, $f_k \to \infty$ in $D \backslash \{a_1, \ldots, a_q\}$.

Again we choose $\varepsilon < \min_{i \neq j} |a_i - a_j|$ and $\delta < \varepsilon/4K$. We also choose $R > \max_j |a_j| + \varepsilon$ and find that if k is sufficiently large, then $|f_k(z)| > R$ for $|z - a_j| = \delta$ and $j = 1, \ldots, q$. On the other hand, since no subsequence of (f_k) is normal at a_j, we have $|f_k(z)| < R$ for some $z \in D(a_j, \delta)$ is k is large. Thus $f_k^{-1}(D(0, R))$ has a component $U_j \subset D(a_j, \delta)$ for $j = 1, \ldots, q$, provided that k is large enough. Clearly, $(f_k|_{U_j}, U_j, D(0, R))$ is a polynomial-like map.

By Lemma 2.2, there exist two simple islands over $D(a_1, \varepsilon)$ or $D(a_2, \varepsilon)$ contained in U_1, and two further simple islands over $D(a_1, \varepsilon)$ or $D(a_2, \varepsilon)$ contained in U_2. Without loss of generality, we may assume that two of these four simple islands are over $D(a_1, \varepsilon)$, that is, we have two simple islands V_1, W_1 over $D(a_1, \varepsilon)$ which are contained in $U_1 \cup U_2$. Using Lemma 2.2, we now find for $j = 2, 3, \ldots, n - 1$ domains $V_j, W_j \subset U_2$ which are simple islands over V_{j-1} or W_{j-1}, and finally domains $V_n, W_n \subset U_1$ which are simple islands over V_{n-1} or W_{n-1}. Then f^n maps V_n bijectively onto $D(a_1, \varepsilon)$. Lemma 2.3 shows that f^n has a fixed point $\xi \in V_n$ with $|(f^n)'(\xi)| > K$. Moreover, $f^j(\xi) \in V_{n-j} \cup W_{n-j} \subset U_2 \subset D(a_2, \delta)$ for $j = 1, 2, \ldots, n - 2$; and this implies that ξ is in fact a periodic point of period n, provided that $n \geq 3$.

We now consider the case that $n = 2$. If V_1, W_1 are both contained in U_2, then the above argument also works for $n = 2$; that is, the fixed point $\xi \in V_2 \subset U_1$ of f^2 constructed above satisfies $f(\xi) \in V_1 \cup W_1 \subset U_2$ and thus is not a fixed point of f, but a periodic point of period 2. Next we consider the case that V_1, W_1 are both contained in U_1. Since V_1 is a simple island over $D(a_1, \varepsilon)$ and $W_1 \subset U_1 \subset D(a_1, \varepsilon)$, we see that V_1 contains a simple island X over W_1. Since f^2 maps X bijectively onto $D(a_1, \varepsilon)$, Lemma 2.2 again yields a fixed point $\xi \in X$ of f^2 with $|(f^2)'(\xi)| > K$. Since $f(\xi) \in W_1$ and $X \cap W_1 \subset V_1 \cap W_1 = \emptyset$, we have $f(\xi) \neq \xi$, so that ξ has period 2. Finally, we consider the case that both U_1 and U_2 contain only one island over $D(a_1, \varepsilon)$, say $V_1 \subset U_1$ and $W_1 \subset U_2$. Then $U_1 \cup U_2$ also contains two simple islands X_1, Y_1 over $D(a_2, \varepsilon)$. We may assume that one of them is contained in U_1 and one is contained in U_2, say $X_1 \subset U_1$ and $Y_1 \subset U_2$, since otherwise we are in one of the situations already considered. Now X_1 contains a simple island Z over W_1, and thus f^2 maps Z bijectively onto $D(a_1, \varepsilon)$. As before, we deduce from Lemma 2.2 that f has a periodic point $\xi \in Z$ of period 2 with $|(f^2)'(\xi)| > K$.

We have thus obtained a contradiction to the assumption that $q \geq 2$ in all cases.

4. Proof of Theorem 1.2

The basic idea is essentially the same as in [**3**, §4]. Let (c_k) be a sequence tending to ∞ such that $(f(c_k))$ is bounded. Define $f_k : \mathbb{C} \to \mathbb{C}$ by $f_k(z) = f(c_k z)/c_k$. Then $f_k(0) \to 0$ and $f_k(1) \to 0$.

Suppose now that the conclusion does not hold. Then there exists $K > 1$ such that $|(f^n)'(\xi)| \leq K$ for all periodic points ξ of period n of f. We note that if ξ

is a periodic point of f_k of period n, then $c_k \xi$ is a periodic point of f of the same period, and $(f_k^n)'(\xi) = (f^n)'(c_k \xi)$. Thus the sequence (f_k) is quasinormal of order 1 by Theorem 1.1. Since (f_k) is bounded at 0 and 1, Lemma 2.6 yields that (f_k) is normal.

On the other hand, it is not difficult to see that (f_k) is not normal at 0. For example, this follows since $M(r, f) := \max_{|z|=r} |f(z)| \geq r^2$ for sufficiently large r so that
$$M\left(\frac{1}{\sqrt{|c_k|}}, f_k\right) = \frac{M(\sqrt{|c_k|}, f)}{|c_k|} \geq 1$$
for large k, while $f_k(0) \to 0$. Thus no subsequence of (f_k) can converge in a neighborhood of 0.

5. Proof of Theorem 1.3

The proof of Theorem 1.3 is based on the following lemma due to X. C. Pang and L. Zalcman [17, Lemma 2].

LEMMA 5.1. *Let $\mathcal{F}$ be a family of functions meromorphic on the unit disc, all of whose zeros have multiplicity at least k, and suppose that there exists $A \geq 1$ such that $|g^{(k)}(\xi)| \leq A$ whenever $g(\xi) = 0$, $g \in \mathcal{F}$. Then if $\mathcal{F}$ is not normal there exist, for each $0 \leq \alpha \leq k$, a number $r \in (0,1)$, points $z_j \in D(0,r)$, functions $g_j \in \mathcal{F}$ and positive numbers ρ_j tending to zero such that*
$$\frac{g_j(z_j + \rho_j z)}{\rho_j^\alpha} \to G(z)$$
locally uniformly, where G is a nonconstant meromorphic function on $\mathbb{C}$ such that the spherical derivative $G^\#$ of G satisfies $G^\#(z) \leq G^\#(0) = kA + 1$ for all $z \in \mathbb{C}$.

We only need the case $k = 1$ of Lemma 5.1. This special case can also be found in [16, Lemma 2].

We also require the following result of Essén and Wu [11, Theorem 4].

LEMMA 5.2. *Let f be a polynomial of degree at least 2 and let $n \geq 2$. Then f^n has a repelling fixed point.*

PROOF OF THEOREM 1.3. We denote by $\mathcal{F}(D, n, K)$ the family of all functions f holomorphic in D such that $|(f^n)'(\xi)| \leq K$ whenever $f^n(\xi) = \xi$. Note that this implies that $|f'(\xi)| \leq \sqrt[n]{K}$ whenever $f(\xi) = \xi$. We fix $n \geq 2$ and suppose that there does not exist $K > 1$ such that $\mathcal{F}(D, n, K)$ is normal, in order to seek a contradiction. We may assume that D is the unit disk.

For $m \in \mathbb{N}$, we choose a non-normal sequence (f_j) in $\mathcal{F}(D, n, 1 + 1/m)$. With $g_j(z) := f_j(z) - z$ we find that if $g_j(\xi) = 0$, then $f_j(\xi) = \xi$ and thus
$$|g_j'(\xi)| \leq |f_j'(\xi)| + 1 \leq \sqrt[n]{1 + \frac{1}{m}} + 1 \leq 3.$$

Clearly, the sequence (g_j) is also not normal. Applying Lemma 5.1 for $\alpha = k = 1$ and $A = 3$, we may assume, passing to a subsequence if necessary, that there exist $z_j \in D$ and $\rho_j > 0$ such that $g_j(z_j + \rho_j z)/\rho_j \to G_m(z)$ for some entire function G_m satisfying $G_m^\#(z) \leq G_m^\#(0) = 4$ for all $z \in \mathbb{C}$. With $L_j(z) = z_j + \rho_j z$, we find that
$$h_j(z) := L_j^{-1}(f_j(L_j(z))) = \frac{f_j(z_j + \rho_j z) - z_j}{\rho_j} = \frac{g_j(z_j + \rho_j z)}{\rho_j} + z \to G_m(z) + z.$$

With $F_m(z) := G_m(z) + z$, we thus have $h_j(z) \to F_m(z)$ as $j \to \infty$. It follows that $h_j^n(z) \to F_m^n(z)$. The assumption that $f_j \in \mathcal{F}(D, n, 1 + 1/m)$ implies that $h_j \in \mathcal{F}(L_j^{-1}(D), n, 1 + 1/m)$; that is, $|(h_j^n)'(\xi)| \leq 1 + 1/m$ whenever $h_j^n(\xi) = \xi$. We deduce that $F_m \in \mathcal{F}(\mathbb{C}, n, 1 + 1/m)$. Since $G_m^\#(z) \leq G_m^\#(0) = 4$ for all $z \in \mathbb{C}$, the sequence (G_m) is normal. Hence (F_m) is normal and we may assume without loss of generality that $F_m \to F$ for some non-constant entire function F. Since $F_m \in \mathcal{F}(\mathbb{C}, n, 1 + 1/m)$, we conclude that $F \in \mathcal{F}(\mathbb{C}, n, 1)$; that is, F is an entire function such that F^n has no repelling fixed point. It follows from Lemma 5.2 and Theorem D (or Theorem 1.2) that F is a polynomial of degree 1 at most. Next we note that $|F_m'(0)| \geq |G_m'(0)| - 1 \geq G_m^\#(0) - 1 = 3$, and thus $|F'(0)| \geq 3$. Hence F has the form $F(z) = az + b$ where $|a| \geq 3$. Taking $\xi := b/(1-a)$, we obtain $F(\xi) = \xi$ and $|F'(\xi)| = |a| \geq 3$, which contradicts $F \in \mathcal{F}(\mathbb{C}, n, 1)$. □

References

[1] L. V. Ahlfors, *Zur Theorie der Überlagerungsflächen*, Acta Math. **65** (1935), 157–194, and *Collected Papers*, Birkhäuser, Boston, Basel, Stuttgart, 1982, Vol. I, pp. 214–251.

[2] I. N. Baker, *Repulsive fixpoints of entire functions*, Math. Z. **104** (1968), 252–256.

[3] D. Bargmann and W. Bergweiler, *Periodic points and normal families*, Proc. Amer. Math. Soc. **129** (2001), 2881–2888.

[4] W. Bergweiler, *Periodic points of entire functions: proof of a conjecture of Baker*, Complex Variables Theory Appl. **17** (1991), 57–72.

[5] ———, *A new proof of the Ahlfors five islands theorem*, J. Analyse Math. **76** (1998), 337–347.

[6] ———, *The role of the Ahlfors five islands theorem in complex dynamics*, Conform. Geom. Dyn. **4** (2000), 22–34.

[7] L. Carleson and T. W. Gamelin, *Complex Dynamics*, Springer, New York, Berlin, Heidelberg, 1993.

[8] C.-T. Chuang, *Normal families of meromorphic functions*, World Scientific, Singapore, 1993.

[9] A. Douady and J. H. Hubbard, *On the dynamics of polynomial-like mappings*, Ann. Sci. École Norm. Sup. (4) **18** (1985), 287–343.

[10] M. Essén and S. Wu, *Fix-points and a normal family of analytic functions*, Complex Variables Theory Appl. **37** (1998), 171–178.

[11] ———, *Repulsive fixpoints of analytic functions with applications to complex dynamics*, J. London Math. Soc. (2) **62** (2000), 139–149.

[12] P. Fatou, *Sur l'itération des fonctions transcendantes entières*, Acta Math. **47** (1926), 337–360.

[13] W. K. Hayman, *Meromorphic Functions*, Clarendon Press, Oxford, 1964.

[14] P. Montel, *Leçons sur les familles normales des fonctions analytiques et leurs applications*, Gauthier-Villars, Paris, 1927.

[15] R. Nevanlinna, *Eindeutige analytische Funktionen*, Springer, Berlin, Göttingen, Heidelberg, 1953.

[16] X. C. Pang, *Shared values and normal families*, Analysis **22** (2002), 175–182.

[17] X. C. Pang and L. Zalcman, *Normal families and shared values*, Bull. London Math. Soc. **32** (2000), 325-331.

[18] P. C. Rosenbloom, *L'itération des fonctions entières*, C. R. Acad. Sci. Paris **227** (1948), 382–383.

[19] J. L. Schiff, *Normal Families*, Springer, New York, Berlin, Heidelberg, 1993.

[20] N. Steinmetz, *Rational Iteration*, de Gruyter, Berlin, New York, 1993.

[21] L. Yang, *Some recent results and problems in the theory of value-distribution*, in Proceedings of the Symposium on Value Distribution Theory in Several Complex Variables, (W. Stoll, ed.), Univ. of Notre Dame Press, Notre Dame Math. Lect. 12 (1992), 157–171.

[22] L. Zalcman, *A heuristic principle in complex function theory*, Amer. Math. Monthly **82** (1975), 813–817.

[23] ———, *Normal families: new perspectives*, Bull. Amer. Math. Soc. (N.S.) **35** (1998), 215–230.

MATHEMATISCHES SEMINAR DER CHRISTIAN–ALBRECHTS–UNIVERSITÄT ZU KIEL, LUDEWIG–MEYN–STRASSE 4, D–24098 KIEL, GERMANY
 E-mail address: bergweiler@math.uni-kiel.de

Local Center Conditions for the Abel Equation and Cyclicity of its Zero Solution

M. Blinov, M. Briskin, and Y. Yomdin

Dedicated to Larry Zalcman on the occasion of his sixtieth birthday

ABSTRACT. An Abel differential equation $y' = p(x)y^2 + q(x)y^3$ is said to have a center at a pair of complex numbers (a, b) if $y(a) = y(b)$ for any solution $y(x)$ (with initial value $y(a)$ small enough). Let p, q be polynomials and let $P = \int p$, $Q = \int q$. We say P and Q satisfy the "polynomial composition condition" if there exist polynomials $\tilde{P}, \tilde{Q}$ and W such that $P(x) = \tilde{P}(W(x))$, $Q(x) = \tilde{Q}(W(x))$, and $W(a) = W(b)$. The main result of this paper is that for a fixed polynomial p (satisfying some minor genericity restrictions) and for a fixed degree d, there exists $\epsilon(p, d) > 0$ such that for any polynomial q of degree d with norm (of q) at most $\epsilon(p, d)$, the Abel equation above has a center if and only if the polynomial composition condition is satisfied. Based on this, we also provide an upper bound for the cyclicity of the zero solution of the Abel equation, i.e., for the maximal number of periodic solutions which can appear in a small perturbation of the zero solution.

1. Introduction

Consider the system of differential equations

(1.1) $$\begin{cases} \dot{x} = -y + F(x, y) \\ \dot{y} = x + G(x, y), \end{cases}$$

where $F(x, y)$ and $G(x, y)$ vanish at the origin with their first derivatives. The system (1.1) has a center at the origin if all its trajectories starting in a sufficiently small neighborhood of the origin are closed. The classical *center-focus problem* is to find conditions on F and G necessary and sufficient for the system (1.1) to have a center at the origin.

This problem, together with the closely related second part of Hilbert's 16-th Problem (asking for the maximal possible number of isolated closed trajectories of (1.1) with $F(x, y)$ and $G(x, y)$ polynomials of a given degree), has until now resisted all attacks. Many deep partial results have been obtained (see [4, 5, 22, 33]), but

---

2000 *Mathematics Subject Classification.* Primary 34C07, 34C25.
This research was supported by ISF Grant No. 264/02 and by BSF Grant No. 2002243.

general center conditions are not known even for $F(x,y)$ and $G(x,y)$ polynomials of degree 3.

The classical approach to the center-focus problem is to analyze the conditions on the parameters of the system (1.1) provided by the vanishing of the first several obstructions to the existence of a center. If as a result one can show the existence of a first integral of (1.1), then the system has a center and no further analysis of the obstructions is necessary. The problem is that already for $F(x,y)$ and $G(x,y)$ polynomials in x and y of degree 3, the obstructions analyzed till now do not necessarily imply the existence of a first integral of any known type.

An alternative approach to both the center-focus problem and Hilbert's 16-th Problem is provided by the study of the perturbed integrable situations (in particular, perturbed Hamiltonian vector fields). See [4, 5, 22]. The investigation of the perturbation version (or of the infinitesimal version) of the above problems has led to many important results, in particular, to a serious progress in understanding the analytic structure of Abelian integrals (see [22] and references there).

In the present paper, we consider a certain variant of the center-focus problem (closely related to the original one) – the center-focus problem for the Abel differential equation

$$(1.2) \qquad \frac{dy}{dx} = p(x)y^2 + q(x)y^3.$$

This problem is to provide necessary and sufficient conditions on p, q and $a, b \in \mathbb{C}$ for all solutions $y(x)$ of (1.2) to satisfy $y(a) = y(b)$. The Abel equation version of the center-focus problem has been studied in [1, 2, 3, 17, 18, 19, 23] (see also [34]) and in many other publications. As shown in [11] and subsequent papers, it suggests important technical simplifications and opens interesting connections with classical analysis and algebra. The Abel equation versions of both the center-focus problem and Hilbert's 16-th Problem apparently reflect the main difficulties of the original classical ones.

The investigation of the infinitesimal center-focus problem for the Abel equation was begun in [11] and continued in subsequent publications of the same authors, in [7, 17] and elsewhere. It has turned out to be essentially a problem in analysis, related to the classical moment problem, on the one hand, and to the composition algebra of univariate polynomials, on the other. By now, a reasonable understanding of the infinitesimal center-focus problem for the Abel equation has been achieved, especially after the recent results of [24, 25, 26, 27, 28].

The main goal of the present paper is to apply the results of this infinitesimal analysis to the center-focus problem itself (at least, locally). This is done by a comparison of the (nonlinear) *center equations* with their linear parts.

Let us recall that, just as in the classical case, so also for the Abel equation the center conditions in the space of the parameters of the problem are given by an *infinite* set of "obstructions," i.e., of certain polynomial equations in the parameters ("center equations" – see Section 3 below). Hence, by Hilbert's basis theorem, the center conditions are in fact provided by a certain *finite* number N of center equations. Formally, one can say that the center-focus problem is just to find this number N. A more constructive approach would be to try to understand the structure of the center equations and thus to produce meaningful necessary and sufficient conditions for the center (obtaining on the way also a bound for N).

This last approach is taken in the present paper. The infinitesimal analysis of the center-focus problem for the Abel differential equation leads to the *moment vanishing condition* and to the *moment equations* which are the linear parts of the center equations. The moment equations imply (usually) the *composition condition*, which in our setting is the main (and the only) integrability condition. In particular, the composition condition implies the existence of a center. Moreover, it implies the vanishing of each nonlinear term in the center equations. We translate these analytic and algebraic information into the information about the algebro-geometric structure of the center equations. Finally, the analysis of this algebro-geometric structure implies our main results: local center conditions for the Abel equation, a description of the Bautin ideal and upper bounds for the "cyclicity" of the zero solution (i.e., for the number of the periodic solutions which can appear in a small perturbation of the zero one).

2. Statement of the main results

Consider the Abel differential equation

(2.1) $$\frac{dy}{dx} = p(x)y^2 + q(x)y^3$$

with $p(x) = P'(x)$ and $q(x) = Q'(x)$ polynomials in $x \in \mathbb{C}$.

DEFINITION 2.1. Let a pair of complex numbers (a,b) be given. The solution $y(x)$ of (2.1) is called periodic at (a,b) if $y(a) = y(b)$. Equation (2.1) is said to have a center at (a,b) if $y(a) = y(b)$ for any solution $y(x)$ with the initial value $y(a)$ small enough.

For small initial values $y(a)$, the solutions $y(x)$ of (2.1) are regular in any fixed disk around $0 \in \mathbb{C}$. Hence we do not need to specify in Definition 2.1 the continuation path from a to b for the solutions $y(x)$.

DEFINITION 2.2. Polynomials $P(x)$ and $Q(x)$ satisfy the polynomial composition condition (PCC) at $(a,b) \in \mathbb{C}$ if there exist polynomials $\tilde{P}(w), \tilde{Q}(w)$ and $W(x)$ with $W(a) = W(b)$ such that

(2.2) $$P(x) = \tilde{P}(W(x)), \; Q(x) = \tilde{Q}(W(x)).$$

The polynomial composition condition for the polynomials P and Q implies that the Abel equation (2.1) with $p(x) = P'(x)$ and $q(x) = Q'(x)$ has a center at a, b. This simple but basic fact follows via a change of variables $w = W(x)$ in (2.1), which closes the integration contour, while the coefficients of the transformed equation still remain polynomials. Hence, for small initial values, the solutions $\tilde{y}(w)$ of the transformed equation do not ramify. The solutions $y(x)$ of (2.1) are expressed as $y(x) = \tilde{y}(W(x))$; and since $W(a) = W(b)$, we have $y(a) = y(b)$ for any solution $y(x)$ of (2.1). A proof of a similar implication for iterated integrals is given in Proposition 3.4 below. See also [**11**].

For given P, Q, the condition (PCC) can be effectively verified by algebraic calculations.

A composition condition similar to (PCC) was introduced for a trigonometric Abel equation in [**1, 2**]. The condition (PCC) was introduced and intensively studied in [**11, 12, 13, 14, 6, 7, 17, 35**] There is growing evidence supporting the major role played by the (PCC) (and in general, by the polynomial composition algebra; see [**29**]) in the structure of the center conditions for the polynomial Abel equation. In particular, we have no counterexamples to the following "Composition Conjecture."

COMPOSITION CONJECTURE. *The Abel equation on the interval $[a,b]$ with polynomials p, q has a center if and only if (PCC) holds for $P = \int p$, $Q = \int q$.*

This conjecture has been verified for polynomials of small degree and in many special cases in [11, 12, 13, 14, 6, 7, 17, 35].

In this paper, we take a "non-symmetric" approach to the center-focus problem for the Abel equation. Namely, we assume the polynomial p in (2.1) to be fixed while only the degree d of q is fixed, and the polynomial q is allowed to vary in the space V_d of all the complex polynomials of degree at most d. We show that under a certain genericity assumption on the fixed p *locally with respect to* $q \in V_d$, the center condition for (2.1) and (PCC) are equivalent.

The assumption on the fixed polynomial p which is central for the method used in the present paper is given in Definition 2.3 below. We stress that this restriction can presumably be eliminated by considering higher order perturbations.

Consider the "one-sided" moments

$$(2.3) \qquad m_k = \int_a^b P^k(x)q(x)dx, \ k = 0, 1, \ldots.$$

The vanishing of the moments m_k (which we call a "moment condition") is also implied by (PCC) (for the same reasons as above and as in Proposition 3.4 below. See also [11]).

Observe that the validity of the moment condition does not depend on the choice of the constant terms in P and Q. This is immediate for Q, since only $q = Q'$ enters into the expression (2.3). The moments for $P + c$ can be linearly expressed in terms of the moments for P and vice versa, so this is true also for P. In what follows, we mostly assume that $P(a) = Q(a) = 0$.

DEFINITION 2.3. A polynomial P is called *definite* (with respect to $a, b \in \mathbb{C}$) if for any polynomial q, the vanishing of the one-sided moments m_k, $0 \leq k < \infty$, implies (and hence is equivalent to) (PCC) for P and $Q = \int q$.

All polynomials P up to degree 5 are definite. In the space V_l of polynomials P of fixed degree $l \geq 6$, non-definite polynomials belong to a certain proper algebraic subset. More specifically, all indecomposable P are definite (for every $a \neq b$), as well as all P with $P'(a) \neq 0$, $P'(b) \neq 0$. The Chebyshev polynomial T_6 is not definite with respect to $a = -\sqrt{3}/2$, $b = \sqrt{3}/2$ ([24]). In the Addendum below, we present a survey of the recent results of [8, 9, 10, 11, 12, 13, 17, 25, 26, 27, 31, 28] describing several classes of definite polynomials.

The following theorem is the first main result of this paper.

THEOREM 2.1. *Let the points a, b and the polynomial $p = P'$ be fixed, with P definite with respect to respect to a, b. Let the maximal degree d of the polynomials q be fixed. There exists $\epsilon = \epsilon(p, d, a, b) > 0$ depending only on p, a, b and d such that for any q of degree d with $\| Q \| \leq \epsilon$, the equation $y' = p(x)y^2 + q(x)y^3$ has a center at a, b if and only if P and $Q = \int q$ satisfy the polynomial composition condition (PCC).*

The proof of Theorem 2.1 is given in Section 4 below.

REMARK. In a recent paper [35], an interesting analysis of the center-focus problem for the Abel differential equation similar to our approach is presented. In particular, Theorem 5.6 of [35] is essentially a special case of our Theorem 3.1 for p of degree 1. Although formally the statement of Theorem 5.6 of [35] is weaker (it does not guarantee the uniformity of the "locality size" with respect to the polynomials q of a fixed degree), we believe that the proof of Theorem 5.6 of [35] essentially provides the uniform bound.

The approach of the present paper allows us also to compute the local Bautin ideal I of (2.1), i.e., the ideal in the ring of holomorphic functions on the ball B_ϵ of radius ϵ in V_d generated by all the Taylor coefficients $v_k(q)$ of the Poincaré first return mapping G on a, b. (See Section 3 below for the definition of G.) We can also compute the Bautin index $b(P, d, a, b)$, which is the minimal number of v_k which generate I.

In a similar way, we can define the stabilization index of the moment equations or the Bautin moment index $N(P, d, a, b)$.

DEFINITION 2.4. For P a definite polynomial on a, b and for any natural d, the Bautin moment index $N(P, d, a, b)$ is the minimal number of moments $m_k = \int_a^b P^k(x) q(x) dx$ whose vanishing implies for any $q \in V_d$ that (PCC) is satisfied by P and $Q = \int q$.

The existence (finiteness) of $N(P, d, a, b)$ follows from the stabilization of the decreasing sequence of linear subspaces L_j defined by the vanishing of the moments m_k, $0 \leq k < j$, in the space V_d. A natural conjecture is that $N(P, d, a, b)$ depends only on the *degree* of P and on d.

The following theorem is the second main result of this paper.

THEOREM 2.2. *Let $p = P'$ be fixed, with P definite. The local Bautin ideal I of the Abel equation (2.1) on a, b coincides with the ideal J generated by the moments $m_i(q)$. It is, in fact, generated by $v_2, \ldots, v_{N+3}$ or by $m_0, m_1, \ldots, m_N$, where $N = N(P, d, a, b)$ is the Bautin moment index. In particular, the Bautin index $b(P, d, a, b)$ is equal to $N(P, d, a, b) + 3$.*

Theorem 2.2 implies an explicit bound on the *cyclicity* of the zero solution $y(x) \equiv 0$ of the Abel equation (2.1), i.e., on the number of periodic solutions $y(x)$ of (2.1) (those with $y(a) = y(b)$) which can bifurcate from the zero solution.

Let a definite P and $\deg q = d$ be fixed and let $\epsilon = \epsilon(P, d) > 0$ be as defined in Theorem 2.1. The following theorem, which we prove in Section 4 below, is the third main result of this paper.

THEOREM 2.3. *There exists $\delta = \delta(P, d) > 0$ such that for any q with $\| q \| \leq \epsilon/2$, the number of solutions y of the Abel equation (2.1) satisfying $y(a) = y(b)$ and $|y(a)| \leq \delta$ does not exceed $N(P, d, a, b) + 3$.*

3. Poincaré mapping and center equations

The *Poincaré first return mapping* $G(y)$ of the Abel equation (2.1) at $a, b \in \mathbb{C}$ associates to each $y = y_a$ the value $G(y) = y(b)$ at the point b of the solution $y(x)$ of (2.1) satisfying $y(a) = y_a$ at the point a. For $y = y_a$ sufficiently small $G(y) = y(b)$

does not depend on the continuation path from a to b, so $G(y)$ is a regular function for y near zero and is given by a convergent power series

$$(3.1) \qquad G(y) = y + \sum_{k=2}^{\infty} v_k(\lambda, a, b) y^k,$$

where $\lambda = (\lambda_1, \lambda_2, \dots)$ is the (finite) set of the coefficients of p, q.

The solution $y(x)$ of (2.1) is periodic at (a, b) if and only if $G(y(a)) = y(a)$. The equation (2.1) has a center at (a, b) if and only if $G(y) \equiv y$. Therefore, we have the following simple but basic fact.

PROPOSITION 3.1. *The Abel equation (2.1) has a center at a, b if and only if the infinite sequence of equations*

$$(3.2) \qquad v_k(\lambda, a, b) = 0, \quad k = 2, \dots$$

is satisfied.

We call equations (3.2) the *center equation* and the set $\mathcal{C}$ of the parameters λ satisfying (3.2) the *center set*.

It is convenient to "free" the endpoint b in (3.1): denoting by $G(y, x)$ the Poincaré first return mapping $G(y)$ at a, x, we obtain a convergent Taylor representation which can be used to express both the Poincaré mapping (when we fix x) and the solutions of (2.1) (when we fix y):

$$(3.3) \qquad G(y, x) = y + \sum_{k=2}^{\infty} v_k(\lambda, a, x) y^k.$$

One shows easily (by substituting expansion (3.3) into equation (2.1)) that $v_k(x) = v_k(\lambda, a, x)$ satisfy the recurrence relation

$$(3.4) \qquad \begin{cases} v_0(x) \equiv 0 \\ v_1(x) \equiv 1 \\ v_n(0) = 0 \quad \text{and} \\ v_n'(x) = p(x) \sum_{i+j=n} v_i(x) v_j(x) + q(x) \sum_{i+j+k=n} v_i(x) v_j(x) v_k(x), \ n \geq 2. \end{cases}$$

An immediate consequence is the following result.

PROPOSITION 3.2. *The Taylor coefficients $v_k(\lambda, a, x)$ are polynomials in a, x and in λ. In particular, for $x = b$, the coefficients $v_k(\lambda) = v_k(\lambda, a, b)$ are polynomials in the parameters a, b, λ.*

PROOF. This follows from the recurrence relation (3.4) via induction on k. □

So, in fact, the center equations (3.2) are polynomial equations. By the Hilbert basis theorem, the center set is defined by a finite subsystem of (3.2) and, in particular, is an algebraic subset of the space of parameters.

It was shown in [11] that the recurrence relation (3.4) can be linearized in the following sense. Consider the inverse Poincaré mapping G^{-1} associating to the end value $y(x) = y_x$ of each solution y of (2.1) its initial value $y(a) = y_a$. We have a Taylor expansion

$$(3.5) \qquad y_a = G^{-1}(y_x) = y_x + \sum_{k=2}^{\infty} \psi_k(\lambda, a, x) y_x^k.$$

In particular, for $x = b$, we get the inverse to the Poincaré mapping G at a, b. Hence the center condition $y(a) \equiv y(b)$ is equivalent to another infinite system of polynomial equations in λ:

(3.6) $$\psi_k(\lambda, a, b) = \psi_k(\lambda) = 0, \quad k = 2, \ldots.$$

One can show (see [**11**]) that for each $k = 2, \ldots$, the ideals $I_k = \{v_2(\lambda), \ldots, v_k(\lambda)\}$ and $I'_k = \{\psi_2(\lambda), \ldots, \psi_k(\lambda)\}$ in the ring of polynomials in λ coincide.

It was shown in [**11**] that for a fixed λ, the Taylor coefficients $\psi_k(x) = \psi_k(\lambda, a, x)$ satisfy a *linear* recurrence relation

(3.7) $$\begin{cases} \psi_0(x) \equiv 0 \\ \psi_1(x) \equiv 1 \\ \psi_n(0) = 0 \quad \text{and} \\ \psi'_n(x) = -(n-1)\psi_{n-1}(x)p(x) - (n-2)\psi_{n-2}(x)q(x), \; n \geq 2. \end{cases}$$

Now one can see that each $\psi_k(\lambda, a, b) = \psi_k(\lambda)$ can be written as a sum of iterated integrals: each summand has the form $Const \cdot \int q \int p \ldots \int p \int q$ (the order and the number of the integrands p and q varies). More precisely, the iterated integrals entering the polynomials $\psi_k(\lambda)$ are given by

(3.8) $$I_\alpha = \int_a^b h_{\alpha_1}(x_1)dx_1 \left(\int_a^{x_1} h_{\alpha_2}(x_2)dx_2 \ldots \left(\int_a^{x_{s-1}} h_{\alpha_s}(x_s)dx_s \right) \ldots \right).$$

Here the α are multi-indices $\alpha = (\alpha_1, \ldots, \alpha_s)$ with $\alpha_j = 1$ or 2, and $h_1 = p, h_2 = q$.

Formally integrating recurrence relation (3.7), we can obtain in a combinatorial way the symbolic expressions for ψ_k through the sums of the iterated integrals (3.8). The first few of these expressions for ψ_k are as follows:

$\psi_0 \equiv 0$

$\psi_1 \equiv 1$

$\psi_2 = -\int p = I_1$

$\psi_3 = 2\int p \int p - \int q = 2I_{11} - I_2$

$\psi_4 = -6\int p \int p \int p + 3 \int p \int q + 2 \int q \int p = -6I_{111} + 3I_{12} + 2I_{21}$

$\psi_5 = 24I_{1111} - 12I_{112} - 8I_{121} - 6I_{211} + 3I_{22}$

$\psi_6 = -120I_{11111} + 60I_{1112} + 40I_{1121} + 30I_{1211} - 15I_{122} + 24I_{2111} - 12I_{212} - 8I_{221}$

The basic combinatorial structure of the symbolic expressions for ψ_k produced via the recurrence relation (3.7) is given by the following proposition.

PROPOSITION 3.3. *For each $k \geq 2$, the Poincaré coefficient ψ_k is given as an integer linear combination of iterated integrals of p and q*

(3.9) $$\psi_k = \sum n_\alpha I_\alpha,$$

with the sum running over all the multi-indices $\alpha = (\alpha_1, \ldots, \alpha_s)$ for which $\sum_1^s \alpha_j = k - 1$. The number of the terms in the expression for ψ_k is the $(k-1)$-st

Fibonacci number. The integer coefficients n_α are given as the products

$$(3.10) \qquad n_\alpha = (-1)^s \prod_{r=1}^{s}\left(k - \sum_{j=1}^{r}\alpha_j\right).$$

PROOF. By induction. Assuming that the result is true for $k < m$, we apply the recurrence relation (3.7) and represent the terms ψ_{k-1} and ψ_{k-2} according to the expression (3.9). Integrating the right hand side of (3.7), we obtain ψ_k as the integer sum of the new iterated integrals, each one containing exactly one integrand more than before the integration. These new iterated integrals can be naturally split into the two groups corresponding to the two terms on the right hand side of the recurrence relation (3.7). In the first group, the new integrand on the left is p and in the second group it is q. Hence the multi-indices α in these two groups are mutually different, and these multi-indices cover together all the α with $\sum_1^s \alpha_j = k$. This proves also that the number of the terms in the expression for ψ_k is the $(k-1)$-st Fibonacci number. The formula (3.10) for the coefficients n_α follows immediately by induction from (3.7). □

REMARK. Another derivation of the iterated integrals form of the center equations has been obtained in [16] by a completely different method.

As was mentioned above, we assume in this paper that in the Abel equation (2.1)

$$y' = p(x)y^2 + q(x)y^3,$$

the polynomial p is fixed, while q is considered as a variable polynomial belonging to the space V_d of all the univariate polynomials of a given degree d. So let us denote by $\mu = (\mu_0, \ldots, \mu_d)$ the coefficients of the polynomial q. The parameters μ form a part of the complete set λ of parameters of the Abel equation (2.1). In the setting where the polynomial p is fixed we can consider the expressions $\psi_k = \psi_k(\lambda)$ introduced above as the functions $\psi_k(\mu)$ of the variables μ only.

COROLLARY 3.1. *For each $k \geq 2$, the Poincaré coefficient $\psi_k(\mu)$ is a polynomial in μ of degree $[\frac{k-1}{2}]$.*

PROOF. The iterated integrals I_α are polynomials in μ of degree equal to the number of the appearances of q in the integral. By Proposition 3.3, the maximal number of the appearances of q in the integrals of the sum (3.9) is $[\frac{k-1}{2}]$. □

Proposition 3.3 also provides the following information about the structure of the polynomials $\psi_k(\mu)$.

COROLLARY 3.2. *For each $k \geq 0$, the term of degree 0 in μ in the polynomial $\psi_k(\mu)$ is $I_{1\ldots 1}$ with the coefficient $(-1)^k k!$. The term of degree 1 is given by the integer linear combination of the iterated integrals I_α with exactly one appearance of q.*

Explicit analysis of the symbolic expressions for ψ_k is not easy. Integration by parts can be used to simplify them but ultimately it leads to a word problem which has been only partly analyzed (and only for the recurrence relation (3.2)) in [18, 19, 3].

However, some of the iterated integrals above containing more than one appearance of both p and q cannot be reduced to the one-sided or double moments by symbolic operations (including integration by parts). This follows, in particular, from the example (given in [**7**]) of the Abel equation (2.1) with the elliptic functions as coefficients, for which all the double moments vanish while the center equations are not satisfied.

In the sequel, we always work with center equations given by the system (3.6). Assuming, as usual, that $P(a) = Q(a) = 0$ and simplifying the subsequent equations via the preceding ones we obtain the following explicit form for the first seven center equations in (3.6) (see, e.g., [**14**]):

$$0 = \psi_2(b) = -P(b)$$
$$0 = \psi_3(b) = -m_0 = -Q(b)$$
$$0 = \psi_4(b) = -m_1$$
$$0 = \psi_5(b) = -m_2$$
$$0 = \psi_6(b) = -m_3 - \frac{1}{2}\int_a^b pQ^2$$
$$0 = \psi_7(b) = -m_4 - 2\int_a^b PpQ^2$$
$$0 = \psi_8(b) = -m_5 - \int_a^b \frac{1}{2}Q^3 p + 23P^3 Qq - 77 \int_a^b P^2(t)q(t)dt \int_a^t Pq.$$

The terms m_k appearing in these equations are the one-sided moments (2.3): $m_k = \int_a^b P^k(x)q(x)dx$. Observe that the first of these equations implies $P(b) = P(a) = 0$, and the second implies $Q(b) = Q(a) = 0$.

The form of these initial center equations suggests some important general patterns, which can be proved by a combination of integration by parts and combinatorial analysis. In particular, the iterated integrals where the integrand q appears exactly once can be transformed via integration by parts to moment form. By Corollary 3.2, we see that the terms linear in q of the center equations are indeed the moments m_k.

THEOREM 3.1. *In each ψ_k, $k \geq 2$, the sum of all the iterated integrals containing exactly one integrand q (i.e., the part of the polynomial $\psi_k(\mu)$ linear in μ) is equal to the one-sided moment $m_{k-3} = \int_a^b P^{k-3}(x)q(x)dx$, taken with the coefficient -1.*

The proof of this result is given in [**15**] (see Proposition 4.1).

In this paper, we use only one additional fact concerning the structure of the center equations. It is given by the following proposition.

PROPOSITION 3.4. *If the polynomials P and Q satisfy* (PCC) *on a, b, then for $p = P'$ and $q = Q'$ all the iterated integrals I_α on a, b vanish. In particular,* (PCC) *implies the vanishing of each of the terms in the center equations (3.6).*

PROOF. Under the factorization $P(x) = \tilde{P}(W(x))$, $Q(x) = \tilde{Q}(W(x))$ provided by (PCC), we can make a change of the independent variable $x \to w = W(x)$ in the iterated integrals. We get

$$(3.11) \qquad I_\alpha = \int_{W(a)}^{W(b)} h_{\alpha_1}(w_1)dw_1 \int_{W(a)}^{w_1} h_{\alpha_2}(w_2)dw_2 \ldots \int_{W(a)}^{w_{s-1}} h_{\alpha_s}(w_s)dw_s.$$

Here $h_{\alpha_j}(w) = \tilde{p}(w) = \tilde{P}'(w)$ for $\alpha_j = 1$ and $h_{\alpha_j}(w) = \tilde{q}(w) = \tilde{Q}'(w)$ for $\alpha_j = 2$. Now since $\tilde{P}'(w)$ and $\tilde{Q}'(w)$ are polynomials, all the subsequent integrands in (3.11) are polynomials. But by the conditions, we have $W(a) = W(b)$; and the most exterior integral must be zero, being the integral of a certain polynomial over a closed contour. $\square$

4. Proof of Main Results

To prove the local coincidence of the center and the composition conditions, we have to translate the information on the center equations obtained in the previous section into algebro-geometric properties of these equations. We use the following result, which is essentially a version of the Nakayama Lemma in commutative algebra (see, for example, [21, chapter 4, Lemma 3.4]), adapted to our situation.

LEMMA 4.1. *Let $f_1, \ldots, f_m$ be polynomials in n complex variables. Let $f_i = f_i^1 + f_i^2$, $i = 1, \ldots, m$, with all f_i^1 homogeneous of degree d_1 and f_i^2 having all terms of degrees greater than d_1.*

Let $C = \{f_1 = 0, \ldots, f_m = 0\}$, $C^1 = \{f_1^1 = 0, \ldots, f_m^1 = 0\}$. Assume, in addition, that $f_1^1, \ldots, f_m^1$ generate the ideal I_1 of the set C^1 and that each f_i^2 vanishes on C^1.

Then there exists $\epsilon > 0$ such that for the ball B_ϵ in $\mathbb{C}^m$,

1. *$C \cap B_\epsilon = C^1 \cap B_\epsilon$;*
2. *in the ring of holomorphic functions on B_ϵ, the ideals $I = \{f_1, \ldots, f_m\}$ and $I_1 = \{f_1^1, \ldots, f_m^1\}$ coincide.*

PROOF. Since f_i^2 vanish on C^1, they belong to the ideal I_1 of the set C^1. By assumption, I_1 is generated by f_i^1. Hence we have

$$(4.1) \qquad f_i^2 = \sum_{j=1}^m a_{ij} f_j^1$$

with certain polynomials a_{ij}.

We can assume that $a_{ij}(0) = 0$. Indeed, since f_j^1 are homogeneous of degree d_1, while all the terms in f_i^2 have degrees strictly greater than d_1, we can omit the free terms of a_{ij} in (4.1) and the equality still remains valid.

From $f_i = f_i^1 + f_i^2$ and from (4.1) we get

$$(4.2) \qquad \begin{pmatrix} f_1 \\ \vdots \\ f_m \end{pmatrix} = \tilde{A} \begin{pmatrix} f_1^1 \\ \vdots \\ f_m^1 \end{pmatrix},$$

where $\tilde{A} = (a_{ij}) + Id$. Since $a_{ij}(0) = 0$, $\tilde{A}(0) = Id$; hence $\tilde{A}$ is invertible in a neighborhood of the origin in $\mathbb{C}^m$, in particular, in a certain ball B_ϵ, $\epsilon > 0$. (Of course, ϵ depends on the polynomials $f_1, \ldots, f_m$ and on their decomposition $f_i = f_i^1 + f_i^2$).

We get

$$(4.3) \qquad \begin{pmatrix} f_1^1 \\ \vdots \\ f_m^1 \end{pmatrix} = \tilde{A}^{-1} \begin{pmatrix} f_1 \\ \vdots \\ f_m \end{pmatrix}.$$

Taken together, (4.2) and (4.3) prove both the conclusions of Lemma 4.1: the coincidence of the ideals I and I_1 and of their zero sets C and C^1 inside B_ϵ. This completes the proof. □

REMARK. The assumptions of Lemma 4.1 concerning the degrees of f_i^1 and f_i^2 and that f_i^1 generate the ideal of their zero set C^1 are essential, as the following examples show.

EXAMPLE 1. Let $f^1(x_1, x_2) = x_1$, $f^2(x_1, x_2) = (-1 + x_2)x_1$. Then $C_1 = \{x_1 = 0\}$, but $f = f_1^1 + f_2^2 = x_1 x_2$, and $C = \{x_1 = 0\} \cup \{x_2 = 0\}$. This example illustrates the importance of "separation of degrees" of f^1 and f^2.

EXAMPLE 2. Let $f = f^1 + f^2 = y^3 + y^2 x^2$. Here $C_1 = \{y = 0\}$, while $C = \{y^2(y + x^2) = 0\}$ has two components at 0. Here y^3 does not generate the ideal of $\{y = 0\}$.

PROOF OF THEOREMS 2.1 AND 2.2. Let us fix the points $a, b \in \mathbb{C}$, a definite polynomial $p = P'$, and the degree d of the polynomial q. As above, V_d denotes the space of complex polynomials q of degree d. We denote by $\mathcal{C} \in V_d$ the center set of the Abel equation (2.1), i.e., the set of $q \in V_d$ for which (2.1) has a center at a, b. We denote by $\mathcal{L} \subset V_d$ the composition linear subspace, consisting of those $q = Q' \in V_d$ for which P and Q satisfy (PCC).

The center set $\mathcal{C} \in V_d$ is defined in the space V_d by the infinite system of polynomial equations (3.6): $\psi_k(\mu) = 0$, $k = 2, \ldots$.

Let $N = N(P, d, a, b)$ be the Bautin moment index of the definite polynomial P on a, b. We apply Lemma 4.1 to the first $N + 3$ equations of the system (3.6). So $f_i(\mu) = \psi_i(\mu)$ and $f_i^1(\mu)$ are chosen to be the linear parts $m_{i-3}(\mu)$ of $\psi_i(\mu)$ while $f_i^2(\mu)$ contain all the non-linear in μ terms of $\psi_i(\mu)$, $i = 2, \ldots, N + 3$. Clearly, in this case, the assumptions of Lemma 4.1 concerning the degrees of f_i^1 and f_i^2 and that f_i^1 generate the ideal of their zero set C^1 are satisfied.

By definition of the Bautin moment index $N = N(P, d, a, b)$, the zero set C^1 of the equations $f_i^1(\mu) = m_{i-3}(\mu) = 0$ for $i = 2, \ldots, N + 3$, is the composition linear subspace $\mathcal{L} \subset V_d$. By Proposition 3.4, the nonlinear parts $f_i^2(\mu)$ vanish on $\mathcal{L}$. Hence the last condition of Lemma 4.1 is also satisfied. We conclude that there exists $\epsilon > 0$ such that for the ball B_ϵ in $V_d = \mathbb{C}^{d+1}$, the following statements hold.

1. $C_{N+3} \cap B_\epsilon = \mathcal{L} \cap B_\epsilon$, where C_{N+3} is the zero set of the first $N + 3$ center equations $\psi_k(\mu) = 0$, $k = 2, \ldots, N + 3$.

2. In the ring of holomorphic functions on B_ϵ, the ideals $I'_{N+3} = \{\psi_2, \ldots, \psi_{N+3}\}$ and $I_N^1 = \{m_0, \ldots, m_N\}$ coincide.

Observe that the radius ϵ of the ball B_ϵ in V_d in which the above conclusions are valid depends only on the equations $\psi_k(\mu) = 0$, $k = 2, \ldots, N$ and on their decomposition into the linear and the nonlinear parts. In other words, $\epsilon = \epsilon(P, d, a, b)$ depends only on the fixed definite polynomial P, the degree d, and the points a, b.

It remains to observe that for each $k \geq N + 4$ the polynomial $\psi_k(\mu)$ belongs to the ideal $I_N^1 = \{m_0, \ldots, m_N\}$ and hence also to $I'_{N+3} = \{\psi_2, \ldots, \psi_{N+3}\}$. Indeed, by Proposition 3.4 $\psi_k(\mu)$ vanishes on the composition subspace $\mathcal{L}$. Since I_N^1 is the ideal of $\mathcal{L}$, we obtain $\psi_k(\mu) \in I_N^1 = I'_{N+3}$. Therefore, the ideal I'_{N+3} coincides with the local Bautin ideal $I = \{\psi_2, \ldots, \psi_{N+3}, \psi_{N+4}, \ldots\}$. In particular, this implies that the Bautin index $b(P, d, a, b)$ is at most $N+3$. On the other hand, $b(P, d, a, b)$ cannot be smaller than $N + 3$. Indeed, if the ideal $I = I'_{N+3}$ were generated by a number of polynomials ψ_j smaller than $N + 3$, we could invert the proof of Lemma 4.1 to conclude that a number of moments m_j smaller than N generate the ideal of $\mathcal{L}$ – which contradicts the definition of the Bautin moment index. Therefore the Bautin index $b(P, d, a, b)$ is equal to the Bautin moment index $N(P, d, a, b)$ plus 3.

From the equality of the local ideals $I = I'_{N+3} = I^1_N$, we see that their zero sets inside the ball B_ϵ coincide:

$$C_{N+3} \cap B_\epsilon = \mathcal{C} \cap B_\epsilon = \mathcal{L} \cap B_\epsilon,$$

where, as above, $\mathcal{C}$ is the center set of (2.1), i.e., the set of zeros of the Bautin ideal I. This completes the proofs of Theorems 2.1 and 2.2. □

REMARK. In fact, each of the polynomials $\psi_k(\mu)$ belongs to the ideal generated by the moments $m_1, \ldots, m_N$ in the *global* ring of polynomials in μ. However, the opposite inclusion is valid in only in the domain where we can invert the corresponding matrix $\tilde{A}$ (in particular, in B_ϵ). In geometric terms, the result of Theorem 2.1 does not exclude the possibility that the center set $\mathcal{C}$ of (2.1) contains components other than the composition set $\mathcal{L}$. However, these components may appear only "far away" from the origin.

PROOF OF THEOREM 2.3. This theorem follows directly from Theorem 2.2 and Theorem 2.3.9 of [20]. Formally, the results of [20] are stated for the Bautin ideal and the Bautin index defined in the global ring of polynomials in μ, while Theorem 2.2 concerns the Bautin index defined in the ring of functions on the ball $B_r \subset V_d$. However, for $\mu \in B_{r/2}$, all the estimates of [20] remain valid via the analytic version of the "effective division theorem". □

REMARK. The bounds of [20] are usually far from being realistic, mostly because of the worst case estimates used in Hironaka's effective division algorithm. We believe that for many analytic families arising in relation to algebraic differential equations (including the most mysterious one – the Poincaré mapping) the "blind" application of the division algorithm can be replaced by a detailed study of the algebraic properties of the Taylor coefficients $a_k(\lambda)$. In particular, for the "moment generating function" $H(y) = \int_a^b \frac{q(x)dx}{1-yP(x)}$, this was done in [15]. In the situation considered in the present paper, one can replace a general division algorithm by an improved version of the *linear division theorem* of [15] combined with an accurate computation of the Bautin moment index $N(P, d, a, b)$, with an estimate of the non-degeneracy of the moment equations, and with a bounding of the norm of the center equations $\psi_k(\mu) = 0$. We plan to present these results separately.

5. Computing the Bautin moment index in examples

For P of degree two, a convenient method for the analysis of the one-sided moments has been suggested in [6]. It is based on a representation of Q via the basis of the ring of polynomials of x considered as a module over the polynomials of P. In this basis (and for P of degree two), the moment equations have a very simple form; and the matrix representation of these equations can be explicitly analyzed. As a result we can compute explicitly the Bautin moment index $N(P, d, a, b)$ for P of degree two with respect to the two zeros $a, b \in \mathbb{C}$ of P. We present this computation below. However, for P of higher degree, the matrices become much more complicated and only partial results can be obtained by this method.

In [8, 9, 12, 13, 14], an algebraic method for the analysis of the moments vanishing has been developed. This method introduces rather delicate algebraic techniques which relate moments of different orders. Recently, this method was extended in [**10**] to produce quantitative information on moments "near-vanishing". In particular, the following result is obtained in [**10**]. For a given P and $q = Q'$, define the *moment polynomials* $m_k(x)$ by

$$(5.1) \qquad m_k(x) = \int_a^x P^k(t)q(t)dt,$$

where a is one of the roots of P. Observe that the zero moment polynomial $m_0(x)$ is equal to $Q(x)$.

We define the *generalized moment vanishing condition* by requiring that all moment polynomials $m_k(x)$ vanish at certain fixed zeros $x_1, ..., x_l$ of P, $x_1 = a$. The *generalized composition condition* is that $P(x) = \tilde{P}(W(x))$ and $Q(x) = \tilde{Q}(W(x))$, where $\tilde{P}(0) = 0, \tilde{Q}(0) = 0$, and $W(x)$ vanishes at $x_1, ..., x_l$.

THEOREM 5.1. *Let P be a polynomial of degree m. Fix l different zeros $x_1, ..., x_l$, $x_1 = a$, of P with $2l \geq m + 1$. Then for any Q of degree d, the vanishing of $N(P, d) = [(d - l)/(2l - m)] + 1$ moments $m_k(x)$ at the points $x_1, ..., x_l$ implies the generalized composition condition, which, in this case, takes the form*

$$P(x) = W^n(x), \ Q(x) = \tilde{Q}(W(x)),$$

where $W(x)$ is a certain polynomial vanishing at all the roots of P.

If not all the above moments vanish, then the deviation of Q from the generalized composition condition can be estimated through the maximum of the absolute values of $m_k(x_j)$, $k = 0, 1, ..., N(P, d)$, $j = 1, ..., l$.

In particular, this theorem allows us to bound explicitly the Bautin moment index $N(P, d, a, b)$ for any P of degree three with respect to any two of its zeros $a, b \in \mathbb{C}$: it does not exceed $d - 1$. This provides also an explicit bound for the "locality size" in the above computations. We plan to present the corresponding results separately.

So let us fix a polynomial $P(x)$ of degree 2. We can always assume that one of the roots of $P(x)$ is zero and so $P(x) = x(x - b)$, $b \neq 0$. The following theorem provides a description of the moment vanishing conditions for any polynomial Q.

THEOREM 5.2. *Let $P(x) = x(x - b)$, $b \neq 0$. Let Q be a polynomial of degree d and let $\nu = [d/2] + 1$. Let $m_j = m_j(b)$ be defined by (5.1). Then*

1) *If for some k we have $m_k = m_{k+1} = \ldots = m_{k+\nu} = 0$, then there exists a polynomial $\tilde{Q}$ such that $Q(x) = \tilde{Q}(P(x))$. In this case, all the moments m_j vanish.*

2) *For any number $r < \nu$ of the equations $m_{j_s} = 0, s = 1, \ldots, r$ (not necessarily consecutive), there exists a polynomial Q of degree d for which all these equations are satisfied, and which cannot be represented as $Q(x) = \tilde{Q}(P(x))$.*

In particular, the Bautin moment index $N(P, d, 0, b)$ equals $[d/2] + 1$.

PROOF. The proof of Theorem 5.2 consists of several steps. First, using the polynomial P and its derivative P' we can construct a basis for the space of all polynomials in x of a given degree. Indeed, the polynomials $P(x)^k$ have degree $2k$, and the polynomials $P(x)^k P'(x)$ have degree $2k + 1$, respectively. Therefore, all these polynomials are linearly independent; and $P(x)^k$, $P(x)^k P'(x)$, $k = 0, 1, \ldots, l$, form the basis for the space V_{2l+1} of all polynomials $r(x)$ of degree at most $2l+1$. The same polynomials except $P(x)^l P'(x)$ form a basis for V_{2l}.

Thus, any polynomial $r(x)$ of the degree $2l+1$ can be uniquely written in the form

$$(5.2) \quad r = P^l(\alpha_l P' + \beta_l) + P^{l-1}(\alpha_{l-1} P' + \beta_{l-1}) + \ldots + (\alpha_0 P' + \beta_0).$$

We have the following simple proposition.

PROPOSITION 5.1. *The polynomial $R(x)$ has the form $R(x) = \tilde{R}(P(x))$ if and only if $\beta_j = 0$ for $j = 0, \ldots, l$ in the representation (5.2) of its derivative $r(x) = R'(x)$.*

PROOF. If $\beta_j = 0$ for $j = 0, \ldots, l$, then integrating the representation (5.2) yields

$$(5.3) \quad R = \left(\frac{\alpha_l}{l+1}\right) P^{l+1} + \left(\frac{\alpha_{l-1}}{l}\right) P^l + \cdots + \alpha_0 P + \delta,$$

or $R(x) = \tilde{R}(P(x))$ with $\tilde{R}(z) = \left(\frac{\alpha_l}{l+1}\right) z^{l+1} + \cdots + \delta$. Conversely, if $R(x) = \tilde{R}(P(x))$, then differentiating this expression yields a representation (5.2) for $r(x) = R'(x)$ in which only the terms $P(x)^k P'(x)$ have nonzero coefficients. □

REMARK. The ring $\mathcal{R}$ of polynomials in x is a module over the polynomials in P. The representation (5.2) shows that as a module, $\mathcal{R}$ has exactly two generators: 1 and P'.

We return now to the proof of Theorem 5.2. Represent $q(x) \in V_d$ according to (5.2):

$$(5.4) \quad q(x) = P^m(\alpha_m P' + \beta_m) + P^{m-1}(\alpha_{m-1} P' + \beta_{m-1}) + \ldots + (\alpha_0 P' + \beta_0).$$

For any j, we get

$$(5.5) \quad m_j = \int_0^b P^j q = \beta_m \int_0^b P^{m+j} + \beta_{m-1} \int_0^a P^{m+j-1} + \ldots + \beta_0 \int_0^b P^j,$$

since all the integrals containing P' vanish on the interval $[0, b]$, whose endpoints are zeros of P. Defining the constants ω_j by $\omega_j = \int_0^b P(t)^j dt$, we obtain the following proposition.

PROPOSITION 5.2. *In the coordinates α_k, β_k of (5.4) the equation $m_j = 0$ takes the form $\sum_{k=0}^m \omega_{j+k} \beta_k = 0$.*

Therefore, the equations $m_{j_1} = 0, \ldots, m_{j_n} = 0$ can be rewritten as the system

$$\begin{cases} \beta_m \omega_{m+j_1} + \beta_{m-1} \omega_{m+j_1-1} + \cdots + \beta_0 \omega_{j_1} = 0 \\ \beta_m \omega_{m+j_2} + \beta_{m-1} \omega_{m+j_2-1} + \cdots + \beta_0 \omega_{j_2} = 0 \\ \vdots \qquad \vdots \qquad \ddots \qquad \vdots \qquad \vdots \\ \beta_m \omega_{m+j_{n+1}} + \beta_{m-1} \omega_{m+j_{n+1}-1} + \cdots + \beta_0 \omega_{j_{n+1}} = 0 \end{cases},$$

i.e.,

$$\begin{pmatrix} \omega_{j_1} & \omega_{j_1+1} & \cdots & \omega_{j_1+m} \\ \omega_{j_2} & \omega_{j_2+1} & \cdots & \omega_{j_2+m} \\ \vdots & \vdots & \ddots & \vdots \\ \omega_{j_{n+1}} & \omega_{j_{n+1}+1} & \cdots & \omega_{j_{n+1}+m} \end{pmatrix} \begin{pmatrix} \beta_0 \\ \beta_1 \\ \vdots \\ \beta_m \end{pmatrix} = 0.$$

For $n < m$, this system always has a nonzero solution. According to Proposition 5.1, for the corresponding polynomial $q \in V_d$, its primitive $Q = \int q$ cannot be represented as $Q(x) = \tilde{Q}(P(x))$. This proves the second part of Theorem 5.2.

To prove the first part of the theorem, let us denote by $D_{k,m}$ the determinant of the system of m *consecutive moment equations*, i.e., the determinant

$$D_{k,m} = \det \begin{vmatrix} \omega_k & \omega_{k+1} & \cdots & \omega_{k+m} \\ \vdots & \vdots & \ddots & \vdots \\ \omega_{k+m} & \omega_{k+m+1} & \cdots & \omega_{k+2m} \end{vmatrix}.$$

PROPOSITION 5.3. *For any natural k and m, $D_{k,m} \neq 0$.*

PROOF. Consider the scalar product $< f, g > = < f, g >_k = \int_0^a P(x)^k f(x) g(x) dx$ on the space of square integrable functions on $[0, b]$. (Without loss of generality, we can assume $b < 0$, so $P(x)$ is positive on $[0, b]$). Then

$$D_{k,m} = \det \begin{vmatrix} < P^0, P^0 > & < P^0, P^1 > & \cdots & < P^0, P^m > \\ < P^1, P^0 > & < P^1, P^1 > & \cdots & < P^1, P^m > \\ \vdots & \vdots & \ddots & \vdots \\ < P^m, P^0 > & < P^m, P^1 > & \cdots & < P^m, P^m > \end{vmatrix}$$

is the Gram determinant of the set of linearly independent vectors $P^j(x) = (x(x-b))^j$, and hence is non-zero. $\square$

Now we can complete the proof of Theorem 5.2. By Proposition 5.3, the vanishing of m consecutive moments $m_k, m_{k+1}, \ldots, m_{k+m}$ implies that in the representation (5.4) of q, all the coefficients β_j must be zero. By Proposition 5.1, this implies that $Q(x) = \tilde{Q}(P(x))$, which in turn implies the vanishing of all the moments m_j. $\square$

REMARK. One can find the values of ω_k explicitly: we have

$$\omega_k = \int_0^b (x(x-b))^k dx = (-1)^k \frac{b^{2k+1} k!}{2^k (2k+1)!!}.$$

Estimating the volume spanned by the vectors $P^j(x)$ with respect to the scalar product $< f, g >_k$, we can obtain an explicit lower bound for the determinants $D_{k,m}$. We do not use these bounds in the present paper. In [15], some explicit bounds for the locality size and for periodic solutions of the Abel equation are obtained via the methods of [10].

6. Addendum: Definite polynomials

At present, we do not have a complete description of definite polynomials. However, some rather wide classes of definite polynomials have been recently specified. Let us present these classes briefly and sketch some relations between them.

6.1. Simple end-points. If the polynomial P has simple zeros at a and b, then it is definite on $[a, b]$. This was shown in [17]. Other proofs can be found in [25, 28].

6.2. Indecomposability.
Any indecomposable polynomial P (one not possessing a representation $P(x) = R(S(x))$ with degrees of both R and S greater than 1) is definite on each interval $[a, b]$ with $P(a) = P(b)$. This fact is proved in [25]. In particular, each P of prime degree is indecomposable and hence definite on each interval $[a, b]$ with $P(a) = P(b)$.

6.3. Simple zeros not at the end-points.
If all the zeros of P, except possibly a, b are simple, then the polynomial P is definite on $[a, b]$. This is shown in [27]. Another result of [27] is the following: if for any critical value c of P except possibly 0, the preimage $P^{-1}(c)$ contains exactly one critical point of P, then P is definite on any $[a, b]$ with $P(a) = P(b) = 0$.

6.4. Real polynomials.
Let P be a polynomial with real coefficients and let $a, b \in \mathbb{R}$. If all the real zeros of P in the open interval (a, b) are simple, then P is definite on $[a, b]$. This is shown in [28]. The techniques presented in [14], Section 4.2, allow one to prove the following result. Let P be a polynomial with real coefficients and let $a, b \in \mathbb{R}$, $P(a) = P(b) = 0$. Assume that the multiplicity of each of the roots of P on the closed interval $[a, b]$ is odd. Then P is definite on $[a, b]$. (We plan to present the proof of this and other results for real polynomials separately.)

6.5. Geometry of $P([a, b])$.
The following results are proved in [28]. Let P be a complex polynomial, $P(a) = P(b) = 0$. Assume that there exists a path $\Gamma \subset \mathbb{C}$ joining a and b such that the curve $\gamma = P(\Gamma)$ has only transversal self-intersections and that the point $0 = P(a) = P(b)$ is on the boundary of the exterior domain with respect to the closed curve γ. Then $P(x)$ is definite on $[a, b]$.

The following criterion can be verified explicitly in many important examples. Let P be a complex polynomial with $P(a) = P(b) = 0$, $a, b \in \mathbb{C}$. Let Γ be a piecewise-analytic curve in $\mathbb{C}$ joining a and b, and let $\gamma = P(\Gamma)$. Assume that the open part $\gamma \setminus 0$ is contained in an open set Ω with piecewise-analytic boundary, and assume that 0 belongs to the exterior boundary of Ω. Then $P(x)$ is definite on $[a, b]$. In particular, this happens for any $P(x) = (x - a)(b - x)P_1(x)$, with $P_1(x) = \sum_{k=0}^{n} a_k x^k$, $a_k \in \mathbb{C}$, if the convex hull CH of the coefficients a_k does not contain $0 \in \mathbb{C}$. (In this case, Ω is an open cone $\alpha < \text{Arg}(z) < \beta$ containing the closed cone with the vertex at $0 \in \mathbb{C}$ generated by CH).

6.6. Recursive representation.
As was mentioned above, in [8, 9, 10, 12, 13, 14], an algebraic method for the analysis of the vanishing moments has been developed, which relates between them the moments of different orders. This method provides a general setting of the moment problem which is in some aspects more natural than the one used in the present paper: the moments vanish not only at two points a, b, but at a possibly larger number of roots of P. Theorem 5.1 above presents some initial results in this direction. The notion of a definite polynomial can be extended to this general setting, and some classes of "generalized definite polynomials" can be described. We plan to present these results separately. As far as the setting of the present paper is concerned, the recursive representation method shows that each polynomial P of degree at most three is definite. Another conclusion is that a polynomial P of degree d having at a, b zeros of total multiplicity d or $d - 1$ is definite (this also follows from 6.3).

6.7. Bernstein Classes. The method of [**31**] combines the study of the integration operator with the "Bernstein Class" approach of [**32**]. The result is that any P is definite on $[a, b]$ for a, b different from a certain finite number of points (which, in general, do *not* coincide with the critical values of P as in subsections 6.1–6.3 above).

It was mentioned above that the Chebyshev polynomial $T_6(x)$ is not definite on the interval $[\frac{-\sqrt{3}}{2}, \frac{\sqrt{3}}{2}]$. The results above show that each polynomial P of degree at most 5 is definite. Indeed, for $\deg P \leq 2$, this follows from 6.1 (and was shown also in Section 5 above). For $\deg P = 2, 3, 5$, the polynomial P is indecomposable and hence definite on any a, b with $P(a) = P(b)$. Finally, for $\deg P = 4$, either the roots of P at a, b are simple or the remaining roots are simple. Hence, P is definite by 6.1 or 6.3, respectively. This fact can also be proved by the method of 6.6.

References

1. M.A.M. Alwash and N.G. Lloyd, *Non-autonomous equations related to polynomial two-dimensional systems*, Proc. Royal Soc. Edinburgh Sect. A **105** (1987), 129–152.
2. M.A.M. Alwash, *On a condition for a centre of cubic non-autonomous equations*, Proc. Royal Soc. Edinburgh Sect. A **113** (1989), 289-291.
3. M.A.M. Alwash, *Word problems and the centers of Abel differential equations*, Ann. Differential Equations **11** (1995), 392-396.
4. V.I. Arnold, *Problems on singularities and dynamical systems*, in: Developments in Mathematics: The Moscow School (V.I. Arnold, M. Monastyrsky, eds.), Chapman & Hall, London, 1993, pp. 251-274.
5. V.I. Arnold and Yu. Il'yashenko, *Ordinary Differential Equations*, in: Dynamical Systems I: Ordinary Differential Equations and Smooth Dynamical Systems (D.V. Anosov, V.I. Arnold, eds.), Encyclopaedia of Mathematical Sciences **1**, Springer, Berlin, 1988.
6. M. Blinov, *Some Computations Around the Center Problem, Related to the Composition Algebra of Univariate Polynomials*, M.Sc. Thesis, Weizmann Institute of Science, 1997.
7. M. Blinov, *Center and Composition Conditions for Abel Equation*, Ph.D. Thesis, Weizmann Institute of Science, 2002.
8. M. Briskin, *Infinitesimal center-focus problem, generalized moments and composition of polynomials*, Funct. Differ. Equ. **6** (1999), 47-53.
9. M. Briskin, *Algebra of generalized moments and composition of polynomials*, Proc. Int. Conference "Dynamical Systems: Stability, Control, Optimization", Minsk, 2000.
10. M. Briskin, *Recursive moments representation and quantitative moment problem*, in preparation.
11. M. Briskin, J.-P. Francoise, and Y. Yomdin, *Center conditions, compositions of polynomials and moments on algebraic curve*, Ergodic Theory Dynam. Syst. **19** (1999), 1201-1220.
12. M. Briskin, J.-P. Francoise, and Y. Yomdin, *Center condition II: Parametric and model center problems*, Israel J. Math. **118** (2000), 61-82.
13. M. Briskin, J.-P. Francoise, and Y. Yomdin, *Center condition III: Parametric and model center problems*, Israel J. Math. **118** (2000), 83-108.
14. M. Briskin, J.-P. Francoise, and Y. Yomdin, *Generalized moments, center-focus conditions and compositions of polynomials*, in: Operator Theory, System Theory and Related Topics, Oper. Theory Adv. Appl. **123** (2001), 161-185.
15. M. Briskin and Y. Yomdin, *Tangential Hilbert problem for Abel equation*, Moscow Math. J., to appear.
16. A. Brudnyi, *On the center problem for ordinary differential equations*, preprint, 2003.
17. C. Christopher, *Abel equations: composition conjectures and the model problem*, Bull. Lond. Math. Soc. 32 (2000), 332-338.
18. J. Devlin, *Word problems related to periodic solutions of a nonautonomous system*, Math. Proc. Cambridge Philos Soc. **108** (1990), 127-151.
19. J. Devlin, *Word problems related to derivatives of the displacement map*, Math. Proc. Cambridge Philos. Soc. **110** (1991), 569-579.

20. J.-P. Francoise and Y. Yomdin, *Bernstein inequality and applications to analytic geometry and differential equations*, J. Funct. Anal. **146** (1997), 185-205.
21. V. Golubitsky and V. Guillemin, *Stable Mappings and their Singularities*, Springer-Verlag, New York-Heidelberg, 1973.
22. Yu. Ilyashenko, *Centennial history of Hilbert's 16th problem*, Bull. Amer. Math. Soc. (N.S.) **39** (2002), 301-354.
23. A. Lins Neto, *On the number of solutions of the equation $\frac{dx}{dt} = \sum_{j=0}^{n} a_j(t)x^j$, $0 \leq t \leq 1$, for which $x(0) = x(1)$*, Invent. Math. **59** (1980), 67-76.
24. F. Pakovich, *A counterexample to the composition conjecture*, Proc. Amer. Math. Soc. **130** (2002), 3747-3749.
25. F. Pakovich, *On the polynomial moment problem*, Math. Res. Lett. **10** (2003), 401-410.
26. F. Pakovich, *On polynomials orthogonal to all degrees of a Chebyshev polynomial on a segment*, Israel J. Math. **142** (2004), 273-284.
27. F. Pakovich, *On polynomials orthogonal to all powers of a given polynomial on a segment*, http://arXiv.org/abs/math.CA/0408019.
28. F. Pakovich, N. Roytvarf, and Y. Yomdin, *Cauchy type integrals of algebraic functions and a tangential center-focus problem for Abel equations*, Israel J. Math., to appear.
29. J. Ritt, *Prime and composite polynomials*, Trans. Amer. Math. Soc. **23** (1922), 51–66.
30. R. Roussarie, *Bifurcation of Planar Vector Fields and Hilbert's Sixteenth Problem*, Birkhäuser Verlag, Basel, 1998.
31. N. Roytvarf, *Generalized moments, composition of polynomials and Bernstein classes*, in: Entire Functions in Modern Analysis. B.Ya. Levin Memorial Volume, Israel Math. Conf. Proc. **15** (2001), 339-355.
32. N. Roytvarf and Y. Yomdin, *Bernstein classes*, Ann. Inst. Fourier **47** (1997), 825-858.
33. K.S. Sibirsky, *Introduction to the Algebraic Theory of Invariants of Differential Equations*, Manchester University Press, Manchester, 1988.
34. S. Smale, *Mathematical problems for the next century*, Math. Intelligencer **20** (1998), no.2, 7-15.
35. Yang Lijun and Tang Yun, *Some new results on Abel Equations*, J. Math. Anal. Appl. **261** (2001), 100-112.

THEORETICAL BIOLOGY AND BIOPHYSICS GROUP, LOS ALAMOS NATIONAL LABORATORY, LOS ALAMOS, NM, 87545, U.S.A.
E-mail address: `mblinov@atlas.lanl.gov`

JERUSALEM COLLEGE OF ENGINEERING, RAMAT BET HAKEREM, P.O.B. 3566, JERUSALEM 91035, ISRAEL
E-mail address: `briskin@server.jce.ac.il`

DEPARTMENT OF MATHEMATICS, THE WEIZMANN INSTITUTE OF SCIENCE, REHOVOT 76100, ISRAEL
E-mail address: `yosef.yomdin@weizmann.ac.il`

Univalent Functions Starlike with Respect to a Boundary Point

D. Bshouty and A. Lyzzaik

On the occasion of Larry Zalcman's 60th birthday

ABSTRACT. The aim of this paper is to give a new proof for an already established necessary condition for a univalent function to be starlike with respect to a boundary point. The proof sheds additional light on the relationship between the analytic and geometric characterizations of these functions and is in the spirit of the recently established sufficient condition.

1. Introduction

A domain Ω of the complex plane $\mathbb{C}$, with $0 \in \overline{\Omega}$, is *starlike with respect to the origin* if for every point $w \in \Omega$, the line segment $(0, w] = \{tw : 0 < t \leq 1\}$ lies in Ω. A univalent function f from the open unit disc $\mathbb{D}$ onto a domain Ω starlike with respect to the origin is called *starlike with respect to the origin;* here we do *not* necessarily assume $f(0) = 0$. If $0 \in \partial\Omega$, then f is called *starlike with respect to the (boundary point at the) origin;* the class of all such functions is denoted by $\mathcal{S}_0^*$.

An analytic function f on $\mathbb{D}$ is said to have the *asymptotic value* $a \in \mathbb{C} \cup \{\infty\}$ at a point $\zeta \in \partial\mathbb{D}$ if there exists a Jordan arc Γ that ends at ζ and lies in $\mathbb{D}$ except for ζ such that
$$f(z) \to a \quad \text{as} \quad z \in \zeta, \quad z \in \Gamma.$$
Also, f is said to have the *angular limit* $a \in \mathbb{C} \cup \{\infty\}$ at ζ if
$$f(z) \to a \quad \text{as} \quad z \to \zeta, \quad z \in A$$
for every Stolz angle $A = \{z \in \mathbb{D} : |\arg(1 - \overline{\zeta}z)| < \pi/2 - \delta\}$ at ζ, where $0 < \delta < \pi/2$. For these notions, see [6].

Univalent functions starlike with respect to a boundary point were first introduced and investigated in 1981 by M.S. Robertson [8] and were further studied by A. Lyzzaik [5], A. Lyzzaik and A. Lecko [4], H. Silverman and E. M. Silvia [9], P. Todorov [10], M. Elin, S. Reich and D. Shoikhet [1], and M. Elin and D. Shoikhet [2]. In [8], the following two classes of univalent functions which are assumed to be nonvanishing in the unit disc were defined.

2000 *Mathematics Subject Classification.* Primary 30C45.
Key words and phrases. Univalent functions, starlike functions.

(i) The class $\mathcal{G}$ of univalent functions f of $\mathbb{D}$ that satisfy $f(0) = 1$ and
$$\Re\left\{2z\frac{f'(z)}{f(z)} + \frac{1+z}{1-z}\right\} > 0, \quad (z \in \mathbb{D}).$$

(ii) The class $\mathcal{G}^*$ of univalent functions f of $\mathbb{D}$ that satisfy $f(0) = 1$, $\lim_{r \to 1^-} f(r) = 0$, $f(\mathbb{D})$ is starlike with respect to the origin, and $\Re\{e^{i\alpha} f(z)\} > 0$ for some real α and all $z \in \mathbb{D}$.

Further, Robertson proved that $\mathcal{G} \subset \mathcal{G}^*$ and conjectured that $\mathcal{G}^* \subset \mathcal{G}$. The conjecture was resolved positively by the second author in [5], where a short proof of the former set-inclusion was also given.

The close relationship between the classes $\mathcal{S}_0^*$ and $\mathcal{G}$ (or $\mathcal{G}^*$) is evident and entails a complete analytic characterization of $\mathcal{S}_0^*$ [4]. A different analytic characterization of $\mathcal{S}_0^*$ was recently established and is formulated in the next theorem. Let $\mathcal{B}$ be the class of all analytic functions from $\mathbb{D}$ into itself, and for $\alpha > 0$, let $\mathcal{B}(\alpha)$ be the subclass of $\mathcal{B}$ consisting of all functions ω such that the angular limit of $(1 - \omega(z))/(1 - z)$ at 1 is α.

THEOREM 1. *Let f be an analytic function of $\mathbb{D}$ with angular limit zero at 1. A necessary and sufficient condition for f to belong to $\mathcal{S}_0^*$ is that there exists $\omega \in \mathcal{B}(\alpha)$, $\alpha \in (0, 1]$, such that*

(1) $$-(1-z)^2 \frac{f'(z)}{f(z)} = 4\frac{1 - \omega(z)}{1 + \omega(z)}, \quad z \in \mathbb{D}.$$

The necessity and sufficiency were shown, respectively, by A. Lecko [3, Theorem 3.2] and A. Lecko and A. Lyzzaik [4, Main Theorem]. However, there were two shortcomings. First, the two proofs are different in spirit; and second, the necessity proof falls short of showing that $2\alpha\pi$ is the size of the vertex angle of the smallest wedge containing $f(\mathbb{D})$ and for which the vertex is the origin, a fact which follows at once from the sufficiency proof.

The object of this note is to give a proof of the following result in the spirit of the sufficiency part in [4].

THEOREM 2. *Let $f \in \mathcal{S}_0^*$ have angular limit zero at 1, and let $2\alpha\pi$, $\alpha \in (0, 1]$, be the size of the vertex angle of the smallest wedge containing $f(\mathbb{D})$ whose vertex is the origin. Then there exists $\omega \in \mathcal{B}(\alpha)$ such that (1) holds.*

2. Proof of Theorem 2

Assume without loss of generality that $f(0) = 1$. Let $G = f(\mathbb{D})$ and $G_n = G \cup D_n$, $n = 1, 2, \cdots$, where $D_n = \{z : |z| < 1/n.\}$ Then each G_n is a starlike domain with respect to the origin. Let $f_n : \mathbb{D} \to G_n$ be the Riemann mapping of G_n satisfying $f_n(0) = 1$ and $\arg f_n'(0) = \arg f'(0)$. Since G_n converges to G with respect to 1 in the sense of Carathéodory kernel convergence [6, p.28], $f_n \to f$ locally uniformly in $\mathbb{D}$. If $z_n = f_n^{-1}(0)$, then the function

$$g_n(z) = f_n\left(\frac{z + z_n}{1 + \overline{z}_n z}\right), \quad z \in \mathbb{D},$$

is starlike with respect to the origin and satisfies $g_n(0) = 0$. It follows that $\Re\{zg_n'(z)/g_n(z)\} > 0$ for $z \in \mathbb{D}$. Letting $\zeta = (z + z_n)/(1 + \overline{z}_n z)$ and then replacing

ζ by z, we have
$$\Re\left\{(z-z_n)(1-\overline{z}_n z)\frac{f_n'(z)}{f_n(z)}\right\} > 0, \quad z \in \mathbb{D}.$$

But $\lim_{n\to\infty} z_n = 1$ by the proof of part (a) of the main theorem in [**5**]; hence the local uniform convergence of the sequences $\{f_n\}$ and $\{f_n'\}$ and the maximum principle yield
$$\Re\left\{-(1-z)^2\frac{f'(z)}{f(z)}\right\} > 0, \quad z \in \mathbb{D},$$

and (1) holds for some $\omega \in \mathcal{B}$.

It remains to show $\omega \in \mathcal{B}(\alpha)$ for some $\alpha \in (0,1]$. There exists β such that the function $h = e^{i\beta} f^{1/(2\alpha)}$, where $f^{1/(2\alpha)}(0) = 1$, belongs to $\mathcal{S}_0^*$, has positive real part, has angular limit 0 at 1, and satisfies

(2) $$H = \{\tau h(z) : \tau > 0, z \in \mathbb{D}\},$$

where H denotes the right half-plane. Let γ be the half-open Jordan arc parameterized by $\gamma(t) = h^{-1}(1-t)$, $0 \leq t < 1$; note that $\lim_{t\to 1^-} \gamma(t) = 1$. For $0 \leq t < 1$, let
$$h_t(z) = (1-t)^{-1} h\left(\frac{z+\gamma(t)}{1+\overline{\gamma}(t)z}\right), \quad z \in \mathbb{D}.$$

Then $h_t(0) = 1$ and $h_t'(0) = (1-t)^{-1}(1-|\gamma(t)|^2) h' \circ \gamma(t)$. There exists $0 < \delta < \pi/2$ such that the circular sector $\{w : |\arg w| < \delta, |w| < 1\}$ lies in $h(\mathbb{D})$. Let $d_h(\gamma(t))$ be the Euclidean distance between $1-t$ and $\partial h(\mathbb{D})$. Then for t sufficiently close to 1,
$$(1-t)\sin\delta < d_h(\gamma(t)) \leq (1-t),$$

and, by [**6**, Corollary 1.4],
$$\frac{1}{4}(1-|\gamma(t)|^2) h' \circ \gamma(t) \leq d_h(\gamma(t)) \leq (1-|\gamma(t)|^2) h' \circ \gamma(t).$$

Hence $\sin\delta < |h_t'(0)| \leq 4$ and there exists a univalent function k such that $h_{t_n} \to k$ converges locally uniformly in $\mathbb{D}$ for some $t_n \to 1^-$. Since by (2), $h_{t_n}(\mathbb{D}) \to H$ as $t_n \to 1^-$, $k(\mathbb{D}) = H$ by the Carathéodory kernel theorem. Further, $k(0) = 1$ since $h_{t_n}(0) = 1$, and $k(-1) = \infty$ since $h_{t_n}(-\gamma(t_n)) = 1/(1-t_n)$. Thus $k(z) = (1-z)/(1+z)$; consequently, $h_t \to k$ converges as $t \to 1^-$ locally uniformly in $\mathbb{D}$. It follows that, with $s = (z+\gamma(t))/(1+\overline{\gamma}(t)z)$,
$$\frac{d}{dz}\log\left\{\frac{1+z}{1-z}h_t(z)\right\} = \frac{2}{1-z^2} + \frac{1+\gamma(t)}{(1+\overline{\gamma}(t)z)(1-z)}(1-s)\frac{h'(s)}{h(s)} \to 0$$

as $t \to 1^-$ locally uniformly in $\mathbb{D}$. In particular, for $z = 0$ we have $s = \gamma(t)$ and
$$\lim_{t\to 1^-}(\gamma(t)-1)\frac{h'(\gamma(t))}{h(\gamma(t))} = \lim_{t\to 1^-}\frac{2}{1+\gamma(t)} = 1;$$

that is, $(z-1)h'(z)/h(z)$ has the asymptotic value 1 at 1. But the latter function is normal [**6**, Lemma 9.2] as it maps $\mathbb{D}$ into $\mathbb{C} \setminus \{z : z \leq 0\}$ since
$$\left|\arg\left\{(z-1)\frac{h'(z)}{h(z)}\right\}\right| < \left|\arg\left\{-(z-1)^2\frac{h'(z)}{h(z)}\right\}\right| + |\arg(1-z)| < \pi.$$

Thus, by [**6**, Corollary 9.2], the angular limit of $(z-1)h'(z)/h(z)$ at 1 is 1. Hence the angular limit of $(z-1)f'(z)/f(z)$ at 1 is 2α, since

$$(z-1)\frac{f'(z)}{f(z)} = 2\alpha(z-1)\frac{h'(z)}{h(z)}.$$

Writing (1) as

$$\frac{1-\omega(z)}{1-z} = \frac{1}{4}(1+\omega(z))(z-1)\frac{f'(z)}{f(z)}$$

shows that the angular limits of ω and $(1-\omega(z))/(1-z)$ at 1 are 1 and α, respectively. Therefore $\omega \in \mathcal{B}_\alpha$, and the proof is complete.

REMARK 1. Another proof of Theorem 2 follows by appealing to M. Elin, S. Reich and D. Shoikhet [**1**, Theorem 7] and M. Elin and D. Shoikhet [**2**] to conclude that the angular limit of $(z-1)f'(z)/f(z)$ at 1 is 2α and then writing (1) as

$$\frac{1-\omega(z)}{1-z} = \frac{1+\omega(z)}{4}\frac{(z-1)f'(z)}{f(z)}$$

to obtain that $\omega \in \mathcal{B}_\alpha$.

We conclude the paper with restating Theorem 1 as follows.

THEOREM 3. *Let f be an analytic function of $\mathbb{D}$ with angular limit zero at 1. A necessary and sufficient condition for f to belong to $\mathcal{S}_0^*$ is that there exist $\omega \in \mathcal{B}(\alpha)$, $\alpha \in (0,1]$, such that (1) holds. In this case, $2\alpha\pi$ is the size of the vertex angle of the smallest wedge containing $f(\mathbb{D})$ whose vertex is the origin.*

REMARK 2. By the Julia-Wolf lemma [**6**, p. 306], $\omega \in \mathcal{B}(\alpha)$ if and only if

$$\frac{1+\omega(z)}{1-\omega(z)} = \frac{1}{\alpha}\frac{1+z}{1-z} + 4p(z),$$

where p is a function of positive real part in $\mathbb{D}$ such that the angular limit of $(z-1)p(z)$ at 1 is zero. In this case, (1) can be written as

$$-\frac{1}{(1-z)^2}\frac{f(z)}{f'(z)} - \frac{1}{4\alpha}\frac{1+z}{1-z} = p(z), \quad z \in \mathbb{D}.$$

Using this equation instead of (1) in the aforementioned theorems entails different reformulations of these theorems.

References

[1] M. Elin, S. Reich and D. Shoikhet, *Dynamics of inequalities in geometric function theory*, J. Inequal. Appl. **6** (2001), 651-664.

[2] M. Elin and D. Shoikhet, *Dynamics extension of the Julia-Wolf Carathéodory Theorem*, Dynam. Systems Appl. **10** (2001), 421-437.

[3] A. Lecko, *On the class of functions starlike with respect to a boundary point*, J. Math. Anal. Appl. **261** (2001), 649-664.

[4] A. Lecko and A. Lyzzaik, *A note on univalent functions starlike with respect to a boundary point*, J. Math. Anal. Appl. **282** (2003), 846-851.

[5] A. Lyzzaik, *On a conjecture of M. S. Robertson*, Proc. Amer. Math. Soc. **91** (1984), 108-110.

[6] Ch. Pommerenke, *Univalent Functions*, Vandenhoeck & Ruprecht, Göttingen, 1975.

[7] Ch. Pommerenke, *Boundary Behaviour of Conformal Maps*, Springer-Verlag, Berlin, 1992.

[8] M. S. Robertson, *Univalent functions starlike with respect to a boundary point*, J. Math. Anal. Appl. **81** (1981), 327-345.

[9] H. Silverman and E. M. Silvia, *Subclasses of univalent functions starlike with respect to a boundary point*, Houston J. Math. **16** (1990), 289-299.

[10] P. Todorov, *On the univalent functions starlike with respect to a boundary point*, Proc. Amer. Math. Soc. **97** (1986), 602-604.

DEPARTMENT OF MATHEMATICS, TECHNION, HAIFA 32000, ISRAEL
E-mail address: `daoud@tx.technion.ac.il`

DEPARTMENT OF MATHEMATICS, AMERICAN UNIVERSITY OF BEIRUT, BEIRUT, LEBANON
E-mail address: `lyzzaik@aub.edu.lb`

On the Geometry Induced by a Grusin Operator

Ovidiu Calin, Der-Chen Chang, Peter Greiner, and Yakar Kannai

Dedicated to Professor Lawrence Zalcman on his sixtieth birthday

ABSTRACT. Let $X_1 = \partial_x$ and $X_2 = x\partial_y$ be two vector fields on $\mathbf{R}^2$. We study the geometric and analytic properties of the Grusin operator $\frac{1}{2}(X_1^2 + X_2^2)$. Using Hamilton-Jacobi theory, we obtain the number of geodesics connecting any two points on $\mathbf{R}^2$. We also construct a complex action function whose critical points give the lengths of those geodesics.

1. Introduction

We are interested in the geometric and analytic properties of the Grusin operator

$$\Delta_G = \frac{1}{2}(X_1^2 + X_2^2), \tag{1.1}$$

where the vector fields X_1 and X_2 in $\mathbf{R}^2$ are given by

$$X_1 = \frac{\partial}{\partial x}, \qquad X_2 = x\frac{\partial}{\partial y}. \tag{1.2}$$

Note that X_1 and X_2 are linearly independent everywhere except on the y-axis, where X_2 vanishes. Consequently, the operator Δ_G is elliptic except on the y-axis. On the other hand, $[X_1, X_2] = \frac{\partial}{\partial y}$ on the y-axis, so Chow's theorem [6] holds, and every two points on the x,y-plane can be connected by a piecewise differentiable horizontal curve. A horizontal curve is a curve whose tangents are linear combinations of X_1 and X_2. Grusin [8] shows that

$$\begin{aligned} P &= (\partial_x - ix\partial_y)(\partial_x + ix\partial_y) + ic\partial_y \\ &= (\partial_x^2 + x^2\partial_y^2) + i(c+1)\partial_x \end{aligned}$$

2000 *Mathematics Subject Classification.* Primary 53C17; Secondary 34K10, 35H20.

Key words and phrases. Grusin operator, geodesic, Hamilton-Jacobi theory.

The second author was partially supported by a William Fulbright Research grant and a competitive research grant at Georgetown University and the third author was partially supported by NSERC Grant OGP0003017.

is hypoelliptic if and only if c is not an even integer. Hence for $c = -1$, the equation $\partial_x^2 + x^2 \partial_y^2 = \Delta_G$ is hypoelliptic (see also Hörmander [9], Nagel-Stein [10] and Nirenberg-Trèves [11]).

Here we use Hamilton-Jacobi theory of bicharacteristics to study the geometry induced by the operator Δ_G (see [1], [2], [3], [5] and [7]). We obtain all geodesics between any two points in $\mathbf{R}^2$ and calculate the lengths of those geodesics. Every neighborhood of each point P on the y-axis contains points which are connected to P by at least one, and some by an infinite number of geodesics. This is similar to results obtained by Beals-Gaveau-Greiner in [4] for the sub-Laplacian on the Heisenberg group $\mathbf{H}_1$. However, there are differences. In particular, given $P \in \mathbf{H}_1$, the set of points conjugate to P is a straight line through P. There are no conjugate points in our case. We know that Δ_G is elliptic off the y-axis; consequently, it induces a Riemannian geometry, which manifests itself in the fact that if P is not on the y-axis, then in some sufficiently small neighborhood U of P, every point of U is connected by a unique geodesic to P.

Given two points $P(\mathbf{x}_0, \mathbf{y}_0)$ and $Q(\mathbf{x}_1, \mathbf{y}_1)$, $\mathbf{x}_1 \neq 0$, we construct a complex distance function $f(\mathbf{x}_0, \mathbf{y}_0, \mathbf{x}_1, \mathbf{y}_1, \tau)$ for each point τ of the characteristic variety of Δ_G. The number of critical points of f agrees with the number of geodesics connecting P and Q. Furthermore, the values of f at these critical points agree with the lengths of these geodesics. More precisely,

$$f(\mathbf{x}_0, \mathbf{y}_0, \mathbf{x}_1, \mathbf{y}_1, \tau_m) = \frac{1}{2} \ell_m^2,$$

where

$$\frac{\partial f}{\partial \tau}(\mathbf{x}_0, \mathbf{y}_0, \mathbf{x}_1, \mathbf{y}_1, \tau_m) = 0,$$

$m = 1, \ldots, N$, and ℓ_m is the length of that particular geodesic connecting P and Q.

Part of this article is based on a lecture presented by the second author during the Second International Conference on Complex Analysis and Dynamical Systems which was held on June 9-12, 2003 at Nahariya, Israel. The second author thanks the organizing committee, especially Professors Lawrence Zalcman, Mark Agranovsky and Lavi Karp, for their warm hospitality during his visit to Israel.

2. Hamiltonian mechanics and geometry induced by a Grusin operator

The Hamiltonian of Δ_G is

$$(2.3) \qquad H(x, y, \xi, \theta) = \frac{1}{2}\left(\xi^2 + x^2 \theta^2\right),$$

where the variables (ξ, θ) are the momenta associated with the coordinates (x, y).

The *geodesics* between the point $P(\mathbf{x}_0, \mathbf{y}_0)$ and the point $Q(\mathbf{x}_1, \mathbf{y}_1)$ are the projections on the (x, y)-plane of the solutions of the Hamiltonian system of equations

$$(2.4) \qquad \begin{cases} \dot{x} = H_\xi = \xi \\ \dot{y} = H_\theta = \theta x^2 \\ \dot{\xi} = -H_x = -\theta^2 x \\ \dot{\theta} = -H_y, \end{cases}$$

with the boundary conditions

$$(2.5) \qquad x(0) = \mathbf{x}_0, \ y(0) = \mathbf{y}_0, \ x(1) = \mathbf{x}_1, \ y(1) = \mathbf{y}_1.$$

The Hamiltonian system is invariant with respect to the symmetries
$$(x, y; \theta) \to (-x, y; \theta), \quad (x, y; \theta) \to (x, -y; -\theta).$$
These symmetries send geodesics into geodesics. Without loss of generality, we study only the case $\mathbf{x}_1 > 0$, $\mathbf{y}_1 - \mathbf{y}_0 > 0$, $\theta > 0$.

Since $H_y = 0$, it follows that the momentum θ is a constant, which can be considered as a Lagrange multiplier. Let $V(x) = \frac{1}{2}\theta^2 x^2$. Then $\ddot{x} = -V'(x)$, which can be written also as the conservation of energy law
$$\frac{1}{2}\dot{x}^2 + V(x) = E \quad \text{or} \quad \dot{x}^2 + \theta^2 x^2 = 2E,$$
where E is the constant of total energy. The energy E depends on the boundary conditions $\mathbf{x}_0$, $\mathbf{x}_1$, $\mathbf{y}_1$, $\mathbf{y}_0$ and the constant θ. The second equation of the Hamiltonian system (2.4) yields
$$\dot{y}^2 = \theta^2 x^4 \implies \frac{\dot{y}^2}{x^2} = \theta^2 x^2.$$
Using the law of conservation of energy, we have
$$\dot{x}^2 + \frac{\dot{y}^2}{x^2} = \dot{x}^2 + \theta^2 x^2 = 2E(\text{constant}).$$
Let $\mathcal{C}(s) = (x(s), y(s))$ be a curve with $s \in [0, 1]$. The velocity is
$$\dot{\mathcal{C}}(s) = (\dot{x}(s), \dot{y}(s)) = \dot{x}(s)\partial_x + \dot{y}(s)\partial_y = \dot{x}(s)X_1 + \frac{\dot{y}(s)}{x(s)}X_2.$$
The length of the curve $\mathcal{C}(s)$ is
$$(2.6) \qquad \ell(\mathcal{C}) = \int_0^1 \left[\dot{x}^2(s) + \frac{\dot{y}^2(s)}{x^2(s)}\right]^{1/2} ds = \int_0^1 \sqrt{2E}\, ds = \sqrt{2E}.$$
In the rest of this section, we find explicit formulas for the geodesics between the points P and Q.

• *Case $\theta = 0$.* We have the following theorem.

THEOREM 2.1. *For any two points $P(\mathbf{x}_0, \mathbf{y}_0)$ and $Q(\mathbf{x}_1, \mathbf{y}_0)$ on the same horizontal line $y = \mathbf{y}_0$, there is a single geodesic*
$$(2.7) \qquad x(s) = s(\mathbf{x}_1 - \mathbf{x}_0) + \mathbf{x}_0, \qquad y(s) = \mathbf{y}_0, \qquad s \in [0, 1]$$
connecting them.

PROOF. Since $\theta = 0$, we have from the Hamiltonian system, $\dot{y}(s) = 0$, $\dot{x}(s) = $ constant. Hence the geodesic should have the form (2.7). Moreover, from the second equation of (2.4), one has $\dot{y} = \theta x^2$. If $\theta \neq 0$, then $\dot{y}$ is either strictly positive or strictly negative. It follows that y is either strictly increasing or strictly decreasing. Hence a geodesic starting from the point $P(\mathbf{x}_0, \mathbf{y}_0)$ never returns to the line $y = \mathbf{y}_0$. Thus, in order to connect the points $P(\mathbf{x}_0, \mathbf{y}_0)$ and $Q(\mathbf{x}_1, \mathbf{y}_0)$, the momentum θ must be zero. □

- *Case $\theta \neq 0$.* The analysis is now split into three situations. Let us start with the case when $(\mathbf{x}_0, \mathbf{y}_0) = (0, \mathbf{y}_0)$ and $(\mathbf{x}_1, \mathbf{y}_1) = (0, \mathbf{y}_1)$, i.e., the geodesic starts from a point on the y-axis and ends at a point on the y-axis. From the Hamiltonian system, $x(s)$ satisfies the boundary equation

(2.8)
$$\begin{cases} \ddot{x}(s) = -\theta^2 x(s) \\ x(0) = 0, \; x(1) = 0, \\ \theta = \text{constant}, \end{cases}$$

and $y(s)$ satisfies

(2.9)
$$y(s) = \mathbf{y}_0 + \theta \int_0^s x^2(u)\, du.$$

From (2.8), it is easy to see that the general solution for $x(s)$ is $A\cos\theta s + B\sin\theta s$. Since $x(0) = 0$, $A = 0$. Because $x(1) = 0$, $B\sin\theta = 0$, which implies that $\theta_m = m\pi$ for $m = 1, 2, 3, \ldots$. Therefore, $x(s) = B\sin m\pi s$. It follows that

$$y(s) = \mathbf{y}_0 + \theta B^2 \int_0^s \sin^2 m\pi u\, du = \mathbf{y}_0 + \frac{B^2}{4}\left[2m\pi s - \sin 2m\pi s\right].$$

The boundary condition $y(1) = \mathbf{y}_1$ gives

$$\mathbf{y}_1 - \mathbf{y}_0 = \frac{m\pi B^2}{2} \quad\Rightarrow\quad B = \sqrt{\frac{2(\mathbf{y}_1 - \mathbf{y}_0)}{m\pi}}.$$

Thus we have the following theorem.

THEOREM 2.2. *Given points $P(0, \mathbf{y}_0)$ and $Q(0, \mathbf{y}_1)$, there are infinitely many geodesics connecting P and Q. The parametric equations of the geodesics are given by*

(2.10)
$$\begin{cases} x_m(s) = \sqrt{\dfrac{2(\mathbf{y}_1 - \mathbf{y}_0)}{m\pi}}\, \sin m\pi s, \\ y_m(s) = \mathbf{y}_0 + (\mathbf{y}_1 - \mathbf{y}_0)\left(s - \dfrac{\sin 2m\pi s}{2m\pi}\right), \; m = 1, 2, 3, \ldots \end{cases}$$

and $\ell_m^2 = 2m\pi(\mathbf{y}_1 - \mathbf{y}_0)$, where ℓ_m is the length of the m-th geodesic. For each length ℓ_m, there are two geodesics connecting the points P and Q.

PROOF. It remains to calculate lengths of the geodesics. From (2.6), we know that the length of a geodesic $\ell(\mathcal{C}) = \sqrt{2E_m}$, where E_m satisfies the equation $\dot{x}_m^2 + \theta_m^2 x_m^2 = 2E_m$. Since $\dot{x}_m(s) = \sqrt{\frac{2(\mathbf{y}_1 - \mathbf{y}_0)}{m\pi}}\, m\pi \cos m\pi s$, one has

$$\begin{aligned} \ell_m^2 &= 2E_m = 2m\pi(\mathbf{y}_1 - \mathbf{y}_0)\cos^2 m\pi s + m^2\pi^2 \cdot \frac{2(\mathbf{y}_1 - \mathbf{y}_0)}{m\pi}\sin^2 m\pi s \\ &= 2m\pi(\mathbf{y}_1 - \mathbf{y}_0). \end{aligned}$$

If $x_m(s)$ is a solution, then $-x_m(s)$ is also a solution because of the Lagrangian symmetries. This completes the proof of the theorem. □

The geodesics $(x_m(s), y_m(s))$ for $m = 1, 2, 3, 4, 5, 6$ are given in Figure 1. One may note that $\lim_{m \to \infty} \ell_m = +\infty$.

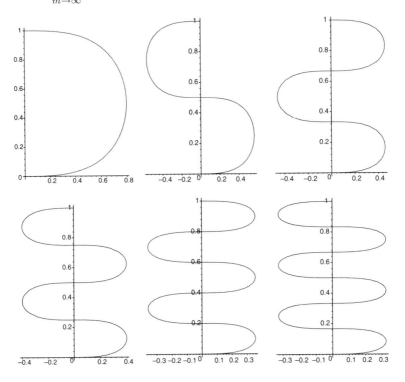

FIGURE 1. The geodesics $(x_m(s), y_m(s))$ between the origin and the point $(0, \mathbf{y})$ for $m = 1, \ldots, 6$

2.1. Geodesics ending outside of the y-axis. Let us continue to analyze the situation when $\theta \neq 0$. Now we have to solve the system (2.8) with the boundary condition
$$x(0) = \mathbf{x}_0, \quad y(0) = \mathbf{y}_0, \quad x(1) = \mathbf{x}_1, \quad y(1) = \mathbf{y}_1.$$
Again, $x(s) = A \cos \theta s + B \sin \theta s$. Since $x(0) = \mathbf{x}_0 = A$, we have $x(1) = \mathbf{x}_0 \cos \theta + B \sin \theta = \mathbf{x}_1$. When $\sin \theta \neq 0$, it follows that

(2.11) $$B = \frac{\mathbf{x}_1 - \mathbf{x}_0 \cos \theta}{\sin \theta} \quad \text{and} \quad x(s) = \mathbf{x}_0 \cos \theta s + \frac{\mathbf{x}_1 - \mathbf{x}_0 \cos \theta}{\sin \theta} \sin \theta s.$$

Plugging (2.11) into $\dot{y}(s) = \theta x^2(s)$, we have

$$\begin{aligned}
y(s) &= \mathbf{y}_0 + \theta \int_0^s x^2(u) \, du \\
&= \mathbf{y}_0 + \theta \int_0^s \left[\mathbf{x}_0 \cos \theta u + \frac{\mathbf{x}_1 - \mathbf{x}_0 \cos \theta}{\sin \theta} \sin \theta u \right]^2 du \\
&= \mathbf{y}_0 + \theta \mathbf{x}_0^2 \int_0^s \cos^2 \theta u \, du + \theta \left(\frac{\mathbf{x}_1 - \mathbf{x}_0 \cos \theta}{\sin \theta} \right)^2 \int_0^s \sin^2 \theta u \, du \\
&\quad + \frac{2\theta \mathbf{x}_0 [\mathbf{x}_1 - \mathbf{x}_0 \cos \theta]}{\sin \theta} \int_0^s \cos \theta u \sin \theta u \, du.
\end{aligned}$$

Then
$$y(s) = \mathbf{y}_0 + \frac{\mathbf{x}_0^2[2\theta s + \sin 2\theta s]}{4} + \left(\frac{\mathbf{x}_1 - \mathbf{x}_0 \cos\theta}{\sin\theta}\right)^2 \frac{2\theta s - \sin 2\theta s}{4}$$
$$(2.12) \qquad + \frac{\mathbf{x}_0[\mathbf{x}_1 - \mathbf{x}_0 \cos\theta]}{\sin\theta} \sin^2\theta s.$$

Since $y(1) = \mathbf{y}_1$, we have
$$\begin{aligned}
\mathbf{y}_1 &= \mathbf{y}_0 + [\mathbf{x}_1 - \mathbf{x}_0\cos\theta]^2 \frac{\mu(\theta)}{2} + \frac{\mathbf{x}_0^2[2\theta + \sin 2\theta]}{4} + \mathbf{x}_0[\mathbf{x}_1 - \mathbf{x}_0\cos\theta]\sin\theta \\
&= \mathbf{y}_0 + \frac{\mathbf{x}_1^2 \mu(\theta)}{2} + \mathbf{x}_0^2 \left\{ \frac{1}{2}\mu(\theta)\cos^2\theta + \frac{1}{4}[2\theta + \sin 2\theta] - \cos\theta\sin\theta \right\} \\
&\quad + \mathbf{x}_0\mathbf{x}_1 \Big[\sin\theta - \mu(\theta)\cos\theta\Big] \\
&= \mathbf{y}_0 + \frac{\mathbf{x}_1^2 \mu(\theta)}{2} + \mathbf{x}_0^2 \left\{ \frac{1}{2}\mu(\theta)\cos^2\theta + \frac{2\theta - 2\sin\theta\cos\theta}{4\sin^2 2\theta}\sin^2 2\theta \right\} \\
&\quad + \mathbf{x}_0\mathbf{x}_1 \Big[\sin\theta - \mu(\theta)\cos\theta\Big] \\
&= \mathbf{y}_0 + \frac{\mathbf{x}_1^2\mu(\theta)}{2} + \frac{\mathbf{x}_0^2\mu(\theta)}{2}\Big[\cos^2\theta + \sin^2\theta\Big] + \mathbf{x}_0\mathbf{x}_1\Big[\sin\theta - \mu(\theta)\cos\theta\Big] \\
&= \mathbf{y}_0 + \frac{(\mathbf{x}_0^2 + \mathbf{x}_1^2)}{2}\mu(\theta) + \mathbf{x}_0\mathbf{x}_1\Big[\sin\theta - \mu(\theta)\cos\theta\Big],
\end{aligned}$$
where
$$(2.13) \qquad \mu(z) = \frac{z}{\sin^2 z} - \cot z.$$
It follows that
$$\frac{2(\mathbf{y}_1 - \mathbf{y}_0)}{\mathbf{x}_0^2 + \mathbf{x}_1^2} = \mu(\theta) + \frac{2\mathbf{x}_0\mathbf{x}_1}{\mathbf{x}_0^2 + \mathbf{x}_1^2}\Big[\sin\theta - \mu(\theta)\cos\theta\Big].$$
In particular, when $\mathbf{x}_0 = 0$ and $\mathbf{x}_1 \neq 0$, one has
$$(2.14) \qquad \frac{2(\mathbf{y}_1 - \mathbf{y}_0)}{\mathbf{x}_1^2} = \mu(\theta).$$

The function $\mu(z)$, introduced by Gaveau, was first studied by Beals, Gaveau, Greiner in [4]. We require the following lemma.

LEMMA 2.3. *The function μ defined by (2.13) is a monotone increasing diffeomorphism of the interval $(-\pi, \pi)$ onto* **R**. *On each interval $(m\pi, (m+1)\pi)$, $m = 1, 2, \ldots$, μ has a unique critical point c_m. On this interval, μ decreases strictly from $+\infty$ to $\mu(c_m)$ and then increases strictly from $\mu(c_m)$ to $+\infty$. Moreover,*
$$\mu(c_m) + \pi < \mu(c_{m+1}), \; m = 1, 2, \ldots$$

The graph of $\mu(x)$ is given in Figure 2, where we take $\mathbf{y}_0 = 0$.

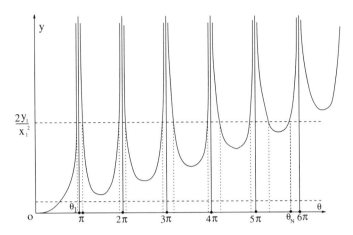

FIGURE 2. The graphs of $\mu(\theta)$ and $y = \dfrac{2\mathbf{y}_1}{\mathbf{x}_1^2}$

Let $c_1 > $ be the first critical point. If $\dfrac{2(\mathbf{y}_1 - \mathbf{y}_0)}{\mathbf{x}_1^2} < \mu(c_1)$, then (2.14) has only one solution θ_1 (see Figure 2). We have the following theorem.

THEOREM 2.4. *Given a point $Q(\mathbf{x}_1, \mathbf{y}_1)$ with $\mathbf{x}_1 \neq 0$, there are only finitely many geodesics joining the point $P(0, \mathbf{y}_0)$ and the point Q. Let $\theta_1, \theta_2, \ldots, \theta_N$ be the solutions of the equation*

$$\frac{2(\mathbf{y}_1 - \mathbf{y}_0)}{\mathbf{x}_1^2} = \mu(\theta).$$

Then the equations of the geodesics are

$$x_m(s) = \frac{\sin \theta_m s}{\sin \theta_m} \mathbf{x}_1,$$

$$y_m(s) = \mathbf{y}_0 + \frac{\mathbf{x}_1^2}{2 \sin^2 \theta_m}\left(\theta_m s - \frac{1}{2}\sin 2\theta_m s\right), \quad m = 1, 2, \ldots, N.$$

The lengths of these geodesics are

$$\ell_m^2 = \nu(\theta_m)\big[(\mathbf{y}_1 - \mathbf{y}_0) + \mathbf{x}_1^2\big],$$

where

$$\nu(z) = \frac{2z^2}{z - \sin z \cos z + 2\sin^2 z}.$$

PROOF. The first part of the theorem follows from formulas (2.11) and (2.12). For the second part, we have by (2.14),

$$(\mathbf{y}_1 - \mathbf{y}_0) + \mathbf{x}_1^2 = \left(\frac{1}{2}\mu(\theta) + 1\right)\mathbf{x}_1^2$$

and hence

(2.15) $$\mathbf{x}_1^2 = \frac{1}{1 + \mu(\theta)/2}\big((\mathbf{y}_1 - \mathbf{y}_0) + \mathbf{x}_1^2\big).$$

Using (2.6), (2.11) and (2.15), we obtain

$$\begin{aligned}
\ell_m^2 &= 2E_m = (\dot{x}_m)^2 + \theta^2 x_m^2 \\
&= \left(\frac{\mathbf{x}_1 \theta_m}{\sin\theta_m}\cos\theta_m s\right)^2 + \theta_m^2 \frac{\mathbf{x}_1^2}{\sin^2\theta_m}\sin^2\theta_m s = \frac{\theta_m^2 \mathbf{x}_1^2}{\sin^2\theta_m} \\
&= \frac{\theta_m^2}{\sin^2\theta_m} \cdot \frac{1}{1+\mu(\theta_m)/2}\left[(\mathbf{y}_1-\mathbf{y}_0)+\mathbf{x}_1^2\right] \\
&= \frac{2\theta_m^2}{\theta_m - \sin\theta_m \cos\theta_m + 2\sin^2\theta_m}\left[(\mathbf{y}_1-\mathbf{y}_0)+\mathbf{x}_1^2\right] \\
&= \nu(\theta_m)\left[(\mathbf{y}_1-\mathbf{y}_0)+\mathbf{x}_1^2\right].
\end{aligned}$$

The proof of the theorem is complete. □

The graph of the function ν is given in Figure 3.

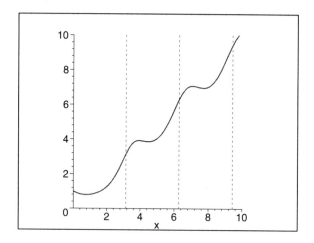

FIGURE 3. The graph of $\nu(x)$

Assume that $\mathbf{x}_0 \mathbf{x}_1 \neq 0$. Then we have to look at the equation

(2.16) $$\frac{2(\mathbf{y}_1-\mathbf{y}_0)}{\mathbf{x}_0^2+\mathbf{x}_1^2} = \mu(\theta) + \frac{2\mathbf{x}_0\mathbf{x}_1}{\mathbf{x}_0^2+\mathbf{x}_1^2}\left[\sin\theta - \mu(\theta)\cos\theta\right].$$

Let $a = \dfrac{2\mathbf{x}_0\mathbf{x}_1}{\mathbf{x}_0^2+\mathbf{x}_1^2}$. There are two cases.

• $a > 0$. This means that the points P and Q are located on the same side of the y-axis. Equation (2.16) becomes

$$\begin{aligned}
\frac{2(\mathbf{y}_1-\mathbf{y}_0)}{\mathbf{x}_0^2+\mathbf{x}_1^2} &= \mu(\theta)(1-a\cos\theta) \\
&\quad + a\sin\theta \\
&= \mu(\theta)\left[1-a+a(1-\cos\theta)\right] + a\sin\theta \\
&= \mu(\theta)\left[1-a+2a\sin^2\theta/2\right] + a\sin\theta \\
&= (1-a)\mu(\theta) + a\frac{(2\sin^2\frac{\theta}{2})(\theta-\cos\theta\sin\theta)+\sin^3\theta}{\sin^2\theta}.
\end{aligned}$$

We consider the term $(\sin^2 \frac{\theta}{2})[\theta - \cos\theta \sin\theta] + \sin^3 \theta$ separately. One has

$$\begin{aligned}
(2\sin^2 \tfrac{\theta}{2})&[\theta - \cos\theta \sin\theta] + \sin^3 \theta \\
&= 2\theta \sin^2 \tfrac{\theta}{2} + \left(1 - 2\sin^2 \tfrac{\theta}{2}\right) \sin\theta \cos\theta - \sin\theta\cos\theta + \sin^3\theta \\
&= 2\theta \sin^2 \tfrac{\theta}{2} + \cos^2\theta \sin\theta + \sin^3\theta - \sin\theta\cos\theta \\
&= 2\theta \sin^2 \tfrac{\theta}{2} + \sin\theta(1 - \cos\theta) \\
&= 2\theta \sin^2 \tfrac{\theta}{2} + 2\sin\theta \sin^2 \tfrac{\theta}{2} \\
&= 2\theta \sin^2 \tfrac{\theta}{2} + 4\sin^3 \tfrac{\theta}{2} \cos \tfrac{\theta}{2}.
\end{aligned}$$

Thus

$$\begin{aligned}
\frac{2(\mathbf{y}_1 - \mathbf{y}_0)}{\mathbf{x}_0^2 + \mathbf{x}_1^2} &= (1-a)\mu(\theta) + a \frac{2\theta \sin^2 \tfrac{\theta}{2} + 4\sin^3 \tfrac{\theta}{2} \cos \tfrac{\theta}{2}}{4\sin^2 \tfrac{\theta}{2} \cos^2 \tfrac{\theta}{2}} \\
&= (1-a)\mu(\theta) + a\left\{\tfrac{\theta/2}{\cos^2 \tfrac{\theta}{2}} + \tan \tfrac{\theta}{2}\right\} \\
&= (1-a)\mu(\theta) + a\tilde{\mu}\left(\tfrac{\theta}{2}\right),
\end{aligned}$$

where

$$\tilde{\mu}(x) = \frac{x}{\cos^2 x} + \tan x.$$

- $a < 0$. This means that the points P and Q are located on different sides of the y-axis. The equation (2.16) becomes

$$\begin{aligned}
\frac{2(\mathbf{y}_1 - \mathbf{y}_0)}{\mathbf{x}_0^2 + \mathbf{x}_1^2} &= \mu(\theta)\big[1 + |a|\cos\theta\big] - |a|\sin\theta \\
&= \mu(\theta)\big[1 - |a| + |a|(1+\cos\theta)\big] - |a|\sin\theta \\
&= \mu(\theta)\left[1 - |a| + 2|a|\cos^2 \tfrac{\theta}{2}\right] - |a|\sin\theta \\
&= (1-|a|)\mu(\theta) + |a|\frac{2\cos^2 \tfrac{\theta}{2}[\theta - \cos\theta\sin\theta] - \sin^3\theta}{\sin^2\theta}.
\end{aligned}$$

Similar to the case $a > 0$, one has

$$\begin{aligned}
(2\cos^2 \tfrac{\theta}{2})&[\theta - \cos\theta\sin\theta] - \sin^3\theta \\
&= 2\theta \cos^2 \tfrac{\theta}{2} - \left(2\cos^2 \tfrac{\theta}{2} - 1\right)\sin\theta\cos\theta - \sin\theta\cos\theta - \sin^3\theta \\
&= 2\theta \cos^2 \tfrac{\theta}{2} - \cos^2\theta \sin\theta - \sin^3\theta - \sin\theta\cos\theta \\
&= 2\theta \cos^2 \tfrac{\theta}{2} - \sin\theta(1 + \cos\theta) \\
&= 2\theta \cos^2 \tfrac{\theta}{2} - 2\sin\theta \cos^2 \tfrac{\theta}{2} \\
&= 2\theta \cos^2 \tfrac{\theta}{2} - 4\cos^3 \tfrac{\theta}{2} \sin \tfrac{\theta}{2}.
\end{aligned}$$

Then

$$\begin{aligned}
\frac{2(\mathbf{y}_1 - \mathbf{y}_0)}{\mathbf{x}_0^2 + \mathbf{x}_1^2} &= (1+a)\mu(\theta) - a\frac{2\theta \cos^2 \tfrac{\theta}{2} - 4\cos^3 \tfrac{\theta}{2} \sin \tfrac{\theta}{2}}{4\sin^2 \tfrac{\theta}{2} \cos^2 \tfrac{\theta}{2}} \\
&= (1+a)\mu(\theta) - a\left\{\frac{\tfrac{\theta}{2}}{\sin^2 \tfrac{\theta}{2}} - \cot \tfrac{\theta}{2}\right\} \\
&= (1+a)\mu(\theta) - a\mu\left(\tfrac{\theta}{2}\right).
\end{aligned}$$

It remains to analyze the functions $(1+a)\mu(\theta) - a\mu(\frac{\theta}{2})$ and $(1-a)\mu(\theta) + a\tilde{\mu}(\frac{\theta}{2})$. By Lemma 2.3, the function $\mu(x)$ has vertical asymptotes at $m\pi$ and a single minimum point on each interval $\bigl(m\pi, (m+1)\pi\bigr)$, $m \geq 1$. The function $\mu(x/2)$ looks very similar but has vertical asymptotes at $2m\pi$ and a single minimum point on each interval $\bigl(2m\pi, (2m+1)\pi\bigr)$, $m \geq 1$. Let

$$F(z) = (1-a)\mu(z) + a\mu(z/2), \quad a \in [0,1].$$

The function $F(z)$ has vertical asymptotes at $m\pi$ and, as a convex combination of two strictly convex functions, is strictly convex ($F''(z) > 0$). Hence $F(z)$ has an unique critical point on each interval $\bigl(m\pi, (m+1)\pi\bigr)$, $m > 1$, which must be a minimum point. We call this point c_m and claim that

$$\lim_{m \to \infty} F(c_m) = \infty,$$

i.e., the critical values of F go to infinity.

Indeed, let

$$\delta_{2m} = \{\min \mu(z); z \in \bigl(2m\pi, (2m+1)\pi\bigr)\},$$
$$\delta'_{2m} = \{\min \mu(z/2); z \in \bigl(2m\pi, (2m+2)\pi\bigr)\}.$$

It is known that

(2.17) $$\lim_{m \to \infty} \delta_{2m} = \lim_{m \to \infty} \delta'_{2m} = +\infty.$$

Then for $z \in \bigl(2m\pi, (2m+1)\pi\bigr)$

$$F(z) \geq (1-a) \min_{[2m\pi,(2m+1)\pi]} \mu(z) + a \min_{[2m\pi,(2m+1)\pi]} \mu(z/2)$$

$$\geq (1-a)\delta_{2m} + a\delta'_{2m} \geq \min\{\delta_{2m}, \delta'_{2m}\}.$$

Hence

$$\min_{[2m\pi,(2m+1)\pi]} F(z) \geq \min\{\delta_{2m}, \delta'_{2m}\}.$$

Equation (2.17) yields

$$\lim_{m \to \infty} \min_{[2m\pi,(2m+1)\pi]} F(z) = \infty.$$

In a similar way, we can show that for $m > 1$,

(2.18) $$\lim_{m \to \infty} \min_{[(2m-1)\pi, 2m\pi]} F(z) = \infty.$$

We now turn our attention to the function

$$\tilde{\mu}(z) = \frac{z}{\cos^2 z} + \tan z, \quad z \geq 0.$$

LEMMA 2.5. *The function $\tilde{\mu}(z)$ has vertical asymptotes at $m\pi + \frac{\pi}{2}$, $m = 1, 2, 3 \ldots$. On each interval $\bigl(m\pi + \frac{\pi}{2}, (m+1)\pi + \frac{\pi}{2}\bigr)$, $\tilde{\mu}(z)$ has an unique critical point $\tilde{c}_m$, which is a minimum. We have*

(2.19) $$\tilde{\mu}(\tilde{c}_m) \geq \pi\left(m + \frac{1}{2}\right),$$

and hence $\lim_{m \to \infty} \tilde{\mu}(\tilde{c}_m) = +\infty$.

PROOF. **Step 1:** We prove the decomposition

(2.20)
$$\tilde{\mu}\left(z + \frac{\pi}{2}\right) = \mu(z) + \frac{\pi}{2}\frac{1}{\sin^2 z}.$$

Indeed,

$$\mu\left(z - \frac{\pi}{2}\right) = \frac{z - \frac{\pi}{2}}{\sin^2(z - \frac{\pi}{2})} - \cot\left(z - \frac{\pi}{2}\right) = \frac{z - \frac{\pi}{2}}{\sin^2(z - \frac{\pi}{2})} - \frac{\cos(z - \frac{\pi}{2})}{\sin(z - \frac{\pi}{2})}$$

$$= \frac{z - \frac{\pi}{2}}{\cos^2 z} - \frac{\sin z}{\cos z} = \left(\frac{z}{\cos^2 z} + \tan z\right) - \frac{\pi}{2}\frac{1}{\cos^2 z}$$

$$= \tilde{\mu}(z) - \frac{\pi}{2}\frac{1}{\cos^2 z}.$$

Hence
$$\tilde{\mu}(z) = \mu\left(z - \frac{\pi}{2}\right) + \frac{\pi}{2}\frac{1}{\cos^2 z}.$$

Putting $z := z + \frac{\pi}{2}$ yields (2.20).

Step 2: Both functions $\mu(z)$ and $\frac{\pi}{2}\frac{1}{\sin^2 z}$ have asymptotes at $m\pi$, $m = 1, 2, 3 \ldots$; hence, by Step 1, $\tilde{\mu}(z)$ has asymptotes at $m\pi + \frac{\pi}{2}$.

Step 3: The function $z \to \frac{\pi}{2}\frac{1}{\sin^2 z}$ is periodic and on each interval $(m\pi, (m+1)\pi)$ has a minimum value equal to $\pi/2$.

Step 4: The first derivative of the function $\tilde{\mu}(z)$ is

$$\tilde{\mu}'(z) = \frac{2\cos z + 2z \sin z}{\cos^2 z}.$$

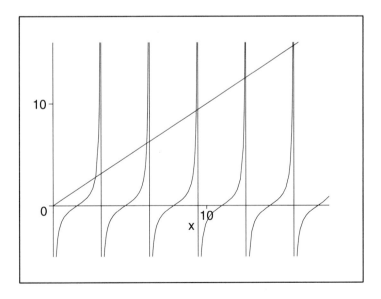

FIGURE 4. The positive solutions of the equation $x = -\cot x$

It follows that $\tilde{\mu}(z)$ has a unique critical point $\tilde{c}_m$ on each interval $(m\pi + \pi/2, (m+1)\pi + \pi/2)$ since it satisfies the equation $\tilde{c}_m = -\cot \tilde{c}_m$; see Figure 4. From Step 1, the function $\tilde{\mu}(z)$ is a convex sum of two strictly convex functions. Therefore, $\tilde{\mu}''(z) > 0$ on each interval $(m\pi + \pi/2, (m+1)\pi + \pi/2)$. Hence, all critical points are minimum points.

Step 5: It remains to prove (2.19). Let
$$\delta_m := \min\{\mu(z) : z \in [m\pi, (m+1)\pi]\}.$$
It is known that $\delta_m > m\pi$ [**5**, p. 739, (1.34)]. Let $z \in [m\pi + \pi/2, (m+1)\pi + \pi/2]$. By Step 1, we have
$$\begin{aligned}
\tilde{\mu}(z) &= \mu\left(z - \frac{\pi}{2}\right) + \frac{\pi}{2}\frac{1}{\cos^2 z} \\
&\geq \mu\left(z - \frac{\pi}{2}\right) + \frac{\pi}{2} \\
&\geq \delta_m + \frac{\pi}{2} \\
&\geq m\pi + \frac{\pi}{2}.
\end{aligned}$$
Hence
$$\tilde{\mu}(\tilde{c}_m) \geq m\pi + \frac{\pi}{2} \quad \text{and} \quad \lim_{m \to \infty} \tilde{\mu}(\tilde{c}_m) = +\infty.$$
$\square$

Summarizing the above discussion, we have the following theorem.

THEOREM 2.6. *Assume that $\mathbf{x}_0\mathbf{x}_1 \neq 0$ with $\mathbf{x}_0 \neq \pm\mathbf{x}_1$. The number of geodesics connecting the points $P(\mathbf{x}_0, \mathbf{y}_0)$ and $Q(\mathbf{x}_1, \mathbf{y}_1)$ is finite. Their lengths are given by*

$$\begin{aligned}
\ell_m^2 &= \frac{\theta_m^2}{\sin^2 \theta_m}\left[\mathbf{x}_0^2 + \mathbf{x}_1^2 - 2\mathbf{x}_0\mathbf{x}_1 \cos\theta_m\right] \\
&= \nu_a(\theta_m)\left[\mathbf{y}_1 - \mathbf{y}_0 + (\mathbf{x}_1 - \mathbf{x}_0)^2\right], \qquad m = 1, 2, \ldots, N,
\end{aligned} \tag{2.21}$$

where

$$\nu_a(\theta) = \begin{cases} \dfrac{2(1 - a\cos\theta)}{(1-a)(\mu(\theta) + 2) + a\tilde{\mu}(\theta/2)} \cdot \dfrac{\theta^2}{\sin^2\theta} & \text{if } a = \dfrac{2\mathbf{x}_0\mathbf{x}_1}{\mathbf{x}_0^2 + \mathbf{x}_1^2} > 0, \\[2mm] \dfrac{2(1 - a\cos\theta)}{(1+a)\mu(\theta) - a(\mu(\theta/2) + 2) + 2} \cdot \dfrac{\theta^2}{\sin^2\theta} & \text{if } a = \dfrac{2\mathbf{x}_0\mathbf{x}_1}{\mathbf{x}_0^2 + \mathbf{x}_1^2} < 0. \end{cases} \tag{2.22}$$

Here θ_m are solutions of

$$\frac{2(\mathbf{y}_1 - \mathbf{y}_0)}{\mathbf{x}_0^2 + \mathbf{x}_1^2} = (1-a)\mu(\theta) + a\tilde{\mu}\theta/2, \quad \text{if} \quad a > 0, \tag{2.23}$$

or

$$\frac{2(\mathbf{y}_1 - \mathbf{y}_0)}{\mathbf{x}_0^2 + \mathbf{x}_1^2} = (1+a)\mu(\theta) - a\mu(\theta/2), \quad \text{if} \quad a < 0. \tag{2.24}$$

PROOF. From (2.18) and Lemma 2.5, we have
$$\lim_{m \to \infty} \min_{\theta \in [(2m-1)\pi, 2m\pi]} (1+a)\mu(\theta) - a\mu(\theta/2) = \infty$$
and
$$\lim_{m \to \infty} \min_{\theta \in [(2m+1)\pi, (2m+3)\pi]} (1-a)\mu(\theta) + a\tilde{\mu}(\theta/2) = \infty.$$

It is easy to conclude that the number of solutions of the equations

$$\frac{2(\mathbf{y}_1 - \mathbf{y}_0)}{\mathbf{x}_0^2 + \mathbf{x}_1^2} = (1+a)\mu(\theta) - a\mu(\theta/2)$$

and

$$\frac{2(\mathbf{y}_1 - \mathbf{y}_0)}{\mathbf{x}_0^2 + \mathbf{x}_1^2} = (1-a)\mu(\theta) + a\tilde{\mu}(\theta/2)$$

is finite if and only if $\mathbf{x}_0^2 + \mathbf{x}_1^2 \neq 0$. Denote the solutions by $\theta_1, \theta_2, ..., \theta_N$. Using (2.6) and (2.11), we have

$$\begin{aligned}
\ell_m^2 &= 2E_m = (\dot{x}_m(s))^2 + \theta_m^2 x_m^2(s) \\
&= \left[-\mathbf{x}_0 \theta_m \sin\theta_m s + \theta_m \frac{\mathbf{x}_1 - \mathbf{x}_0 \cos\theta_m}{\sin\theta_m} \cos\theta_m s \right]^2 \\
&\quad + \theta_m^2 \left[\mathbf{x}_0 \cos\theta_m s + \frac{\mathbf{x}_1 - \mathbf{x}_0 \cos\theta_m}{\sin\theta_m} \sin\theta_m s \right]^2 \\
&= \frac{\theta_m^2}{\sin^2 \theta_m} \left[\mathbf{x}_0^2 \sin^2\theta_m + \mathbf{x}_1^2 - 2\mathbf{x}_0 \mathbf{x}_1 \cos\theta_m + \mathbf{x}_0^2 \cos^2\theta_m \right],
\end{aligned}$$

which gives

(2.25) $$E_m = \frac{\theta_m^2}{2}\left[\mathbf{x}_0^2 + \left(\frac{\mathbf{x}_1 - \mathbf{x}_0 \cos\theta_m}{\sin\theta_m}\right)^2\right] = \frac{\theta_m^2}{2}\left[\frac{\mathbf{x}_0^2 + \mathbf{x}_1^2 - 2\mathbf{x}_0\mathbf{x}_1 \cos\theta_m}{\sin^2\theta_m}\right]$$

and

$$\ell_m^2 = \frac{\theta_m^2}{\sin^2\theta_m}\left[\mathbf{x}_0^2 + \mathbf{x}_1^2 - 2\mathbf{x}_0\mathbf{x}_1 \cos\theta_m\right].$$

Now we are in a position to prove (2.21). There are two cases.
Case $a > 0$.

$$\begin{aligned}
\ell_m^2 &= \frac{\theta_m^2}{\sin^2\theta_m}(\mathbf{x}_0^2 + \mathbf{x}_1^2 - 2\mathbf{x}_0\mathbf{x}_1 \cos\theta_m) \\
&= \frac{\theta_m^2}{\sin^2\theta_m}(\mathbf{x}_0^2 + \mathbf{x}_1^2)(1 - a\cos\theta_m).
\end{aligned}$$ (2.26)

Since θ_m is the m-th solution of the equation

$$\frac{2(\mathbf{y}_1 - \mathbf{y}_0)}{\mathbf{x}_0^2 + \mathbf{x}_1^2} = \widetilde{F}(a,\theta) := (1-a)\mu(\theta) + a\tilde{\mu}(\theta/2),$$

we have

$$\begin{aligned}
\mathbf{y}_1 - \mathbf{y}_0 + (\mathbf{x}_1 - \mathbf{x}_0)^2 &= \mathbf{y}_1 - \mathbf{y}_0 + \mathbf{x}_1^2 + \mathbf{x}_0^2 - 2\mathbf{x}_0\mathbf{x}_1 \\
&= (\mathbf{x}_0^2 + \mathbf{x}_1^2)\left(\frac{\mathbf{y}_1 - \mathbf{y}_0}{\mathbf{x}_0^2 + \mathbf{x}_1^2} + 1 - a\right) \\
&= (\mathbf{x}_0^2 + \mathbf{x}_1^2)\left(\frac{1}{2}\widetilde{F}(a,\theta) + 1 - a\right)
\end{aligned}$$

and hence

(2.27) $$\mathbf{x}_0^2 + \mathbf{x}_1^2 = \frac{\mathbf{y}_1 - \mathbf{y}_0 + (\mathbf{x}_1 - \mathbf{x}_0)^2}{\frac{1}{2}\widetilde{F}(a,\theta) + 1 - a}.$$

Substituting in (2.26) yields

$$\begin{aligned}
\ell_m^2 &= \frac{\theta_m^2}{\sin^2\theta_m}(1 - a\cos\theta_m)(\mathbf{x}_0^2 + \mathbf{x}_1^2) \\
&= \frac{\theta_m^2}{\sin^2\theta_m}(1 - a\cos\theta_m)\frac{\mathbf{y}_1 - \mathbf{y}_0 + (\mathbf{x}_1 - \mathbf{x}_0)^2}{\frac{1}{2}\widetilde{F}(a,\theta) + 1 - a} \\
&= \frac{2\theta_m^2}{\sin^2\theta_m}\frac{1 - a\cos\theta_m}{\widetilde{F}(a,\theta) + 2(1-a)}[\mathbf{y}_1 - \mathbf{y}_0 + (\mathbf{x}_1 - \mathbf{x}_0)^2] \\
&= \frac{2\theta_m^2}{\sin^2\theta_m}\frac{1 - a\cos\theta_m}{(1-a)(2 + \mu(\theta_m)) + a\tilde{\mu}(\theta_m/2)}[\mathbf{y}_1 - \mathbf{y}_0 + (\mathbf{x}_1 - \mathbf{x}_0)^2] \\
&= \nu_a(\theta_m)[\mathbf{y}_1 - \mathbf{y}_0 + (\mathbf{x}_1 - \mathbf{x}_0)^2].
\end{aligned}$$

Case $a < 0$. In this case, θ_m is a solution of the equation

$$\frac{2(\mathbf{y}_1 - \mathbf{y}_0)}{\mathbf{x}_0^2 + \mathbf{x}_1^2} = F(a,\theta) := (1+a)\mu(\theta) - a\mu(\theta/2).$$

In a way similar to that above, we arrive at the following analogue of equation (2.27)

$$\mathbf{x}_0^2 + \mathbf{x}_1^2 = \frac{\mathbf{y}_1 - \mathbf{y}_0 + (\mathbf{x}_1 - \mathbf{x}_0)^2}{\frac{1}{2}F(a,\theta) + 1 - a}.$$

Therefore,

$$\begin{aligned}
\ell_m^2 &= \frac{\theta_m^2}{\sin^2\theta_m}(\mathbf{x}_0^2 + \mathbf{x}_1^2 - 2\mathbf{x}_0\mathbf{x}_1\cos\theta_m) \\
&= \frac{\theta_m^2}{\sin^2\theta_m}(1 - a\cos\theta_m)(\mathbf{x}_0^2 + \mathbf{x}_1^2) \\
&= \frac{\theta_m^2}{\sin^2\theta_m}(1 - a\cos\theta_m)\frac{\mathbf{y}_1 - \mathbf{y}_0 + (\mathbf{x}_1 - \mathbf{x}_0)^2}{\frac{1}{2}F(a,\theta) + 1 - a} \\
&= \frac{2\theta_m^2}{\sin^2\theta_m}\frac{1 - a\cos\theta_m}{(1+a)\mu(\theta_m) - a\big(\mu(\theta_m/2) + 2\big) + 2}[\mathbf{y}_1 - \mathbf{y}_0 + (\mathbf{x}_1 - \mathbf{x}_0)^2] \\
&= \nu_a(\theta_m)[\mathbf{y}_1 - \mathbf{y}_0 + (\mathbf{x}_1 - \mathbf{x}_0)^2].
\end{aligned}$$

The proof of the theorem is complete. $\square$

When $\sin\theta = 0$, then $\theta = m\pi$ for some $m \in \mathbf{N}$. This yields

$$\mathbf{x}_1 = \mathbf{x}_0\cos\theta + B\sin\theta \Rightarrow \mathbf{x}_1 = \pm\mathbf{x}_0.$$

It follows that

$$y(s) = \mathbf{y}_0 + \frac{m\pi}{2}\left[\mathbf{x}_0^2\left(s + \frac{\sin 2m\pi s}{2m\pi}\right) + B^2\left(s - \frac{\sin 2m\pi s}{2m\pi}\right) + 2\mathbf{x}_0 B\sin^2 m\pi s\right].$$

Hence

$$\mathbf{y}_1 = \mathbf{y}_0 + \frac{m\pi}{2}(\mathbf{x}_0^2 + B^2) \Leftrightarrow B^2 = \frac{2(\mathbf{y}_1 - \mathbf{y}_0)}{m\pi} - \mathbf{x}_0^2.$$

The above makes sense only when

(2.28) $$\frac{2(\mathbf{y}_1 - \mathbf{y}_0)}{m\pi} \geq \mathbf{x}_0^2.$$

Therefore, with given $\mathbf{y}_0$, $\mathbf{y}_1$, and $\mathbf{x}_0$, there are only finitely many m's such that (2.28) holds. There is no extra geodesic when $2(\mathbf{y}_1 - \mathbf{y}_0) < \mathbf{x}_0^2 \pi$. In case that $2(\mathbf{y}_1 - \mathbf{y}_0) \geq \mathbf{x}_0^2 \pi$, we may find $m = 1, \ldots, N$ such that

$$B_m = \sqrt{\frac{2(\mathbf{y}_1 - \mathbf{y}_0)}{m\pi} - \mathbf{x}_0^2}.$$

In this case, the parametric equations of the geodesics have the form

$$x_m(s) = \mathbf{x}_0 \cos m\pi s + B_m \sin m\pi s,$$

$$y_m(s) = \mathbf{y}_0 + \frac{m\pi}{2}\left[\mathbf{x}_0^2\left(s + \frac{\sin 2m\pi s}{2m\pi}\right) + B_m^2\left(s - \frac{\sin 2m\pi s}{2m\pi}\right) + 2\mathbf{x}_0 B_m \sin^2 m\pi s\right].$$

The length of the corresponding geodesic satisfies

$$\ell_m^2 = m^2\pi^2[\mathbf{x}_0^2 + B_m^2] = 2m\pi(\mathbf{y}_1 - \mathbf{y}_0), \quad m = 1, \ldots, N.$$

Therefore, when $\mathbf{x}_0 = \pm\mathbf{x}_1 \neq \mathbf{0}$, those θ satisfying either (2.23) or (2.24) give finitely many geodesics. They all have different lengths. Furthermore, if (2.28) holds, then there are finitely many extra geodesics connecting the points P and Q.

THEOREM 2.7. *Set $a = \frac{\mathbf{x}_0 \mathbf{x}_1}{\mathbf{x}_0^2 + \mathbf{x}_1^2} \neq 0$. If $a > 0$, then the m-th geodesic connecting $(\mathbf{x}_0, \mathbf{y}_0)$ and $(\mathbf{x}_1, \mathbf{y}_1)$ intersects the y-axis $2\left[\frac{m+2}{4}\right]$ times. If $a < 0$, then this geodesic intersects the y-axis $2\left[\frac{m}{4}\right] + 1$ times.*

PROOF. Assume $\mathbf{x}_0 > 0$ and $\mathbf{x}_1 \neq 0$. Set

(2.29) $$\frac{\mathbf{x}_0 \sin\theta}{\mathbf{x}_0 \cos\theta - \mathbf{x}_1} = \psi(\theta).$$

By (2.11) and (2.29),

(2.30) $$x(s) = 0 \iff \tan\theta s = \psi(\theta).$$

We are looking for solutions of (2.30) in the interval $s \in (0,1)$ or $\theta s \in (0,\theta)$ where $\theta = \theta_m$ is a solution of either (2.23) or (2.24). Note that θ_m is not an integral multiple of π. Consider first the case $a > 0$, i.e., $\mathbf{x}_1 > 0$. Without loss of generality, one may assume $\mathbf{x}_0 \leq \mathbf{x}_1$. We distinguish four cases.

(i) $2n\pi < \theta \leq 2n\pi + \frac{\pi}{2}$. Then $\sin\theta > 0$, $\cos\theta \geq 0$ and $\psi(\theta) < 0$. Hence no solution exists with $\theta s > 2n\pi$. However, every period of the tangent function contains one solution. Hence there exist $2n$ solutions in $\theta s \in (0, 2n\pi)$.

(ii) $2n\pi + \frac{\pi}{2} < \theta \leq (2n+1)\pi$. Then

$$\sin\theta > 0, \ \cos\theta < 0, \ |\psi(\theta)| < |\tan\theta| \implies \tan\theta < \psi(\theta) < 0.$$

We cannot have a solution θs with $\theta s \in (2n\pi + \frac{\pi}{2}, \theta)$. We also know that $\tan\theta s > 0$ in the interval $[2n\pi, 2n\pi + \frac{\pi}{2})$. However, we do have $2n$ solutions when $\theta s \in (0, 2n\pi)$.

(iii) $(2n+1)\pi < \theta \leq (2n+1)\pi + \frac{\pi}{2}$. Then $\sin\theta < 0$, $\cos\theta \leq 0$ and $\psi(\theta) > 0$. We have now, besides the $2n$ solutions with $\theta s \in (0, 2n\pi)$, two additional solutions: one in the interval $\theta s \in [2n\pi, 2n\pi + \frac{\pi}{2})$, and one satisfying $(2n+1)\pi < \theta s < \theta$, as $\psi(\theta) < \tan\theta$. Therefore, the total number of solutions is $2n + 2 = 2(n+1)$.

(iv) $(2n+1)\pi + \frac{\pi}{2} < \theta < (2n+2)\pi$. Then $\sin\theta < 0$, $\cos\theta > 0$ and $\psi(\theta) > 0$. Hence we have $2n + 2$ solutions: one for each interval where $\tan\theta s > 0$.

Consider now the case $a < 0$ and $\mathbf{x}_1 < 0$. We again distinguish four cases.

(i) $2n\pi < \theta \leq 2n\pi + \frac{\pi}{2}$. Then $0 < \psi(\theta) < \tan\theta$. We have $2n$ solutions with $\theta s \in (0, 2n\pi)$, and an additional one when $\theta s \in (2n\pi, \theta)$. Therefore, we have $2n+1$ solutions.

(ii) $2n\pi + \frac{\pi}{2} < \theta < (2n+1)\pi$. Then $\cos\theta < 0$ and $\tan\theta < 0$. There are three subcases.

(a) $\mathbf{x}_0 \cos\theta - \mathbf{x}_1 > 0$. Then $\psi(\theta) > 0$, and we have $2n+1$ solutions with $\theta s \in (0, 2n + \frac{\pi}{2})$.

(b) $\mathbf{x}_0 \cos\theta - \mathbf{x}_1 = 0$. Then $\psi(\theta) = \infty$ and $\theta s = k\pi + \pi/2$ for $0 < k \leq 2n$. Hence we have $2n+1$ solutions.

(c) $\mathbf{x}_0 \cos\theta - \mathbf{x}_1 < 0$. Then $\psi(\theta) < 0$. Note that in this subcase $|\mathbf{x}_0 \cos\theta - \mathbf{x}_1| < |\mathbf{x}_0 \cos\theta|$ and $\sin\theta > 0$. Hence $\psi(\theta) < \tan\theta < 0$, so that, besides the $2n$ solutions with $\theta s \in (0, 2n\pi)$, there is also a solution with $\theta s \in (2n\pi + \frac{\pi}{2}, \theta)$.

(iii) $(2n+1)\pi < \theta < (2n+1)\pi + \frac{\pi}{2}$. Here $\sin\theta < 0$, $\cos\theta > 0$, and we consider three subcases.

(a) $\mathbf{x}_0 \cos\theta - \mathbf{x}_1 > 0$. Then $\psi(\theta) < 0$, and we have $2n+1$ solutions with $\theta s \in (0, (2n+1)\pi)$.

(b) $\mathbf{x}_0 \cos\theta - \mathbf{x}_1 = 0$. Then $\psi(\theta) = \infty$, and we proceed as in subcase (b) of (ii).

(c) $\mathbf{x}_0 \cos\theta - \mathbf{x}_1 < 0$. Then $\psi(\theta) > 0$. We have $2n+1$ solutions with $\theta s \in (0, (2n+1)\pi)$. Moreover,

$$\left|\cos\theta - \frac{\mathbf{x}_1}{\mathbf{x}_0}\right| < |\cos\theta| \Rightarrow |\psi(\theta)| > |\tan\theta| \Rightarrow \psi(\theta) > \tan\theta;$$

so there is no solution with $\theta s \in ((2n+1)\pi, \theta)$, and we have $2n+1$ solutions.

(iv) $(2n+1)\pi + \frac{\pi}{2} < \theta < (2n+2)\pi$. Here $\sin\theta < 0$, $\cos\theta > 0$, and $\psi(\theta) < 0$. Also,

$$|\psi(\theta)| < |\tan\theta| \Rightarrow \tan\theta < \psi(\theta) < 0;$$

and we have no solution with $\theta s \in \left((2n+1)\pi + \frac{\pi}{2}, \theta\right)$, but we do have $2n+1$ solutions with $\theta s \in \left(0, (2n+1)\pi\right)$.

Recall that $\theta_1 \in [0, \pi)$, $\theta_2, \theta_3 \in (\pi, 2\pi)$,... etc. We see that if $a > 0$, then the number of intersections with the y-axis is $2\left[\frac{m+2}{4}\right]$. If $a < 0$, then it is $2\left[\frac{m}{4}\right] + 1$. The proof of the theorem is complete. $\square$

As a consequence, we have the following corollary.

COROLLARY 2.8. *If $\mathbf{x}_0 = 0$ and $\mathbf{x}_1 \neq 0$, then the m-th geodesic connecting $(0, \mathbf{y}_0)$ and $(\mathbf{x}_1, \mathbf{y}_1)$ intersects the y-axis $\left[\frac{m}{2}\right]$ times.*

PROOF. When $\mathbf{x}_0 = 0$, (2.30) reduces to

$$x(s) = 0 \iff \sin\theta_m s = 0, \quad s \in [0, 1],$$

where θ_m satisfies the equation $\frac{2(\mathbf{y}_1 - \mathbf{y}_0)}{\mathbf{x}_1^2} = \mu(\theta_m)$. From Lemma 2.3, we know that $\theta_1 \in [0, \pi)$, $\theta_2, \theta_3 \in (\pi, 2\pi)$,... etc. Now the result follows immediately. $\square$

3. Complex Hamiltonian mechanics

In this section, we define the *complex geodesics* and compute the complex action. We show how the complex action can be used to obtain the length of the real geodesics. Before we calculate the complex action function, we compute the real action function. Let $\mathcal{C}(s) = (x(s), y(s))$ be a curve in $\mathbf{R}^2$ with $s \in [0, \tau]$ and let θ be a real parameter. The real action S is defined as an integral of the Lagrangian along the bicharacteristics

$$S(x, y, \tau; \theta) = \int_0^\tau (\dot{x}\xi + \dot{y}\theta - H)\, ds = \theta(\mathbf{y}_1 - \mathbf{y}_0) + \int_0^\tau (\dot{x}\xi - H)\, ds.$$

It is known that S is a solution of the Hamilton-Jacobi equation

(3.31) $$\frac{\partial S}{\partial \tau} + H\left(x, y, \frac{\partial S}{\partial x}, \frac{\partial S}{\partial y}\right) = 0.$$

A classical result provides the momenta as derivatives of the action

(3.32) $$\frac{\partial S}{\partial x} = \xi, \qquad \frac{\partial S}{\partial y} = \theta,$$

where $(x(s), y(s), \xi(s), \theta(s))$ is a solution of the Hamiltonian system. Since

$$\frac{d}{ds} H(x(s), y(s), \xi(s), \theta(s)) = \frac{\partial}{\partial s} H(x(s), y(s), \xi(s), \theta(s)) = 0,$$

and by (3.32),

$$H\left(x(s), y(s), \frac{\partial S}{\partial x}, \frac{\partial S}{\partial y}\right) = H(x(s), y(s), \xi(s), \theta(s)) = \frac{1}{2}(\xi^2 + x^2\theta^2) = E,$$

the energy E is constant along the bicharacteristics. Because H is homogeneous of degree 2 with respect of (ξ, θ),

$$S = \int_0^\tau \left(\xi H_\xi + \theta H_\theta - H\right) ds = \int_0^\tau (2H - H)\, ds = \tau E.$$

Formula (2.25) with $\mathbf{x}_0 = 0$ yields

$$E = \frac{\theta^2 \mathbf{x}_1^2}{2 \sin^2 \theta \tau},$$

where $x(\tau) = \mathbf{x}_1$, $y(0) = 0$ and $y(\tau) = \mathbf{y}_1 - \mathbf{y}_0$. Hence the classical action, with $\mathbf{x}_0 = 0$, is

(3.33) $$S(\mathbf{x}_1, \mathbf{y}_1, \tau; \theta) = \frac{\tau \theta^2 \mathbf{x}_1^2}{2 \sin^2 \theta \tau}.$$

We note that S is parametrized by θ, the solutions of the equation (2.14). For $\tau\theta = 0$, the action S can be extended by continuity with the value $\dfrac{\mathbf{x}_1^2}{2\tau}$. Formula (3.33) works only for $\mathbf{x}_1 \neq 0$. When $\mathbf{x}_1 = 0$, $\theta\tau = m\pi$ and Theorem 2.2 yields

$$S = \tau E = \tau \frac{m\pi(\mathbf{y}_1 - \mathbf{y}_0)}{\tau} = m\pi(\mathbf{y}_1 - \mathbf{y}_0), \quad m = 1, 2, 3, \ldots.$$

The next result overcomes the flaw above, providing a single formula for the action.

PROPOSITION 3.1. *The real action with* $\mathbf{x}_0 = 0$ *is given by*

$$S(\mathbf{x}_1, \mathbf{y}_1, \tau; \theta) = \theta\left(\frac{1}{2}\mathbf{x}_1^2 \cot\theta\tau + \mathbf{y}_1 - \mathbf{y}_0\right), \tag{3.34}$$

where θ is a solution of equation (2.14).

PROOF.

$$\begin{aligned}
S &= \frac{\tau\theta^2 \mathbf{x}_1^2}{2\sin^2\theta\tau} = -\theta\left((\mathbf{y}_1 - \mathbf{y}_0) - \frac{\tau\theta\mathbf{x}_1^2}{2\sin^2\theta\tau} - (\mathbf{y}_1 - \mathbf{y}_0)\right) \\
&= -\theta\left(\frac{1}{2}\mathbf{x}_1^2\mu(\theta\tau) - \frac{\tau\theta\mathbf{x}_1^2}{2\sin^2\theta\tau} - (\mathbf{y}_1 - \mathbf{y}_0)\right) \\
&= -\theta\left(\mathbf{x}_1^2\left(\frac{1}{2}\mu(\theta\tau) - \frac{\tau\theta}{2\sin^2\theta\tau}\right) - (\mathbf{y}_1 - \mathbf{y}_0)\right) \\
&= -\theta\left(\frac{1}{2}\mathbf{x}_1^2\left(\frac{\theta\tau}{\sin^2\theta\tau} - \cot\theta\tau - \frac{\theta\tau}{\sin^2\theta\tau}\right) - (\mathbf{y}_1 - \mathbf{y}_0)\right) \\
&= \theta\left(\frac{1}{2}\mathbf{x}_1^2 \cot\theta\tau + \mathbf{y}_1 - \mathbf{y}_0\right).
\end{aligned}$$

□

3.1. The complex action from the y-axis.
Now let us turn to the complex action function.

DEFINITION 3.2. A complex geodesic is the projection on (x, y)-space of a solution of the Hamiltonian system

$$\dot{x} = H_\xi, \qquad \dot{y} = H_\theta,$$

with non-standard boundary conditions

$$x(0) = \mathbf{x}_0, \qquad \theta = \theta_0 = -i, \qquad x(\tau) = \mathbf{x}_1, \qquad y(\tau) = \mathbf{y}_1 - \mathbf{y}_0,$$

and Hamiltonian given in (2.3).

Note the boundary conditions are real, i.e., $\mathbf{x}_0, \mathbf{x}_1, \mathbf{y}_1 - \mathbf{y}_0 \in \mathbf{R}$, while the geodesic components between the end-points might be complex. Observe that the condition $y(0) = 0$ has been replaced by a condition which fixes the momentum $\theta = -i$.

The *complex action* is defined as

$$\begin{aligned}
g(\mathbf{x}_1, \mathbf{y}_0, \mathbf{y}_1, \tau) &= \int_0^\tau (\dot{x}\xi - H)\, ds - i(\mathbf{y}_1 - \mathbf{y}_0) \\
&= \int_0^\tau (\dot{x}\xi + (-i)\dot{y} - H)\, ds - iy(0) \\
&= S(\mathbf{x}_1, \mathbf{y}_0, \mathbf{y}_1, \tau; -i) - iy(0).
\end{aligned}$$

Hence the relation between the real and the complex action is

$$S(\mathbf{x}_1, \mathbf{y}_0, \mathbf{y}_1, \tau; -i) = g(\mathbf{x}_1, \mathbf{y}_0, \mathbf{y}_1, \tau) + iy(0). \tag{3.35}$$

PROPOSITION 3.3. *The complex action with* $\mathbf{x}_0 = 0$ *is given by*

$$(3.36) \qquad g(\mathbf{x}_1, \mathbf{y}_0, \mathbf{y}_1, \tau) = -i(\mathbf{y}_1 - \mathbf{y}_0) + \frac{1}{2}\mathbf{x}_1^2 \coth \tau.$$

PROOF. Using $\cot(-i\tau) = i \coth \tau$, and substituting $\theta = -i$ in formula (3.34) with the new boundary conditions yields

$$\begin{aligned} S(\mathbf{x}_1, \mathbf{y}_1, \tau; -i) &= -i\left(\frac{1}{2}\mathbf{x}_1^2 \cot(-i\tau) + (\mathbf{y}_1 - \mathbf{y}_0) - y(0)\right) \\ &= -i(\mathbf{y}_1 - \mathbf{y}_0) + iy(0) + \frac{1}{2}\mathbf{x}_1^2 \coth \tau. \end{aligned}$$

Comparing with (3.35) yields (3.36). $\square$

Let

$$(3.37) \qquad f(\mathbf{x}_1, \mathbf{y}_0, \mathbf{y}_1, \tau) = \tau g(\mathbf{x}_1, \mathbf{y}_0, \mathbf{y}_1, \tau) = -i(\mathbf{y}_1 - \mathbf{y}_0)\tau + \frac{1}{2}\mathbf{x}_1^2 \tau \coth \tau$$

be the *modified complex action*. In the following, we show that the critical points τ_c of $f(\mathbf{x}_1, \mathbf{y}_0, \mathbf{y}_1, \tau)$ play an important role in finding the length of geodesics joining the origin and the point $(\mathbf{x}_1, \mathbf{y}_1 - \mathbf{y}_0)$.

THEOREM 3.4. *Let* $\mathbf{x}_1 \neq 0$. *The function* $f(\mathbf{x}_1, \mathbf{y}_0, \mathbf{y}_1, \tau)$ *has a finite number of critical points* $\tau_{c,1}, \ldots, \tau_{c,N}$. *These critical points are purely imaginary. The lengths of the geodesics joining the origin and the point* $(\mathbf{x}_1, \mathbf{y}_1 - \mathbf{y}_0)$ *are given by*

$$\ell_j^2 = \nu(-i\tau_{c,j})\big((\mathbf{y}_1 - \mathbf{y}_0) + \mathbf{x}_1^2\big), \quad j = 1, \ldots N.$$

PROOF. From the Hamilton-Jacobi equation and equation (3.35), we have

$$\begin{aligned} \frac{\partial g}{\partial \tau} &= -E = -\frac{S(\mathbf{x}_1, \mathbf{y}_0, \mathbf{y}_1, \tau; -i)}{\tau} \\ &= -\frac{g(\mathbf{x}_1, \mathbf{y}_0, \mathbf{y}_1, \tau) + iy(0)}{\tau}. \end{aligned}$$

Then

$$\begin{aligned} \frac{\partial f}{\partial \tau} &= \frac{\partial (\tau g)}{\partial \tau} = g + \tau \frac{\partial g}{\partial \tau} \\ &= g + \tau \frac{-(g + iy(0))}{\tau} = -iy(0). \end{aligned}$$

Using

$$\frac{2[(\mathbf{y}_1 - \mathbf{y}_0) - y(0)]}{\mathbf{x}_1^2} = \mu(-i\tau),$$

we see that if τ_c is a critical point, i.e., $\frac{\partial f}{\partial \tau}(\tau_c) = 0$, then τ_c satisfies the equation

$$\frac{2(\mathbf{y}_1 - \mathbf{y}_0)}{\mathbf{x}_1^2} = \mu(-i\tau_c) = \mu\left(-i\frac{\tau_c}{\tau}\tau\right) = \mu(\zeta\tau).$$

Let $\zeta_1, \ldots, \zeta_N$ be the solutions of the above equation, see Figure 2. Applying Theorem 2.4 with $y(0) = 0$, we obtain

$$\ell_j^2 = \nu(\zeta_j \tau)((\mathbf{y}_1 - \mathbf{y}_0) + \mathbf{x}_1^2) = \nu(-i\tau_{c,j})((\mathbf{y}_1 - \mathbf{y}_0) + \mathbf{x}_1^2).$$

$\square$

3.2. The complex action outside the y-axis.
We follow the same steps as in the previous section. First we find the real action starting from $(\mathbf{x}_0, 0)$ which ends at $(\mathbf{x}_1, \mathbf{y}_1 - \mathbf{y}_0)$.

THEOREM 3.5. *The real action is given by*

$$S(\mathbf{x}_0, \mathbf{y}_0, \mathbf{x}_1, \mathbf{y}_1, \tau; \theta) = \theta\left[(\mathbf{y}_1 - \mathbf{y}_0) + \frac{1}{2}(\mathbf{x}_1^2 + \mathbf{x}_0^2)\cot\theta\tau - \mathbf{x}_1\mathbf{x}_0 \cos\theta\tau\right].$$

PROOF. First, the equation

$$\frac{2(\mathbf{y}_1 - \mathbf{y}_0)}{\mathbf{x}_0^2 + \mathbf{x}_1^2} = \mu(\theta\tau) + \frac{2\mathbf{x}_0\mathbf{x}_1}{\mathbf{x}_0^2 + \mathbf{x}_1^2}\left[\sin\theta\tau - \mu(\theta\tau)\cos\theta\tau\right]$$

can be rewritten as

$$(3.38) \quad \frac{1}{2}(\mathbf{x}_1^2 + \mathbf{x}_0^2)\mu(\theta\tau) = \mathbf{y}_1 - \mathbf{y}_0 - \mathbf{x}_1\mathbf{x}_0\big(\sin\theta\tau - \mu(\theta\tau)\cos\theta\tau\big).$$

We also have

$$(3.39) \quad \frac{\theta\tau}{2\sin^2\theta\tau}(\mathbf{x}_1^2 + \mathbf{x}_0^2) = \frac{1}{2}(\mathbf{x}_1^2 + \mathbf{x}_0^2)\mu(\theta\tau) + \frac{1}{2}(\mathbf{x}_1^2 + \mathbf{x}_0^2)\cot\theta\tau.$$

By Theorem 2.6, we have

$$\begin{aligned}
S &= \tau E = \tau\frac{\theta^2}{2}\left[\mathbf{x}_0^2 + \left(\frac{\mathbf{x}_1 - \mathbf{x}_0\cos\theta\tau}{\sin\theta\tau}\right)^2\right]\\
&= \theta\frac{\theta\tau}{2\sin^2\theta\tau}\left(\mathbf{x}_0^2 + \mathbf{x}_1^2 - 2\mathbf{x}_1\mathbf{x}_0\cos\theta\tau\right)\\
&= \theta\left[\frac{\theta\tau}{2\sin^2\theta\tau}(\mathbf{x}_0^2 + \mathbf{x}_1^2) - \mathbf{x}_1\mathbf{x}_0\frac{\theta\tau}{\sin^2\theta\tau}\cos\theta\tau\right]\\
&= \theta\left[\frac{1}{2}(\mathbf{x}_1^2 + \mathbf{x}_0^2)\mu(\theta\tau) + \frac{1}{2}(\mathbf{x}_1^2 + \mathbf{x}_0^2)\cot\theta\tau - \mathbf{x}_1\mathbf{x}_0\frac{\theta\tau}{\sin^2\theta\tau}\cos\theta\tau\right]\\
&= \theta\Bigg[\mathbf{y}_1 - \mathbf{y}_0 - \mathbf{x}_1\mathbf{x}_0\big(\sin\theta\tau - \mu(\theta\tau)\cos\theta\tau\big)\\
&\qquad + \frac{1}{2}(\mathbf{x}_1^2 + \mathbf{x}_0^2)\cot\theta\tau - \mathbf{x}_1\mathbf{x}_0\frac{\theta\tau}{\sin^2\theta\tau}\cos\theta\tau\Bigg]\\
&= \theta\Bigg[\mathbf{y}_1 - \mathbf{y}_0 + \frac{1}{2}(\mathbf{x}_1^2 + \mathbf{x}_0^2)\cot\theta\tau\\
&\qquad - \mathbf{x}_1\mathbf{x}_0\left(\frac{\theta\tau}{\sin^2\theta\tau}\cos\theta\tau + \sin\theta\tau - \mu(\theta\tau)\cos\theta\tau\right)\Bigg],
\end{aligned}$$

where we have used relations (3.38) and (3.39). A direct computation shows

$$\frac{\theta\tau}{\sin^2\theta\tau}\cos\theta\tau + \sin\theta\tau - \mu(\theta\tau)\cos\theta\tau$$
$$= \frac{\theta\tau}{\sin^2\theta\tau}\cos\theta\tau + \bigl(\tan\theta\tau - \mu(\theta\tau)\bigr)\cos\theta\tau$$
$$= \left(\frac{\theta\tau}{\sin^2\theta\tau} + \tan\theta\tau - \mu(\theta\tau)\right)\cos\theta\tau$$
$$= \Bigl(\tan\theta\tau + \cot\theta\tau\Bigr)\cos\theta\tau$$
$$= \frac{1}{\sin\theta\tau\cos\theta\tau}\cos\theta\tau = \frac{1}{\sin\theta\tau} = \csc\theta\tau.$$

Substituting in the above formula for the action, we obtain

$$S = \theta\Bigl[\mathbf{y}_1 - \mathbf{y}_0 + \frac{1}{2}(\mathbf{x}_1^2 + \mathbf{x}_0^2)\cot\theta\tau - \mathbf{x}_1\mathbf{x}_0\csc\theta\tau\Bigr].$$

□

Note that for $\mathbf{x}_0 = 0$, the formula above reduces to (3.34).

THEOREM 3.6. *The complex action starting outside of the y-axis is*

(3.40) $\qquad g(\mathbf{x}_0, \mathbf{y}_0, \mathbf{x}_1, \mathbf{y}_1, \tau) = -i(\mathbf{y}_1 - \mathbf{y}_0) + \frac{1}{2}(\mathbf{x}_1^2 + \mathbf{x}_0^2)\coth\tau - \frac{\mathbf{x}_1\mathbf{x}_0}{\sinh\tau}.$

PROOF. By (3.35) and Theorem 3.5,

$$g(\mathbf{x}_0, \mathbf{y}_0, \mathbf{x}_1, \mathbf{y}_1, \tau)$$
$$= S(\mathbf{x}_0, \mathbf{y}_0, \mathbf{x}_1, \mathbf{y}_1, \tau; -i) - iy(0)$$
$$= -i(\mathbf{y}_1 - \mathbf{y}_0) + iy(0) - i\frac{1}{2}\cot(-\tau)(\mathbf{x}_1^2 + \mathbf{x}_0^2) + i\mathbf{x}_1\mathbf{x}_0\csc(-i\tau) - iy(0)$$
$$= -i(\mathbf{y}_1 - \mathbf{y}_0) + \frac{1}{2}(\mathbf{x}_1^2 + \mathbf{x}_0^2)\coth\tau - \mathbf{x}_1\mathbf{x}_0\frac{1}{\sinh\tau}.$$

□

For $\mathbf{x}_0 = 0$, this last formula reduces to (3.36). Similarly, we may define the modified complex action in the case $\mathbf{x}_0 \neq 0$ as

$$f(\mathbf{x}_0, \mathbf{y}_0, \mathbf{x}_1, \mathbf{y}_1, \tau) = \tau g(\mathbf{x}_0, \mathbf{y}_0, \mathbf{x}_1, \mathbf{y}_1, \tau)$$
$$= -i\tau(\mathbf{y}_1 - \mathbf{y}_0) + \frac{1}{2}\tau(\mathbf{x}_1^2 + \mathbf{x}_0^2)\coth\tau - \tau\frac{\mathbf{x}_1\mathbf{x}_0}{\sinh\tau}.$$

The following is a generalization of Theorem 3.4.

THEOREM 3.7. *Let $\mathbf{x}_1 \neq 0$. The function $f(\mathbf{x}_0, \mathbf{y}_0, \mathbf{x}_1, \mathbf{y}_1, \tau)$ has a finite number of critical points $\tau_{c,1}, \ldots, \tau_{c,N}$, all of which are purely imaginary. The lengths ℓ_m of the geodesics joining the points $(\mathbf{x}_0, 0)$ and $(\mathbf{x}_1, \mathbf{y}_1 - \mathbf{y}_0)$ are given by*

$$\ell_m^2 = \nu_a(-i\tau_{c,m})\Bigl[(\mathbf{y}_1 - \mathbf{y}_0) + (\mathbf{x}_1 - \mathbf{x}_0)^2\Bigr], \quad m = 1, \ldots N,$$

with ν_a as in (2.22).

PROOF. The first part of the proof is the same as in Theorem 3.4 and is based on general properties of the action. From the Hamilton-Jacobi equation and equation (3.35),

$$\frac{\partial g}{\partial \tau} = -E = -\frac{S(\mathbf{x}_0, \mathbf{y}_0, \mathbf{x}_1, \mathbf{y}_1, \tau; -i)}{\tau}$$
$$= -\frac{g(\mathbf{x}_0, \mathbf{y}_0, \mathbf{x}_1, \mathbf{y}_1, \tau) + iy(0)}{\tau}.$$

Then

$$\frac{\partial f}{\partial \tau} = \frac{\partial(\tau g)}{\partial \tau} = g + \tau \frac{\partial g}{\partial \tau}$$
$$= g + \tau \frac{-(g + iy(0))}{\tau} = -iy(0).$$

Putting $\theta = -i$ yields

$$\frac{2[(\mathbf{y}_1 - \mathbf{y}_0) - y(0)]}{\mathbf{x}_1^2 + \mathbf{x}_0^2} = \begin{cases} \widetilde{F}(a, \theta\tau) = \widetilde{F}(a, -i\tau) & \text{if } a > 0, \\ F(a, \theta\tau) = F(a, -i\tau) & \text{if } a < 0. \end{cases}$$

If $\frac{\partial f}{\partial \tau}(\tau_c) = 0$ then τ_c satisfies the equation

$$\frac{2(\mathbf{y}_1 - \mathbf{y}_0)}{\mathbf{x}_1^2 + \mathbf{x}_0^2} = \begin{cases} \widetilde{F}(a, -i\tau_c) & \text{if } a > 0, \\ F(a, -i\tau_c) & \text{if } a < 0. \end{cases}$$

Let $\zeta_1, \ldots, \zeta_N$ be the solutions of the above equation. Applying Theorem 2.6, we see that the lengths are

$$\ell_m^2 = \nu_a(-i\tau_{c,m})\Big[(\mathbf{y}_1 - \mathbf{y}_0) + (\mathbf{x}_1 - \mathbf{x}_0)^2\Big].$$

$\square$

THEOREM 3.8. *The value of the modified complex action at the critical points yields the lengths of the geodesics*

$$f(\mathbf{x}_0, \mathbf{y}_0, \mathbf{x}_1, \mathbf{y}_1, \tau_{c,m}(\mathbf{x}_0, \mathbf{y}_0, \mathbf{x}_1, \mathbf{y}_1)) = \frac{1}{2}\ell_m^2,$$

where θ_m is a solution of either (2.23) *or* (2.24).

PROOF. From Theorem 2.6, the lengths of the geodesics between $(\mathbf{x}_0, 0)$ and $(\mathbf{x}_1, \mathbf{y}_1 - \mathbf{y}_0)$ are given by

$$\ell_m^2 = \frac{\theta_m^2}{\sin^2 \theta_m}\Big[\mathbf{x}_0^2 + \mathbf{x}_1^2 - 2\mathbf{x}_0\mathbf{x}_1 \cos \theta_m\Big], \quad m = 1, \ldots, N,$$

where θ_m, $m = 1, \ldots, N$, is a solution of either (2.23) or (2.24). By the proof of Theorem 3.5, the real action is given by

$$S(\mathbf{x}_0, \mathbf{y}_0, \mathbf{x}_1, \mathbf{y}_1, \tau; \theta) = \frac{\theta^2 \tau}{2 \sin^2 \theta\tau}\Big[\mathbf{x}_0^2 + \mathbf{x}_1^2 - 2\mathbf{x}_0\mathbf{x}_1 \cos \theta\tau\Big]$$

and the complex action is

$$g(\mathbf{x}_0, \mathbf{y}_0, \mathbf{x}_1, \mathbf{y}_1, \tau) = S(\mathbf{x}_0, \mathbf{y}_0, \mathbf{x}_1, \mathbf{y}_1, \tau; -i) - iy(0)$$
$$= \frac{(-i)^2 \tau}{2 \sin^2(-i\tau)}\Big[\mathbf{x}_0^2 + \mathbf{x}_1^2 - 2\mathbf{x}_0\mathbf{x}_1 \cos(-i\tau)\Big] - iy(0).$$

Thus the modified complex action evaluated at the critical points is

$$\begin{aligned}
&f(\mathbf{x}_0,\mathbf{y}_0,\mathbf{x}_1,\mathbf{y}_1,\tau_{c,m}(\mathbf{x}_0,\mathbf{y}_0,\mathbf{x}_1,\mathbf{y}_1))\\
&= \tau_{c,m} g(\mathbf{x}_0,\mathbf{y}_0,\mathbf{x}_1,\mathbf{y}_1,\tau_{c,m}(\mathbf{x}_0,\mathbf{y}_0,\mathbf{x}_1,\mathbf{y}_1))\\
&= \frac{(-i\tau_{c,m})^2}{2\sin^2(-i\tau_{c,m})}\left[\mathbf{x}_0^2+\mathbf{x}_1^2-2\mathbf{x}_0\mathbf{x}_1\cos(-i\tau_{c,m})\right] - i\tau_{c,m} y(0)\\
&= \frac{1}{2}\frac{\theta_m^2}{\sin^2\theta_m}\left[\mathbf{x}_0^2+\mathbf{x}_1^2-2\mathbf{x}_0\mathbf{x}_1\cos\theta_m\right]\\
&= \frac{1}{2}\ell_m^2,
\end{aligned}$$

since $0 = \frac{\partial f}{\partial \tau}(\tau_c) = -iy(0)$. $\square$

References

[1] R. Beals, B. Gaveau and P.C. Greiner, *Complex Hamiltonian mechanics and parametrices for subelliptic Laplacians, I, II, III*, Bull. Sci. Math. **121** (1997), 1-36, 97-149, 195-259.

[2] R. Beals, B. Gaveau and P.C. Greiner, *On a geometric formula for the fundamental solution of subelliptic Laplacians*, Math. Nachr. **181** (1996) 81-163.

[3] R. Beals, B. Gaveau and P.C. Greiner, *Green's functions for some highly degenerate elliptic operators*, J. Funct. Anal. **165** (1999), 407-429.

[4] R. Beals, B. Gaveau and P.C. Greiner, *Hamilton-Jacobi theory and the heat kernel on Heisenberg groups*, J. Math. Pures Appl. **79** (2000), 633-689.

[5] R. Beals, B. Gaveau, P.C. Greiner and Y. Kannai, *Exact fundamental solutions for a class of degenerate elliptic operators*, Comm. Partial Differential Equations **24** (1999), 719-742.

[6] W.L. Chow, *Über Systeme von linearen partiellen Differentialgleichungen erster Ordnung*, Math. Ann. **117** (1939), 98-105.

[7] P.C. Greiner, D. Holcman and Y. Kannai, *Wave kernels related to second-order operators*, Duke Math. J. **114** (2002), 329-386.

[8] V.V. Grusin, *On a class of hypoelliptic operators*, Mat. Sb. (N.S.) **83** (**125**) (1970), 456-473.

[9] L. Hörmander, *Hypoelliptic second order differential equations*, Acta Math. **119** (1967), 141-171.

[10] A. Nagel and E.M. Stein, *Lectures on Pseudodifferential Operators: Regularity Theorems and Applications to Nonelliptic Problems*, Princeton University Press, Princeton, New Jersey, 1979.

[11] L. Nirenberg and F. Trèves, *On local solvability of linear partial differential equations, Part I, Necessary conditions*, Comm. Pure Appl. Math. **23** (1970), 1-38; *On local solvability of linear partial differential equations, Part II, Sufficient conditions*, ibid. **23** (1970), 459-509; Correction, ibid. **24** (1971), 279-288.

DEPARTMENT OF MATHEMATICS, EASTERN MICHIGAN UNIVERSITY, YPSILANTI, MI 48197, U.S.A.
  E-mail address: ocalin@emunix.emich.edu

DEPARTMENT OF MATHEMATICS, GEORGETOWN UNIVERSITY, WASHINGTON, DC 20057-0001, U.S.A.
  E-mail address: chang@math.georgetown.edu

DEPARTMENT OF MATHEMATICS, UNIVERSITY OF TORONTO, TORONTO, ON M5S 3G3, CANADA
  E-mail address: greiner@math.toronto.edu

DEPARTMENT OF MATHEMATICS, THE WEIZMANN INSTITUTE OF SCIENCES, REHOVOT 76100, ISRAEL
  E-mail address: kannai@wisdom.weizmann.ac.il

The Kœnigs Embedding Problem for Operator Affine Mappings

Mark Elin and Victor Khatskevich

Dedicated to Lawrence Zalcman on his 60th birthday

ABSTRACT. The Kœnigs embedding problem for affine mappings F of the unit operator ball is studied. We establish the embeddability of the discrete semigroup of iterates $\{F^n\}$, $n = 1, 2, \ldots$, into a one-parameter continuous semigroup $\{F^t\}_{t \geq 0}$ when the fixed point of F is located in a neighborhood of zero as well as when it is a boundary point.

Let D be a domain in a Banach space X and F be a holomorphic self-mapping of D. The problem of embedding a discrete semigroup of iterates $\{F^n\}_{n \in \mathbb{N}}$ into a continuous one-parameter semigroup $\{F^t\}_{t \in \mathbb{R}^+}$ of self-mappings such that $F_1 = F$ is called the Kœnigs Embedding Problem (briefly KEProblem). If the discrete-time semigroup $\{F^n\}_{n \in \mathbb{N}}$ is embeddable into $\{F^t\}_{t \in \mathbb{R}^+}$, then we say that F has the Kœnigs Embedding Property (KEProperty for brevity). In [10], this problem was considered when $D = \Delta$ is the open unit disk in the complex plane $\mathbb{C}$ (see also [2]).

Let $\mathcal{K}$ be the open unit ball of the space $X = L(\mathfrak{H}_1, \mathfrak{H}_2)$ of all linear bounded operators acting from one Hilbert space $\mathfrak{H}_1$ to another, $\mathfrak{H}_2$. Consider a linear fractional mapping (LFM for brevity) $F = F_A : \mathcal{K} \mapsto X$ defined by

$$(1) \qquad F_A(K) = (A_{21} + A_{22}K)(A_{11} + A_{12}K)^{-1}, \quad K \in \mathcal{K},$$

where $A_{ij} \in L(\mathfrak{H}_i, \mathfrak{H}_j)$, $i, j = 1, 2$, A_{11} is invertible and $\|A_{11}^{-1} A_{22}\| < 1$. If $F_A(\mathcal{K}) \subseteq \mathcal{K}$, we call F_A a linear fractional transform (LFT for brevity). In [4, 5], the KEProblem was studied in the case when $D = \mathcal{K}$ and $F = F_A$ is an LFT. The existence of an interior fixed point $F_A(K_0) = K_0$, $K_0 \in \mathcal{K}$, was supposed. It was shown that to solve the KEProblem, it is enough to solve it in the particular case $K_0 = 0$. The latter equality means that $A_{21} = 0$ since $0 = F_A(0) = A_{21} A_{11}^{-1}$, i.e., the operator matrix $A = (A_{ij})_{i,j=1}^2$ is upper triangular and

$$(2) \qquad F_A(K) = A_{22} K (A_{11} + A_{12} K)^{-1}.$$

In the present paper, we continue the investigations of [4, 5]. We consider the dual situation, namely, we study LFT F_B of the form (1) such that the matrix

---

2000 *Mathematics Subject Classification.* Primary 46G20, 47H20; Secondary 47A60, 47H06.

$B = (B_{ij})_{i,j=1}^2$ is lower triangular: $B_{12} = 0$ and
(3) $$F_B(K) = (B_{21} + B_{22}K)B_{11}^{-1},$$
that is, the LFT F_B is affine. Note that F_B has a fixed point $S_0 \in \overline{\mathcal{K}}$ (see, for example, [1]).

Below we need the following facts from operator theory in Krein spaces. Let $\mathfrak{H} = \mathfrak{H}_1 \oplus \mathfrak{H}_2$ be a Krein space (see, for example, [1]) with canonical orthoprojections P_1 and P_2 such that $P_i^* = P_i$, $P_i\mathfrak{H} = \mathfrak{H}_i$, $i = 1, 2$, $P_1 + P_2 = I$. As usual, we let $\mathfrak{F}_+ = \{x \in \mathfrak{H} : \|x_1\| \geq \|x_2\|\}$, $\mathfrak{F}_- = \{x \in \mathfrak{H} : \|x_1\| \geq \|x_2\|\}$, where $x_i = P_i x$, $x \in \mathfrak{H}$, $i = 1, 2$. An indefinite metric on $\mathfrak{H}$ is introduced via
$$[x, y] = (Jx, y), \quad x, y \in \mathfrak{H},$$
where $(\cdot, \cdot)$ is the usual scalar product on $\mathfrak{H}$ and $J = P_1 - P_2$, so that $J = J^* = J^{-1}$.

Denote by $\mathfrak{M}_+$ the set of all maximal nonnegative subspaces of $\mathfrak{H}$, that is, $\mathfrak{L} \in \mathfrak{M}_+$ if and only if $\mathfrak{L} \subset \mathfrak{F}_+$ and $\mathfrak{L}$ is not strictly contained in any other subspace $\widetilde{\mathfrak{L}}$ such that $\widetilde{\mathfrak{L}} \subset \mathfrak{F}_+$. There is a one-to-one correspondence between $\mathfrak{M}_+$ and the closed unit ball $\overline{\mathcal{K}}$ of the space $L(\mathfrak{H}_1, \mathfrak{H}_2)$, defined by the formula
$$\mathfrak{L} \in \mathfrak{M}_+ \quad \text{if and only if} \quad \mathfrak{L} = (P_1 + K)(\mathfrak{H}_1), \quad K \in \overline{\mathcal{K}},$$
where K is uniquely defined.

A bounded linear operator A with domain $D(A) = \mathfrak{H}$ is called a plus-operator if $A\mathfrak{F}_+ \subset \mathfrak{F}_+$. A plus-operator A is said to be strict if, in addition,
$$\mu(A) = \inf_{[x,x]>0} \frac{[Ax, Ax]}{[x, x]} > 0.$$

It is known (see, for example, [8] and [1]) that for a non-strict plus-operator A, $\operatorname{Im}(A) \subset \mathfrak{F}_+$, where $\operatorname{Im}(A)$ is the range of A. A strict plus-operator is called bistrict if its adjoint operator A^* is also a strict plus-operator.

For any operator $T \in L(\mathfrak{H}) = L(\mathfrak{H}, \mathfrak{H})$, consider its block-matrix
$$T = (T_{ij})_{i,j=1}^2,$$
where $T_{ij} = P_i T P_j \in L(\mathfrak{H}_j, \mathfrak{H}_i)$, $i, j = 1, 2$.

PROPOSITION 1 ([8, 1]). *A strict plus-operator A is bistrict if and only if A_{11} is invertible.*

Let us turn to LFM's. When is an LFM F_B an LFT? That is, when is $F_B(\mathcal{K}) \subseteq \mathcal{K}$? A number of necessary and sufficient conditions can be found in [5, 6].

We give the following criterion for an affine LFM F_B to be an LFT.

PROPOSITION 2. *An affine LFM F_B is an LFT if and only if $B = (B_{ij})_{i,j=1}^2$ is a bistrict plus-operator satisfying*
   (i) *there exists $Q \in \mathcal{K}$ such that $B_{21} = QB_{11}$;*
   (ii) $2\operatorname{Re} B_{21}^* B_{22} K + K^* B_{22}^* B_{22} K \leq B_{11}^*(I - Q^*Q)B_{11}$, $K \in \mathcal{K}$.

PROOF. Necessity. First suppose that F_B is non-constant. Let $x \in \mathfrak{F}_+$. Then there exists $K_+ \in \mathcal{K}$ such that $x \in L_+ = (P_1 + K_+)\mathfrak{H}_1 \in \mathfrak{M}_+$. Straightforward calculation shows that $Bx \in BL = (P_1 + F_B(K_+))\mathfrak{H}_1$, that is, $Bx \in \mathfrak{F}_+$, and B is a plus-operator. Since B_{11} is invertible, B is a bistrict plus-operator (see Proposition 1 above). We have $F_B(0) = B_{21}B_{11}^{-1}$; hence we get (i). The fact that

B is a plus-operator is equivalent to $\|(B_{21} + B_{22}K)x\| \leq \|B_{11}x\|$ for each $x \in \mathfrak{H}_1$ and each $K \in \mathcal{K}$. Rewriting the latter inequality as

$$(B_{21} + B_{22}K)^*(B_{21} + B_{22}K) \leq B_{11}^* B_{11}$$

and replacing B_{21} by QB_{11}, we get (ii).

Sufficiency. Straightforward.

Now let F_B be constant, that is, $F_B(K) = Q \in \mathcal{K}$, $K \in \mathcal{K}$. Then $B_{21} = QB_{11}$, $B_{22} = 0$ and $B = \begin{pmatrix} B_{11} & 0 \\ QB_{11} & 0 \end{pmatrix}$. Evidently, B is a bistrict plus-operator, and (i) and (ii) are satisfied. □

Note that (ii) can be rewritten in the form

(ii′) $2\operatorname{Re} B_{21}^* B_{22} K + |B_{22}K|^2 \leq \left| (I - Q^*Q)^{\frac{1}{2}} B_{11} \right|^2$, $K \in \mathcal{K}$.

(Recall that for $T \in L(\mathfrak{H})$, $|T| = (T^*T)^{\frac{1}{2}}$.)

As is well-known, for a given linear operator $A \in L(\mathfrak{H})$ whose spectrum $\sigma(A)$ does not separate zero and infinity, one can define the operator $\log A \in L(\mathfrak{H})$ by using the Riesz–Dunford integral and choosing an analytic branch of the function $\log z$. Furthermore, for $t \geq 0$, the operator $A^t := e^{t \log A}$ is well-defined. If in addition, $\|A\| \leq 1$, then by [3, Lemma 2.1.1], $\operatorname{Re} \log A \leq 0$, and consequently $\|A^t\| \leq 1$. This fact will be used in the sequel.

Note also that the closed ball $\overline{\mathcal{K}}$ is compact in the weak operator topology. The mapping (3) is evidently continuous in this topology, hence it has a (not necessarily unique) fixed point $S_0 \in \overline{\mathcal{K}}$. Using this fixed point, we can see that $B_{21} = (I - B_{22}S_0)$.

In the present paper, we study the KEProblem using the following notation.

DEFINITION 1. Given $B_{22} \in L(\mathfrak{H}_2)$ such that the spectrum $\sigma(B_{22})$ does not separate zero and infinity, we denote by $\mathfrak{S} = \mathfrak{S}(B_{22})$ the subset of $\overline{\mathcal{K}}$ such that for all $S_0 \in \mathfrak{S}$, the LFT F_B with $B = \begin{pmatrix} I & 0 \\ B_{21} & B_{22} \end{pmatrix}$, $B_{21} = (I - B_{22})S_0$, has the KEProperty.

If $S_0 \in \mathfrak{S}$, then the semigroup $\{F^t\}_{t \geq 0}$,

(4) $$F^t = F_T^{-1} \circ F_{\widetilde{B}^t} \circ F_T,$$

is a solution of the KEProblem for F_B; here $\widetilde{B} = \operatorname{diag} B$ and

(5) $$F_T(K) = T_{22}(K - S_0)T_{11}^{-1}$$

with invertible operators $T_{11} \in L(\mathfrak{H}_1)$ and $T_{22} \in L(\mathfrak{H}_2)$ and T_{22} commutes with B_{22}.

In the trivial case $S_0 = 0$, both F_B and $F_{\widetilde{B}^t}$ are evidently linear. We exhibit a family of operators such that $\mathfrak{S}(B_{22}) \neq \{0\}$ for all B_{22} of this family. Also we describe a set of operators such that for all B_{22} from this set there exists $S_0 \in \mathfrak{S}(B_{22})$, $\|S_0\| = 1$, for which a solution of the KEProblem to F_B is of the form (4).

As in [4, 5], we use the Schröder equation

(6) $$F_T \circ F_B = F_B'(0) \circ F_T.$$

to construct an LFM F_T which is a biholomorphic similarity between F_B and $F_B'(0)$.

Let F_B be a LFT of the form (3). In this case, the Fréchet derivative $F'_B(K)$, $K \in \mathcal{K}$, can be identified with $F_{\widetilde{B}}$, where $\widetilde{B} = \operatorname{diag} B$. Consequently, (6) is equivalent to the equation

$$F_T \circ F_B = F_{\widetilde{B}} \circ F_T. \tag{7}$$

THEOREM 1. *Let F_B be an LFT of the form (3). Then for any invertible operators T_{11} and T_{22} which commute with B_{11} and B_{22} respectively, the LFM defined by formula (5) in which S_0 satisfies*

$$B_{22} S_0 - S_0 B_{11} = B_{21}, \tag{8}$$

is a solution to the Schröder equation (7).

Conversely, any affine mapping F_T satisfying (7) has the form (5), where the invertible operators $T_{11} \in L(\mathfrak{H}_1)$ and $T_{22} \in L(\mathfrak{H}_2)$ scalar commute with B_{11} and B_{22}, respectively: $T_{ii} B_{ii} = \lambda B_{ii} T_{ii}$, $i = 1, 2$ for some number $\lambda \neq 0$.

PROOF. Sufficiency. Consider the operator $T = \begin{pmatrix} T_{11} & 0 \\ T_{22} S_0 & T_{22} \end{pmatrix}$, where T_{11}, T_{22} and S_0 are as in the statement of the theorem. Consider the corresponding LFM F_T of the form (5). By the chain rule, to prove the equality (7), we have to show that

$$TB = \widetilde{B} T. \tag{9}$$

The latter equality can be checked directly using (8).

Necessity. By [7, Theorem 3.1], it follows from (7) that there exists $\lambda \in \mathbb{C}$ such that

$$TB = \lambda \widetilde{B} T. \tag{10}$$

The latter means that the operators T_{11} and T_{22} scalar commute with B_{11} and B_{22}, respectively. The proof is complete. □

Suppose now that the spectrum $\sigma(\widetilde{B})$ of the operator $\widetilde{B}$ does not separate zero and infinity. It follows by Theorem 1 that the family $\{F^t\}_{t \geq 0}$, where

$$F^t = F_T^{-1} \circ F_{\widetilde{B}^t} \circ F_T,$$

is a semigroup of affine mappings acting on the whole space $L(\mathfrak{H}_1, \mathfrak{H}_2)$ which is a solution to the KEProblem in this space.

Example. Let $F(z) = \frac{i}{4}\left(z - \frac{1}{\sqrt{2}}\right) + \frac{1}{\sqrt{2}}$ be an affine mapping of the complex plane $\mathbb{C}^1$ related to the operator $B = \begin{pmatrix} 1 & 0 \\ \frac{4-i}{4\sqrt{2}} & \frac{i}{4} \end{pmatrix}$. There are infinitely many semigroups $\{F^t_k\}_{t \geq 0}$, $k \in \mathbb{Z}$, acting on $\mathbb{C}^1$ such that F is embedded into any one of them. Namely,

$$F^t_k(z) = \frac{e^{it\left(\frac{\pi}{2} + 2\pi k\right)}}{4^t} \left(z - \frac{1}{\sqrt{2}}\right) + \frac{1}{\sqrt{2}}.$$

At the same time, it is easy to check that the given function F is a self-mapping of the open unit disk Δ, but no semigroup $\{F^t_k\}_{t \geq 0}$ is a semigroup of self-mappings of Δ. For instance, if $t = 0.5$, then no function

$$F^{0.5}_k(z) = \frac{e^{i\left(\frac{\pi}{4} + \pi k\right)}}{2} \left(z - \frac{1}{\sqrt{2}}\right) + \frac{1}{\sqrt{2}}, \quad k \in \mathbb{Z},$$

is a self-mapping of Δ.

Below in Theorems 2 and 3, we impose some additional restrictions on B ensuring that $\{F^t\}_{t\geq 0}$ is a solution of the KEProblem for F_B in the ball $\mathcal{K}$.

THEOREM 2. *Let $B_{22} \in L(\mathfrak{H}_2)$. Suppose that the spectrum $\sigma(B_{22})$ does not separate zero and infinity and $\|B_{22}\| \leq 1$. Set*

$$\delta := \inf_{t \geq 0} \frac{1 - \|B_{22}^t\|}{\|I - B_{22}^t\|}, \tag{11}$$

and let $\delta\overline{\mathcal{K}} := \{K \in L(\mathfrak{H}_1, \mathfrak{H}_2) : \|K\| \leq \delta\}$. Then
  a) $B = \begin{pmatrix} I & 0 \\ (I - B_{22})S_0 & B_{22} \end{pmatrix}$ is a plus-operator for each $S_0 \in \delta\overline{\mathcal{K}}$;
  b) $\delta\overline{\mathcal{K}} \subset \mathfrak{S}(B_{22})$, i.e., the corresponding LFT F_B has the KEProperty.

Moreover, in this case, the point $S_0 \in \delta\overline{\mathcal{K}}$ is a fixed point of the semigroup $\{F^t\}_{t \in \mathbb{R}^+}$ of self-mappings of the ball $\mathcal{K}$ corresponding to the semigroup of linear operators B^t.

PROOF. To prove the first assertion, we have to show that for any positive vector $x = (x_1, x_2) \in \mathfrak{H}_1 \oplus \mathfrak{H}_2$, $x_i \in \mathfrak{H}_i$, the vector Bx is also positive, i.e., the inequality $\|x_1\| \geq \|x_2\|$ implies $\|x_1\| \geq \|(I - B_{22})S_0 x_1 + B_{22} x_2\|$. Indeed, by (11), we have $\|S_0\| \leq \frac{1 - \|B_{22}\|}{\|I - B_{22}\|}$ and consequently

$$\|(I - B_{22})S_0 x_1 + B_{22} x_2\| \leq \|(I - B_{22})S_0 x_1\| + \|B_{22} x_2\|$$
$$\leq (1 - \|B_{22}\|)\|x_1\| + \|B_{22}\|\|x_2\| \leq \|x_1\|.$$

To prove the second assertion, note that under our assumptions,

$$B^t = \begin{pmatrix} I & 0 \\ (I - B_{22})S_0 & B_{22} \end{pmatrix}^t = \begin{pmatrix} I & 0 \\ (I - B_{22}^t)S_0 & B_{22}^t \end{pmatrix}.$$

Thus the LFM F_{B^t} generated by B^t is actually affine and has the form

$$F_{B^t}(Z) = (I - B_{22}^t)S_0 + B_{22}^t Z.$$

It is clear that $F_{B^t}(S_0) = S_0$. Moreover, $F_{B^t} \circ F_{B^s} = F_{B^{t+s}}$. To complete the proof, we just estimate $F_{B^t}(Z)$ for any $Z \in \mathcal{K}$:

$$\|F_{B^t}(Z)\| \leq \|(I - B_{22}^t)S_0\| + \|B_{22}^t Z\| \leq (1 - \|B_{22}^t\|) + \|B_{22}^t\| \cdot \|Z\| < 1.$$

So F_B is embeddable into $\{F_{B^t}\}$. □

The following assertion is an immediate consequence of Theorem 2.

COROLLARY 1. *Let $B_{22} \in L(\mathfrak{H}_2)$ be such that*

$$cI \leq B_{22} \leq dI, \tag{12}$$

where $0 < c \leq d < 1$, and $\|S_0\| \leq \frac{\ln d}{\ln c}$. Then $\sigma(B_{22})$ does not separate zero and infinity and $S_0 \in \mathfrak{S}(B_{22})$, i.e., the LFT F_B with $B = \begin{pmatrix} I & 0 \\ (I - B_{22})S_0 & B_{22} \end{pmatrix}$ has the KEProperty.

PROOF. The assertion follows by calculation:
$$\delta = \inf_{t\geq 0} \frac{1-\|B_{22}^t\|}{\|I-B_{22}^t\|} \geq \inf_{t\geq 0} \frac{1-d^t}{1-c^t} = \frac{\ln d}{\ln c}.$$
□

It is clear that $\delta \leq 1$ and in various cases $\delta < 1$ (for example, in the case when $B_{22} > 0$, $B_{22} \neq cI$, $c \in [0,1)$). Below, in Theorem 3, we describe a rather general situation by which the set $\mathfrak{S}$ contains boundary points of the unit ball $\mathcal{K}$. To proceed, we need the following auxiliary lemma, which is close to [**11**, Proposition 3.1].

LEMMA 1. *Let $A \in L(\mathfrak{H})$ be a linear operator on a Hilbert space $\mathfrak{H}$ for which the number $\lambda = \|A\|$ is an eigenvalue. Denote by E_λ and by $E_\perp$ the eigenspace of λ and the orthogonal subspace to E_λ, respectively. Then the operator A has a diagonal representation with respect to the decomposition $\mathfrak{H} = E_\lambda \oplus E_\perp$, namely*

(13) $$A = \lambda \begin{pmatrix} I & 0 \\ 0 & C \end{pmatrix},$$

where $\|C\| \leq 1$.

PROOF. Write $A = \begin{pmatrix} A_1 & A_2 \\ A_3 & A_4 \end{pmatrix}$. First note that if $x = (x_1, 0)$, $x_1 \in E_\lambda$, then $Ax = \lambda x$. Hence $A_1 = \lambda I_{E_\lambda}$ and $A_3 = 0$. Further, for any $x = (x_1, x_2) \in \mathfrak{H}$, where $x_1 \in E_\lambda$, $x_2 \in E_\perp$, we have
$$\|Ax\|^2 = \|\lambda x_1 + A_2 x_2\|^2 + \|A_4 x_2\|^2 \leq \|A\|^2 \left(\|x_1\|^2 + \|x_2\|^2\right)$$
or, equivalently,
$$\|A_2 x_2\|^2 + \|A_4 x_2\|^2 - \|A\|^2 \|x_2\|^2 \leq -2\|A\| \operatorname{Re}\langle x_1, A_2 x_2\rangle.$$
Since the left-hand side does not depend on x_1, we conclude that the latter inequality cannot hold when $\operatorname{Re}\langle x_1, A_2 x_2\rangle \neq 0$. So $A_2 = 0$. The lemma is proved. □

THEOREM 3. *Let $B_{22} \in L(\mathfrak{H}_2)$ be a linear operator for which $\lambda = \|B_{22}\|$ is an eigenvalue and the spectrum $\sigma(B_{22})$ does not separate zero and infinity. Suppose that $S_0 \in \overline{\mathcal{K}}$ and $\operatorname{Im}(S_0) \subset E_\lambda$, where E_λ is the eigenspace of λ. Then $S_0 \in \mathfrak{S}(B_{22})$, i.e., the LFT F_B has the KEProperty.*

Moreover, in this case, the point $S_0 \in \overline{\mathcal{K}}$ is a fixed point of the semigroup $\{F^t\}_{t\in\mathbb{R}^+}$ of self-mappings of the ball $\mathcal{K}$ related to the semigroup of linear operators B^t.

REMARK. The conditions of the theorem hold, for example, in the following cases.

1) $B_{22} \in L(\mathfrak{H}_2)$ is a uniformly positive operator for which $\lambda = \|B_{22}\|$ is an eigenvalue;

2) $\mathfrak{H}_2$ is a separable space and B_{22} has diagonal form
$$B_{22} = \begin{pmatrix} \lambda_1 & 0 & 0 & \cdots \\ 0 & \lambda_2 & 0 & \cdots \\ 0 & 0 & \lambda_3 & \cdots \\ \cdots & \cdots & \cdots & \cdots \end{pmatrix},$$
where $|\lambda_1| \geq |\lambda_i|$ for all i and the set $\overline{\{\lambda_i\}}$ does not separate zero and infinity.

PROOF. First note that by Lemma 1, the operator B_{22} has the form
$$B_{22} = \lambda \begin{pmatrix} I & 0 \\ 0 & C \end{pmatrix}$$
with $\|C\| \leq 1$, and its spectrum $\sigma(C)$ does not separate zero and infinity. Therefore, all powers C^t and B_{22}^t, $t \geq 0$, are well-defined. Moreover, $\|C^t\| \leq 1$ and $B_{22}^t = \lambda^t \begin{pmatrix} I & 0 \\ 0 & C^t \end{pmatrix}$. Also, it is easy to check that for each S_0, we have $B^t = \begin{pmatrix} I & 0 \\ (I - B_{22}^t)S_0 & B_{22}^t \end{pmatrix}$.

Exactly as in the proof of Theorem 2, we have to show that the inequality $\|x_1\| \geq \|x_2\|$ implies $\|x_1\| \geq \|(I - B_{22}^t)S_0 x_1 + B_{22}^t x_2\|$ for all $t \geq 0$.

Indeed, by our assumption $S_0 x_1 \in E_\lambda$. By the spectral mapping theorem (see, for example, [9]) the number $\lambda^t = \|B_{22}\|^t$ is an eigenvalue of the operator B_{22}^t for each $t \geq 0$. Thus
$$(I - B_{22}^t)S_0 x_1 = (1 - \|B_{22}\|^t)S_0 x_1.$$
So
$$\|(I - B_{22}^t)S_0 x_1 + B_{22}^t x_2\| \leq \|S_0 x_1\|(1 - \|B_{22}\|^t) + \|B_{22}^t x_2\|$$
$$\leq \|S_0\|\|x_1\|(1 - \|B_{22}\|^t) + \|B_{22}\|^t\|x_2\| < \|x_1\|,$$
which completes the proof. $\square$

Finally, we show the convexity of the set $\mathfrak{S}$.

PROPOSITION 3. *Let $B_{22} \in L(\mathfrak{H}_2)$ be a given linear operator whose spectrum $\sigma(B_{22})$ does not separate zero and infinity. Then $\mathfrak{S}(B_{22})$ is a convex subset of $\overline{\mathcal{K}}$.*

PROOF. If $S_1, S_2 \in \mathfrak{S}$, then for any given $Z \in \mathcal{K}$, the two points $(I - B_{22}^t)S_1 + B_{22}^t Z$ and $(I - B_{22}^t)S_2 + B_{22}^t Z$ also belong to $\mathcal{K}$. Writing $S_0 = \lambda S_1 + (1 - \lambda)S_2$, we see that
$$(I - B_{22}^t)S_0 + B_{22}^t Z =$$
$$= \lambda \left((I - B_{22}^t)S_1 + B_{22}^t Z\right) + (1 - \lambda)\left((I - B_{22}^t)S_2 + B_{22}^t Z\right) \in \mathcal{K}.$$
Since the point $Z \in \mathcal{K}$ is arbitrary, the proposition is proved. $\square$

REMARK. Combining the results of Theorems 2 and 3 and using Proposition 3, we see that in the setting of Theorem 3, the set $\mathfrak{S}(B_{22})$ contains the convex hull of the ball $\delta\overline{\mathcal{K}}$ of the space $L(\mathfrak{H}_1, \mathfrak{H}_2)$ and of the closed unit ball of some subspace of $L(\mathfrak{H}_1, \mathfrak{H}_2)$.

References

[1] T. Ja. Azizov and I. S. Iohvidov, *Foundations of Operator Theory in Spaces with Indefinite Metric*, Nauka, Moscow, 1986.

[2] M. Elin, V. Goryainov, S. Reich and D. Shoikhet, *Fractional iteration and functional equations for functions analytic in the unit disk*, Comput. Methods Funct. Theory **2** (2002), 353–366.

[3] M. Elin, L. A. Harris, S. Reich and D. Shoikhet, *Evolution equations and geometric function theory in J^*-algebras*, J. Nonlinear Convex Anal. **3** (2002), 81–121.

[4] V. Khatskevich, S. Reich and D. Shoikhet, *Schröder's functional equation and the Koenigs embedding property*, Nonlinear Anal. **47** (2001), 3977–3988.

[5] V. Khatskevich, S. Reich and D. Shoikhet, *Abel-Schröder equations for linear fractional mappings and the Koenigs embedding problem*, Acta Sci. Math. **69** (2003), 67–98.

[6] V. Khatskevich and V. Senderov, *Abel-Schröder type equations for maps of operator balls*, Funct. Differ. Equ. **10** (2003), 239–258.

[7] V. Khatskevich, V. Senderov and V. Shulman, *On operator matrices generating linear fractional maps of operator balls*, Contemp. Math. **364** (2004), 93–102.

[8] M. G. Krein and Yu. L. Smuljan, *Plus-operators in a space with an indefinite metric*, Amer. Math. Soc. Transl. **85** (1969), 93–113.

[9] W. Rudin, *Functional Analysis*, McGraw-Hill, New York, 1973.

[10] D. Shoikhet, *Semigroups in Geometrical Function Theory*, Kluwer, Dordrecht, 2001.

[11] B. Sz.-Nagy and Č. Foiaš, *Harmonic Analysis of Operators on Hilbert Space*, American Elsevier, New York, 1970.

DEPARTMENT OF MATHEMATICS, ORT BRAUDE COLLEGE, 21982 KARMIEL, ISRAEL
E-mail address: `elin_mark@hotmail.com`

DEPARTMENT OF MATHEMATICS, ORT BRAUDE COLLEGE, 21982 KARMIEL, ISRAEL
E-mail address: `victor_kh@hotmail.com`

On an Arithmetical Function II

Hershel M. Farkas

This paper is dedicated to my friend and colleague Larry Zalcman on the occasion of his 60th birthday

1. Introduction

A classical result which goes back to Jacobi gives a formula for the number of representations of a positive integer as a sum of two squares. Jacobi's result and formula is proved by the following identity in the theory of elliptic (theta) functions:

$$\left(\sum_{n=-\infty}^{\infty} z^{n^2}\right)^2 = 1 + 4\sum_{n=1}^{\infty} \frac{z^{4n+1}}{1-z^{4n+1}} - \frac{z^{4n+3}}{1-z^{4n+3}}.$$

A new proof of this result has recently been discovered by Y. Godin and appears in his HU thesis. The important point for us is the interpretation of the power series on the right, which we now write as

$$1 + 4\sum_{n=1}^{\infty} \delta_{4,2}(n) z^n.$$

The interpretation is that the coefficient $\delta_{4,2}(n)$ is the difference between the number of divisors of n congruent to 1 mod 4 and those congruent to -1 mod 4.

Jacobi's result can be reformulated as the answer to the following question: what is the number of solutions of the diophantine equation

$$x^2 + y^2 = N?$$

When asked this way, we can generalize and ask the more general question about the number of solutions of the diophantine equation

$$x^2 + ky^2 = N.$$

These types of questions have been treated by many people, the most famous perhaps being Dirichlet (1840). As an application of his identities between theta constants, Godin rederives Jacobi's formula and also solves the case of $k = 2$. Godin and the current author then also solved the problem when $k = 3$ via a theta identity.

This paper is not about these problems but rather about some generalizations of them. Jacobi also derived a formula for the number of representations as a sum

---

2000 *Mathematics Subject Classification.* Primary 11D09, 33E05.

of 4 squares. His proof was again a consequence of an identity in the theory of elliptic (theta) functions.

We can however once again rephrase Jacobi and inquire as to the number of solutions of the diophantine equation

$$x^2 + y^2 + z^2 + w^2 = N.$$

Once again, the generalization as to the number of solutions to the diophantine equation

$$x^2 + ky^2 + z^2 + kw^2 = N$$

is immediate. I mean the question is immediate, not the solution.

The question is immediate because the number of solutions to

$$x^2 + y^2 + z^2 + w^2 = N$$

is simply the coefficient of x^N in the power series expansion of $(\sum_{n=-\infty}^{\infty} z^{n^2})^4$, which is the square of the one we dealt with previously. Hence, if we denote the number of solutions to the diophantine equation $x^2 + ky^2 = N$ by $A(n)$, and let

$$F(z) = 1 + \sum_{n=1}^{\infty} A(n) z^n,$$

then the number of solutions to the diophantine equation

$$x^2 + ky^2 + z^2 + kw^2 = n$$

is given by the coefficient of x^n in the power series expansion of $F^2(z)$.

Since we have solved the problem for $A(n)$ when $k = 1, 2, 3$, we should have a solution to the next problem also. This is indeed the case. I shall show how the Jacobi 4-square theorem follows from the 2-square theorem and also the results for $k = 2, 3$.

In a previous paper [1], we pointed out how the introduction of the function $\delta(n)$, defined as the difference between the number of divisors of n congruent to 1 mod 3 and those congruent to 2 mod 3, leads to interesting relations between the function $\delta(n)$ and the classical function $\sigma(n)$, the sum of the divisors of the positive integer n. Here we shall show formally that similar results are true for primes greater than 3 and lead to criteria for primality and solutions of diophantine equations.

We assume the reader is familiar with the previous paper and freely assume all the material there demonstrated. We consider here a more general function $\delta_{k,l}(n)$, where k is a prime greater than 3 and l is an odd number $1 \leq l \leq k - 2$. In fact, k need not be prime or even odd. If k is even, it must be at least 4 and then l must be an even number with $2 \leq l \leq k - 2$.

DEFINITION 1. $\delta_{k,l}(n)$ denotes the difference between the number of divisors of n congruent to $\frac{k-l}{2}$ and those congruent to $\frac{k+l}{2}$ mod k.

Thus our previous function in this setting is $\delta_{3,1}(n)$. We saw that the latter was a nonnegative function and also multiplicative. The same is no longer necessarily true for these functions with $k > 3$.

2. Theta functions and $\delta_{k,l}(n)$

It is a simple consequence of the Jacobi triple product formula that letting $x = e^{\pi i \tau}, z = e^{2\pi i \zeta}$, we have

$$\theta\begin{bmatrix} \frac{l}{k} \\ 1 \end{bmatrix}(\zeta,\tau) = e^{\frac{\pi i l}{2k}} x^{\frac{l^2}{4k^2}} z^{\frac{l}{2k}} \prod_{n=0}^{\infty}(1-x^{2n+2})(1-x^{2n+1+\frac{l}{k}}z)(1-x^{2n+1-\frac{l}{k}}/z).$$

If we change parameters and let now $w^{\frac{k}{2}} = x$, we find

$$\theta\begin{bmatrix} \frac{l}{k} \\ 1 \end{bmatrix}(\zeta,\tau) = e^{\frac{\pi i l}{2k}} w^{\frac{l^2}{8k}} z^{\frac{l}{2k}} \prod_{n=0}^{\infty}(1-w^{k(n+1)})(1-w^{kn+\frac{k+l}{2}}z)(1-w^{kn+\frac{k-l}{2}}/z).$$

We now differentiate logarithmically with respect to ζ and obtain

$$\frac{\theta'\begin{bmatrix} \frac{l}{k} \\ 1 \end{bmatrix}(\zeta,\tau)}{\theta\begin{bmatrix} \frac{l}{k} \\ 1 \end{bmatrix}(\zeta,\tau)} = 2\pi i \left[\frac{l}{2k} + \sum_{n=0}^{\infty} \frac{w^{kn+\frac{k-l}{2}}/z}{1-w^{kn+\frac{k-l}{2}}/z} - \frac{w^{kn+\frac{k+l}{2}}z}{1-w^{kn+\frac{k+l}{2}}z}\right].$$

We now evaluate at $\zeta = 0$ to obtain

$$\frac{\theta'\begin{bmatrix} \frac{l}{k} \\ 1 \end{bmatrix}(0,\tau)}{\theta\begin{bmatrix} \frac{l}{k} \\ 1 \end{bmatrix}(0,\tau)} = 2\pi i \left[\frac{l}{2k} + \sum_{n=0}^{\infty} \frac{w^{kn+\frac{k-l}{2}}}{1-w^{kn+\frac{k-l}{2}}} - \frac{w^{kn+\frac{k+l}{2}}}{1-w^{kn+\frac{k+l}{2}}}\right].$$

The above expression is clearly equal to

$$2\pi i \left[\frac{l}{2k} + \sum_{n=\frac{k-l}{2}}^{\infty} \delta_{k,l}(n) w^n\right].$$

Differentiating again with respect to ζ and evaluating once more at $\zeta = 0$, we find

$$\left(\frac{\theta'\begin{bmatrix} \frac{l}{k} \\ 1 \end{bmatrix}(0,\tau)}{\theta\begin{bmatrix} \frac{l}{k} \\ 1 \end{bmatrix}(0,\tau)}\right)'$$

$$= -(2\pi i)^2 \left[\sum_{n=0}^{\infty} \frac{w^{kn+\frac{k-l}{2}}}{1-w^{kn+\frac{k-l}{2}}} + \frac{w^{kn+\frac{k+l}{2}}}{1-w^{kn+\frac{k+l}{2}}} + \frac{w^{2kn+k+l}}{(1-w^{kn+\frac{k+l}{2}})^2} + \frac{w^{2kn+k-l}}{(1-w^{kn+\frac{k-l}{2}})^2}\right].$$

We now write the power series of the right hand side as $\sum_{n=\frac{k-l}{2}}^{\infty} a_{k,l}(n) w^n$ and easily check that

$$a_{k,l}(n) = \sum_{d|n} \frac{n}{d},$$

where the sum runs over all d which are congruent to either $\frac{k-l}{2}$ or $\frac{k+l}{2}$ mod k. We immediately note that

$$\left(\frac{\theta'\begin{bmatrix} \frac{l}{k} \\ 1 \end{bmatrix}(0,\tau)}{\theta\begin{bmatrix} \frac{l}{k} \\ 1 \end{bmatrix}(0,\tau)}\right)' = \frac{\theta''\begin{bmatrix} \frac{l}{k} \\ 1 \end{bmatrix}(0,\tau)}{\theta\begin{bmatrix} \frac{l}{k} \\ 1 \end{bmatrix}(0,\tau)} - \left(\frac{\theta'\begin{bmatrix} \frac{l}{k} \\ 1 \end{bmatrix}(0,\tau)}{\theta\begin{bmatrix} \frac{l}{k} \\ 1 \end{bmatrix}(0,\tau)}\right)^2;$$

by the heat equations satisfied by the theta function, the right hand side of the previous line can be rewritten as

$$\frac{(4\pi i)d/d\tau\theta\begin{bmatrix}\frac{l}{k}\\1\end{bmatrix}(0,\tau)}{\theta\begin{bmatrix}\frac{l}{k}\\1\end{bmatrix}(0,\tau)} - \left(\frac{\theta'\begin{bmatrix}\frac{l}{k}\\1\end{bmatrix}(0,\tau)}{\theta\begin{bmatrix}\frac{l}{k}\\1\end{bmatrix}(0,\tau)}\right)^2.$$

The second term has already been computed above. We observe that

$$\frac{d/d\tau\theta\begin{bmatrix}\frac{l}{k}\\1\end{bmatrix}(0,\tau)}{\theta\begin{bmatrix}\frac{l}{k}\\1\end{bmatrix}(0,\tau)}$$

$$= \frac{2\pi i}{k}\left[\frac{l^2}{8k} - \sum_{n=0}^{\infty}\frac{k(n+1)w^{k(n+1)}}{1-w^{k(n+1)}} + \frac{(kn+\frac{k+l}{2})w^{kn+\frac{k+l}{2}}}{1-w^{kn+\frac{k+l}{2}}} + \frac{(kn+\frac{k-l}{2})w^{kn+\frac{k-l}{2}}}{1-w^{kn+\frac{k-l}{2}}}\right]$$

$$= \frac{2\pi i}{k}\left[\frac{l^2}{8k} - \sum_{n=\frac{k-l}{2}}^{\infty}\sigma'_{k,l}(n)w^n\right],$$

where $\sigma'_{k,l}(n) = \sum_{d|n} d$ and d runs over the divisors of n which are congruent to $0, \frac{k+l}{2}, \frac{k-l}{2}$ mod k. The above computation leads to the following theorem.

THEOREM 1. *We have the following identity between power series:*

$$\sum_{n=\frac{k-l}{2}}^{\infty} a_{k,l}(n)w^n = \frac{2}{k}\sum_{n=\frac{k-l}{2}}^{\infty}\sigma'_{k,l}(n)w^n + \frac{l}{k}\sum_{n=\frac{k-l}{2}}^{\infty}\delta_{k,l}(n)w^n + \left(\sum_{n=\frac{k-l}{2}}^{\infty}\delta_{k,l}(n)w^n\right)^2.$$

Therefore, for all $n \geq \frac{k-l}{2}$,

$$ka_{k,l}(n) - 2\sigma'_{k,l}(n) = l\delta_{k,l}(n) + k\sum_{j=1}^{n-1}\delta_{k,l}(j)\delta_{k,l}(n-j).$$

The above theorem has the following corollary.

COROLLARY 1. *Let* $l = k-2$ *and let* n *be a prime at least 3. Then*

a) *If* $n \equiv \pm 1$ mod k, $a_{k,l}(n) = \sigma'_{k,l}(n) = \sigma(n)$ *and* $\delta_{k,l}(n) = \begin{cases} 2 & n \equiv 1 \\ 0 & n \equiv -1 \end{cases}$.

Thus

$$(k-2)\sigma(n) = \begin{cases} 2l + k\sum_{j=1}^{n-1}\delta_{k,l}(j)\delta_{k,l}(n-j) & n \equiv 1 \\ k\sum_{j=1}^{n-1}\delta_{k,l}(j)\delta_{k,l}(n-j) & n \equiv -1 \end{cases}.$$

b) *If* n *is not equivalent to* ± 1 mod k *or not equal to* k *(this case is empty when* $k = 3$*), then* $a_{k,l}(n) = n, \sigma'_{k,l}(n) = 1$ *and* $\delta_{k,l}(n) = 1$, *so that*

$$kn - 2 = l + k\sum_{j=1}^{n-1}\delta_{k,l}(j)\delta_{k,l}(n-j).$$

c) If $n = k$, then $a_{k,l}(n) = k$, $\sigma'_{k,l}(n) = k+1$, $\delta_{k,l}(n) = 1$, so

$$k^2 - 2(k+1) = l + k \sum_{j=1}^{k-1} \delta_{k,l}(j)\delta_{k,l}(k-j).$$

If $1 \leq l \leq k-4$, so that now $k \geq 5$,

a') If $n \equiv \frac{k-l}{2}, \frac{k+l}{2}$, $a_{k,l}(n) = 1$, $\sigma'_{k,l}(n) = \begin{cases} n & n = \frac{k-l}{2} \\ n & n = \frac{k+l}{2} \end{cases}$,

$\delta_{k,l}(n) = \begin{cases} 1 & n = \frac{k-l}{2} \\ -1 & n = \frac{k+l}{2} \end{cases}$. Thus, if $n \equiv \frac{k-l}{2}$, we have

$$k - 2n = l + k \sum_{j=1}^{n-1} \delta_{k,l}(j)\delta_{k,l}(n-j),$$

while if $n \equiv \frac{k+l}{2}$, we have

$$k - 2n = -l + k \sum_{j=1}^{n-1} \delta_{k,l}(j)\delta_{k,l}(n-j).$$

b') If n is not equivalent to $\frac{k-l}{2}, \frac{k+l}{2}, 0 \bmod k$, then once again we have

$$\sum_{j=1}^{n-1} \delta_{k,l}(j)\delta_{k,l}(n-j) = 0.$$

c') If $n = k$, then $a_{k,l}(k) = 0, \sigma'_{k,l}(k) = k, \delta_{k,l}(k) = 0$, and we have

$$-2 = \sum_{j=1}^{k-1} \delta_{k,l}(j)\delta_{k,l}(k-j).$$

The above corollary gives us primality conditions which may turn out to be useful. We do not pursue this here but rather take this opportunity to show how these ideas can be used to discuss some other questions.

3. Classical Results of Jacobi

Our first example will be Jacobi's 4-square theorem. We recall that it is a well-known result of Jacobi that

$$\left(\sum_{n=-\infty}^{\infty} w^{n^2}\right)^2 = 1 + 4 \sum_{n=0}^{\infty} \left(\frac{w^{4n+1}}{1 - w^{4n+1}} - \frac{w^{4n+3}}{1 - w^{4n+3}}\right).$$

A consequence of this result is the fact that the number of representations of the positive integer N as a sum of 2 squares is given by $4\delta_{4,2}(N)$. At any rate, we see the number of representations of N as a sum of 4 squares is given by $8\delta_{4,2}(N) + 16 \sum_{j=1}^{N-1} \delta_{4,2}(j)\delta_{4,2}(N-j)$; therefore, in order to get the 4-square theorem, we need only compute the above. Thus we have

$$\left(\sum_{n=-\infty}^{\infty} w^{n^2}\right)^4 = 1 + 8 \sum_{n=1}^{\infty} \delta_{4,2}(n)w^n + 16 \left(\sum_{n=1}^{\infty} \delta_{4,2}(n)w^n\right)^2.$$

Therefore the number of representations of N as a sum of 4 squares ($N \geq 2$) is given by
$$8\delta_{4,2}(N) + 16 \sum_{j=1}^{N-1} \delta_{4,2}(j)\delta_{4,2}(N-j).$$
By the theorem, we can rewrite this as
$$8\delta_{4,2}(N) + 4(4a_{4,2}(N) - 2\sigma'_{4,2}(N) - 2\delta_{4,2}(N)),$$
which is equal to
$$8(2a_{4,2}(N) - \sigma'_{4,2}(N)).$$
Clearly $a_{4,2}(N) = \sigma'_{4,2}(N) = \sigma(N)$ whenever N is odd, so in this case we obtain immediately Jacobi's result $8\sigma(N)$. If N is even, say $N = 2^r M$ with M odd, we have now
$$a_{4,2}(N) = 2^r \sigma(M), \sigma'_{4,2}(N) = \sigma(2^r M) - 2\sigma(M),$$
so that once again we get
$$8(2a_{4,2}(N) - \sigma'_{4,2}(N)) = 8\sigma(2M),$$
which is the same as the sum of the divisors of N neglecting those divisors which are congruent to 0 mod 4. Once again Jacobi's result.

The point of the previous remarks is to show that the 4-square theorem is really an elementary exercise which follows from the 2-square theorem after one has Theorem 1 at one's disposal. There are some other results of the same type, to which we now turn.

4. Combinatorial Problems

Jacobi's 2-square theorem is just one of a host of problems one can consider. For example, in place of considering the diophantine equation which Jacobi considered
$$x^2 + y^2 = N,$$
we could consider the diophantine equation
$$x^2 + py^2 = N$$
for any prime p. We can consider this equation even for non-primes. In fact, Dirichlet considered the problem
$$ax^2 + bxy + cy^2 = N,$$
with a general quadratic form replacing the expression $x^2 + py^2$. We basically stick with the above except to make some further remarks. Y. Godin rediscovered the fact that the number of solutions of this equation for $p = 2$ is given by $2(\delta_{8,2}(N) + \delta_{8,6}(N))$. In other words, he showed that the number of representations of N as a sum of a square and twice a square is given by the above formula. His proof is not difficult and depends on the classical formulas for the product of two theta functions. We do not give the proof here but rather wish to show how this result and our theorem above allow us to deduce properties of the power series expansion given by
$$\left(1 + 2\sum_{n=1}^{\infty}(\delta_{8,2}(n) + \delta_{8,6}(n))w^n\right)^2,$$

whose coefficients represent the number of solutions to the diophantine equation
$$x^2 + y^2 + 2(z^2 + w^2) = N.$$

THEOREM 2. *Let*
$$\left(1 + 2\sum_{n=1}^{\infty}(\delta_{8,2}(n) + \delta_{8,6}(n))w^n\right)^2 = 1 + \sum_{n=1}^{\infty} A_n w^n$$

Then

(1) *if N is odd, $A_N/4 = \sigma(N)$;*

(2) *if N is even and $N = 2^r M$ with M odd, $A_N/4 = \begin{cases} 2\sigma(M) & r = 1 \\ 6\sigma(M) & r \geq 2 \end{cases}$.*

PROOF. Since the left hand side of the of the equation in the theorem is $\theta^2\begin{bmatrix} 0 \\ 0 \end{bmatrix}(0,\tau)\theta^2\begin{bmatrix} 0 \\ 0 \end{bmatrix}(0,2\tau)$, which is the same as

$$\left(1 + 4\sum_{n=1}^{\infty}\delta_{4,2}(n)w^n\right)\left(1 + 4\sum_{n=1}^{\infty}\delta_{4,2}(n)w^{2n}\right),$$

we can write the left hand side as

$$1 + 4\sum_{n=1}^{\infty}(\delta_{4,2}(n) + \delta_{4,2}(n/2))w^n + 16\sum_{n=2}^{\infty}\left(\sum_{j=1}^{n-1}\delta_{4,2}(j/2)\delta_{4,2}(n-j)\right)w^n.$$

It thus follows that for $n \geq 2$, we have

$$A_n = 4(\delta_{4,2}(n) + \delta_{4,2}(n/2)) + 16\sum_{j=1}^{n-1}\delta_{4,2}(j/2)\delta_{4,2}(n-j).$$

If n is odd, we can replace this by

$$4(\delta_{4,2}(n)) + 16\sum_{k=1}^{\frac{n-1}{2}}\delta_{4,2}(k)\delta_{4,2}(n-2k).$$

Since $\delta_{4,2}(k) = \delta_{4,2}(2k)$, we may replace the last line by

$$4(\delta_{4,2}(n)) + 16\sum_{k=1}^{\frac{n-1}{2}}\delta_{4,2}(2k)\delta_{4,2}(n-2k);$$

and this is just

$$4(\delta_{4,2}(n)) + 16 \cdot \frac{1}{2}\sum_{j=1}^{n-1}\delta_{4,2}(j)\delta_{4,2}(n-j).$$

We can now use our theorem to get the last line equal to

$$4(\delta_{4,2}(n)) + 2 \cdot (4a_{4,2}(n) - 2\sigma'_{4,2}(n) - 2\delta_{4,2}(n)).$$

Since we assumed n is odd, we had $\delta_{4,2}(n/2) = 0$. Because n is odd, $a_{4,2}(n) = \sigma'_{4,2}(n) = \sigma(n)$. Therefore, when n is odd, we get $4\sigma(n)$ as stated in the theorem.

If n is even and $n = 2^r M$, we must proceed slightly differently. Return to the line preceding the statement "If n is odd"; when n is even, the previous line can be replaced by

$$4(\delta_{4,2}(n) + \delta_{4,2}(n/2)) + 16 \sum_{k=1}^{\frac{n-2}{2}} \delta_{4,2}(k)\delta_{4,2}(n-2k).$$

Once again, since $\delta_{4,2}(k) = \delta_{4,2}(2k)$, we may replace the last line by

$$8(\delta_{4,2}(n)) + 16 \sum_{k=1}^{\frac{n-2}{2}} \delta_{4,2}(2k)\delta_{4,2}\left(\frac{n-2k}{2}\right),$$

which is the same as

$$8(\delta_{4,2}(n)) + 16 \sum_{k=1}^{n/2-1} \delta(k)\delta(n/2-k).$$

We again use the theorem to rewrite the last line as

$$8\delta_{4,2}(n) + 4(4a_{4,2}(n/2) - 2\sigma'_{4,2}(n/2) - 2\delta_{4,2}(n/2)).$$

Now write $n = 2^r M$ and consider the cases $r = 1$ and $r \geq 2$ separately. Since for $r \geq 2$, $\sigma'_{4,2}(2^{r-1}M) = \sigma(2^{r-1}M) - \sigma(2M)$, it is clear that when $r = 1$, we have $A_n = 8\sigma(n)$ and when $r \geq 2$, we have $A_n = 24\sigma(n)$. This concludes the proof of the theorem. $\square$

Another example is obtained by considering the case $p = 3$. The author and Y. Godin [2] have shown that the number of solutions to the diophantine equation

$$x^2 + 3y^2 = N$$

is given by $2\delta_{3,1}(N)$ when N is odd and by $6\delta_{3,1}(N)$ when N is even. By the same technique as above, we now prove

THEOREM 3. *The number of solutions to the diophantine equation*

$$x^2 + y^2 + 3(z^2 + w^2) = N = 2^r M, (2, M) = 1$$

is given by

1) $\sigma''(M)(2^{r+3} - 12)$ *when* $r \geq 2$
2) $4\sigma''(M)$ *when* $r = 0, 1$

where $\sigma''(M)$ denotes the sum of the divisors of M not congruent to $0 \mod 3$.

PROOF. The proof consists of determining the coefficients of the power series

$$\left(1 + \sum_{n=1}^{\infty} 2\delta_{3,1}(2n-1)x^{2n-1} + 6\delta_{3,1}(2n)x^{2n}\right)^2.$$

We write this as

$$\sum_{n=0}^{\infty} A_n x^n$$

with $A_0 = 1$ and readily see that

$$A_n = \sum_{k=0}^{n} \alpha(k)\alpha(n-k)$$

where $\alpha(k) = \begin{cases} 6\delta_{3,1}(k) & k \equiv 0 \bmod 2 \\ 2\delta_{3,1}(k) & k \equiv 1 \bmod 2 \end{cases}$, where $\alpha(0) = 1$.

We begin with the easier case of n odd. In this case, we have

$$A_n = 4\delta_{3,1}(n) + 12 \sum_{k=1}^{n-1} \delta_{3,1}(k)\delta_{3,1}(n-k).$$

By our theorem, we can rewrite this as

$$A_n = 4(\delta_{3,1}(n) + 3 \sum_{k=1}^{n-1} \delta_{3,1}(k)\delta_{3,1}(n-k) = 4(3a_{3,1}(n) - 2\sigma'_{3,1}(n)).$$

Assume now that $n = 3^j M$ with $(3, M) = 1$. Then $a_{3,1}(n) = 3^j \sigma(M)$ and $\sigma'_{3,1}(n) = \sigma(n)$. Hence the last expression is just

$$4(3^{j+1}\sigma(M) - 2\sigma(3^j M)) = 4\sigma(M) = 4\sigma''(n).$$

If n is even, the argument is a bit more complicated. We now assume $n = 2^r M$ with $(2, M) = 1$ and also that $r \geq 2$. In this case,

$$A_n = A_{2^r M} = 12\delta_{3,1}(2^r M) + 4 \sum_{j=1}^{2^{r-1}M} \delta_{3,1}(2j-1)\delta_{3,1}(2^r M - (2j-1))$$

$$+ 36 \sum_{j=1}^{2^{r-1}M-1} \delta_{3,1}(2j)\delta_{3,1}(2^r M - 2j)$$

$$= 12\delta_{3,1}(2^r M) + 4 \sum_{j=1}^{2^r M-1} \delta_{3,1}(j)\delta_{3,1}(2^r M - j) + 32 \sum_{j=1}^{2^{r-1}M-1} \delta_{3,1}(2j)\delta_{3,1}(2^r M - 2j).$$

In order to simplify what follows, we assume also that $n \equiv 1, 2 \bmod 3$. It should be clear from the case above with n odd what happens if $n \equiv 0 \bmod 3$.

The last expression, by our theorem, is the same as

$$12\delta_{3,1}(2^r M) + \frac{4}{3}(\sigma(2^r M) - \delta_{3,1}(2^r M)) + 32 \sum_{j=1}^{2^{r-1}M-1} \delta_{3,1}(2j)\delta_{3,1}(2^r M - 2j).$$

In the sum above, we can just sum over even j, for otherwise each term vanishes. This is because $\delta_{3,1}(2M) = 0$ when $(2, M) = 1$. Hence we have

$$12\delta_{3,1}(2^r M) + \frac{4}{3}(\sigma(2^r M) - \delta_{3,1}(2^r M)) + 32 \sum_{j=1}^{2^{r-2}M-1} \delta_{3,1}(4j)\delta_{3,1}(2^r M - 4j).$$

Here is where we use the fact that $r \geq 2$. We can now factor out a 4 and use the elementary fact that in all cases $\delta_{3,1}(4k) = \delta_{3,1}(k)$ to obtain

$$12\delta_{3,1}(2^r M) + \frac{4}{3}(\sigma(2^r M) - \delta_{3,1}(2^r M)) + \frac{32}{3}(\sigma(2^{r-2}M) - \delta_{3,1}(2^{r-2}M)).$$

The above is clearly

$$12\delta_{3,1}(2^r M) + \frac{4}{3}(2^{r+1} - 1)\sigma(M) + \frac{32}{3}(2^{r-1} - 1)\sigma(M) - 12\delta_{3,1}(2^{r-2}M)$$
$$= \sigma(M)(2^{r+3} - 12).$$

In the event that $M \equiv 0 \bmod 3$, the term $\sigma(M)$ becomes $\sigma''(M)$. The above used the condition $r \geq 2$. If $r = 1$, then the second sum in the expression for A_n vanishes and the answer is just $4\sigma''(M)$. $\square$

It seems worth mentioning the following facts. It is an easy exercise to show that the number of solutions to
$$x^2 + y^2 + 3(z^2 + w^2) = 3^r M$$
with $(3, m) = 1$ is the same as the number of solutions to
$$x^2 + y^2 + 3(z^2 + w^2) = M.$$
This explains perhaps why σ'' replaces σ. It is, however, not clear to me why from an elementary point of view it is true that
$$A_{2^r} - A_{2^{r-1}} = 2^{r+2}.$$
However, this is a consequence of our result.

References

[1] H.M. Farkas, *On an arithmetical function*, Ramanujan J., to appear.
[2] H.M. Farkas, Y. Godin, *Logarithmic derivatives of theta functions*, Israel J. Math., to appear.

INSTITUTE OF MATHEMATICS, THE HEBREW UNIVERSITY OF JERUSALEM, JERUSALEM 91904, ISRAEL

E-mail address: `farkas@math.huji.ac.il`

A Glance at Wiman-Valiron Theory

P. C. Fenton

To Lawrence Zalcman, with respect and affection

ABSTRACT. A sketch of Wiman-Valiron theory is given from the point of view of the Kövari-Hayman comparison method.

1. Introduction

W.K. Hayman's survey of Wiman-Valiron theory [6] draws on work of Clunie [1], [2], Kövari [7] and others and includes results of his own. It is the most complete available account of the Kövari-Hayman comparison method. Thorough as it is, however, it may be somewhat daunting; the present note is intended for the reader who wants no more than a glance at the basic results. No attempt is made at completeness, but what is done is done completely. The main inequalities are derived in a general setting, incorporating developments since Hayman's article appeared and are used to obtain information about the asymptotic behaviour of the derivatives of entire functions. A weaker result than the one established here is the basis of Laine's discussion of entire solutions of differential equations ([8], p. 51).

2. Maximum term and central index

Since, for an entire function $f(z) = \sum_{n=0}^{\infty} a_n z^n$, the general term of the series tends to zero for each z, there is a *maximum term*:

$$\mu(r) = \mu(r, f) = \max_{n \geq 0} |a_n| r^n \qquad (r = |z|).$$

The value of n for which the maximum is attained is the *central index*, $N(r) = N(r, f)$. If there is ambiguity, so that $\mu(r)$ is attained for several values of n, then $N(r)$ is taken to be the largest one. For $n < N$, the ratio $|a_N| r^N / |a_n| r^n$ increases with r, and thus $N(r)$ is non-decreasing, and constant in the half-open subintervals $[r_{j-1}, r_j)$ of some partition of $[0, \infty)$:

(2.1) $$0 = r_0 < r_1 < r_2 < \ldots.$$

2000 *Mathematics Subject Classification.* Primary 30D20.

The functions μ and N are the main tools of Wiman-Valiron theory. They are connected by the identity

$$\frac{d}{dr}(\log \mu(r)) = \frac{N(r)}{r}, \tag{2.2}$$

which holds in the open subintervals of the partition (2.1); equivalently,

$$\log \mu(r') - \log \mu(r) = \int_r^{r'} \frac{N(t)}{t}\, dt, \quad r' \geq r > 0. \tag{2.3}$$

If $a_0 \neq 0$, $N(t) = 0$ for all small t, and (2.3) can be extended to allow $r = 0$. The *maximum modulus* of f, $M(r) = M(r, f) = \max_{|z|=r} |f(z)|$, is sometimes included with μ and N. A consequence of Cauchy's inequalities is that

$$\mu(r, f) \leq M(r, f), \quad r \geq 0. \tag{2.4}$$

As an illustration, consider the entire function

$$\Phi(z) = \sum_{n=0}^{\infty} n^{-n/\sigma} z^n, \tag{2.5}$$

where σ is a positive constant, $n^{-n/\sigma}$ being interpreted as 1 when $n = 0$. The central index $N = N(r)$ maximizes $-(n/\sigma) \log n + n \log r$ and so, ignoring for the moment that N is an integer, $N = e^{-1} r^\sigma$. To be precise, $N = e^{-1} r^\sigma + \eta$, where $|\eta| < 1$, but with

$$\rho_N = (eN)^{1/\sigma}, \tag{2.6}$$

we have $N(\rho_N) = N$ for all non-negative integers N. When $r = \rho_N$, the relative significance of the terms in (2.5) is measured by the ratios

$$\frac{n^{-n/\sigma} \rho_N^n}{\mu(\rho_N, \Phi)} = \exp\left(-\sigma^{-1} \int_N^n \frac{n-t}{t}\, dt\right). \tag{2.7}$$

Using (2.7) it can be shown ([6], p. 330) that

$$M(\rho_N, \Phi) = \Phi(\rho_N) = (1 + o(1))\mu(\rho_N, \Phi)\sqrt{2\pi \sigma N}, \tag{2.8}$$

which neatly connects M, μ and N. Since $\mu(\rho_N, \Phi) = N^{-N/\sigma} \rho_N^N = e^{N/\sigma}$, we have $N = \sigma \log \mu(\rho_N, \Phi)$, and N can be eliminated from (2.8) to give

$$M(\rho_N, \Phi) = (\sigma + o(1))\mu(\rho_N, \Phi)\sqrt{2\pi \log \mu(\rho_N, \Phi)}. \tag{2.9}$$

These are specimen results for a special function; but, as we shall see, they can be extended to functions that are, in a certain sense, comparable to Φ.

3. Comparing functions

In the comparison method, an entire function f is compared to a *fully indexed* power series F (which may have finite radius of convergence), F being fully indexed in this context if its central index takes every possible value. Evidently, F is fully indexed if and only if we can find, for every $n \geq 0$, a positive number R_n such that (i) $N(R_n, F) = n$, and (ii) the maximum term at R_n is unique. The function Φ of (2.5) is fully indexed, and the numbers ρ_n defined by (2.6) are suitable choices for R_n, at least for $n \geq 1$ ($\rho_0 = 0$, which is not allowed, but we could take, for example, $R_0 = 1/2$).

The aim is to obtain estimates for the terms of f similar to (2.7), but with the right hand side replaced by something depending on F. Success depends on f and F having similarities in their growth. The main result is

THEOREM 3.1. [5] *Suppose that $F(z) = \sum_{n=0}^{\infty} A_n z^n$ is a fully indexed power series, and that R_n is a sequence of positive numbers such that $N(R_n, F) = n$ for $n = 0, 1, 2, \ldots$, and the maximum term of F at R_n is unique. Given an entire function $f(z) = \sum_{n=0}^{\infty} a_n z^n$ and a number $K > 1$,*

$$\text{(3.1)} \qquad \frac{|a_n| r^n}{\mu(r, f)} \leq \frac{|A_n| R_N^n}{\mu(R_N, F)}, \quad 0 < n \leq KN,$$

$$\text{(3.2)} \qquad \frac{|a_n| r^n}{\mu(r, f)} \leq \max\left\{ \frac{|A_n| R_N^n}{\mu(R_N, F)}, \left(\frac{R_N}{R_{[KN]}}\right)^{(1-K^{-1})n} \right\}, \quad n > KN,$$

for all positive values of r outside a set E of intervals such that, for all $r' > 0$,

$$\text{(3.3)} \qquad \int_{E \cap (0, r')} t^{-1} dt \leq \log(R_{[KN']}/R_0).$$

Here $N = N(r, f)$, $N' = N(r', f)$ and $[\ ]$ denotes the integral part.

The left hand side of (3.3) is the logarithmic measure of $E \cap (0, r')$, written here as $\operatorname{logmeas}(E \cap (0, r'))$.

Theorem 3.1 is completely general in that no assumptions are made about f. Nothing comes from nothing however, and the conclusions (3.1) and (3.2) may be vacuous if the right hand side of (3.3) is about $\log r'$, which happens if f and F are ill-matched. (The ratio of the right hand side of (3.3) to $\log r'$ is a kind of measure of the comparability of f and F up to r'.)

The proof is elementary. Given $r' > 0$ and t satisfying $0 \leq t \leq r'/R_{[KN']}$, suppose that ν is the index of the largest term in the set

$$\left\{ \frac{|a_n|}{|A_n|} t^n : 0 \leq n \leq KN' \right\}$$

(taking the largest index if there is ambiguity). Then

$$\text{(3.4)} \qquad \frac{|a_n|}{|a_\nu|} t^{n-\nu} \leq \frac{|A_n|}{|A_\nu|}, \quad 0 \leq n \leq KN';$$

multiplying both sides by R_ν and writing $r = tR_\nu$, we have

$$\text{(3.5)} \qquad \frac{|a_n|}{|a_\nu|} r^{n-\nu} \leq \frac{|A_n|}{|A_\nu|} R_\nu^{n-\nu}, \quad 0 \leq n \leq KN'.$$

The right hand side of (3.5) is less than 1 for $n \neq \nu$, from the definition of R_ν; and since $r = tR_\nu \leq r' R_\nu / R_{[KN']} \leq r'$, we have $N(r, f) \leq N(r', f) = N'$. Thus $\nu = N(r, f)$, and then (3.1) follows from (3.5). For (3.2) note that for $KN < n \leq KN'$, (3.2) follows from (3.5), while for $n > KN'$,

$$|a_n| r^n = |a_n| r'^n (r/r')^n \leq |a_{N'}| r'^{N'} (r/r')^n = |a_{N'}| r'^{N'} (r/r')^{n-N'}$$
$$\leq \mu(r, f)(R_N/R_{[KN']})^{n-N'} \leq \mu(r, f)(R_N/R_{[KN]})^{n-N'}$$
$$\leq \mu(r, f)(R_N/R_{[KN]})^{(1-K^{-1})n}.$$

It remains to establish (3.3). Suppose that, for $0 \leq t \leq r'/R_{[KN']}$, the index $\nu = \nu(t)$ assumes the values $\nu_1 < \cdots < \nu_q$, the jumps occurring at $t_1, \ldots, t_{q-1}$, where $0 < t_1 < \cdots < t_{q-1} \le r'/R_{[KN']}$. We have shown that (3.1) and (3.2) hold in the intervals

$$[0, t_1 R_{\nu_1}), [t_1 R_{\nu_2}, t_2 R_{\nu_2}), \ldots, [t_{q-1} R_{\nu_q}, r' R_{\nu_q}/R_{[KN']}].$$

The contributions of the complementary intervals

(3.6) $\qquad [t_1 R_{\nu_1}, t_1 R_{\nu_2}), \ldots, [t_{q-1} R_{\nu_{q-1}}, t_{q-1} R_{\nu_q}), (r' R_{\nu_q}/R_{[KN']}, r']$

to the left hand side of (3.3) are

$$\log(R_{\nu_2}/R_{\nu_1}), \ldots, \log(R_{\nu_q}/R_{\nu_{q-1}}), \log(R_{[KN']}/R_{\nu_q}),$$

and (3.3) follows.

4. Comparison with Φ

As we have seen, Φ of (2.5) is fully indexed and the numbers ρ_n given by (2.6) are suitable choices for R_n in Theorem 3.1, for $n \ge 1$. Applying Theorem 3.1 with $F = \Phi$ and $K = 2$, we obtain a non-vacuous conclusion if the estimate (3.3) is non-trivial, i.e., if

$$\log(\rho_{[KN']}/R_0) = \sigma^{-1} \log N(r', f) + O(1) \ll \log r'.$$

This holds for certain arbitrarily large r' if $\liminf_{r\to\infty} \log N(r, f)/\log r < \sigma$, which is guaranteed if λ, the lower order of f, satisfies

$$\lambda \equiv \liminf_{r\to\infty} \frac{\log \log M(r, f)}{\log r} < \sigma,$$

since, from (2.3) and (2.4),

$$N(r, f) \le \int_r^{er} \frac{N(t, f)}{t} dt \le \log \mu(er, f) \le \log M(er, f).$$

Thus

(4.1) $\qquad \dfrac{|a_n| r^n}{\mu(r, f)} \le \dfrac{n^{-n/\sigma} \rho_N^n}{\mu(\rho_N, \Phi)}, \quad 0 \le n \le 2N,$

(4.2) $\qquad \dfrac{|a_n| r^n}{\mu(r, f)} \le \max\left\{ \dfrac{n^{-n/\sigma} \rho_N^n}{\mu(\rho_N, \Phi)}, 2^{-n/(2\sigma)} \right\}, \quad n > 2N,$

for all r outside a set E such that

(4.3) $\qquad \liminf_{r'\to\infty} \dfrac{\log \operatorname{meas} E \cap [0, r']}{\log r'} \le \lambda/\sigma.$

The left hand side of (4.3) is the *lower logarithmic density* of E. Since

$$\sum_{n=2N+1}^{\infty} 2^{-n/(2\sigma)} = o(1)$$

as $N \to \infty$, it follows from (4.1) and (4.2) that

(4.4) $\qquad M(r, f)/\mu(r, f) \le \sum_{n=0}^{\infty} |a_n| r^n/\mu(r, f) \le \Phi(\rho_N)/\mu(\rho_N, \Phi) + o(1),$

as $r \to \infty$ outside E. Taking account of (2.8), we have

THEOREM 4.1. [**4**] *If f is an entire function of finite lower order λ, then given $\sigma > \lambda$,*

(4.5) $$M(r,f) \leq (1 + o(1))\mu(r,f)\sqrt{2\pi\sigma N(r,f)}$$

as $r \to \infty$ outside a set of lower logarithmic density at most λ/σ.

This can be taken a little further. For the purposes of (4.5), we may assume without loss of generality that $a_0 = 1$; and then (4.1) gives, when $n = 0$, $1/\mu(r,f) \leq 1/\mu(\rho_N, \Phi)$. From this, (2.9) and (4.4), we obtain

THEOREM 4.2. [**4**] *If f is an entire function of finite lower order λ, then given $\sigma > \lambda$,*

(4.6) $$M(r,f) \leq (\sigma + o(1))\mu(r,f)\sqrt{2\pi \log \mu(r,f)}$$

as $r \to \infty$ outside a set of lower logarithmic density at most λ/σ.

In applying Theorem 3.1, it might have been assumed that the order rather than the lower order of f was finite, the order being

$$\limsup_{r \to \infty} \frac{\log \log M(r,f)}{\log r};$$

and then versions of Theorems 4.1 and 4.2 would follow in which lower order and lower logarithmic density are replaced by order and upper logarithmic density, the upper logarithmic density of a set E being

$$\limsup_{r' \to \infty} \frac{\log \operatorname{meas} E \cap [0, r']}{\log r'}.$$

5. A universal fully indexed function

Kövari [**7**] observed that $F(z) = \sum_{n=0}^{\infty} A_n z^n$ is fully indexed if

(5.1) $$A_n = \exp\left(\int_0^n \alpha(t)dt\right), \quad n = 0, 1, 2, \ldots,$$

where α is strictly decreasing on $[0, \infty)$. In fact, with R_n defined by

(5.2) $$R_n = \exp(-\alpha(n)), \quad n = 0, 1, 2, \ldots,$$

we have, for $N \geq 0$,

(5.3) $$\frac{A_n}{A_N} R_N^{n-N} = \exp\left(\int_N^n (\alpha(t) - \alpha(N))dt\right) < 1, \quad \text{if } n \neq N,$$

so that $N = N(R_N, F)$. Thus the central index takes all values, and the maximum term of F at R_N is unique, for all N.

(It is sometimes convenient to widen the class of functions α, and allow $\alpha(0) = +\infty$, provided that α is integrable on $(0,t)$ for every positive t. Then (5.3) holds as before, except for the case $N = 0$, for R_0 given by (5.2) is now 0. But it is clear that there is some positive number R_0 such that $1 = A_0 > A_n R_0^n$, for all $n \geq 1$, so that $N(R_0, F) = 0$, and again F is fully indexed. The function Φ of (2.5) corresponds to $\alpha(t) = -\sigma^{-1}(1 + \log t)$.)

Each such F has an associated class of comparable entire functions, in the sense that the inequalities of Theorem 3.1 are valid over a non-trivial set. Every entire function is comparable to $\Psi(z) = \sum_{n=0}^{\infty} B_n z^n$, where

$$(5.4) \qquad B_n = \exp\left(\int_0^n \beta(t) dt\right), \quad n = 0, 1, 2, \ldots,$$

and $\beta(t) = (\log(t+2))^{-1}$; indeed, since $\exp(-\beta(n))$ is bounded, the right hand side of (3.3) is bounded also. (There are evidently many other such functions.) From Theorem 3.1 with $K = 2$, and with

$$(5.5) \qquad \tau_n = \exp(-\beta(n)), \quad n = 0, 1, 2, \ldots,$$

we have

given an entire function $f(z) = \sum_{n=0}^{\infty} a_n z^n$,

$$(5.6) \qquad \frac{|a_n| r^n}{\mu(r, f)} \leq \frac{B_n \tau_N^n}{\mu(\tau_N, \psi)}, \quad 0 \leq n \leq 2N,$$

$$(5.7) \qquad \frac{|a_n| r^n}{\mu(r, f)} \leq \max\left\{\frac{B_n \tau_N^n}{\mu(\tau_N, \psi)}, (\tau_N/\tau_{2N})^{n/2}\right\}, \quad n > 2N,$$

for all r outside a set of finite logarithmic measure, where $N = N(r, f)$.

Inequalities for $M(r, f)$ in the same vein as those of Theorems 4.1 and 4.2 (but not as good) follow from these estimates, but we use (5.6) and (5.7) instead to obtain information about the derivatives of f at points where $|f|$ is reasonably large. Inequalities (5.6) and (5.7) provide bounds for the terms of the series that are far from the maximum term. Where these terms are negligible, it is the remaining polynomial, consisting of terms grouped around the maximum term, that dominates and links the behaviour of the derivatives of f to that of f itself.

The remainder of this section is concerned with the question: how much of the head and tail of the series for f can be discarded without significant loss, at points where $|f(z)|$ is large compared with $M(|z|, f)$? We focus on transcendental functions, since for polynomials the result we prove (Theorem 6.1 below) is sufficiently plain.

LEMMA 5.1. (cf. [6], p. 327) *Let q be a non-negative integer, η a positive number and k_N the smallest positive integer such that*

$$(5.8) \qquad k_N \geq \sqrt{8(q + 1 + \eta)N} (\log 2N)^3, \quad N = 1, 2, 3, \ldots.$$

Then given a transcendental entire function $f(z) = \sum_{n=0}^{\infty} a_n z^n$,

$$(5.9) \qquad \sum_{|n-N|>k_N} n^q |a_n| r^n = o(N^{-\eta} \mu(r, f)),$$

as $r \to \infty$ outside a set of finite logarithmic measure, where $N = N(r, f)$.

From (5.6) and (5.7), it is enough to show that

$$(5.10) \qquad \sum_{|n-N|>k_N} n^q B_n \tau_N^n = o(N^{-\eta} \mu(\tau_N, \psi))$$

and

(5.11) $$\sum_{n=2N+1} n^q (\tau_N/\tau_{2N})^{n/2} = o(N^{-\eta}).$$

From (5.4) and (5.5) we have, after integrating by parts,

$$B_n \tau_N{}^n / \mu(\tau_N, \psi)) = \exp\left(-\int_N^n \frac{n-t}{(t+2)(\log(t+2))^2} dt\right).$$

For $k_N < |n - N| \leq N$,

$$\int_N^n \frac{n-t}{(t+2)(\log(t+2))^2} dt \geq \frac{(n-N)^2}{4(N+1)(\log(2N+2))^2} > \frac{k_N{}^2}{8N(\log 2N)^2},$$

for all large N, while for $n > 2N$,

$$\int_N^n \frac{n-t}{(t+2)(\log(t+2))^2} dt \geq \int_N^{3N/2} \frac{n-t}{(t+2)(\log(t+2))^2} dt \geq \frac{Cn}{(\log N)^2},$$

for all large N. Here and below, C denotes an absolute constant, not necessarily the same at each occurrence. It follows that

$$\sum_{|n-N|>k_N} n^q B_n \tau_N^n / \mu(\tau_N, \psi)$$

$$\leq (2N)^{q+1} e^{-k_N^2/(8N(\log 2N)^2)} + \sum_{n=2N+1}^{\infty} n^q e^{-Cn/(\log N)^2}$$

$$\leq (2N)^{-\eta} + \sum_{n=2N+1}^{\infty} e^{-Cn/(\log N)^2}$$

$$= (2N)^{-\eta} + e^{-C(2N+1)/(\log N)^2} / (1 - e^{-C/(\log N)^2})$$

(5.12) $$= O(N^{-\eta}),$$

as $N \to \infty$, in view of (5.8). This establishes (5.10). Finally ([**3**], Lemma 9),

$$\sum_{n=2N+1}^{\infty} n^q t^n \leq C_0 \frac{N^q t^{2N}}{1-t} (1 + (2N(1-t))^{-q-1}),$$

where C_0 depends only on q; and (5.11) follows with $t = \tau_N/\tau_{2N} = \exp(-(C + o(1))/(\log N)^2)$.

6. Behaviour of the derivatives

According to the last result we shall prove, $f(z)$ behaves locally like its maximum term, the monomial z^N, on a significant set of z.

THEOREM 6.1. (cf. [**6**], p. 341) *Suppose that $f(z) = \sum_{n=0}^{\infty} a_n z^n$ is a transcendental entire function, and that z is such that*

(6.1) $$|f(z)| \geq N^{-\gamma} M(|z|, f),$$

where $N = N(|z|, f)$ and γ is a constant satisfying $0 < \gamma < 1/2$. Then given any non-negative integer q,

(6.2) $$f^{(q)}(z) = (1 + o(1)) \left(\frac{N}{z}\right)^q f(z),$$

uniformly as $|z| \to \infty$ outside a set of finite logarithmic measure.

From Lemma 5.1 with $\eta = 1/2$,

$$\sum_{|n-N|>k_N} n^q |a_n| r^n = o(N^{-1/2} \mu(r,f)), \tag{6.3}$$

as $r \to \infty$ outside a set of finite logarithmic measure, where $N = N(r,f)$ and k_N is the smallest positive integer satisfying (5.8). For whatever z the left hand side of (6.3) is negligible, so are the corresponding terms of the series for $f(z)$. The behaviour of $f(z)$ is then determined by the central portion of the series, that is, by $\sum_{|n-N|\le k_N} a_n z^n = z^{N-k_N} P_N(z)$, P_N being a polynomial of degree at most $2k_N + 1$. Indeed, if z is such that (6.1) holds, then

$$f(z) = (1 + o(1)) z^{N-k_N} P_N(z) \tag{6.4}$$

(from which it follows incidentally that

$$M(r,f) = O(N^\gamma z^{N-k_N} P_N(z)), \tag{6.5}$$

which will be useful in a moment). The same is true if z is such that $P_N(z)$ is large, and in particular,

$$\max_{|z|=r} |P_N(z)| = (1 + o(1)) r^{k_N - N} M(r,f). \tag{6.6}$$

Having made these observations, let us consider the derivatives of f:

$$f^{(q)}(z) = \frac{d^q}{dz^q} (z^{N-k_N} P_N(z)) + O\left(\sum_{|n-N|>k_N} n^q |a_n| r^{n-q} \right). \tag{6.7}$$

From (6.3) and (6.5), we have

$$\sum_{|n-N|>k_N} n^q |a_n| r^{n-q} = o\left(N^{\gamma - 1/2} z^{N-k_N-q} P_N(z) \right), \tag{6.8}$$

while

$$\frac{d^q}{dz^q} \left(z^{N-k_N} P_N(z) \right)$$
$$= \sum_{j=0}^{q} C_j^q (N-k_N)\ldots(N-k_N-q+j+1) z^{N-k_N-q+j} P_N^{(j)}(z)$$
$$= (1+o(1)) N^q z^{N-k_N-q} P_N(z) + O\left(\sum_{j=1}^{q} N^{q-j} z^{N-k_N-q+j} P_N^{(j)}(z) \right). \tag{6.9}$$

To proceed, we need estimates for the derivatives of P_N.

LEMMA 6.1. ([6], p. 337) *Let P be a polynomial of degree m and j a positive integer. For any $r > 0$,*

$$\max_{|z|=r} |P^{(j)}(z)| \le ej! (m/r)^j \max_{|z|=r} |P(z)|. \tag{6.10}$$

To see this, write $M = \max_{|\zeta|=r} |P(\zeta)|$ and note that since $P(z)/z^m$ is analytic in $|z| \ge r$, including the point at infinity, $|P(z)| \le M(|z|/r)^m$ for $|z| \ge r$, by the maximum principle. Applying Cauchy's inequalities over the disc with centre z and radius r/m, where $|z| = r$, we have

$$|P^{(j)}(z)| \le j! \left(\frac{m}{r}\right)^j \max_{|\zeta|=r(1+1/m)} |P(\zeta)| \le Mj! \left(\frac{m}{r}\right)^j \left(1+\frac{1}{m}\right)^m \le Mej! \left(\frac{m}{r}\right)^j,$$

which is (6.10).

From (6.6) and Lemma 6.1, we obtain

(6.11) $$|P_N^{(j)}(z)| = O\left(k_N^j r^{k_N-N-j} M(r,f)\right),$$

so the general term of the sum on the right of (6.9) is

(6.12) $$O\left(N^{q-j} k_N^j r^{-q} M(r,f)\right) = O\left(N^{q-j+\gamma} k_N^j z^{N-k_N-q} P_N(z)\right),$$

taking account of (6.5). Given the definition of k_N, this is asymptotically largest when j is as small as possible (that is, when $j=1$), in which case it is

$$O\left(N^{q-1+\gamma} k_N z^{N-k_N-q} P_N(z)\right).$$

This is negligible compared with the first term on the right of (6.9), since $N^{-1+\gamma} k_N = o(1)$ as $N \to \infty$, γ being less than $1/2$. The left hand side of (6.8) is similarly negligible; thus, from (6.7), (6.9) and (6.4), we have

$$f^{(q)}(z)) = (1+o(1)) N^q z^{N-k_N-q} P_N(z) = (1+o(1))\left(\frac{N}{z}\right)^q f(z),$$

which proves Theorem 6.1.

References

1. J. Clunie, *The determination of an integral function of finite order from its Taylor series*, J. London Math. Soc. **28** (1953), 58–66.
2. J. Clunie, *On the determination of an integral function from its Taylor Series*, J. London Math. Soc. **30** (1955), 32–42.
3. C.C. Davis and P.C. Fenton, *The real part of entire functions*, Michigan Math. J. **43** (1996), 475-494.
4. P.C. Fenton, *Some results of Wiman-Valiron type for integral functions of finite lower order*, Ann. of Math. **103** (1976), 237-252.
5. Fenton P.C., *A note on the Wiman-Valiron method*, Proc. Edinburgh Math. Soc. (2) **37** (1994), 53–56.
6. W.K. Hayman, *The local growth of power series: a survey of the Wiman-Valiron method*, Canad. Math. Bull. **17** (1974), 317–358.
7. T. Kövari, *On the Borel exceptional values of lacunary integral functions*, J. Analyse Math. **9** (1961), 71–109.
8. I. Laine, *Nevanlinna Theory and Complex Differential Equations*, Walter de Gruyter & Co., Berlin, 1993.

DEPARTMENT OF MATHEMATICS, UNIVERSITY OF OTAGO, DUNEDIN, NEW ZEALAND
E-mail address: pfenton@maths.otago.ac.nz

Billiards in an Ellipse

Leopold Flatto

Dedicated to Larry Zalcman on his sixtieth birthday

ABSTRACT. The billiard problem was introduced by G. Birkhoff. For elliptic tables, the problem is integrable. Using notions from Riemann surfaces, we derive the invariant measure associated with the problem. The measure is used to obtain a complete description of billiard trajectories and their associated dynamics.

1. Billiards in an ellipse

The billiard problem was formulated by G. Birkhoff [**B**, p. 169] in his investigations of certain dynamical systems. A point particle moves in the interior of a domain bounded by a smooth convex curve C. The point moves along a straight line with constant velocity and is perfectly reflected at the boundary, i.e., the angle of incidence equals the angle of reflection. The problem is to describe all trajectories.

In general, it seems impossible to give a complete solution to this problem. However, when C is an ellipse, the problem has a rather elegant solution, which we now describe.

We first dispose of two special cases. (i) The trajectory passes through a focus of C. In this case, the consecutive line segments making up the trajectory pass alternately through the two foci of C (a proof is given [**C-R**, p. 333]). (ii) The trajectory moves along the minor axis of C, and reverses direction upon each contact with C.

In all other cases, we have

THEOREM 1.1. *The lines containing the segments of a billiard trajectory in an ellipse C are either all tangent to another ellipse confocal with C or are all tangent to a hyperbola confocal with C.*

A proof of Theorem 1.1 is given in [**Si**, p. 86]. We refer to the two possibilities as cases 1 and 2. Let F_1, F_2 be the foci of C and F_1F_2 the line segment joining them. In case 1, the trajectory does not meet F_1F_2, and in case 2 all segments of

---

2000 *Mathematics Subject Classification.* Primary 37D40, 37D50; Secondary 33E05.

the trajectory meet F_1F_2. The ellipses and hyperbolas of Theorem 1.1 are called caustics of the trajectories.

In appropriate Cartesian coordinates x, y, the equations of confocal ellipses and hyperbolas are given by

$$\text{(1.1)} \qquad \frac{x^2}{a^2 - \lambda} + \frac{y^2}{b^2 - \lambda} = 1,$$

where $0 < b < a$. For $\lambda < b^2$, the curve is an ellipse; and for $b^2 < \lambda < a^2$, it is a hyperbola. The foci are $(\pm\sqrt{a^2 - b^2}, 0)$. We assume that the ellipse C, inside which the trajectories move, corresponds to $\lambda = 0$. The caustic, for a given λ, is denoted by D_λ.

We also use the parametric equations for C given by

$$\text{(1.2)} \qquad x = a\cos\theta, \qquad y = b\sin\theta, \qquad -\pi \leq \theta \leq \pi.$$

2. The maps T and η

We proceed to give further information on billiard trajectories in C with the same caustic D. In particular, we obtain a criterion as to when these trajectories are periodic. To this end, we define two homeomorphisms, T and η, both related to billiards. In § 4, we discuss the dynamics of T and η, that is, the behavior of their iterates. The dynamics are applied to describe billiard trajectories with the same caustic. The proofs of the results stated in § 4 are outlined in § 6.

T: This map is defined only for case 1. Let C, D be two confocal ellipses, with D inside C. For any point $p \in C$, choose a line through p which is tangent to D and which varies continuously with p (there are two such choices). The line meets C again in a point $T(p)$ distinct from p. Then $p \mapsto T(p)$ is an orientation-preserving homeomorphism of C. The map T relates to billiards, as a trajectory moving counterclockwise and meeting C at p meets C again at $T(p)$.

η: This map is defined for both cases 1 and 2. Let

$$\mathfrak{M} = \{(p, \xi) : p \in C, \xi \in D^*, p \in \xi\},$$

where D^* is the dual of D, that is, the set of lines tangent to D. For $(p, \xi) \in \mathfrak{M}$, let $\eta(p, \xi) = (p', \xi')$, where p, p' are the points of C contained in ξ, and ξ, ξ' are the lines of D^* containing p'. The map η is a homeomorphism[1] of $\mathfrak{M}$; it relates to billiards, as a trajectory leaving C at p and moving along ξ meets C again at p' and then moves along ξ'.

REMARK. The map η can also be described as the composition $\tau \circ \sigma$, where σ and τ are the two involutions of $\mathfrak{M}$ given by

$$\sigma(p, \xi) = (p', \xi); \quad \tau(p, \xi) = (p, \xi').$$

Here p, p' are the points of C contained in ξ, and ξ, ξ' are the lines through p tangent to D. This description of η is used in § 6.

For $(p, \xi) \in \mathfrak{M}$, let $\pi(p, \xi) = p$. We call π the projection map from $\mathfrak{M}$ to C, and (p, ξ) is said to be over p.

---

[1]The topology on $\mathfrak{M}$ is the one it obtains as a subset of $E_2 \times P_2$, where E_2 is the real plane and P_2 is the real projective plane.

3. Description of $\mathfrak{M}$

The set $\mathfrak{M}$ is the union of two disjoint simple closed curves $\mathfrak{M}^1$ and $\mathfrak{M}^2$, described as follows.

Case 1: For $p \in C$, let $\xi_1(p)$ and $\xi_2(p)$ be the lines joining p to $T(p)$ and $T^{-1}(p)$, respectively. Then $\mathfrak{M}^i := \{(p, \xi_i(p)) : p \in C\}$, for $i = 1, 2$. Both $\mathfrak{M}^1$ and $\mathfrak{M}^2$ lie over C.

Case 2: Let C_1, C_2 be the two arcs of C contained between the two branches of the hyperbola D. For $p \in C_1 \cup C_2$, let $\xi_1(p)$ and $\xi_2(p)$ be the two lines through p which are tangent to D, with the segment of $\xi_1(p)$ inside C to the right of the segment of $\xi_2(p)$ inside C ($\xi_1(p) \neq \xi_2(p)$, except when p is an endpoint of C_1 or C_2). Then $\mathfrak{M}^i = \mathfrak{M}^{i1} \cup \mathfrak{M}^{i2}$, where $\mathfrak{M}^{ij} := \{(p, \xi_j(p)) : p \in C_i\}$, for $i, j = 1, 2$; and $\mathfrak{M}^1$ and $\mathfrak{M}^2$ lie over C_1 and C_2, respectively.

For case 1, the curves $\mathfrak{M}^1$ and $\mathfrak{M}^2$ are both invariant under η. The parametrizations and orientations of $\mathfrak{M}^1$ and $\mathfrak{M}^2$ are derived in an obvious manner from those of C.

For case 2, η interchanges $\mathfrak{M}^1$ and $\mathfrak{M}^2$, so both $\mathfrak{M}^1$ and $\mathfrak{M}^2$ are invariant under η^2. We parametrize and orient the arcs $\mathfrak{M}^{ij}$.

Parametrization. Let $p(\theta) = (a\cos\theta, b\sin\theta)$, and let $p(\theta_0)$ be the right end point of C_2. A calculation gives that, when $D = D_\lambda$,

$$(3.1) \qquad \theta_0 = \arccos\sqrt{\frac{a^2 - \lambda}{a^2 - b^2}}, \quad 0 < \theta_0 < \pi/2.$$

For $i, j = 1, 2$, $\mathfrak{M}^{ij}$ consists of the points $(p(\theta), \xi_j(p(\theta)))$ with $\theta_0 - \pi \leq \theta \leq -\theta_0$ for $i = 1$, and $\theta_0 \leq \theta \leq \pi - \theta_0$ for $i = 2$.

Orientation. For $i = 1, 2$, we orient $\mathfrak{M}^{i1}$ and $\mathfrak{M}^{i2}$ so that $\pi(\mathfrak{M}^{i1})$ moves along C_i from left to right and $\pi(\mathfrak{M}^{i2})$ moves along C_i from right to left. The orientation of $\mathfrak{M}^i$ derives from those of $\mathfrak{M}^{i1}$ and $\mathfrak{M}^{i2}$.

4. Dynamics of T and η

To describe the dynamics of T and η, we introduce the concepts of topological conjugacy, rotation number and invariant measure. We first describe the

Dynamics of T: Let K be a circle of unit circumference and R_α a rotation of K by angle $2\pi\alpha$, $0 \leq \alpha < 1$. Both C and K are endowed with the counterclockwise orientation.

THEOREM 4.1. *There exists an orientation preserving homeomorphism ψ from C to K such that*
$$\psi \circ T \circ \psi^{-1} = R_\alpha$$
for some $0 \leq \alpha < 1$.

The map ψ is called a topological conjugacy from T to R_α, and T and R_α are said to be topologically conjugate. We call α the rotation number of T and denote it by $\theta(T)$. The geometric meaning of $\theta(T)$ is as follows. For $p \in C$, let $\omega(n, p)$ be the number of times $p, Tp, \ldots, T^n p$ winds around C in the counterclockwise direction. Then

$$(4.1) \qquad \lim_{n \to \infty} \frac{\omega(n, p)}{n} = \theta(T),$$

the limit being independent of p.

Let λ denote Lebesgue measure on K, and let

(4.2) $$\mu(E) := \lambda(\psi E),$$

for any Borel set E of C.

Now λ is R_α-invariant, that is, $\lambda(F) = \lambda(R_\alpha F)$, for any Borel set of K. It follows from (4.2) that μ is T-invariant.

In Theorem 4.2, T is the homeomorphism of C determined by C and D_λ.

THEOREM 4.2. *i) The T-invariant measure of C is given by*

(4.3) $$d\mu = c \frac{d\theta}{\sqrt{1 - k^2 \cos^2 \theta}},$$

where c is the normalizing constant chosen so that $\mu(C) = 1$, and

(4.4) $$k = \sqrt{\frac{a^2 - b^2}{a^2 - \lambda}} < 1$$

is the eccentricity of D_λ.

ii) The rotation number of T is given by

(4.5) $$\theta(T) = \int_0^{\theta_1} \frac{d\theta}{\sqrt{1 - k^2 \cos^2 \theta}} \bigg/ \int_0^\pi \frac{d\theta}{\sqrt{1 - k^2 \cos^2 \theta}},$$

where

(4.6) $$\theta_1 = \arccos \frac{\sqrt{a^2 - \lambda}}{a}, \quad 0 < \theta_1 < \pi/2.$$

REMARKS. 1. (4.5) follows from (4.3). For we have

(4.7) $$\alpha = \lambda[q, R_\alpha q] \quad \text{for all } q \in K,$$

where $[q, R_\alpha q]$, is the arc of K going clockwise from q to $R_\alpha q$. From (4.2) and (4.7), we then get

(4.8) $$\theta(T) = \mu[p, Tp], \quad \text{for all } p \in C,$$

where $[p, Tp]$ is the arc of C going counterclockwise from p to Tp. Then (4.8) becomes (4.5), when p is chosen as the lower intersection of the vertical line $x = \sqrt{a^2 - \lambda}$ with C.

2. The measure μ given by (4.3) can also be interpreted as a measure on $\mathfrak{M}^1$ and $\mathfrak{M}^2$. As such, μ is also η_i-invariant, where η_i is the restriction of η to $\mathfrak{M}^i$. For we have

$$\pi_1 \circ \eta_1 = T \circ \pi_1, \quad \pi_2 \circ \eta_2 = T^{-1} \circ \pi_2,$$

where π_i is the restriction of π to $\mathfrak{M}^i$. From these conjugacy relations, we obtain that the following three properties of μ are equivalent: μ is an invariant measure for T, η_1 and η_2.

The significance of the invariant measure and rotation number for the dynamics of T is given in Theorem 4.3.

THEOREM 4.3. *i) If $\theta(T) = m/n$, where m and n are positive relatively prime integers, then T^n is the identity map. The T-orbit (set of T-iterates) of $p \in C$ consists of the distinct points $p, Tp, ... T^{n-1}p$, which wind m times around C.*

ii) If $\theta(T)$ is irrational, then for any $p \in C$ and any arc A of C, we have

$$\lim_{n \to \infty} \frac{|\{p, Tp, ..., T^{n-1}p\} \cap A|}{n} = \mu(A),$$

where $|\cdot|$ denotes set cardinality.

The proof of Theorem 4.3 follows from the corresponding statement for R_α. In this case, i) is obvious, and ii) is the Kronecker–Weyl theorem [**H-W**, Theorem 345]. Passing from R_α to T by conjugation, we obtain Theorem 4.3.

Dynamics of η. For case 1, the dynamics of η follow from that of T. For if $P = (p, \xi) \in \mathfrak{M}$, then $\eta^n(P)$ lies over $T^n p$ when $P \in \mathfrak{M}^1$ and over $T^{-n}(p)$ when $P \in \mathfrak{M}^2$.

For case 2, it suffices to consider the dynamics of the restrictions of η^2 to $\mathfrak{M}^1$ and $\mathfrak{M}^2$. Denote these by η_1^2 and η_2^2.

THEOREM 4.4. *i) For $i = 1, 2$, η_i^2 is an orientation-preserving homeomorphism of $\mathfrak{M}^1$ and is conjugate to a rotation of a circle.*

ii) η_i^2 has the invariant measure (4.3), where the constant c is now chosen so that $\mu(\mathfrak{M}_i) = 1$.

iii) η_1^2, η_2^2 have the some rotation number

$$(4.9) \qquad \theta(\eta_i^2) = 2 \int_{\theta_0}^{\theta_1} \frac{d\theta}{\sqrt{1 - k^2 \cos^2 \theta}} \Big/ \int_{\theta_0}^{\pi - \theta_0} \frac{d\theta}{\sqrt{1 - k^2 \cos^2 \theta}},$$

where θ_0 and θ_1 are given by (3.1) and (4.6).

From Theorem 4.4, we obtain the dynamics of η_i^2. They are given by Theorem 4.3, with C and T replaced by $\mathfrak{M}_i$ and η_i^2.

5. Billiard trajectories with the same caustic

The following theorem is essentially a reformulation of Theorem 4.3, as applied to billiard trajectories. In case 1, we may assume, by symmetry, that the trajectories move counterclockwise around C.

THEOREM 5.1. *Let D_λ be an ellipse and T the homeomorphism of C determined by C and D_λ.*

i) If $\theta(T) = m/n$, where m and n are positive relatively prime integers, then all trajectories in C with caustic D_λ are closed n-gons winding m times around C.

ii) If $\theta(T)$ is irrational, then all trajectories in C with caustic D_λ are aperiodic. Let $p_1, \ldots, p_n, \ldots$ be the successive points in which such a trajectory meets C. Then for any arc A of C,

$$\lim_{n \to \infty} \frac{|\{p_1, \ldots, p_n\} \cap A|}{n} = \mu(A),$$

where μ is the measure given by (4.3).

If D_λ is a hyperbola, then we use Theorem 4.4 to obtain an analogue of Theorem 5.1. In this case, when $\theta(\eta_i^2) = \frac{m}{n}$, the trajectories are $2n$-gons with n vertices on C_1 and n vertices on C_2.

REMARK. Formulas (4.5), (4.9) can be used to show that $\theta(T)$ is a continuous strictly increasing function of λ and $\theta(\eta_i^2)$ a continuous strictly decreasing function of λ. It follows that trajectories with caustic D_λ are periodic for only a countable number of λ.

6. Invariant measure

We derive the invariant measure (4.3) by complex analysis. This requires that we complexify C and D, which means that the variables x, y in the equations (1.1) for C and D now assume complex values. Then $\mathfrak{M}$ becomes a Riemann surface, and η becomes an automorphism (conformal map) of $\mathfrak{M}$. The real counterparts of $C, D, \ldots$ are now denoted by $C_R, D_R, \ldots$.

The derivation is divided into the following steps.

1) We obtain an algebraic equation for $\mathfrak{M}$.
2) We describe the covering map from the complex plane $\mathbb{C}$ to $\mathfrak{M}$.
3) The map η is lifted to a map $\tilde{\eta}$, which is a translation of $\mathbb{C}$.
4) The invariant measure for η_R is recovered from the invariant measure for the restriction of $\tilde{\eta}$ to $\phi^{-1}(\mathfrak{M}_R)$.

We carry out the details for case 1. The derivation for case 2 is similar and so omitted.

1) *Equation for* $\mathfrak{M}$. Let $\alpha = \sqrt{a^2 - \lambda}$, $\beta = \sqrt{b^2 - \lambda}$. Then C and D have the respective parametric equations

$$(6.1) \qquad x = a\frac{1-t^2}{1+t^2}, \quad y = \frac{2bt}{1-t^2};$$

$$(6.2) \qquad x = \alpha\frac{1-u^2}{1+u^2}, \quad y = \frac{2\beta u}{1-u^2},$$

where t, u assume all complex values.

The tangent line to D at (x_0, y_0) has equation

$$(6.3) \qquad 0 = \frac{x_0}{\alpha^2}(x - x_0) + \frac{y_0}{\beta^2}(y - y_0) = \frac{x_0 x}{\alpha^2} + \frac{y_0 y}{\beta^2} - 1.$$

Writing x_0, y_0 for x, y in (6.2), we conclude from (6.1)–(6.3), after slightly tedious but straightforward manipulations, that $\mathfrak{M}$ is defined by either of the equations

$$(6.4) \qquad u = \frac{-2Bt + \sqrt{A^2-1}\sqrt{\Delta(t)}}{(A-1)t^2 - (A+1)}, \quad t = \frac{-2Bu + \sqrt{A^2-1}\sqrt{\Delta(u)}}{(A-1)u^2 - (A+1)},$$

where $A = a/\alpha$, $B = b/\beta$,

$$\Delta(t) = (t^2 + m^2)(t^2 + 1/m^2); \quad m = \sqrt{\frac{1+k}{1-k}}.$$

In terms of t and u, C and D become respectively the t- and u-spheres. We may picture $\mathfrak{M}$ as a 2-sheeted Riemann surface over either the t- or u-spheres, with branch points at $t, u = \pm i/m, \pm im$. By the Riemann–Hurwitz formula, $\mathfrak{M}$ has genus one, i.e., is a torus.

Furthermore, C_R becomes the real t-axis, and $\mathfrak{M}_R^1, \mathfrak{M}_R^2$ become the two curves in $\mathfrak{M}$ lying over the real t-axis.

2) *Covering map of* $\mathfrak{M}$. Since $\mathfrak{M}$ is a torus, its universal covering space is the complex plane $\mathbb{C}$. The covering map $\phi(s) = (t(s), u(s))$ from $\mathbb{C}$ to $\mathfrak{M}$ can be obtained explicitly by means of the Schwarz–Christoffel map from a half-plane to a rectangle (for details, see [**Sp**, p. 36]). The map ϕ has the following properties.

a) $t(s), u(s)$ are elliptic functions with periods $\omega_1, i\omega_2$, where $\omega_1, \omega_2 > 0$.
b) $\phi(s_1) = \phi(s_2)$ if and only if $s_1 - s_2 = m\omega_1 + in\omega_2$, m and n arbitrary integers.

c) Let $J_1 = \{s : 0 \leq s < \omega_1\}$, $J_2 = \{\frac{i\omega_2}{2} + s : 0 \leq s < \omega_1\}$. Then ϕ bijects J_i to $\mathfrak{M}_R^i$, $i = 1, 2$. For $0 < s < \infty$, $t(s)$ ($t(\frac{i\omega_2}{2} + s)$) is a continuous strictly increasing (decreasing) function with range $(-\infty, \infty)$.

d) $t(s)$ satisfies the differential equation

$$\left(\frac{dt}{ds}\right)^2 = \Delta(t). \tag{6.5}$$

We remark that from b) and c), we obtain the set relations

$$\phi^{-1}(\mathfrak{M}_R^1) = \bigcup_k l_k, \quad \phi^{-1}(\mathfrak{M}_R^2) = \bigcup_k l_k',$$

where

$$l_k = \{s : \operatorname{Im} s = k\omega_2\}; \quad l_k' = \{s : \operatorname{Im} s = (k + 1/2)\omega_2\}, \quad k = 0, \pm 1, \ldots.$$

3) *Lift of η.* We show that η is an automorphism of $\mathfrak{M}$ and it has a lift $\widetilde{\eta}$ given by

$$\widetilde{\eta}(s) = s + b \tag{6.6}$$

for $s \in \mathbb{C}$ and some $b \in (0, \omega_1)$.

We have $\eta = \tau \circ \sigma$. In terms of t and u, σ and τ become the sheet interchanges of $\mathfrak{M}$, viewed as a cover of the u- and t-spheres, respectively. As such, σ and τ are both automorphisms of $\mathfrak{M}$. Hence η is also an automorphism of $\mathfrak{M}$. Lift σ, τ to automorphisms $\widetilde{\sigma}, \widetilde{\tau}$ of $\mathbb{C}$. Then

$$\widetilde{\sigma}(s) = \alpha s + \beta, \quad \widetilde{\tau}(s) = \gamma s + \delta \tag{6.7}$$

for some constants $\alpha, \beta, \gamma, \delta$. Since σ^2 is the identity map, we have

$$\widetilde{\sigma}^2(s) = \alpha^2 s + (\alpha + 1)\beta \equiv s \pmod{\Lambda} \quad \text{for all } s \in \mathbb{C}. \tag{6.8}$$

Hence $\alpha^2 = 1$. If $\alpha = 1$, then σ would have no fixed points. But σ has four fixed points, namely the four branch points at $u = \pm i/m$, $\pm im$. Thus $\alpha = -1$ and, similarly, $\gamma = -1$.

Then η has the lift $\widetilde{\tau} \circ \widetilde{\sigma}(s) = s + \delta - \beta$. Since η maps $\mathfrak{M}_R^1$ to $\mathfrak{M}_R^1$, we conclude from the properties of ϕ that $\delta - \beta = b + \omega$, for some $b \in (0, \omega_1)$ and some $\omega \in \Lambda$. But then $\widetilde{\eta}(s) = s + b$ is also a lift of η.

4. *Invariant measure for η_R.* For $i = 1, 2$, let η_i denote the restriction of η_R to $\mathfrak{M}_R^i$. We show that (4.3) is an invariant measure for η_1 and hence also an invariant measure for η_2. To this end, we first establish the conjugacy relation (6.9).

Let ϕ_1 be the restriction of ϕ to J_1. For $s \in J_1$, let $R_b(s) = s \oplus b$, where b is defined as in (6.6) and $\oplus$ denotes addition mod ω_1. We view J_1 has a circle of length ω_1. Then ϕ_1 is a homeomorphism from J_1 to $\mathfrak{M}_R^1$, and R_b is a rotation of J_1 (by angle $\frac{b}{\omega_1}$). We have

$$\phi_1 \circ R_b = \eta_1 \circ \phi_1, \tag{6.9}$$

which follows from $\widetilde{\eta}(s) = s + b$ being a lift of η and $\phi(s + \omega_1) = \phi(s)$.

The Lebesgue measure ds is R_b-invariant; and from (6.9), we get that

$$d\mu := \frac{ds}{dt} dt, \quad -\infty < t < \infty, \tag{6.10}$$

is η_1-invariant. By (6.5),

$$\text{(6.11)} \qquad \frac{ds}{dt} = \frac{1}{\sqrt{(t^2+m^2)(t^2+1/m^2)}}, \quad -\infty < t < \infty.$$

Now use $m^2 = \frac{1+k}{1-k}$ and the change of variable $t = \tan\frac{\theta}{2}$, which is obtained from (6.1) and (6.2). Then (6.10) and (6.11) give

$$d\mu = \sqrt{1-k^2}\frac{d\theta}{\sqrt{1-k^2\cos^2\theta}},$$

which is (4.3) except for a multiplicative constant.

References

[B]   G. Birkhoff, *Dynamical Systems*, Amer. Math. Soc., Providence, R.I., 1927.
[C-R] R. Courant and H. Robbins, *What is Mathematics?*, Oxford University Press, New York, 1944.
[H-W] G.H. Hardy and E.M. Wright, *An Introduction to the Theory of Numbers*, Clarendon Press, Oxford, 1945.
[K]   J. King, *Three problems in search of a measure*, Amer. Math. Monthly **101** (1994), 609-628.
[Si]   Y. Sinai, *Introduction to Ergodic Theory*, Princeton University Press, 1976.
[Sp]   G. Springer, *Introduction to Riemann Surfaces*, Addison Wesley, Reading, Mass., 1957.

3116 ARLINGTON AVENUE, BRONX, NY 10463, U.S.A.
E-mail address: zeflatto@jtsa.edu

(p,q,r)–Kleinian Groups and the Margulis Constant

F. W. Gehring and G. J. Martin

Dedicated to Larry Zalcman on the occasion of his 60th birthday

ABSTRACT. A (p,q,r)–Kleinian group is a Kleinian group generated by three rotations of orders p,q and r whose axes of rotation are coplanar. The *Margulis constant* for Kleinian groups is the smallest constant c_M such that for each discrete group G and each point x in the upper half space $\mathbb{H}^3$, the group generated by the elements in G which move x less than distance c_M is elementary. We show here that the sharp bound for this constant is $c = 0.230989\ldots$ for (p,q,r)–Kleinian groups. With our earlier work it follows that the Margulis constant is achieved in a Kleinian group generated by three elliptic elements of order 2. We give an example in this case to show that $c_M \leq 0.132409\ldots$. We conjecture this value to be the Margulis constant. We also determine some other useful universal constraints concerning (p,q,r)–Kleinian groups which are necessary in determining sharp co–volume bounds for general Kleinian groups.

1. Introduction

In this paper, we study Kleinian groups generated by three rotations of orders p, q and r whose axes are coplanar. That is, they lie in a common hyperbolic plane in $\mathbb{H}^3$. We call such a group a (p,q,r)–Kleinian group when the generating elements have orders p, q and r. These groups are natural generalizations of triangle groups and also of the index 2 orientation preserving subgroups of the group generated by reflection in the faces of a hyperbolic tetrahedron. In particular, the (p,q,r)–Kleinian groups properly contain both of these classes. These groups also appear as extremals for certain interesting problems in the theory of Kleinian groups. Apart from the triangle groups and reflection subgroups, there are a number of interesting infinite families of web groups. These are the groups whose limit set is a Sierpinski gasket, that is, $\overline{\mathbb{C}}$ minus a certain infinite family of disjoint disks such that what remains has no interior. The stabilizers of components of the ordinary set, the union of the disks removed, are triangle groups.

2000 *Mathematics Subject Classification.* Primary 30F40, 30D50.

Research supported in part by grants from the N.Z. Marsden Fund, the U.S. National Science Foundation and the Volkswagen–Stiftung (RiP–program of the Mathematisches Forschungsinstitut at Oberwolfach).

When the axes of the rotations bound a hyperbolic triangle and the group is discrete, these groups have a presentation of the form

(1) $$\langle a, b, c : a^p = b^q = c^r = (ab)^\ell = (ac)^m = (bc)^n \rangle.$$

When such a group Γ is discrete, the sum

$$\Sigma = \frac{1}{\ell} + \frac{1}{m} + \frac{1}{n}$$

determines whether Γ is compact ($\Sigma > 1$), Γ is noncompact but of finite covolume ($\Sigma = 1$) or Γ is a web group ($\Sigma < 1$). If $p = q = r = 2$, we have in fact a $\mathbb{Z}_2$–extension of a triangle group.

In this paper, we identify all discrete (p, q, r) groups and certain related groups in geometrically important situations, namely when the axes bound a hyperbolic triangle. We then identify certain geometric constants associated with these groups. These include the Margulis constant and the minimal distances between points fixed by spherical triangle subgroups. The latter, together with the results of [6], constitute an essential part of our program to determine the Kleinian group of minimal co–volume. The former extend the results of our paper [5].

The Margulis constant c_M for Kleinian groups is the smallest positive constant c with the following property. For each discrete group Γ and each point x in the upper half space $\mathbb{H}^3$, the group generated by the elements in Γ which move x less than distance c is elementary, that is, virtually abelian.

Since Kleinian groups act as universal covering groups for hyperbolic 3–folds, there is also a geometric interpretation of the Margulis constant. For hyperbolic 3–folds $\mathcal{Q}^3$, c_M is the largest constant c such that for all $x_0 \in \mathcal{Q}^3$, the subgroup of the (orbifold) fundamental group generated by all loops based at x_0 and of length less than c is virtually abelian, cyclic if $\mathcal{Q}^3$ is a manifold.

A fundamental step in Margulis' proof of the rigidity theorem [13, 10] is to show that $c_M > 0$, actually for the larger class of orbifolds covered by rank 1 symmetric spaces. It is well understood that the problem of determining the Margulis constant reduces to the problem of calculating the constant for all two generator groups and for certain three generator groups. See [9, 12, 18] for the two dimensional case. In [5], we determine this constant for certain two generator groups, namely those with a parabolic or elliptic generator. Culler and Shalen have given lower bounds on the Margulis constant for certain two generator torsion free groups. See [3].

In this paper, we identify all the relevant three generator discrete groups and calculate the associated constant, except for those groups generated by three elements of order 2. It follows from the results established here and in [5] that the Margulis constant is attained in a Kleinian group generated by three elements of order 2. We give the conjectured extremal example with a constant of $c_M = 0.132409...$.

We begin with some basic definitions and results about elementary discrete groups.

DEFINITION. A *Kleinian* group Γ is a discrete nonelementary subgroup of $Isom^+(\mathbb{H}^3)$, the group of orientation preserving isometries of hyperbolic 3–space. In this setting, *nonelementary* means that Γ does not contain an abelian subgroup of finite index.

We denote the hyperbolic metric of $\mathbb{H}^3$ by $\rho(x,y)$ and view hyperbolic 3-space $\mathbb{H}^3$ as the upper half space of $\mathbb{R}^3$,
$$\mathbb{H}^3 = \{(x_1, x_2, x_3) \in \mathbb{R}^3 : x_3 > 0\}.$$

Other than the identity, there are three types of orientation preserving isometries of $\mathbb{H}^3$.

- *Parabolic*: f is conjugate to the translation $z \to z + 1$.
- *Elliptic*: f is conjugate to the rotation $z \to \lambda z$, $|\lambda| = 1$.
- *Loxodromic*: f is conjugate to the dilation $z \to \lambda z$, $|\lambda| \neq 1$.

Loxodromic and elliptic transformations have two fixed points on the Riemann sphere $\overline{\mathbb{C}} = \partial \mathbb{H}^3$. For such a transformation g, we can therefore define the *axis* of g to be the closed hyperbolic line joining these two fixed points. We denote the axis of g by $ax(g)$. If g is elliptic, then every point on $ax(g)$ is fixed by g. If g is loxodromic, then $ax(g)$ is only setwise fixed. A discrete subgroup of $Isom^+(\mathbb{H}^3)$ which stabilizes a point of $\mathbb{H}^3$ is easily seen to be finite and therefore elementary, [**2**].

For a Kleinian group $\Gamma = \langle g_1, g_2, g_3, \ldots \rangle$, we define the *Margulis constant* for Γ to be
$$m_\Gamma = \inf_{x \in \mathbb{H}^3} \sup_{i \geq 1} \{\rho(g_i(x), x)\}. \tag{2}$$

Hence the Margulis constant for a group Γ depends on the particular choice of generators for Γ. The Margulis constant for Kleinian groups is
$$c_M = \inf_\Gamma m_\Gamma, \tag{3}$$
where the infimum is over all Kleinian groups Γ.

The following is an immediate consequence of the above definitions.

LEMMA 1.1. *If* $\Gamma = \langle g_1, g_2, g_3, \ldots \rangle$ *and* $\Gamma' = \langle h_1, h_2, h_3, \ldots \rangle$ *where for all* $i \geq 1$, $h_i \in \{g_1, g_2, g_3, \ldots\}$, *then*
$$m_{\Gamma'} \leq m_\Gamma. \tag{4}$$

Thus, in order to bound the Margulis constant for a given Kleinian group, we may choose a suitable subcollection of generators and estimate the Margulis constant for the Kleinian group they generate. In particular, if there are two or three generators that generate a nonelementary subgroup, then the Margulis constant will be bounded below by the constant for such subgroups.

We recall next the following useful classification theorems for the elementary subgroups of a Kleinian group [**2, 14**].

THEOREM 1.1. *Let Γ be a Kleinian group and F a finite subgroup. Then F is isomorhic to one of the following groups: C_p, the cyclic group of order $p \geq 2$; D_p, the dihedral group of order $2p \geq 4$; A_4, the tetrahedral group; S_4, the octahedral group; or A_5, the icosahedral group.*

We then have the following well-known corollary.

COROLLARY 1.1. *Let $\Gamma = \langle f, g \rangle$ be a discrete group with $ax(f) \neq ax(g)$.*
- *If f is loxodromic and g is not of order 2, then G is nonelementary.*
- *If f and g are elliptic not both of order two and if $ax(f) \cap ax(g) = \emptyset$, then Γ is nonelementary.*

- If f is parabolic and if g has order 5 or order at least 7 or does not share the fixed point of f, then Γ is nonelementary.

The following theorem determines the three generator groups we must study.

THEOREM 1.2. *Let $\Gamma = \langle g_1, g_2, g_3, \ldots \rangle$ be a Kleinian group. Then there are indices $i_1 < i_2$ such that $\Gamma' = \langle g_{i_1}, g_{i_2} \rangle$ is a Kleinian group or there are indices $i_1 < i_2 < i_3$ such that $\Gamma' = \langle g_{i_1}, g_{i_2}, g_{i_3} \rangle$ is a Kleinian group.*

The case of three generators is necessary only if $g_{i_1}, g_{i_2}, g_{i_3}$ are elliptic elements of finite order $2 \leq p \leq q \leq r \leq 6$ and (p,q,r) is one of the following triples:

$(2,2,k)$, $2 \leq k \leq 6$, $\quad (2,3,k)$, $3 \leq k \leq 6$, $\quad (2,k,k)$, $4 \leq k \leq 6$,

$(3,3,k)$, $3 \leq k \leq 6$, $\quad (3,k,k)$, $4 \leq k \leq 6$, $\quad (k,k,k)$, $4 \leq k \leq 6$.

In all but the case $(2,2,k)$, the elliptic axes are coplanar and bound a hyperbolic triangle.

PROOF. It is not difficult to establish the reduction to the two or three generator case. See [9] for the details.

If three generators are in fact necessary, then each pair of them must generate an elementary group. It is clear from the classification of the elementary discrete groups that no g_{i_j} can be parabolic, loxodromic or elliptic of order at least 7.

If two elliptic elements generate an elementary group, Corollary 1.1 implies their axes must meet unless they both have order 2. Suppose that $q \geq 3$. Then the axes of the generators must pairwise intersect, possibly on the Riemann sphere; hence they are coplanar and bound a hyperbolic triangle. Each vertex of this triangle is stabilized by an elementary group. Using the classification of the elementary discrete groups once more, we obtain the stated restrictions on (p,q,r). □

The labeling (p,q,r) and the hyperbolic triangle obtained when the axes are coplanar determine the Kleinian group completely. We thus want metric restrictions on this hyperbolic triangle. However, first we show that the groups we are interested in embed in reflection groups.

2. The associated reflection group

The following result is of considerable help in deciding when a possible triple (p,q,r) and the associated hyperbolic triangle arise from a discrete group.

THEOREM 2.1. *Let f,g,h be elliptic Möbius transformations whose axes are coplanar. Then there are reflections σ_i in hyperplanes Π_i, $i = 0,1,2,3$, such that the following hold.*

(1) *$\langle f,g,h \rangle$ is a subgroup of index 2 in $\langle \sigma_0, \sigma_1, \sigma_2, \sigma_3 \rangle$; hence both groups are simultaneously discrete or non–discrete.*

(2) *Let $\theta_{i,j}$ denote the dihedral angle between Π_i and Π_j, $0 \leq i \neq j \leq 3$ and let $\theta_{i,j} = 0$ if they do not meet. Then $\theta_{0,1} = \pi/p$, $\theta_{0,2} = \pi/q$ and $\theta_{0,3} = \pi/r$. If*

$$\theta_{1,2} + \theta_{1,3} + \theta_{2,3} \geq \pi,$$

then the hyperplanes Π_i bound a hyperbolic tetrahedron.

PROOF. We may assume that f,g,h are primitive elliptic elements of order p,q,r, respectively. Let Π_0 be the plane in which the axes of f,g,h lie and let σ_0 be reflection in Π_0. Π_0 separates $\mathbb{H}^3$ into two components. Fix one such component

U. Let Π_1 be the plane having dihedral angle π/p with Π_0, chosen so that the acute angle is formed in U and $\Pi_0 \cap \Pi_1$ is the axis of f. Do the same for g and h. Then $\sigma_0\sigma_1 = f$ or f^{-1}; in the latter case, we replace f with f^{-1}. We again do the same for $\sigma_0\sigma_2 = g$ and $\sigma_0\sigma_3 = h$. Next we see that $\sigma_1\sigma_2 = \sigma_1\sigma_0\sigma_0\sigma_2 = fg^{-1}$. Arguing similarly, we conclude that $\langle f, g, h \rangle$ has index 2 in the reflection group; thus the first part of the theorem is proved. The second part is a well-known fact from hyperbolic geometry [16]. $\square$

The Poincaré theorem [14] can be used to obtain a presentation for the reflection group and then for the subgroup $\langle f, g, h \rangle$ when it is discrete. We shall see that the group is discrete if and only if the dihedral angles of intersection of the hyperplanes are submultiples of π together with a number of exceptions, in fact infinitely many, but all of a similar pattern. From this, we obtain a presentation of the reflection group:

$$\begin{aligned}\langle \sigma_0, \sigma_1, \sigma_2, \sigma_3, \sigma_4 : \sigma_i^2 &= (\sigma_0\sigma_1)^p = (\sigma_0\sigma_2)^q = (\sigma_0\sigma_3)^r \\ &= (\sigma_1\sigma_3)^\ell = (\sigma_1\sigma_3)^m = (\sigma_1\sigma_2)^n = 1 \rangle.\end{aligned}$$

The associated (p, q, r)–Kleinian group is the index two subgroup generated by the products and has presentation

(5) $$\langle f, g, h : f^p = g^q = h^r = (fh)^\ell = (gh)^m = (fg)^n = 1 \rangle.$$

3. Angles

In this section, we recall a few facts about spherical and euclidean triangle groups that we need. We simply tabulate the results. The verification of these tables is a rather lengthy calculation in spherical trigonometry.

LEMMA 3.1. *Let $\theta_{2,3}$ and $\theta_{3,3}$ denote, respectively, the angles subtended at the origin between the axes of order 2 and 3 and the axes of order 3 of a spherical $(2, 3, 3)$–triangle group. Then*

(6) $$\cos(\theta_{2,3}) = 1/\sqrt{3}, \qquad \cos(\theta_{3,3}) = 1/3.$$

LEMMA 3.2. *Let $\phi_{2,3}$, $\phi_{2,4}$ and $\phi_{3,4}$ denote, respectively, the angles subtended at the origin between the axes of order 2 and 3, the axes of order 2 and 4 and the axes of order 3 and 4 of a spherical $(2, 3, 4)$–triangle. Then*

(7) $$\cos(\phi_{2,3}) = \sqrt{2/3}, \quad \cos(\phi_{2,4}) = 1/\sqrt{2}, \quad \cos(\phi_{3,4}) = 1/\sqrt{3}.$$

LEMMA 3.3. *Let $\psi_{2,3}$, $\psi_{2,5}$ and $\psi_{3,5}$ denote, respectively, the angles subtended at the origin between the axes of order 2 and 3, the axes of order 2 and 5 and the axes of order 3 and 5 of a spherical $(2, 3, 5)$–triangle. Then*

(8) $$\begin{cases}\cos(\psi(2,3)) = 2\cos(\pi/5)/\sqrt{3}, \\ \cos(\psi(2,5)) = 1/2\sin(\pi/5), \\ \cos(\psi(3,5)) = \cot(\pi/5)/\sqrt{3}.\end{cases}$$

From these elementary observations and some spherical trigonometry, we obtain the following tables, which list the possible angles, up to four decimal places, at which elliptic axes in a Kleinian group can meet. We also include the dihedral angle ψ opposite this angle in the spherical triangle formed when the adjacent vertices are π/p and π/q.

Table 1 - Angles: 2, m axes

m	$\sin(\theta)$	θ	ψ	Group
3	$\sqrt{\frac{2}{3}}$	.9553	$\pi/3$	A_4
3	$-\sqrt{\frac{2}{3}}$	2.1462	$2\pi/3$	A_4
3	$\sqrt{\frac{1}{3}}$	.6154	$\pi/4$	S_4
3	$-\sqrt{\frac{1}{3}}$	2.5261	$3\pi/4$	S_4
3	$\sqrt{\frac{3-\sqrt{5}}{6}}$	.3648	$\pi/5$	A_5
3	$-\sqrt{\frac{3-\sqrt{5}}{6}}$	2.7767	$4\pi/5$	A_5
3	$\sqrt{\frac{3+\sqrt{5}}{6}}$	1.2059	$2\pi/5$	A_5
3	$-\sqrt{\frac{3+\sqrt{5}}{6}}$	1.9356	$3\pi/5$	A_5
3	1	$\pi/2$	$\pi/2$	D_3
4	$\frac{1}{\sqrt{2}}$	$\pi/4$	$\pi/3$	S_4
4	$\frac{1}{\sqrt{2}}$	$3\pi/4$	$2\pi/3$	S_4
4	1	$\pi/2$	$\pi/2$	D_4
5	$\sqrt{\frac{5-\sqrt{5}}{10}}$	.5535	$\pi/3$	A_5
5	$-\sqrt{\frac{5-\sqrt{5}}{10}}$	2.5880	$2\pi/3$	A_5
5	$\sqrt{\frac{5+\sqrt{5}}{10}}$	1.0172	$2\pi/5$	A_5
5	$-\sqrt{\frac{5+\sqrt{5}}{10}}$	2.1243	$3\pi/5$	A_5
5	1	$\pi/2$	$\pi/2$	D_5

Table 2 - Angles: 3, m axes

m	$\sin(\theta)$	θ	ψ	Group
3	$\frac{2}{3}$	.7297	$2\pi/5$	A_5
3	$-\frac{2}{3}$	2.4118	$3\pi/5$	A_5
3	$\frac{2\sqrt{2}}{3}$	1.2309	$\pi/2$	A_4
3	$-\frac{2\sqrt{2}}{3}$	1.9106	$2\pi/3$	A_4
4	$\sqrt{\frac{2}{3}}$	.9553	$\pi/2$	S_4
4	$-\sqrt{\frac{2}{3}}$	2.1462	$3\pi/4$	S_4
5	$\sqrt{\frac{10-2\sqrt{5}}{15}}$	.6523	$\pi/2$	A_5
5	$-\sqrt{\frac{10-2\sqrt{5}}{15}}$	2.4892	$4\pi/5$	A_5
5	$\sqrt{\frac{10+2\sqrt{5}}{15}}$	1.3820	$3\pi/5$	A_5
5	$-\sqrt{\frac{10+2\sqrt{5}}{15}}$	1.7595	$2\pi/3$	A_5

We include, in addition, the following remarks.

(1) The angle between intersecting axes of ellipitcs of order 4 in a discrete group is always either 0 when they meet on the Riemann sphere or $\pi/2$. In the first case, the dihedral angle ψ as above is $\pi/2$, while in the second case it is $2\pi/3$.

(2) The angle between intersecting axes of elliptics of order 5 in a discrete group is either $\arcsin(2/\sqrt{5}) = 1.107149\ldots$ or its complement $\arcsin(-2/\sqrt{5}) = 2.03444\ldots$. In the former case, the dihedral angle ψ is $2\pi/3$. In the latter case, it is $4\pi/5$.

(3) The axes of elliptics both of order 2 can intersect at an angle $k\pi/m$ for any k and $n \geq 2$. In this case, the dihedral angle ψ is $k\pi/m$.

(4) The axes of elliptics of order p and q, $p \leq q$, in a discrete group meet on the sphere at infinity, i.e., meeting with angle 0, if and only if

$$(p,q) \in \{(2,2),(2,3),(2,4),(2,6),(3,3),(3,6),(4,4),(6,6)\}.$$

In each case, the dihedral angle ψ is $\pi - \pi/p - \pi/q$.

4. Admissible Triangles and Discreteness

DEFINITION. We say that a hyperbolic triangle in $\mathbb{H}^3$ is an *admissible (p,q,r)-triangle* if the triangle is formed from the axes of elliptic elements f, g, h of order p, q, r respectively and if $\langle f, g, h \rangle$ is a discrete group.

We want to see what sort of admissible hyperbolic triangles our axes can bound. There are two obvious necessary conditions. The first is that the sum of the (interior) angles of intersection of the axes is smaller than π, for we have a hyperbolic triangle of positive area. The second is that all the interior angles must be chosen from those of our list or in the subsequent remarks.

These two conditions readily reduce our list of possibilities to manageable proportions. However, we seek to identify all such groups, and so we must decide when a candidate generates a discrete group.

The third condition we apply concerns the sum of the dihedral angles at each vertex (opposite the angle of intersection) formed in the associated reflection group. Let Σ be the sum of dihedral angles opposite each vertex. If $\Sigma \geq \pi$, then the three hyperplanes (those other than the hyperplane containing our triangle) of the associated reflection group meet at a finite point and, together with the hyperplane containing our triangle, bound a finite volume hyperbolic tetrahedron. Our (p, q, r)–group is of index 2 (if either group is discrete). Notice that there are more reflection groups than those appearing in the standard lists, since we do not require that the dihedral angles be submultiples of π. We have tabulated the reflection groups appearing in §12.

If $\Sigma < \pi$, then the three planes subtend a hyperbolic triangle on the sphere at infinity and therefore give rise to a reflection group in the sides of a hyperbolic triangle. This group must be discrete, and this requirement places restrictions on the angles. Of course, if all angles are submultiples of π, then the group is certainly discrete. However, it is shown by Knapp [11] that in a discrete reflection group in the sides of a hyperbolic triangle, if one angle is of the form $2\pi/p$, p odd, then the other two angles are of the form $\pi/n, \pi/n$ or $\pi/2, \pi/p$ (with a further exception for the $(2, 3, 7)$–triangle group which will not concern us). If the angle is of the form $3k\pi/p$, then the other two angles are $\pi/3, \pi/p$. Finally, if an angle is of the form $4\pi/p$, then the other two angles are $\pi/p, \pi/p$. No other rational multiples of π occur. Knapp's theorem is more or less restated in §4.4 below concerning admissible $(2, 2, 2)$–triangles. This theorem gives us a powerful tool for limiting the possible dihedral angles.

There are two basic cases to consider. The first case occurs when all the dihedral angles are submultiples of π and the second when one or more such angles is not. In the first case, the group is discrete by the Poincaré theorem [14]. Thus we obtain necessary and sufficient conditions for discreteness. It is a fairly routine matter to run through the possible angles and find all the discrete groups. These are tabulated in §11. We have indicated the Margulis constant for these groups and also the associated tetrahedral reflection group when the group has finite covolume. From the tables of covolumes of these reflection groups, one can obtain the covolume of our (p, q, r)–Kleinian group by doubling this value. We have also indicated the structure of the stabilizers of vertices of our hyperbolic triangle. These are important for our application in [6]. When the group is a web group, we further indicate the triangle group which is a component stabilizer.

From the tables we deduce the following theorem.

THEOREM 4.1. *Let Γ be a (p, q, r)–Kleinian group, with $3 \leq p \leq q \leq r$. Then Γ is the index two orientation preserving subgroup of the group of reflections in a hyperbolic tetrahedron.*

Note that there are nonclassical reflection groups here.

4.1. Dihedral angles not submultiples of π. In this case, we proceed as above to identify the triangles involved. We can use Knapp's theorem to eliminate a number of cases as well as the simple observation that if the dihedral sum $\Sigma \geq \pi$, then the stabilizer of this vertex must be one of the elementary groups. Thus, for instance, there can be no angles of the form $m\pi/4$ and $n\pi/5$ which meet.

If we have a candidate satisfying all these conditions, we must still decide if it is discrete. We do this in the following way. Let Γ be a (p,q,r)-Kleinian group and suppose the dihedral angle between two hyperplanes Π_1 and Π_2 at some edge, say $p-q$, with angle θ_{pq}, is of the form $k\pi/m$ (usually $k=2$). Let σ_i denote the reflection in the hyperplane Π_i, $i=1,2$. Then in $\langle \sigma_1, \sigma_2 \rangle$ we find reflections in hyperplanes meeting Π_1 and Π_2 at all the dihedral angles $m\pi/n$, $m=1,2,\ldots,n$. At least one of these hyperplanes intersects the interior of our triangle. This intersection is the axis of an elliptic element of some order, say s; and thus we have found two new groups of the form (p,r,s) and (q,r,s), each of which must be discrete. We do not have $s=2$ unless the tetrahedron is symmetric. Observe that this process provides k such groups, all having angle of intersection at the edge $p-q$ of the form θ_{pq}/k. The hyperbolic area of the new triangle has decreased, so this process of subdivision must stop if the group is discrete and at one of the groups whose dihedral angles are all submultiples of π. In practice, we find the subdivision process stops at the first stage and the group is either discrete as an index two subgroup of a group with dihedral angles all submultiples of π, or the angles of intersection of the axes s and r are not permissible in a discrete group. Sometimes we must subdivide 3 times. This subdivision exhibits some interesting inclusions among the (p,q,r)-groups. The resulting discrete (p,q,r)-Kleinian groups are tabulated in §11.

Incidentally, this process identifies for us all discrete reflection groups in the faces of a hyperbolic tetrahedron such that at one face all the dihedral angles are submultiples of π. These appear in §13.

We say a word here about one interesting case. The ideas here are similar to those used in other cases as well.

Let Γ be a group associated with a $(2,2,p)$-triangle, $p \geq 3$, and suppose the dihedral angle at the $2-p$ vertices is twice a submultiple of π. If we split the angle as above, we find a group with three coplanar axes, one of which has order p, and all meeting at one point. The axis splitting the initial angle cannot have order 2, for this would conjugate an order 2 axis onto an order p axis.

To see that this leads to a contradiction, suppose first $p=3$. Then we would have two order 6 axes meeting necessarily at infinity; and so the original dihedral angle between the 2-3 vertices would be $\pi/6$, a submultiple. The same thing would happen if $p=4$, while $p=5$ or 6 is clearly not possible. Therefore, the splitting axis has order $3 \leq q \leq 6$. A quick look through our tables of possible angle intersection shows few cases where a $p-q$ intersection angle is half a $2-p$ intersection angle. There are none when $p=3$ other than 0 which gives a submultiple at intersection. The same is true when $p=4,5$ and 6.

Next, if the dihedral angle at a $2-p$ vertex is a triple or quadruple submultiple of π, we can split the angle and reduce to the previous case by looking at a triple of succesive axes.

Thus, only the case that the angle of intersection at the $2-3$ vertex is a non–submultiple of π is of interest. This angle must be of the form $k\pi/n$, $k \geq 2$. We can split this angle to find $(k-1)$ axes of order 2 lying in our triangle and meeting the p-axis and the $2-3$ vertex. From the first part of our argument, none of the angles of intersection between one of these $(k-1)$ axes can give rise to a dihedral angle which is not a submultiple of π. This applies to the complementary angle of intersection as well. Therefore, the angle of intersection has the property that both it and its complement yield dihedral angles which are submultiples of π. The only angle with this property is $\pi/2$. Thus $k = 2$. Hence we have found two copies of a $(2, 2, p)$–triangle, whose interior angles are π/n, $\pi/2$, θ, where θ is a $2 - p$ angle with a corresponding dihedral angle which is a submultiple of π. Of course, these are the discrete $(2, 2, p)$ groups with all dihedral angles submultiples of π. This is how we obtain our list.

4.2. Admissible $(2, 2, 2)$-triangles. A $(2, 2, 2)$–Kleinian group when restricted to the hyperbolic plane containing the axes is actually a reflection group in the sides of a hyperbolic triangle and thus is an index 2 supergroup of a Fuchsian triangle group. The results of Knapp [11] can then be used to determine all possible triples of angles. Each such angle is a rational multiple of π and, in fact, with few exceptions a submultiple of π.

THEOREM 4.2. *Let Γ be a $(2,2,2)$–Kleinian group. Then the angles of intersection of the axes are of the following form:*

(1) π/ℓ, π/m, π/n with $\frac{1}{\ell} + \frac{1}{m} + \frac{1}{n} < 1$;
(2) $2\pi/\ell$, π/m, π/m with $\frac{1}{\ell} + \frac{1}{m} < \frac{1}{2}$;
(3) $2\pi/\ell$, $\pi/2$, π/ℓ with $\ell \geq 7$;
(4) $3\pi/\ell$, $\pi/3$, π/ℓ with $\ell \geq 7$;
(5) $4\pi/\ell$, π/ℓ, π/ℓ with $\ell \geq 7$;
(6) $2\pi/7$, $\pi/3$, $\pi/7$.

The reader can verify that in a group generated by three coplanar axes of order 2, the Margulis constant is twice the radius of the maximal inscribed disk of the triangle. The following formula for the radius r of the inscribed disk in a hyperbolic triangle with angles α, β, γ follows from Theorem 7.14.2 in [2]:

$$\tanh^2(r) = \frac{\cos^2 \alpha + \cos^2 \beta + \cos^2 \gamma + 2\cos\alpha\cos\beta\cos\gamma - 1}{2(1 + \cos\alpha)(1 + \cos\beta)(1 + \cos\gamma)}. \tag{9}$$

The reader can then fill in the details for the proof of the following result. Note that there is no assumption that the axes actually form a triangle in the plane they span. Once the problem is reduced to this case, it is relatively straightforward.

THEOREM 4.3. *Let Γ be a $(2,2,2)$–Kleinian group. Then*

$$m_\Gamma \geq 2\tanh^{-1}\sqrt{\frac{\cos^2 \pi/3 + \cos^2 \pi/7 - 1}{2(1 + \cos \pi/3)(1 + \cos \pi/7)}} = 0.2088\ldots \tag{10}$$

This bound is sharp and uniquely achieved by the reflection group of the $(2, 3, 7)$ triangle.

SKETCH OF PROOF. Let $\Gamma = \langle f, g, h \rangle$ where each element is of order 2. We consider the hyperbolic plane spanned by these axes. There is a unique disk tangent to each of the three axes. The diameter of this disk is the Margulis constant for Γ. We may suppose that the radius of this disk is less than $0.1044\ldots$ and that the disk is centered at the origin. The angle at the origin subtended by each of these axes exceeds $2\pi/3$, so a pair of them meet, say those of f and g. This point of intersection is fixed by the elliptic fg. We replace g by a conjugate (also denoted g) of the form $(fg)^n g(fg)^{-n}$ so that the axes of f and g meet at a smallest possible submultiple of π, say π/p. Consider the new group $\Gamma_1 = \langle f, g, h \rangle$ generated by three elliptics of order 2. It is easy to see that the Margulis constant for Γ_1 is smaller than that of Γ (with the correct choice of conjugate), as the radius of the inscribed disk must have decreased.

Move the point of intersection of the axes of f and g to the origin. The axis of h is tangent to our (new) inscribed disk of radius at most 0.1044, thus the distance from the origin to the axis of h is at most

$$d = \operatorname{arcsinh}\left(\frac{\sinh(0.1044\ldots)}{\sin \pi/2p}\right) + 0.1044\ldots$$

Then the angle subtended by the axis of h is at least α with

$$\sin \alpha/2 = 1/\cosh(d) > \sin \pi/2p$$

for $p \geq 2$. So the axis of h meets either f or g and as before we may assume this angle of intersection is a smallest possible submultiple of π, say π/q, and moreover that $1/p + 1/q < 1/2$ (otherwise Γ was elementary).

We are now in the situation that one axis meets the other two at submultiples of π. If these axes do not already bound a triangle of finite area (and therefore all angles are submultiples of π by Theorem 4.2), then we can push a pair of axes together keeping the angles π/p and π/q fixed until they become tangent. As we do this, the groups remain discrete and the Margulis constant goes down. The remainder of the proof consists in verifying that the $(2,3,7)$ triangle has the smallest inscribed radius. See [2]. □

5. The Margulis constant for groups generated by an admissible triangle

In this section, we show how the Margulis constant for each group in the following tables is calculated.

We say that a group Γ is generated by an admissible triangle if $\Gamma = \langle f_1, f_2, f_3 \rangle$, f_i are all elliptic of order less than 6 and the axes α_i of f_i form an admissible triangle. In view of the results above, we want to bound the Margulis constant for the groups generated by an admissible triangle. The Margulis constant for such a group $\Gamma = \langle f_1, f_2, f_3 \rangle$ is simply

$$m_\Gamma = \inf\{\rho(x, f_i(x)) : x \in \mathbb{H}^3, i = 1, 2, 3\},$$

where ρ is the hyperbolic metric of $\mathbb{H}^3$ of constant curvature equal to -1. A moment's thought should convince the reader that the infimum is actually a minimum and that (by convexity) the minimum is attained at a point x_0 lying in the plane spanned by the three axes interior to the region bounded by the triangle and such that

(11) $$\rho(x_0, f_1(x_0)) = \rho(x_0, f_2(x_0)) = \rho(x_0, f_3(x_0)).$$

Since the plane spanned by the three axes is totally geodesic, we may use the metric geometry in that plane. Let x_0 be the point in question and let

$$d_i = \rho(x_0, ax(f_i)).$$

We also abuse notation and identify d_i with the geodesic segment from x_0 perpendicular to $ax(f_i)$ (of length d_i). We denote the vertex opposite the axis of f_i by v_i.

We suppose for the moment that all the angles are positive. We require the result when one, two or possibly all three angles are zero, so we discuss the asymptotics later.

One can calculate the edge lengths ℓ_i (of the edge opposite v_i) via the formula

$$\cosh(\ell_2) = \frac{\cos(\theta_1)\cos(\theta_3) + \cos(\theta_2)}{\sin(\theta_1)\sin(\theta_3)} \tag{12}$$

and its obvious symmetries; see §7.12 in [**2**]. Let f_i be of order n_i. We may suppose that f_i is primitive. Then for each i, we calculate that

$$m_\Gamma = \rho(x_0, f_i(x_0)) = 2\operatorname{arcsinh}\left(\sin(\pi/n_i)\sinh(d_i)\right). \tag{13}$$

Next, let r be the geodesic segment from x_0 to v_1 (we continue to abuse notation and denote its length by r as well). This segment bisects the angle θ_1 into two angles $\theta_{1,2}$ from r to α_2 and $\theta_{1,3}$ from r to α_3. From the law of hyperbolic sines, we find

$$\sinh(r) = \frac{\sinh(d_2)}{\sin(\theta_{1,2})} = \frac{\sinh(d_3)}{\sin(\theta_{1,3})}. \tag{14}$$

Since

$$\theta_{1,2} + \theta_{1,3} = \theta_1$$

we obtain

$$(\sin(\theta_1)\cot(\theta_{1,2}) - \cos(\theta_1))\sinh(d_2) = \sinh(d_3). \tag{15}$$

Then, substituting

$$\sinh(d_i) = \frac{\sinh(m_\Gamma/2)}{\sin(\pi/n_i)}, \tag{16}$$

we conclude that

$$\sin(\pi/n_2) = \sin(\pi/n_3)(\sin(\theta_1)\cot(\theta_{1,2}) - \cos(\theta_1)) \tag{17}$$

or, equivalently,

$$\cot(\theta_{1,2}) = \frac{\sin(\pi/n_2)/\sin(\pi/n_3) + \cos(\theta_1)}{\sin(\theta_1)}. \tag{18}$$

By symmetry,

$$\cot(\theta_{3,2}) = \frac{\sin(\pi/n_2)/\sin(\pi/n_1) + \cos(\theta_3)}{\sin(\theta_3)}. \tag{19}$$

The two angles $\theta_{1,2}$, $\theta_{3,2}$ and the length ℓ_2 are enough to determine d_2 and then to determine m_Γ. This is done as follows.

Consider the triangle determined by x_0 and ℓ_2. Let s and t denote the other two edges, the angle between s and ℓ_2 being $\theta_{1,2}$ and the angle between t and ℓ_2 being $\theta_{3,2}$. Let η denote the angle between s and t. Then, since d_2 is perpendicular to ℓ_2, we have $\sinh(t)\sin(\theta_{3,2}) = \sinh(d_2)$ and $\sinh(t)\sin(\eta) = \sinh(\ell_2)\sin(\theta_{1,2})$. Thus

$$\sinh(d_2) = \frac{\sin(\theta_{1,2})\sin(\theta_{3,2})\sinh(\ell_2)}{\sin(\eta)}. \tag{20}$$

We also have from above

$$\cos(\eta) = \sin(\theta_{1,2})\sin(\theta_{3,2})\cosh(\ell_2) - \cos(\theta_{1,2})\cos(\theta_{3,2}). \tag{21}$$

Putting this together and rearranging terms, we find $\sinh^2\left(\frac{m_\Gamma}{2}\right)$ is equal to

$$\frac{\sin^2(\pi/n_2)(\cosh^2(\ell_2) - 1)}{1 - \cosh^2(\ell_2) + \cot^2(\theta_{1,2}) + \cot^2(\theta_{3,2}) + 2\cot(\theta_{1,2})\cot(\theta_{3,2})\cosh(\ell_2)}, \tag{22}$$

which gives m_Γ in terms of the three angles in view of our formulas above.

It is clear that the above formula remains valid if $\theta_2 = 0$, since $\cot(\theta_{1,2})$ and $\cot(\theta_{1,3})$ are not defined in terms of θ_2 and $\cosh(\ell_2)$ is finite. We first consider what happens as $\theta_1 \to 0$. To do this, we can clear the singular terms by multiplying top and bottom by $\sin^2(\theta_1)\sin^2(\theta_3)$ and setting $\theta_2 = 0$. After doing this, we find the numerator of (22) has the form

(23) $$N = \sin^2(\pi/n_2) N_1,$$

where

(24) $$N_1 = (\cos(\theta_1) + \cos(\theta_3))^2,$$

while the denominator of (22) is given by $D = -N_1 + D_1 - D_2$, where

(25) $$D_1 = \left(\frac{\sin(\pi/n_2)}{\sin(\pi/n_3)} + \frac{\sin(\pi/n_2)}{\sin(\pi/n_1)} + \cos(\theta_1) + \cos(\theta_3)\right)^2$$

and

(26) $$D_2 = \left(\frac{\sin(\pi/n_2)}{\sin(\pi/n_3)}\cos(\theta_1) - \frac{\sin(\pi/n_2)}{\sin(\pi/n_1)}\cos(\theta_3)\right)^2.$$

We can now put $\theta_1 = 0$ and find that when there are two infinite vertices, the Margulis constant can be calculated from the ratio N/D, where

$$N = \sin^2(\pi/n_2)(1 + \cos(\theta_3))^2$$

and

$$D = -(1 + \cos(\theta_3))^2 + \sin^2(\theta_3)\left(\frac{\sin(\pi/n_2)}{\sin(\pi/n_3)} + 1\right)^2$$
$$+ \ 2(1 + \cos(\theta_3))\left(\frac{\sin(\pi/n_2)}{\sin(\pi/n_3)} + 1\right)\left(\frac{\sin(\pi/n_2)}{\sin(\pi/n_1)} + \cos(\theta_3)\right).$$

Then

(27) $$\sinh^2\left(\frac{m_\Gamma}{2}\right) = \frac{N}{D},$$

while if all the vertices are infinite, we obtain the symmetric formula

(28) $$\sinh^2\left(\frac{m_\Gamma}{2}\right) = \frac{\sin(\pi/n_1)\sin(\pi/n_2)\sin(\pi/n_3)}{\sin(\pi/n_1) + \sin(\pi/n_2) + \sin(\pi/n_3)}.$$

6. Margulis constant for the $(2,2,n)$ case

In this section, we deal with the case of two elliptic axes of order 2 meeting a third axis of order $n \geq 3$ not only in the case of a $(2,2,n)$–Kleinian group, but also in the case that the three axes may not be coplanar. In this latter case, we are unfortunately unable to give explicit computations of the constant. Instead, we just give a lower bound which is sharp in some instances.

Thus let α be the axis of order n and β_1, β_2 the two axes of order two. Then β_i meets α at vertex v_i with angle θ_i. It is possible that β_1 does not meet β_2 at all. We use a discreteness criteria from our earlier work concerning the distance between the fixed points of spherical triangle subgroups of a discrete group. Such a bound gives a number ρ between v_1 and v_2. Now we ignore discreteness criteria to move the situation to a geometrically extreme situation. It is clear that the Margulis constant will go down if we swing the two order 2 axes into the same plane. (To see this, consider the plane containing α and the point where the Margulis constant is attained. The constant goes down as we move the β_i closer to this plane). Thus we can assume that α, β_1 and β_2 are coplanar. It is clear that the point attaining the Margulis constant in this case lies in the triangular region bounded by these three lines (the region itself need not be bounded, or even of finite area). If neither θ_1 nor θ_2 is $\pi/2$, we make an additional simplification. We replace β_2 by the hyperbolic line in the plane containing the three axes which is perpendicular to α and bisects v_1 and v_2. We relabel this line as β_2 and declare it to be an elliptic axis of order 2 and set $\theta = \theta_1$. The point here is that we may not have arrived at a configuration which is possible for a discrete group, but we can calculate the "Margulis constant" giving a bound for the case in the initial configuration. Actually the extremal configurations arise in similar situations as above, so our estimates are not too bad. The bound on ρ is the minimal distance between the various type of spherical triangle group points on a common axis of order n as identified in [6]. After our reduction, the distance between the vertices is at least $\rho/2$. This enables us to calculate the angle between the two order 2 axes if they meet. This gives us the angle ψ since

(29) $$\cos(\psi) = \sin(\theta_1)\cosh(\rho/2).$$

If the two axes do not meet, we move them closer in the plane until the angle ψ is zero. We can then use our earlier formulas to obtain a lower bound for the Margulis constant.

Table 3 - Margulis Constants for (2,2,n)–nontriangles

Order	Vertex	ρ	θ	ψ	Constant
3	A_4	0.6931	0.95531	$\pi/6$	.230989
3	A_5	1.736	0.36486	$\pi/3$	.276814
3	S_4	1.059	0.61547	.8496	.23167
3	A_5	1.736	1.20593	0	.302572
4	S_4	1.059	$\pi/4$	.62894	.28693
5	A_5	1.382	.55357	.85474	.26088
5	A_5	1.382	1.01772	0	.420531
6	$\Delta_{(2,3,6)}$	∞	0	0	.594241

7. Three elliptics of order 2

In this section, we show that a Kleinian group generated by two loxodromic transformations can be identified as a subgroup of index at most two in a Kleinian group generated by three elliptics of order 2 and that this supergroup has a smaller Margulis constant.

THEOREM 7.1. *Let $\Gamma = \langle f, g \rangle$ be a Kleinian group generated by two loxodromic transformations. Then there are three elliptic transformations ϕ_1, ϕ_2, ϕ_3, each of order two, generating a Kleinian group $\Gamma' = \langle \phi_1, \phi_2, \phi_3 \rangle$ which contains Γ with index at most two and*

(30) $$m'_\Gamma < m_\Gamma.$$

PROOF. Let $\alpha = ax(g)$ and $\beta = ax(g)$, and let γ be the hyperbolic line perpendicular to α and β. Such a line γ exists because Γ is discrete, so $\alpha \cap \beta$ is either empty or a finite point of hyperbolic space. Let ϕ_1 be the half turn whose axis is γ. There is a half turn ϕ_2 whose axis is perpendicular to α and such that $\phi_2\phi_1 = f$. Similarly, there is ϕ_3 with $\phi_3\phi_1 = g$. It is easy to see that $\langle \phi_1, \phi_2, \phi_3 \rangle$ contains Γ with index at most two, and therefore Γ' is Kleinian.

Next, the point $x_0 \in \mathbb{H}^3$ such that $m_\Gamma = \rho(x_0, f(x_0)) = \rho(x_0, g(x_0))$ lies on the line γ between the two axes. To see this, note that the set of all points y such that for a given constant δ we have $\rho(y, f(y)) = \delta$ forms a hyperbolic cylinder about α, a fact easily observed if one conjugates f so that its fixed points are $0, \infty \in \overline{\mathbb{C}}$. These cylinders bound geodesically convex regions, and the two cylinders $\{y : \rho(y, f(y)) = m_\Gamma\}$ and $\{y : \rho(y, g(y)) = m_\Gamma\}$ meet at a unique point of γ. Now as $x_0 \in \gamma$, we have

$$\rho(x_0, \phi_2(x_0)) = 0 \tag{31}$$
$$\rho(x_0, \phi_2(x_0)) = \rho(x_0, \phi_2\phi_1(x_0)) = \rho(x_0, f(x_0)) = m_\Gamma \tag{32}$$
$$\rho(x_0, \phi_3(x_0)) = \rho(x_0, \phi_3\phi_1(x_0)) = \rho(x_0, g(x_0)) = m_\Gamma, \tag{33}$$

so of course $m'_\Gamma \leq m_\Gamma$. That strict inequality holds follows since for a Kleinian group generated by three elliptics of order 2, the Margulis constant is the diameter of the smallest hyperbolic ball which is tangent to all three axes. □

In fact, Lemma 2.7 of [5] can be used to find the Margulis constant for any group generated by a pair of loxodromic transformations in terms of the traces of the elements and their commutators. However, the solution involves solving highly nonlinear equations and is only practical numerically. Inequalities for Kleinian groups, such as Jørgensen's inequality, can be used together with these formulae to give lower bounds for the Margulis constant in the general case. However, at present our methods are far from sharp. We shall return to this problem elsewhere.

Further note that, as limiting cases of the above, it is easily seen that any two generator Kleinian group lies with index at most two in a Kleinian group generated by three elliptics of order 2.

8. An Example

In this section, we give an example of a Kleinian group generated by three elliptic elements of order 2 with a Margulis constant which we conjecture to be smallest possible among all Kleinian groups. The group in question arises from an arithmetic Kleinian group (a $\mathbb{Z}_2$–extension of the $3-5-3$ Coxeter group, Γ_1 in our list). This group has the smallest co–volume among all arithmetic groups and is conjectured to be of minimal co–volume among all Kleinian groups. Unfortunately, the calculation is rather messy, involving the numerical solution of a number of simultaneous nonlinear equations. We give the example and leave the computational details to the reader.

Consider the hyperbolic tetrahedron with vertices A, B, C, D and dihedral angles at the edges as below:

- dihedral angle at $\overline{AB}$ is $\pi/3$,
- dihedral angle at $\overline{AC}$ is $\pi/2$,
- dihedral angle at $\overline{AD}$ is $\pi/2$,
- dihedral angle at $\overline{BC}$ is $\pi/5$,
- dihedral angle at $\overline{BD}$ is $\pi/2$,
- dihedral angle at $\overline{CD}$ is $\pi/3$.

Let ℓ be the hyperbolic line which is perpendicular to $\overline{AD}$ and $\overline{BC}$. We note first that the angle $\angle ADB = \pi/3$ and that ℓ and $\overline{AD}$ meet at right angles. Also, the angle between the plane spanned by AD and BD and the plane spanned by AD and ℓ is $\pi/4$.

The group we want is the group generated by the three half turns ϕ_1, ϕ_2, ϕ_3 whose axes are $\overline{AD}$, $\overline{BD}$ and ℓ, respectively. That this group is discrete is easily seen from the $\mathbb{Z}_2$ symmetry of the Coxeter tetrahedron about ℓ and the fact that the reflection group in the faces of this tetrahedron is discrete.

We want the radius of the unique ball tangent to the axes of ϕ_i, $i = 1, 2, 3$. To find this, we subdivide the tetrahedron about the center of this ball and use elementary hyperbolic trigonometry and symmetry to produce a number of different equations for r, which we then solve numerically. We obtain the following result.

THEOREM 8.1. *The Margulis constant for the group* $\Gamma = \langle \phi_1, \phi_2, \phi_3 \rangle$ *is*

$$(34) \qquad m_\Gamma = 0.1324\ldots$$

We conjecture this value to be the Margulis constant for Kleinian groups.

9. Summary

Here we briefly summarize what we know about the Margulis constant from the results of [**5**] and the results herein after one examines the values of the Margulis constant in our tables.

THEOREM 9.1. *Let* $\Gamma = \langle f_1, f_2, \ldots \rangle$ *be a Kleinian group.*
  (1) *If some f_i is parabolic, then $m_\Gamma \geq 0.189\ldots$.*
  (2) *If some f_i is elliptic of order $p \geq 3$ and the axis of f_i does not meet the axis of some f_j, then $m_\Gamma \geq 0.189\ldots$.*
  (3) *If $\langle f_i, f_j \rangle$ is elementary for all i, j and the f_i are not all of order 2, then $m_\Gamma \geq 0.2309\ldots$.*
  (4) *If there are 3 f_i's each of order two and with coplanar axes, then $m_\Gamma \geq 0.2088\ldots$.*

Each estimate is sharp.

Of course, one of the most interesting cases, namely that of a Kleinian group generated by a pair of loxodromic transformations, is not covered in this result. As we have noted, the Margulis constant in this situation is the radius of the smallest hyperbolic ball tangent to each of the axes. This leads to the following conjecture.

CONJECTURE. *Let f, g, h be elliptic elements each of order 2 whose axes are all tangent to a hyperbolic ball of radius $r < m_0 = 0.1324\ldots$. Then if $\langle f, g, h \rangle$ is nonelementary, it is not discrete.*

10. Distance between spherical fixed points

An important part of our program to identify the minimal volume hyperbolic 3–orbifold (equivalently the minimal covolume Kleinian group) concerns proving various universal geometric constraints on the geometry of the singular locus of such an object. This singular locus is a trivalent graph. The vertices of this graph arise from spherical triangle subgroups (including the dihedral groups) of the uniformizing Kleinian group. We relate the geometric complexity of this graph to the covolume of the associated Kleinian group by finding precisely invariant tubular neighbourhoods of vertices and edges; see [**6**, **8**]. The radius of a maximal embedded ball about a vertex of the singular graph is half the distance between a point in $\mathbb{H}^3$ fixed by the associated finite subgroup and its nearest translate. This nearest translate is also stabilized by a finite subgroup. Thus we are presented with the problem of determining how close different points stabilized by finite subgroups can be. Universal lower bounds for this distance provide covolume estimates in an obvious fashion. Notice that it also bounds from below the length of the edges in the singular graph connecting two vertices. Our (p, q, r)-Kleinian groups give a number of potential minima for the distance between points stabilized by finite spherical triangle subgroups; we record them below. The minimal distances between points stabilized by icosahedral points were determined by Derevnin and Mednykh [4] and do occur in our lists. However, for sharp volume estimates a little more is needed. Roughly speaking, the spectrum of such distances initially comes in a discrete part and then a continuous part; we need to know the first few distances in the discrete part.

Here we record the possible distances that occur in the (p, q, r)-Kleinian groups. In [6] it is determined which of these values are sharp.

The tables contain the distances between the various spherical triangle points. We list the first 5 distances, or include as a last entry the closest distance on an axis not already represented. In each case, the finite groups share a common axis, and we indicate the order of that axis.

Table 4 - A_4-A_4

Common	distance	Group
3	0.6931	Γ_{27}
3	0.7691	Γ_4
3	0.9290	Γ_5
3	1.0050	Γ_{10}
2	1.0612	Γ_{32}

Table 5 - A_4-S_4

Common	distance	Group
3	1.0148	Γ_4
3	1.3169	Γ_{14}
3	1.4336	$(2,3,3)$
2	1.4379	Γ_4
3	1.4927	$(2,3,3)$

Table 6 - A_4-A_5

Common	distance	Group
3	1.2264	Γ_9
3	1.6266	Γ_5
2	1.7611	Γ_5
3	1.8798	$(2,3,3)$
3	1.9833	$(2,3,3)$

Table 7 - S_4-S_4

Common	distance	Group
4	1.0612	Γ_4
4	1.1283	Γ_4
3	1.3169	Γ_{25}
4	1.3169	Γ_{25}
2	1.7000	Γ_6

Table 8 - S_4-A_5

Common	distance	Group
3	1.2264	Γ_2
3	1.9833	$(2,2,3)$
3	2.1327	Γ_7
2	2.2731	Γ_7
3	2.3486	$(2,3,4)$

Table 9 - A_5-A_5

Common	distance	Group
5	1.3825	Γ_1
5	1.6169	Γ_9
3	1.9028	Γ_3
5	2.0444	Γ_5
2	2.8264	Γ_8

We also record some infinite sets of distances (which accumulate on the continuous part of the distance spectrum).

Table 10 - distances

Groups	Common	distances	(p,q,r)
$A_4 - A_4$	3	$d_n = \operatorname{arccosh}(\frac{1}{2}(1 + 3\cos(2\pi/n)))$, $n \geq 7$	$(2,2,3)$
$A_4 - S_4$	3	$d_n = \operatorname{arccosh}(1 + 3\cos(\pi/n)/\sqrt{2}))$, $n \geq 3$	$(2,2,3)$
$A_4 - A_5$	3	$d_n = \operatorname{arccosh}(\frac{1+\sqrt{5}+6\cos(\pi/n)}{2\sqrt{3-\sqrt{5}}})$, $n \geq 3$	$(2,2,3)$
$S_4 - S_4$	3	$d_n = \operatorname{arccosh}(2 + 3\cos(2\pi/n)))$, $n \geq 7$	$(2,2,3)$
$S_4 - S_4$	4	$d_n = \operatorname{arccosh}(1 + 2\cos(2\pi/n)))$, $n \geq 7$	$(2,2,4)$
$S_4 - A_5$	3	$d_n = \operatorname{arccosh}(\frac{\sqrt{2}+\sqrt{10}+6\cos(\pi/n)}{\sqrt{6-2\sqrt{5}}})$, $n \geq 3$	$(2,2,3)$
$A_5 - A_5$	3	$d_n = \operatorname{arccosh}(3(3 + \sqrt{5})\cos^2(\pi/n) - 1)$, $n \geq 7$	$(2,2,3)$
$A_5 - A_5$	5	$d_n = \operatorname{arccosh}(\frac{(5-\sqrt{5})\cos(2\pi/n)+2}{3-\sqrt{5}})$, $n \geq 7$	$(2,2,3)$

11. Discrete (p,q,r)–Kleinian groups

11.1. Dihedral angles submultiples of π.

Table 11 - (p,q,r)–triangles ($p \geq 3$). Dihedral angles submultiples of π

#	m_Γ	(p,q,r)	Angle	Angle	Angle	Σ dihedral	Discrete	Vertex Structure
1	0.9624	(3,3,3)	0	0	0	π	Γ_{31}	$\Delta_{(3,3,3)}, \Delta_{(3,3,3)}, \Delta_{(3,3,3)}$
2	0.7564	(3,3,3)	0	0	1.230	$7\pi/6$	Γ_{32}	$A_4, \Delta_{(3,3,3)}, \Delta_{(3,3,3)}$
3	0.4435	(3,3,3)	0	1.230	1.230	$4\pi/3$	Γ_{27}	$A_4, A_4, \Delta_{(3,3,3)}$
4	0.6329	(3,3,4)	0	.955	.955	$4\pi/3$	Γ_{28}	$\Delta_{(3,3,3)}, S_4, S_4$
5	0.7362	(3,3,5)	0	.652	.652	$4\pi/5$	Γ_{29}	$\Delta_{(3,3,3)}, A_5, A_5$
6	0.3796	(3,3,5)	1.230	.652	.652	$3\pi/2$	Γ_9	A_4, A_5, A_5
7	0.7983	(3,3,6)	0	0	0	$4\pi/3$	Γ_{30}	$\Delta_{(3,3,3)}, \Delta_{(2,3,6)}, \Delta_{(2,3,6)}$
8	0.6349	(3,3,6)	1.230	0	0	$3\pi/2$	Γ_{24}	$A_4, \Delta_{(2,3,6)}, \Delta_{(2,3,6)}$
9	0.7953	(3,4,4)	.955	.955	0	$3\pi/2$	Γ_{25}	$\Delta_{(2,4,4)}, \Delta_{(2,4,4)}, \Delta_{(2,4,4)}$
10	0.5777	(4,4,4)	0	0	0	$3\pi/2$	Γ_{26}	$\Delta_{(2,4,4)}, \Delta_{(2,4,4)}, \Delta_{(2,4,4)}$

Table 12 - (2,3,3)–triangles. Dihedral angles submultiples of π

#	m_Γ	$3-3$	$2-3$	$2-3$	dihedral angles	discrete	Vertex Structure
1	1.0050	0	0	0	$\pi/3, \pi/6, \pi/6$	$\Delta_{(3,3,6)}$	$\Delta_{(3,3,3)}, \Delta_{(2,3,6)}, \Delta_{(2,3,6)}$
2	0.9841	0	.364	0	$\pi/3, \pi/5, \pi/6$	$\Delta_{(3,5,6)}$	$\Delta_{(3,3,3)}, A_5, \Delta_{(2,3,6)}$
3	0.9465	0	.615	0	$\pi/3, \pi/4, \pi/6$	$\Delta_{(3,4,6)}$	$\Delta_{(3,3,3)}, S_4, \Delta_{(2,3,6)}$
4	0.8689	0	.955	0	$\pi/3, \pi/3, \pi/6$	$\Delta_{(3,3,6)}$	$\Delta_{(3,3,3)}, A_4, \Delta_{(2,3,6)}$
5	0.6687	0	$\pi/2$	0	$\pi/3, \pi/2, \pi/6$	Γ_{30}	$\Delta_{(3,3,3)}, D_3, \Delta_{(2,3,6)}$
6	0.9624	0	.364	.364	$\pi/3, \pi/5, \pi/5$	$\Delta_{(3,5,5)}$	$\Delta_{(3,3,3)}, A_5, A_5$
7	0.9233	0	.615	.364	$\pi/3, \pi/4, \pi/5$	$\Delta_{(3,4,5)}$	$\Delta_{(3,3,3)}, S_4, A_5$
8	0.8423	0	.955	.364	$\pi/3, \pi/3, \pi/5$	$\Delta_{(3,3,5)}$	$\Delta_{(3,3,3)}, A_4, A_5$
9	0.6309	0	$\pi/2$	.364	$\pi/3, \pi/2, \pi/5$	Γ_{29}	$\Delta_{(3,3,3)}, D_3, A_5$
10	0.8813	0	.615	.615	$\pi/3, \pi/4, \pi/4$	$\Delta_{(3,4,4)}$	$\Delta_{(3,3,3)}, S_4, S_4$
11	0.7941	0	.955	.615	$\pi/3, \pi/3, \pi/4$	$\Delta_{(3,3,4)}$	$\Delta_{(3,3,3)}, A_4, S_4$
12	0.5622	0	$\pi/2$	.615	$\pi/3, \pi/2, \pi/4$	Γ_{28}	$\Delta_{(3,3,3)}, D_3, S_4$
13	0.6931	0	.955	.955	$\pi/3, \pi/3, \pi/3$	Γ_{32}	$\Delta_{(3,3,3)}, A_4, A_4$
14	0.4170	0	$\pi/2$	.955	$\pi/3, \pi/2, \pi/3$	Γ_{27}	$\Delta_{(3,3,3)}, D_3, A_4$
15	0.7872	1.230	0	0	$\pi/2, \pi/6, \pi/6$	$\Delta_{(2,6,6)}$	$A_4, \Delta_{(2,3,6)}, \Delta_{(2,3,6)}$
16	0.7673	1.230	.364	0	$\pi/2, \pi/5, \pi/6$	$\Delta_{(2,5,6)}$	$A_4, A_5, \Delta_{(2,3,6)}$
17	0.7123	1.230	.615	0	$\pi/2, \pi/4, \pi/6$	$\Delta_{(2,4,6)}$	$A_4, S_4, \Delta_{(2,3,6)}$
18	0.5961	1.230	.955	0	$\pi/2, \pi/3, \pi/6$	Γ_{10}	$A_4, A_4, \Delta_{(2,3,6)}$
19	0.2671	1.230	$\pi/2$	0	$\pi/2, \pi/2, \pi/6$	Γ_{17}	$A_4, D_3, \Delta_{(2,3,6)}$
20	0.7344	1.230	.364	.364	$\pi/2, \pi/5, \pi/5$	$\Delta_{(2,5,5)}$	A_4, A_5, A_5
21	0.6743	1.230	.615	.364	$\pi/2, \pi/4, \pi/5$	$\Delta_{(2,4,5)}$	A_4, S_4, A_5
22	0.5438	1.230	.955	.364	$\pi/2, \pi/3, \pi/5$	Γ_5	A_4, A_4, A_5
23	0.6034	1.230	.615	.615	$\pi/2, \pi/4, \pi/4$	Γ_{14}	A_4, S_4, S_4
24	0.4397	1.230	.955	.615	$\pi/2, \pi/3, \pi/4$	Γ_4	A_4, A_4, S_4

Now we turn to the $(2,3,4)$–triangles.

Table 13 - (2,3,4)–triangles. Dihedral angles submultiples of π

#	m_Γ	$3-4$	$2-4$	$2-3$	dihedral angles	discrete	Vertex Structure
1	0.8040	.955	0	0	$\pi/2, \pi/4, \pi/6$	$\Delta_{(2,4,6)}$	$S_4, \Delta_{(2,3,6)}, \Delta_{(2,3,6)}$
2	0.7955	.955	0	.364	$\pi/2, \pi/4, \pi/5$	$\Delta_{(2,4,5)}$	$S_4, A_5, \Delta_{(2,4,4)}$
3	0.7478	.955	0	.615	$\pi/2, \pi/4, \pi/4$	Γ_{15}	$S_4, S_4, \Delta_{(2,4,4)}$
4	0.6494	.955	0	.955	$\pi/2, \pi/4, \pi/3$	Γ_{14}	$S_4, A_4, \Delta_{(2,4,4)}$
5	0.3887	.955	0	$\pi/2$	$\pi/2, \pi/4, \pi/2$	Γ_{19}	$S_4, D_3, \Delta_{(2,4,4)}$
6	0.7138	.955	$\pi/4$	0	$\pi/2, \pi/3, \pi/6$	Γ_{11}	$S_4, S_4, \Delta_{(2,3,6)}$
7	0.6792	.955	$\pi/4$	.364	$\pi/2, \pi/3, \pi/5$	Γ_7	S_4, S_4, A_5
8	0.6155	.955	$\pi/4$	.615	$\pi/2, \pi/3, \pi/4$	Γ_6	S_4, S_4, S_4
9	0.4749	.955	$\pi/4$	.955	$\pi/2, \pi/3, \pi/3$	Γ_4	S_4, S_4, A_4
10	0.4098	.955	$\pi/2$	0	$\pi/2, \pi/2, \pi/6$	Γ_{20}	$S_4, D_4, \Delta_{(2,3,6)}$
11	0.3231	.955	$\pi/2$	0.364	$\pi/2, \pi/2, \pi/5$	Γ_2	S_4, D_4, A_5

Table 14 - (2,3,5)–triangles. Dihedral angles submultiples of π

#	m_Γ	$3-5$	$2-5$	$2-3$	dihedral angles	discrete	Vertex Structure
1	0.7855	.652	.553	0	$\pi/2, \pi/3, \pi/6$	Γ_{12}	$A_5, A_5, \Delta_{(2,3,6)}$
2	0.7589	.653	.553	.364	$\pi/2, \pi/3, \pi/5$	Γ_8	A_5, A_5, A_5
3	0.7115	.652	.553	.615	$\pi/2, \pi/3, \pi/4$	Γ_7	A_5, A_5, S_4
4	0.6147	.652	.553	.955	$\pi/2, \pi/3, \pi/3$	Γ_5	A_5, A_5, A_4
5	0.3541	.652	.553	$\pi/2$	$\pi/2, \pi/3, \pi/2$	Γ_1	A_5, A_5, D_3
6	0.5014	.652	$\pi/2$	0	$\pi/2, \pi/2, \pi/6$	Γ_{21}	$A_5, D_5, \Delta_{(2,3,6)}$
7	0.4481	.652	$\pi/2$	.364	$\pi/2, \pi/2, \pi/5$	Γ_3	A_5, D_5, A_5
8	0.3456	.652	$\pi/2$	.615	$\pi/2, \pi/2, \pi/4$	Γ_2	A_5, D_5, S_4

Table 15 - (2,3,6)–triangles. Dihedral angles submultiples of π

#	m_Γ	$2-3$	$2-6$	$3-6$	dihedral angles	discrete	Vertex Structure
1	0.8314	0	0	0	$\pi/6, \pi/3, \pi/2$	Γ_{13}	$\Delta_{(2,3,6)}, \Delta_{(2,3,6)}, \Delta_{(2,3,6)}$
2	0.8152	.364	0	0	$\pi/5, \pi/3, \pi/2$	Γ_{12}	$A_5, \Delta_{(2,3,6)}, \Delta_{(2,3,6)}$
3	0.7860	.615	0	0	$\pi/4, \pi/3, \pi/2$	Γ_{11}	$S_4, \Delta_{(2,3,6)}, \Delta_{(2,3,6)}$
4	0.7249	.955	0	0	$\pi/3, \pi/3, \pi/2$	Γ_{10}	$A_4, \Delta_{(2,3,6)}, \Delta_{(2,3,6)}$
5	0.5636	$\pi/2$	0	0	$\pi/2, \pi/3, \pi/2$	Γ_{18}	$D_3, \Delta_{(2,3,6)}, \Delta_{(2,3,6)}$
6	0.5120	0	$\pi/2$	0	$\pi/6, \pi/2, \pi/2$	Γ_{22}	$D_6, \Delta_{(2,3,6)}, \Delta_{(2,3,6)}$
7	0.6572	.364	$\pi/2$	0	$\pi/5, \pi/2, \pi/2$	Γ_{21}	$D_6, \Delta_{(2,3,6)}, A_5$
8	0.5879	.615	$\pi/2$	0	$\pi/4, \pi/2, \pi/2$	Γ_{20}	$D_6, \Delta_{(2,3,6)}, S_4$
9	0.4393	.955	$\pi/2$	0	$\pi/3, \pi/2, \pi/2$	Γ_{17}	$D_6, \Delta_{(2,3,6)}, A_4$

Table 16 - (2,4,4)–triangles. Dihedral angles submultiples of π

#	m_Γ	$2-4$	$2-4$	$4-4$	dihedral angles	discrete	Vertex Structure
1	0.8813	0	0	0	$\pi/4, \pi/4, \pi/2$	$\Gamma(16)$	$\Delta_{(2,4,4)}, \Delta_{(2,4,4)}, \Delta_{(2,4,4)}$
2	0.5777	$\pi/2$	0	0	$\pi/3, \pi/4, \pi/2$	Γ_{23}	$D_4, \Delta_{(2,4,4)}, \Delta_{(2,4,4)}$
3	0.7953	$\pi/4$	0	0	$\pi/3, \pi/4, \pi/2$	Γ_{15}	$S_4, \Delta_{(2,4,4)}, \Delta_{(2,4,4)}$
4	0.4232	$\pi/2$	$\pi/4$	0	$\pi/3, \pi/3, \pi/2$	Γ_{19}	$D_4, S_4, \Delta_{(2,4,4)}$
5	0.6931	$\pi/4$	$\pi/4$	0	$\pi/3, \pi/3, \pi/2$	Γ_{14}	$S_4, S_4, \Delta_{(2,4,4)}$

Table 17 - (2,2,3)–triangles. Dihedral angles submultiples of π

#	m_Γ	$2-3$	$2-3$	$2-2$	dihedral angles	discrete	Vertex Structure
1	0.7095	0	0	$\pi/n\ (2 \leq n \leq \infty)$	$\pi/6, \pi/6, \pi/n$	$\Delta_{(n,6,6)}$	$\Delta_{(2,3,6)}, \Delta_{(2,3,6)}, D_n$
2	0.6694	0	.364	$\pi/n\ (2 \leq n \leq \infty)$	$\pi/6, \pi/5, \pi/n$	$\Delta_{(5,6,n)}$	$\Delta_{(2,3,6)}, A_5, D_n$
3	0.5966	0	.615	$\pi/n\ (2 \leq n \leq \infty)$	$\pi/6, \pi/4, \pi/n$	$\Delta_{(4,6,n)}$	$\Delta_{(2,3,6)}, S_4, D_n$
4	0.4429	0	.955	$\pi/2$	$\pi/6, \pi/3, \pi/2$	Γ_{24}	$\Delta_{(2,3,6)}, A_4, D_2$
5	0.6941	0	.955	$\pi/n\ (3 \leq n \leq \infty)$	$\pi/6, \pi/3, \pi/n$	$\Delta_{(3,n,6)}$	$\Delta_{(2,3,6)}, A_4, D_n$
6	0.3908	0	$\pi/2$	$\pi/3$	$\pi/6, \pi/2, \pi/3$	Γ_{18}	$\Delta_{(2,3,6)}, D_3, D_3$
7	0.5274	0	$\pi/2$	$\pi/n\ (4 \leq n \leq \infty)$	$\pi/6, \pi/2, \pi/n$	$\Delta_{(2,6,n)}$	$\Delta_{(2,3,6)}, D_3, D_n$
8	0.6146	.364	.364	$\pi/n\ (2 \leq n \leq \infty)$	$\pi/5, \pi/5, \pi/n$	$\Delta_{(5,5,n)}$	A_5, A_5, D_n
9	0.5349	.364	.615	$\pi/n\ (2 \leq n \leq \infty)$	$\pi/5, \pi/4, \pi/n$	$\Delta_{(4,5,n)}$	A_5, S_4, D_n
10	0.3490	.364	.955	$\pi/2$	$\pi/5, \pi/3, \pi/n$	Γ_9	A_5, A_4, D_2
11	0.6496	.364	.955	$\pi/n\ (3 \leq n \leq \infty)$	$\pi/5, \pi/3, \pi/n$	$\Delta_{(3,5,n)}$	A_5, A_4, D_n
12	0.2768	.364	$\pi/2$	$\pi/3$	$\pi/5, \pi/2, \pi/3$	Γ_1	A_5, D_3, D_3
13	0.4610	.364	$\pi/2$	$\pi/n\ (4 \leq n \leq \infty)$	$\pi/5, \pi/2, \pi/n$	$\Delta_{(2,5,n)}$	A_5, D_3, D_n
14	0.4299	.615	.615	$\pi/2$	$\pi/4, \pi/4, \pi/2$	Γ_{25}	S_4, S_4, D_2
15	0.7048	.615	.615	$\pi/n\ (3 \leq n \leq \infty)$	$\pi/4, \pi/4, \pi/n$	$\Delta_{(4,4,n)}$	S_4, S_4, D_n
16	0.5659	.615	.955	$\pi/n\ (3 \leq n \leq \infty)$	$\pi/4, \pi/3, \pi/n$	$\Delta_{(3,4,n)}$	S_4, A_4, D_n
17	0.3157	.615	$\pi/2$	$\pi/4$	$\pi/4, \pi/2, \pi/4$	Γ_{19}	S_4, D_3, D_4
18	0.4305	.615	$\pi/2$	$\pi/n\ (5 \leq n \leq \infty)$	$\pi/4, \pi/2, \pi/n$	$\Delta_{(2,4,n)}$	S_4, D_3, D_n
19	0.3517	.955	.955	$\pi/3$	$\pi/3, \pi/3, \pi/3$	Γ_{27}	A_4, A_4, D_3
20	0.5372	.955	.955	$\pi/n\ (4 \leq n \leq \infty)$	$\pi/3, \pi/3, \pi/n$	$\Delta_{(3,3,n)}$	A_4, A_4, D_n
21	0.2309	.955	$\pi/2$	$\pi/6$	$\pi/3, \pi/2, \pi/6$	Γ_{17}	A_4, D_3, D_6
22	0.30345	.955	$\pi/2$	$\pi/n\ (7 \leq n \leq \infty)$	$\pi/3, \pi/2, \pi/n$	$\Delta_{(2,3,n)}$	A_4, D_3, D_n

Table 18 - (2,2,4)–triangles. Dihedral angles submultiples of π

#	m_Γ	$2-4$	$2-4$	$2-2$	dihedral angles	discrete	Vertex Structure
1	0.5777	0	0	$\pi/2$	$\pi/4, \pi/4, \pi/2$	Γ_{26}	$\Delta_{(2,4,4)}, \Delta_{(2,4,4)}, D_2$
2	0.7337	0	0	$\pi/n\ (3 \leq n \leq \infty)$	$\pi/4, \pi/4, \pi/n$	$\Delta_{(4,4,n)}$	$\Delta_{(2,4,4)}, \Delta_{(2,4,4)}, D_n$
3	0.4738	0	$\pi/4$	$\pi/2$	$\pi/4, \pi/3, \pi/2$	Γ_{25}	$\Delta_{(2,4,4)}, S_4, D_2$
4	0.6931	0	$\pi/4$	$\pi/n\ (3 \leq n \leq \infty)$	$\pi/4, \pi/3, \pi/n$	$\Delta_{(3,4,n)}$	$\Delta_{(2,4,4)}, S_4, D_n$
5	0.3574	0	$\pi/2$	$\pi/3$	$\pi/4, \pi/2, \pi/3$	Γ_{19}	$\Delta_{(2,4,4)}, D_4, D_3$
6	0.4877	0	$\pi/2$	$\pi/4$	$\pi/4, \pi/2, \pi/4$	Γ_{23}	$\Delta_{(2,4,4)}, D_4, D_4$
7	0.5504	0	$\pi/2$	$\pi/n\ (5 \leq n \leq \infty)$	$\pi/4, \pi/2, \pi/n$	$\Delta_{(2,4,n)}$	$\Delta_{(2,4,4)}, D_4, D_n$
8	0.5283	$\pi/4$	$\pi/4$	$\pi/3$	$\pi/3, \pi/3, \pi/3$	Γ_{28}	S_4, S_4, D_3
9	0.6441	$\pi/4$	$\pi/4$	$\pi/n\ (4 \leq n \leq \infty)$	$\pi/3, \pi/3, \pi/n$	$\Delta_{(3,3,n)}$	S_4, S_4, D_n
10	0.2874	$\pi/4$	$\pi/2$	$\pi/5$	$\pi/3, \pi/2, \pi/5$	Γ_2	S_4, D_4, D_5
11	0.3640	$\pi/4$	$\pi/2$	$\pi/6$	$\pi/3, \pi/2, \pi/6$	Γ_{20}	S_4, D_4, D_6
12	0.4055	$\pi/4$	$\pi/2$	$\pi/n\ (7 \leq n \leq \infty)$	$\pi/3, \pi/2, \pi/n$	$\Delta_{(2,3,n)}$	S_4, D_4, D_n

Table 19 - (2,2,5)–triangles. Dihedral angles submultiples of π

#	m_Γ	$2-5$	$2-5$	$2-2$	dihedral angles	discrete	Vertex Structure
1	0.4064	.55357	.55357	$\pi/2$	$\pi/3, \pi/3, \pi/2$	Γ_9	A_5, A_5, D_2
2	0.6300	.55357	.55357	$\pi/3$	$\pi/3, \pi/3, \pi/3$	Γ_{29}	A_5, A_5, D_3
3	0.7157	.55357	.55357	$\pi/n,\ (4 \leq n \leq \infty)$	$\pi/3, \pi/3, \pi/n$	$\Delta(3,3,n)$	A_5, A_5, D_n
4	0.3126	.55357	$\pi/2$	$\pi/4$	$\pi/3, \pi/2, \pi/4$	Γ_2	A_5, D_5, D_4
5	0.4042	.55357	$\pi/2$	$\pi/5$	$\pi/3, \pi/2, \pi/5$	Γ_3	A_5, D_5, D_5
6	0.4515	.55357	$\pi/2$	$\pi/6$	$\pi/3, \pi/2, \pi/6$	Γ_{21}	A_5, D_5, D_6
7	0.4798	.55357	$\pi/2$	$\pi/n,\ (7 \leq n \leq \infty)$	$\pi/3, \pi/2, \pi/n$	$\Delta(2,3,n)$	A_5, D_5, D_n

Table 20 - (2,2,6)–triangles. Dihedral angles submultiples of π

#	m_Γ	$2-6$	$2-6$	$2-2$	dihedral angles	discrete	Vertex Structure
1	0.5942	0	0	$\pi/2$	$\pi/3,\pi/3,\pi/2$	Γ_{24}	$\Delta_{(2,3,6)},\Delta_{(2,3,6)},D_2$
2	0.7389	0	0	$\pi/3$	$\pi/3,\pi/3,\pi/3$	Γ_{30}	$\Delta_{(2,3,6)},\Delta_{(2,3,6)},D_3$
3	0.7935	0	0	$\pi/n,\ 4\le n\le\infty$	$\pi/3,\pi/3,\pi/n$	$\Delta_{(3,3,n)}$	$\Delta_{(2,3,6)},\Delta_{(2,3,6)},D_n$
4	0.3003	0	$\pi/2$	$\pi/3$	$\pi/3,\pi/2,\pi/3$	Γ_{17}	$\Delta_{(2,3,6)},D_6,D_3$
5	0.4187	0	$\pi/2$	$\pi/4$	$\pi/3,\pi/2,\pi/4$	Γ_{20}	$\Delta_{(2,3,6)},D_6,D_4$
6	0.4781	0	$\pi/2$	$\pi/5$	$\pi/3,\pi/2,\pi/5$	Γ_{21}	$\Delta_{(2,3,6)},D_6,D_5$
7	0.5120	0	$\pi/2$	$\pi/6$	$\pi/3,\pi/2,\pi/6$	Γ_{22}	$\Delta_{(2,3,6)},D_6,D_6$
8	0.5330	0	$\pi/2$	$\pi/n,\ 7\le n\le\infty$	$\pi/3,\pi/2,\pi/n$	$\Delta_{(2,3,n)}$	$\Delta_{(2,3,6)},D_6,D_n$

11.2. Dihedral angles not submultiples of π. In this section, we record those groups which arise from reflection groups whose dihedral angles are not submultiples of π. There is no need to record the Margulis constant for these groups as each one of them is a subgroup of one of the (p,q,r)–Kleinian groups found in the previous section. This group is found by refining the triangle and therefore decreasing the Margulis constant. We indicate the related subgroup from the above tables in each case.

Table 21 - (p,q,r)–triangles $(p\ge 3)$. Dihedral angles not submultiples of π

#	(p,q,r)	Angle	Angle	Angle	Σ dihedral	group	Vertex Structure
1*	(3,3,3)	0.7297	0.7297	0.7297	$6\pi/5$	$6\times(2,2,3)_{12}$	A_5,A_5,A_5
2*	(3,3,3)	0	0	0.7297	$16\pi/15$	$2\times(2,3,3)_9$	$A_5,\Delta_{(3,3,3)},\Delta_{(3,3,3)}$
3*	(3,3,3)	0	0	1.9106	$4\pi/3$	$2\times(2,3,3)_{14}$	$A_4,\Delta_{(3,3,3)},\Delta_{(3,3,3)}$
4*	(3,3,4)	0.7297	.955	.955	$16\pi/15$	$2\times(2,3,4)_{11}$	$\Delta_{(3,3,3)},S_4,S_4$
5*	(3,3,5)	0.7297	.652	.652	$7\pi/5$	$2\times(2,3,5)_7$	A_5,A_5,A_5
6*	(3,3,6)	0.7297	0	0	$7\pi/5$	$2\times(2,3,6)_7$	$A_5,\Delta_{(2,3,6)},\Delta_{(2,3,6)}$
7*	(3,3,6)	1.9106	0	0	$5\pi/3$	$2\times(2,3,6)_9$	$A_4,\Delta_{(2,3,6)},\Delta_{(2,3,6)}$
8*	(3,5,5)	1.10719	.652	.652	$5\pi/3$	$2\times(2,3,5)_2$	A_5,A_5,A_5
9*	(3,6,6)	0	0	0	$5\pi/3$	$2\times(2,3,6)_5$	A_5,A_5,A_5
10*	(6,6,6)	0	0	0	2π	$6\times(2,2,6)_2$	A_5,A_5,A_5

Table 22 - (2,3,3)–triangles. Dihedral angles not submultiples of π

#	$3-3$	$2-3$	$2-3$	dihedral angles	group	Vertex Structure
1*	0.7297	0	0	$2\pi/5,\pi/6,\pi/6$	$2\times(2,2,3)_2$	$A_5,\Delta_{(2,3,6)},\Delta_{(2,3,6)}$
2*	0.7297	0.364	0.364	$2\pi/5,\pi/5,\pi/5$	$2\times(2,2,3)_8$	A_5,A_5,A_5
3*	0.7297	0.615	0.615	$2\pi/5,\pi/4,\pi/4$	$2\times(2,2,3)_9$	A_5,S_4,S_4
4*	0.7297	0.955	0.955	$2\pi/5,\pi/3,\pi/3$	$2\times(2,2,3)_{10}$	A_4,A_4,A_5
5*	1.9106	0	0	$2\pi/3,\pi/6,\pi/6$	$2\times(2,2,3)_4$	$A_4,\Delta_{(2,3,6)},\Delta_{(2,3,6)}$
6*	1.9106	.364	.364	$2\pi/3,\pi/5,\pi/5$	$2\times(2,2,3)_{10}$	A_4,A_5,A_5

Table 23 - (2,p,q)–triangles. Dihedral angles not submultiples of π

#	$4-4$	$2-4$	$2-4$	dihedral angles	group	Vertex Structure
1*	$\pi/2$	0	0	$2\pi/3,\pi/4,\pi/4$	$2\times(2,2,4)_3$	$S_4,\Delta_{(2,4,4)},\Delta_{(2,4,4)}$
#	$5-5$	$2-5$	$2-5$			
1*	1.107149	0.5535	0.5535	$2\pi/3,\pi/3,\pi/3$	$2\times(2,2,5)_1$	A_5,A_5,A_5
#	$6-6$	$2-6$	$2-6$			
1*	0	0	0	$2\pi/3,\pi/3,\pi/3$	$2\times(2,2,6)_1$,	$\Delta(2,3,6),\Delta_{(2,3,6)},\Delta_{(2,3,6)}$
2*	0	0	$\pi/2$	$2\pi/3,\pi/3,\pi/2$	$3\times(2,2,6)_2$,	$\Delta(2,3,6),\Delta_{(2,3,6)},D_6$

Table 24 - (2,2,3)–triangles. Dihedral angles not submultiples of π

#	$2-3$	$2-3$	$2-2$	dihedral angles	group	Vertex Structure
1*	0	0	$2\pi/3$	$\pi/6,\pi/6,2\pi/3$	$2\times(2,2,3)_6$	$\Delta_{(2,3,6)},\Delta_{(2,3,6)},D_3$
2*	0	0	$2\pi/n\ (5\le n\le\infty)$	$\pi/6,\pi/6,2\pi/n$	$\Delta_{(2,6,n)}$	$\Delta_{(2,3,6)},\Delta_{(2,3,6)},D_n$
3*	.364	.364	$2\pi/3$	$\pi/5,\pi/5,2\pi/3$	$2\times(2,2,3)_{12}$	A_5,A_5,D_3
4*	.364	.364	$2\pi/n\ (2\le n\le\infty)$	$\pi/5,\pi/5,2\pi/n$	$\Delta_{(5,5,n)}$	A_5,A_5,D_n
5*	.615	.615	$2\pi/n\ (5\le n\le\infty)$	$\pi/4,\pi/4,2\pi/n$	$\Delta_{(2,4,n)}$	S_4,S_4,D_n
6*	.955	.955	$2\pi/n\ (7\le n\le\infty)$	$\pi/3,\pi/3,2\pi/n$	$\Delta_{(2,3,n)}$	A_4,A_4,D_n

Table 25 - (2,2,4)–triangles. Dihedral angles not submultiples of π

#	$2-4$	$2-4$	$2-2$	dihedral angles	group	Vertex Structure
1*	0	0	$2\pi/3$	$\pi/4,\pi/4,2\pi/3$	$2\times(2,2,4)_5$	$\Delta_{(2,4,4)},\Delta_{(2,4,4)},D_3$
2*	0	0	$2\pi/n\ (5\le n\le\infty)$	$\pi/4,\pi/4,2\pi/n$	$\Delta_{(2,4,n)}$	$\Delta_{(2,4,4)},\Delta_{(2,4,4)},D_n$
3*	$\pi/4$	$\pi/4$	$2\pi/5$	$\pi/3,\pi/3,2\pi/5$	$2\times(2,2,4)_{10}$	S_4,S_4,D_5
4*	$\pi/4$	$\pi/4$	$2\pi/n\ (7\le n\le\infty)$	$\pi/3,\pi/3,2\pi/n$	$\Delta_{(2,3,n)}$	S_4,S_4,D_n

Table 26 - (2,2,5)–triangles. Dihedral angles not submultiples of π

#	$2-5$	$2-5$	$2-2$	dihedral angles	group	Vertex Structure
1*	.55357	.55357	$2\pi/5,$	$\pi/3,\pi/3,2\pi/5$	$2\times(2,2,5)_5$	A_5,A_5,D_5
2*	.55357	.55357	$2\pi/n,\ (7\le n\le\infty)$	$\pi/3,\pi/3,2\pi/n$	$\Delta(2,3,n)$	A_5,A_5,D_n

Table 27 - (2,2,6)–triangles. Dihedral angles not submultiples of π

#	2 − 6	2 − 6	2 − 2	dihedral angles	group	Vertex Structure
1*	0	0	$2\pi/3$	$\pi/3, \pi/3, 2\pi/3$	$2 \times (2,2,6)_4$	$\Delta_{(2,3,6)}, \Delta_{(2,3,6)}, D_3$
2*	0	0	$2\pi/5$	$\pi/3, \pi/3, 2\pi/5$	$2 \times (2,2,6)_6$	$\Delta_{(2,3,6)}, \Delta_{(2,3,6)}, D_5$
3*	0	0	$2\pi/n,\ 7 \leq n \leq \infty$	$\pi/3, \pi/3, 2\pi/n$	$\Delta_{(2,3,n)}$	$\Delta_{(2,3,6)}, \Delta_{(2,3,6)}, D_n$

12. Appendix: The reflection groups

Here, for the convenience of the reader, we collect together the tetrahedral reflection groups which occur and tabulate them so that they can be identified in our previous tables. The first table contains the classical reflection groups with all dihedral angles submultiples of π. We have extracted from [15] estimates of the covolumes of the reflection groups. The second table contains the additional reflection groups for which the dihedral angles at one face are all submultiples of π.

Table 28 - Reflection groups. Angles submultiples of π

Γ_i	$\angle\overline{AB}$	$\angle\overline{BC}$	$\angle\overline{CD}$	$\angle\overline{AC}$	$\angle\overline{AD}$	$\angle\overline{BD}$	covolume
1	$\pi/3$	$\pi/2$	$\pi/3$	$\pi/2$	$\pi/5$	$\pi/2$	0.039050
2	$\pi/5$	$\pi/2$	$\pi/4$	$\pi/2$	$\pi/3$	$\pi/2$	0.035885
3	$\pi/5$	$\pi/2$	$\pi/5$	$\pi/2$	$\pi/3$	$\pi/2$	0.093326
4	$\pi/3$	$\pi/3$	$\pi/3$	$\pi/2$	$\pi/4$	$\pi/2$	0.085770
5	$\pi/3$	$\pi/3$	$\pi/3$	$\pi/2$	$\pi/5$	$\pi/2$	0.205289
6	$\pi/3$	$\pi/4$	$\pi/3$	$\pi/2$	$\pi/4$	$\pi/2$	0.222229
7	$\pi/3$	$\pi/5$	$\pi/3$	$\pi/2$	$\pi/4$	$\pi/2$	0.358653
8	$\pi/3$	$\pi/5$	$\pi/3$	$\pi/2$	$\pi/5$	$\pi/2$	0.502131
9	$\pi/2$	$\pi/3$	$\pi/3$	$\pi/5$	$\pi/3$	$\pi/2$	0.071770
10	$\pi/3$	$\pi/3$	$\pi/6$	$\pi/2$	$\pi/3$	$\pi/2$	0.364107
11	$\pi/4$	$\pi/3$	$\pi/6$	$\pi/2$	$\pi/3$	$\pi/2$	0.525840
12	$\pi/5$	$\pi/3$	$\pi/6$	$\pi/2$	$\pi/3$	$\pi/2$	0.672986
13	$\pi/6$	$\pi/3$	$\pi/6$	$\pi/2$	$\pi/3$	$\pi/2$	0.845785
14	$\pi/3$	$\pi/3$	$\pi/4$	$\pi/2$	$\pi/4$	$\pi/2$	0.305322
15	$\pi/3$	$\pi/4$	$\pi/4$	$\pi/2$	$\pi/4$	$\pi/2$	0.556282
16	$\pi/4$	$\pi/4$	$\pi/4$	$\pi/2$	$\pi/4$	$\pi/2$	0.915965
17	$\pi/6$	$\pi/2$	$\pi/3$	$\pi/2$	$\pi/3$	$\pi/2$	0.042289
18	$\pi/3$	$\pi/2$	$\pi/3$	$\pi/2$	$\pi/6$	$\pi/2$	0.169157
19	$\pi/4$	$\pi/2$	$\pi/3$	$\pi/2$	$\pi/4$	$\pi/2$	0.076330
20	$\pi/6$	$\pi/2$	$\pi/4$	$\pi/2$	$\pi/3$	$\pi/2$	0.105723
21	$\pi/6$	$\pi/2$	$\pi/5$	$\pi/2$	$\pi/3$	$\pi/2$	0.171502
22	$\pi/6$	$\pi/2$	$\pi/6$	$\pi/2$	$\pi/3$	$\pi/2$	0.253735
23	$\pi/4$	$\pi/2$	$\pi/4$	$\pi/2$	$\pi/4$	$\pi/2$	0.228991
24	$\pi/2$	$\pi/2$	$\pi/3$	$\pi/6$	$\pi/3$	$\pi/2$	0.211446
25	$\pi/2$	$\pi/2$	$\pi/3$	$\pi/4$	$\pi/4$	$\pi/2$	0.152661
26	$\pi/2$	$\pi/2$	$\pi/4$	$\pi/4$	$\pi/4$	$\pi/2$	0.457983
27	$\pi/3$	$\pi/2$	$\pi/3$	$\pi/3$	$\pi/3$	$\pi/2$	0.084578
28	$\pi/3$	$\pi/2$	$\pi/4$	$\pi/3$	$\pi/3$	$\pi/2$	0.211446
29	$\pi/3$	$\pi/2$	$\pi/5$	$\pi/3$	$\pi/3$	$\pi/2$	0.343003
30	$\pi/3$	$\pi/2$	$\pi/6$	$\pi/3$	$\pi/3$	$\pi/2$	0.507471
31	$\pi/3$	$\pi/3$	$\pi/3$	$\pi/3$	$\pi/3$	$\pi/3$	1.014941
32	$\pi/3$	$\pi/3$	$\pi/3$	$\pi/3$	$\pi/3$	$\pi/2$	0.422892

Table 29 - Reflection groups. Non–submultiples of π

Γ_i^*	$\angle AB$	$\angle AC$	$\angle AD$	$\angle BC$	$\angle BD$	$\angle CD$
1^*	$2\pi/5$	$2\pi/5$	$2\pi/5$	$\pi/3$	$\pi/3$	$\pi/3$
2^*	$2\pi/5$	$\pi/3$	$\pi/3$	$\pi/3$	$\pi/3$	$\pi/3$
3^*	$2\pi/5$	$\pi/3$	$\pi/3$	$\pi/3$	$\pi/3$	$\pi/4$
4^*	$2\pi/5$	$\pi/2$	$\pi/2$	$\pi/3$	$\pi/3$	$\pi/5$
5^*	$2\pi/5$	$\pi/2$	$\pi/2$	$\pi/3$	$\pi/3$	$\pi/6$
6^*	$2\pi/5$	$\pi/6$	$\pi/6$	$\pi/3$	$\pi/3$	$\pi/2$
7^*	$2\pi/5$	$\pi/5$	$\pi/5$	$\pi/3$	$\pi/3$	$\pi/2$
8^*	$2\pi/5$	$\pi/4$	$\pi/4$	$\pi/3$	$\pi/3$	$\pi/2$
9^*	$2\pi/5$	$\pi/3$	$\pi/3$	$\pi/3$	$\pi/3$	$\pi/2$
10^*	$2\pi/5$	$\pi/3$	$\pi/3$	$\pi/2$	$\pi/2$	$\pi/4$
11^*	$2\pi/3$	$\pi/3$	$\pi/3$	$\pi/3$	$\pi/3$	$\pi/3$
12^*	$2\pi/3$	$\pi/2$	$\pi/2$	$\pi/3$	$\pi/3$	$\pi/6$
13^*	$2\pi/3$	$\pi/2$	$\pi/2$	$\pi/5$	$\pi/5$	$\pi/3$
14^*	$2\pi/3$	$\pi/2$	$\pi/2$	$\pi/6$	$\pi/6$	$\pi/3$
15^*	$2\pi/3$	$2\pi/3$	$2\pi/3$	$\pi/6$	$\pi/6$	$\pi/6$
16^*	$2\pi/3$	$\pi/6$	$\pi/6$	$\pi/3$	$\pi/3$	$\pi/2$
17^*	$2\pi/3$	$\pi/5$	$\pi/5$	$\pi/3$	$\pi/3$	$\pi/2$
18^*	$2\pi/3$	$\pi/4$	$\pi/4$	$\pi/4$	$\pi/4$	$\pi/2$
19^*	$2\pi/3$	$\pi/4$	$\pi/4$	$\pi/2$	$\pi/2$	$\pi/4$

References

[1] B. N. Apanasov, *A universal property of Kleinian groups in the hyperbolic metric*, Dokl. Akad. Nauk SSSR **225** (1975), 1418-1421.

[2] A. Beardon, *The Geometry of Discrete Groups*, Springer–Verlag, 1983.

[3] M. Culler and P. Shalen, *Paradoxical decompositions, 2-generator Kleinian groups and volumes of hyperbolic 3-manifolds*, J. Amer. Math. Soc. **5** (1992), 231–288.

[4] D. A. Derevnin and A.D. Mednykh, *Geometric properties of discrete groups acting with fixed points in Lobachevsky space*, Soviet Math. Dokl. **37** (1988), 614–617.

[5] F. W. Gehring and G. J. Martin, *On the Margulis constant for Kleinian groups I*, Ann. Acad. Sci. Fenn. Math. **21** (1996), 439–462.

[6] F. W. Gehring and G. J. Martin, *Minimal covolume hyperbolic lattices I*, to appear.

[7] F. W. Gehring and G. J. Martin, *Commutators, collars and the geometry of Möbius groups*, J. Anal. Math. **63** (1994), 175–219.

[8] F. W. Gehring and G. J. Martin, *On the minimal volume hyperbolic 3-orbifold* Math. Research Letters **1** (1994), 107–114.

[9] T. Inada, *An elementary proof of a theorem of Margulis for Kleinian groups*, Math. J. Okayama Univ. **30** (1988), 177–186.

[10] D. A. Každan and G. A. Margulis, *A proof of Selberg's conjecture*, Math. USSR–Sbornik **4** (1968), 147–152.

[11] A. W. Knapp, *Doubly generated Fuchsian groups*, Michigan Math. J. **15** (1968), 289–304.

[12] A. Marden, *Universal properties of Fuchsian groups in the Poincaré metric*, Discontinuous Groups and Riemann Surfaces, Princeton Univ. Press, 1974, pp. 315–339.

[13] G. A. Margulis, *Discrete groups of motions on manifolds of nonpositive curvature*, Proceedings of the International Congress of Mathematicians (Vancouver, B.C., 1974), Vol. 2, 1975, pp. 21–34.

[14] B. Maskit, *Kleinian Groups*, Springer–Verlag, 1987.

[15] R. Meyerhoff, *A lower bound for the volume of hyperbolic 3-orbifolds*, Duke Math. J. **57** (1988), 185–203.

[16] J.G. Ratcliffe, *Foundations of Hyperbolic Manifolds*, Springer–Verlag, 1994.

[17] E. B. Vinberg, *Hyperbolic reflection groups*, Russian Math. Surveys **40** (1985), 31–75.

[18] A. Yamada, *On Marden's universal constant of Fuchsian groups*, Kodai Math. J. **4** (1981), 266–277.

UNIVERSITY OF MICHIGAN, ANN ARBOR, MI 48109, U.S.A.
E-mail address: `fgehring@umich.edu`

UNIVERSITY OF AUCKLAND, AUCKLAND, NEW ZEALAND
E-mail address: `martin@math.auckland.ac.nz`

Holomorphic Extendibility and the Argument Principle

Josip Globevnik

Dedicated to the memory of Herb Alexander

ABSTRACT. The paper gives the following characterization of the disc algebra in terms of the argument principle: A continuous function f on the unit circle $b\Delta$ extends holomorphically through the unit disc if and only if for each polynomial P such that $f + P \neq 0$ on $b\Delta$, the change of argument of $f + P$ around $b\Delta$ is nonnegative.

1. Introduction and the main result

This paper deals with the problem of characterizing the continuous functions on the unit circle which extend holomorphically into the unit disc, in terms of the argument principle. Generalizing some results of H. Alexander and J. Wermer [**AW**], E.L. Stout [**S**] obtained a characterization of continuous functions on boundaries of certain domains D in $\mathbb{C}^n$, $n \geq 1$, which extend holomorphically through D, in terms of a generalized argument principle. In the special case of Δ, the open unit disc in $\mathbb{C}$, a version of his result is

THEOREM 1.1. [**S**] *A continuous function f on $b\Delta$ extends holomorphically through Δ if and only if*

(1.1) $\quad\begin{cases} \text{for each polynomial } Q \text{ of two complex variables such} \\ \text{that } Q(z, f(z)) \neq 0 \ (z \in b\Delta), \text{ the change of argument of} \\ \text{the function } z \mapsto Q(z, f(z)) \text{ around } b\Delta \text{ is nonnegative.} \end{cases}$

J. Wermer [**W**] showed that it suffices to assume (1.1) only for polynomials of the form $Q(z, w) = w + P(z)$ provided that f is smooth and asked whether the same holds for continuous functions. In the present paper we prove that this indeed is the case.

2000 *Mathematics Subject Classification.* Primary 30E25.
This work was supported in part by the Ministry of Education, Science and Sport of Slovenia through research program Analysis and Geometry, Contract No. P1-0291.

THEOREM 1.2. *A continuous function f on $b\Delta$ extends holomorphically through Δ if and only if*

(1.2) $\quad\begin{cases} \text{for each polynomial } P \text{ such that } f + P \neq 0 \text{ on } b\Delta, \text{ the} \\ \text{change of argument of } f + P \text{ around } b\Delta \text{ is nonnegative.} \end{cases}$

Note that the only if part is an obvious consequence of the argument principle. In fact, if f admits a holomorphic extension $\tilde{f}$, then the change of argument of $f + P$ around $b\Delta$ equals 2π times the number of zeros of $\tilde{f} + P$ in Δ.

2. The Morera condition

LEMMA 2.1. *Let f be a continuous function on $b\Delta$ which satisfies (1.2). Then*

$$\int_{b\Delta} f(z)\,dz = 0.$$

PROOF. Suppose that $\int_{b\Delta} f(z)\,dz \neq 0$. With no loss of generality assume that

$$\frac{1}{2\pi i} \int_{b\Delta} f(z)\,dz = 1.$$

Then $z \mapsto zf(z) - 1$ is a continuous function on $b\Delta$ with zero average. Since polynomials in z and $\bar{z}$ are dense in $C(b\Delta)$, it follows that there are polynomials P and Q and a continuous function g on $b\Delta$ such that

(2.1) $\qquad\qquad\qquad |g(z)| \leq 1/2 \quad (z \in b\Delta)$

and

$$zf(z) - 1 = zP(z) + \overline{zQ(z)} + g(z) \quad (z \in b\Delta).$$

It follows that

$$z[f(z) - P(z) - Q(z)] \in 1 + g(z) + i\mathbb{R} \quad (z \in b\Delta)$$

which, by (2.1), implies that

$$\frac{1}{2} \leq \Im\bigl(z[f(z) - P(z) - Q(z)]\bigr) \leq \frac{3}{2} \quad (z \in b\Delta),$$

so the change of argument of $z \mapsto z[f(z) - P(z) - Q(z)]$ around $b\Delta$ equals zero. Thus, $f - P - Q \neq 0$ on $b\Delta$; and the change of argument of $f - P - Q$ around $b\Delta$ equals -2π, contradicting the assumption that f satisfies (1.2). This completes the proof. □

Before proceeding to the proof of Theorem 1.2, observe that Lemma 2.1 provides a simple alternative proof of Theorem 1.1; in fact, it provides a proof of the following corollary, which sharpens Theorem 1.1.

COROLLARY 2.1. *A continuous function f on $b\Delta$ extends holomorphically through Δ if and only if for each nonnegative integer n and each polynomial P such that $z^n f(z) + P(z) \neq 0$ $(z \in b\Delta)$, the change of argument of the function $z \mapsto z^n f(z) + P(z)$ around $b\Delta$ is nonnegative.*

PROOF. If f satisfies the condition, then Lemma 2.1 implies that

$$\int_{b\Delta} z^n f(z)\,dz = 0$$

for each nonnegative integer n. It is well-known that this implies that f extends holomorphically through Δ. This completes the proof. □

3. Proof of Theorem 1.2

LEMMA 3.1. *Suppose that f is a continuous function on $b\Delta$ that satisfies* (1.2). *Then for each $w \in \mathbb{C} \setminus \overline{\Delta}$, the function $z \mapsto f(z)/(z-w)$ satisfies* (1.2).

PROOF. Suppose that $w \in \mathbb{C} \setminus \overline{\Delta}$ and assume that P is a polynomial such that
$$\frac{f(z)}{z-w} + P(z) \neq 0 \quad (z \in b\Delta).$$
Then
$$f(z) + (z-w)P(z) \neq 0 \quad (z \in b\Delta);$$
and since f satisfies (1.2), it follows that the change of argument of the function $z \mapsto f(z) + (z-w)P(z)$ around $b\Delta$ is nonnegative. Since
$$\frac{f(z)}{z-w} + P(z) = \frac{1}{z-w}\left[f(z) + (z-w)P(z)\right] \quad (z \in b\Delta)$$
and the change of argument of $z \mapsto 1/(z-w)$ around $b\Delta$ is zero, it follows that the change of argument of
$$(3.1) \qquad z \mapsto \frac{f(z)}{z-w} + P(z)$$
around $b\Delta$ is nonnegative. This completes the proof. $\square$

PROOF OF THEOREM 1.2. Suppose that f satisfies (1.2). By Lemma 3.1, for each $w \in \mathbb{C} \setminus \overline{\Delta}$ the function (3.1) satisfies (1.2). By Lemma 2.1, it follows that
$$(3.2) \qquad \int_{b\Delta} \frac{f(z)}{z-w} dz = 0 \quad (w \in \mathbb{C} \setminus \overline{\Delta}).$$
It is well-known that (3.2) implies that f extends holomorphically through Δ. This completes the proof. $\square$

4. An example

Theorem 1.2 gives a simple characterization of continuous functions on $b\Delta$ which extend holomorphically to Δ in terms of the argument principle. One can ask whether one can further simplify this characterization. J.Wermer [**W**] showed that in Theorem 1.2, it is not enough to assume (1.2) for polynomials of the form $P(z) = az + b$, $a, b, \in \mathbb{C}$. In this section, we sharpen this by showing that for Theorem 1.2 to hold, (1.2) must hold for polynomials of arbitrarily large degree.

PROPOSITION 4.1. *For every $n_0 \in \mathbb{N}$, there is a continuous function f on $b\Delta$ such that whenever P is a polynomial of degree not exceeding n_0 such that $f + P \neq 0$ on $b\Delta$, then the change of argument of $f + P$ around $b\Delta$ is nonnegative, yet f does not extend holomorphically through Δ.*

PROOF. Let $n_0 \in \mathbb{N}$ and let $n \in \mathbb{N}$, $n \geq n_0 + 1$. Set
$$f(z) = z^n + \frac{1}{2z} \quad (z \in b\Delta).$$
We claim that for every polynomial P of degree not exceeding n_0 such that $f+P \neq 0$ on $b\Delta$, the change of argument of $f+P$ around $b\Delta$ is nonnegative. Assume, contrary to our claim, that there exists a polynomial P, $\deg(P) \leq n_0$, such that $f + P$ has no zero on $b\Delta$ but the change of argument of $f + P$ around $b\Delta$ is negative.

Since $f + P$ is rational with a single pole in Δ, the argument principle implies that the change of argument of $f + P$ around $b\Delta$ equals $2\pi(\nu - 1)$, where ν is the number of zeros of $f + P$ on Δ. By our assumption, $2\pi(\nu - 1) < 0$, so $f + P$ has no zero on Δ. The zeros of $f + P$ are the zeros of

$$(4.1) \qquad z^{n+1} + zP(z) + \frac{1}{2}.$$

Since $n \geq n_0 + 1$ and the degree of P does not exceed n_0, it follows that the leading term in (4.1) is z^{n+1}. Since the constant term in (4.1) is $1/2$, it follows that the product of zeros of (4.1) equals $1/2$, which implies that at least one of the zeros of (4.1) is contained in Δ. Thus at least one of the zeros of $f + P$ is contained in Δ, a contradiction. This completes the proof. □

5. Holomorphic extendibility to finite Riemann surfaces

Theorem 1.1 has yet another, less elementary but even shorter proof using Wermer's maximality theorem. Suppose that f is a continuous function on $b\Delta$ which satisfies (1.1) and which does not extend holomorphically through Δ. By Wermer's maximality theorem [**H**], the polynomials in z and f are dense in $C(b\Delta)$. In particular, there is a polynomial Q such that

$$|Q(z, f(z)) - \bar{z}| < \frac{1}{2} \qquad (z \in b\Delta).$$

Obviously $Q(z, f(z)) \neq 0$ ($z \in b\Delta$), and the change of argument of $z \mapsto Q(z, f(z))$ around $b\Delta$ equals the change of argument of $z \mapsto \bar{z}$ around $b\Delta$, which equals -2π. Thus, the change of argument of $z \mapsto Q(z, f(z))$ around $b\Delta$ is negative, which contradicts the fact that f satisfies (1.1) and so completes the proof of Theorem 1.1. This proof of Stout's theorem was the first which the author found. It was only after a careful reading of Cohen's proof of Wermer's maximality theorem [**C**] that the author found the proof of Lemma 2.1, which gives a more elementary proof of Stout's theorem. J. Wermer has kindly pointed out that the preceding proof generalizes to finite Riemann surfaces.

THEOREM 5.1. *Let X be a finite Riemann surface with boundary Γ. Let A be the algebra of all continuous functions on Γ which extend holomorphically through X. A continuous function f on Γ extends holomorphically through X if and only if for every polynomial P with coefficients in A such that $P(f) \neq 0$ on Γ, the change of argument of $P(f)$ along Γ is nonnegative.*

PROOF. Suppose that $f \in C(\Gamma)$ has the property that the change of argument of $P(f)$ along Γ is nonnegative whenever P is a polynomial with coefficients in A such that $P(f) \neq 0$ on Γ. Suppose that f does not extend holomorphically through X. By the maximality of A in $C(\Gamma)$ [**R**], the functions of the form $P(f)$, where P is a polynomial with coefficients in A, are dense in $C(\Gamma)$. Choose $g \in C(\Gamma)$ such that $|g| = 1$ on Γ and the change of argument of g along Γ equals -2π. There is a polynomial P with coefficients in A such that $|P(f)(z) - g(z)| < 1/2$ ($z \in \Gamma$). Since $|g| = 1$ on Γ, the change of argument of $P(f)$ along Γ equals the change of argument of g along Γ. So the change of argument of $P(f)$ along Γ is negative, contradicting the hypothesis. This proves that f extends holomorphically through X. The only if part follows from the argument principle. This completes the proof. □

Mark Agranovsky observed that a substantially longer argument in the proof of Theorem 1.1 in the original version of the paper can be replaced by Lemma 3.1, which is due to him. The author thanks him for his kind permission to include Lemma 3.1 in the final version of the paper. The author also thanks John Wermer for pointing out that Theorem 5.1 follows from the maximality theorem.

Finally, the author thanks Larry Zalcman for the kind invitation to include this paper in the proceedings of the Nahariya Conference.

References

[AW] H. Alexander and J. Wermer, *Linking numbers and boundaries of varieties*, Ann. of Math. (2) **151** (2000), 125-150.

[C] P.J. Cohen, *A note on constructive methods in Banach algebras*, Proc. Amer. Math. Soc. **12** (1961), 159-163.

[H] K. Hoffman, *Banach Spaces of Analytic Functions*, Prentice-Hall, Englewood Cliffs, N.J., 1962.

[R] H. Royden, *The boundary values of analytic and harmonic functions*, Math. Z. **78** (1962), 1-24.

[S] E.L. Stout, *Boundary values and mapping degree*, Michigan Math. J. **47** (2000), 353-368.

[W] J. Wermer, *The argument principle and boundaries of analytic varieties*, Recent Advances in Operator Theory and Related Topics (Szeged, 1999), Birkhäuser, Basel, 2001, pp. 639-659.

INSTITUTE OF MATHEMATICS, PHYSICS AND MECHANICS, UNIVERSITY OF LJUBLJANA, LJUBLJANA, SLOVENIA

E-mail address: `josip.globevnik@fmf.uni-lj.si`

Homeomorphisms with Finite Mean Dilatations

Anatoly Golberg

Dedicated to Professor Lawrence Zalcman on the occasion of his 60th birthday

In geometric function theory, quasiconformal homeomorphisms form a natural interpolating class of mappings between the classes of bilipschitz mappings and general homeomorphisms. Under q-quasiconformal mappings, the n-module of a curve family can change by a factor of at most q. All properties of quasiconformal mappings can be obtained from this simple inequality.

In this paper, we consider homeomorphisms whose dilatations are bounded in a certain integral sense. The resulting notion generalizes quasiconformal mappings, mappings quasiconformal in the mean, etc. The main tool for investigation relies on the method of p-moduli of curve families and surface families and involves more general inequalities than quasi-invariance.

1. Quasiconformal Dilatations

Let $A: \mathbb{R}^n \to \mathbb{R}^n$ be a linear bijection. The numbers

$$H_I(A) = \frac{|\det A|}{l^n(A)}, \quad H_O(A) = \frac{L^n(A)}{|\det A|}, \quad H(x,f) = \frac{L(A)}{l(A)},$$

are called the *inner*, the *outer* and the *linear* dilatations of A, respectively. Here

$$l(A) = \min_{|h|=1} |Ah|, \quad L(A) = \max_{|h|=1} |Ah|,$$

and $\det A$ is the determinant of A (see, e.g., [18]).

Obviously, all three dilatations are not less than 1. They have the following geometric interpretation. The image of the unit ball B^n under A is an ellipsoid $E(A)$. Let $B_I(A)$ and $B_O(A)$ be the inscribed and the circumscribed balls of $E(A)$, respectively.

Then

$$H_I(A) = \frac{mE(A)}{mB_I(A)}, \quad H_O(A) = \frac{mB_O(A)}{mE(A)},$$

and $H(A)$ is the ratio of the greatest and the smallest semi-axis of $E(A)$. Here $mD = m_n D$ denotes the n-dimensional Lebesgue measure of a set D.

2000 *Mathematics Subject Classification.* Primary 30C65.

Let $\lambda_1 \geq \lambda_2 \geq \ldots \geq \lambda_n$ be the semi-axes of $E(A)$. Then
$$L(A) = \lambda_1, \quad l(A) = \lambda_n, \quad |\det A| = \lambda_1 \cdot \ldots \cdot \lambda_n,$$
and we can also write
$$H_I(A) = \frac{\lambda_1 \cdot \ldots \cdot \lambda_{n-1}}{\lambda_n^{n-1}}, \quad H_O(A) = \frac{\lambda_1^{n-1}}{\lambda_2 \cdot \ldots \cdot \lambda_n}, \quad H(A) = \frac{\lambda_1}{\lambda_n}.$$

If $n = 2$, then $H_I(A) = H_O(A) = H(A)$. In the general case, we have the relations

(1)
$$H(A) \leq \min(H_I(A), H_O(A)) \leq H^{n/2}(A)$$
$$\leq \max(H_I(A), H_O(A)) \leq H^{n-1}(A).$$

Let G and G^* be two bounded domains in $\mathbb{R}^n$, $n \geq 2$, and let the mapping $f : G \to G^*$ be differentiable at the point $x \in G$. This means there exists a linear mapping $f'(x) : \mathbb{R}^n \to \mathbb{R}^n$, called the (strong) derivative of the mapping f at x, such that
$$f(x+h) = f(x) + f'(x)h + \omega(x,h)|h|,$$
where $\omega(x,h) \to 0$ as $h \to 0$.

We denote
$$H_I(x,f) = H_I(f'(x)), \quad H_O(x,f) = H_O(f'(x)),$$
and
$$L(x,f) = L(f'(x)), \quad l(x,f) = l(f'(x)), \quad J(x,f) = J(f'(x)).$$

PROPOSITION. *Let $f : G \to G^*$ be a q-quasiconformal homeomorphism. Then*
 (i) *f is ACL (absolutely continuous on lines);*
 (ii) *$f \in W^1_{n,loc}(G)$ (Sobolev class);*
 (iii) *for almost every $x \in G$,*
$$H_I(x,f) \leq q, \quad H_O(x,f) \leq q.$$

F.W. Gehring [2] proved that $f \in W^1_{p,loc}(G)$ for $p \in [n, n+c)$, where $c \leq n/(q^{1/(n-1)} - 1)$. In the planar case, this fact was first discovered by B. Bojarski [1].

2. Integrable bounded dilatations

The various classes of spatial mappings quasiconformal in the mean have been studied intensively for more than 40 years (see, e.g., [9]). For the contemporary development of this direction, we refer to [13].

Now we consider the quantities
$$H_{I,\alpha}(A) = \frac{|J(A)|}{l^\alpha(A)}, \quad H_{O,\alpha}(A) = \frac{L^\alpha(A)}{|J(A)|}$$
with $\alpha \geq 1$.

For a homeomorphism f which is differentiable a.e. in G and real numbers α, β such that $1 \leq \alpha < \beta < \infty$, we introduce the integrals

(2) $$HI_{\alpha,\beta}(f) = \int_G H_{I,\alpha}^{\frac{\beta}{\beta-\alpha}}(x,f)\,dx, \quad HO_{\alpha,\beta}(f) = \int_G H_{O,\beta}^{\frac{\alpha}{\beta-\alpha}}(x,f)\,dx,$$

where $H_{I,\alpha}(x,f) = H_{I,\alpha}(f'(x))$, $H_{O,\beta}(x,f) = H_{O,\beta}(f'(x))$. We call these the *inner* and *outer* mean dilatations of the mapping f in G, respectively.

The characteristics (2) were introduced in [**11**]. See also [**4**] and [**19**].

For fixed real numbers $\alpha, \beta, \gamma, \delta$ such that $1 \le \alpha < \beta < \infty$, $1 \le \gamma < \delta < \infty$, we define the *class $B(G)$* of such homeomorphisms $f : G \to G^*$ which satisfy:

(iv) f and f^{-1} are ACL-homeomorphisms,
(v) f and f^{-1} are differentiable with Jacobians $J(x, f) \ne 0$ and $J(y, f^{-1}) \ne 0$ a.e. in G and G^*, respectively,
(vi) the inner and the outer mean dilatations $HI_{\alpha,\beta}(f)$ and $HO_{\gamma,\delta}(f)$ are finite.

We call $B(G)$ the *class of mappings with finite mean dilatations*. The following theorem describes some of the basic differential properties of mappings in the class $B(G)$.

Recall that a homeomorphism $h : G \to G^*$ satisfies the N-property if $mD = 0$ implies $mf(D) = 0$ for each $D \subset G$.

THEOREM 1. *Suppose that $\alpha, \beta, \gamma, \delta$ are fixed real numbers such that $n - 1 \le \alpha < \beta < \infty$, $n - 1 \le \gamma < \delta < \infty$. Then the mappings of $B(G)$ belong to the Sobolev class $W^1_{p,loc}(G)$, while $f^{-1} \in W^1_{q,loc}(G^*)$, with $p = \max(\gamma, \beta/(\beta - n + 1))$ and $q = \max(\alpha, \delta/(\delta - n + 1))$.*

PROOF. It follows from the assumption that the mappings f and f^{-1} satisfy the N-property in the domains G and G^*, respectively. This property allows us to apply the standard rule for a change of variables under integration. Then the Hölder inequality and condition (vi) yield

$$\int_G L^\gamma(x, f) \, dx = \int_G \left[\left(\frac{L^\delta(x,f)}{|J(x,f)|}\right)^{\frac{\gamma}{\delta-\gamma}}\right]^{\frac{\delta-\gamma}{\delta}} |J(x,f)|^{\frac{\gamma}{\delta}} dx \le HO_{\gamma,\delta}^{\frac{\delta-\gamma}{\delta}}(f)(mG^*)^{\frac{\gamma}{\delta}} < \infty,$$

$$\int_G L^{\frac{\beta}{\beta-n+1}}(x, f) \, dx \le \int_G \left(\frac{|J(x,f)|}{l^{n-1}(x,f)}\right)^{\frac{\beta}{\beta-n+1}} dx = HI_{n-1,\beta}(f)$$

$$\le HI_{\alpha,\beta}^{\frac{(\alpha-n+1)\beta}{\alpha(\beta-n+1)}}(f)(mG^*)^{\frac{(\beta-\alpha)(n-1)}{\alpha(\beta-n+1)}} < \infty.$$

The conditions (iv)-(vi) provide that the mappings $f \in B(G)$ satisfy

$$HI_{\alpha,\beta}(f^{-1}) = HO_{\alpha,\beta}(f), \quad HO_{\alpha,\beta}(f^{-1}) = HI_{\alpha,\beta}(f);$$

therefore, if we denote the class $B(G)$ by $B(G, G^*, \alpha, \beta, \gamma, \delta)$, we have $f^{-1} \in B(G^*, G, \gamma, \delta, \alpha, \beta)$. This completes the proof of the theorem. $\square$

The relations (1) show that in the classical case of quasiconformal mappings, their dilatations are simultaneously either finite or infinity. However, this is not true for the mean dilatations. The following example shows that each of the mean dilatations $HI_{\alpha,\beta}(f)$ and $H_{\gamma,\delta}(f)$ can be unbounded, independently of restrictions on any other dilatation.

EXAMPLE. Let

$$G = \{x = (x_1, \ldots, x_n) : 0 < x_k < 1, k = 1, \ldots, n\}$$

and

$$g(x) = \left(x_1, \ldots, x_{n-1}, \frac{x_n^{1-c}}{1-c}\right), \quad 0 < c < 1.$$

Then the image $g(G)$ is the domain
$$G^* = \{y = (y_1, \ldots, y_n) : 0 < y_k < 1, k = 1, \ldots, n-1, 0 < y_n < 1/(1-c)\}.$$
It is easily to verify that g is differentiable in G and
$$l(x,g) = 1, \quad L(x,g) = J(x,g) = x_n^{-c} > 1.$$
Thus
$$H_{I,\alpha}(x,g) = \frac{J(x,g)}{l^\alpha(x,g)} = x_n^{-c}, \quad H_{O,\delta}(x,g) = \frac{L^\delta(x,g)}{J(x,g)} = x_n^{-c(\delta-1)},$$

$$HI_{\alpha,\beta}(g) = \int_G H_{I,\alpha}^{\frac{\beta}{\beta-\alpha}}(x,g)\,dx = \int_0^1 x_n^{-\frac{c\beta}{\beta-\alpha}}\,dx_n,$$

$$HO_{\gamma,\delta}(g) = \int_G H_{O,\delta}^{\frac{\gamma}{\delta-\gamma}}(x,g)\,dx = \int_0^1 x_n^{-\frac{c(\delta-1)\gamma}{\delta-\gamma}}\,dx_n,$$

One concludes from this that
$$HI_{\alpha,\beta}(g) < \infty \iff 0 < c < 1 - \alpha/\beta,$$
$$HI_{\alpha,\beta}(g) = \infty \iff 1 - \alpha/\beta \le c < 1,$$
$$HO_{\gamma,\delta}(g) < \infty \iff 0 < c < 1 - (\gamma-1)\delta/(\delta-1)\gamma,$$
$$HO_{\gamma,\delta}(g) = \infty \iff 1 - (\gamma-1)\delta/(\delta-1)\gamma \le c < 1.$$

The above example shows also that the class $B(G)$ is much wider than the class of quasiconformal mappings. For example, the mapping g belongs to $B(G)$ if
$$c \le \min\{1 - \alpha/\beta, 1 - (\gamma-1)\delta/(\delta-1)\gamma\},$$
but g is not q-quasiconformal for any c.

3. p-moduli of the families of k-dimensional surfaces

Let $\mathcal{S}^k$ be a family of k-dimensional surfaces $\mathcal{S}$ in $\mathbb{R}^n$, $1 \le k \le n-1$. $\mathcal{S}$ is a k-dimensional surface if $\mathcal{S} : D_s \to \mathbb{R}^k$ is a continuous image of the closed domain $D_s \subset \mathbb{R}^k$.

The p-modulus of $\mathcal{S}^k$ is defined as
$$M_p(\mathcal{S}^k) = \inf \int_{\mathbb{R}^n} \rho^p\,dx, \quad p \ge 1,$$
where the infimum is taken over all Borel measurable functions $\rho \ge 0$ such that
$$\int_S \rho^k\,d\sigma_k \ge 1$$
for every $\mathcal{S} \in \mathcal{S}^k$. We call each such ρ an *admissible function* for $\mathcal{S}^k$.

The following proposition characterizes quasiconformality in terms of the p-moduli of $\mathcal{S}_k$ (see, [**16**]).

A homeomorphism f of a domain $G \subset \overline{\mathbb{R}^n}$ is q-quasiconformal, $1 \leq q < \infty$, if for any family $\mathcal{S}^k$, $1 \leq k \leq n-1$, of k-dimensional surfaces in G the double inequality

$$q^{k-n} M_n(\mathcal{S}^k) \leq M_n(f(\mathcal{S}^k)) \leq q^{n-k} M_n(\mathcal{S}^k) \tag{3}$$

holds.

The following theorem generalizes the double inequality (3).

THEOREM 2. *Let $f : G \to G^*$ be a homeomorphism satisfying*
 (vii) *f and f^{-1} are ACL;*
 (viii) *f and f^{-1} are differentiable a.e. in G and G^*, respectively;*
 (ix) *the Jacobians $J(x, f)$ and $J(y, f^{-1})$ do not vanish a.e. in G and G^*, respectively.*

Then for every quadruple of fixed values $\alpha, \beta, \gamma, \delta$ such that $1 \leq \alpha < \beta < \infty$, $1 \leq \gamma < \delta < \infty$ and for any ring domain $D \subset G$, the inequalities

$$M_\alpha^\beta(\mathcal{S}_k^*) \leq HI_{\alpha,\beta}^{\beta-\alpha}(f) M_\beta^\alpha(\mathcal{S}_k), \tag{4}$$

$$M_\gamma^\delta(\mathcal{S}_k) \leq HO_{\gamma,\delta}^{\delta-\gamma}(f) M_\delta^\gamma(\mathcal{S}_k^*), \tag{5}$$

hold, where $\mathcal{S}_k^* = f(\mathcal{S}_k)$.

PROOF. Let $D \subset G$ be a ring domain and let ρ be an admissible function for $\mathcal{S}_k$. Denote by $\mu_k(x)$ the minimal distortion of k-dimensional measures at x under f, i. e.,

$$\mu_k(x) = \lambda_n \cdot \lambda_{n-1} \cdot \ldots \cdot \lambda_{n-k+1}.$$

Note that $\mu_k(x) \geq l^k(x, f)$ for a.e. $x \in G$. Define in $D^* = f(D)$ the function

$$\rho^*(y) = \frac{\rho(x)}{\left[\mu_k(x)\right]^{1/k}},$$

where $x = f^{-1}(y)$. It is easy to check that $\rho^*(y)$ is an admissible function for $\mathcal{S}_k^*$.

Since $d\sigma_k^* \geq \mu_k(x) d\sigma_k$, we have

$$\int_{\mathcal{S}^*} \rho^{*k}(y) d\sigma_k^* \geq \int_{\mathcal{S}} \rho^k(x) d\sigma_k \geq 1$$

for every surface $\mathcal{S}^* \in \mathcal{S}_k^*$.

We conclude from (vii)–(ix) that f and f^{-1} satisfy the N-property in G and G^*, respectively, and

$$H_{I,\alpha}(x, f) = H_{O,\alpha}(y, f^{-1})$$

for a.e. $x \in G$ and $y \in G^*$. Applying Hölder's inequality and the properties of f and f^{-1}, we obtain

$$\int_{D^*} \rho^{*\alpha}(y) \, dy = \int_D \frac{\rho^\alpha(x)}{\left[\mu_k(x)\right]^{\alpha/k}} |J(x, f)| \, dx \leq \int_D \rho^\alpha(x) \frac{|J(x, f)|}{l^\alpha(x, f)} \, dx$$

$$\leq \left(\int_D \rho^\beta(x) \, dx\right)^{\alpha/\beta} \left(\int_D H_{I,\alpha}^{\frac{\beta}{\beta-\alpha}}(x, f) \, dx\right)^{(\beta-\alpha)/\beta}.$$

Taking the infima over all such $\rho(x)$ yields (4). Interchanging $f : G \to G^*$ and $f^{-1} : G^* \to G$ in (4), we obtain the inequality (5). $\square$

4. p-moduli of the families of curves and of $(n-1)$-dimensional surfaces

A ring domain $D \subset \mathbb{R}^n$ is a finite domain whose complement consists of two components C_0 and C_1. Setting $F_0 = \partial C_0$ and $F_1 = \partial C_1$, we obtain two boundary components of D. For definiteness, let us assume that $\infty \in C_1$.

We say that a curve γ *joins the boundary components in D* if γ lies in D, except for its endpoints, one of which lies on F_0 and the second on F_1. A compact set Σ is said to *separate the boundary components of D* if $\Sigma \subset D$ and if C_0 and C_1 are located in the different components of $\widehat{\mathbb{R}} \setminus \Sigma$. Denote by Γ_D the family of all locally rectifiable curves γ which join the boundary components of D and by Σ_D the family of all compact piecewise smooth $(n-1)$-dimensional surfaces Σ which separate the boundary components of D. For each quantity V associated with D such as subset of D or a family of sets contained in D, we let V^* denote its image under f.

Now we introduce new classes of homeomorphisms. These classes depend on the values of parameters $\alpha, \beta, \gamma, \delta$ and some set functions.

Let Φ be a finite nonnegative function in G defined for open subsets E of G such that
$$\sum_{k=1}^{m} \Phi(E_k) \leq \Phi(E)$$
for any finite collection $\{E_k\}_{k=1}^{m}$ of nonintersecting open sets $E_k \subset E$. We denote the class of all such set functions Φ by $\mathcal{F}$.

We fix the numbers $\alpha, \beta, \gamma, \delta$ satisfying
$$n-1 \leq \alpha < \beta < \infty, \quad n-1 \leq \gamma < \delta < \infty,$$
and assume that there exists a nonempty family of homeomorphisms $f : G \to G^*$ such that there are two set functions $\Phi, \Psi \in \mathcal{F}$ not depending on f so that for each ring domain $D \subset G$ the inequalities

(6) $$M_\alpha^\beta(\Sigma_D^*) \leq \Phi^{\beta-\alpha}(D) M_\beta^\alpha(\Sigma_D),$$

(7) $$M_\gamma^\delta(\Sigma_D) \leq \Psi^{\delta-\gamma}(D) M_\delta^\gamma(\Sigma_D^*),$$

hold. The class of such homeomorphisms will be denoted by $\mathcal{MS}(G)$ (in fact, it depends also on $\alpha, \beta, \gamma, \delta$).

We shall need the following theorem from [5].

THEOREM 3. *Let*
$$n-1 < \alpha < \beta < \infty \quad \text{and} \quad n-1 < \gamma < \delta < \infty.$$
Then every mapping $f \in \mathcal{MS}(G)$ admits the following properties:
(a) f is ACL in G;
(b) f^{-1} is ACL in G^*;
(c) $f \in W_{a,loc}^1(G)$, $a = \beta/(\beta - n + 1)$;
(d) $f^{-1} \in W_{b,loc}^1(G^*)$, $b = \delta/(\delta - n + 1)$.

It is not hard to see that if $\beta \leq n$, then $\beta/(\beta - n + 1) \geq n$ and if $\beta \geq n$, then $\beta/(\beta - n + 1) \leq n$.

Now we introduce the mapping class $\mathcal{MJ}(G)$, using p-moduli of the families of joining curves. Fix the numbers p, q, s, t which satisfy
$$1 \leq p < q < \infty, \quad 1 \leq s < t < \infty.$$

Suppose that there exists a (nonempty) family of homeomorphisms $f : G \to G^*$ such that for every ring domain $D \subset G$,

(8) $$M_s^t(\Gamma_D^*) \leq \Theta^{t-s}(D) M_t^s(\Gamma_D),$$

(9) $$M_p^q(\Gamma_D) \leq \Pi^{q-p}(D) M_q^p(\Gamma_D^*),$$

where Θ and Π are two given set functions in $\mathcal{F}$ not depending from f.

The properties of the mappings satisfying inequalities (8) and (9) in the equivalent terms of p-capacity were investigated in [11] for $n-1 < s < t \leq n$ and $n-1 < p < q \leq n$ and in [19] for the wider bounds $n-1 < s < t < \infty$ and $n-1 < p < q < \infty$. It was proved in [11] that f and f^{-1} are ACL and belong to $W^1_{t/(t-n+1),loc}$ and $W^1_{q/(q-n+1),loc}$, respectively. In [19], the inequality (9) was extended to the mappings of Carnot groups; and using only this inequality, the authors have established various properties of such mappings. One of these states that if a given mapping belongs to $W^1_{p,loc}$, then the inverse mapping belongs to $W^1_{q/(q-n+1),loc}$. The case $\beta = \delta = n$ has been explicitly studied in [8].

The relations between the p-capacities and the α-moduli of families of separating sets was obtained by W.P. Ziemer [20]. For the equality of p-capacity and p-moduli of families of joining curves, we refer to [7]. Ziemer has considered the condition

$$\int_s \rho \, d\sigma_{n-1} \geq 1$$

and established that

$$M_p(\Gamma_D) = M^{1-p}_{\frac{p}{p-1}}(\Sigma_D).$$

In our case

$$M_p(\Gamma_D) = M^{1-p}_{\frac{p(n-1)}{p-1}}(\Sigma_D).$$

It is important to observe that

$$1 < p < n \iff n < \frac{p(n-1)}{p-1} < \infty,$$

$$p = n \iff \frac{p(n-1)}{p-1} = n,$$

$$n < p < \infty \iff 1 < \frac{p(n-1)}{p-1} < n.$$

Suppose that

$$\alpha = \frac{q(n-1)}{q-1}, \quad \beta = \frac{p(n-1)}{p-1}, \quad \gamma = \frac{t(n-1)}{t-1}, \quad \delta = \frac{s(n-1)}{s-1}.$$

It is easy to verify that the inequalities (6) and (9) are equivalent when $\Pi(D) = \Phi(D)$. The same assertion is true for (7) and (8) if $\Theta(D) = \Psi(D)$. The above conclusions result in

THEOREM 4. *Let*

$$n-1 < \alpha < \beta \leq n \quad \text{and} \quad n-1 < \gamma < \delta \leq n$$

or

$$n \leq \alpha < \beta < \frac{(n-1)^2}{n-2} \quad \text{and} \quad n \leq \gamma < \delta < \frac{(n-1)^2}{n-2}.$$

Then every mapping $f \in \mathcal{MS}(G)$ admits the following properties:

(a') f is ACL in G;
(b') f^{-1} is ACL in G^*;
(c') $f \in W^1_{a,loc}(G)$ with $a = \max\left(\gamma, \beta/(\beta - n + 1)\right)$;
(d') $f^{-1} \in W^1_{b,loc}(G^*)$ with $b = \max\left(\alpha, \delta/(\delta - n + 1)\right)$.

REMARKS. 1. To get the properties (a') and (b'), it suffices to apply only one in the inequality pairs (6), (7) or (8), (9).
2. The properties (c') and (d') yield that the both mappings f and f^{-1} have the N-property (cf. [**10**]).

5. Quasiconformal and quasiisometric mappings

In the case in which the set functions Φ and Ψ are of more special type, we obtain a new characterization of quasiconformality and of quasi-isometry.

THEOREM 5. *A homeomorphism $f : G \to G^*$ is K-quasiconformal if and only if there exists a constant K, $1 \leq K < \infty$, such that for any ring domain $D \subset G$, the inequalities*

$$(10) \qquad M_p^n(\Sigma_D^*) \leq K^{\frac{p}{n-1}} \left(mD^*\right)^{n-p} M_n^p(\Sigma_D),$$

$$(11) \qquad M_q^n(\Sigma_D) \leq K^{\frac{q}{n-1}} \left(mD\right)^{n-q} M_n^q(\Sigma_D^*),$$

hold for $n - 1 < p < q \leq n$ or the inequalities

$$(12) \qquad M_n^p(\Sigma_D^*) \leq K^{\frac{p}{n-1}} \left(mD\right)^{p-n} M_p^n(\Sigma_D),$$

$$(13) \qquad M_n^q(\Sigma_D) \leq K^{\frac{q}{n-1}} \left(mD^*\right)^{q-n} M_q^n(\Sigma_D^*),$$

hold for $n \leq p < q < (n-1)^2/(n-2)$.

SKETCH OF PROOF. The case of relations (12)–(13) follows in fact from the results of [**12**]. Thus it remains only to prove the theorem for inequalities (10)–(11). These inequalities can be obtained from (3) applying Hölder's inequality to $(n-1)$-dimensional surfaces. The proof of sufficiency is accomplished in three stages following to the classical work [**14**] (see also [**8**]).

First we prove that f is ACL. Let $\Theta(V) = mV$ and Q be an n-dimensional interval in G. Then the set function Θ belongs to the class $\mathcal{F}$. Write $Q = Q_0 \times J$, where Q_0 is an $(n-1)$-dimensional interval in $\mathbb{R}^{n-1}$ and J is an open segment of the axis x_n.

Using the notations of [**15**], we can write $\Theta(T, Q) = \Theta(T \times J)$. The function $\Theta(T, Q)$ also belong to the class $\mathcal{F}$ for the Borel sets $T \subset Q_0$.

Fix $z \in Q_0$ so that $\overline{\Theta}'(z, Q) < \infty$, and let $\Delta_1, \ldots, \Delta_k$ be the disjoint closed subintervals of the segment $J_z = \{z\} \times J$. Put $C_{0,i} = \Delta_i + r\overline{B}^n$ and $A_i = \Delta_i + 2rB^n$, where $\overline{B}^n$ is the closure of $B^n = B^n(x, r)$. The positive number r is chosen so that the domains $D_i = A_i \setminus C_{o,i}$ are disjoint and $D_i \subset Q$. Using the estimates of p-moduli of Σ_D, we obtain

$$\sum_{i=1}^k d(A_i^*) \leq C_1 \left(\overline{\Theta}'(z, Q)\right)^{\frac{1}{n}} \left(\sum_{i=1}^k m_1 \Delta_i\right)^{\frac{n-1}{n}},$$

where the constant C_1 depends only on p, n and K. Thus f is ACL.

In the second step, we show that a given homeomorphism f belongs to $W^1_{n,loc}$. For $\tilde{x}, x \in G$, $\tilde{x} \neq x$, define
$$k(x) = \limsup_{\tilde{x} \to x} \frac{|f(\tilde{x}) - f(x)|}{|\tilde{x} - x|}.$$
Consider for a point $x \in D$ the spherical ring $D_r(x) = \{y : r < |x - y| < 2r\}$, choosing $r > 0$ so that $D_r(x) \subset G$. It follows form the estimate
$$k(x) \leq \limsup_{r \to 0} \frac{d(A_r^*)}{r} \leq C_2\big(\overline{\Theta}'(x)\big)^{\frac{1}{n}},$$
that $k(x) < \infty$ a.e. in G. Here C_2 depends only on n and K. Applying now the Stepanov theorem [**17**], one concludes that f is differentiable a.e. in G. Moreover, for an arbitrary Borel set $V \subset G$, we have
$$\int_V k^n(x)\,dx \leq C_2 \int_V \overline{\Theta}'(x)\,dx \leq C_2 mV < \infty.$$

Finally, from (10) and (11), we obtain
$$H_O(x, f) \leq K, \qquad H_I(x, f) \leq K,$$
respectively. This completes the proof of Theorem 5. □

Replacing (10)–(11) and (12)–(13) by suitable inequalities for α-moduli with $\alpha \neq n$ (given explicitly), one obtains quasi-isometry for the mapping.

Recall that a homeomorphism $f : G \to G^*$ is called *quasi-isometric* if for any $x, z \in G$ and $y, t \in G^*$, the inequalities
$$\limsup_{z \to x} \frac{|f(x) - f(z)|}{|x - z|} \leq M, \quad \limsup_{t \to y} \frac{|f^{-1}(y) - f^{-1}(t)|}{|y - t|} \leq M,$$
hold, with a constant M depending only on G and G^* (see [**6**]).

THEOREM 6. *Let $f : G \to G^*$ be a homeomorphism, then the following conditions are equivalent:*
1^0. *f is quasi-isometric;*
2^0. *for fixed real numbers $\alpha, \beta, \gamma, \delta$ such that*
$$n - 1 < \alpha < \beta < n \quad \text{and} \quad n - 1 < \gamma < \delta < n$$
or
$$n < \alpha < \beta < (n-1)^2/(n-2) \quad \text{and} \quad n < \gamma < \delta < (n-1)^2/(n-2),$$
there exists a constant K such that for any ring domain $D \subset G$, the inequalities
$$M_\alpha^\beta(\Sigma_D^*) \leq K^\beta (mD)^{\beta - \alpha} M_\beta^\alpha(\Sigma_D),$$
$$M_\gamma^\delta(\Sigma_D) \leq K^\delta (mD^*)^{\delta - \gamma} M_\delta^\gamma(\Sigma_D^*)$$
hold.

The quasi-invariance of p-moduli of curve or surface families is a characteristic property of quasi-isometry (see, e.g., [**6**]). This quasi-invariance is represented by a double inequality. The implication $1^0 \Rightarrow 2^0$ follows from this inequality by applying Hölder's inequality. The inverse implication $2^0 \Rightarrow 1^0$ is proved using the estimates which are given in [**3**].

References

[1] B. Bojarski, *Homeomorphic solutions of Beltrami systems*, Dokl. Akad. Nauk SSSR **102** (1955), 661–664.

[2] F. W. Gehring, *The L^p - integrability of the partial derivatives of a quasiconformal mapping*, Acta Math. **130** (1973), 265–277.

[3] F. W. Gehring, *Lipschitz mappings and the p-capacity of rings in n-space*, Advances in the Theory of Riemann Surfaces, Princeton University Press, 1971, pp. 175-193.

[4] A. Golberg, *Some classes of plane topological mappings with generalized first derivatives*, Ukrainian Math. J. **44** (1992), 1016–1018.

[5] A. Golberg, *Distortions of p-moduli of separating sets under mappings with finite integral dilatations*, Bull. Soc. Sci. Lettres Łódź Sér. Rech. Déform. **40** (2003), 41-51.

[6] V. M. Gol'dshtein and Yu. G. Reshetnyak, *Quasiconformal Mappings and Sobolev Spaces*, Kluwer Academic Publishers, 1990.

[7] J. Hesse, *A p-extremal length and p-capacity*, Ark. Mat. **13** (1975), 131–144.

[8] V. I. Kruglikov, *Capacities of condensors and quasiconformal in the mean mappings in space*, Mat. Sb. **130** (1986), no. 2, 185–206.

[9] S. L. Krushkal, *On mean quasiconformal mappings*, Doklady Akad. Nauk SSSR **157** (1964), 517–519.

[10] S. L. Krushkal, *The absolute continuity and differentiability of certain classes of maps of many-dimensional regions*, Sibirsk. Mat. Z. **6** (1965), 692–696.

[11] V. S. Kud'yavin, *A characteristic property of a class of n-dimensional homeomorphisms*, Dokl. Akad. Nauk Ukrain. SSR Ser. A **1990**, no. 3, 7–9.

[12] V. S. Kud'yavin, *Quasiconformal mappings and α-moduli of families of curves*, Dokl. Akad. Nauk Ukrainy **1992**, no. 7, 11–13.

[13] O. Martio, V. Ryazanov, U. Srebro, and E. Yakubov, *Mappings with finite length distortion*, J. Anal. Math., to appear.

[14] O. Martio, S. Rickman, and J. Väisälä, *Definitions for quasiregular mappings*, Ann. Acad. Sci. Fenn. Ser. A I Math. **448** (1969).

[15] T. Rado and P. Reichelderfer, *Continuous Transformations in Analysis*, Springer-Verlag, 1955.

[16] B. V. Shabat, *The modulus method in space*, Soviet Math. Dokl. **130** (1960), 1210–1213.

[17] V. V. Stepanov, *Sur les conditions de l'existence de la différentielle total*, Mat. Sb. **30** (1924), 487–489.

[18] J. Väisälä, *Lectures on n-dimensional Quasiconformal Mappings*, Springer-Verlag, 1971.

[19] S. Vodop'yanov and A. Ukhlov, *Sobolev spaces and (P,Q)-quasiconformal mappings of Carnot groups*, Siberian Math. J. **39** (1998), 665–682.

[20] W. P. Ziemer, *Extremal length and p-capacity*, Michigan Math. J. **16** (1969), 43–51.

DEPARTMENT OF MATHEMATICS, BAR-ILAN UNIVERSITY, 52900 RAMAT-GAN, ISRAEL
E-mail address: `golbera@macs.biu.ac.il`

On a Connection Between the Number of Poles of a Meromorphic Function and the Number of Zeros of its Derivatives

Anatolii Gol'dberg

Dedicated to Professor Lawrence Zalcman on his sixtieth birthday

ABSTRACT. Let f be a meromorphic function in the finite plane. We prove an estimate of the function $\overline{N}(r, f)$ from above by using a sum of the functions $N(r, 1/f^{(k)})$ for certain sets of $k \in \mathbb{N}$ without restrictions on the multiplicity of poles.

Let f be a meromorphic function in the finite plane. We make use of the standard notations of Nevanlinna theory [4], [5]. We also denote by $Q(r, f)$ any quantity for which

$$Q(r, f) = O(\log r) + O(\log T(r, f))$$

when $r \to \infty$ outside a set $E \subset \mathbb{R}_+$ such that $\operatorname{mes} E < \infty$.

Denote the set of poles of f by $P(f)$ and its set of zeros by $Z(f)$. Let a_1, a_2, a_3 be distinct complex numbers. The Second Main Theorem of value distribution theory implies the inequality

(1) $\quad N(r, f) < N(r, 1/(f - a_1)) + N(r, 1/(f - a_2)) + N(r, 1/(f - a_3)) + Q(r, f).$

Applying (1) to $1/(f - a)$, where $a \in \mathbb{C}$ is chosen so that $N(r, a, f) > T(r, f) + Q(r, f)$, we may replace $N(r, f)$ by $T(r, f)$ in the left hand side of (1) (the possibility of such a choice of a follows from [3]). So inequality (1) is equivalent to the Second Main Theorem in its essential features.

Beginning from the result of Milloux [12] (see also [4, Ch. III, § 2, Theorem 2.4]), the inequality (1) has been generalized by the use of derivatives of f in the right hand side of (1). Milloux himself proved the inequality

$$T(r, f) \leq \overline{N}(r, f) + N(r, 1/(f - a_1)) + \overline{N}(r, 1/(f^{(k)} - a_2)) + Q(r, f)$$

for $k \in \mathbb{N}$. Subsequently, many other authors obtained other results in this direction ([5, Theorems 3.2, 3.5], [4, Ch. III, §2, Theorem 2.6], [4, Ch. III, §2, Theorem 2.4], [1]-[3], [6]-[10], [13], [14].

2000 *Mathematics Subject Classification.* Primary 30D35.

In all these papers except for [2] and [3], $N(r, f)$ is estimated from above by a sum of functions $N(r, 1/(f^{(k)} - a))$ for various values of a. In [3], Frank and Weissenborn proved (under an additional assumption) the important inequality

$$\overline{N}(r, f) < N(r, 1/f'') + o(T(r, f))$$

when $r \to \infty$ outside a set $E \subset \mathbb{R}_+$ such that mes $E < \infty$. This inequality has interesting applications. The inequality of Frank-Weissenborn is proved under the additional assumption that all poles of f are simple.

In this note, we prove an estimate for $N(r, f)$ from above in terms of $N(r, 1/f^{(k)})$ for certain sets of $k \in \mathbb{N}$ without any restriction on the multiplicity of the poles.

Let f have a pole of order p at a point z_0. We call the pole *marked* if the Laurent expansion of f in a neighborhood of z_0 has the form

$$a_{-p}(z - z_0)^{-p} + \sum_{j=2}^{\infty} a_{-p+j}(z - z_0)^{-p+j}, \quad a_{-p} \neq 0.$$

It is obvious that if a pole of a function f is simple, then the same pole of its derivative is marked. If a pole of f is marked, then the same pole of all derivatives of f is marked. If z_0 is a pole of f and $f(z_0 + z)$ is either even or odd, then the pole z_0 of the function f and all its derivatives is marked.

THEOREM 1. *Let $n_1 < n_2 < n_3$ be positive integers. If $f^{(n_1)}$ has all of its poles marked, then*

(2) $$\overline{N}(r, f) \leq \overline{N}(r, 1/f^{(n_1)}) + \overline{N}(r, 1/f^{(n_2)}) + \overline{N}(r, 1/f^{(n_3)}) + Q(r, f).$$

For the proof, we require the following

LEMMA. *If f has a marked pole $z_0 \in \mathbb{C}$ of order p, then in a neighborhood of z_0,*

$$\frac{f'(z)}{f(z)} = \frac{-p}{z - z_0} + h(z),$$

where h is a function analytic at z_0 and $h(z_0) = 0$.

PROOF. Without loss of generality, we suppose that $z_0 = 0$. Then $a_{-p} \neq 0$ and $a_{-p+1} = 0$ imply

$$f(z) = \frac{a_{-p} + z^2 g(z)}{z^p},$$

where $g(z)$ is a function analytic at $z = 0$. So

$$\frac{f'(z)}{f(z)} = \frac{-p}{z} + \frac{z(2g(z) + zg'(z))}{a_{-p} + z^2 g(z)}.$$

Clearly,

$$h(z) = \frac{z(2g(z) + zg'(z))}{a_{-p} + z^2 g(z)}$$

is analytic near 0 and $h(0) = 0$.

PROOF OF THEOREM 1. Consider the function

$$A_f = \frac{f^{(n_1+1)}}{f^{(n_1)}} + s_2 \frac{f^{(n_2+1)}}{f^{(n_2)}} + s_3 \frac{f^{(n_3+1)}}{f^{(n_3)}},$$

where s_2, s_3 are real numbers defined by

$$s_2 = \frac{n_1 - n_3}{n_3 - n_2}, s_3 = \frac{n_2 - n_1}{n_3 - n_2}.$$

By the Lemma of the Logarithmic Derivative,

(3) $$m(r, A_f) = Q(r, f).$$

It is obvious that all poles of $f^{(n_2)}$ and $f^{(n_3)}$ are marked. If z_0 is a pole of f, then (by the Lemma), A_f is analytic at z_0 and $A_f(z_0) = 0$. Thus $P(f) \subseteq Z(A_f)$ and $\overline{N}(r, f) \leq N(r, 1/A_f)$. Obviously, all poles of A_f are simple,

$$P(A_f) \subseteq Z(f^{(n_1)}) \cup Z(f^{(n_2)}) \cup Z(f^{(n_3)})$$

and

(4) $$N(r, A_f) \leq \overline{N}(r, 1/f^{(n_1)}) + \overline{N}(r, 1/f^{(n_2)}) + \overline{N}(r, 1/f^{(n_3)}).$$

It follows from (3) and (4) that

(5) $$\begin{aligned}T(r, A_f) &\leq N(r, A_f) + Q(r, f) \\ &\leq \overline{N}(r, 1/f^{(n_1)}) + \overline{N}(r, 1/f^{(n_2)}) + \overline{N}(r, 1/f^{(n_3)}) + Q(r, f).\end{aligned}$$

On the other hand,

(6) $$T(r, A_f) \geq N(r, 1/A_f) + O(1) \geq \overline{N}(r, f) + O(1).$$

The inequality (2) now follows from (5) and (6). $\square$

COROLLARY. *If all poles of f' are marked, then*

$$\overline{N}(r, f) \leq \overline{N}(r, 1/f') + \overline{N}(r, 1/f'') + \overline{N}(r, 1/f''') + Q(r, f).$$

The formulation of the following theorem does not involve the notion of marked pole, but the concept plays a role in the proof.

Denote by $W(f_1, f_2)$ the Wronskian of f_1 and f_2.

THEOREM 2. *For any meromorphic function f,*

(7) $$\begin{aligned}\overline{N}(r, f) &\leq \overline{N}(r, 1/f) + \overline{N}(r, 1/f') \\ &\quad + \overline{N}(r, 1/W(f, f')) + \overline{N}(r, 1/W(f', f'')) + Q(r, f).\end{aligned}$$

PROOF. All poles of f'/f are simple. Thus all poles of $(f'/f)'$ are marked and have second order. The same is true for the function $(f''/f')'$.

Set $B_f = (f'/f)''/(f'/f)'$, $C_f = B_f - B_{f'}$. If z_0 is a pole of f, then by the Lemma, the function $C_f(z)$ is analytic at z_0 and $C_f(z_0) = 0$. Thus $P(f) \subseteq Z(C_f)$ and

(8) $$\overline{N}(r, f) \leq \overline{N}(r, 1/C_f).$$

Let $\overline{n}(z, f) = 0$ if f is analytic at z and $\overline{n}(z, f) = 1$ if f has a pole at z; in this latter case, denote by $n(z, f)$ the order of the pole of f at z.

Suppose $z \notin P(f)$. Then

$$\begin{aligned}n(z, B_f) &= \overline{n}(z, (f'/f)') + \overline{n}(z, 1/(f'/f)') \\ &= \overline{n}(z, f'/f) + \overline{n}(z, f^2/W(f, f')) \leq \overline{n}(z, 1/f) + \overline{n}(z, 1/W(f, f')).\end{aligned}$$

Substituting f' for f, we obtain

$$n(z, B_{f'}) \leq \overline{n}(z, 1/f') + \overline{n}(z, 1/W(f', f'')),$$

and hence

$$n(z, C_f) \leq \overline{n}(z, 1/f') + \overline{n}(z, 1/W(f', f'')) + \overline{n}(z, 1/f) + \overline{n}(z, 1/W(f, f'))$$

because all poles of C_f are simple.

Summing the last inequality over all $z \in \{z : |z| \leq r\} \setminus Z(C_f)$, we obtain
$$n(r, C_f) \leq \overline{n}(r, 1/f') + \overline{n}(r, 1/W(f', f'')) + \overline{n}(r, 1/f) + \overline{n}(r, 1/W(f, f'))$$
and
(9) $\quad N(r, C_f) \leq \overline{N}(r, 1/f') + \overline{N}(r, 1/W(f', f'')) + \overline{N}(r, 1/f) + \overline{N}(r, 1/W(f, f')).$

By the Lemma of the Logarithmic Derivative, $m(r, C_f) = Q(r, f)$; thus
(10) $\qquad \overline{N}(r, 1/C_f) \leq T(r, C_f) + O(1) \leq N(r, C_f) + Q(r, f).$

Applying (8), (10) and (9) consecutively, we obtain (7). $\square$

References

[1] C.T. Chuang, *On theorems of Frank-Weissenborn and Hayman - Miles*, Acta Scientiarum Universitatis Pekinensis **31** (1995), 282-290.

[2] G. Frank and W. Hennekemper, *Einige Ergebnisse über die Werteverteilung meromorpher Funktionen und ihrer Ableitungen*, Resultate Math. **4** (1981), 39-54.

[3] G. Frank and G. Weissenborn, *Rational deficient functions of meromorphic functions*, Bull. London Math. Soc. **18** (1986), 29-33.

[4] A.A. Gol'dberg and I.V. Ostrovskii, *Distribution of Values of Meromorphic Functions*, Nauka, Moscow, 1970.

[5] W.K. Hayman, *Meromorphic Functions*, Clarendon Press, Oxford, 1964.

[6] K.L. Hiong, *Nouvelle démonstration et amélioration d'une inégalité de M. Milloux*, Bull. Sci. Math. (2) **79** (1955), 135-160.

[7] K.L. Hiong, *Sur la limitation de $T(r, f)$ sans intervention des pôles*, Bull. Sci. Math. (2) **80** (1956), 175-190.

[8] K.L. Hiong, *Sur les fonctions méromorphes en rapport avec leurs dérivées*, Sci. Sinica **7** (1958), 661-685.

[9] X.H. Hua, *On a problem of Hayman*, Kodai Math. J. **13** (1990), 386-390.

[10] X.H. Hua and C.T. Chuang, *On a conjecture of Hayman*, Acta Math. Sinica (N.S.) **7** (1991), 119-126.

[11] J. Littlewood, *On the exceptional values of power series*, J. London Math. Soc. **5** (1930), 82-87.

[12] H. Milloux, *Les dérivées des fonctions méromorphes et la théorie des defauts*, Ann.Sci. École Norm. Sup. (3) **63** (1947), 289-316.

[13] L. Yang, *Precise fundamental inequalities and sum of deficiencies*, Sci. China Ser. A **34** (1991), 157-165

[14] Y.F. Wang, *On Mues' Conjecture and Picard values*, Sci. China Ser. A **36** (1993), 28-35.

DEPARTMENT OF MATHEMATICS AND STATISTICS, BAR-ILAN UNIVERSITY, 52900 RAMAT-GAN, ISRAEL

The General Solution of the Loewner Differential Equation on the Unit Ball in $\mathbb{C}^n$

Ian Graham, Gabriela Kohr, and John A. Pfaltzgraff

To Professor Lawrence Zalcman on the occasion of his sixtieth birthday

ABSTRACT. In this paper, we generalize to several complex variables a theorem of Becker concerning the form of arbitrary solutions to the Loewner differential equation. In particular, the form of univalent solutions, which need not be unique in higher dimensions, is determined.

1. Introduction and preliminaries

The study of subordination chains in several complex variables was begun by Pfaltzgraff [**Pf1**], [**Pf2**] in the 1970's. The existence and regularity theory of the associated differential equations has been considered by several authors, and applications have been given to the characterization of subclasses of biholomorphic mappings, univalence criteria, coefficient estimates, and quasiconformal extensions (see [**Ch**], [**GrHaKo**], [**GrKo1**], [**GrKo2**], [**GrKoKo1**], [**GrKoKo2**], [**HaKo1**], [**HaKo2**], [**Ko1**], [**Pf1**], [**Pf2**], [**Pf3**], [**PfSu**]). The existence theorems for the Loewner differential equations in several complex variables were recently improved in [**GrHaKo**] as a consequence of the compactness of the analogue of the Carathéodory class.

In this paper, we consider questions of uniqueness, or rather nonuniqueness, associated to the Loewner equation in higher dimensions. Of course, the initial value problem for the transition mapping of a Loewner chain has a unique solution. However, the partial differential equation for Loewner chains with a given transition mapping does not. Indeed, if $f(z,t)$ is a Loewner chain, and if $\Phi : \mathbb{C}^n \to \mathbb{C}^n$ is a normalized biholomorphic mapping (not necessarily surjective), then $(\Phi \circ f)(z,t)$ is another Loewner chain with the same transition mapping (and a solution of the same partial differential equation). The question we consider is to what extent all

---

2000 *Mathematics Subject Classification.* Primary 32H02, Secondary 30C45.

Key words and phrases. biholomorphic mapping, kernel convergence, Loewner differential equation, Loewner chain, subordination, subordination chain.

The first author was partially supported by the Natural Sciences and Engineering Research Council of Canada under Grant A9221.

solutions of the Loewner differential equation have this form. The corresponding question in one variable was considered by Becker [**Be1**], [**Be2**], [**Be3**] (see also [**Pom1**]). Our arguments are similar; but in higher dimensions they lead to different conclusions, because a normalized biholomorphic mapping of $\mathbb{C}^n$, $n \geq 2$, need not be the identity.

Let $\mathbb{C}^n$ denote the space of n complex variables $z = (z_1, \ldots, z_n)$ with the Euclidean inner product $\langle \cdot, \cdot \rangle$ and the Euclidean norm $\|z\| = \langle z, z \rangle^{1/2}$. Let $B_r = \{z \in \mathbb{C}^n : \|z\| < r\}$ and let $B = B_1$. In the case of one variable, B_r is denoted by U_r and U_1 by U. The topological closure of a subset A of $\mathbb{C}^n$ is denoted by $\overline{A}$. If $\Omega \subset \mathbb{C}^n$ is an open set, $H(\Omega)$ is the set of holomorphic mappings from Ω into $\mathbb{C}^n$, with the topology of locally uniform convergence (or uniform convergence on compact subsets). Denote by $L(\mathbb{C}^n, \mathbb{C}^m)$ the space of continuous linear operators from $\mathbb{C}^n$ into $\mathbb{C}^m$ with the standard operator norm and by I the identity in $L(\mathbb{C}^n, \mathbb{C}^n)$.

We say that $f \in H(B)$ is normalized if $f(0) = 0$ and $Df(0) = I$. Also for $f \in H(B)$, we denote by $D^k f(z)$ the k-th Fréchet derivative of f at $z \in B$ and write
$$D^k f(z)(w^k) = D^k f(z)(\underbrace{w, \ldots, w}_{k-times}), \quad w \in \mathbb{C}^n.$$

Let $S(B)$ be the set of normalized biholomorphic mappings on the unit ball B.

If $f, g \in H(B)$, we say that f is subordinate to g and write $f \prec g$ if there exists a Schwarz mapping v (i.e., $v \in H(B)$, $v(0) = 0$ and $\|v(z)\| < 1$, $z \in B$) such that $f(z) = g(v(z))$ for $z \in B$. If g is biholomorphic on B, this condition is equivalent to requiring that $f(0) = 0$ and $f(B) \subseteq g(B)$.

DEFINITION 1.1. A mapping $f : B \times [0, \infty) \to \mathbb{C}^n$ is called a Loewner chain if it satisfies the following conditions:
 (i) $f(\cdot, t)$ is biholomorphic on B, $f(0, t) = 0$ and $Df(0, t) = e^t I$ for $t \geq 0$;
 (ii) $f(\cdot, s) \prec f(\cdot, t)$ whenever $0 \leq s \leq t < \infty$.

We mention that the second condition and the fact that $f(\cdot, t)$ is biholomorphic, $t \geq 0$, imply that there is a unique biholomorphic Schwarz mapping $v = v(z, s, t)$, called the transition mapping associated to $f(z, t)$, such that

(1.1) $\qquad f(z, s) = f(v(z, s, t), t), \quad z \in B, \ 0 \leq s \leq t < \infty.$

Moreover, the normalization of $f(z, t)$ implies the normalization
$$Dv(0, s, t) = e^{s-t} I, \quad 0 \leq s \leq t < \infty,$$
for the transition mapping.

If $f : B \times [0, \infty) \to \mathbb{C}^n$ is a mapping which is holomorphic on B for fixed t (not necessarily biholomorphic) and satisfies the second condition in Definition 1.1, then we say that $f(z, t)$ is a subordination chain. Thus $f(z, t)$ is a Loewner chain if and only if it is a subordination chain which is biholomorphic on B for fixed t and satisfies the normalization $f(0, t) = 0$ and $Df(0, t) = e^t I$ for $t \geq 0$.

A key role in the study of the Loewner equation in higher dimensions is played by the n-dimensional version of the Carathéodory set
$$\mathcal{M} = \Big\{ h \in H(B) : h(0) = 0, \ Dh(0) = I, \ \mathrm{Re}\, \langle h(z), z \rangle > 0, \ z \in B \setminus \{0\} \Big\}.$$

It is well-known that in the case of one variable this set is compact. Recently Graham, Hamada and Kohr [**GrHaKo**] have established the same result in the case of several complex variables. We have

LEMMA 1.2. *Let $h \in \mathcal{M}$. Then for each $r \in (0,1)$, there is a positive constant $M = M(r) \leq 4r/(1-r)^2$, which is independent of h, such that $\|h(z)\| \leq M(r)$ for $\|z\| \leq r$.*

The basic existence theorem for the Loewner differential equation, due to Pfaltzgraff in its original version (see [**Pf1**, Theorem 2.1]) can be stated as follows (cf. [**Pf1**, Theorem 2.1], [**Por**, Theorems 2 and 3], and [**GrHaKo**]). Boundedness assumptions on the mapping $h(z,t)$ can be omitted because of Lemma 1.2.

LEMMA 1.3. *Let $h : B \times [0,\infty) \to \mathbb{C}^n$ be a mapping which satisfies the following assumptions:*
(i) $h(\cdot, t) \in \mathcal{M}$, $t \geq 0$;
(ii) $h(z, \cdot)$ is measurable on $[0,\infty)$, $z \in B$.
Then there exists a unique solution $\varphi(t) = \varphi(z,s,t)$ of the initial value problem

$$(1.2) \qquad \frac{\partial \varphi}{\partial t} = -h(\varphi, t) \text{ a.e. } t \geq s, \quad \varphi(s) = z.$$

The mapping $\varphi(z,s,t) = e^{s-t}z + \cdots$ is a biholomorphic Schwarz mapping and is Lipschitz continuous in $t \in [s,\infty)$ locally uniformly with respect to $z \in B$. In addition,

$$(1.3) \qquad \lim_{t \to \infty} e^t \varphi(z,s,t) = f(z,s)$$

exists locally uniformly on B for each $s \geq 0$, $f(\cdot, s)$ is biholomorphic on B and

$$(1.4) \qquad f(z,s) = f(\varphi(z,s,t),t), \quad z \in B, \ t \geq s \geq 0.$$

Thus $f(z,t)$ is a Loewner chain. Moreover, $f(z,\cdot)$ is locally Lipschitz continuous on $[0,\infty)$ locally uniformly with respect to $z \in B$ and

$$(1.5) \qquad \frac{\partial f}{\partial t}(z,t) = Df(z,t)h(z,t) \text{ a.e. } t \geq 0, \ \forall \, z \in B.$$

The Loewner chain given by (1.3) may be called the *canonical solution* of the Loewner differential equation (1.5). The result below is a direct consequence of [**Pf1**, Lemma 2.2] and the relation (1.3).

LEMMA 1.4. *Let $f(z,t)$ be the Loewner chain given by (1.3). Then $\{e^{-t}f(z,t)\}_{t \geq 0}$ is a normal family on B and*

$$(1.6) \qquad e^s \frac{\|z\|}{(1+\|z\|)^2} \leq \|f(z,s)\| \leq e^s \frac{\|z\|}{(1-\|z\|)^2}, \quad z \in B, \ s \geq 0.$$

We remark that the solution $\varphi(z,s,t)$ of (1.2) satisfies the following semigroup property (see [**Pf1**], [**GrKo2**]):

$$(1.7) \qquad \varphi(z,s,\tau) = \varphi(\varphi(z,s,t),t,\tau), \quad z \in B, \ 0 \leq s \leq t \leq \tau < \infty.$$

Of course, this property is also satisfied by the transition mapping of any Loewner chain (see [**GrKoKo2**]).

Next, we recall that the canonical solution of the Loewner differential equation may be characterized among all solutions by a normality condition (essentially a growth condition in the t variable). This result is due to Pfaltzgraff [**Pf1**, Theorem 2.3]. Poreda [**Por**, Theorem 6] obtained a similar result for subordination chains in Banach spaces with stronger regularity assumptions in t. Recently, Hamada and Kohr [**HaKo3**] have obtained an analogous result for subordination chains in reflexive Banach spaces with similar regularity assumptions as in finite dimensions.

LEMMA 1.5. *Let $f : B \times [0, \infty) \to \mathbb{C}^n$ be a mapping such that $f(\cdot, t) \in H(B)$, $f(0, t) = 0$, $Df(0, t) = e^t I$, $t \geq 0$, and $f(z, \cdot)$ is locally absolutely continuous on $[0, \infty)$ locally uniformly with respect to $z \in B$. Let $h : B \times [0, \infty)$ satisfy the assumptions (i) and (ii) in Lemma* 1.3. *Assume that*

$$(1.8) \qquad \frac{\partial f}{\partial t}(z, t) = Df(z, t) h(z, t) \ a.e. \ t \geq 0, \ \forall \ z \in B.$$

Assume furthermore that there is an increasing sequence $\{t_m\}_{m \in \mathbb{N}}$ such that $t_m \geq 0$, $t_m \to \infty$, and

$$(1.9) \qquad \lim_{m \to \infty} e^{-t_m} f(z, t_m) = F(z)$$

locally uniformly on B. Then $f(z, t)$ is a Loewner chain; and for each $s \geq 0$,

$$\lim_{t \to \infty} e^t \varphi(z, s, t) = f(z, s)$$

locally uniformly on B, where $\varphi(t) = \varphi(z, s, t)$ is the solution of the initial value problem (1.2).

We next mention the following properties which are satisfied by the solution of the initial value problem (1.2) (see [**Pf1**] and also [**GrKo2**]).

LEMMA 1.6. *Let $\varphi = \varphi(z, s, t)$ be the solution of the initial value problem* (1.2). *Then for each $r \in (0, 1)$, there is $L = L(r) > 0$ such that*

$$(1.10) \qquad \|\varphi(z, s, t_1) - \varphi(z, s, t_2)\| \leq L(r)|t_1 - t_2|, \quad \|z\| \leq r, \ t_1, t_2 \in [s, \infty).$$

Moreover,

$$(1.11) \qquad e^t \|\varphi(z, s, t)\| / (1 - \|\varphi(z, s, t)\|)^2 \leq e^s \|z\| / (1 - \|z\|)^2$$

and

$$(1.12) \qquad e^s \|z\| / (1 + \|z\|)^2 \leq e^t \|\varphi(z, s, t)\| / (1 + \|\varphi(z, s, t)\|)^2,$$

for all $z \in B$ and $0 \leq s \leq t < \infty$.

Finally, we conclude this section with a result which is an analogue of the well-known Carathéodory kernel convergence theorem, on the convergence of conformal mappings of one complex variable, for biholomorphic mappings which satisfy a growth result (see [**Ko2**]). To this end, we need to recall the following notions.

DEFINITION 1.7. Let $\{G_k\}_{k \in \mathbb{N}}$ be a sequence of domains in $\mathbb{C}^n$ such that $0 \in G_k$, $k \in \mathbb{N}$. If 0 is an interior point of $\bigcap_{k \in \mathbb{N}} G_k$, define the kernel G of $\{G_k\}_{k \in \mathbb{N}}$ to be the largest domain in $\mathbb{C}^n$ which contains 0, such that if K is a compact subset of G, then K is contained in all but finitely many of G_k. If 0 is not an interior point of $\bigcap_{k \in \mathbb{N}} G_k$, define the kernel to be $\{0\}$.

We say that $\{G_k\}_{k \in \mathbb{N}}$ kernel converges to G, and write $G_k \to G$, if each subsequence of $\{G_k\}_{k \in \mathbb{N}}$ has the same kernel G.

Next, let $S^c(B)$ be a compact subset of $S(B)$. We have (see [**Ko2**])

LEMMA 1.8. *Let $\{f_k\}_{k \in \mathbb{N}}$ be a sequence of biholomorphic mappings on B such that $f_k(0) = 0$ and $Df_k(0) = \alpha_k I$, where $\alpha_k > 0$ for $k \in \mathbb{N}$. Assume that $f_k/\alpha_k \in S^c(B)$, $k \in \mathbb{N}$. Also let $G_k = f_k(B)$, $k \in \mathbb{N}$, and let G be the kernel of $\{G_k\}_{k \in \mathbb{N}}$. Then $\{f_k\}_{k \in \mathbb{N}}$ converges locally uniformly on B to a mapping f if and only if $G_k \to G \neq \mathbb{C}^n$. In the case of convergence, either $f \equiv 0$ and $G = \{0\}$, or else f is biholomorphic on B, $f/\alpha \in S^c(B)$ where $\alpha = \lim_{k \to \infty} \alpha_k$, and $f(B) = G$. In the latter case, $f_k^{-1} \to f^{-1}$ locally uniformly on G.*

We now proceed to show how an arbitrary solution of the Loewner differential equation (1.8) can be expressed in terms of the canonical solution. In the case of one variable, this question was considered by Becker [**Be1**]-[**Be3**]; his methods can be applied to the several variable case.

2. Solutions of the generalized Loewner differential equation

The main result of this section is a generalization of a one-variable theorem of Becker (see [**Be1**, Satz 2]). In one variable, it is useful to consider solutions of the Loewner differential equation which, for fixed t, are holomorphic on a punctured disc (rather than a disc) centered at 0. However, in higher dimensions point singularities of holomorphic functions are removable, so we assume that the solutions are holomorphic at 0 for fixed t. As in [**Be1**]-[**Be3**], we allow the radius of the ball on which the solution is initially defined in z to vary with t; this is potentially useful for applications. The mapping $h(z,t)$ must of course be defined on $B \times [0, \infty)$. Recall that a Fatou-Bieberbach map in $\mathbb{C}^n$, $n \geq 2$, is a biholomorphic mapping of $\mathbb{C}^n$ onto a proper subset of $\mathbb{C}^n$.

THEOREM 2.1. *Let $h(z,t)$ satisfy assumptions (i) and (ii) of Lemma 1.3 and let $f(z,t)$ be given by (1.3). Let $g(z,t)$ be a mapping such that for each $t \geq 0$, $g(\cdot, t) \in H(B_{r(t)})$, where $r(t) \in (0,1]$ and*

$$(2.1) \qquad \limsup_{t \to \infty} e^t r(t) = \infty.$$

Assume that there exist positive functions ρ and δ on $[0, \infty)$ such that $\rho(t) < 1$, $t \geq 0$, and for each $t_0 \geq 0$, the following conditions hold:
  (a) *$r(t) \geq \rho(t_0)$ for $t \in E_{\delta(t_0)} = [t_0 - \delta(t_0), t_0 + \delta(t_0)] \cap [0, \infty)$ (thus $g(\cdot, t)$ is holomorphic on $B_{\rho(t_0)}$ for $t \in E_{\delta(t_0)}$);*
  (b) *$g(z, \cdot)$ is absolutely continuous on $E_{\delta(t_0)}$ for $z \in B_{\rho(t_0)}$, and for almost all $t \in E_{\delta(t_0)}$,*

$$(2.2) \qquad \frac{\partial g}{\partial t}(z,t) = Dg(z,t) h(z,t), \quad z \in B_{\rho(t_0)}.$$

Then $g(z,t)$ extends to a subordination chain on $B \times [0, \infty)$, again denoted by $g(z,t)$, and there exists a holomorphic mapping $\Phi : \mathbb{C}^n \to \mathbb{C}^n$ such that

$$(2.3) \qquad g(z,t) = \Phi(f(z,t)), \quad z \in B, \; t \geq 0.$$

Moreover, $g(z,t)$ is a univalent subordination chain if and only if Φ is univalent, i.e., an automorphism of $\mathbb{C}^n$ or a Fatou-Bieberbach map.

The proof of this result is based on the following lemmas.

LEMMA 2.2. *Fix $s \geq 0$ and $z \in B$. Let $\varphi(t) = \varphi(z, s, t)$ be the unique solution of (1.2), and let $g(z,t)$ satisfy the assumptions of Theorem 2.1. Then for each $t_0 \geq s$,*

(2.4) $\qquad g(\varphi(z,s,t),t) = g(\varphi(z,s,t_0),t_0), \quad z \in B_{\rho(t_0)}, \ t \in E_{\delta_1(t_0)},$

where $E_{\delta_1(t_0)} = [t_0 - \delta(t_0), t_0 + \delta(t_0)] \cap [s, \infty)$.

PROOF. First note that $E_{\delta_1(t_0)} \subseteq E_{\delta(t_0)}$ since $s \geq 0$. Therefore, both mappings $g(\varphi(\cdot, s, t), t), \ t \in E_{\delta_1(t_0)}$, and $g(\varphi(\cdot, s, t_0), t_0)$ are well-defined and holomorphic on $B_{\rho(t_0)}$.

Let $\varphi_{s,t}(z) = \varphi(z, s, t)$ for $z \in B$ and $t \geq s \geq 0$. Since $g = g(z,t)$ is continuous on $B_{\rho(t_0)} \times E_{\delta(t_0)}$ by the assumptions (a) and (b), it follows that for each $\rho \in (0, \rho(t_0))$, there is some $K = K(\rho, t_0) > 0$ such that

$$\|g(z,t)\| \leq K, \quad z \in \overline{B}_\rho, \ t \in E_{\delta(t_0)}.$$

Moreover, by the Cauchy integral formula for vector valued holomorphic functions, it is easy to deduce for each $\rho \in (0, \rho(t_0))$ that there exists $K' = K'(\rho, t_0) > 0$ such that

(2.5) $\qquad \|Dg(z,t)\| \leq K', \quad z \in \overline{B}_\rho, \ t \in E_{\rho(t_0)}.$

Consequently, for each $\rho \in (0, \rho(t_0))$, we obtain

(2.6) $\qquad \|g(z,t) - g(w,t)\| \leq K'\|z - w\|, \quad z, w \in \overline{B}_\rho, \ t \in E_{\delta(t_0)}.$

On the other hand, taking into account the relations (2.2) and (2.5) and Lemma 1.2, we deduce for each $\rho \in (0, \rho(t_0))$ that there is some $K^* = K^*(\rho, t_0)$ such that

$$\left\|\frac{\partial g}{\partial t}(z,t)\right\| \leq K^* \text{ a.e. } t \in E_{\delta(t_0)}, \ \forall \ z \in \overline{B}_\rho.$$

Since $g(z, \cdot)$ is absolutely continuous on $E_{\delta(t_0)}$, in view of the above relation we obtain that

(2.7) $\qquad \|g(z,t_1) - g(z,t_2)\| \leq K^*|t_1 - t_2|, \quad \|z\| \leq \rho < \rho(t_0), \ t_1, t_2 \in E_{\delta(t_0)}.$

Next, fix $\rho \in (0, \rho(t_0))$ and let $q(z,t) = g(\varphi_{s,t}(z), t)$ for $z \in \overline{B}_\rho$ and $t \in E_{\delta_1(t_0)}$. According to the relations (2.6), (2.7) and (1.10), we have

$\|q(z,t_1) - q(z,t_2)\|$
$\quad \leq \|g(\varphi_{s,t_1}(z), t_1) - g(\varphi_{s,t_1}(z), t_2)\| + \|g(\varphi_{s,t_1}(z), t_2) - g(\varphi_{s,t_2}(z), t_2)\|$
$\quad \leq K^*|t_1 - t_2| + K'\|\varphi_{s,t_1}(z) - \varphi_{s,t_2}(z)\|$
$\quad \leq [K^* + K'L(r)]|t_1 - t_2| \leq K^{**}|t_1 - t_2|, \quad z \in \overline{B}_\rho, \ t_1, t_2 \in E_{\delta_1(t_0)}.$

Hence for each $z \in \overline{B}_\rho$, $q(z,t)$ is Lipschitz continuous with respect to $t \in E_{\delta_1(t_0)}$ and

$$\frac{\partial q}{\partial t}(z,t) = Dg(\varphi_{s,t}(z), t)\frac{\partial \varphi_{s,t}}{\partial t}(z) + \frac{\partial g}{\partial t}(\varphi_{s,t}(z), t)$$
$$= -Dg(\varphi_{s,t}(z), t)h(\varphi_{s,t}(z), t) + Dg(\varphi_{s,t}(z), t)h(\varphi_{s,t}(z,t), t) = 0$$

for almost all $t \in E_{\delta_1(t_0)}$. Here we have used (1.2) and (2.2). Since $q(z, \cdot)$ is absolutely continuous on $E_{\delta_1(t_0)}$, we conclude that $q(z, \cdot)$ is constant on $E_{\delta_1(t_0)}$; thus

$$g(\varphi(z,s,t),t) = g(\varphi(z,s,t_0),t_0), \quad t \in E_{\delta_1(t_0)}, \ z \in \overline{B}_\rho.$$

By the identity theorem for holomorphic mappings, we obtain (2.4), as desired. $\square$

LEMMA 2.3. *Let $g(z,t)$ satisfy the assumptions of Theorem 2.1. Then for each $t \geq 0$, $g(\cdot, t)$ can be extended to a holomorphic mapping on B, again denoted by $g(\cdot, t)$, and*

(2.8) $$g(z, s) = g(\varphi(z, s, t), t), \quad z \in B, \ 0 \leq s \leq t < \infty,$$

i.e., $g(z,t)$ is a subordination chain.

PROOF. Fix $s \geq 0$ and $r \in (0,1)$. Let $g_t(z) = g(z,t)$. By (2.1), there is a sequence $\{t_m\}_{m \in \mathbb{N}}$ such that $t_m \geq 0$, $t_m \to \infty$ and $e^{t_m} r(t_m) \to \infty$ as $m \to \infty$. Moreover, in view of (1.11), we have

$$\|\varphi_{s,t}(z)\| \leq e^{s-t} \frac{r}{(1-r)^2} = \frac{e^s r}{(1-r)^2} \left[\frac{r(t)}{e^t r(t)} \right], \quad z \in \overline{B}_r, \ t \geq s;$$

thus there exists $m_0 = m_0(r, s) \in \mathbb{N}$ such that

$$\varphi_{s,t_m}(B_r) \subseteq B_{r(t_m)} \text{ for } m \geq m_0.$$

Consequently, the mapping $g_{t_m} \circ \varphi_{s,t_m}$ is well-defined and holomorphic on B_r for $m \geq m_0$. From Lemma 2.2, for each $t_0 \in [s, t_m]$, there is a neighbourhood $W(t_0)$ of t_0 such that

$$g(\varphi(z, s, t), t) = g(\varphi(z, s, t_0), t_0), \quad t \in W(t_0) \cap [s, \infty), \ z \in B_{\rho(t_0)}.$$

On the other hand, the compact interval $[s, t_m]$ can be covered by a finite number of intervals $W(\tau_1), \ldots, W(\tau_N)$ such that

$$g(\varphi(z, s, t), t) = g(\varphi(z, s, \tau_\nu), \tau_\nu), \quad t \in W(\tau_\nu) \cap [s, \infty), \ z \in B_{\rho(\tau_\nu)},$$

for $\nu = 1, \ldots, N$. Let $\rho_m = \min\{\rho(\tau_\nu) : \nu = 1, \ldots, N\}$ and

$$A = \{t \in [s, t_m] : g(\varphi(z, s, t), t) = g(z, s), \ z \in B_{\rho_m}\};$$

we show that A is a nonempty, open and closed subset of $[s, t_m]$, and hence $A = [s, t_m]$. Indeed, A is nonempty since $s \in A$. To see that A is closed, let $t^* \in \overline{A}$; then $t^* \in [s, t_m]$ and there exist $\nu \in \{1, \ldots, N\}$ and a neighbourhood $W(\tau_\nu)$ of t^* such that $t^* \in W(\tau_\nu)$. Hence

$$g(\varphi(z, s, t), t) = g(\varphi(z, s, \tau_\nu), \tau_\nu) = g(\varphi(z, s, t^*), t^*), \ t \in W(\tau_\nu) \cap [s, \infty).$$

On the other hand, since $t^* \in \overline{A}$, $W(\tau_\nu) \cap A \neq \emptyset$, and thus there exists $\widetilde{t} \in W(\tau_\nu) \cap A$. Then

$$g(\varphi(z, s, \widetilde{t}), \widetilde{t}) = g(\varphi(z, s, \tau_\nu), \tau_\nu) = g(z, s).$$

Combining the above arguments, we obtain $g(\varphi(z, s, t^*), t^*) = g(z, s)$, which yields $t^* \in A$.

We now show that A is open in $[s, t_m]$. Let $u \in A$. Then $u \in W(\tau_\nu) \cap [s, \infty)$ for some $\nu \in \{1, \ldots, N\}$. Hence

$$g(\varphi(z, s, u), u) = g(\varphi(z, s, t), t) = g(\varphi(z, s, \tau_\nu), \tau_\nu), \quad t \in W(\tau_\nu) \cap [s, \infty).$$

But $u \in A$, and thus

$$g(z, s) = g(\varphi(z, s, u), u).$$

Consequently,

$$g(z, s) = g(\varphi(z, s, u), u) = g(\varphi(z, s, t), t), \quad t \in W(\tau_\nu) \cap [s, \infty),$$

and therefore $W(\tau_\nu) \cap [s, \infty)$ is a neighbourhood of u which is contained in A.

We have proved that $A = [s, t_m]$, as claimed. Hence

(2.9) $$g(\varphi(z, s, t), t) = g(z, s), \quad z \in B_{\rho_m}, \ s \leq t \leq t_m.$$

Moreover, since $g_{t_m} \circ \varphi_{s,t_m}$ is holomorphic on B_r for $m \geq m_0$, we deduce in view of (2.9) that g_s can be extended holomorphically to B_r. Since r and s are arbitrary, we conclude that g_s can be extended holomorphically to the whole ball B, and according to (2.9) and the identity theorem for holomorphic mappings, we obtain
$$g(\varphi(z,s,t),t) = g(z,s), \quad z \in B, \ s \leq t \leq t_m, \ m = 1, 2, \ldots.$$
Letting $m \to \infty$ in the above relation, we deduce (2.8). □

We are now able to prove the main result of this paper.

PROOF OF THEOREM 2.1. First, we write the relation (2.8) as
$$(2.10) \qquad g_s(z) = g_t(e^{-t}e^t \varphi_{s,t}(z)), \quad z \in B, \ 0 \leq s \leq t < \infty.$$

Let $\psi_{s,t}(z) = e^t \varphi_{s,t}(z)$ for $z \in B$ and $0 \leq s \leq t < \infty$. In view of (1.3), we have for each $s \geq 0$
$$(2.11) \qquad \lim_{t \to \infty} \psi_{s,t}(z) = f(z,s)$$
locally uniformly on B. Moreover, $f_s(z) = f(z,s)$ satisfies
$$e^s \frac{\|z\|}{(1+\|z\|)^2} \leq \|f_s(z)\| \leq e^s \frac{\|z\|}{(1-\|z\|)^2}, \quad z \in B, \ s \geq 0,$$
by (1.6); thus for each $m = 1, 2, \ldots$, there exists $s_m \geq 0$ such that $f_{s_m}(B) \supseteq \overline{B}_m$. Consequently, taking into account this relation and (2.11), we deduce for each $m = 1, 2, \ldots$ that there exists $t_m \geq s_m$ such that
$$\psi_{s_m,t}(B) \supseteq \overline{B}_m, \quad t \geq t_m.$$

Since $\{\psi_{s_m,t}\}_{t \geq t_m}$ is a family of biholomorphic mappings such that $\psi_{s_m,t}(0) = 0$, $D\psi_{s_m,t}(0) = e^{s_m} I$, $t \geq t_m$, and by (1.11) and (1.12),
$$\frac{\|z\|}{(1+\|z\|)^2} \leq e^{-s_m} \|\psi_{s_m,t}(z)\| \leq \frac{\|z\|}{(1-\|z\|)^2}, \quad z \in B, \ t \geq t_m,$$
we obtain according to Lemma 1.8
$$(2.12) \qquad \lim_{t \to \infty} \psi_{s_m,t}^{-1}(w) = f_{s_m}^{-1}(w);$$
and the above limit holds locally uniformly on B_m for each fixed $m = 1, 2, \ldots$. In view of (2.10), we have
$$g_t(e^{-t}w) = g_{s_m}(\psi_{s_m,t}^{-1}(w)), \quad w \in \overline{B}_m, \ t \geq t_m;$$
and since $g_{s_m} \in H(B)$ and $\psi_{s_m,t}^{-1} \to f_{s_m}^{-1}$ locally uniformly on B_m as $t \to \infty$ by (2.12), we conclude from the above relations that $\{g_t(e^{-t}w)\}_{t \geq t_m}$ converges locally uniformly on B_m as $t \to \infty$ to a mapping Φ_m which, in view of Weierstrass' theorem, is holomorphic on B_m. Since m is arbitrary, we deduce that there is a mapping $\Phi \in H(\mathbb{C}^n)$ such that $g_t(e^{-t}w) \to \Phi$ locally uniformly on $\mathbb{C}^n$ as $t \to \infty$. Clearly Φ is the holomorphic extension to $\mathbb{C}^n$ of Φ_m, $m = 1, 2, \ldots$, according to the identity theorem for holomorphic mappings. Letting $t \to \infty$ in (2.10) and using (2.11), we conclude that
$$g_s(z) = \Phi(f_s(z)), \quad z \in B, \ s \geq 0,$$
and thus obtain (2.3), as desired.

Finally, we note that the estimate (1.6) implies that $\bigcup_{s\geq 0} f_s(B) = \mathbb{C}^n$. This easily implies that $g(z,t)$ is a univalent subordination chain if and only if Φ is univalent. This completes the proof. $\square$

Taking into account Lemma 1.5, we obtain the following equivalent version of Theorem 2.1.

THEOREM 2.4. *Let $f : B \times [0,\infty) \to \mathbb{C}^n$ be a mapping such that $f(\cdot,t) \in H(B)$, $f(0,t) = 0$, $Df(0,t) = e^t I$, $t \geq 0$, and $f(z,\cdot)$ is locally absolutely continuous on $[0,\infty)$ locally uniformly with respect to $z \in B$. Assume that $h(z,t)$ satisfies the conditions (i) and (ii) in Lemma 1.3 and $f(z,t)$ satisfies the Loewner differential equation (1.5). Assume further that there exists a sequence $\{t_m\}_{m\in\mathbb{N}}$ such that $t_m > 0$, $t_m \to \infty$ as $m \to \infty$, and*

$$\lim_{m\to\infty} e^{-t_m} f(z,t_m) = F(z)$$

locally uniformly on B. Let $g(z,t)$ satisfy the assumptions of Theorem 2.1. Then $g(z,t)$ can be extended to a subordination chain, again denoted by $g(z,t)$, and there is a mapping $\Phi \in H(\mathbb{C}^n)$ such that $g(z,t) = \Phi(f(z,t))$ for $z \in B$ and $t \geq 0$. Conversely, if $g(z,t) = \Phi(f(z,t))$, $z \in B$, $t \geq 0$, where Φ is a holomorphic maping of $\mathbb{C}^n$ into $\mathbb{C}^n$, then $g(z,t)$ is a subordination chain which satisfies the same Loewner differential equation. Moreover, $g(z,t)$ is a univalent subordination chain if and only if Φ is univalent, i.e., an automorphism of $\mathbb{C}^n$ or a Fatou-Bieberbach map.

We next present the following particular cases of Theorem 2.1.

COROLLARY 2.5. *Let $r \in (0,1]$ and $g = g(z,t) : B_r \times [0,\infty) \to \mathbb{C}^n$ be a mapping such that $g(\cdot,t) \in H(B_r)$, $t \geq 0$ and $g(z,\cdot)$ is locally absolutely continuous on $[0,\infty)$ locally uniformly with respect to $z \in B_r$. Let $h = h(z,t) : B \times [0,\infty) \to \mathbb{C}^n$ satisfy assumptions (i) and (ii) of Lemma 1.3 and let $f(z,t)$ be given by (1.3). Assume that $g(z,t)$ satisfies the Loewner differential equation*

$$\frac{\partial g}{\partial t}(z,t) = Dg(z,t)h(z,t) \quad a.e. \ t \geq 0, \ \forall \ z \in B_r.$$

Then $g(z,t)$ can be extended to a subordination chain, again denoted by $g(z,t)$, and there exists a holomorphic mapping Φ of $\mathbb{C}^n$ into $\mathbb{C}^n$ such that $g(z,t) = \Phi(f(z,t))$ for $z \in B$ and $0 \leq t < \infty$.

PROOF. It suffices to consider $r(t) \equiv r$ in Theorem 2.1. $\square$

We next prove that $g(z,t)$ given in Corollary 2.5 coincides with $f(z,t)$ if $\{e^{-t}g(z,t)\}_{t\geq 0}$ is a normal family on B_r. Indeed, we have (see [**GrHaKo**], [**GrKoKo2**]).

COROLLARY 2.6. *Let $r \in (0,1]$, $h(z,t)$ and $g(z,t)$ satisfy the assumptions of Corollary 2.5. Assume that $\{e^{-t}g(z,t)\}_{t\geq 0}$ is a normal family on B_r, $g(0,t) = 0$ and $Dg(0,t) = e^t I$, $t \geq 0$. Then $g(z,t)$ can be extended to a Loewner chain, again denoted by $g(z,t)$, and $g(z,t) = f(z,t)$, $z \in B$, $t \geq 0$.*

PROOF. In [**GrHaKo**] (see also the proof of [**Pf1**, Theorem 2.3]), it is shown that

$$\lim_{t\to\infty} e^t \varphi(z,s,t) = g(z,s)$$

locally uniformly on B for each $s \geq 0$, where $\varphi(z,s,t)$ is the unique solution of (1.2). Thus $f(z,s) = g(z,s)$, $z \in B$, $s \geq 0$, according to Lemma 1.3. $\square$

We next give another consequence of Theorem 2.1 in terms of the coefficients of $g(z,t)$. This generalizes a result of Becker [**Be1**, Lemma 2]. Note that the condition (2.13) holds if the extended subordination chain $g(z,t)$ is such that $\{e^{-t}g(z,t)\}_{t\geq 0}$ is a normal family.

COROLLARY 2.7. *Let $g(z,t)$ and $h(z,t)$ satisfy the assumptions of Theorem 2.1. Also let $f(z,t)$ be given by (1.3) and let*
$$c_k(t) = \frac{1}{k!}D^k g_t(0), \quad t \geq 0, \ k \geq 0.$$

Assume that

(2.13) $$\liminf_{t\to\infty} e^{-kt}\|c_k(t)\| = 0, \quad k \geq 2.$$

Then $g(z,t) = c_0(0) + c_1(0)(f(z,t))$, $z \in B$, $t \geq 0$.

PROOF. By Theorem 2.1, $g(z,t)$ extends to a subordination chain, again denoted by $g(z,t)$. Let $\Phi \in H(B)$ satisfy

(2.14) $$g(z,t) = \Phi(f(z,t)), \quad z \in B, \ t \geq 0.$$

Since Φ is holomorphic on $\mathbb{C}^n$, it can be expanded in a power series
$$\Phi(w) = A_0 + A_1(w) + \cdots + A_k(w^k) + \cdots, \quad w \in \mathbb{C}^n,$$
where
$$A_k(w^k) = \frac{1}{k!}D^k\Phi(0)(\underbrace{w,\ldots,w}_{k-times}), \quad w \in \mathbb{C}^n.$$

We show that $A_k \equiv 0$ for $k \geq 2$. To this end, let
$$a_k(t) = \frac{1}{k!}D^k f_t(0), \quad t \geq 0, \ k \geq 2.$$

Then $a_0(t) = 0$, $a_1(t) = e^t I$, $t \geq 0$, and by (2.14), we obtain
$$\sum_{k=0}^{\infty} c_k(t)(z^k) = \Phi\left(e^t z + \sum_{k=2}^{\infty} a_k(t)(z^k)\right) = \sum_{k=0}^{\infty} A_k(w^k), \quad z \in B,$$

where $w = e^t z + \sum_{j=2}^{\infty} a_j(t)(z^j)$. Consequently, $c_0(t) = A_0$, $c_1(t) = e^t A_1$, $t \geq 0$, and

(2.15) $$c_k(t) = A_1 \circ a_k(t) + e^{kt} A_k, \quad k \geq 2, \ t \geq 0.$$

On the other hand, in view of Lemma 1.4 and the Cauchy integral formula for vector valued holomorphic functions, for each $k \geq 2$, there exists $C_k > 0$ such that

(2.16) $$\|a_k(t)\| \leq C_k e^t, \quad t \geq 0.$$

Indeed, since
$$a_k(t)(w) = \frac{1}{2\pi i}\int_{|\zeta|=r}\frac{f(\zeta w, t)}{\zeta^{k+1}}d\zeta, \quad \|w\| = 1,$$

for any $r \in (0,1)$, and $\|f(z,t)\| \leq e^t r/(1-r)^2$, $\|z\| = r$, $t \geq 0$, we have
$$\|a_k(t)\| \leq e^t \min_{r\in(0,1)} \frac{1}{r^{k-1}(1-r)^2} \leq e^t \left[\frac{e(k+1)}{2}\right]^2.$$

Then (2.16) holds with $C_k = e^2(k+1)^2/4$.

Next, fix $k \geq 2$. Multiplying both sides of (2.15) by e^{-kt} and using (2.16), we obtain
$$\|e^{-kt}c_k(t) - A_k\| \leq \|A_1\|e^{-kt}\|a_k(t)\| \leq C_k e^{(1-k)t}\|A_1\|, \quad t \geq 0.$$

According to (2.13), there exists a sequence $\{t_m\}_{m\in\mathbb{N}}$ such that $t_m > 0$, $t_m \to \infty$ and $e^{-kt_m}\|c_k(t_m)\| \to 0$ as $m \to \infty$. Finally, letting $m \to \infty$ in the inequality
$$\|e^{-kt_m}c_k(t_m) - A_k\| \leq C_k e^{(1-k)t_m}\|A_1\|,$$
we obtain $A_k \equiv 0$, as claimed. Consequently,
$$g(z,t) = A_0 + A_1(f(z,t)) = c_0(0) + c_1(0)(f(z,t)), \quad z \in B, \ t \geq 0,$$
as desired. $\square$

REMARK 2.8. Let $f(z,t)$ be a Loewner chain. Recently, in [**GrHaKo**] (see also [**GrKoKo1**]), it has been proved that there exists a mapping $h = h(z,t)$ such that $h(\cdot, t) \in \mathcal{M}$ for $t \geq 0$, $h(z, \cdot)$ is measurable on $[0, \infty)$ for $z \in B$, and

(2.17) $$\frac{\partial f}{\partial t}(z,t) = Df(z,t)h(z,t) \text{ a.e. } t \geq 0, \ \forall \ z \in B.$$

Moreover, if $\{e^{-t}f(z,t)\}_{t\geq 0}$ is a normal family on B, then for each $s \geq 0$,
$$f(z,s) = \lim_{t\to\infty} e^t v(z,s,t)$$
locally uniformly on B, where $v(t) = v(z,s,t)$ is the unique solution of the initial value problem
$$\frac{\partial v}{\partial t} = -h(v,t) \text{ a.e. } t \geq s, \ v(s) = z,$$
for $z \in B$ and $s \geq 0$.

Combining Theorem 2.1 and Remark 2.8, we obtain the following connection between a Loewner chain and a subordination chain, both satisfying the same differential equation (compare with [**Pom1**, Satz 5]):

COROLLARY 2.9. *Let $f(z,t)$ be a Loewner chain such that $\{e^{-t}f(z,t)\}_{t\geq 0}$ is a normal family on B. Let $h(z,t)$ be the mapping given by (2.17). Also let $g = g(z,t) : B \times [0,\infty) \to \mathbb{C}^n$ be a mapping such that $g(\cdot, t) \in H(B)$, $g(0,t) = 0$, $Dg(0,t) = e^t I$, for $t \geq 0$, and $g(z,\cdot)$ is locally absolutely continuous on $[0,\infty)$ locally uniformly with respect to $z \in B$. Assume that*
$$\frac{\partial g}{\partial t}(z,t) = Dg(z,t)h(z,t) \text{ a.e. } t \geq 0, \ \forall \ z \in B.$$

Then there is a normalized holomorphic mapping Φ of $\mathbb{C}^n$ into $\mathbb{C}^n$ such that $g(z,t) = \Phi(f(z,t))$ for $z \in B$ and $t \geq 0$. Moreover, $g(z,t)$ is a univalent subordination chain if and only if Φ is univalent, i.e., an automorphism of $\mathbb{C}^n$ or a Fatou-Bieberbach map.

Acknowledgements. Some of the research for this paper was carried out in August, 2003, while the second and third authors visited the Department of Mathematics of the University of Toronto. G. Kohr and J. Pfaltzgraff express their gratitude to the members of this department for their hospitality during this visit.

References

[Be1]   J. Becker, *Löwnersche Differentialgleichung und Schlichtheitskriterien*, Math. Ann. **202** (1973), 321-335.

[Be2]   J. Becker, *Über die Lösungsstruktur einer Differentialgleichung in der konformen Abbildung*, J. Reine Angew. Math. **285** (1976), 66-74.

[Be3]   J. Becker, *Conformal mappings with quasiconformal extensions*, in: Aspects of Contemporary Complex Analysis, Academic Press, London-New York, 1980, pp.37-77.

[Ch]   M. Chuaqui, *Applications of subordination chains to starlike mappings in $\mathbb{C}^n$*, Pacific J. Math. **168** (1995), 33-48.

[GrHaKo]   I. Graham, H. Hamada and G. Kohr, *Parametric representation of univalent mappings in several complex variables*, Canad. J. Math. **54** (2002), 324-351.

[GrHaKoSu]   I. Graham, H. Hamada, G. Kohr and T. Suffridge, *Extension operators for locally univalent mappings*, Michigan Math. J. **50** (2002), 37-55.

[GrKo1]   I. Graham and G. Kohr, *An extension theorem and subclasses of univalent mappings in several complex variables*, Complex Variables **47** (2002), 59-72.

[GrKo2]   I. Graham and G. Kohr, *Geometric Function Theory in One and Higher Dimensions*, Marcel Dekker Inc., New York, 2003.

[GrKoKo1]   I. Graham, G. Kohr and M. Kohr, *Loewner chains and parametric representation in several complex variables*, J. Math. Anal. Appl. **281** (2003), 425-438.

[GrKoKo2]   I. Graham, G. Kohr and M. Kohr, *Basic properties of Loewner chains in several complex variables*, Proceedings of a Conference on Geometric Function Theory in Several Complex Variables held in Hefei, China, (C. FitzGerald and S. Gong, eds.), World Scientific Publishing Co., to appear.

[Ha]   H. Hamada, *Starlike mappings on bounded balanced pseudoconvex domains with C^1-plurisubharmonic defining functions*, Pacific J. Math. **194** (2000), 359-371.

[HaKo1]   H. Hamada and G. Kohr, *The growth theorem and quasiconformal extension of strongly spirallike mappings of type α*, Complex Variables **44** (2001), 281-297.

[HaKo2]   H. Hamada and G. Kohr, *Loewner chains and quasiconformal extension of holomorphic mappings*, Ann. Polon. Math. **81** (2003), 85-100.

[HaKo3]   H. Hamada and G. Kohr, *Loewner chains and the Loewner differential equation in reflexive complex Banach spaces*, Rev. Roum. Math. Pures Appl., to appear.

[Ko1]   G. Kohr, *Using the method of Loewner chains to introduce some subclasses of biholomorphic mappings in $\mathbb{C}^n$*, Rev. Roumaine Math. Pures Appl. **46** (2001), 743-760.

[Ko2]   G. Kohr, *Kernel convergence and biholomorphic mappings in several complex variables*, Int. J. Math. Math. Sci. **67** (2003), 4229-4239.

[Pf1]   J.A. Pfaltzgraff, *Subordination chains and univalence of holomorphic mappings in $\mathbb{C}^n$*, Math. Ann. **210** (1974), 55-68.

[Pf2]   J.A. Pfaltzgraff, *Subordination chains and quasiconformal extension of holomorphic maps in $\mathbb{C}^n$*, Ann. Acad. Sci. Fenn. Ser. A I Math **1** (1975), 13-25.

[Pf3]   J.A. Pfaltzgraff, *Loewner theory in $\mathbb{C}^n$*, Abstract of papers presented to AMS, **11** (66) (1990), 46.

[PfSu]   J.A. Pfaltzgraff and T.J. Suffridge, *Close-to-starlike holomorphic functions of several variables*, Pacific J. Math. **57** (1975), 271-279.

[Pom1]   C. Pommerenke, *Über die Subordination analytischer Funktionen*, J. Reine Angew. Math. **218** (1965), 159-173.

[Pom2]   C. Pommerenke, *Univalent Functions*, Vandenhoeck & Ruprecht, Göttingen, 1975.

[Por]   T. Poreda, *On the univalent subordination chains of holomorphic mappings in Banach spaces*, Comment. Math. Prace Mat. **28** (1989), 295-304.

[Su]   T.J. Suffridge, *Starlike and convex maps in Banach spaces*, Pacific J. Math. **46** (1973), 575-589.

Department of Mathematics, University of Toronto, Toronto, Ontario M5S 3G3, Canada
E-mail address: graham@math.toronto.edu

Faculty of Mathematics and Computer Science, Babeş-Bolyai University, 1 M. Kogălniceanu Str., 3400 Cluj-Napoca, Romania
E-mail address: gkohr@math.ubbcluj.ro

Department of Mathematics, CB 3250, University of North Carolina, Chapel Hill, NC 27599-3250, U.S.A.
E-mail address: jap@math.unc.edu

On the Zeros of a q-Bessel Function

W. K. Hayman

Dedicated to Larry Zalcman on his 60th birthday

1. Introduction

Suppose that a and q are non-zero complex constants such that $|q| < 1$ and $aq^\nu \neq 1$ for $\nu \in \mathbb{N}$. Let

(1.1) $$f(z) = \sum_0^\infty a_n z^n$$

be a solution of the functional equation

(1.2) $$(a - qz)f(q^2 z) - (1 + a)f(qz) + f(z) = 0.$$

Then, writing as usual $(a;q)_n = \prod_{k=0}^{n-1}(1 - aq^k)$, we have

(1.3) $$a_n = a_0 q^{n^2} / \prod_{\nu=1}^n \{(1 - q^\nu)(1 - aq^\nu)\} = \frac{a_0 q^{n^2}}{(q;q)_n (aq;q)_n}.$$

We recall the definition (see Ismail [4]) of the q-Bessel function:

$$J_\nu^{(2)}(z;q) = \frac{(q^{\nu+1};q)_\infty}{(q;q)_\infty} \sum_{n=0}^\infty \frac{(-1)^n q^{n(n+\nu)}}{(q;q)_n (q^{\nu+1};q)_n} \left(\frac{z}{2}\right)^{2n+\nu}.$$

Thus, with $a_0 = 1$ and $a = q^\nu$, we have

$$J_\nu^{(2)}(z;q) = \frac{(q^{\nu+1};q)_\infty}{(q;q)_\infty} \left(\frac{z}{2}\right)^\nu f\left(-\frac{q^\nu z^2}{4}\right);$$

and the zeros ζ_n of $J_\nu^{(2)}(z;q)$ and z_n of $f(z)$ are related by

$$z_n = -\frac{q^\nu \zeta_n^2}{4}.$$

Mourad Ismail [5] has conjectured that in this and more general cases, the zeros z_n of $f(z)$ arranged in order of nondecreasing moduli satisfy

(1.4) $$z_n = c_1 q^{-n/\alpha} + c_2 q^{\beta n}[1 + O(q^{\gamma n})], \quad \text{as} \quad n \to \infty,$$

2000 *Mathematics Subject Classification.* Primary 30D05.

for some constants c_1, c_2, α, β, γ. In this paper, we prove the following.

THEOREM 1. *With the above hypotheses, if k is a positive integer, we have an asymptotic expansion*

$$(1.5) \quad z_n = -q^{(1-2n)} \left\{ 1 + \sum_{\nu=1}^{k} b_\nu q^{n\nu} + O\left(q^{(k+1)n}\right) \right\} \quad \text{as} \quad n \to \infty.$$

Here the constants b_ν depend on a and q and

$$(1.6) \quad b_1 = -\frac{(1+a)}{(1-q)} \prod_{m=1}^{\infty} \left(\frac{1-q^{2m-1}}{1-q^{2m}}\right)^2.$$

In a recent paper by W. Bergweiler and the author [3], more general equations than (1.2) were considered and estimates for the coefficients and the zeros of the solutions were established. In the present situation, Theorem 2 of [3] yields

LEMMA 1. *We have*

$$(1.7) \quad a_n = a_0 q^{n^2} \left(A + O\left(q^n\right)\right)$$

and

$$(1.8) \quad z_n = A' q^{-2n} \left(1 + O(q^n)\right),$$

where A and A' are constants (which depend on a and q).

Thus Theorem 1 yields more precise results, but in a special case.

Bruce Berndt has informed me that Theorem 1 contains in the special case $a = 0$ the third identity on page 57 of Ramanujan's lost notebook. This identity reads

$$(1.9) \quad \sum_{0}^{\infty} \frac{z^n q^{n^2}}{(q;q)_n} = \prod_{n=1}^{\infty} \left(1 + \frac{zq^{2n-1}}{1 - q^n y_1 - q^{2n} y_2 - q^{3n} y_3 - \ldots} \right),$$

where

$$y_1 = \frac{1}{(1-q)\psi^2(q)} = \frac{1}{(1-q)} \left\{ \prod_{n=1}^{\infty} \frac{(1-q^{2m-1})}{(1-q^{2m})} \right\}^2,$$

$$y_2 = 0, \quad y_3 = \frac{(q+q^3)}{(1-q)(1-q^2)(1-q^3)\psi^2(q)} - \frac{\sum_0^\infty \frac{(2n+1)q^{2n+1}}{1-q^{2n+1}}}{(1-q)^3 \psi^6(q)}$$

and

$$y_4 = y_1 y_3.$$

Here

$$\psi(q) = \sum_{n=0}^{\infty} q^{n(n+1)/2} = \prod_{n=1}^{\infty} \left(\frac{1-q^{2n}}{1-q^{2n-1}}\right).$$

Andrews ([1, Theorem 6.2], see also Andrews and Berndt ([2]), proves this result by showing that, at least if $0 < q < \frac{1}{4}$, the zeros z_n are given exactly for $n = 1, 2, \ldots$ by the convergent series

$$(1.10) \quad z_n = -q^{(1-2n)} \left(1 - \sum_{\nu=1}^{\infty} y_\nu q^{\nu n} \right).$$

Thus, in this case, (1.5) may be replaced by the stronger (1.10) with $b_\nu = -y_\nu$. This leads Mourad Ismail and me to conjecture that a similar conclusion holds in general for all n if q is complex and $|q|$ is sufficiently small, and for $n > n_0(q)$, whenever $|q| < 1$. Professor Berndt has verified that $b_\nu = -y_\nu$ for $1 \leq \nu \leq 4$, if $a = 0$.

In fact, since $f(z)$ is an entire function of genus zero, we have by Hadamard's product expansion,

$$f(z) = f(0) \prod_{n=1}^{\infty} \left(1 - \frac{z}{z_n}\right),$$

where z_n are the zeros of $f(z)$. Substituting $f(0) = 1$ and z_n from (1.5) with $a = 0$, we obtain (1.9). The identity (1.9) is proved by Andrews and Berndt [2] by a different method involving orthogonal polynomials.

2. Preliminary Estimates

As in [3], we obtain our results by comparing $f(z)$ with the theta function [3, p. 66],

(2.1) $$\theta(z) = \theta(z, q) = \sum_{n=-\infty}^{+\infty} q^{n^2} z^n = \omega \prod_{n=1}^{\infty} \left(1 + \frac{q^{2n-1}}{z}\right)\left(1 + q^{2n-1} z\right),$$

where

(2.2) $$\omega(q) = \prod_{n=1}^{\infty} \left(1 - q^{2n}\right).$$

We also need the entire function

(2.3) $$\theta_0(z) = \sum_{0}^{\infty} q^{n^2} z^n.$$

Then if $|z| = r \geq 1$,

(2.4) $$|\theta(z) - \theta_0(z)| \leq \sum_{1}^{\infty} \frac{|q^{n^2}|}{r^n} \leq \frac{1}{r} \sum_{1}^{\infty} |q|^{n^2} \to 0, \quad \text{as} \quad r \to \infty.$$

We put together the other properties of $\theta(z)$ which we need in

LEMMA 2. *Suppose that $r = |z| = q^{2-\tau-2\nu}, 0 \leq \tau < 2$, i.e., $q^{2-2\nu} \leq |z| < q^{-2\nu}$, where $\nu \in \mathbb{N}_0$. Then [3, p. 69] we have (2.4) and*

(2.5) $$\log |\theta(z)| = \left(\frac{(\log r)^2}{4 \log (1/|q|)}\right) + \log |1 + q^{2\nu-1} z| + O(1)$$

uniformly as $z \to \infty$.
    Also, if $\alpha \in \mathbb{Z}$, we have from the series in (2.1)

(2.6) $$\theta(z) = q^{\alpha^2} z^{-\alpha} \theta(q^{-2\alpha} z).$$

We next compare $f(z)$ with $\theta(z)$. In what follows, we assume without loss of generality that $a_0 = 1$ in (1.3).

LEMMA 3. *Given a positive integer k, we have*

$$(2.7) \qquad a_n = K_0 q^{n^2}(1 - \varepsilon_n),$$

where

$$(2.8) \qquad 1 - \varepsilon_n = \sum_{m=0}^{k} d_m q^{mn} + O\left(|q|^{(k+1)n}\right), \text{ as } n \to \infty.$$

Here K_0 and the d_m are constants depending only on a, q;

$$(2.9) \qquad d_0 = 1, \; d_1 = \frac{-(1+a)q}{1-q}, \; d_2 = \frac{q^2 \{a + q(1 + a + a^2)\}}{(1-q)(1-q^2)}, \ldots$$

Hence we have for $|z| = r$, as $r \to \infty$, while $|1 + q^{2\nu-1}z| > \delta > 0$, for every ν,

$$(2.10) \qquad f(z) = K_0 \sum_{m=0}^{k} d_m \theta(q^m z, q) + O(\theta(r|q|^{k+1}, |q|)).$$

We have by (1.3),

$$(2.11) \qquad a_n = \frac{q^{n^2} \Pi_{\nu=n+1}^{\infty}(1-q^\nu)(1-aq^\nu)}{\Pi_{\nu=1}^{\infty}\{(1-q^\nu)(1-aq^\nu)\}} = K_0 q^{n^2} \exp(-\eta_n)$$

say, where

$$(2.12) \qquad K_0^{-1} = \prod_{\nu=1}^{\infty}(1-q^\nu)(1-aq^\nu).$$

We write $\mu = \max(1, [a])$ and suppose that n is so large that $\mu|q|^{n+1} < \frac{1}{2}$. Let k be a positive integer. Then

$$\eta_n = \sum_{\nu=n+1}^{\infty}\left\{\log\left(\frac{1}{1-q^\nu}\right) + \log\left(\frac{1}{1-aq^\nu}\right)\right\}$$

$$= \sum_{\nu=n+1}^{\infty}\left\{\sum_{m=1}^{k}\frac{q^{\nu m} + a^m q^{\nu m}}{m} + \varepsilon_{k,\nu}\right\}$$

where

$$|\varepsilon_{k\nu}| < \frac{2\mu^{k+1}|q|^{\nu(k+1)}}{1 - \mu|q^\nu|} < 4\mu^{k+1}|q|^{\nu(k+1)}.$$

We now reverse the order of summation. Note that

$$\sum_{\nu=n+1}^{\infty} \frac{q^{\nu m} + a^m q^{\nu m}}{m} = \frac{q^{m(n+1)}(1+a^m)}{m(1-q^m)} = f_m q^{m(n+1)},$$

say. Also,

$$\left|\sum_{\nu=n+1}^{\infty} \varepsilon_{k,\nu}\right| \leqslant \sum_{n+1}^{\infty} 4\mu^{k+1}|q|^{(k+1)\nu} = \frac{4\mu^{k+1}|q|^{(k+1)(n+1)}}{1-|q|}.$$

Thus

$$(2.13) \qquad \eta_n = \sum_{m=1}^{k} f_m q^{m(n+1)} + O(|q|^{(k+1)(n+1)}),$$

where

(2.14) $$f_m = \frac{1+a^m}{m(1-q^m)}.$$

Since $|q| < 1$, $q^{m(n+1)} \to 0$ as $n \to \infty$ for $1 \leqslant m \leqslant k+1$ and so does η_n. We now write

(2.15) $$(1-\varepsilon_n) = \exp(-\eta_n) = \sum_0^\infty \frac{(-\eta_n)^p}{p!}.$$

We substitute for η_n from (2.13) and (2.15) and obtain (2.8). In fact, the terms in η_n^p contribute only to the error term in (2.8) if $p > k$. Taking $k = 2$, we obtain

$$\varepsilon_n = \eta_n - \frac{\eta_n^2}{2} + O(|q|^{3(n+1)})$$

$$= f_1 q^{n+1} + \left(f_2 - \frac{1}{2}f_1^2\right) q^{2(n+1)} + O\left(q^{3(n+1)}\right)$$

(2.16) $$= \frac{1+a}{1-q} q^{n+1} + \left\{\frac{1+a^2}{2(1-q^2)} - \frac{1}{2}\frac{(1+a)^2}{(1-q)^2}\right\} q^{2(n+1)} + O\left(q^{3(n+1)}\right).$$

This yields (2.9).

It remains to deduce (2.10). We note that for fixed m, we have as $z \to \infty$

$$\sum_{n=0}^\infty q^{n^2} q^{mn} z^n$$

$$= \sum_{n=0}^\infty (q^m z)^n q^{n^2} = \theta_0(q^m z, q) = \theta(q^m z, q) + o(1)$$

by (2.4). Also, by Lemma 2,

$$|\theta_0(z,q)| \leqslant \theta(|z|, |q|) = \exp\left\{\frac{(\log|z|)^2}{4\log|1/q|} + O(1)\right\}.$$

Now (2.8) yields (2.10).

Remark. We note that our method extends to more general functions $f(z)$ whose coefficients a_n are given by

$$a_n = a_0 q^{n^2} / \prod_{p=1}^{p_0} (c_p q; q)_n^{t_p},$$

where c_p, t_p are real or complex constants such that $c_p q^k \neq 1$, $k \in \mathbb{N}$. We need only replace f_m in (2.14) and the subsequent analysis by

$$f_m = \sum_{p=1}^{p_0} \frac{t_p c_p^m}{m(1-q^m)}.$$

3. A local reduction

In order to solve the equation $f(z) = 0$ when z is large, it is convenient to reduce the problem to a compact part of the plane. To do this, we use (2.6).

Suppose that, with the notation of Lemma 2, we set $\alpha = -\nu$. Then
$$\theta(z) = q^{\nu^2} z^\nu \, \theta(q^{2\nu} z).$$
We apply this result with $q^m z$ instead of z and obtain
$$\theta(q^m z) = q^{\nu^2}(q^m z)^\nu \theta(q^{2\nu+m} z).$$
We now write

(3.1) $$z = -q^{1-2\nu}(1+Z), \; zq^m = -q^{m+1-2\nu}(1+Z),$$

so that
$$\begin{aligned}\theta(zq^m) &= \theta\left(-q^{m+1-2\nu}(1+Z)\right) \\ &= q^{\nu^2}\left\{-q^{m+1-2\nu}(1+Z)\right\}^\nu \theta(-q^{2\nu+m+1-2\nu}(1+Z)) \\ &= (-1)^\nu q^{-\nu^2+(m+1)\nu}(1+Z)^\nu \, \theta(-q^{m+1}(1+Z)) \\ (3.2) \quad &= (-1)^\nu q^{\nu-\nu^2}(1+Z)^\nu \left\{q^{m\nu}\theta\left(-q^{m+1}(1+Z)\right)\right\}.\end{aligned}$$

Also, by (3.1) $r = |z| = |q|^{1-2\nu}|1+Z|$. Thus (2.6) yields
$$\begin{aligned}\theta(r|q|^{k+1}, |q|) &= |q|^{\nu^2}(|q|^{k+2-2\nu}|1+Z|)^\nu \theta(|q|^{k+2-2\nu+2\nu}|1+Z|) \\ (3.3) \quad &= |q|^{\nu-\nu^2}|1+Z|^\nu |q|^{(k+1)\nu} \, \theta(|q|^{k+2}|1+Z|, |q|).\end{aligned}$$

We make the substitution (3.1) and write

(3.4) $$f(z) = f(\{-q^{1-2\nu}(1+Z)\}) = K_0(-1)^\nu q^{\nu-\nu^2}(1+Z)^\nu F(Z).$$

Then $F(Z)$ is analytic in Z for fixed ν, and the equation $f(z) = 0$ reduces by (2.10) and (3.1) to (3.4) to

(3.5) $$F(Z) = \sum_{m=0}^{k} d_m \, q^{m\nu} \, \theta\left(-q^{m+1}(1+Z)\right) + O(|q|^{(k+1)\nu}) = 0.$$

We may expand $\theta\left(-q^{m+1} - q^{m+1}Z\right)$ as a power series in Z near $Z = 0$ with coefficients depending on m and q. We also recall (3.1) and write

(3.6) $$W = q^\nu, \; F(Z) = F(W, Z).$$

Then the equation (3.5) becomes

(3.7) $$F(W, Z) = 0,$$

where

(3.8) $$F(W, Z) = \sum_{m \geq 0, n \geq 0} \sum_{m+n \leq k} a_{m,n} W^m Z^n + O\left(|Z|^{k+1} + |W|^{k+1}\right),$$

and

(3.9) $$a_{m,n} = \frac{d_m(-q^{m+1})^n \, \theta^{(n)}(-q^{m+1})}{n!}.$$

Here d_m is defined by (2.8) and (2.13) to (2.15). Further, $F(W, Z)$ is analytic in Z for fixed W; and the constant implicit in O is uniform in some bicylinder $|z| < \delta, |W| < \delta$.

Also,

(3.10) $\quad a_{00} = d_0\theta(-q) = 0$

(3.11) $\quad a_{0,1} = d_0(-q)\theta'(-q) = -q\theta'(-q) \neq 0,$

since $\theta(z)$ has a simple zero at $z = q$ by (2.1) and $d_0 = 1$ by (2.9). Further,

$$a_{10} = d_1\theta(-q^2) = \frac{-(1+a)q}{1-q}\theta(-q^2) \neq 0;$$

$$a_{02} = \frac{q^2\theta''(-q)}{2}, \quad a_{20} = d_2\theta(-q^3) = 0,$$

$$a_{11} = \frac{(1+a)q^3}{(1-q)}\theta'(-q^2).$$

(3.12)

4. An inversion formula

We need the following inversion formula, which may have independent interest.

LEMMA 4. *Suppose that $F(W, Z)$ is analytic in Z for fixed W in some bicylinder $|Z| < \delta_0, |W| < \delta_0$ and satisfies (3.8) there, where $a_{00} = 0$, $a_{01} \neq 0$. Then for sufficiently small δ_1 there exists δ_2, $0 < \delta_2 < \delta_1$, such that if $0 < |W| < \delta_2$, the equation (3.7) has exactly one root $Z = Z(W)$ in $|Z| < \delta_1$. Further*

(4.1) $$Z(W) = \sum_{n=1}^{k} b_n W^n + O(|W|^{k+1}) \quad as \quad W \to 0.$$

Here

(4.2) $$b_1 = -\frac{a_{10}}{a_{01}}, \quad b_2 = (-a_{02}a_{10}^2 + a_{01}a_{10}a_{11} - a_{20}a_{01}^2)/a_{01}^3,$$

and more generally $a_{01}^{2n-1}b_n$ is a homogeneous polynomial of degree $2n - 1$, with integer coefficients, in the a_{ij} with $0 \leq i + j \leq n$.

COROLLARY 1. *For large ν, the equation $f(z) = 0$ has a unique zero z_ν near $-q^{1-2\nu}$ such that for a fixed positive integer k, we have*

(4.3) $$z_\nu = -q^{1-2\nu}\left(1 + \sum_{n=1}^{k} b_n q^{\nu n} + O\left(|q|^{\nu(k+1)}\right)\right),$$

where
(4.4)
$$b_1 = -\frac{1+a}{1-q}\frac{\theta(-q^2)}{\theta'(-q)}, \quad b_2 = \frac{q(1+a)^2\theta''(-q)\theta(-q^2)^2}{2(1-q)^2\{\theta'(-q)\}^3} - \frac{(1+a)^2q^2}{(1-q)^2}\frac{\theta(-q^2)\theta'(-q^2)}{\theta'(-q)^2},$$

etc.

To deduce (1.6), note that by (2.1)

$$\theta(-q^2) = \omega(q)\prod_{n=1}^{\infty}(1-q^{2n-3})(1-q^{2n+1})$$

$$= \left\{\omega(q)(1-\frac{1}{q})/(1-q)\right\}\prod_{n=1}^{\infty}(1-q^{2n-1})^2 = -\frac{\omega(q)}{q}\prod_{n=1}^{\infty}(1-q^{2n-1})^2.$$

Also,

$$\theta'(-q) = \lim_{z \to -q} \frac{\theta(z)}{z+q} = \omega(q) \frac{1-q^2}{-q} \prod_{n=2}^{\infty}(1-q^{2n-2})(1-q^{2n})$$

$$= -\frac{\omega(q)}{q} \prod_{n=1}^{\infty}(1-q^{2n})^2.$$

Thus (1.6) follows from (4.4).

We start by proving that if (3.8) holds with $k=1$ and $a_{00}=0$, $a_{01} \neq 0$, then the equation (3.3) has a unique root $Z = Z(W)$ in a fixed small neighbourhood of $Z=0$, if W is small. Further $Z(W) = O(|W|)$ as $W \to 0$.

Clearly, if (3.8) holds for some value of k, then (3.8) is also true for any k' such that $1 \leqslant k' < k$. So we assume that $k=1$. Thus there exists δ_0 such that for $|Z| \leqslant \delta_0, |W| \leqslant \delta_0$,

$$F(W,Z) = a_{10}W + a_{01}Z + \varepsilon(W,Z),$$

where

(4.5) $$|\varepsilon(W,Z)| < C(|Z|^2 + |W|^2),$$

and C is a positive constant.

We now choose δ_1 so small that

(4.6) $$0 < \delta_1 \leqslant \min\left\{1, \delta_0, \frac{|a_{01}|}{4C}\right\}.$$

This is possible since $a_{01} \neq 0$. Next we choose

(4.7) $$\delta_2 = \frac{\delta_1}{3} \min(1, |a_{01}/a_{10}|).$$

Thus $0 < \delta_2 < \delta_1$. Suppose now that $|W| \leqslant \delta_2, |Z| = \delta_1$. We write $f(Z) = a_{01}Z$, $g(Z) = a_{10}W + \varepsilon(W,Z)$. Then

$$\begin{aligned}|g(Z)| &\leqslant |a_{10}W| + |\varepsilon(W,Z)| \leqslant \delta_2 |a_{10}| + 2C\delta_1^2 \\ &< \frac{1}{2}\delta_1|a_{01}| + 2C\delta_1^2 \\ &\leqslant \frac{1}{2}\delta_1|a_{01}| + \frac{1}{2}\delta_1|a_{01}| = \delta_1|a_{01}|,\end{aligned}$$

while $|f(z)| = |a_{01}|\delta_1$. Thus, by Rouché's Theorem, $f(W,Z) = f(Z) + g(Z)$ has as many zeros in $|Z| < \delta_1$ as $f(Z)$, i.e., exactly one.

Let us denote the root of $f(W,Z) = 0$ in $|Z| < \delta_1$ for a given W by $Z = Z(W)$. The above argument shows that $Z(W)$ exists and is unique if

(4.8) $$|W| < \frac{1}{3}\min\left(1, \delta_0, \frac{|a_{01}|}{4C}\right) \min\left(1, \frac{|a_{01}|}{|a_{10}|}\right).$$

Further, we have in this case,

(4.9) $$|Z(W)| \leqslant 3|W| \max\left(1, \frac{|a_{10}|}{|a_{01}|}\right).$$

For we may choose $\delta_2 = |W|$; then
$$\delta_1 = 3\delta_2 \max\left(1, \frac{|a_{10}|}{|a_{01}|}\right)$$
in this case, which is equivalent to (4.7). Also, by (4.8), (4.6) is satisfied. Thus $f(W, Z) = 0$ has a root in $|Z| < \delta_1$, which gives (4.9). Qualitatively, we may say that $Z(W)$ exists and is unique in a fixed neighbourhood of $W = 0$ and

(4.10) $$Z(W) = O(W) \text{ as } W \to 0.$$

This is the case $k = 0$ of (4.1). We now proceed by induction on k. Suppose that $k \geqslant 1$ and that (4.1) holds for $k - 1$, where the sum in (4.1) is taken to be zero if $k = 0$. Thus

(4.11) $$Z(W) = \sum_{n=1}^{k-1} b_n W^n + b_k(W) W^k,$$

where $b_k(W)$ remains bounded as $W \to 0$. We substitute (4.11) in (3.8) and use (3.7). We deduce from (4.11) that if $2 \leqslant n \leqslant k$,

(4.12) $$Z^n = \sum_{\mu=n}^{k} b_{\mu,n} W^\mu + O(W^{k+1}),$$

where the $b_{\mu,n}$ are polynomials of degree n with integer coefficients in b_1 to $b_{\mu+1}$. Using (3.7), (3.8), (4.11) and (4.12), we obtain

(4.13) $$\sum_{m=1}^{k} a_{m0} W^m + \left(\sum_{n=1}^{k-1} b_n W^n + b_k(W) W^k\right)\left(\sum_{m=0}^{k-1} a_{m1} W^m\right)$$
$$+ \sum_{n=2}^{k}\left\{\sum_{m=0}^{k-n} a_{mn} W^m \sum_{\mu=n}^{k-m} b_{\mu,n} W^\mu\right\} + O(|W|^{k+1}) = 0.$$

We can write (4.13) in the form

(4.14) $$P_k(W) + a_{01} b_k(W) W^k + O(W^{k+1}) = 0,$$

where $P_k(W)$ is a polynomial of degree k.

Suppose that $a_l W^l$ is the term of lowest degree in $P_k(W)$. Then if $l < k$, we obtain a contradiction from (4.14), as $W \to 0$. Thus $l = k$. Now (4.14) yields
$$W^k \{a_k + a_{01} b_k(W) + O(W)\} = 0$$
or
$$b_k(W) = \frac{-a_k}{a_{01}} + O(W),$$
and (4.11) yields (4.1) with

(4.15) $$b_k = \frac{-a_k}{a_{01}}.$$

To complete the proof of Lemma 4, we investigate more closely the constants b_n. We may assume that $n = k$. The polynomial $P_k(W) = a_k W^k$ can be calculated from (4.13). This yields

(4.16) $$a_k = a_{k0} + \sum_{n=1}^{k-1} b_n a_{k-n,1} + \sum_{n=2}^{k} \sum_{m=0}^{k-n} a_{m,n} b_{k-m,n}.$$

The quantities $a_{m,n}$ are given in (3.9), the b_n for $n < k$ are as in (4.11) and the $b_{\mu,n}$ are calculated from the b_n by (4.12). Then b_k is given by (4.15) and (4.16). In particular,

$$a_{01}b_1 = -a_1 = -a_{10}, -a_{01}b_2 = a_2 = a_{20} + a_{11}b_1 + a_{02}b_2^2 = a_{20} + a_{11}b_1 + a_{02}b_1^2$$

$$b_2 = \left\{ \frac{-a_{20}a_{01}^2 + a_{01}a_{10}a_{11} - a_{02}a_{10}^2}{a_{01}^3} \right\},$$

as stated in Lemma 4. Also, if b_1 to b_{k-1} are polynomials in the a_{ij} divided by a_{01}^{2k-3}, then $b_{k-m,n}$ is such a polynomial divided by at most a_{01}^{2k-2}, since $m \geqslant 0$ and $n \geqslant 2$, and hence so are a_k and $a_{01}b_k$. Thus, by induction, $a_{01}^{2k-1}b_k$ is a polynomial in the $a_{m,n}$ with $m \geqslant 0, n \geqslant 0, m+n \leqslant k$. Since (3.7) and (4.1) are unaltered if all the $a_{m,n}$ are multiplied by a constant λ, the polynomial must be homogeneous of degree $2k - 1$. This proves Lemma 4.

It remains to prove the Corollary. We set $W = q^\nu$ and write $Z(W) = Z_\nu$. Then $F(W, Z_\nu) = 0$, and so (3.1) yields

$$f\left\{-q^{1-2\nu}(1 + Z_\nu)\right\} = 0.$$

Thus, for large ν, there is a unique z_ν close to $-q^{1-2\nu}$ such that $f(z_\nu) = 0$. Further,

$$z_\nu = -q^{1-2\nu}(1 + Z_\nu) = -q^{1-2\nu}\left\{1 + \sum_{n=1}^{k} b_n q^{\nu n} + O\{q^{\nu(k+1)}\}\right\},$$

where the b_n are as in (4.1). This gives (4.3) and (1.5).

It remains to verify (4.4). To do this, we use (4.2) to express b_1 and b_2 in terms of the a_{mn}. We express a_{mn} in terms of $\theta(z)$ by (3.10) to (3.12). This yields (4.4) and the Corollary.

5. Completion of the proof of Theorem 1

It follows from the Corollary to Lemma 4 that, if δ is a small positive number and n is large, $f(z)$ has a unique zero z_n in the disk

$$\left|z + q^{1-2n}\right| < \delta \left|q^{1-2n}\right|.$$

It remains to prove that there are no other zeros. It follows from (2.5) that if

(5.1) $$\left|q^{2-2n}\right| \leqslant |z| \leqslant \left|q^{-2n}\right|$$

and

$$\left|z + q^{1-2n}\right| \geqslant \delta|q|^{1-2n},$$

then

$$\log|\theta(z)| = \frac{(\log|z|)^2}{4\log 1/|q|} + O(1) \to \infty.$$

Using (2.10) with $k = 0$, we deduce that in the same region, we have for large n

$$|f(z)| > K\exp\frac{(\log|z|)^2}{4\log 1/|q|},$$

where K is a positive constant. Thus the zero z_n satisfying (1.5) is the only zero in the annulus (5.1) when n is large. Suppose that n is a large positive integer and that $f(z)$ has exactly $n + p$ zeros in

$$|z| < R_n = |q|^{-2n}.$$

Since $f(z)$ has exactly one zero z_{n+1} in $R_n \leqslant |z| < R_{n+1}$, the same statement holds, when n is replaced by $n+1$, and so for all large n. We use Jensen's formula to show that $p = 0$

Let $\nu(r)$ be the number of zeros of $f(z)$ in $|z| \leqslant r$. It follows that if n is large, we have
$$\nu(r) = n + p \text{ for } |z_n| \leqslant r < |z_{n+1}|.$$
Hence
$$\int_{|z_n|}^{|z_{n+1}|} \nu(r) \frac{dr}{r} = (n+p) \log \left|\frac{z_{n+1}}{z_n}\right| = 2(n+p) \left\{ \log \frac{1}{|q|} + O(q^n) \right\}$$
by (1.5). Also, since $f(0) \neq 0$, $\nu(r) = 0$ for small positive r. Thus
$$\int_0^{|z_n|} \nu(t) \frac{dt}{t} = 2 \log \left|\frac{1}{q}\right| \sum_{m=1}^{n-1} (m+p) + O(1)$$
$$= 2 \log \left|\frac{1}{q}\right| \left\{ \frac{n(n-1)}{2} + pn + O(1) \right\}.$$

Also, if $R = R_n$, we have
$$N(R) = \int_0^R \frac{\nu(t) dt}{t} = \int_0^{|z_n|} \frac{\nu(t) dt}{t} + (n+p) \log \frac{R}{|z_n|}$$
$$= \log \left|\frac{1}{q}\right| \left\{ (n(n-1) + 2pn + n + p + O(1) \right\}$$
(5.2) $$= \log \left|\frac{1}{q}\right| \left\{ n^2 + 2pn + O(1) \right\}.$$

On the other hand, by (2.5) and (2.10) with $k = 0$, we have on $|z| = R$,
$$\log |f(z)| = \frac{(\log R)^2}{4 \log |1/q|} + O(1) = n^2 \log \left|\frac{1}{q}\right| + O(1).$$

Now Jensen's formula gives
$$\frac{1}{2\pi} \int_0^{2\pi} \log |f(Re^{i\theta})| \, d\theta = n^2 \left\{ \log \left|\frac{1}{q}\right| + O(1) \right\}$$
(5.3) $$= N(R) + \log |f(0)| = \log \left|\frac{1}{q}\right| \left\{ n^2 + 2pn + O(1) \right\}.$$

Hence $p = 0$, and $f(z)$ has precisely n zeros in $|z| < |q|^{-2n}$.

It follows that for large n, the zero z_n which satisfies (1.5) is precisely the n-th zero in order of increasing moduli. This completes the proof of Theorem 1.

Acknowledgements. The author is grateful to Mourad Ismail for suggesting the problem and for subsequent advice and help, to Bruce Berndt for valuable conversations and to Amy Bensted for typing a very illegible manuscript.

References

[1] G.E Andrews, *Ramanujan's "Lost" Notebook VIII. The entire Rogers-Ramanujan function*, Adv. Math., to appear.
[2] G.E. Andrews and B.C. Berndt, *Ramanujan's Lost Notebooks, Part I*, Springer Verlag, New York, to appear.
[3] Walter Bergweiler and Walter Hayman, *Zeros of solutions of a functional equation*, Comput. Methods Funct. Theory **3** (2003), 55-78.
[4] M.E.H. Ismail, *The zeros of basic Bessel functions, the functions $J_{\nu+ax}(x)$ and the associated orthogonal polynomials*, J. Math. Anal. Appl. **86** (1982), 1-19.
[5] Mourad Ismail, letter to the author.

DEPARTMENT OF MATHEMATICS, IMPERIAL COLLEGE LONDON, LONDON SW7 2AZ, U.K.

Entire Functions with No Unbounded Fatou Components

A. Hinkkanen

Dedicated to Lawrence Zalcman on his 60th birthday

ABSTRACT. Let f be a transcendental entire function of order less than $1/2$. We introduce a condition on the regularity of growth of f and show that it implies that every component of the Fatou set of f is bounded.

1. Introduction and results

Let f be a transcendental entire function. We consider the question, raised by I.N. Baker in 1981 [4], of under what circumstances each component of the Fatou set of f is bounded. First we review the relevant definitions.

The Fatou set $\mathcal{F}(f)$ of f is the set of those points z in the complex plane $\mathbb{C}$ that have a neighborhood U such that the family $\{f^n|U : n \geq 1\}$ of restrictions of the iterates f^n of f to U is a normal family. Here $f^1 = f$, and $f^n = f \circ f^{n-1}$ for $n \geq 2$. (Some authors write $f^{\circ n}$ for the iterates of f.) The Julia set $\mathcal{J}(f)$ of f is the complement of the Fatou set, thus $\mathcal{J}(f) = \mathbb{C} \setminus \mathcal{F}(f)$. The set $\mathcal{F}(f)$ is trivially open while, as shown by Fatou and by Julia, $\mathcal{J}(f)$ is a non-empty perfect set. If $\mathcal{J}(f) \neq \mathbb{C}$, then $\mathcal{J}(f)$ is nowhere dense in $\mathbb{C}$. We have $\mathcal{F}(f^n) = \mathcal{F}(f)$ and $\mathcal{J}(f^n) = \mathcal{J}(f)$ for each $n \geq 1$.

We say that a set E is (forward) invariant under f if $f(E) \subset E$, and backward invariant under f if $f^{-1}(E) \subset E$. We say that E is completely invariant under f if E is both invariant and backward invariant under f. Both $\mathcal{F}(f)$ and $\mathcal{J}(f)$ are completely invariant under f.

The fundamental results in complex dynamics for rational and entire functions can be found in the original papers of Fatou [10, 11, 12] and Julia [15] and in the books of Beardon [7], Carleson and Gamelin [9], Milnor [16], and Steinmetz [19].

We study the components of $\mathcal{F}(f)$. These have been classified as follows. Let D be a component of $\mathcal{F}(f)$. Since $\mathcal{F}(f)$ is invariant under f, there exists, for each $n \geq 1$, a component D_n of $\mathcal{F}(f)$ such that $f^n(D) \subset D_n$. In fact, Herring [13] has shown that $D_n \setminus f^n(D)$ can contain at most one point. If all the components D_n are disjoint, then D is called a wandering domain. If there is a smallest positive

---

2000 *Mathematics Subject Classification.* Primary 30D05, 37F10.

This material is based upon work supported by the National Science Foundation under Grant No. 0200752.

integer p such that $D_p = D$ then D is periodic of period p. In particular, if $p = 1$, then D is called invariant, in accordance with our general definition of invariant sets. Otherwise, D is called preperiodic; and then $D_n = D_{n+p}$ for some minimal $n, p \geq 1$, while for all $p \geq 1$ we have $D_p \neq D$.

Since $\mathcal{F}(f^p) = \mathcal{F}(f)$, a periodic domain of period p is an invariant domain for f^p. Invariant components D of $\mathcal{F}(f)$ can be of the following five types.

There may be a fixed point α of f in D such that $\lim_{n \to \infty} f^n(z) = \alpha$, locally uniformly for $z \in D$. Then the multiplier $\lambda = f'(\alpha)$ satisfies $|\lambda| < 1$. If $\lambda = 0$, we call α a superattracting fixed point of f and D a superattracting domain. If $0 < |\lambda| < 1$, we call α an attracting fixed point of f and D an attracting domain.

There may be a fixed point α of f on ∂D such that $\lim_{n \to \infty} f^n(z) = \alpha$, locally uniformly for $z \in D$. Then $f'(\alpha) = 1$. We call α a parabolic fixed point of f and D a parabolic domain.

It may be that D is simply connected, containing a fixed point α of f such that if φ is a conformal mapping of D onto the unit disk with $\varphi(\alpha) = 0$, then $(\varphi \circ f \circ \varphi^{-1})(z) \equiv e^{2\pi i \beta} z$ for some real irrational number β. In this case, D is called a Siegel disk with centrum α.

Finally, it may be that $\lim_{n \to \infty} f^n(z) = \infty$, locally uniformly for $z \in D$. Then D is called a Baker domain.

We use the following notation for the maximum and minimum modulus of f:
$$M(r, f) = \max\{|f(z)| : |z| = r\}, \qquad m(r, f) = \min\{|f(z)| : |z| = r\}.$$
Recall that the order $\rho(f)$ and lower order $\lambda(f)$ of f are defined by
$$\rho(f) = \limsup_{r \to \infty} \frac{\log \log M(r, f)}{\log r}, \qquad \lambda(f) = \liminf_{r \to \infty} \frac{\log \log M(r, f)}{\log r}.$$
If $0 < \rho(f) = \rho < +\infty$, we define the type of f by
$$\tau(f) = \limsup_{r \to \infty} \frac{\log M(r, f)}{r^\rho}.$$
If $\tau(f) = 0$, we say that f is of minimal type. If $0 < \tau(f) < +\infty$, we say that f is of mean type. If $\tau(f) = +\infty$, we say that f is of maximal type.

Baker [4] asked whether every component of $\mathcal{F}(f)$ is bounded if the growth of f is sufficiently small. The best possible growth condition in terms of order would be of order $1/2$, minimal type at most, as shown by the following example due to Baker [4, p. 489]. For any sufficiently large positive real number a, the function $f(z) = z^{-1/2} \sin \sqrt{z} + z + a$ is of order $1/2$, mean type, and has an unbounded component D of $\mathcal{F}(f)$ containing a segment $[x_0, \infty)$ of the positive real axis, such that $f^n(z) \to \infty$ as $n \to \infty$, locally uniformly in D.

Baker obtained the following result [4, Theorem 1, p. 484, Lemma 7, p. 490].

THEOREM A. *Let f be a transcendental entire function of growth not exceeding order $1/2$, minimal type. Let D be a component of $\mathcal{F}(f)$ such that either*
  (i) *all limit functions of convergent subsequences of iterates of f are finite in D; or*
  (ii) *there is a component U of $\mathcal{F}(f)$ and an integer $m \geq 0$ such that $f^m(D) \subset U$ and $f(U) \subset U$, and furthermore $f^n(z) \to \infty$ as $n \to \infty$, locally uniformly for $z \in U$.*

Then D is bounded.

This fundamental result of Baker has subsequently been improved by several authors. However, many of the methods used have been modifications of Baker's original arguments.

We investigate what Theorem A actually states. It is helpful to recall the $\cos\pi\rho$-theorem in the form given in [**6**, p. 294].

If $E \subset [1,\infty)$, the lower logarithmic density of the set E is defined by

$$\underline{\log \text{ dens }} E = \liminf_{R\to\infty} \frac{1}{\log R} \int_{E\cap(1,R)} \frac{dt}{t}.$$

THEOREM B. *Let f be a transcendental entire function of order $\rho < 1/2$, and suppose that $\rho < \alpha < 1/2$. Then if*

$$E = \{\, r \geq 1 \,:\, \log m(r,f) > (\cos\pi\alpha) \log M(r,f)\,\},$$

we have

$$\underline{\log \text{ dens }} E \geq 1 - \frac{\rho}{\alpha}.$$

In particular, $\limsup_{r\to\infty} m(r,f) = \infty$.

If f is of order $1/2$, minimal type, then $m(r,f)$ is unbounded.

The following example illustrates what Theorem B actually implies. When $\rho(f) < 1/2$, set $\alpha = (1+2\rho)/4$,

$$\beta = \cos\left(\pi\frac{1+2\rho}{4}\right), \qquad \sigma = \frac{2}{1-2\rho}.$$

Now it is easily seen that there is a number R_0 such that for each $R \geq R_0$, we have

$$\log m(r,f) > \beta \log M(r,f)$$

for some r with $R \leq r \leq R^\sigma$.

If the component D of $\mathcal{F}(f)$ is unbounded and f is as in Theorem A, then by Theorem B, the minimum modulus $m(r,f)$ is unbounded, so that $f(D)$ is unbounded. By induction, $f^n(D)$ is unbounded for each $n \geq 1$. So, if D is not a wandering domain, we may assume that D is periodic, that is, $f^p(D) \subset D$ for some $p \geq 1$. Now if all limit functions of convergent subsequences of iterates of f are finite in D, then the conclusion of Theorem A follows from [**4**, Lemma 7, p. 490]. Otherwise, if D is not a wandering domain, so that we may assume that $f^p(D) \subset D$, we note that $f^n(z) \to \infty$ in D (otherwise all limit functions are finite by the classification of periodic domains), that is, D is a Baker domain. When D is invariant (that is, $p = 1$ here), the conclusion of Theorem A follows from [**4**, Theorem 1, p. 484]. An extension of Theorem A(ii) to the case $p \geq 2$ was obtained by Stallard [**17**, Theorem 3A, p. 49] in her PhD thesis. A different proof was given by Anderson and the author [**1**, Theorem 1, p. 3244]. Combining the results mentioned so far, we get the following.

THEOREM C. *Let f be a transcendental entire function of order less than $1/2$. Then every preperiodic or periodic component of $\mathcal{F}(f)$ is bounded.*

Theorem C leaves the possibility that a component D of $\mathcal{F}(f)$ (for $\rho(f) < 1/2$) is unbounded only if D is a wandering domain in which the constant infinity is among the limit functions of convergent subsequences of iterates of f. When considering wandering domains, one can assume that they are simply connected, since by another (later) result due to Baker [**5**, Theorem 3.1, p. 565], if f is any transcendental entire function such that $\mathcal{F}(f)$ has a multiply connected component, then each component of $\mathcal{F}(f)$ is bounded. (Furthermore, each multiply connected component of $\mathcal{F}(f)$ if a wandering domain on which the iterates of f tend to infinity, and, of course, is a bounded domain.)

Concerning simply connected wandering domains D in which the constant infinity is among the limit functions of convergent subsequences of iterates of f, Baker [4, Theorem 3.1, p. 565] proved that D is bounded if

$$\log M(r, f) = O((\log r)^p)$$

as $r \to \infty$, where $1 < p < 3$. Stallard [18, Theorem B, p. 43] improved this sufficient condition to

$$\log \log M(r, f) = O\left(\frac{(\log r)^{1/2}}{(\log \log r)^\varepsilon}\right)$$

for some $\varepsilon > 0$. Thus these growth conditions imply that all components of $\mathcal{F}(f)$ are bounded. These were the first conditions to cover all components, including wandering domains.

Stallard [8, Theorem C, p. 44] also proved that every component of $\mathcal{F}(f)$ is bounded provided that f is of order less than $1/2$ and

$$\frac{\log M(2r, f)}{\log M(r, f)} \to c \qquad \text{as } r \to \infty,$$

where $c \geq 1$ is a finite constant that depends on f.

Further results have been obtained by various authors under additional assumptions that limit not only the order of growth but also the regularity of the growth of the function. The last result due to Stallard mentioned above may be viewed as such a result. Anderson and the author [1, Theorem 2, p. 3245] obtained the following sufficient condition. Their method of proof was based on the notion of "self-sustaining spread". We set $\varphi(x) = \log M(e^x, f)$ so that, by the Hadamard three-circles theorem, $\varphi(x)$ is an increasing convex function of x. The function $\varphi'(x)$ may fail to exist at a countable set of points. At such points, $\varphi'(x)$ is understood to be the right-hand derivative.

THEOREM D. *Let f be a transcendental entire function of order $< 1/2$ such that for some positive constant c,*

$$\frac{\varphi'(x)}{\varphi(x)} \geq \frac{1+c}{x}$$

for all sufficiently large x, where $\varphi(x) = \log M(e^x, f)$. Then every component of $\mathcal{F}(f)$ is bounded.

The condition of Theorem D should be compared to the condition

$$\varphi(x+1) \sim c\varphi(x) \qquad \text{as } x \to \infty,$$

which is essentially Stallard's regularity condition.

As the authors of [1] remark, the assumption of Theorem D is a condition of growth as well as regularity and is equivalent to the condition

$$\liminf_{r \to \infty} \frac{d \log \log M(r, f)}{d(\log \log r)} > 1,$$

which requires a certain minimal rate of growth at all times and implies, in particular, that

$$\log M(r, f) > (\log r)^{1+\delta}$$

for some $\delta > 0$ and all $r > r_0(\delta)$. Therefore, it should not be surprising, and indeed is easily seen, that a transcendental entire function of order $< 1/2$ and positive lower order satisfies the assumptions of Theorem D. Thus, the difficulty of using the method of self-sustaining spread to prove that all Fatou components of a transcendental entire function of order $< 1/2$ are bounded arises from the possible presence of large annuli which are almost zero free, so that the growth of the function there is like that of a polynomial.

Subsequently, Hua and Yang [14] proved the following result.

THEOREM E. *Let f be a transcendental entire function. If $\lambda(f) < 1/2$, then every (pre)periodic component of $\mathcal{F}(f)$ is bounded. If $\lambda(f) < 1/2$ and for each $m > 1$,*
$$\log M(r^m, f) \geq m^2 \log M(r, f)$$
for all sufficiently large r, then every component of $\mathcal{F}(f)$ is bounded.

Note that if there is a large annulus $\mathcal{A}$ centered at the origin in which f has no zeros, then $|f(z)|$ grows there like $|z|^p$ for some positive integer p depending on $\mathcal{A}$, so that $\log M(r, f)/\log r$ is close to p there. If $m > 1$ and $\mathcal{A}$ is large enough to contain circles of radii r and r^m centered at the origin, then $\log M(r^m, f)/\log r$ is close to mp while $m^2 \log M(r, f)/\log r$ is close to $m^2 p$. This violates the requirement of Theorem E for this r, regardless of how large p is. Thus, if there is a sequence of such annuli (in which case the corresponding numbers p tend to infinity), the assumption of Theorem E does not hold.

Zheng [20] refined Baker's original method and obtained the following result where, at least for (pre)periodic components, there is no upper bound for the growth of f.

THEOREM F. *If*
$$\limsup_{r \to \infty} \frac{m(r, f)}{r} = \infty,$$
then $\mathcal{F}(f)$ has no unbounded preperiodic or periodic components. Also, for a function of order less than $1/2$, any wandering Fatou component U containing a point z such that $\log^+ \log^+ |f^n(z)| = O(n)$ as $n \to \infty$ must be bounded.

In this paper, we prove the following result.

THEOREM 1. *Let f be a transcendental entire function of order $< 1/2$. Suppose that there exist positive numbers R_0, L, δ, and C with $R_0 > e$, $M(R_0, f) > e$, $L > 1$, and $0 < \delta \leq 1$ such that for every $r > R_0$ there exists $t \in (r, r^L]$ with*

(1.1) $$\frac{\log m(t, f)}{\log M(r, f)} \geq L\left(1 - \frac{C}{(\log r)^\delta}\right).$$

Then all components of the Fatou set $\mathcal{F}(f)$ are bounded.

QUESTION. Does every transcendental entire function f of order $< 1/2$ satisfy the assumptions of Theorem 1?

The point of the condition of Theorem 1 is the following. Let f be a transcendental entire function of order $< 1/2$. Theorem B and the discussion after it show that in a range where R grows to a suitable power of R, we are guaranteed to find a radius t such that $|f|$ is uniformly large on the circle of radius t centered at the origin. The proof of Theorem 1 shows how a condition of that nature (we use (1.1) instead) allows us to show that if there is a considerable spread of moduli, in the sense of (2.2) below, in a compact subset of a hypothetical unbounded wandering domain of f, then the spread will propagate to the images under all the iterates of f. The novelty of the condition of Theorem 1 is that, unlike any of its predecessors, it allows this spread to be reduced. Nonetheless, it turns out that if we start with a sufficiently large spread, and in an unbounded domain we may start with as large a spread as we like, we are able to keep the propagated spreads large enough even if they are reduced from the original value.

Even though we are allowed to use any $L > 1$ in Theorem 1 and may, in particular, take L close to 1 if that works, it seems likely, particularly if the order of f is close to $1/2$, that we may need to take L very large, depending on $\rho(f)$. If we are not close to or inside an annulus containing very few zeros of f, it would seem plausible that the condition (1.1) should be easy to satisfy, with a wide margin, by taking t to be a value arising from the consequences of Theorem B. This is because then $\log m(t,f)/\log M(t,f)$ is greater than a fixed constant, while $\log M(t,f)/\log M(r,f)$ should be quite large. So there should be a potential problem at most if we are in an annulus where f behaves like a polynomial. But in that case, we should be able to take t close to r^L; and then the three numbers $\log m(t,f)$, $\log M(t,f)$, and $L\log M(r,f)$, should be close together. There may be some error term required to estimate $\log m(t,f)/(L\log M(r,f))$ from below, but (1.1) allows for such a term. For this reason, it seems sensible to ask whether every transcendental entire function f of order $< 1/2$ satisfies the assumptions of Theorem 1 and, perhaps, even to conjecture that this is so.

2. Proof of Theorem 1

To prove Theorem 1, let f satisfy the assumptions of Theorem 1 and set
$$C_1 = C + \log^+ M(1,f) \geq C > 0.$$
Recall that $\log^+ x = \max\{0, \log x\}$ for $x \geq 0$. Choose $C_2 > 2$ so large that
$$\frac{1}{L} < \prod_{n=1}^{\infty} \left(1 - \frac{1}{C_2^{n\delta}}\right).$$

If $\mathcal{F}(f)$ has an unbounded component D, then by Theorem C, the component D must be a wandering domain; and then, as mentioned, all components of $\mathcal{F}(f)$ must be simply connected.

To get a contradiction, suppose that D is an unbounded wandering domain of f. Choose $R' > 1$ so that

(2.1) $$r < r^{C_2} < M(r,f)$$

for all $r \geq R'$. Choose $R_1 = \max\{R_0, R', 2\exp(C_1^{1/\delta}C_2)\}$. Let K be a compact connected subset of D containing points z_0 and w_0 with

(2.2) $$|w_0| > |z_0| > R_1$$

and
$$\frac{\log|w_0|}{\log|z_0|} > L^2.$$

Set $K_n = f^n(K)$, so that K_n is a compact connected subset of the component D_n of $\mathcal{F}(f)$ containing $f^n(D)$. We seek to prove that for each $n \geq 1$, there are points

$z_n, w_n \in K_n$ with
$$|w_n| > |z_n| > R_1$$
and
$$\frac{\log |w_n|}{\log |z_n|} > L^2 \prod_{k=1}^{n}\left(1 - \frac{1}{C_2^{k\delta}}\right) > L.$$

Since K is connected and (2.2) holds, there is $\zeta_0 \in K$ with $|w_0| = |\zeta_0|^L$. Thus $|\zeta_0| > |z_0|$. By (1.1), there is $t \in (|\zeta_0|, |w_0|]$ with
$$\frac{\log m(t,f)}{\log M(|\zeta_0|, f)} \geq L\left(1 - \frac{C}{(\log |\zeta_0|)^\delta}\right).$$

We have
$$|f(z_0)| \leq M(|z_0|, f).$$

Take any point $u_0 \in K$ with $|u_0| = t$. This is possible since K is connected. We have
$$\frac{\log |f(u_0)|}{\log M(|z_0|, f)} \geq \frac{\log m(t,f)}{\log M(|z_0|, f)} = \frac{\log m(t,f)}{\log M(|\zeta_0|, f)} \frac{\log M(|\zeta_0|, f)}{\log M(|z_0|, f)}.$$

We next find a lower bound for
$$\frac{\log M(|\zeta_0|, f)}{\log M(|z_0|, f)}.$$

If $1 < r_1 < r_2$ and $x_j = \log r_j$ for $j = 1, 2$, and if $\varphi(x) = \log M(e^x, f)$, then φ is an increasing convex function of x for real x, and with $r_1 = |z_0|$ and $r_2 = |\zeta_0|$, we have
$$\frac{\log M(|\zeta_0|, f)}{\log M(|z_0|, f)} = \frac{\varphi(x_2)}{\varphi(x_1)}.$$

Since φ is convex, we have
$$\varphi(x_1) \leq \frac{x_2 - x_1}{x_2} \varphi(0) + \frac{x_1}{x_2} \varphi(x_2),$$
so that
$$\varphi(x_2) \geq \frac{x_2}{x_1} \varphi(x_1) - \frac{x_2 - x_1}{x_1} \varphi(0);$$
hence
$$\frac{\varphi(x_2)}{\varphi(x_1)} \geq \frac{x_2}{x_1} - \left(\frac{x_2}{x_1} - 1\right)\frac{\varphi(0)}{\varphi(x_1)}.$$

If $\varphi(0) \leq 0$, we get
$$\frac{\varphi(x_2)}{\varphi(x_1)} \geq \frac{x_2}{x_1}.$$

If $\varphi(0) > 0$, we have
$$\frac{\varphi(x_2)}{\varphi(x_1)} \geq \frac{x_2}{x_1}\left\{1 - \left(1 - \frac{x_1}{x_2}\right)\frac{\varphi(0)}{\varphi(x_1)}\right\} \geq \frac{x_2}{x_1}\left\{1 - \frac{\varphi(0)}{\varphi(x_1)}\right\}.$$

In particular,
$$\frac{\log M(|\zeta_0|, f)}{\log M(|z_0|, f)} = \frac{\varphi(x_2)}{\varphi(x_1)} \geq \frac{\log |\zeta_0|}{\log |z_0|}\left\{1 - \frac{\log M(1, f)}{\log M(|z_0|, f)}\right\}.$$

We conclude that in all cases
$$\frac{\log|f(u_0)|}{\log M(|z_0|,f)} \geq L\frac{\log|\zeta_0|}{\log|z_0|}\left(1 - \frac{C}{(\log|\zeta_0|)^\delta}\right)\left(1 - \frac{\log^+ M(1,f)}{\log M(|z_0|,f)}\right)$$
$$= \frac{\log|w_0|}{\log|z_0|}\left(1 - \frac{C}{(\log|\zeta_0|)^\delta}\right)\left(1 - \frac{\log^+ M(1,f)}{\log M(|z_0|,f)}\right)$$
$$\geq \frac{\log|w_0|}{\log|z_0|}\left(1 - \frac{C}{(\log|\zeta_0|)^\delta} - \frac{\log^+ M(1,f)}{\log M(|z_0|,f)}\right)$$
$$\geq \frac{\log|w_0|}{\log|z_0|}\left(1 - \frac{C}{(\log|z_0|)^\delta} - \frac{\log^+ M(1,f)}{\log M(|z_0|,f)}\right)$$
$$\geq \frac{\log|w_0|}{\log|z_0|}\left(1 - \frac{C + \log^+ M(1,f)}{(\log|z_0|)^\delta}\right)$$
since $\log M(|z_0|,f) \geq \log|z_0| \geq (\log|z_0|)^\delta$. So, whether $\varphi(0) \leq 0$ or $\varphi(0) > 0$, we have
$$\frac{\log|f(u_0)|}{\log M(|z_0|,f)} \geq L^2\left(1 - \frac{C_1}{(\log|z_0|)^\delta}\right) > L^2\left(1 - \frac{1}{C_2^\delta}\right) > L > 1,$$
since $(\log|z_0|)^\delta > (\log R_1)^\delta \geq C_1 C_2^\delta$.

Since $|f(z_0)| \leq M(|z_0|,f)$ and $f(K) = K_1$ is connected, the set K_1 contains a point z_1 with $|z_1| = M(|z_0|,f)$. We set $f(u_0) = w_1$ and note that
$$\frac{\log|w_1|}{\log|z_1|} > L^2\left(1 - \frac{1}{C_2^\delta}\right) > L.$$

LEMMA 1. *Suppose that $n \geq 1$ and that for all m with $1 \leq m \leq n$, there exist $z_m, w_m \in K_m$ with*
$$|w_m| > |z_m| > R_1,$$
$$|z_m| \geq |z_{m-1}|^{C_2} \geq |z_0|^{C_2^m}$$
for $1 \leq m \leq n$, and
$$\kappa_m \equiv \frac{\log|w_m|}{\log|z_m|} > L^2 \prod_{k=1}^m \left(1 - \frac{1}{C_2^{k\delta}}\right) > L.$$
Then K_{n+1} contains points z_{n+1} and w_{n+1} such that
$$|w_{n+1}| > |z_{n+1}| > R_1,$$
$$|z_{n+1}| \geq |z_n|^{C_2} \geq |z_0|^{C_2^{n+1}},$$
and
$$\kappa_{n+1} \equiv \frac{\log|w_{n+1}|}{\log|z_{n+1}|} > L^2 \prod_{k=1}^{n+1}\left(1 - \frac{1}{C_2^{k\delta}}\right) > L.$$

PROOF OF LEMMA 1. Since K_n is connected and
$$\frac{\log|w_n|}{\log|z_n|} > L,$$
there is $\zeta_n \in K_n$ with
$$\frac{\log|w_n|}{\log|\zeta_n|} = L.$$

Now by (1.1), find $t \in (|\zeta_n|, |w_n|] = (|\zeta_n|, |\zeta_n|^L]$ with

(2.3) $$\frac{\log m(t,f)}{\log M(|\zeta_n|, f)} \geq L\left(1 - \frac{C}{(\log |\zeta_n|)^\delta}\right) > 1.$$

Then choose $u_n \in K_n$ with $|u_n| = t$. Note that $|z_n| < |\zeta_n|$. We have, as before,

$$\frac{\log |f(u_n)|}{\log M(|z_n|, f)} \geq \frac{\log m(t,f)}{\log M(|\zeta_n|, f)} \frac{\log M(|\zeta_n|, f)}{\log M(|z_n|, f)}$$
$$> L\left(1 - \frac{C}{(\log |\zeta_n|)^\delta}\right) \frac{\log |\zeta_n|}{\log |z_n|}\left(1 - \frac{\log^+ M(1,f)}{\log M(|z_n|, f)}\right)$$
$$\geq \frac{\log |w_n|}{\log |z_n|}\left(1 - \frac{C}{(\log |\zeta_n|)^\delta} - \frac{\log^+ M(1,f)}{\log M(|z_n|, f)}\right)$$
$$\geq \frac{\log |w_n|}{\log |z_n|}\left(1 - \frac{C_1}{(\log |z_n|)^\delta}\right).$$

Now by our induction assumption, we have
$$\log |z_n| \geq C_2^n \log |z_0|,$$

so that
$$\frac{C_1}{(\log |z_n|)^\delta} \leq \frac{1}{C_2^{n\delta}} \frac{C_1}{(\log |z_0|)^\delta} \leq \frac{1}{C_2^{(n+1)\delta}}.$$

Choose $z_{n+1} \in K_{n+1}$ with $|z_{n+1}| = M(|z_n|, f)$. This is possible since K_{n+1} is connected, $f(z_n) \in K_{n+1}$, $|f(z_n)| \leq M(|z_n|, f)$, and $f(u_n) \in K_{n+1}$ while by (2.3),
$$|f(u_n)| \geq m(t, f) \geq M(|\zeta_n|, f) > M(|z_n|, f).$$

We get, with $w_{n+1} = f(u_n)$, that
$$\kappa_{n+1} = \frac{\log |w_{n+1}|}{\log |z_{n+1}|} > \kappa_n\left(1 - \frac{1}{C_2^{(n+1)\delta}}\right) > L^2 \prod_{k=1}^{n+1}\left(1 - \frac{1}{C_2^{k\delta}}\right) > L.$$

Also, $|z_{n+1}| = M(|z_n|, f) \geq |z_n|^{C_2} > |z_n| > R_1$ by (2.1). This completes the proof of Lemma 1. □

We continue with the proof of Theorem 1. We have previously shown that the hypothesis of Lemma 1 holds for $n = 1$. Hence induction on n together with Lemma 1 shows that for every $n \geq 1$, there are $z_n, w_n \in K_n$ with
$$\frac{\log |w_n|}{\log |z_n|} > L.$$

However, it was proved in [**1**, p. 3251] that for every $C_0 > 1$, there exists $j_0 \geq 1$ such that for all $j \geq j_0$ and for all $z, w \in K$, we have

(2.4) $$\frac{\log |f^j(z)|}{\log |f^j(w)|} \leq C_0.$$

Taking C_0 with $1 < C_0 < L$ and choosing the appropriate j_0, we obtain a contradiction as soon as $n \geq j_0$. This completes the proof of Theorem 1.

REMARK. The proof of Theorem 1 should be based on both dynamics and on special properties of transcendental entire functions of order $< 1/2$. Dynamics has been used only via (2.4). A longer treatment would otherwise have been necessary, but we were able to quote a result from [**1**]. The rest of the proof depended on the assumptions of Theorem 1 and on general properties of entire functions. The assumption that $\rho(f) < 1/2$ does not seem to appear much in the proof but was, in fact, used to deduce that we only need to consider unbounded simply connected wandering domains.

References

[1] J.M. Anderson and A. Hinkkanen, *Unbounded domains of normality*, Proc. Amer. Math. Soc. **126** (1998), 3243–3252.
[2] I.N. Baker, *Zusammensetzungen ganzer Funktionen*, Math. Z. **69** (1958), 121–163.
[3] I.N. Baker, *Fixpoints and iterates of entire functions*, Math. Z. **71** (1959), 146–153.
[4] I.N. Baker, *The iteration of polynomials and transcendental entire functions*, J. Austral. Math. Soc. Ser. A **30** (1981), 483–495.
[5] I.N. Baker, *Wandering domains in the iteration of entire functions*, Proc. London Math. Soc. (3) **49** (1984), 563–576.
[6] P.D. Barry, *On a theorem of Besicovitch*, Quart. J. Math. Oxford Ser. 2 **14** (1963), 292–302.
[7] A.F. Beardon, *Iteration of Rational Functions*, Springer, New York, 1991.
[8] W. Bergweiler, *The iteration of meromorphic functions*, Bull. Amer. Math. Soc. (N.S.) **29** (1993), 151–188.
[9] L. Carleson and T. Gamelin, *Complex Dynamics*, Springer, New York, 1993.
[10] P. Fatou, *Sur les équations fonctionnelles*, Bull. Soc. Math. France **47** (1919), 161–271.
[11] P. Fatou, *Sur les équations fonctionnelles*, Bull. Soc. Math. France **48** (1920), 33–94, 208–314.
[12] P. Fatou, *Sur l'itération des fonctions transcendantes entières*, Acta Math. **47** (1926), 337–370.
[13] M. E. Herring, *Mapping properties of Fatou components*, Ann. Acad. Sci. Fenn. Math. **23** (1998), 263–274.
[14] Xinhou Hua and Chung-Chun Yang, *Fatou components of entire functions of small growth*, Ergodic Theory Dynam. Systems **19** (1999), 1281–1293.
[15] G. Julia, *Mémoire sur l'itération des fonctions rationnelles*, J. Math. Pures Appl. **8** (1918), 47–245.
[16] J. Milnor, *Dynamics in One Complex Variable. Introductory Lectures*, Vieweg, Braunschweig, 1999.
[17] G. Stallard, *Some Problems in the Iteration of Meromorphic Functions*, PhD thesis, Imperial College, London, 1991.
[18] G. Stallard, *The iteration of entire functions of small growth*, Math. Proc. Camb. Phil. Soc. **114** (1993), 43–55.
[19] N. Steinmetz, *Rational Iteration. Complex Analytic Dynamical Systems*, de Gruyter, Berlin, 1993.
[20] Jian-Hua Zheng, *Unbounded domains of normality of entire functions of small growth*, Math. Proc. Cambridge Philos. Soc. **128** (2000), 355–361.

DEPARTMENT OF MATHEMATICS, UNIVERSITY OF ILLINOIS AT URBANA-CHAMPAIGN, URBANA, IL 61801, U.S.A.

E-mail address: aimo@math.uiuc.edu

A Note on a Theorem of J. Globevnik

Dmitry Khavinson

Affectionately dedicated to Professor Zalcman on the occasion of his 60th birthday

In this note, we offer a short proof of Globevnik's theorem [**Gl1**]. Denote the unit disk by **D** and the unit circle by **T**.

THEOREM 1. *If for any function g in the disk-algebra $A := A(\mathbf{D})$ such that $f + g \neq 0$ on $\mathbf{T}$ $\Delta_\mathbf{T} \arg(f+g) \geq 0$, then f extends analytically to $\mathbf{D}$.*

PROOF. The proof is by contradiction. Assume that f does not extend analytically to $\mathbf{D}$; then $\operatorname{dist}(f, A) = d > 0$, where the distance is understood in the $C(\mathbf{T})$-metric. Let $\{f_n\}$ be a sequence of rational functions with poles off $\mathbf{T}$ which converges uniformly to f on $\mathbf{T}$; we can assume that $\operatorname{dist}(f_n, A) = d_n > d/2$. Let g_n be the best $A(\mathbf{D})$ approximation to f_n. We make use of the theory of linear extremal problems in spaces of analytic functions, for which see [**Kh**] or [**Du**, Chapter 8]. It follows from this theory that g_n is analytic across $\mathbf{T}$ and $|g_n - f_n| = d_n > d/2 > 0$ everywhere on $\mathbf{T}$. Moreover, since f_n is continuous on $\mathbf{T}$, there exists an extremal function F_n for the dual problem

$$\max \left| \int_\mathbf{T} f_n h \, dz \right|,$$

where h ranges over the the unit ball of the Hardy space H^1.

Now F_n is also analytic across $\mathbf{T}$ and

(1) $$F_n(f_n - g_n)dz = |F_n| \, d_n \, |dz| \geq 0 \quad \text{on} \quad \mathbf{T}.$$

Since $\Delta_\mathbf{T} \arg F_n \geq 0$ and $\Delta_\mathbf{T} \arg dz = 2\pi$, (1) implies that

(2) $$\Delta_\mathbf{T} \arg(f_n - g_n) \leq -2\pi$$

for all sufficiently large n.

The rest is standard. Assume that $|f - f_n| < \varepsilon < d/2$ on $\mathbf{T}$ so that $|f - g_n| = |f_n - g_n - (f_n - f)| \geq d_n - \varepsilon > 0$ for all n. Then

$$\Delta_\mathbf{T} \arg(f - g_n) = \Delta_\mathbf{T} \arg\left[(f_n - g_n)\left(1 - \frac{f_n - f}{f_n - g_n}\right)\right] = \Delta_\mathbf{T} \arg(f_n - g_n) \leq -2\pi,$$

which contradicts the assumption of the Theorem. This completes the proof. □

2000 *Mathematics Subject Classification*. Primary 30E25.

The author gratefully acknowledges support by the National Science Foundation.

REMARKS. (i) The proof could be shortened further by noticing that the function $f_n - g_n$ is *badly approximable*; cf. [**Ga**, p. 177]. Then, Poreda's theorem yields (2).

(ii) This argument does not extend directly to multiply connected domains, the focus of Globevnik's paper [**Gl2**]. There, $\Delta \arg dz = 2 - k$, where k is the number of the boundary components; the latter number is nonpositive for all $k \geq 2$.

References

[Du]   P. Duren, *Theory of H^p Spaces*, Academic Press, 1970.
[Ga]   J. Garnett, *Bounded Analytic Functions*, Academic Press, 1983.
[Gl1]   J. Globevnik, *Holomorphic extendibility and the argument principle*, this volume, 171-175.
[Gl2]   J. Globevnik, *The argument principle and holomorphic extendibility*, J. Analyse Math. **94** (2004), 385-395.
[Kh]   S.Ya. Khavinson, *Two Papers on Extremal Problems in Complex Analysis*, Amer. Math. Soc. Transl. Ser.2 **129**, 1986.

DEPARTMENT OF MATHEMATICAL SCIENCES, UNIVERSITY OF ARKANSAS, FAYETTEVILLE, AR 72701, U.S.A.
   E-mail address: `dmitry@uark.edu`

Contemporary Mathematics
Volume **382**, 2005

Behaviour of a Dynamical System Far from its Equilibrium

F. C. Klebaner

To Professor Lawrence Zalcman, with best wishes for many more years of productive research

> ABSTRACT. The Lotka-Volterra model is a nonlinear system of differential equations which has periodic solutions. Rothe (1985) used complex analysis to obtain results on the period of the system far from from its equilibrium as well as asymptotics of times the system spends in different regions. We use Chebyshev's inequality to show that far from equilibrium most of the time one of the coordinates is exponentially small. It is an interesting question whether either of the above methods generalizes for other systems.

1. Introduction

The Lotka-Volterra system ([**2**] and [**6**]) is a nonlinear system of differential equations

(1.1) $$x'_t = \mathfrak{a} x_t - \mathfrak{b} x_t y_t, \quad y'_t = -\mathfrak{c} y_t + \mathfrak{d} x_t y_t,$$

with initial conditions $x_0 > 0, y_0 > 0$ and positive coefficients $\mathfrak{a}$, $\mathfrak{b}$, $\mathfrak{c}$, $\mathfrak{d}$. It models behaviour of a predator-prey system (in terms of the prey and predator densities x_t and y_t), and we are interested in positive solutions.

The system has one equilibrium point in R^{2+}, $x^* = \frac{\mathfrak{c}}{\mathfrak{d}}$, $y^* = \frac{\mathfrak{a}}{\mathfrak{b}}$, and possesses a first integral

$$r(x, y) = \mathfrak{d} x - \mathfrak{c} \ln x + \mathfrak{b} y - \mathfrak{a} \ln y;$$

that is, for any t,

$$r(x_t, y_t) = r(x_0, y_0) = \text{const.} = r.$$

It is not hard to see that solutions are periodic with period $T = T(r)$.

The Lotka-Volterra system and its various generalizations have been studied extensively and there is a large amount of literature on the subject.

Behaviour of dynamical systems near the equilibrium is obtained by linearization and is well-known. Here we are interested in the behaviour of the system far from its equilibrium.

Dependence of the period on r was studied in [**3**], [**4**], [**5**], and [**7**].

1991 *Mathematics Subject Classification.* Primary 34F05, 34E10; Secondary 60H10.
The author was supported in part by an ARC Grant.

In [1], a system (1.1) with randomly perturbed coefficients was studied; and, in the process, a result on the behaviour far from its equilibrium was obtained.

The purpose of this note is to call attention to this result and the result of Rothe (1985) with a view to possible generalization to other systems.

2. Behaviour far from equilibrium

THEOREM 2.1 (Rothe [3]). *As* $r \to \infty$, $T_{++}(r) = O(\frac{\ln r}{r})$, $T_{+-}(r) = O(\ln r)$, $T_{--}(r) = O(r)$, *where* T_{ij}, $(i,j \in \{+,-\})$ *is the time spent by the orbit in the quadrants of the plane with the center at the equilibrium.*

PROOF. The change of variables $x = e^p$, $y = e^q$, transforms $r(x, y)$ into the Hamiltonian $H(p, q)$

$$H(p, q) = \mathfrak{d} e^p - \mathfrak{c} p + \mathfrak{b} e^q - \mathfrak{a} q,$$

and the system (1.1) into the Hamiltonian system (2.1)

$$(2.1) \qquad \frac{dp}{dt} = -\frac{\partial H}{\partial q}, \quad \frac{dq}{dt} = \frac{\partial H}{\partial p}.$$

For simplicity take $\mathfrak{a} = \mathfrak{b} = \lambda$, $\mathfrak{d} = \mathfrak{c} = \mu$, then the equilibrium point becomes $(0,0)$ in the (p,q) plane. Let $Z(u) = \int\int e^{-uH(p,q)} dp\, dq$. Calculate it in two ways: directly and then changing variables. This gives the Laplace transform of $T(r)$:

$$Z(u) = z(u\mu)z(u\lambda),$$

with

$$z(s) = e^{s(1-\ln s)} \Gamma(s).$$

It is easy to see that the change of variables $(p, q) \to (t, r)$ given by (2.1) has Jacobian -1. Thus

$$Z(u) = \int\int e^{-uH(p,q)} dp\, dq = \int_0^\infty \int_0^{T(r)} e^{-ur} dt\, dr = \int_0^\infty e^{-ur} T(r)\, dr.$$

Furthermore, the function $z(s)$ is itself a Laplace transform, a fact that allows us to represent the period $T(r)$ as a convolution. Indeed,

$$z(s) = \int_0^\infty e^{-sr} \tau(r)\, dr,$$

where $\tau(r) = \tau_-(r) + \tau_+(r)$, with $\tau_\pm = \frac{1}{h'(p_\pm)}$, and $p_\pm$ are the positive and the negative solutions of $h(p) = e^p - p - 1 = r$. We have

$$T_{ij} = \frac{1}{\mu} \int_0^r \tau_i(a/\lambda) \tau_j((r-a)/\mu)\, da.$$

Inversion of Laplace transforms gives the result. For details, see [3]. $\square$

THEOREM 2.2 (Khasminskii & Klebaner [1]). *There exists a constant* $C > 0$ *such that for large values of* r *and any* $\epsilon > 0$, *solutions of system (1.1) on* $[0, T(r)]$ *satisfy*

$$\frac{1}{T(r)} \text{Leb}\{t : x(t) < e^{-r/(2\mathfrak{c}) + r^{\epsilon+1/2}/(2\mathfrak{c})} \text{ or } y(t) < e^{-r/(2\mathfrak{a}) + r^{\epsilon+1/2}/(2\mathfrak{a})}\} > 1 - Cr^{-2\epsilon},$$

where $\text{Leb}(A)$ *denotes the Lebesgue measure of* A.

PROOF. For a fixed r, let $P^r(dt) = dt/T(r)$ be the uniform probability measure on $[0, T(r)]$. Denote by bar the expectation with respect to P^r, i.e., for a function h

$$\overline{h(x,y)}(r) = \frac{1}{T(r)} \int_0^{T(r)} h(x(t), y(t)) dt.$$

Moments of the solutions of (1.1) can be found and then used to obtain the desired bound.

Using the periodicity of x_t and y_t, separating variables and integrating, we obtain

$$\overline{y} = \mathfrak{a}/\mathfrak{b}, \quad \overline{x} = \mathfrak{c}/\mathfrak{d}.$$

Using this and integrating directly, we see that x and y are uncorrelated:

$$\overline{xy} - (\overline{x})(\overline{y}) = 0.$$

Let

$$\gamma = \gamma(t, r) = r(x(t), y(t)) - (\mathfrak{d}x(t) - \mathfrak{c}) - (\mathfrak{b}y(t) - \mathfrak{a});$$

then $\overline{\gamma(t, r)} = r$. It is possible to show that the variance of γ grows linearly with r. Here the bound

$$Var(\gamma) = \overline{\gamma^2} - (\overline{\gamma})^2 \leq Cr$$

for some constant $C > 0$ suffices. Applying Chebyshev's inequality to γ, we have

$$P(|\gamma - r| > r^{1/2+\epsilon}) \leq \frac{Var(\gamma)}{r^{1+2\epsilon}} < \frac{C}{r^{2\epsilon}},$$

which yields

$$P(|\gamma - r| \leq r^{1/2+\epsilon}) \geq 1 - \frac{C}{r^{2\epsilon}}.$$

As $\gamma - r = -\mathfrak{c} \ln x(t) - \mathfrak{a} \ln y(t)$, the result follows. See [1] for details. $\square$

The above methods rely on the specific structure of the Lotka-Volterra system. It seems that the method of obtaining bounds in terms of integrals over trajectories can be generalized to other systems.

However, it remains an interesting question whether it is possible to generalize the above methods for other systems.

References

[1] R.Z. Khasminskii and F.C. Klebaner, *Long term behavior of solutions of the Lotka-Volterra system under small random perturbations*, Ann. Appl. Prob. **11** (2001), 952–963.
[2] A.J. Lotka, *Elements of Physical Biology*, William and Wilkins, Baltimore, 1925.
[3] F. Rothe, *Thermodynamics, real and complex periods of the Volterra model*, Z. Angew. Math. Phys. **36** (1985), 395-421.
[4] F. Rothe, *The periods of the Volterra-Lotka system*, J. Reine Angew. Math. **355** (1985), 129-138.
[5] R. Schaaf, *Global behaviour of solution of branches for some Neumann problems depending on one or several parameters*, J. Reine Angew. Math. **346** (1984), 1-31.
[6] V. Volterra, *Variazioni e fluttuazioni del numero d'individui in specie d' animali conviventi*, Mem. Acad. Lincei 2 (1926), 31-113.
[7] J. Waldvogel, *The period in the Lotka-Volterra system is monotonic*, J. Math. Anal. Appl. **114** (1986), 178-184.

SCHOOL OF MATHEMATICAL SCIENCES, MONASH UNIVERSITY, VICTORIA 3800, AUSTRALIA
E-mail address: fima.klebaner@sci.monash.edu.au

A Tauberian Theorem for Laplace Transforms with Pseudofunction Boundary Behavior

Jaap Korevaar

To my young friend Larry Zalcman on his sixtieth birthday

ABSTRACT. The prime number theorem provided the chief impulse for complex Tauberian theory, in which the boundary behavior of a transform in the complex plane plays a crucial role. We consider Laplace transforms of bounded functions. Our Tauberian theorem does not allow first-order poles on the imaginary axis; but any milder singularities, characterized by pseudofunction boundary behavior, are permissible. In this context, we obtain a useful Tauberian theorem by exploiting Newman's 'contour method'.

1. Introduction

In 1980, Donald Newman [20] published a beautiful proof for the prime number theorem (PNT) by complex analysis. His vehicle was an old theorem of Ingham [10] involving Dirichlet series, for which he found a clever proof by contour integration. The method is easily adapted to give Theorem 1.1 for Laplace transforms below; cf. the author's paper [15] and Zagier [24]. (Preprints of these papers circulated shortly after Newman's article appeared.) The contour method has recently been used in numerous articles motivated by operator theory; see, for example, Allan, O'Farrell and Ransford [1], Arendt and Batty [2], Batty [4], and the book by Arendt, Batty, Hieber and Neubrander [3].

If one is interested only in a quick proof of the PNT, the following result will suffice.

THEOREM 1.1. *Let $a(\cdot)$ be (measurable and) bounded on $[0, \infty)$, so that the Laplace transform*

$$(1.1) \qquad f(z) = \mathcal{L}a(z) = \int_0^\infty a(t) e^{-zt} dt, \quad z = x + iy,$$

2000 *Mathematics Subject Classification.* Primary 40E05; Secondary 11M45, 11N05, 42A38, 44A10, 46F20.

Key words and phrases. Distributions, Fourier transform, Laplace transform, prime number theorem, pseudofunctions, Tauberian theory.

is well-defined and analytic throughout the open half-plane $\{x = \operatorname{Re} z > 0\}$. Suppose that $f(z)$ has an analytic extension to the open interval $(-iB, iB)$ of the imaginary axis. Then

$$(1.2) \qquad \limsup_{T \to \infty} \left| \int_0^T a(t)dt - f(0) \right| \leq \frac{2M}{B}, \quad \text{where } M = \sup_{t>0} |a(t)|.$$

COROLLARY 1.2. *If $a(\cdot)$ is bounded and $f = \mathcal{L}a$ extends analytically to every point of the imaginary axis, the improper integral*

$$(1.3) \qquad \int_0^{\infty-} a(t)dt = \lim_{T \to \infty} \int_0^T a(t)dt \ \text{ exists and equals } f(0).$$

Here the 'Tauberian condition' that $a(\cdot)$ be bounded can (in the real case) be replaced by boundedness from below. However, this makes the proof more complicated; cf. [16] (Section 9). In Section 2 we sketch how to deduce the PNT.

Theorem 1.1 and Corollary 1.2 are contained in results of Karamata [11] (Theorem B) and Ingham [10] (Theorem III), which were obtained by Wiener's method [23]. They did not require that $f(z)$ can be extended analytically to every point of the imaginary axis, but could get by with weaker boundary conditions. The aim of the present paper is to reduce the boundary requirements in Theorem 1.1 to a minimum.

THEOREM 1.3. *Let $a(\cdot)$ be bounded on $[0, \infty)$, so that the Laplace transform $f(z) = \mathcal{L}a(z)$, $z = x + iy$ is analytic for $x = \operatorname{Re} z > 0$. Suppose that $f(x)$ tends to a limit $f(0)$ as $x \searrow 0$ and that the quotient*

$$(1.4) \qquad q(x+iy) = \frac{f(x+iy) - f(x)}{iy}, \quad x > 0,$$

converges in distributional sense to a pseudofunction $q(iy)$ on the interval $\{-B < y < B\}$ as $x \searrow 0$. Then one has inequality (1.2).

Known sufficient conditions for (1.2) are uniform or L^1 convergence of $q(x+iy)$ to a limit function $q(iy)$ on $(-B, B)$. The distributional conditions in the Theorem require two things:

(i) (convergence condition) that

$$\int_{\mathbb{R}} q(x+iy)\phi(y)dy \ \text{ should tend to a limit } \ <q(iy), \phi(y)>$$

for every C^∞ function ϕ with support in $(-B, B)$;

(ii) (pseudofunction condition) that $q(iy)$ be the restriction to $(-B, B)$ of the distributional Fourier transform of a bounded function which tends to zero at $\pm\infty$. Cf. Sections 4 and 5 below.

We remark that related distributional conditions received inadequate treatment in [16] (Theorem 14.6). General background material on Tauberian theory can be found in the forthcoming book [17].

2. From Corollary 1.2 to the Prime Number Theorem

Background material in number theory may be found in many books; classics are Landau [18] and Hardy and Wright [7].

To obtain the PNT from Corollary 1.2 one may take $a(t)$ equal to

$$(2.1) \qquad b(t) = \frac{\psi(e^t) - [e^t]}{e^t} = e^{-t} \sum_{1 \le n \le e^t} (\Lambda(n) - 1),$$

where $\psi(v) = \sum_{n \le v} \Lambda(n)$ is Chebyshev's function. The symbol $\Lambda(\cdot)$ stands for von Mangoldt's function, which is given by the Dirichlet series

$$\sum_{n=1}^{\infty} \frac{\Lambda(n)}{n^w} = -\frac{d}{dw} \log \zeta(w) = \frac{d}{dw} \sum_{p \text{ prime}} \log(1 - p^{-w}) = \sum_{p \text{ prime}} \frac{p^{-w} \log p}{1 - p^{-w}}$$

when $\operatorname{Re} w > 1$. It is elementary that $\psi(v) = \mathcal{O}(v)$, so that $b(\cdot)$ is bounded. For $\operatorname{Re} z > 0$,

$$(2.2) \qquad \begin{aligned} g(z) &= \mathcal{L}b(z) = \int_0^\infty \{\psi(e^t) - [e^t]\} e^{-(z+1)t} dt \\ &= \int_1^\infty \{\psi(v) - [v]\} v^{-z-2} dv = \frac{1}{z+1} \int_{1-}^\infty v^{-z-1} d\{\psi(v) - [v]\} \\ &= \frac{1}{z+1} \sum_1^\infty \frac{\Lambda(n) - 1}{n^{z+1}} = \frac{1}{z+1} \left(-\frac{\zeta'(z+1)}{\zeta(z+1)} - \zeta(z+1) \right). \end{aligned}$$

The function $g(z)$ is analytic at every point of the line $\{\operatorname{Re} z = 0\}$. Indeed, $\zeta(w)$ is free of zeros on the line $\{\operatorname{Re} w = 1\}$ and the poles of $-(\zeta'/\zeta)(w)$ and $\zeta(w)$ at the point $w = 1$ cancel each other. Conclusion:

$$(2.3) \qquad \int_0^{\infty-} b(t) dt = \int_1^{\infty-} \frac{\psi(v) - [v]}{v^2} dv = g(0).$$

By the monotonicity of ψ, this readily gives

$$\psi(v) \sim v \text{ as } v \to \infty \quad \text{and} \quad \sum_{n=1}^\infty \frac{\Lambda(n) - 1}{n} = g(0).$$

The relation $\psi(v) \sim v$ is equivalent to the PNT:

$$\pi(v) \sim \frac{v}{\log v} \quad \text{as } v \to \infty.$$

3. An Auxiliary Result

We will prove Theorem 1.1 but begin with a useful preliminary form.

PROPOSITION 3.1. *Let $\sup_{t>0} |a(t)| = M < \infty$ and let the Laplace transform*

$$(3.1) \qquad f(z) = \mathcal{L}a(z), \quad z = x + iy, \quad x > 0,$$

have an analytic extension to a neighborhood of the segment $[-iR, iR]$, where $R > 0$. Then for every number $T > 0$,

$$(3.2) \qquad \left| \int_0^T a(t) dt - f(0) \right| \\ \le \frac{2M}{R} + \frac{|f(0)|}{eRT} + \frac{1}{2\pi} \left| \int_{-R}^R \{f(iy) - f(0)\} \left(\frac{1}{iy} + \frac{iy}{R^2} \right) e^{iTy} dy \right|.$$

PROOF. Define

(3.3) $$f_T(z) = \int_0^T a(t)e^{-zt}dt.$$

(i) One begins with some simple estimates. For $x = \operatorname{Re} z > 0$,

(3.4) $$|f_T(z) - f(z)| = \left|\int_T^\infty a(t)e^{-zt}dt\right| \leq M\int_T^\infty e^{-xt}dt = \frac{M}{x}e^{-Tx}.$$

Similarly, for $x = \operatorname{Re} z < 0$,

(3.5) $$|f_T(z)| = \left|\int_0^T a(t)e^{-zt}dt\right| \leq \int_0^T Me^{-xt}dt < \frac{M}{|x|}e^{-Tx}.$$

(ii) Let Γ be the positively oriented circle $C(0, R) = \{|z| = R\}$. We let Γ_1 be the part of Γ in the half-plane $\{x = \operatorname{Re} z > 0\}$, Γ_2 the part in the half-plane $\{x < 0\}$. Finally, let σ be the oriented segment of the imaginary axis from $+iR$ to $-iR$ (Figure 1). Observe that for $z = x + iy \in \Gamma$, one has

(3.6) $$\frac{1}{z} + \frac{z}{R^2} = \frac{2x}{R^2}.$$

By the hypotheses, the quotient $\{f(z) - f(0)\}/z$ is analytic on the segment σ. Observe also that $f_T(z)$ is analytic throughout the complex plane. Formulas (3.4)–(3.6) motivate the following ingenious application of Cauchy's theorem and Cauchy's formula due to Newman:

(3.7) $$\begin{aligned}0 &= \frac{1}{2\pi i}\int_{\Gamma_1+\sigma}\frac{f(z)-f(0)}{z}dz \\ &= \frac{1}{2\pi i}\int_{\Gamma_1+\sigma}\{f(z)-f(0)\}e^{Tz}\left(\frac{1}{z}+\frac{z}{R^2}\right)dz,\end{aligned}$$

(3.8) $$\begin{aligned}f_T(0) - f(0) &= \frac{1}{2\pi i}\int_\Gamma\frac{f_T(z)-f(0)}{z}dz \\ &= \frac{1}{2\pi i}\int_\Gamma\{f_T(z)-f(0)\}e^{Tz}\left(\frac{1}{z}+\frac{z}{R^2}\right)dz.\end{aligned}$$

Subtracting (3.7) from (3.8) and rearranging the result, one obtains the formula

(3.9) $$\begin{aligned}f_T(0) - f(0) &= \frac{1}{2\pi i}\int_{\Gamma_1}\{f_T(z)-f(z)\}e^{Tz}\left(\frac{1}{z}+\frac{z}{R^2}\right)dz \\ &+ \frac{1}{2\pi i}\int_{\Gamma_2}\{f_T(z)-f(0)\}e^{Tz}\left(\frac{1}{z}+\frac{z}{R^2}\right)dz \\ &- \frac{1}{2\pi i}\int_\sigma\{f(z)-f(0)\}e^{Tz}\left(\frac{1}{z}+\frac{z}{R^2}\right)dz \\ &= I_1 + I_2 + I_3,\end{aligned}$$

say.

(iii) By (3.4) and (3.6) for $z \in \Gamma_1$, the integrand $f^*(z)$ in I_1 can be estimated as follows:

$$|f^*(z)| = \left|\{f_T(z)-f(z)\}e^{Tz}\left(\frac{1}{z}+\frac{z}{R^2}\right)\right| \leq \frac{M}{x}e^{-Tx}e^{Tx}\frac{2x}{R^2} = \frac{2M}{R^2}.$$

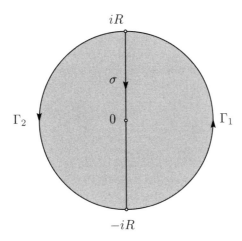

FIGURE 1. The paths of integration

Thus

(3.10) $$|I_1| \leq \frac{1}{2\pi} \int_{\Gamma_1} |f^*(z)||dz| \leq \frac{1}{2\pi} \frac{2M}{R^2} \pi R = \frac{M}{R}.$$

For $z \in \Gamma_2$, where $|x|e^{Tx} \leq 1/(eT)$, formulas (3.5) and (3.6) imply the estimate

(3.11) $$|I_2| = \left| \frac{1}{2\pi i} \int_{\Gamma_2} \{f_T(z) - f(0)\} e^{Tz} \left(\frac{1}{z} + \frac{z}{R^2} \right) dz \right| \leq \frac{M}{R} + \frac{|f(0)|}{eRT}.$$

Combination of (3.3) and (3.9)–(3.11) gives (3.2). □

DERIVATION OF THEOREM 1.1. Let a and $f = \mathcal{L}a$ satisfy the hypotheses of Theorem 1.1. Then we can apply Proposition 3.1 for any $R \in (0, B)$. For the proof of (1.2), one has to show that for any number $\varepsilon > 0$, we can choose T_0 so large that the left-hand side of (3.2) is bounded by $2(M/B) + \varepsilon$ for all $T \geq T_0$. To this end, choose R so close to B that $2M/R < 2(M/B) + \varepsilon/2$. In order to deal with the final term in (3.2), or with

(3.12) $$I_3 = \frac{1}{2\pi} \int_{-R}^{R} \{f(iy) - f(0)\} \left(\frac{1}{iy} + \frac{iy}{R^2} \right) e^{iTy} dy,$$

one may apply integration by parts: $e^{iTy} dy = de^{iTy}/(iT)$, etc., or one may use the Riemann–Lebesgue lemma. Either method will show that for our R,

(3.13) $$I_3 = I_3(R, T) \to 0 \quad \text{as } T \to \infty.$$

We now determine T_0 so large that

$$\frac{|f(0)|}{eRT} + |I_3| < \varepsilon/2, \quad \forall T \geq T_0.$$

Then by (3.2)

$$\left| \int_0^T a(t) dt - f(0) \right| \leq \frac{2M}{B} + \varepsilon, \quad \forall T \geq T_0.$$

□

4. Pseudofunction Boundary Behavior

The preceding results may be refined with the aid of a distributional approach. Motivated by operator theory, Katznelson and Tzafriri [13] used pseudofunctions to strengthen the following theorem of Fatou [5], [6]:

THEOREM 4.1. *Let the function*

$$(4.1) \qquad g(z) = \sum_{n=0}^{\infty} a_n z^n, \quad |z| < 1,$$

have an analytic continuation to (a neighborhood of) the point $z = 1$ on the unit circle $C(0,1)$. Suppose that the coefficients satisfy the 'Tauberian condition' $a_n \to 0$ as $n \to \infty$. Then the series $\sum_{n=0}^{\infty} a_n$ converges to $g(1)$.

The condition of analyticity at the point $z = 1$ can be relaxed in various ways. The most notable refinements in this direction are due to M. Riesz and Ingham (cf. [10], [19]); another refinement is mentioned below.

The condition $a_n \to 0$ is the signature of *pseudofunction* boundary behavior. In Fatou's theorem, and for real a_n, it can be replaced by the one-sided condition $\liminf a_n \geq 0$; cf. [14], [16]. A 2π-periodic distribution $G(t) = \sum_{n \in \mathbb{Z}} c_n e^{int}$ is called a pseudofunction if $c_n \to 0$ as $n \to \pm\infty$. The latter condition first appeared in Riemann's localization principle [21], which Fatou used in the proof of his theorem. (A careful discussion of the localization principle may be found in [25], item (5.7) in Chapter 9.)

Let $g(z)$ as in (4.1) be any function analytic in the unit disc. Among other things, Katznelson and Tzafriri proved that pseudofunction boundary behavior of g on $C(0,1) \setminus \{z = 1\}$, together with boundedness of the sequence $\{s_n = \sum_{k=1}^{n} a_k\}$, implies that $a_n \to 0$. Their method can be used also for further relaxation of the analyticity condition at the point 1. Knowing that $a_n \to 0$, it is enough for convergence of $\sum a_n$ if g in (4.1) is 'weakly regular' at the point 1 in the following sense. For some constant which may be called $g(1)$, the quotient

$$\frac{g(z) - g(1)}{z - 1}$$

has pseudofunction boundary behavior at the point $z = 1$ (more precisely, in some angle $|\arg z| < \delta$); cf. [16], [17].

LAPLACE AND FOURIER TRANSFORMS. Our aim is to prove an extension of Theorem 1.1 involving pseudofunction boundary behavior of the Laplace transform $f(z) = \mathcal{L}a(z)$. We begin with some general remarks on tempered distributions, that is, continuous linear functionals F on the Schwartz space $\mathcal{S}$. The 'testing functions' $\phi \in \mathcal{S}$ are the C^∞ functions with compact support and their limits under a certain family of seminorms which define convergence in $\mathcal{S}$. The tempered distributions F on $\mathbb{R}$ form the dual space $\mathcal{S}'$ of $\mathcal{S}$. Applying F to ϕ one obtains a bilinear functional, denoted by $<F, \phi>$. Locally integrable functions $F_x(y)$ of at most polynomial growth on $\{-\infty < y < \infty\}$ converge to a tempered distribution $F(y)$ as $x \searrow 0$ if

$$\int_{\mathbb{R}} F_x(y)\phi(y)dy \to\, <F(y), \phi(y)>$$

for every function $\phi \in \mathcal{S}$.

The Fourier transform $\hat{F}$ of $F \in \mathcal{S}'$ is defined by the relation

$$<\hat{F}, \phi> = <F, \hat{\phi}>$$

for all testing functions ϕ. Fourier transformation on $\mathcal{S}'$ is continuous, one to one and onto; cf. [22], [8]. If G is the Fourier transform of F, then F is equal to $1/(2\pi)$ times the reflected Fourier transform of G.

A tempered distribution F on $\mathbb{R}$ is called a *pseudomeasure* if it is the Fourier transform of a bounded (measurable) function $b(\cdot)$; it is called a *pseudofunction* if it is the Fourier transform of a bounded function which tends to zero at $\pm\infty$. Reference: Katznelson [12] (Section 6.4). By Fourier inversion and the Riemann–Lebesgue theorem, every function in $L^1(\mathbb{R})$ is a pseudofunction. A nontrivial example of a pseudomeasure on $\mathbb{R}$ is the distribution

$$\frac{1}{y - i0} = \lim_{x \searrow 0} \frac{i}{x + iy} = \lim_{x \searrow 0} i \int_0^\infty e^{-xt} e^{-iyt} dt.$$

It is the Fourier transform of i times the Heaviside function, $1_+(t)$. Other examples are the Dirac measure and the principal-value distribution, p.v. $(1/y)$. In the case of boundary singularities, and roughly speaking, first order poles correspond to pseudomeasures, slightly milder singularities to pseudofunctions.

Every pseudomeasure or pseudofunction $F = \hat{b}$ on $\mathbb{R}$ can be represented in the form

$$(4.2) \qquad F(y) = \lim_{x \searrow 0} \int_\mathbb{R} e^{-x|t|} b(t) e^{-iyt} dt.$$

Indeed, for bounded $b(\cdot)$ and $x \searrow 0$, the product $e^{-x|t|}b(t)$ tends to $b(t)$ in $\mathcal{S}'$ and Fourier transformation is continuous. This formula can be used to justify formal inversion of the order of integration in some situations. An important consequence is a *Riemann–Lebesgue lemma* for pseudofunctions F.

LEMMA 4.2. *For any pseudofunction F on $\mathbb{R}$ and any testing function ϕ,*

$$(4.3) \qquad <F(y), \phi(y) e^{iTy}> \to 0 \quad as \quad T \to \pm\infty.$$

Indeed, by representation (4.2),

$$<F(y), \phi(y)e^{iTy}> = \lim_{x \searrow 0} <\int_\mathbb{R} e^{-x|t|} b(t) e^{-iyt} dt, \phi(y) e^{iTy}>$$

$$= \lim_{x \searrow 0} \int_\mathbb{R} e^{-x|t|} b(t) dt \int_\mathbb{R} e^{-i(t-T)y} \phi(y) dy$$

$$= \int_\mathbb{R} b(t) \hat{\phi}(t - T) dt,$$

and the final integral tends to zero as $T \to \pm\infty$.

PRODUCTS. Let F be a pseudomeasure or pseudofunction as in (4.2) and let ϕ be a testing function. Computing the Fourier transform of $F(y)\phi(y)$, one finds that this product is the Fourier transform of the convolution

$$\int_\mathbb{R} b(v - u) \hat{\phi}(u) / (2\pi) du.$$

For any other function Φ whose Fourier transform $\hat{\Phi}(u)$ is $\mathcal{O}\{1/(u^2+1)\}$, the product $F\Phi$ may be *defined* as the Fourier transform of

(4.4) $$b^*(v) = \int_{\mathbb{R}} b(v-u)\hat{\Phi}(u)/(2\pi)du.$$

With F, the product $F\Phi$ is again a pseudomeasure or pseudofunction.

5. Proof of Theorem 1.3

Let $a(\cdot)$ and $f = \mathcal{L}a$ satisfy the hypotheses of the Theorem. It is convenient to set $a(t) = 0$ for $t < 0$. Denoting $\sup_{t>0}|a(t)|$ by M, taking $\varepsilon > 0$ and $0 < R < B$, we now apply Proposition 3.1 to $a(t)e^{-\varepsilon t}$ and $f(\varepsilon + z)$ instead of $a(t)$ and $f(z)$. Thus we obtain the inequality

(5.1) $$\left| \int_0^T a(t)e^{-\varepsilon t} dt - f(\varepsilon) \right|$$
$$\leq \frac{2M}{R} + \frac{|f(\varepsilon)|}{eRT} + \frac{1}{2\pi}\left| \int_{-R}^{R} \{f(\varepsilon+iy) - f(\varepsilon)\}\left(\frac{1}{iy} + \frac{iy}{R^2}\right)e^{iTy}dy \right|.$$

To treat the final integral, we set

(5.2) $$\{f(\varepsilon+iy) - f(\varepsilon)\}\left(\frac{1}{iy} + \frac{iy}{R^2}\right) = g_\varepsilon(y).$$

Let χ_R denote the characteristic function of the interval $[-R,R]$. For any number $\lambda > 0$ we let τ_λ denote a 'trapezoidal' testing function, that is, a C^∞ function which is equal to 1 on $[-\lambda,\lambda]$ and equal to 0 outside a suitable neighborhood of $[-\lambda,\lambda]$. The last integral in (5.1) may then be written in distributional notation as

(5.3) $$I(T,\varepsilon) = <g_\varepsilon(y)\tau_R(y)\chi_R(y), e^{iTy}\tau_R(y)>.$$

Here we take the support of τ_R inside $(-B,B)$. Then by the hypotheses, $g_\varepsilon(y)\tau_R(y)$ tends to the pseudofunction

$$g_0(y)\tau_R(y) = q(iy)(1 - y^2/R^2)\tau_R(y)$$

as $\varepsilon \searrow 0$; cf. (1.4). The question is whether the integral $I(T,\varepsilon)$ tends to the formal limit $I(T,0)$. Multiplication by the cut-off function $\chi_R(y)$ in (5.3) may cause problems!

One can get around this difficulty by splitting the integral $I(T,\varepsilon)$. Choosing a trapezoidal function τ_μ with support in $(-R,R)$, we first consider the relation

(5.4) $$<g_\varepsilon(y)\tau_\mu(y), e^{iTy}\tau_R(y)> \to <g_0(y)\tau_\mu(y), e^{iTy}\tau_R(y)> \quad \text{as } \varepsilon \searrow 0.$$

By our Riemann–Lebesgue Lemma 4.2, the final expression tends to zero as $T \to \infty$.

Looking at (5.3), we now consider the 'inner product'

(5.5) $$<g_\varepsilon(y)\tau_R(y)\{1-\tau_\mu(y)\}\chi_R(y), e^{iTy}\tau_R(y)>.$$

As $\varepsilon \searrow 0$, the part of this expression which comes from $f(\varepsilon)$ tends to a trigonometric integral of an integrable function,

$$\int_{-R}^{R} f(0)\left(\frac{1}{iy} + \frac{iy}{R^2}\right)\{1-\tau_\mu(y)\}e^{iTy}dy.$$

The latter tends to zero as $T \to \infty$. From here on, we focus on the constituent of the first factor in (5.5) which involves $f(\varepsilon + iy)$:

(5.6) $$f(\varepsilon + iy)\tau_R(y) \cdot \left(\frac{1}{iy} + \frac{iy}{R^2}\right)\{1 - \tau_\mu(y)\}\chi_R(y).$$

The functions $f(\varepsilon + iy)$ tend to the pseudomeasure $f(iy) = \hat{a}(y)$ as $\varepsilon \searrow 0$; and by the hypothesis about the quotient in (1.4), the restriction of $f(iy)$ to $(-B, B)$ [more precisely, to the testing functions with support in $(-B, B)$] is equal to a pseudofunction. Hence the product $f(iy)\tau_R(y)$, which by (4.4) is the Fourier transform of

$$\frac{1}{2\pi}\int_\mathbb{R} a(v-u)\hat{\tau}_R(u)du,$$

is a pseudofunction.

The functions $f(\varepsilon+iy)\tau_R(y)$ are the Fourier transforms of the functions $a(t)e^{-\varepsilon t}$, which form a uniformly bounded family. The factor

$$\Phi(y) = \left(\frac{1}{iy} + \frac{iy}{R^2}\right)\{1 - \tau_\mu(y)\}\chi_R(y),$$

which vanishes for $|y| \leq \mu$ and $|y| \geq R$, has Fourier transform $\hat{\Phi}(t) = \mathcal{O}\{1/(t^2+1)\}$. It follows that the functions in (5.6) are distributionally convergent. The limit $f(iy)\tau_R(y)\Phi(y)$ is a pseudofunction; cf. (4.4). The same is then true for the limit

$$g_0(y)\tau_R(y)\{1 - \tau_\mu(y)\}\chi_R(y) = \lim_{\varepsilon \searrow 0} g_\varepsilon(y)\tau_R(y)\{1 - \tau_\mu(y)\}\chi_R(y)$$

of the functions in the first member of (5.5). Combining the results, one concludes that the limit $I(T, 0)$ of $I(T, \varepsilon)$ can be written as an inner product

$$I(T, 0) = <H(y), e^{iTy}\tau_R(y)>$$

involving a pseudofunction H, so that $I(T, 0) \to 0$ as $T \to \infty$.

To complete the proof of Theorem 1.3, we return to inequality (5.1). Letting ε go to zero, one finds that

(5.7) $$\left|\int_0^T a(t)dt - f(0)\right| \leq \frac{2M}{R} + \frac{|f(0)|}{eRT} + \frac{1}{2\pi}|I(T,0)|.$$

Finally, taking T large and R close to B, one obtains the desired inequality (1.2).

REMARK 5.1. Related considerations show that one can introduce pseudofunction boundary behavior in the statement of the Wiener–Ikehara theorem [9], [23]. One thus obtains

THEOREM 5.2. *Let $S(t)$ vanish for $t < 0$, be nondecreasing, continuous from the right and such that the Laplace–Stieltjes transform*

(5.8) $$f(z) = \mathcal{L}dS(z) = \int_{0-}^\infty e^{-zt}dS(t) = z\int_0^\infty S(t)e^{-zt}dt, \quad z = x + iy,$$

exists for $\operatorname{Re} z = x > 1$. *Suppose that for some constant A, the analytic function*

(5.9) $$g(x+iy) = f(x+iy) - \frac{A}{x+iy-1}, \quad x > 1,$$

has a distributional limit $g(1+iy)$ as $x \searrow 1$, which on every finite interval $\{-B < y < B\}$ coincides with a pseudofunction $g_B(1+iy)$. Then

(5.10) $$e^{-t}S(t) \to A \quad as \ t \to \infty.$$

References

[1] G.R. Allan, A.G. O'Farrell, and T.J. Ransford, *A Tauberian theorem arising in operator theory*, Bull. London Math. Soc. **19** (1987), 537–545.
[2] W. Arendt and C.J.K. Batty, *Tauberian theorems and stability of one-parameter semigroups*, Trans. Amer. Math. Soc. **306** (1988), 837–852.
[3] W. Arendt, C.J.K. Batty, M. Hieber, and F. Neubrander, *Vector-Valued Laplace Transforms and Cauchy Problems*, Birkhäuser, Basel, 2001.
[4] C.J.K. Batty, *Tauberian theorems for the Laplace–Stieltjes transform*, Trans. Amer. Math. Soc. **322** (1990), 783–804.
[5] P. Fatou, *Sur quelques théorèmes de Riemann*, C. R. Acad. Sci. Paris **140** (1905), 569–570.
[6] P. Fatou, *Séries trigonométriques et séries de Taylor*, Acta Math. **30** (1906), 335–400.
[7] G.H. Hardy and E.M. Wright, *An Introduction to the Theory of Numbers*, fifth edition, Clarendon Press, Oxford, 1979. (First edition 1938.)
[8] L. Hörmander, *The Analysis of Linear Partial Differential Operators*, vol. 1, Springer, Berlin, 1983.
[9] S. Ikehara, *An extension of Landau's theorem in the analytic theory of numbers*, J. Math. and Phys. M.I.T. **10** (1931), 1–12.
[10] A.E. Ingham, *On Wiener's method in Tauberian theorems*, Proc. London Math. Soc. (2) **38** (1935), 458–480.
[11] J. Karamata, *Weiterführung der N. Wienerschen Methode*, Math. Z. **38** (1934), 701–708.
[12] Y. Katznelson, *An Introduction to Harmonic Analysis*, Wiley, New York, 1968.
[13] Y. Katznelson and L. Tzafriri, *On power bounded operators*, J. Funct. Anal. **68** (1986), 313–328.
[14] J. Korevaar, *Another numerical Tauberian theorem for power series*, Nederl. Akad. Wetensch. Proc. Ser. A. **57** = Indag. Math. **16** (1954), 46–56.
[15] J. Korevaar, *On Newman's quick way to the prime number theorem*, Math. Intelligencer **4** (1982), 108–115.
[16] J. Korevaar, *A century of complex Tauberian theory*, Bull. Amer. Math. Soc. (N.S.) **39** (2002), 475–531.
[17] J. Korevaar, *Tauberian Theory*, Springer, Berlin, 2004.
[18] E. Landau, *Handbuch der Lehre von der Verteilung der Primzahlen I, II*, Teubner, Leipzig, 1909. (Second edition with an appendix by P. T. Bateman, Chelsea Publ. Co., New York, 1953.)
[19] E. Landau and D. Gaier, *Darstellung und Begründung einiger neuerer Ergebnisse der Funktionentheorie*, third enlarged edition, Springer, Berlin, 1986. (First and second editions 1916, 1929 by E. Landau.)
[20] D.J. Newman, *Simple analytic proof of the prime number theorem*, Amer. Math. Monthly **87** (1980), 693–696.
[21] B. Riemann, *Gesammelte mathematische Werke und wissenschaftlicher Nachlass*, Teubner, Leipzig, 1892. (Reprinted by Dover Publ., New York, 1953.)
[22] L. Schwartz, *Théorie des distributions I, II*, Hermann, Paris. (First edition 1950/51.)
[23] N. Wiener, *Tauberian theorems*, Ann. of Math. (2) **33** (1932), 1–100.
[24] D. Zagier, *Newman's short proof of the prime number theorem*, Amer. Math. Monthly **104** (1997), 705–708.
[25] A. Zygmund, *Trigonometric Series*, Cambridge Univ. Press, second edition, 1959.

DEPARTMENT OF MATHEMATICS, UNIVERSITY OF AMSTERDAM, PLANTAGE MUIDERGRACHT 24, 1018 TV, AMSTERDAM, NETHERLANDS
E-mail address: `korevaar@science.uva.nl`

The Schwarzian Derivative and Complex Finsler Metrics

Samuel L. Krushkal

Dedicated to Larry Zalcman on the occasion of his 60th birthday

ABSTRACT. In this paper, we discuss the extent to which a conformal map f of the disk onto a Jordan domain $D \subset \widehat{\mathbb{C}}$ controls the geometric and conformal invariants of the complementary domain $D^* = \widehat{\mathbb{C}} \setminus \overline{D}$. By controlling, we mean the sharp quantitative distortion of those invariants. The answer is given in terms of the Schwarzian derivative of the Riemann mapping function f and involves its Grunsky coefficients.

1. Introduction. Main theorem

1.1. In this paper, we address the question of to what extent the conformal map f of the disk onto a Jordan domain $D \subset \widehat{\mathbb{C}}$ controls the geometric and conformal invariants of the complementary domain $D^* = \widehat{\mathbb{C}} \setminus \overline{D}$. By controlling, we mean the sharp quantitative distortion of those invariants.

The answer is given in terms of the Schwarzian derivative

$$S_f = (f''/f')' - (f''/f')^2/2$$

of the Riemann mapping function f of D and involves its Grunsky coefficients. It is well-known that this invariant differential operator is intrinsically related to the features of univalent functions and provides a connection between these functions and Teichmüller spaces.

This key result is applied to solving some problems concerning the Fredholm eigenvalues of quasidisks, in particular, of domains bounded by piecewise analytic arcs, and the corresponding reflection coefficients. Though the origins of these problems are classical, many open questions remain.

1.2. The Schwarzian derivatives S_f of conformal maps of the disk

$$\Delta^* = \{z \in \widehat{\mathbb{C}} = \mathbb{C} \cup \{\infty\} : |z| > 1\}$$

2000 *Mathematics Subject Classification.* Primary 30C35, 30C62, 30F45, 30F60; Secondary 32G15, 32Q45, 58B20.

Key words and phrases. Univalent function, Schwarzian derivative, extremal quasiconformal map, universal Teichmüller space, Finsler metrics, invariant metrics, Grunsky coefficients, subharmonic function, quasireflection, Fredholm eigenvalues.

belong to the Banach space $\mathbf{B}$ of hyperbolically bounded holomorphic functions in Δ^* with the norm $\|\varphi\|_{\mathbf{B}} = \sup_{\Delta^*}(|z|^2 - 1)^2|\varphi(z)|$. Note that $\varphi(z) = O(|z|^{-4})$ as $z \to \infty$.

For an arbitrary $\varphi \in \mathbf{B}$, the solutions of the corresponding Schwarzian differential equation $S_f = \varphi$ are locally univalent in Δ^*. The classical theorems of Nehari [**Ne**] and of Ahlfors and Weill [**AW**] state, respectively, that the inequality $\|\varphi\|_{\mathbf{B}} \leq 2$ implies the global univalence of solutions f to the differential equation $S_f = \varphi$ on Δ^*, while the strict inequality

(1.1) $$\|S_f\|_{\mathbf{B}} < 2$$

provides the existence of a quasiconformal extension of f onto the whole Riemann sphere $\widehat{\mathbb{C}}$ whose Beltrami coefficient in the unit disk $\Delta = \{z : |z| < 1\}$ is *harmonic* (in the spirit of the Kodaira-Spencer deformation theory), i.e., is of the form

(1.2) $$\nu_\varphi(z) = \frac{1}{2}(1 - |z|^2)^2 \varphi(1/\bar{z}) 1/\bar{z}^4, \quad \varphi = S_f \quad (z \in \Delta);$$

in particular, the curve $L = f(\partial\Delta^*)$ is a quasicircle. (The existence of a quasiconformal extension of f across the unit circle $S^1 = \partial\Delta^*$ satisfying (1.1) is also a direct consequence of the lambda-lemma of Mañé-Sad-Sullivan [**MSS**], which does not provide, however, an explicit form of the Beltrami coefficient of this extension.)

For a given Jordan domain $D \subset \widehat{\mathbb{C}}$, let $A_1(D)$ denote the subspace of $L_1(D)$ formed by holomorphic functions in D; and let

$$A_1^2(D) = \{\psi \in A_1(D) : \psi = \omega^2\};$$

this set consists of the integrable holomorphic functions on D having only zeros of even orders. Put

$$\langle \mu, \psi \rangle_D = \iint_D \mu(z)\psi(z)dxdy, \quad \mu \in L_\infty(D), \ \psi \in L_1(D) \quad (z = x + iy).$$

Let $\lambda_D(z)|dz|$ be the hyperbolic metric on the domain D of the Gaussian curvature -4, i.e.,

$$\lambda_D(z) = h_*(\lambda_\Delta) = |h'(z)|/(1 - |h(z)|^2),$$

where h is any conformal map of D onto Δ. Using this metric, one obtains the complex Banach space $\mathbf{B}(D) = h_*\mathbf{B}(\Delta)$ of hyperbolically bounded holomorphic functions in D, with the norm $\|\omega\|_{\mathbf{B}(D)} = \sup_D \lambda_D^{-2}|\omega(z)|$.

It was shown by Bers [**Ber**] that this space is dual to $A_1(D)$, i.e., each linear functional $l(\psi)$ on $A_1(D)$ is of the form

(1.3) $$l(\psi) = \langle \lambda_D^{-2}\bar{\omega}, \psi \rangle_D$$

for some $\omega \in \mathbf{B}(D)$, and this ω is uniquely determined by l.

1.3. To formulate the main theorem, we need to introduce the Grunsky matrix $\mathcal{G} = (\alpha_{mn})$ of normalized univalent functions as well as some related quantities. This matrix is a powerful tool for many problems involving the methods of geometric function theory.

Without loss of generality, we can restrict ourselves to conformal maps of the disk Δ^* normalized by

(1.4) $$f(z) = z + b_0 + b_1 z^{-1} + b_2 z^{-2} + \ldots,$$

and assume that f does not vanish in Δ^*. The class of such univalent functions is denoted by Σ.

Let $\Sigma(k)$ consist of $f \in \Sigma$ having k-quasiconformal extensions to $\widehat{\mathbb{C}}$. As usual, the quasiconformal maps are regarded as the orientation preserving homeomorphic solutions of the Beltrami equation $\partial_{\bar{z}} w = \mu(z)\, \partial_z w$ in a domain $D \subseteq \widehat{\mathbb{C}} = \mathbb{C} \cup \{\infty\}$ with $\|\mu\|_\infty < 1$, where the derivatives $\partial_{\bar{z}}$ and ∂_z are distributional and belong locally to L_2. The function μ is called the *Beltrami coefficient* of the map w; the quantities $k(w) = \|\mu\|_\infty$ and $K(f) = (1+k(w))(1-k(w))$ are, respectively, its *dilatation* and *maximal dilatation*. The maps with $k(w) \leq k_0$ are called k_0-*quasiconformal*.

To distinguish a map $f^\mu \in \bigcup_k \Sigma(k)$ with a given Beltrami coefficient μ, we add to (1.4) the third normalization condition

(1.5) $$f^\mu(0) = 0,$$

passing to the maps

(1.6) $$f^\mu(z) = z + \frac{1}{2\pi i} \iint\limits_{|\zeta|<1} \partial_{\bar\zeta} f^\mu \left(\frac{1}{\zeta - z} - \frac{1}{\zeta} \right) d\zeta \wedge d\bar\zeta, \quad z \in \mathbb{C}.$$

Then the correspondence $f^\mu \longleftrightarrow S_{f^\mu}$ becomes one-to-one.

The *Grunsky coefficients* of $f \in \Sigma$ are determined from the expansion

$$\log \frac{f(z) - f(\zeta)}{z - \zeta} = -\sum_{m,n=1}^\infty \alpha_{mn} z^{-m} \zeta^{-n}, \quad (z, \zeta) \in (\Delta^*)^2,$$

where the single-valued branch of the logarithmic function which tends to zero as $z = \zeta \to \infty$ is chosen. One defines also the *Grunsky constant* of the map f by

(1.7) $$\varkappa(f) = \sup \left\{ \left| \sum_{m,n=1}^\infty \sqrt{mn}\, \alpha_{mn} x_m x_n \right| : \mathbf{x} = (x_n) \in l^2,\ \|\mathbf{x}\| = \left(\sum_1^\infty |x_n|^2 \right)^{1/2} = 1 \right\}.$$

By [**Gr**], a $\widehat{\mathbb{C}}$-holomorphic map (1.4) is univalent in Δ^* if and only if $\varkappa(f) \leq 1$.

1.4. Our main result is

THEOREM 1.1. *Suppose that $\|\varphi\|_\mathbf{B} < 2$ (consequently, the solutions to the Schwarzian equation $S_f = \varphi$ are conformal maps $f : \Delta^* \to \widehat{\mathbb{C}}$ admitting quasiconformal extensions across the unit circle S^1). Then the following properties are equivalent:*

(a$_1$) *the Schwarzian derivative $\varphi = S_f$ satisfies*

(1.8) $$\sup_{\|\psi\|_{A_1(\Delta)} = 1} |\langle \nu_\varphi, \psi \rangle_\Delta| = \sup_{\|\psi\|_{A_1^2(\Delta)} = 1} |\langle \nu_\varphi, \psi \rangle_\Delta|$$

which means geometrically that the length of the tangent vector to the holomorphic curve $\{t\varphi\} \subset \mathbf{B}$ at the origin is attained on the integrable squares of holomorphic functions;

(b$_1$) *the extremal (minimal) dilatation*

(1.9) $$k_0(f) = \inf\{k(w^\mu) = \|\mu\|_\infty :\ w^\mu|\overline{\Delta}^* = f\}$$

of quasiconformal extensions of f equals the Grunsky constant of f

(1.10) $$k_0(f) = \varkappa(f).$$

The quantity in the left-hand side of (1.8) is equal to Teichmüller's norm of the form (1.2) regarded as a tangent vector to $\mathbf{T}$ at the base point. However, the origins of these conditions are different: (1.8) arose from the representation (1.3) going back to automorphic forms and has an infinitesimal nature, while (1.10) relates to the features of the Grunsky coefficients of the functions with quasiconformal extension.

Theorem 1.1 provides a key to all other results formulated below. The proof of this theorem involves deep results concerning the complex metric geometry of the universal Teichmüller space $\mathbf{T}$ and the Finsler metrics as well as the Grunsky coefficient inequalities.

1.5. The Grunsky matrix $(\alpha_{mn}(f))$ is closely related to the Fredholm eigenvalues of the curve $L = f(S^1)$; and Theorem 1.1 can be reformulated in terms of the least positive eigenvalue ρ_L, defined by

$$\frac{1}{\rho_L} = \sup \frac{|D_{f(\Delta)}(h) - D_{f(\Delta^*)}(h)|}{D_{f(\Delta)}(h) + D_{f(\Delta^*)}(h)}, \quad D_G(h) = \iint_G (\nabla h)^2 dx dy,$$

where the supremum is taken over all functions h which are continuous on $\widehat{\mathbb{C}}$ and harmonic on $\widehat{\mathbb{C}} \setminus L$. When L is smooth, ρ_L is the least positive eigenvalue of the double-layer potential over L. This eigenvalue plays a special role. Its importance follows, for example, from the well-known Kühnau-Schiffer theorem, which gives that the Grunsky constant $\varkappa$ of a conformal map $\Delta^* \to D^*$ and this least eigenvalue are reciprocal:

(1.11) $$\varkappa = 1/\rho_L$$

(cf. [**Ah1**], [**Ku3**], [**Sc**]). Further, every quasicircle $L \subset \widehat{\mathbb{C}}$ admits *quasiconformal reflections*, or *quasireflections*, i.e., orientation reversing quasiconformal self-maps w of $\widehat{\mathbb{C}}$ which keep L point-wise fixed and interchange the interior and exterior of L. One defines the reflection coefficient q_L of the curve L by

$$q_L = \inf\{k(w) : w \text{ quasireflection across } L\}$$

and its *quasiconformal dilatation* $Q_L = (1 + q_L)/(1 - q_L) \geq 1$; then $Q_L = K_L^2 := \inf K(f)$, where f ranges over (orientation preserving) quasiconformal automorphisms f of $\widehat{\mathbb{C}}$ so that $f(S^1) = L$ (see [**Ah2**], [**Ku5**], [**Kr6**]).

Together with these results, Theorem 1.1 implies

COROLLARY 1.2. *For any $f \in \bigcup_k \Sigma(k)$ with $\|\varphi\|_\mathbf{B} < 2$ satisfying (1.8), we have*

(1.12) $$k_0(f) = \varkappa(f) = \frac{1}{\rho_L} = q_L, \quad L = f(S^1).$$

1.6. The next theorem provides an essential improvement of Theorem 1.1 for a special case.

THEOREM 1.3. *Let $\|\varphi\|_\mathbf{B} < 2$ and*

(1.13) $$\varphi(z) = o\left(\frac{1}{(|z|-1)^2}\right) \quad as \quad |z| \to 1.$$

Then the following properties are equivalent.

(a_1) *The function φ satisfies (1.8).*

(b_1) *This function is the Schwarzian derivative of a univalent function f on Δ^* whose extremal quasiconformal extension $\widehat{f}$ to $\widehat{\mathbb{C}}$ is unique and of the Teichmüller-Kühnau type, i.e., its Beltrami coefficient in Δ is defined by a holomorphic quadratic differential $\psi = \omega^2 \in A_1^2(\Delta)$. Moreover, in this case,*

$$\mu_{\widehat{f}}(z) = \varkappa(f)\frac{|\psi(z)|}{\psi(z)} = \varkappa(f)\frac{\overline{\omega(z)}}{\omega(z)} \tag{1.14}$$

with

$$\psi(z) = \frac{1}{\pi}\sum_{m+n=2}^{\infty}\sqrt{mn}\, x_m x_n z^{m+n-2}; \tag{1.15}$$

and this ψ is uniquely determined by φ, letting $\sum_{1}^{\infty}|x_n|^2 = 1$. The image $L = f(S^1)$ is an asymptotically conformal quasicircle.

Recall that a Jordan curve $L \subset \mathbb{C}$ is called *asymptotically conformal* if for any pair of points $a, b \in L$,

$$\max_{z \in L}\frac{|a-z|+|z-b|}{|a-b|} \to 1 \quad \text{as } |a-b| \to 0,$$

where z lies between a and b.

Such curves are quasicircles without corners and can be rather pathological (see, e.g., [**Po2**], p.249). In particular, all C^1-smooth curves are asymptotically conformal.

The assumption (1.13) holds, for example, for φ from the Bergman space $B^p(\Delta^*)$, $p \geq 1$ (i.e., $\|\varphi\|_{L_p(\Delta^*)} < \infty$).

The following consequence of Theorem 1.1 and of Theorem B from Section 2.2 relates to Theorem 1.3.

COROLLARY 1.4. *The equality (1.10) holds for any Schwarzian $\varphi = S_f \in \mathbf{B}$ with $\|\varphi\|_\mathbf{B} < 2$ admitting Teichmüller extension $\widehat{f}^{k|\psi|/\psi}$ of f to Δ whose quadratic differential ψ has only zeros of even orders in Δ.*

Some illustrative examples are constructed in Section 6.

1.7. Remarks on restriction $\|\varphi\|_\mathbf{B} < 2$. The bound $\|\varphi\|_\mathbf{B} < 2$ for Schwarzians in Theorems 1.1 and 1.3 is, of course, not sharp. The exact bound for the admissible range of φ is given explicitly in the proof of Theorem 1.1. It is nothing other than the topological boundary of the connected component of intersection of the complex line $\{t\varphi : t \in \mathbb{C}\} \subset \mathbf{B}$ with the universal Teichmüller space $\mathbf{T}$; this component contains the origin.

In the last section, we show that this intersection can have more than one connected component, even for φ corresponding to polygons.

2. Preliminaries

The main tool exploited in the proof of both Theorems 1.1 and 1.3 relies on the complex metric geometry of the universal Teichmüller space $\mathbf{T}$ and on the technique of Grunsky inequalities. We briefly present here the basic results concerning these subjects.

2.1. Universal Teichmüller space.

The universal Teichmüller space $\mathbf{T}$ is the space of quasisymmetric homeomorphisms h of the unit circle factorized by Möbius transformations. Its topology and real geometry is determined by the Teichmüller metric which naturally arises from extensions of those h to the unit disk. This space also admits the complex structure of a complex Banach manifold by means of the Bers embedding as a bounded subdomain of $\mathbf{B}$.

We shall identify the space $\mathbf{T}$ with this domain. In this model, the points $\psi \in \mathbf{T}$ represent the Schwarzian derivatives S_f of univalent holomorphic functions f in Δ^*, which have quasiconformal extensions to the whole sphere $\widehat{\mathbb{C}}$.

To get the universal Teichmüller space $\mathbf{T}$, consider the Banach ball

$$(2.1) \qquad \operatorname{Belt}(\Delta)_1 = \{\mu \in L_\infty(\mathbb{C}) : \mu|\Delta^* = 0, \ \|\mu\| < 1\}.$$

Each $\mu \in \operatorname{Belt}(\Delta)_1$ defines a conformal structure on the extended complex plane $\widehat{\mathbb{C}}$, i.e., a vector field of infinitesimal ellipses, or equivalently, a class of conformally equivalent Riemannian metrics $ds^2 = \lambda(z)|dz + \mu d\bar{z}|^2$, $\lambda(z) > 0$; these ellipses reduce to circles for $z \in \Delta^*$.

The universal Teichmüller space $\mathbf{T}$ is obtained from the ball (2.1) by the natural identification, letting μ and ν in $\operatorname{Belt}(\Delta)_1$ be equivalent if $w^\mu|S^1 = w^\nu|S^1$, $S^1 = \partial\Delta$. We denote the equivalence classes by $[\mu]$.

We shall deal with the intersections

$$(2.2) \qquad \Omega_\varphi = \{t'\varphi : \ t' \in \mathbb{C}\} \cap \mathbf{T},$$

defined for each $\varphi \in \mathbf{T}$. Each Ω_φ is an open planar set, which (in view of the results of [**Kr3**]) may be disconnected (see Section 9). Thus we distinguish the connected component of Ω_φ containing the basepoint $\psi = \mathbf{0}$ of $\mathbf{T}$ and denote this component by Ω_φ^0.

It follows from Zhuravlev's theorem that *every connected component of the set Ω_φ is simply connected* (see [**KK**], part 1, ch. V; [**Zh**]).

From now on, we identify Ω_φ^0 with the corresponding region in the complex plane $\mathbb{C}$ and write both $\psi = t\varphi \in \Omega_\varphi^0$ and $t \in \Omega_\varphi^0$, when this does not lead to a misunderstanding.

There are certain natural intrinsic complete metrics on the space $\mathbf{T}$. The first one is the *Teichmüller metric*

$$(2.3) \quad \tau_\mathbf{T}(\phi_\mathbf{T}(\mu), \phi_\mathbf{T}(\nu)) = \frac{1}{2}\inf\{\log K\left(w^{\mu_*} \circ (w^{\nu_*})^{-1}\right) : \ \mu_* \in \phi_\mathbf{T}(\mu), \nu_* \in \phi_\mathbf{T}(\nu)\},$$

where $\phi_\mathbf{T}$ is the canonical projection

$$\phi_\mathbf{T}(\mu) = [\mu] : \ \operatorname{Belt}(\Delta)_1 \to \mathbf{T}.$$

This metric is generated by the *Finsler structure on* $\mathbf{T}$ (in fact, on the tangent bundle $\mathcal{T}(\mathbf{T}) = \mathbf{T} \times \mathbf{B}$ of $\mathbf{T}$); this structure is defined by

$$(2.4) \quad \begin{aligned} F_\mathbf{T}(\phi_\mathbf{T}(\mu), \phi'_\mathbf{T}(\mu)\nu) &= \inf\{\left\|\nu_*(1-|\mu|^2)^{-1}\right\|_\infty : \\ &\phi'_\mathbf{T}(\mu)\nu_* = \phi'_\mathbf{T}(\mu)\nu; \ \mu \in \operatorname{Belt}(\Delta)_1; \ \nu, \nu_* \in L_\infty(\mathbb{C})\}. \end{aligned}$$

The *Carathéodory* and *Kobayashi metrics* on $\mathbf{T}$ are, as usually, the smallest and the largest semi-metrics d on $\mathbf{T}$, which are contracted by holomorphic maps $h : \Delta \to \mathbf{T}$. Denote these metrics by $c_\mathbf{T}$ and $d_\mathbf{T}$, respectively; then

$$c_\mathbf{T}(\psi_1, \psi_2) = \sup\{d_\Delta(h(\psi_1), h(\psi_2)) : \ h \in \operatorname{Hol}(\mathbf{T}, \Delta)\},$$

while $d_\mathbf{T}(\psi_1, \psi_2)$ is the largest pseudometric d on $\mathbf{T}$ satisfying

$$d(\psi_1, \psi_2) \leq \inf\{d_\Delta(0,t): h(0) = \psi_1, \text{ and } h(t) = \psi_2 \ \ h \in \text{Hol}(\Delta, \mathbf{T})\},$$

where d_Δ is the hyperbolic Poincaré metric on Δ of Gaussian curvature -4.

The infinitesimal Kobayashi metric (called also the Kobayashi-Royden metric) is defined on the tangent bundle $\mathcal{T}(\mathbf{T})$ of $\mathbf{T}$ by

$$\begin{aligned}
\mathcal{K}_T(\psi, v) &= \inf\{|t|: h \in \text{Hol}(\Delta, \mathbf{T}), h(0) = \psi, dh(0)t = v\} \\
&= \inf\left\{\frac{1}{r}: r > 0, h \in \text{Hol}(\Delta_r, \mathbf{T}), h(0) = \psi, h'(0) = v\right\}.
\end{aligned} \quad (2.5)$$

Here v is a tangent vector at the point $\psi \in \mathbf{T}$ and Δ_r denotes the disk $\{|z| < r\}$ (see, e.g., [**Ko**], [**FV**], [**Kl**]).

The main underlying ingredient in the proofs of Theorems 1.1 and 1.3 is the following important fact established in [**Kr7**], [**Kr8**], [**Kr9**].

THEOREM A. *The infinitesimal Kobayashi metric $\mathcal{K}_\mathbf{T}(\psi, v)$ on the tangent bundle $\mathcal{T}(\mathbf{T})$ of the universal Teichmüller space $\mathbf{T}$ is logarithmically plurisubharmonic in $\psi \in \mathbf{T}$, equals the canonical Finsler structure $F_\mathbf{T}(\psi, v)$ on $\mathcal{T}(\mathbf{T})$ generating the Teichmüller metric of $\mathbf{T}$, and has constant holomorphic sectional curvature $\varkappa_\mathcal{K}(\psi, v) = -4$ on $\mathcal{T}(\mathbf{T})$.*

This result has various applications, from which we present the following two statements closely related to results of this paper.

COROLLARY 2.1. (a) *The Kobayashi metric $d_\mathbf{T}$ of the universal Teichmüller space $\mathbf{T}$ coincides with its Teichmüller metric $\tau_\mathbf{T}$, and*

$$d_\mathbf{T}(\psi_1, \psi_2) = \tau_\mathbf{T}(\psi_1, \psi_2) = \inf d_\Delta(h^{-1}(\psi_1), h^{-1}(\psi_2)), \quad (2.6)$$

where the infimum is taken over all holomorphic maps $h: \Delta \to \mathbf{T}$.

(b) *The Teichmüller metric $\tau_\mathbf{T}(\psi_1, \psi_2)$ is plurisubharmonic separately in each of its arguments; hence,*

$$g_\mathbf{T}(\psi_1, \psi_2) = \log \tanh \tau_\mathbf{T}(\psi_1, \psi_2) = \log k(\psi_1, \psi_2), \quad (2.7)$$

where $g_\mathbf{T}$ is the pluricomplex Green function of $\mathbf{T}$ and k is an extremal Beltrami coefficient defining the Teichmüller distance between the points ψ_1, ψ_2 in $\mathbf{T}$.

Recall that the pluricomplex Green function g_X of a domain X in a complex Banach space E is defined by

$$g_X(x, y) = \sup u_y(x) \quad (x, y \in X),$$

where the supremum is taken over all plurisubharmonic functions $u_y(x): X \to [-\infty, 0)$ such that $u_y(x) = \log \|x - y\|_E + O(1)$ in a neighborhood of the point y (see, e.g., [**Kl**]).

The proof of Theorem A relies on the specific features of the space $\mathbf{T}$ and involves the technique of Grunsky coefficient inequalities. The equality $d_\mathbf{T} = \tau_\mathbf{T}$ follows also from the Royden-Gardiner theorem (see, e.g., [**EKK**], [**GL**], [**Ro1**]).

In fact, one has much more. Namely, by [**Kr7**], [**Kr8**], all invariant Finsler metrics on $\mathbf{T}$ coincide and are, in fact, determined by the Grunsky coefficients of appropriate conformal maps which are naturally associated to the Schwarzians $\varphi \in \mathbf{T}$. However, we do not use this fact here.

2.2. Grunsky coefficients revised.
We now recall the known results concerning the connection between the functional (1.7) and quasiconformal extensions of univalent functions in Δ^*.

First of all, observe that *the Grunsky coefficients $\alpha_{mn}(f) = \alpha_{mn}(\psi)$ depend holomorphically on the Schwarzians $\psi = S_f \in \mathbf{T}$*, because each α_{mn} is a polynomial in the first Taylor coefficients $b_1, \dots, b_p$, $p \leq \min(m,n)$, which are holomorphic functions of ψ.

For any $f \in \Sigma(k)$, we have
$$\varkappa(f) \leq k; \tag{2.8}$$
on the other hand, if $f \in \Sigma$ satisfies (1.7) with some $k < 1$, then it has a quasiconformal extension to $\widehat{\mathbb{C}}$ with a dilatation $k' \geq k$ (see, e.g., [**Ku1**], [**Po1**], [**Zh**]; [**KK**], pp. 82-84).

The following key result characterizes those functions for which the inequality (2.8) is both necessary and sufficient to belong to $\Sigma(k)$, i.e., the value $\varkappa(f)$ coincides with the least (extremal) dilatation $k(f)$ among the possible extensions of f.

THEOREM B. [**Kr2**], [**Kr4**] *The equality*
$$\varkappa(f) = \inf\{\|\mu\|_\infty : w^\mu |\Delta^* = f\} \tag{2.9}$$
holds if and only if the function f is the restriction to Δ^* of a quasiconformal self-map w^{μ_0} of $\widehat{\mathbb{C}}$ with the Beltrami coefficient μ_0 satisfying the condition
$$\sup |\langle \mu_0, \varphi \rangle_\Delta| = \|\mu_0\|_\infty, \tag{2.10}$$
where the supremum is taken over holomorphic functions $\varphi \in A_1^2(\Delta)$ with $\|\varphi\|_{A_1(\Delta)} = 1$.

This theorem reveals the crucial role played by integrable holomorphic functions with zeros of even order. Geometrically, it means that the Carathéodory metric on the holomorphic extremal disk $\Delta_{\mu_0} = \{\phi_{\mathbf{T}}(t\mu_0/\|\mu_0\|) : t \in \Delta\}$ in the space $\mathbf{T}$ coincides with the intrinsic Teichmüller metric of this space.

In particular, for any function $f \in \Sigma(k)$ admitting a k-quasiconformal extension onto $\overline{\Delta}$ with the Beltrami coefficient μ of the form
$$\mu(z) = k|\varphi(z)|/\varphi(z), \quad \varphi \in A_1^2(\Delta) \setminus \{\mathbf{0}\}, \tag{2.11}$$
we have
$$\varkappa(f) = k. \tag{2.12}$$
Moreover, if the boundary quasicircle $L = f(S^1)$ is asymptotically conformal, then (2.12) is both necessary and sufficient for $f \in \Sigma(k)$ to have (2.11)(cf. [**Kr5**], [**Ku3**]). Hence, any function $f^{\mu_0} \in \bigcup_k \Sigma(k)$ with the Beltrami coefficient of the form $\mu_0 = k|\varphi|/\varphi$ with $\varphi \in A_1(\Delta) \setminus A_1^2(\Delta)$ mapping the unit circle onto an asymptotically conformal curve satisfies $\varkappa(f) < k(f)$.

This shows, for example, that *there exist convex domains even with real analytic boundaries for which equality (1.10) does not hold.*

Applying Parseval's equality to the functions $\omega(z) = \sum_0^\infty c_k z^k \in L_2(\Delta)$, one obtains for their squares $\varphi = \omega^2 \in A_1^2(\Delta)$ the representation
$$\varphi(z) = \frac{1}{\pi} \sum_{m+n=2}^\infty \sqrt{mn}\, x_m x_n z^{m+n-2}, \tag{2.13}$$

with $\mathbf{x} = (x_n) \in l^2$, $\|\mathbf{x}\| = \|\omega\|_{L_2}$. In the case of (1.15), we have $\|\mathbf{x}\| = 1$.

2.3. Holomorphic sectional curvature of invariant metrics. Let X be a domain in a complex Banach space E (or a complex Banach manifold modelled on E). One defines for upper semicontinuous functions $u : X \to [-\infty, \infty)$, the *generalized Hessian* $\Delta_v u(x)$ *in a direction* v *at a point* $x \in X$ by

$$\Delta_v u(x) = 4 \liminf_{r \to 0} \frac{1}{r^2} \left\{ \frac{1}{2\pi} \int_0^{2\pi} u(x + vre^{i\theta}) d\theta - u(x) \right\} \quad (r > 0).$$

It is straightforward to verify that for C^2 functions u on the domains in $\mathbb{C}^n$, Δ_v coincides with the usual Hessian of u and that u is plurisubharmonic on X if and only if $\Delta_v u(x) \geq 0$. Hence, at a point x_0 of a local maximum of an upper semicontinuous function u such that $u(x_0) > -\infty$, $\Delta_v u(x_0) \leq 0$ for all v.

In particular, for domains in $\mathbb{C}$, letting $v = 1$, one obtains the *generalized Laplacian*

$$(2.14) \qquad \Delta u(x) = 4 \liminf_{r \to 0} \frac{1}{r^2} \left\{ \frac{1}{2\pi} \int_0^{2\pi} u(x + re^{i\theta}) d\theta - u(x) \right\},$$

which reduces to the usual Laplacian $\Delta = 4 \partial \bar{\partial}$ for C^2 functions.

Let $F_X(x, v)$ be a Finsler structure on $\mathcal{T}(X) \subset X \times E$ (hence $F_X(x, v) \geq 0$), and suppose that this function is upper semicontinuous on $\mathcal{T}(X)$. Consider for a fixed $(x, v) \in \mathcal{T}(X)$ the holomorphic maps $h : \Delta_r \to X$ with $h(0) = x$, $h'(0) = v$ (for suitable $r > 0$). Any such h determines a conformal metric $ds = \lambda_h(t)|dt|$ on the disk Δ_r, with

$$(2.15) \qquad \lambda_h(t) = F_X(h(t), h'(t)).$$

It is an infinitesimal Finsler metric whose density λ_h is upper semicontinuous on Δ_r. If $F_X(x, v)$ is plurisubharmonic (respectively, logarithmically plurisubharmonic) in $x \in X$, the function λ_h becomes subharmonic (respectively, logarithmically subharmonic) on the disk Δ_r.

The structure $F_X(x, v)$ can be regarded as an infinitesimal Finsler metric on $\mathcal{T}(X)$, and its *holomorphic sectional curvature* $\varkappa_F(x, v)$ at (x, v), where $F_X(x, v) > 0$, is defined as the upper bound of Gaussian curvatures of metrics (2.16), using the generalized Laplacian (2.15); namely,

$$(2.16) \qquad \varkappa_F(x, v) = \sup \left. \frac{\Delta \log F_X^2(h(t), h'(t))}{-2 F_X^2(h(t), h'(t))} \right|_{t=0},$$

where the supremum is taken over all holomorphic maps $h : \Delta_r \to X$ with $h(0) = x, h'(0) = v$ and all admissible $r > 0$.

It is well-known that for any complex Banach manifold X which is complete and hyperbolic, its Kobayashi metric $F_\mathcal{K}(x, v) = \mathcal{K}(x, v)$ has holomorphic curvature $\varkappa_\mathcal{K}(x, v) \geq -4$ for all $(x, v) \in \mathcal{T}(X)$. If X is Carathéodory-hyperbolic, the curvature of its infinitesimal Carathéodory metric $\mathcal{C}(x, v)$ satisfies $\varkappa_\mathcal{C}(x, v) \leq -4$ (see [**Bu**], [**Di**], [**Ro2**], [**Su**], [**Wo**], [**Wu**]).

3. Proof of Theorem 1.1

The proof is accomplished in several steps; it relies upon the fact that the Kobayashi metric on **T** has constant holomorphic curvature -4.

First observe that by the properties of holomorphic motions, the solutions to the equation
$$2(w''/w'(z))' - (w''/w')^2 = t\varphi \quad \text{on } \Delta^* \text{ for } t \in \Omega_\varphi^0$$
admit quasiconformal extensions $w_t(z) = w(z,t)$ to $\mathbb{C}$ with Beltrami coefficients $\mu(z,t)$ depending holomorphically on the parameter t. This determines a holomorphic curve
$$t \mapsto \mu(z,t) : \Omega_\varphi^0 \to \mathbf{T}$$
covering Ω_φ^0. Moreover, using the Riemann mapping function $t = g_\varphi(t_*) : \Delta \to \Omega_\varphi^0$ with $g_\varphi(0) = 0$, $g_\varphi'(0) > 0$, one obtains by Schwarz's Lemma,

(3.1) $$\|\mu(z, g(t_*))\|_\infty \leq |t_*|$$

or, equivalently, $K(w_{t_*}) \leq (1+|t_*|)/(1-|t_*|)$; and this estimate cannot be improved in the general case (see, e.g., [**BR**], [**EKK**], [**MSS**], [**Sl**], [**ST**])).

Every holomorphic map $h : \Delta_r \to \mathbf{T}$ generates such a motion.

Step 1: Estimate of curvatures. Fix $(\psi, v) \in \mathcal{T}(\mathbf{T})$ and take the minimizing sequences for the Kobayashi metric at (ψ, v), i.e., a sequence of complex numbers t_n and a sequence of holomorphic maps $h_n \in \text{Hol}(\Delta, \mathbf{T})$ with $h_n(0) = \psi$, $dh_n(0)t_n = v/F_{\mathbf{T}}(\psi, v)$, for which

(3.2) $$\lim_{n \to \infty} |t_n| = \mathcal{K}_{\mathbf{T}}\left(\psi, \frac{v}{F_{\mathbf{T}}(\psi, v)}\right) = F_{\mathbf{T}}\left(\psi, \frac{v}{F_{\mathbf{T}}(\psi, v)}\right) \quad (n = 1, 2, ...).$$

The corresponding Finsler metrics
$$\lambda_{h_n}(t) = F_{\mathbf{T}}(h_n(t), h_n'(t))$$
are logarithmically subharmonic in Δ (because $F_{\mathbf{T}}$ and $\mathcal{K}_{\mathbf{T}}$ are logarithmically plurisubharmonic by Theorem A) and satisfy

(3.3) $$\lambda_{h_n}(t) \leq 1/(1 - |t|^2) \quad \text{for all } t \in \Delta.$$

Consider the upper envelope of these metrics $\tilde{\lambda}_{\psi,v}(t) = \sup_n \lambda_{h_n}(t)$, $t \in \Delta$, and take its upper semi-continuous regularization
$$\lambda_{\psi,v}(t) = \limsup_{t' \to t} \tilde{\lambda}_{\psi,v}(t').$$

This gives a (logarithmically subharmonic) metric on Δ whose curvature $\varkappa_{\lambda_{\psi,v}}$ is well-defined by (2.19), using the generalized Laplacian. By (3.3),

(3.4) $$\lambda_{\psi,v}(t) \leq \lambda_0(t) = 1/(1 - |t|^2)$$

(we have used here the notation λ_0 instead of λ_Δ). Note also that

(3.5) $$\lambda_{\psi,v}(0) = \lim_{n \to \infty} |t_n|$$

and hence
$$\lambda_{\psi,v}(0) = F_{\mathbf{T}}(\psi, v/F_{\mathbf{T}}(\psi, v)).$$

To prove (3.5), observe that the maps h_n generate holomorphic motions $f_n(z,t)$ of Δ^* extending to motions of $\widehat{\mathbb{C}}$. The Beltrami coefficients $\mu_{f_{n,t}}$ of these motions in Δ are controlled by (3.1). This bound yields
$$\lambda_{h_n}(0) = F_{\mathbf{T}}(\psi, v/F_{\mathbf{T}}(\psi, v)) + o(|t|),$$
where the remainder is uniformly estimated by $o(|t|)/|t| \leq \beta(|t|)$ with a function $\beta(|t|) \to 0$ as $|t| \to 0$ depending only on the pair (ψ, v). Therefore, the above

regularization does not decrease the value $\lambda_{\psi,v}(0)$. Together with the continuity of $F_\mathbf{T}(\psi, v)$, this yields (3.5).

We claim that the curvature of $\lambda_{\psi,v}$ satisfies

(3.6) $$\varkappa_{\lambda_{\psi,v}}(t) \geq -4.$$

Indeed, the relations (3.4) and (3.5) imply that the function $\log \frac{\lambda_{\psi,v}}{\lambda_\Delta}$ has a local maximum at $t = 0$; thus its (generalized) Laplacian at this point cannot be positive, i.e.,

$$\Delta\left(\log \frac{\lambda_{\psi,v}}{\lambda_0}\right)(0) = \Delta \log \lambda_{\psi,v}(0) - \Delta \log \lambda_0(0) \leq 0,$$

which yields (3.6).

Moreover, in fact, one has for all t the equality in (3.6) (which also yields the assertion $\varkappa_\mathcal{K} = -4$ of Theorem A). To see this, set $\varkappa_\mathcal{K}^*(\psi, v) = \sup \varkappa_{\lambda_n}(0)$, $\lambda_n := \lambda_{h_n}$, where the supremum is taken over all sequences $\{h_n\}$ and $\{t_n\}$ satisfying (3.2) for a given (ψ, v). Clearly, $\varkappa_\mathcal{K}^*(\psi, v) \leq \varkappa_\mathcal{K}(\psi, v)$. On the other hand, by (3.6),

$$\varkappa_\mathcal{K}^*(\psi, v) \geq \varkappa_{\lambda_{\psi,v}}(t) \geq -4 \quad \text{for all } (\psi, v) \in \mathcal{T}(\mathbf{T}).$$

Hence

(3.7) $$\varkappa_{\lambda_{\psi,v}}(t) = \varkappa_\mathcal{K}^*\left(\psi, \frac{v}{F_\mathbf{T}(\psi, v)}\right) = \varkappa_\mathcal{K}\left(\psi, \frac{v}{F_\mathbf{T}(\psi, v)}\right) = -4.$$

We now define the conformal metric $\lambda_\varphi(t)|dt|$ on the domain Ω_φ^0 as the restriction of the Finsler structure $F_\mathbf{T}$ to this domain, i.e.,

(3.8) $$\lambda_\varphi(t) = F_\mathbf{T}\left(t\varphi, \frac{v}{F_\mathbf{T}(t\varphi, v)}\right) = \mathcal{K}_\mathbf{T}\left(t\varphi, \frac{v}{\mathcal{K}_\mathbf{T}(t\varphi, v)}\right), \quad t \in \Omega_\varphi^0,$$

and get from (3.7) (or from Theorem A) that the (generalized) curvature of this metric also satisfies

(3.9) $$\varkappa_{\lambda_\varphi}(t) = -4 \quad \text{for all } t \in \Omega_\varphi^0.$$

Step 2: Metric associated with the Grunsky functional. The Grunsky coefficients determine the holomorphic maps

(3.10) $$h_\mathbf{x}(\psi) = \sum_{m,n=1}^\infty \sqrt{mn}\, \alpha_{mn}(\psi) x_m x_n : \mathbf{T} \to \Delta$$

parametrized by the points of the sphere $S(l^2) = \{\mathbf{x} = (x_n) \in l^2 : \|\mathbf{x}\| = 1\}$. Applying Schwarz's Lemma to their lifts

$$\widehat{h}_\mathbf{x}(\mu) = h_\mathbf{x} \circ \phi_\mathbf{T} : \text{Belt}(\Delta)_1 \to \Delta$$

yields $|\widehat{h}_\mathbf{x}(\zeta\mu)| \leq |\zeta|$ for $\|\mu\|_\infty = 1$ and $\zeta \in \Delta$. Note that differential of $\widehat{h}_\mathbf{x}$ at zero is given by

(3.11) $$d\widehat{h}_\mathbf{x}(\mathbf{0})\nu = -\pi^{-1} \iint_\Delta \nu(z) \sum_{m+n=2}^\infty \sqrt{mn}\, x_m x_n z^{m+n-2} dx dy,$$

because by (1.6) we have

$$\alpha_{mn}(\phi_\mathbf{T}(\mu)) = -\pi^{-1} \iint_\Delta \mu(z) z^{m+n-2} dx dy + O(\|\nu\|^2), \quad \|\mu\|_\infty \to 0.$$

We now define on Ω_φ^0 the conformal metrics $ds = \lambda_{h_{\mathbf{x}}}(t)|dt|$ as the pull-back of the hyperbolic metric $\lambda_0(t_*)|dt_*|$ in Δ under the maps (3.10) at $\psi = t\varphi$; explicitly,

$$\lambda_{h_{\mathbf{x}}}(t) = \frac{|h'_{\mathbf{x}}(t)|}{1 - |h_{\mathbf{x}}(t)|^2}. \tag{3.12}$$

The curvature of each $\lambda_{h_{\mathbf{x}}}$ equals -4 (at noncritical points of $h_{\mathbf{x}}$). We pass to the upper envelope $\widetilde{\lambda}_{\varkappa}(t) = \sup\{\lambda_{h_{\mathbf{x}}}(t) : \mathbf{x} \in S(l^2)\}$ and, applying upper semicontinuous regularization

$$\lambda_{\varkappa}(t) = \limsup_{t' \to t} \widetilde{\lambda}_{\varkappa}(t'), \tag{3.13}$$

again obtain a logarithmically subharmonic metric on Ω_φ^0 which is continuous at $t = 0$. By (3.11) and (3.12),

$$\lambda_{\varkappa}(t) = \varkappa(t\varphi) + o(t) \quad \text{as } t \to 0, \tag{3.14}$$

where $\varkappa(t\varphi)$ denotes the Grunsky constant of the maps $f \in \bigcup_k \Sigma(k)$ with $S_{f|\Delta^*} = t\varphi$.

Our goal is to establish that the curvature of $\lambda_{\varkappa}$ is at most -4. To this end, we take a maximizing sequence $\{h_{\mathbf{x}^{(p)}}\}$ for $\varkappa(\varphi)$ so that

$$\lim_{p \to \infty} h_{\mathbf{x}^{(p)}}(\varphi) = \varkappa(f),$$

and the restrictions $h_{\mathbf{x}^{(p)}}(t\varphi)$ converge uniformly on the compact subsets of Ω_φ^0 to a holomorphic function $h_0(t) : \Omega_\varphi^0 \to \Delta$ with $h'_0(0) = \varkappa(f)$. This provides a conformal metric

$$\lambda_{h_0}(t) = \frac{|h'_0(t)|}{1 - |h_0(t)|^2} \quad \text{on } \Omega_\varphi^0$$

of constant curvature -4 (again at noncritical points) which is, in view of (3.14) and definition of $\lambda_{\varkappa}$, a *supporting metric* for $\lambda_{\varkappa}$ at 0, i.e., $\lambda_{h_0}(0) = \lambda_{\varkappa}(0)$ and $\lambda_{h_0}(t) \leq \lambda_{\varkappa}(t)$ in a neighborhood of $t = 0$.

Hence, $\log \frac{\lambda_{h_0}}{\lambda_{\varkappa}}$ has a local maximum at 0, and therefore,

$$\Delta \log \frac{\lambda_{h_0}}{\lambda_{\varkappa}}(0) = \Delta \log \lambda_{h_0}(0) - \Delta \log \lambda_{\varkappa}(0) \leq 0,$$

which yields

$$-\frac{\Delta \log \lambda_{\varkappa}(0)}{\lambda_{\varkappa}(0)^2} \leq -\frac{\Delta \log \lambda_{h_0}(0)}{\lambda_{h_0}^2(0)};$$

and the desired inequality $\varkappa_{\lambda_{\varkappa}}(t) \leq -4$ on Ω_φ^0 follows.

Step 3: Comparison of the metrics $\lambda_{\varkappa}$ and λ_{φ}. We use the metric $\lambda_{\varkappa}$ in the case when

$$\lambda_{\varkappa}(0) = \lambda_{\varphi}(0). \tag{3.15}$$

To compare these metrics in the whole domain Ω_φ^0, we apply Minda's strong maximum principle for upper semicontinuous generalized solutions of the differential inequality $\Delta u \geq Ku$.

THEOREM C. [**Mi**] *Suppose Ω is a region in $\mathbb{C}$, $u : \Omega \to [-\infty, +\infty)$ is upper semicontinuous and there is a positive constant K such that the generalized Laplacian $\Delta u(z) \geq Ku(z)$ at any point $z \in \Omega$ with $u(z) > -\infty$. If*

$$\limsup_{z \to \zeta} u(z) \leq 0 \quad \text{for all } \zeta \in \partial\Omega,$$

then either $u(z) < 0$ for all $z \in \Omega$ or else $u(z) = 0$ for all $z \in \Omega$.

This principle, applied to $u = \log \lambda_\varkappa - \log_\varphi$, shows that either $\log \lambda_\varkappa(t) < \log \lambda_\varphi(t)$ or else $\lambda_\varkappa(t) = \log \lambda_\varphi(t)$ on Ω_φ^0; thus, *in the case of equality (3.15), both metrics $\lambda_\varkappa$ and λ_φ must coincide on Ω_φ^0.*

Step 4. Completion of the proof of Theorem 1.1. We may now establish the assertion of Theorem 1.1. For a given $\varphi \in \mathbf{T}$, define the functions $w = f_t(z) \in \bigcup_k \Sigma(k)$ by the Schwarzian equation

$$S_w = t\varphi, \quad t \in \Omega_\varphi^0.$$

Assuming $\|\varphi\|_{\mathbf{B}} < 2$, one can take for $|t| < \|\varphi\|$ the Ahlfors-Weill extensions $\widehat{f}_t$ of f_t onto $\overline{\Delta}$. Explicitly,

$$(3.16) \qquad \widehat{f}_t(z) = f_t(1/\bar{z}) + \frac{(z - 1/\bar{z}) f_t'(1/\bar{z})}{1 - \frac{1}{2}(z - 1/\bar{z}) f_t''(1/\bar{z})/f_t'(1/\bar{z})}, \quad |z| \leq 1,$$

and these maps have in Δ harmonic Beltrami coefficients $t\nu_\varphi(z)$ given by (1.2). For small $|t|$, the extremal Beltrami coefficients for f_t are equal to $t\mu_0$, for some $\mu_0 \in \text{Belt}(\Delta)_1$, and

$$(3.17) \qquad t\nu_\varphi(z) = t\mu_0(z) + \sigma(z, t),$$

where

$$\sigma(z, t) \in A_1(\Delta)^\perp = \{\omega \in L_\infty(\Delta) : <\omega, \psi>_\Delta = 0 \text{ for all } \psi \in A_1(\Delta)\}$$

and $\|\sigma(\cdot, t)\|_\infty \to 0$ as $t \to 0$; in other words, the differentials ν_φ and μ_0 determine the same tangent vector $\phi_{\mathbf{T}}(\nu_\varphi) = \phi_{\mathbf{T}}(\mu_0)$ at the basepoint of $\mathbf{T}$.

Note also that, by a characteristic property of extremal Beltrami differentials,

$$(3.18) \qquad \|\mu_0\|_\infty = \inf\{\|\mu_0 + \sigma\|_\infty : \sigma \in A_1(\Delta)^\perp\}$$

(see, e.g., [**Ah2**], [**GL**], [**Kr1**]). Now, if a given Schwarzian φ satisfies equality (1.8), the relations (3.17) and (3.18) yield

$$F_{\mathbf{T}}(\mathbf{0}, t\varphi) = \sup_{\|\psi\|_{A_1^2(\Delta)}=1} |\langle \nu_\varphi, \psi \rangle_\Delta|;$$

then, by (3.12), (3.14) and Theorem B, we have for $\lambda_\varkappa$ the equality (3.15). Consequently, by step 3, the metric $\lambda_\varkappa$ coincides with λ_φ, i.e., with $F_{\mathbf{T}}(t\varphi, v)$, on the whole domain Ω_φ^0. This means, by definition of $\lambda_\varkappa$, that the functions (3.10) produce the maximizing sequences for the Carathéodory distance $c_{\mathbf{T}}(\mathbf{0}, t\varphi)$ for any t. In particular, this holds for $t = t_0$ corresponding to the point φ itself, which implies, again by Theorem B, equality (1.10).

Conversely, having equality (1.10), one can define on Ω_φ^0 the complex parameter $\tilde{t} = g_\varphi \circ \gamma \circ g_\varphi^{-1}(t)$ with

$$\gamma(t_*) = (t_* - t_{0,*})/(1 - \bar{t}_{0,*} t_*) \quad (t_{0,*} = g_\varphi^{-1}(1)),$$

carrying the initial value t_0 into the new basepoint $\tilde{t} = 0$ on Ω_φ^0. Then equality (1.8) follows by applying the same arguments. This completes the proof of Theorem 1.1. $\square$

4. Proof of Theorem 1.3 and of Corollary 1.2

4.1. Proof of Theorem 1.3. The condition (1.13) is equivalent to

$$\lim_{|z| \to 1+} (|z|^2 - 1)^2 \, |S_f(z)| = 0,$$

which, in turn, is one of the equivalent characterizations of asymptotically conformal boundaries (see, e.g., [**Po2**]). Thus the curve $L = f(\partial \Delta)$ is an asymptotically conformal quasicircle.

The proof consists of two steps. First, when $k(\widehat{f}) = \varkappa(f)$, we can combine this equality with the asymptotic conformality and Strebel's frame mapping criterion involving the dilatations of arbitrary quasiconformal extensions of f into Δ (see [**St**], [**EL**]). This allows us to establish that f has an extremal extension $\widehat{f}$ of Teichmüller type, i.e., with $\mu_{\widehat{f}} = k|\psi|/\psi$, where $\psi \in A_1(\Delta) \setminus \{\mathbf{0}\}$ (cf. [**Kr5**]). The second step involves arguments related to Theorem B which imply that the quadratic differential ψ must be a square of some holomorphic ω in Δ, hence of the form (1.14). □

4.2. Proof of Corollary 1.2. By [**Ah1**], $q_L \geq 1/\rho_L$. On the other hand, $\varkappa \leq q_L \leq k_0(f)$. Combining all these inequalities with equality (1.10), one concludes that in the case $\varkappa(f) = k_0(f)$ all the above relations can occur only as equalities, which gives (1.12). □

5. Generalization of Theorem 1.1

The bound $\|\varphi\| < 2$ (and even the weaker assumption $\varphi \in \Omega_\varphi^0$) is rather restrictive in applications of Theorem 1.1. One can try to drop this condition by using more general complex curves (holomorphic disks) in $\mathbf{T}$ connecting a given Schwarzian $\varphi_0 = S_{f_0}$ with the basepoint $\varphi = \mathbf{0}$. In the case of extremal holomorphic disks, Theorem 1.1 is reduced to Theorem B.

The arguments exploited in the proof of Theorem 1.1 admit a straightforward extension from the flat disks Ω_φ^0 to complex curves in $\mathbf{T}$ without self-intersections, which yields, for example, the following result.

THEOREM 5.1. *Let a Schwarzian $\varphi_0 = S_{f_0} \in \mathbf{T}$, and suppose there exists a holomorphically embedded disk $\Omega = h(\Delta) \subset \mathbf{T}$ with $h(t) = h'(0)t + h''(0)t^2/2 + \ldots$, passing through the point φ_0 and such that $\|h'(0)\|_\mathbf{B} \leq 2$. Then the equality*

$$\sup_{\|\psi\|_{A_1(\Delta)}=1} |\langle \nu_{h'(0)}, \psi \rangle_\Delta| = \sup_{\|\psi\|_{A_1^2(\Delta)}=1} |\langle \nu_{h'(0)}, \psi \rangle_\Delta|$$

(equivalent to (1.8)) is necessary and sufficient for the corresponding normalized univalent function f_0 to have the property (1.9).

The assumption $\|h'(0)\|_\mathbf{B} \leq 2$ ensures the existence of holomorphic maps from the whole unit disk onto the tangent holomorphic disks to Ω at the initial point $h(0) = \mathbf{0}$, which allows us to define the metric $\lambda_\varkappa$ at this point (and on the whole disk Ω). Thereafter, one can apply the arguments from the last steps in the proof of Theorem 1.1.

6. Examples

6.1. Fix an integer $n \geq 3$, and let

$$(6.1) \quad \varphi_n(z) = \epsilon \sum_{j,l=1}^{n} \frac{c_{jl}}{(z-x_j)(z-x_l)} = \epsilon \sum_{j=1}^{n} \frac{c_{jj}}{(z-x_j)^2} + \epsilon \sum_{j=1}^{n} \frac{C_j}{z-x_j}, \quad |\epsilon|=1,$$

with distinct real poles $x_1, \ldots, x_n$ and non-negative coefficients c_{jl} satisfying

$$(6.2) \quad \sum_{j,l=1}^{n} c_{jl} = \frac{1}{2}.$$

It is well-known that any solution of the equation $(w''/w')' - 1/2(w''/w')^2 = \varphi(z)$ with a given holomorphic function φ in a domain $D \subseteq \mathbb{C}$ is locally univalent on D (and vice versa) and is represented as a ratio of two linearly independent solutions of the linear equation $2v'' + \varphi v = 0$.

Such linear equations, even with rational φ, arise in various fields of mathematics and its applications. Their solutions can have both algebraic and transcendental singularities. In the important case of Fuchsian equations (i.e., those with only regular singular points on $\widehat{\mathbb{C}}$), the coefficients C_j must satisfy some additional relations. Our example concerns a special case of such equations.

6.2. Consider the solutions $w_t(z)$ of the equation

$$\left(\frac{w''}{w'}\right)' - \frac{1}{2}\left(\frac{w''}{w'}\right)^2 = t\varphi_n(z)$$

on the lower half-plane $U^* = \{z : \operatorname{Im} z < 0\}$ (for a fixed t, $|t| < 1$), which are homeomorphisms on this half-plane. Every w_t maps U^* conformally onto a quasidisk (either bounded or not), which can be regarded as an analytic polygon P_n^* with vertices $w_t(x_1), \ldots, w_t(x_n)$, whose boundary consists either of n real analytic arcs with nonzero intersection angles or else of arcs of spirals, which are analytic at their interior points. Since

$$\sup_y \frac{-y}{|z-x_j|} = 4\sup_{U^*} y^2|\varphi(z)| = \sup_y \frac{-y}{\sqrt{(x-x_j)^2+y^2}} = 1 \quad (z=x+iy,\ y<0),$$

the assumption (6.2) yields

$$(6.3) \quad \|\varphi_n\|_{\mathbf{B}(U^*)} \leq 4\sum_{j,l} c_{jl} = 2.$$

The harmonic Beltrami coefficients $\nu_{t\varphi_n}$ of the Ahlfors-Weill extensions of w_t into the upper half-plane $U = \{\operatorname{Im} z > 0\}$ assume the form $\nu_{t\varphi_n}(z) = -2ty^2\varphi_n(\bar{z})$, $z \in U$; and by (6.3),

$$(6.4) \quad \|\nu_{t\varphi_n}\|_\infty \leq |t|.$$

We now verify that

$$(6.5) \quad \sup_{\|\psi\|_{A_1(U)}=1} |\langle \nu_{\varphi_n}, \psi \rangle_U| = \sup_{\|\psi\|_{A_1^2(U)}=1} |\langle \nu_{\varphi_n}, \psi \rangle_U| = 2\sum_{j,l} c_{jl},$$

which implies that equality must hold in relations (6.3) and (6.4). First of all, it follows from (6.1) and (6.3) that

$$(6.6) \quad \lim_{y\to\infty} |\nu_{\varphi_n}(x+iy)| = \|\nu_{\varphi_n}\|_\infty = 1.$$

To establish the first equality in (6.5), we use the function $z = g(\zeta) := \cosh \zeta$ to map the half-strip
$$\Pi_+ = \{\zeta = \xi + i\eta : \xi > 0,\ 0 < \eta < 1\}$$
conformally onto U with $g(\infty) = \infty$, and pull-back ν_{φ_n} to the Beltrami coefficient
$$\nu_*(\zeta) := g_*(\nu_{\varphi_n})(\zeta) = (\nu_{\varphi_n} \circ g)(\zeta)\, \overline{g'(\zeta)}/g'(\zeta)$$
on Π_+. Then by (6.6), $\lim_{\xi \to \infty} |\nu_*(\xi + i\eta)| = \|\nu_*\|_\infty = 1$. Note also that in view of the smoothness of g on $\partial \Pi_+ \setminus \{0, i, \infty\}$, there exists the limit function $\nu_*(\zeta_0) = \lim_{\zeta \to \zeta_0 \in \partial \Pi_+} \nu_*(\zeta)$, and

(6.7) $$\nu_*(i\eta) = 0.$$

Let us take the sequence
$$\omega_m(\zeta) = \frac{1}{m} e^{-\zeta/m},\quad \zeta \in \Pi_+\ \ (m = 1, 2, \ldots);$$
all these ω_m belong to $\mathcal{A}_1^2(\Pi_+)$ and $\omega_m(\zeta) \to 0$ uniformly on $\Pi_+ \cap \{|\zeta| < M\}$ for any $M < \infty$. Also, $\|\omega_m\|_{A_1(\Pi_+)} = 1$ (moreover, $\left|\iint_{\Pi_+} \omega_m d\xi d\eta\right| = 1 - O(1/m)$), which shows that $\{\omega_m\}$ is a degenerating sequence for the affine horizontal stretching of Π_+.

We prove that this sequence is degenerating also for ν_*. Indeed,
(6.8)
$$\langle \nu_*, \omega_m \rangle_{\Pi_+} = \frac{1}{m} \iint_{\Pi_+} \nu_*(\zeta) \omega_m(\zeta) d\xi d\eta = \int_0^1 e^{-i\eta/m} d\eta \left(\frac{1}{m} \int_0^\infty \nu_*(\xi + i\eta) e^{-\xi/m} d\xi\right).$$

The inner integral relates to the Laplace transform of ν_* in ξ and can be evaluated explicitly. Integrating by parts and applying (6.7), one obtains
$$\int_0^\infty \frac{\partial \nu_*(\xi + i\eta)}{\partial \xi} e^{-\xi/m} d\xi = \frac{1}{m} \int_0^\infty \nu_*(\xi + i\eta) e^{-\xi/m} d\xi.$$

On the other hand, by Abel's theorem for the Laplace transform, the nontangential limit
$$\lim_{s \to 0} \int_0^\infty \frac{\partial \nu_*(\xi + i\eta)}{\partial \xi} e^{-s\xi} d\xi = \int_0^\infty \frac{\partial \nu_*(\xi + i\eta)}{\partial \xi} d\xi = \nu_*(\infty) - \nu_*(i\eta);$$
hence,
$$\lim_{m \to \infty} \frac{1}{m} \int_0^\infty \nu_*(\xi + i\eta) e^{-\xi/m} d\xi = \nu_*(\infty).$$

Applying Lebesgue's theorem on dominated convergence to the iterated integral on the right-hand side of (6.8), one gets

(6.9) $$\lim_{m \to \infty} |\langle \nu_*, \omega_m \rangle_{\Pi_+}| = \left|\int_0^1 d\eta \lim_{m \to \infty} \frac{1}{m} \int_0^\infty \nu_*(\xi + i\eta) e^{-\xi/m} d\xi\right| = 1.$$

Using the inverse conformal map $\zeta = g^{-1}(z) : \Pi_+ \to U$, i.e., a suitable branch of arccosh z, one obtains the sequence $\{\psi_m = (\omega_m \circ g^{-1})(g')^{-2}\} \subset A_1^2(U)$, which is a degenerating sequence for the initial Beltrami coefficient ν_{φ_n} on U, and by (6.9),

$$\lim_{m \to \infty} |\langle \nu_{\varphi_n}, \psi_m \rangle_U| = \|\nu_{\varphi_n}\|_\infty = 1.$$

It follows from the Hamilton-Krushkal-Reich-Strebel theorem that any $t\nu_{\varphi_n}$ is an extremal Beltrami coefficient in the class of the boundary values $w_t \mid \mathbb{R}$ (see, e.g., [**Ha**], [**Kr1**], [**RS**]). This boundary homeomorphism does not satisfy Strebel's frame mapping criterion, so the maps w_t cannot have Teichmüller extremal extensions into the upper half-plane.

Composing $w_t(z)$ with a fractional linear map $\Delta^* \to U^*$, one obtains maps $f \in \Sigma(k)$ which satisfy (1.7), (1.9) and (1.11).

6.3. The above arguments also work well for appropriate Schwarzians with infinite sets of poles, for example, those of the form

$$\varphi(z) = \epsilon \sum_{j=1}^{\infty} \frac{\alpha_j}{(z - x_j)^2}, \quad \text{Im } z < 0,$$

where $|\epsilon| = 1$, the poles $x_j \in \mathbb{R}$ accumulate to infinity, and the coefficients α_j are positive numbers satisfying $\sum_{1}^{\infty} \alpha_j = 1/2$.

7. Remarks on sufficiency of Grunsky inequalities for k-quasiconformal extension

The important results of Pommerenke and Zuravlev mentioned in Section 2.2 provide that any function $f \in \Sigma$ with $\varkappa(f) < 1$ admits a k'-quasiconformal extensions $\widehat{f}$ to $\widehat{\mathbb{C}}$ with $k(\widehat{f}) \geq k = \varkappa(f)$. Theorem B characterizes the functions for which this equality is both a necessary and sufficient condition for belonging to $\Sigma(k)$.

It is important for applications to have sufficient conditions ensuring such extensions. Reiner Kühnau first raised this question (see, e.g., [**Ku2**], [**Ku4**]); later it was investigated by others, but only a few results have been established.

As a straightforward extension of the above examples, we have the following sufficiency theorem.

THEOREM 7.1. *If the Schwarzian derivative of a function $f \in \Sigma$ satisfies $\|S_f\|_\mathbf{B} < 2$, S_f is regular on $\partial \Delta^*$ excluding a set E_0 of zero linear Lebesgue measure and*

(7.1) $$\lim_{r \to 1} \sup_{1 < |z| < r} (|z|^2 - 1)^2 |S_f(z)| = \|S_f\|_\mathbf{B},$$

then f admits a quasiconformal extension $\widehat{f}$ onto the the whole sphere $\widehat{\mathbb{C}}$ with dilatation $k(\widehat{f}) = \varkappa(f)$.

From the geometric point of view, this theorem describes the situation when the domain Ω_φ^0 represents a geodesic holomorphic disk in $\mathbf{T}$.

PROOF. One can assume that $z = 1$ is not a regular point of S_f and, in view of (7.1), that

$$\lim_{z \to 1} \sup_{1 < |z| < r} (|z|^2 - 1)^2 |S_f(z)| = \|S_f\|_\mathbf{B}$$

as $z \to 1$ in Δ^* in an arbitrary way.

Having Δ mapped conformally onto the half-strip Π_+, using a function $\zeta = g_1(z)$ with $g_1(1) = \infty$, one produces on Π_+ a harmonic Beltrami coefficient as the pull-back of ν_{S_f}. Applying to it the same arguments as in the above example, we have, after returning to the unit disk, that similarly to (6.13),

$$(7.2) \qquad \lim_{m \to \infty} |\langle \nu_{S_f}, \psi_m \rangle_U| = \|\nu_{S_f}\|_\infty,$$

where $\{\psi_m = (\omega_m \circ g_1^{-1})(g_1')^{-2}\}_{m=1}^\infty \subset A_1^2(\Delta)$ is a corresponding degenerating sequence in Δ. Equality (7.2) yields that ν_{S_f} is an extremal Beltrami coefficient among quasiconformal extensions $\widehat{f}$ of the map f and, moreover, satisfies the necessary and sufficient condition of Theorem B for the equality $\varkappa(f) = k(\widehat{f})$. This completes the proof of the theorem. $\square$

Theorem 7.1 can be improved; for example, one can replace in (7.1) the limit by upper limit and drop the assumption of regularity of S_f, adding weaker assumptions.

8. Schwarzians with nonconnected sets Ω_φ

By results of [**Kr3**], for any homeomorphic embedding $\gamma(t) : [0,1] \to \mathbb{C}$, there exist points $\varphi_0 \in \mathbf{T}$ such that the Jordan curve

$$(8.1) \qquad t \mapsto \gamma(t)\varphi_0 : [0,1] \to \mathbf{B}$$

must contain the points $\varphi \in \mathbf{B} \setminus \mathbf{S}$. This means that the functions f on Δ^* with $S_f = \varphi$ are only locally univalent.

The proof of this fact relies on Thurston's theorem on existence of conformally rigid domains which correspond to isolated components of $\mathbf{S} \setminus \mathbf{T}$ ([**Th**]; see also [**As**]).

Take such $\varphi_0 \in \mathbf{T}$. There exists an open neighborhood V of φ_0 in the topology of uniform convergence on compact subsets of Δ^*, such that for any $\varphi \in V$, the curve $\gamma(t)\varphi$ is not contained entirely in $\mathbf{S}$. (Otherwise, the curve (9.1) would be contained in $\mathbf{S}$, since $\mathbf{S}$ is closed in the topology of uniform convergence on compact sets.)

The domain $D_0^* = f_0(\Delta^*)$ determined by φ_0 can be approximated by a sequence of domains $D_n^* = \widehat{\mathbb{C}} \setminus P_n = f_n(\Delta^*)$, $n = 1, 2, \ldots$, where P_n are polygons with vertices on the boundary $f_0(S^1)$. Then D_0^* is the kernel of this sequence, and f_n converge to f_0 uniformly on compact sets in Δ^*; hence, we have the similar weak convergence of the Schwarzians S_{f_n} to S_{f_0}. This yields as in [**Kr3**] that the neighborhood V must meet the sets $\Omega_{S_{f_n}}$ for n sufficiently large; and consequently, these sets also are not connected. In other words, even for linear polygons, the corresponding sets Ω_φ can have more than one connected component.

References

[Ah1]   L.V. Ahlfors, *Remarks on the Neumann-Poincaré equation*, Pacific J. Math. **2** (1952), 271-280.

[Ah2]   L. V. Ahlfors, *Lectures on Quasiconformal Mappings*, Van Nostrand, Princeton, 1966.

[AW]   L.V. Ahlfors and G. Weill, *A uniqueness theorem for Beltrami equations*, Proc. Amer. Math. Soc. **13** (1962), 975-978.

[As]   K. Astala, *Self-similar zippers*, Holomorphic Functions and Moduli, Vol. I, (Berkeley, CA, 1986), Springer, New York, 1988, pp. 61-73.

[Ber]   L. Bers, *Automorphic forms and Poincaré series for infinitely generated Fuchsian groups*, Amer. J. Math. **87** (1965), 196-214.

[BR]   L. Bers and H.L. Royden, *Holomorphic families of injections*, Acta Math. **157** (1986), 259-286.

[Bu]   J. Burbea, *On the Hessian of the Carathéodory metric*, Rocky Mountain J. Math. **8** (1978), 555-559.

[Di]   S. Dineen, *The Schwarz Lemma*, Clarendon Press, Oxford, 1989.

[EKK]   C.J. Earle, I. Kra and S.L. Krushkal, *Holomorphic motions and Teichmüller spaces*, Trans. Amer. Math. Soc. **343** (1994), 927-948.

[EL]   C.J. Earle and and Z. Li, *Isometrically embedded polydisks in Teichmüller spaces*, J. Geom. Anal. **9** (1999), 51-71.

[FV]   T. Franzony and E. Vesentini, *Holomorphic Maps and Invariant Distances*, North-Holland, Amsterdam, 1980.

[GL]   F.P. Gardiner and N. Lakic, *Quasiconformal Teichmüller Theory*, Amer. Math. Soc., Providence, RI, 2000.

[Gr]   H. Grunsky, *Koeffizientenbediengungen für schlicht abbildende meromorphe Funktionen*, Math. Z. **45** (1939), 29-61.

[Ha]   R. Hamilton, *Extremal quasiconformal mappings with prescribed boundary values*, Trans. Amer. Math. Soc. **138** (1969), 399-406.

[Kl]   M. Klimek, *Pluripotential Theory*, Clarendon Press, Oxford, 1991.

[Ko]   S. Kobayayshi, *Hyperbolic Complex Spaces*, Springer, New York, 1998.

[Kr1]   S.L. Krushkal, *Quasiconformal Mappings and Riemann Surfaces*, Wiley, New York, 1979.

[Kr2]   S.L. Krushkal, *On the Grunsky coefficient conditions*, Siberian Math. J. **28** (1987), 104-110.

[Kr3]   S.L. Krushkal, *On the question of the structure of the universal Teichmüller space*, Soviet Math. Dokl. **38** (1989), 435-437.

[Kr4]   S.L. Krushkal, *Grunsky coefficient inequalities, Carathéodory metric and extremal quasiconformal mappings*, Comment. Math. Helv. **64** (1989), 650-660.

[Kr5]   S.L. Krushkal, *On Grunsky conditions, Fredholm eigenvalues and asymptotically conformal curves*, Mitt. Math. Sem. Giessen **228** (1996), 17-23.

[Kr6]   S.L. Krushkal, *Quasiconformal reflections across arbitrary planar sets*, Sci. Ser. A Math. Sci. (N.S.) (2002), 57-62.

[Kr7]   S.L. Krushkal, *Polygonal quasiconformal maps and Grunsky inequalities*, J. Anal. Math. **90** (2003), 175-196.

[Kr8]   S.L. Krushkal, *Grunsky inequalities of higher rank with applications to complex geometry and function theory*, Complex Analysis and Dynamical Systems, Contemp. Math. **364** (2004), 127-153.

[Kr9]   S.L. Krushkal, *Plurisubharmonic features of the Teichmüller metric*, Publ. Inst. Math. (Beograd) (N.S.) **75 (89)** (2004), to appear.

[KK]   S.L. Krushkal and R. Kühnau, *Quasikonforme Abbildungen - neue Methoden und Anwendungen*, Teubner, Leipzig, 1983.

[Ku1]   R. Kühnau, *Verzerrungssätze und Koeffizientenbedingungen vom Grunskyschen Typ für quasikonforme Abbildungen*, Math. Nachr. **48** (1971), 77-105.

[Ku2]   R. Kühnau, *Zur quasikonformen Fortsetzbarkeit schlichter konformer Abbildungen*, Bull. Soc. Sci. Lettres Lódz **24** (1974), no. 6 (1975).

[Ku3]   R. Kühnau, *Zu den Grunskyschen Koeffizientenbedingungen*, Ann. Acad. Sci. Fenn. Ser. A I Math. **6** (1981), 125-130.

[Ku4]   R. Kühnau, *Wann sind die Grunskyschen Koeffizientenbedingungen hinreichend für Q-quasikonforme Fortsetzbarkeit?*, Comment. Math. Helv. **61** (1986), 290-307.

[Ku5]   R. Kühnau, *Möglichst konforme Spiegelung an einer Jordankurve*, Jahresber. Deutsch. Math.-Verein. **90** (1988), 90-109.

[MSS]   R. Mañé, P. Sad and D. Sullivan, *On dynamics of rational maps*, Ann. Sci. Ecol. Norm. Sup. **16** (1983), 193-216.

[Mi]   D. Minda, *The strong form of Ahlfors' lemma*, Rocky Mountain J. Math. **17** (1987), 457-461.

[Ne]   Z. Nehari, *The Schwarzian derivative and schlicht functions*, Bull. Amer. Math. Soc. **55** (1949), 545-551.

[Po1]   Chr. Pommerenke, *Univalent Functions*, Vandenhoeck & Ruprecht, Göttingen, 1975.

[Po2] Chr. Pommerenke, *Boundary Behavior of Conformal Maps*, Springer, Berlin, 1992.
[RS]   E. Reich and K. Strebel, *Extremal quasiconformal mappings with given boundary values*, Contributions to Analysis, Academic Press, New York, 1974, pp. 375-391.
[Ro1] H.L. Royden, *Automorphisms and isometries of Teichmüller space*, Advances in the Theory of Riemann Surfaces, Princeton Univ. Press, Princeton, 1971, pp. 369-383.
[Ro2] H.L. Royden, *Complex Finsler metrics*, Complex Differential Geometry and Nonlinear Differential Equations, Contemp. Math. **49** (1986), 119-124.
[Sc]   M. Schiffer, *Fredholm eigenvalues and Grunsky matrices*, Ann. Polon. Math. **39** (1981), 149-164.
[Sl]   Z. Slodkowski, *Holomorphic motions and polynomial hulls*, Proc. Amer. Math. Soc. **111** (1991), 347-355.
[St]   K. Strebel, *On the existence of extremal Teichmüller mappings*, J. Anal. Math. **30** (1976), 464-480.
[ST]   D. Sullivan and W.P. Thurston, *Extending holomorphic motions*, Acta Math. **157** (1986), 243-257.
[Su]   M. Suzuki, *The intrinsic metrics on the domains in* $\mathbb{C}^n$, Math. Rep. Toyama Univ. **6** (1983), 143-177.
[Te]   O. Teichmüller, *Extremale quasikonforme Abbildungen und quadratische Differentiale*, Abh. Preuss. Akad. Wiss., Math.-Naturw. Kl., 1939 **22** (1940), 1-197.
[Th]   W.P. Thurston, *Zippers and univalent functions*, The Bieberbach Conjecture (West Lafayette, Ind., 1985), Amer. Math. Soc., Providence, RI, 1986, pp. 185-197.
[Wo]   B. Wong, *On the holomorphic curvature of some intrinsic metrics*, Proc. Amer. Math. Soc. **65** (1977), 57-61.
[Wu]   H. Wu, *A remark on holomorphic sectional curvature*, Indiana Univ. Math. J. **22** (1973), 1103-1107.
[Zh]   I. V. Zhuravlev, *Univalent functions and Teichmüller spaces*, Inst. of Mathematics, Novosibirsk, preprint, 1979, pp. 1-23 (Russian).

DEPARTMENT OF MATHEMATICS, BAR-ILAN UNIVERSITY, 52900 RAMAT-GAN, ISRAEL
E-mail address: krushkal@macs.biu.ac.il

On Evaluation of the Cauchy Principal Value of the Singular Cauchy-Szegö Integral in a Ball of $\mathbb{C}^n$

A. M. Kytmanov and S. G. Myslivets

This paper is dedicated to Professor Lawrence Zalcman on his sixtieth birthday.

ABSTRACT. It is shown that in the multidimensional case, the Cauchy principal value of the Cauchy-Szegö integral in the ball is equal to the boundary value of the Cauchy-Szegö integral inside the ball.

Let B be the unit ball of $\mathbb{C}^n$
$$B = \{z : |z| < 1\},$$
where $z = (z_1, z_2, \ldots, z_n)$, $|z|^2 = |z_1|^2 + \ldots + |z_n|^2$, and let $S = \{z : |z| = 1\}$ be its boundary.

Denote by $K(\zeta, z)$ the Cauchy-Szegö kernel for the ball, i.e.,
$$K(\zeta, z) = \frac{(n-1)!}{(2\pi i)^n} \cdot \frac{1}{(1 - \langle \zeta, z \rangle)^n},$$
where $\langle \zeta, z \rangle = \bar\zeta_1 z_1 + \ldots + \bar\zeta_n z_n$ is the Euclidean scalar product in $\mathbb{C}^n$. This can also be written in the form of a matrix product: if z is a column vector then $\langle \zeta, z \rangle = \bar\zeta' \cdot z$, where prime indicates the matrix transpose.

Denote by $\sigma(\zeta)$ the differential form
$$\sigma(\zeta) = \sum_{k=1}^{n} (-1)^{k-1} \bar\zeta_k d\bar\zeta[k] \wedge d\zeta,$$
where $d\zeta = d\zeta_1 \wedge \ldots \wedge d\zeta_n$, and $d\bar\zeta[k]$ is obtained from $d\bar\zeta$ by deletion of $d\bar\zeta_k$. The restriction of the form $\sigma(\zeta)$ to the boundary of the ball coincides (to within a constant) with Lebesgue measure on S.

The following assertion is well-known (see, for example, [1, 2]).

THEOREM 1. *If a function f is holomorphic in the ball B and continuous in the closure $\overline{B}$ (i.e., $g \in \mathcal{O}(B) \cap \mathcal{C}(\overline{B})$), then*

(1) $$f(z) = \int_S f(\zeta) K(\zeta, z) \sigma(\zeta), \quad z \in B.$$

2000 *Mathematics Subject Classification.* Primary 32A25.
Supported by grants RFFI 02-01-00167 and NSh-1212.2003.1.

Moreover, the integral operator defined by (1) gives the orthogonal projection of the Hilbert space $\mathcal{L}^2(S)$ onto the subspace of functions admitting holomorphic continuation from S into B.

The aim of our paper is to obtain the analogue of the Sokhotskij-Plemelj formula for the Cauchy-Szegö integral. Such formulas for the Bochner-Martinelli integral are well-known (see, for example, [**3**, Ch. 1]). They are entirely analogous to the classical Sokhotskij-Plemelj formula for the Cauchy integral in the complex plane. For the Henkin-Ramirez integral in strictly pseudoconvex domains D, similar formulas were given in [**4, 5**]; but there the principal value differ from the Cauchy principal value. For the ball, the Henkin-Ramirez integral coincides with the Cauchy-Szegö integral.

The Cauchy principal value of a multidimensional integral is defined as usual by

$$\text{v.p.} \int_S u(\zeta) K(\zeta,z)\,\sigma(\zeta) = \lim_{\varepsilon \to +0} \int_{S \setminus \{\zeta : |\zeta - z| < \varepsilon\}} u(\zeta) K(\zeta,z)\,\sigma(\zeta), \quad z \in S.$$

For the Henkin-Ramirez integral, the principal value in [**4, 5**] is defined by

$$\text{v.p.h.} \int_S u(\zeta) K(\zeta,z)\,\sigma(\zeta) = \lim_{\varepsilon \to +0} \int_{S \setminus \{\zeta : |1-\langle \zeta,z\rangle| < \varepsilon\}} u(\zeta) K(\zeta,z)\,\sigma(\zeta), \quad z \in S.$$

We show that the Cauchy principal value of the singular Cauchy-Szegö integral of 1 equals 1, not 1/2, as in [**4, 5**]. Thus, there exists an essential difference between the behavior of the Bochner-Martinelli and Cauchy-Szegö integrals in the multidimensional case.

THEOREM 2. *For $n > 1$,*

(2)
$$\text{v.p.} \int_S K(\zeta,z)\,\sigma(\zeta) = 1 \quad \text{for any point} \quad z \in S.$$

PROOF. To begin with, we show that it is sufficient to take the point z to be $(1, 0, \ldots, 0)$.

Let $\zeta = Aw$ be the linear transformation given by a unitary matrix A; then the differential form $\sigma(\zeta)$ is invariant under that transform

$$\sigma(\zeta) = \sigma(Aw) = \sigma(w).$$

In fact, $d\zeta = \det A\, dw$. In addition,

$$\sum_{k=1}^n (-1)^{k-1} \zeta_k d\zeta[k] = \sum_{k=1}^n (-1)^{k-1} \sum_{j=1}^n a_{jk} w_j \sum_{p=1}^n A_{pk} dw[k],$$

where the a_{jk} are the entries of the matrix A, and the A_{pk} are the minors of the matrix A corresponding to the entries a_{pk}. Since

$$\sum_{k=1}^n (-1)^{k-1} a_{jk} A_{pk} = \begin{cases} 0, & j \neq p, \\ (-1)^{p-1} \det A, & j = p, \end{cases}$$

we have

$$\sum_{k=1}^n (-1)^{k-1} \zeta_k d\zeta[k] = \det A \sum_{p=1}^n (-1)^{p-1} w_k dw[k].$$

From this and by properties of the unitary transform, we obtain
$$\sigma(\zeta) = |\det A|^2 \sigma(w) = \sigma(w).$$

Now
$$1 - \langle \zeta, z \rangle = 1 - \langle Aw, z \rangle = 1 - (\overline{Aw})' \cdot z = 1 - \bar{w}' \bar{A}' z$$
$$= 1 - \bar{w}' \cdot (\bar{A}' z) = 1 - \langle w, \bar{A}' z \rangle = 1 - \langle w, Az \rangle.$$

Thus, using the invariance of the sphere S under unitary transformations and the formulas above, we can transform any point $z \in S$ to the point $(1, 0, \ldots, 0)$. In addition, the integrals in the formula (2) do not change (the sets which are omitted from the sphere S are also unitarily invariant).

Thus it is sufficient to evaluate the integrals
$$I_\varepsilon = \frac{(n-1)!}{(2\pi i)^n} \int_{S \setminus U_\varepsilon} \frac{\sigma(\zeta)}{(1-\bar{\zeta}_1)^n},$$

where U_ε is the set $\{\zeta : |\zeta_1 - 1|^2 + |\zeta_2|^2 + \ldots + |\zeta_n|^2 < \varepsilon^2\}$, and then to find the limit of these integrals as $\varepsilon \to +0$. Observe that on S, $|\zeta_1 - 1|^2 + |\zeta_2|^2 + \ldots + |\zeta_n|^2 = |\zeta_1 - 1|^2 + 1 - |\zeta_1|^2 = 2 - 2\operatorname{Re}\zeta_1$, therefore, we can consider the set $\{\operatorname{Re}\zeta_1 > 1 - \varepsilon^2/2\}$ as the set U_ε.

Since $|\zeta|^2 = 1$ on S, we have
$$\sum_{k=1}^n \bar{\zeta}_k d\zeta_k + \sum_{k=1}^n \zeta_k d\bar{\zeta}_k = 0 \quad \text{on} \quad S.$$

Solving for $d\bar{\zeta}_n$ and substituting in $\sigma(\zeta)$, we obtain
$$(3) \qquad \sigma(\zeta) = \frac{(n-1)!}{(2\pi i)^n} \cdot \frac{(-1)^{n-1}}{\zeta_n} \cdot d\bar{\zeta}[n] \wedge d\zeta, \quad \zeta_n \neq 0.$$

Using Fubini's theorem and taking into account the orientation of S, we get
$$I_\varepsilon = \frac{(n-1)!}{(2\pi i)^n} \int_{S \setminus U_\varepsilon} \frac{1}{(1-\bar{\zeta}_1)^n} \cdot \frac{(-1)^{n-1}}{\zeta_n} \cdot d\bar{\zeta}[n] \wedge d\zeta$$
$$(4) \qquad = \frac{(n-1)!}{(2\pi i)^n} \int_{\{|\zeta_1|<1\} \setminus U'_\varepsilon} \frac{d\bar{\zeta}_1 \wedge d\zeta_1}{(1-\bar{\zeta}_1)^n} \int_{\{|\zeta_2|^2+\ldots+|\zeta_n|^2=1-|\zeta_1|^2\}} \frac{d\bar{\zeta}[1,n] \wedge d\zeta[1]}{\zeta_n},$$

where $U'_\varepsilon \subset \mathbb{C}$ is the set $\{\operatorname{Re}\zeta_1 > 1 - \varepsilon^2/2\}$.

Using (3) and Cauchy's formula, we evaluate the interior integral in (4):
$$\int_{\{|\zeta_2|^2+\ldots+|\zeta_n|^2=1-|\zeta_1|^2\}} \frac{d\bar{\zeta}[1,n] \wedge d\zeta[1]}{\zeta_n}$$
$$= \int_{\{|\zeta_2|^2+\ldots+|\zeta_{n-1}|^2<1-|\zeta_1|^2\}} d\bar{\zeta}[1,n] \wedge d\zeta[1,n] \int_{\{|\zeta_n|^2=1-(|\zeta_1|^2+\ldots+|\zeta_{n-1}|^2)\}} \frac{d\zeta_n}{\zeta_n}$$
$$= 2\pi i \int_{\{|\zeta_2|^2+\ldots+|\zeta_{n-1}|^2<1-|\zeta_1|^2\}} d\bar{\zeta}[1,n] \wedge d\zeta[1,n] = \frac{(2\pi i)^{n-1}}{(n-2)!}(1-|\zeta_1|^2)^{n-2}.$$

Here we have used that the volume of the $(2n-4)$-dimensional unit ball is $\dfrac{\pi^{n-2}}{(n-2)!}$.

Thus, we get

$$(5) \qquad I_\varepsilon = \frac{n-1}{2\pi i} \int\limits_{\{|\zeta_1|<1\}\setminus U'_\varepsilon} \frac{(1-|\zeta_1|^2)^{n-2}}{(1-\bar\zeta_1)^n} d\bar\zeta_1 \wedge d\zeta_1.$$

To evaluate I_ε, we introduce the polar coordinates $\zeta_1 = re^{i\varphi}$, where as usual $r > 0$, $0 \leqslant \varphi \leqslant 2\pi$. Then $d\bar\zeta_1 \wedge d\zeta_1 = 2ir\, dr \wedge d\varphi$. Write $I_\varepsilon = I^1_\varepsilon + I^2_\varepsilon$, where

$$I^1_\varepsilon = \frac{n-1}{2\pi i} \int_0^{1-\varepsilon^2/2} 2ir\, dr \int_0^{2\pi} \frac{(1-r^2)^{n-2}}{(1-re^{-i\varphi})^n} d\varphi$$

$$= \frac{n-1}{\pi i} \int_0^{1-\varepsilon^2/2} (1-r^2)^{n-2} dr \int_0^{2\pi} \frac{e^{i(n-1)\varphi}}{(e^{i\varphi}-r)^n} de^{i\varphi}.$$

By Cauchy's formula,

$$\int_0^{2\pi} \frac{e^{i(n-1)\varphi}}{(e^{i\varphi}-r)^n} de^{i\varphi} = 2\pi i;$$

therefore,

$$I^1_\varepsilon = 2(n-1) \int_0^{1-\varepsilon^2/2} (1-r^2)^{n-2} r\, dr = -\int_0^{1-\varepsilon^2/2} d(1-r^2)^{n-1} = 1 - \left(1 - \left(1 - \frac{\varepsilon^2}{2}\right)^2\right)^{n-1},$$

so $I^1_\varepsilon \to 1$ as $\varepsilon \to +0$.

For the second integral, we have

$$I^2_\varepsilon = \frac{n-1}{2\pi i} \int_{1-\varepsilon^2/2}^{1} 2ir\, dr \int_{\arccos \frac{1-\varepsilon^2/2}{r}}^{2\pi - \arccos \frac{1-\varepsilon^2/2}{r}} \frac{(1-r^2)^{n-2}}{(1-re^{-i\varphi})^n} d\varphi =$$

$$= \frac{n-1}{\pi i} \int_{1-\varepsilon^2/2}^{1} (1-r^2)^{n-2} r\, dr \int_{\arccos \frac{1-\varepsilon^2/2}{r}}^{2\pi - \arccos \frac{1-\varepsilon^2/2}{r}} \frac{e^{i(n-1)\varphi}}{(e^{i\varphi}-r)^n} de^{i\varphi}.$$

It is easy to find a primitive of the interior integral. Write $w = e^{i\varphi}$ to get

$$\frac{w^{n-1}}{(w-r)^n} = \frac{((w-r)+r)^{n-1}}{(w-r)^n} = \frac{\sum_{k=0}^{n-1} C_{n-1}^k r^k (w-r)^{n-1-k}}{(w-r)^n} = \sum_{k=0}^{n-1} \frac{C_{n-1}^k r^k}{(w-r)^{k+1}},$$

where C_{n-1}^k are binomial coefficients. It follows that

$$(6) \qquad \int \frac{w^{n-1}}{(w-r)^n} dw = \ln(w-r) - \sum_{k=1}^{n-1} \frac{1}{k} \frac{C_{n-1}^k r^k}{(w-r)^k} + C,$$

where $\ln(w-r)$ is the value of logarithm with cross-cut on the non-negative real axis.

Denote by ζ_r the point on a line $\mathrm{Re}\,\zeta = 1 - \varepsilon^2/2$ with argument φ_r which is equal to $\cos\dfrac{1-\varepsilon^2/2}{r}$. Then for the first term of (6), we have

$$\ln(\bar{w}_r - r) - \ln(w_r - r) = \ln\frac{\bar{w}_r - r}{w_r - r} = \ln\left|\frac{\bar{w}_r - r}{w_r - r}\right| + i\arg\frac{\bar{w}_r - r}{w_r - r} = i\arg\frac{\bar{w}_r - r}{w_r - r}.$$

Hence
$$|\ln(\bar{w}_r - r) - \ln(w_r - r)| \leqslant 4\pi.$$

Therefore,
$$\int_{1-\varepsilon^2/2}^{1} (1-r^2)^{n-1}|\ln(\bar{w}_r - r) - \ln(w_r - r)|\,dr \to 0 \quad \text{as} \quad \varepsilon \to +0.$$

For the terms $\dfrac{1}{k}\dfrac{C_n^k r^k}{(w-r)^k}$ with $k < n-1$, we have the estimate

$$\left|\frac{1-r^2}{w_r - r}\right| \leqslant 1 + r.$$

Thus
$$\left|\frac{1}{k} \cdot \frac{C_{n-1}^k (1-r^2)^{n-2}}{(w_r - r)^k}\right| \leqslant C_1,$$

(for some constant C_1), and the integrals of such functions over the interval $[1-\varepsilon^2/2, 1]$ tend to 0 as $\varepsilon \to +0$ ($k = 1, \ldots, n-2$).

Considering the last term of (6), we have

$$\left|\frac{1}{n-1} \cdot \frac{(1-r^2)^{n-2}}{(w_r - r)^{n-1}}\right| \leqslant C_2 \frac{1}{|w_r - r|} = C_2 \frac{1}{\sqrt{r^2 + \varepsilon^2 - 1}}.$$

Now
$$\int_{1-\varepsilon^2/2}^{1} \frac{dr}{\sqrt{r^2 + \varepsilon^2 - 1}} = \ln(r + \sqrt{r^2 + \varepsilon^2 - 1})\bigg|_{1-\varepsilon^2/2}^{1} = \ln(1+\varepsilon) \to 0 \quad \text{as} \quad \varepsilon \to +0.$$

It follows that $I_\varepsilon^2 \to 0$ as $\varepsilon \to +0$. $\square$

As is shown in [4, 5],
$$\text{v.p.h.} \int_S K(\zeta, z)\,\sigma(\zeta) = \frac{1}{2} \quad \text{for} \quad z \in S.$$

We check this fact in the two-dimensional case. Consider the integral

$$J_\varepsilon = \frac{1}{(2\pi i)^2} \int_{S\setminus\{\zeta: |1-\langle\zeta,z\rangle|<\varepsilon\}} \frac{\sigma(\zeta)}{(1-\langle\zeta,z\rangle)^2}, \quad z \in S.$$

Using the unitary invariance of the set of integration, we may (as in Theorem 2), take the point z to be $(1,0)$, i.e.,

$$J_\varepsilon = \frac{1}{(2\pi i)^2} \int_{S\setminus\{\zeta: |1-\bar\zeta_1|<\varepsilon\}} \frac{\sigma(\zeta)}{(1-\bar\zeta_1)^2}, \quad z \in S.$$

Performing the calculation as in Theorem 2, we reduce the integral J_ε to the form

$$J_\varepsilon = \frac{1}{2\pi i} \int\limits_{\{\zeta_1 : |\zeta_1| < 1,\ |\zeta_1 - 1| \geq \varepsilon\}} \frac{d\bar\zeta_1 \wedge d\zeta_1}{(1 - \bar\zeta_1)^2}.$$

Thus

$$M_\varepsilon = I_\varepsilon - J_\varepsilon = \frac{1}{2\pi i} \int\limits_{S_\varepsilon} \frac{d\bar\zeta_1 \wedge d\zeta_1}{(1 - \bar\zeta_1)^2},$$

where $S_\varepsilon = \{\zeta_1 : |\zeta_1 - 1| < \varepsilon,\ \operatorname{Re}\zeta_1 \leq 1 - \varepsilon^2/2\}$.

To calculate M_ε, we change variables $\zeta_1 = 1 - w = 1 - re^{i\psi}$. Then $d\bar\zeta_1 \wedge d\zeta_1 = 2ir\, dr \wedge d\psi$, and

$$M_\varepsilon = \frac{1}{\pi}\int\limits_{S_\varepsilon} \frac{1}{r} e^{2i\psi}\, dr \wedge d\psi = \frac{1}{\pi} \int\limits_{\varepsilon^2/2}^{\varepsilon} \frac{dr}{r} \int\limits_{-\arcsin\sqrt{1-\frac{\varepsilon^4}{4r^2}}}^{\arcsin\sqrt{1-\frac{\varepsilon^4}{4r^2}}} \cos 2\psi\, d\psi$$

$$= \frac{1}{\pi}\int\limits_{\varepsilon^2/2}^{\varepsilon} \frac{dr}{r}\cdot \sin\left(2\arcsin\sqrt{1 - \frac{\varepsilon^4}{4r^2}}\right) = \frac{1}{\pi}\int\limits_{\varepsilon^2/2}^{\varepsilon}\frac{dr}{r}\cdot\sqrt{1 - \frac{\varepsilon^4}{4r^2}}\cdot\frac{\varepsilon^2}{r}$$

$$= \frac{\varepsilon^2}{2\pi}\int\limits_{\varepsilon^2/2}^{\varepsilon}\frac{\sqrt{4r^2 - \varepsilon^4}}{r^3}\,dr.$$

Changing variables $r = \varepsilon t$ in the last integral, we get

$$M_\varepsilon = \frac{\varepsilon}{2\pi}\int\limits_{\varepsilon/2}^{1}\frac{\sqrt{4t^2 - \varepsilon^2}}{t^3}\,dt = -\frac{\varepsilon}{4\pi}\int\limits_{\varepsilon/2}^{1}\sqrt{4t^2 - \varepsilon^2}\cdot d\frac{1}{t^2}$$

$$= -\frac{\varepsilon\sqrt{4-\varepsilon^2}}{4\pi} + \frac{\varepsilon}{\pi}\int\limits_{\varepsilon/2}^{1}\frac{dt}{t\sqrt{4t^2 - \varepsilon^2}}.$$

Setting $t = 1/x$ in the last integral, we have

$$M_\varepsilon = -\frac{\varepsilon\sqrt{4-\varepsilon^2}}{4\pi} + \frac{\varepsilon}{\pi}\int\limits_{1}^{2/\varepsilon}\frac{dx}{\sqrt{4 - \varepsilon^2 x^2}} = -\frac{\varepsilon\sqrt{4-\varepsilon^2}}{4\pi} + \frac{1}{\pi}\int\limits_{1}^{2/\varepsilon}\frac{dx}{\sqrt{4/\varepsilon^2 - x^2}}$$

$$= -\frac{\varepsilon\sqrt{4-\varepsilon^2}}{4\pi} + \frac{1}{\pi}\arcsin\frac{\varepsilon x}{2}\Big|_1^{2/\varepsilon} = -\frac{\varepsilon\sqrt{4-\varepsilon^2}}{4\pi} + \frac{1}{2} - \frac{1}{\pi}\arcsin\frac{\varepsilon}{2}.$$

Thus, $M_\varepsilon \to 1/2$ as $\varepsilon \to 0$.

Denote by $K^+[u]$ for a function u integrable on S the boundary value of the integral

$$\int\limits_S u(\zeta) K(\zeta, z)\sigma(\zeta)$$

from within the ball B, and by $K_s[u]$ the Cauchy principal value of that integral, i.e.,

$$K_s[u] = \text{v.p.} \int u(\zeta) K(\zeta, z) \sigma(\zeta), \quad z \in S.$$

COROLLARY 1. *Let $n > 1$. Let u satisfy a Hölder condition with exponent α, $0 < \alpha \leqslant 1$ on S. Then $K^+[u]$ extends to a function on S satisfying on S a Hölder condition with some exponent β, $0 < \beta \leqslant 1$; moreover, $K_s[u]$ exists and*

(7) $$K^+[u] = K_s[u].$$

The formula (7) is an analogue of the Sokhotskij-Plemelj formula for the Cauchy-Szegö integral in a ball. It is different from the analogous formula in [**4, 5**], which considers a different principal value of the singular Cauchy-Szegö integral.

PROOF. For points $\zeta, z \in S$, write $d(\zeta, z) = |1 - \langle \zeta, z \rangle|$. As is shown in [**5**],

$$C_1 d(\zeta, z) \leqslant |\zeta - z| \leqslant C_2 \sqrt{d(\zeta, z)}.$$

It follows that a function u, satisfying a Hölder condition with respect to the Euclidean metric $|\zeta - z|$, also satisfies a Hölder condition for the metric $d(\zeta, z)$, but with another Hölder exponent.

Consider the integral

$$\Phi(z) = \int_S (u(\zeta) - u(z)) K(\zeta, z) \sigma(\zeta).$$

As shown in [**4**] (Theorem 1), this integral is a function satisfying a Hölder condition for the metric $d(\zeta, z)$ in the closure of the ball $\overline{B}$. Then, on the one hand,

$$\Phi(z) = K^+[u](z) - u(z), \quad z \in S,$$

while, on the other hand (by Theorem 2),

$$\Phi(z) = K_s[u](z) - u(z).$$

□

Thus, the principal values of the singular Cauchy-Szegö integral depend on the definition. Consider the same problem for the Bochner-Martinelli integral. We show that for it, the principal values v.p and v.p.h coincide. Recall that the Bochner-Martinelli kernel has the form

$$U(\zeta, z) = \frac{(n-1)!}{(2\pi i)^n} \cdot \frac{\sum_{k=1}^n (-1)^{k-1} (\bar{\zeta}_k - \bar{z}_k) d\bar{\zeta}[k] \wedge d\zeta}{|\zeta - z|^{2n}}.$$

It is well-known (see, for example, [**3**, Ch. 1]) that

$$\text{v.p.} \int_S U(\zeta, z) = \frac{1}{2}, \quad z \in S.$$

THEOREM 3. *For any point $z \in S$,*

(8) $$\text{v.p.h.} \int_S U(\zeta, z) = \frac{1}{2}.$$

PROOF. Since the Bochner-Martinelli kernel is invariant under unitary transformations (see [**3**, Ch. 1]), we can assume (as above) that the point $z = z^0 = (1, 0, \ldots, 0)$.

On the sphere S, $|\zeta_1 - 1|^2 + |\zeta_2|^2 + \ldots + |\zeta_n|^2 = |\zeta_1 - 1|^2 + 1 - |\zeta_1|^2 = 2(1 - \operatorname{Re} \zeta_1)$.

Expressing on S the differential $d\bar{\zeta}_n$ through other differentials as in formula (3), we obtain

$$U(\zeta, z^0) = \frac{(n-1)!}{(2\pi i)^n} \cdot \frac{(-1)^{n-1}}{\zeta_n} \cdot \frac{(1-\zeta_1) d\bar{\zeta}[n] \wedge d\zeta}{(2(1-\operatorname{Re}\zeta_1))^n}, \quad \zeta_n \neq 0.$$

Consider the integral

$$B_\varepsilon = \int_{S \setminus \{\zeta : |1-\zeta_1| < \varepsilon\}} U(\zeta, z^0).$$

Performing the calculations (as in Theorem 2), we get

$$B_\varepsilon = \frac{n-1}{2\pi i} \int_{\{\zeta_1 : |\zeta_1| \leqslant 1\} \setminus \{\zeta_1 : |1-\zeta_1| < \varepsilon\}} \frac{(1-\zeta_1)(1-|\zeta_1|^2)^{n-2}}{2^n (1-\operatorname{Re}\zeta_1)^n} d\bar{\zeta}_1 \wedge d\zeta_1.$$

The given integral differs from the integral corresponding to the Cauchy principal value v.p. by the integral

$$N_\varepsilon = \frac{n-1}{2\pi i} \int_{S_\varepsilon} \frac{(1-\zeta_1)(1-|\zeta_1|^2)^{n-2}}{2^n (1-\operatorname{Re}\zeta_1)^n} d\bar{\zeta}_1 \wedge d\zeta_1,$$

where $S_\varepsilon = \{\zeta_1 : |\zeta_1 - 1| < \varepsilon, \operatorname{Re} \zeta_1 \geqslant 1 - \varepsilon^2/2\}$.

Changing variables $\zeta_1 = 1 - w = 1 - re^{i\psi}$, we have $d\bar{\zeta}_1 \wedge d\zeta_1 = 2ir\, dr \wedge d\psi$ and

$$N_\varepsilon = \frac{n-1}{2^n \pi} \int_{S_\varepsilon} \frac{w(1-|1-w|^2)^{n-2} \cdot r\, dr \wedge d\psi}{(\operatorname{Re} w)^n} = \frac{n-1}{2^n \pi} \int_{S_\varepsilon} \frac{(2\cos\psi - r)^{n-2}\, dr \wedge d\psi}{\cos^{n-1}\psi}.$$

Putting $r = \varepsilon R$, we obtain

$$N_\varepsilon = \frac{n-1}{2^{n-1}\pi} \int_{S'_\varepsilon} \frac{\varepsilon(2\cos\psi - \varepsilon R)^{n-2}\, dR \wedge d\psi}{\cos^{n-1}\psi},$$

where $S'_\varepsilon = \{(R, \psi) : 0 \leqslant R \leqslant 1, \psi \geqslant 0, R\cos\psi \geqslant \varepsilon/2\}$.

The integral in the last formula is the sum of terms of the form

$$C_j \int_{S'_\varepsilon} \frac{\varepsilon^{j+1} R^j\, dR \wedge d\psi}{\cos^{j+1}\psi}, \quad 0 \leqslant j \leqslant n-2.$$

On the set S'_ε, we have $\dfrac{\varepsilon}{\cos\psi} \leqslant 2$; hence

$$0 \leqslant \int_{S'_\varepsilon} \frac{\varepsilon^{j+1} R^j\, dR \wedge d\psi}{\cos^{j+1}\psi} \leqslant \varepsilon 2^j \int_{S'_\varepsilon} \frac{dR \wedge d\psi}{\cos\psi}.$$

In what follows,

$$\varepsilon \int_{S'_\varepsilon} \frac{dR \wedge d\psi}{\cos \psi} = \varepsilon \int_{\varepsilon/2}^{1} dR \int_{0}^{\arcsin \sqrt{1-\frac{\varepsilon^2}{4R^2}}} \frac{d\psi}{\cos \psi} = \frac{\varepsilon}{2} \int_{\varepsilon/2}^{1} \ln \frac{1+\sqrt{1-\frac{\varepsilon^2}{4R^2}}}{1-\sqrt{1-\frac{\varepsilon^2}{4R^2}}} dR$$

$$= \varepsilon \int_{\varepsilon/2}^{1} \ln \frac{2R + \sqrt{4R^2 - \varepsilon^2}}{\varepsilon} dR \leqslant \varepsilon \int_{\varepsilon/2}^{1} \ln \frac{4}{\varepsilon} dR \to 0 \quad \text{as} \quad \varepsilon \to +0.$$

Therefore, $N_\varepsilon \to 0$ as $\varepsilon \to +0$. Thus

$$\text{v.p.h.} \int_S U(\zeta, z) = \text{v.p.} \int_S U(\zeta, z) = \frac{1}{2},$$

and formula (8) is proved. □

Thus, the Sokhotskij-Plemelj formulas for the Bochner-Martinelli integral are the same as those given by the Cauchy principal value.

References

[1] G.M. Henkin, *The method of integral representations in complex analysis*, Several Complex Variables I, Encyclopaedia Math. Sci., **7**, Springer-Verlag, 1990, pp. 19–116.
[2] W. Rudin, *Function Theory in the Unit Ball of* $\mathbb{C}^n$, Springer-Verlag, New York–Berlin, 1980.
[3] A.M. Kytmanov, *The Bochner-Martinelli Integral and Its Applications*, Birkhäuser Verlag, Basel, 1995.
[4] W. Alt, *Singuläre Integrale mit gemischten Homogenitäten auf Mannigfaltigkeiten und Anwendungen in der Funktionentheorie*, Math. Z. **137** (1974), 227–256.
[5] N. Kerzman and E.M. Stein, *The Szegö kernel in terms of Cauchy-Fantappiè kernels*, Duke Math. J. **45** (1978), 197–224.

DEPARTMENT OF MATHEMATICS AND COMPUTER SCIENCE, KRASNOYARSK STATE UNIVERSITY, 660041 KRASNOYARSK, RUSSIA
  E-mail address: kytmanov@lan.krasu.ru

DEPARTMENT OF MATHEMATICS AND COMPUTER SCIENCE, KRASNOYARSK STATE UNIVERSITY, 660041 KRASNOYARSK, RUSSIA
  E-mail address: simona@lan.krasu.ru

Boundary Properties of Convex Functions

Adam Lecko

ABSTRACT. The aim of this paper is to apply the method of the Julia lemma to study boundary behaviour of convex functions.

1. Introduction

In the sequence of papers [1], [3-5], [9-13], the Julia lemma was applied as a basic tool to study classes of univalent functions having certain geometrical properties connected with a boundary point (starlikeness, spirallikeness, convexity in one direction). Alternative proofs of well-known classical results or new formulas showed the importance of the Julia lemma as a technique in the geometric theory of univalent functions. This technique allows one to avoid the approximation methods usually used to solve such problems. In [9-13], the Julia lemma was applied directly by the construction of self-mappings of the unit disk $\mathbb{D}$ and a suitable sequence of points in $\mathbb{D}$. In [1], [3-5], observing independently that certain self-mappings of $\mathbb{D}$ form one-parameter continuous semigroups in $\mathbb{D}$, the authors developed the technique of applying dynamical systems and the theory of semigroups to the theory of univalent functions.

In this paper, we indicate how the method of the Julia lemma applied for convexity, that most classical geometrical property of plane domains, encounters certain difficulties. In Theorem 3.1, we prove that convexity is preserved on each horocycle by every convex function. As in our earlier works, we apply the method of the Julia lemma by constructing a suitable self-mapping of $\mathbb{D}$. Unfortunately, the proof of Theorem 3.1 is long and complicated; however, the result can be proved quite easily if we use Robertson's result, as is demonstrated in Theorem 3.3.

2. Preliminaries

1. Let $\mathbb{C}$ be the complex plane and let $\overline{\mathbb{C}} = \mathbb{C} \cup \{\infty\}$. For $z_0 \in \mathbb{C}$ and $r > 0$, let $\mathbb{D}(z_0, r) = \{z \in \mathbb{C} : |z - z_0| < r\}$. Let $\mathbb{D} = \mathbb{D}(0,1)$ and $\mathbb{T} = \partial \mathbb{D}$. Let $\mathbb{B}(\xi, \rho) = \{z \in \mathbb{D} : |(z-\xi)/(1-\overline{\xi}z)| < \tanh \rho\}$, $\xi \in \mathbb{D}$, $\rho > 0$, denote the hyperbolic open disk in $\mathbb{D}$ with hyperbolic center at ξ and hyperbolic radius ρ. For each $k > 0$,

---

2000 *Mathematics Subject Classification.* Primary 30C45.
Key words and phrases. convex functions, Julia lemma, boundary property.

consider the horocycle
$$\mathbb{O}_k = \left\{ z \in \mathbb{D} : \frac{|1-z|^2}{1-|z|^2} < k \right\}.$$
Let $O_k = \partial \mathbb{O}_k \setminus \{1\}$. In the next sections we use the following parametrization of O_k:

(2.1) $$O_k : \quad z = o_k(\theta) = \frac{1+ke^{i\theta}}{1+k}, \qquad \theta \in (0, 2\pi),$$

so the arc O_k is positively oriented.

For $\zeta \subset \mathbb{T}$ and $b \subset (0, \pi/2)$, let $\Delta(\zeta, b) = \{ z \in \mathbb{D} : |\arg(1-\bar{\zeta}z)| < b, |z - \zeta| < \rho \}$, $\rho < 2\cos b$, denote the Stolz angle at ζ. When the angle $2b$ of $\Delta(\zeta, b)$ is not relevant, we write $\Delta(\zeta)$ for short. We also write Δ instead of $\Delta(1)$.

2. Let $\mathcal{A}$ denote the set of all analytic functions in $\mathbb{D}$. The subset of $\mathcal{A}$ of all univalent functions is denoted by $\mathcal{S}$, and the subset of $\mathcal{A}$ of all self-mappings of $\mathbb{D}$ is denoted by $\mathcal{B}$.

For each $\xi \in \mathbb{D}$, let $\varphi_\xi(z) = (z-\xi)/(1-\bar{\xi}z)$, $z \in \mathbb{C} \setminus \{1/\bar{\xi}\}$. We denote the set of all analytic automorphisms of $\mathbb{D}$ by $\mathrm{Aut}(\mathbb{D})$.

3. Let $f \in \mathcal{A}$ and $\zeta \in \mathbb{T}$. By $f_\angle(\zeta) \in \overline{\mathbb{C}}$ we denote the angular limit of f at ζ. If f has an unrestricted limit $v \in \overline{\mathbb{C}}$ at ζ, then we write $f(\zeta)$. Let $f'_\angle(\zeta) \in \overline{\mathbb{C}}$ denote the angular derivative of f at ζ. Let us recall that f is conformal at ζ if $f'_\angle(\zeta)$ exists and is different from zero and infinity.

Assume that the finite radial limit $\lim_{r \to 1^-} f(r\zeta) = v$ exists at $\zeta \in \mathbb{T}$. The Visser-Ostrowski quotient of f at ζ (e.g., [**16**], p. 251) is defined as
$$Q_f(z;\zeta) = \frac{(z-\zeta)f'(z)}{f(z)-v}, \qquad z \in \mathbb{D}.$$

Let $Q_f(z;1) = Q_f(z)$. Recall that f satisfies the Visser-Ostrowski condition at $\zeta \in \mathbb{T}$ if $\lim_{\Delta(\zeta) \ni z \to \zeta} Q_f(z;\zeta) = 1$ for every Stolz angle $\Delta(\zeta)$ (see, e.g., [**16**], p. 252).

4. The Julia lemma ([**8**]; [**18**], pp. 68-72) recalled below is the basis for our considerations.

LEMMA 2.1 (JULIA). *Let $\omega \in \mathcal{B}$. Assume that there exists a sequence (z_n) of points in $\mathbb{D}$ such that*

(2.2) $$\lim_{n \to \infty} z_n = 1, \qquad \lim_{n \to \infty} \omega(z_n) = 1$$

and

(2.3) $$\lim_{n \to \infty} \frac{1-|\omega(z_n)|}{1-|z_n|} = \lambda < \infty.$$

Then $\lambda > 0$ and

(2.4) $$\frac{|1-\omega(z)|^2}{1-|\omega(z)|^2} \leq \lambda \frac{|1-z|^2}{1-|z|^2}, \qquad z \in \mathbb{D},$$

i.e., for every $k > 0$,
$$\omega(\mathbb{O}_k) \subset \mathbb{O}_{\lambda k}.$$
Equality in (2.4) for one $z \in \mathbb{D}$ can occur only for $\omega \in \mathrm{Aut}(\mathbb{D})$.

For $\omega \in \mathcal{B}$ with angular limit $\omega_\angle(\zeta)$ at $\zeta \in \mathbb{T}$, let

$$\Lambda(\zeta) = \sup \left\{ \frac{|\omega_\angle(\zeta) - \omega(z)|^2}{1 - |\omega(z)|^2} \cdot \frac{1 - |z|^2}{|\zeta - z|^2} : z \in \mathbb{D} \right\}.$$

Let $\Lambda = \Lambda(1)$. The lemma below in the another version Julia's lemma, also called the Julia-Wolff lemma ([**16**], p. 82).

LEMMA 2.2. *Let $\omega \in \mathcal{B}$. If $\omega_\angle(\zeta)$ at $\zeta \in \mathbb{T}$ exists and $\omega_\angle(\zeta) \in \mathbb{T}$, then $\omega'_\angle(\zeta)$ exists and*

$$0 < \zeta \frac{\omega'_\angle(\zeta)}{\omega_\angle(\zeta)} = \Lambda(\zeta) \leq \infty.$$

REMARK 2.1. In what follows, we assume that $\zeta = \omega_\angle(\zeta) = 1$. Then from Lemma 2.2, $\omega'_\angle(1)$ exists and

(2.5) $$0 < \omega'_\angle(1) = \Lambda \leq \infty.$$

Moreover,

$$\lim_{\Delta \ni z \to 1} \frac{1 - \omega(z)}{1 - z} = \lim_{\Delta \ni z \to 1} \omega'(z) = \omega'_\angle(1) = \Lambda$$

for every Stolz angle Δ (see [**6**], pp. 42-44).

For our needs, it is convenient to define the following class of functions.

DEFINITION 2.1. Fix $\lambda \in (0, \infty]$. We say that $\omega \in \mathcal{B}$ belongs to the class $\mathcal{B}(\lambda)$ if $\omega_\angle(1) = 1$ and (2.5) holds with $\Lambda = \lambda$.

Let $\mathcal{P}(\lambda)$ denote the class of all functions p of the form

$$p(z) = 4 \frac{1 - \omega(z)}{1 + \omega(z)}, \qquad z \in \mathbb{D},$$

where $\omega \in \mathcal{B}(\lambda)$.

REMARK 2.2. The assumptions of Lemma 2.1 imply that $\omega_\angle(1) = 1$. It then follows from Lemma 2.2 that $\omega'_\angle(1) = \Lambda$. Therefore, every $\omega \in \mathcal{B}$ satisfying (2.2) and (2.3) belongs to $\mathcal{B}(\Lambda)$. Clearly, $\Lambda \leq \lambda$, where λ is given by (2.3).

3. Convex functions

1. Let $\mathcal{K}$ denote the set of all convex domains in $\mathbb{C}$, and let $\mathcal{C}$ denote the class of all convex functions, i.e., all $f \in \mathcal{S}$ such that $f(\mathbb{D}) \in \mathcal{K}$. Clearly, we can assume that $0 \in \Omega$ for every $\Omega \in \mathcal{K}$. By $\mathcal{C}(\xi)$, $\xi \in \mathbb{D}$, we denote the set of all $f \in \mathcal{C}$ such that $f(\xi) = 0$.

If $f \in \mathcal{C}$ is bounded, then $\partial f(\mathbb{D})$ is a Jordan curve since it is convex. Therefore, by the well-known Carathéodory theorem (see, e.g., [**16**], p. 24), f has a continuous injective extension to $\overline{\mathbb{D}}$. If $f(\mathbb{D})$ is unbounded, the same holds if we use the spherical metric instead of the euclidean metric. Thus, for every point $\zeta \in \mathbb{T}$, an unrestricted limit $\lim_{z \to \zeta} f(z) = f(\zeta) \in \overline{\mathbb{C}}$ exists. Let us also recall the fact below needed for our considerations (see [**16**], p. 84).

LEMMA 3.1. *If $f \in \mathcal{C}$ is bounded, then $0 \neq f'_\angle(\zeta) \in \overline{\mathbb{C}}$ exists for all $\zeta \in \mathbb{T}$.*

2. The theorem below shows that for convexity the method of the Julia lemma creates some difficulties.

THEOREM 3.1. *Let* $f \in \mathcal{S}$ *and suppose* $f'_\angle(1) \neq \infty$ *exists. Then* $f \in \mathcal{C}$ *if and only if* $f(\mathbb{O}_k) \in \mathcal{K}$ *for every* $k > 0$.

PROOF. Assume that $f \in \mathcal{C}$. Fix $k > 0$ and let $v_1, v_2 \in f(\mathbb{O}_k)$ be such that $v_1 \neq v_2$. There exist $\xi_1, \xi_2 \in \mathbb{O}_k$ such that $v_1 = f(\xi_1)$ and $v_2 = f(\xi_2)$. Since $\xi_i \in \partial \mathbb{O}_{k_i}$ for $k_i = |1 - \xi_i|^2/(1 - |\xi_i|^2) > 0$, $i = 1, 2$, respectively, we can assume that $k_1 \geq k_2$. This means that $\xi_2 \in \overline{\mathbb{O}}_{k_1}$.

Fix $t \in (0,1)$ and let $w_1 = (1-t)v_1 + tv_2$. Since f is convex, we can find $z_1 \in \mathbb{D}$ such that $w_1 = f(z_1)$. To prove the assertion, we show that $z_1 \in \mathbb{O}_k$, i.e., $w_1 \in f(\mathbb{O}_k)$.

Let $u = \varphi_{-\xi_2} \circ (\kappa \varphi_{\xi_1})$, where $\kappa = \varphi_{\xi_2}(1)/\varphi_{\xi_1}(1) \in \mathbb{T}$. Then $u(\xi_1) = \xi_2$ and $u(1) = 1$. Moreover,

$$(3.1) \qquad u'(1) = \varphi'_{-\xi_2}(\varphi_{\xi_2}(1))\kappa \varphi'_{\xi_1}(1) = \frac{|1-\xi_2|^2}{1-|\xi_2|^2} \cdot \frac{1-|\xi_1|^2}{|1-\xi_1|^2} = \frac{k_2}{k_1} \leq 1.$$

For each $t \in [0,1]$, define

$$(3.2) \qquad \omega_t(z) = f^{-1}((1-t)f(z) + tf(u(z))), \qquad z \in \mathbb{D}.$$

The convexity of $f(\mathbb{D})$ implies that $(1-t)f(z) + tf(u(z)) \in f(\mathbb{D})$ for every $t \in [0,1]$ and $z \in \mathbb{D}$, and the univalence of f shows that ω_t is well-defined for each $t \in [0,1]$. Clearly, $\omega_t(\mathbb{D}) \subset \mathbb{D}$. Since f and u are continuous on $\overline{\mathbb{D}}$, so is each ω_t, i.e., for every $\zeta \in \mathbb{T}$, an unrestricted limit $\omega_t(\zeta) \in \overline{\mathbb{D}}$ exists. We show that ω_t satisfies the assumption of the Julia lemma.

Let $I = [0,1)$. Since u is a Möbius transformation of $\overline{\mathbb{D}}$ onto itself, either $u(I)$ is a circular arc joining $\varphi_{-\xi_2}(-\kappa \xi_1)$ with 1, orthogonal to $\mathbb{T}$ at 1, or $u(I) \subset I$. Clearly, I and $u(I)$ are tangent at 1. Let Δ be any Stolz angle. Thus $I_1 = I \cap \Delta$ and $u(I) \cap \Delta$ are arcs in Δ ending at 1. Since f has finite angular derivative $f'_\angle(1)$, by ([**16**], Proposition 4.7), f' has finite angular limit at 1 equal $f'_\angle(1)$. Hence f' has the asymptotic value $f'_\angle(1)$ along $I \cap \Delta$ and $u(I) \cap \Delta$. Equivalently,

$$(3.3) \qquad \lim_{r \to 1^-} f'(u(r)) = f'_\angle(1).$$

Now we prove that

$$(3.4) \qquad \lim_{r \to 1^-} f'(\omega_t(r)) = f'_\angle(1)$$

for every $t \in [0,1]$, i.e., we prove that f' has the asymptotic value $f'_\angle(1)$ along the curve $\omega_t(I) \cap \Delta$.

The cases $t = 0$ and $t = 1$ reduce to (3.3). We prove (3.4) for $t \in (0,1)$. For each $t \in [0,1]$, let us consider arcs $\Gamma_t = (1-t)f(I) + tf(u(I))$ parametrized by

$$\Gamma_t : w = \phi_t(r) = (1-t)f(r) + tf(u(r)), \qquad r \in [0,1).$$

In particular, $\Gamma_0 = f(I)$ and $\Gamma_1 = f(u(I))$ are parametrized by ϕ_0 and ϕ_1, respectively. Clearly, each Γ_t ends at the origin. In view of (3.1), (3.3) and Lemma 3.1, the one-sided tangent vector to Γ_t at the origin equals

$$\lim_{r \to 1^-} \phi'_t(r) = (1-t)\lim_{r \to 1^-} f'(r) + t \lim_{r \to 1^-}\{f'(u(r))u'(r)\}$$

$$= (1 - t + tu'(1)) f'_\angle(1) = \left(1 - t + t\frac{k_2}{k_1}\right) f'_\angle(1).$$

Since $1-t+tk_2/k_1$ is a positive real number, we see that all arcs Γ_t, $t \in [0,1]$, are mutually tangent at the origin.

Let $\Delta(1,b_1)$ and $\Delta(1,b_2)$, b_1, $b_2 \in (0,\pi/2)$, be two arbitrary Stolz angles such that $\Delta(1,b_1) \subset \Delta(1,b_2)$. Clearly, $\Delta(1,b_1)$ is the Jordan domain bounded by two segments J_1 and J_2 that intersect at 1 and form the angle b_1 with the segment I, and by a circular arc. Since f has the unrestricted limit 0 at 1, it has the asymptotic value 0 along every curve ending at 1. Therefore, $f(J_1)$ and $f(J_2)$ are smooth arcs ending at the origin, and $f(\Delta(1,b_1))$ is a Jordan domain. Since f is conformal at 1, it is isogonal there. Then, by ([**16**], Proposition 4.10), smooth arcs in each Stolz angle are mapped onto smooth arcs, and the angles between arcs are preserved. Since J_1 and J_2 lie in $\Delta(1,b_2)$, the arcs $f(J_1)$ and $f(J_2)$ lying in $f(\Delta(1,b_2))$ form the angle b_1 with the smooth arc $\Gamma_0 = f(I)$ and, consequently, with every smooth arc Γ_t. But, clearly, $f(I) \cap f(\Delta(1,b_1))$ is an arc in $f(\Delta(1,b_1))$, which yields that $\Gamma_t \cap f(\Delta(1,b_1))$ is a nonempty set and is a smooth arc ending at the origin for every $t \in [0,1]$. By ([**16**], Proposition 2.14), each $\gamma_t = f^{-1}(\Gamma_t)$ is a curve in $\mathbb{D}$ ending at 1. Hence each $f^{-1}(\Gamma_t \cap f(\Delta(1,b_1))) = \gamma_t \cap \Delta(1,b_1)$ is a curve in $\Delta(1,b_1)$ ending at 1 and tangent to I there. Thus f' along $\gamma_t \cap \Delta(1,b_1)$ exists and

$$\lim_{r \to 1^-} f'(\omega_t(r)) = \lim_{\omega_t(I) \ni z \to 1} f'(z)$$
$$= \lim_{\gamma_t(I) \ni z \to 1} f'(z) = \lim_{\gamma_t \cap \Delta(1,b_1) \ni z \to 1} f'(z) = f'_\angle(1),$$

which proves (3.4).

Since $(\omega_t)'_\angle(1)$ exists, by Proposition 4.7 ([**16**], p. 79) it equals the asymptotic value ω'_t along I. Thus from (3.2), (3.3) and (3.4), we have

$$(\omega_t)'_\angle(1) = \lim_{r \to 1^-} \omega'_t(r) = (1-t) \lim_{r \to 1^-} \frac{f'(z)}{f'(\omega_t(r))} + tu'(1) \lim_{r \to 1^-} \frac{f'(u(r))}{f'(\omega_t(r))}$$
$$= 1 - t + t\frac{k_2}{k_1} = \Lambda \leq 1$$

for every $t \in [0,1]$.

Hence each ω_t satisfies the assumptions of the Julia lemma, with $\lambda(t) = \Lambda = 1 - t + tk_2/k_1 \leq 1$ for every $t \in (0,1)$. In consequence,

(3.5) $$\omega_t(\mathbb{O}_k) \subset \mathbb{O}_{\lambda(t)k} \subset \mathbb{O}_k$$

for every $k > 0$. Since

$$f(\omega_t(\xi_1)) = (1-t)f(\xi_1) + tf(u(\xi_1)) = (1-t)v_1 + tv_2 = w_1,$$

we have $\omega_t(\xi_1) = z_1$. As $\xi_1 \in \mathbb{O}_k$, from (3.5) we deduce that $z_1 \in \mathbb{O}_k$, so $f(z_1) \in f(\mathbb{O}_k)$. This implies that $f(\mathbb{O}_k) \in \mathcal{K}$.

Since $f(\mathbb{D}) = \bigcup_{k>0} f(\mathbb{O}_k)$, the assumption that $f(\mathbb{O}_k) \in \mathcal{K}$ for every $k > 0$ yields $f(\mathbb{D}) \in \mathcal{K}$ and proves the converse. $\square$

3. Theorem 3.1 can be proved quite easily if we apply the following well-known result due to Robertson [**17**].

THEOREM 3.2. *A convex function maps every disk in $\mathbb{D}$ onto a convex domain.*

Using Theorem 3.2, we have

THEOREM 3.3. *Let $f \in \mathcal{S}$. Then $f \in \mathcal{C}$ if and only if $f(\mathbb{O}_k) \in \mathcal{K}$ for every $k > 0$.*

PROOF. Fix $k > 0$. Let $w_1, w_2 \in f(\mathbb{O}_k)$. Thus $w_1 = f(z_1)$ and $w_2 = f(z_2)$ for some z_1 and z_2 lying in $\mathbb{O}_k$. Clearly, there exists a disk $\mathbb{D}(z, r) \subset \mathbb{O}_k$ such that $\{z_1, z_2\} \subset \mathbb{D}(z, r)$. Since, in view of Theorem 3.2, we have $f(\mathbb{D}(z, r)) \in \mathcal{K}$, $[w_1, w_2] \subset f(\mathbb{D}(z, r)) \subset f(\mathbb{O}_k)$. Hence $f(\mathbb{O}_k) \in \mathcal{K}$. □

The normalization of convex functions plays no role in our considerations. Looking at Theorem 3.2, we see that a convex function maps euclidean and hyperbolic disks in $\mathbb{D}$ onto convex domains independently of its normalization in $\mathbb{D}$. Therefore, by applying Theorem 3.2, the analytic formula for convex functions may depend on each $\xi \in \mathbb{D}$, i.e., may be expressed for the class $\mathcal{C}(\xi)$. Since each $\xi \in \mathbb{D}$ is the center of the family of hyperbolic disks $\mathbb{B}(\xi, \rho)$, $\rho > 0$, $f(\mathbb{D}) = \bigcup_{\rho > 0} f(\mathbb{B}(\xi, \rho))$, and each $f(\mathbb{B}(\xi, \rho))$ is convex, applying Study's result (see, e.g., [**15**], p. 44) to the convex function $f \circ \varphi_{-\xi}$ leads at once to the following result.

THEOREM 3.4. *Fix $\xi \in \mathbb{D}$ and let $f \in \mathcal{A}$ with $f'(\xi) \neq 0$. Then $f \in \mathcal{C}$ if and only if*

$$\operatorname{Re}\left\{1 - 2\bar{\xi}z + |\xi|^2 + (1 - \bar{\xi}z)(z - \xi)\frac{f''(z)}{f'(z)}\right\} > 0, \qquad z \in \mathbb{D}.$$

Clearly, the simplest form appears when $\xi = 0$, i.e., we have Study's characterization of convex functions. The method of proof of Study's theorem may be based on the Schwarz lemma ([**15**], pp. 44-45).

In fact, the result of Theorem 3.3 is known. The remark below was reported to the author by Prof. T. Carroll.

REMARK 3.1. Heins [**7**] raised and solved the problem of finding a necessary and sufficient condition on subsets X of $\mathbb{D}$ for their images under every convex function to be convex domains. Heins proved that this property for convex functions holds only for so called D-convex sets; $X \subset \mathbb{D}$ is called *D-convex* if for any a and b in X the intersection of the two disks, having boundary circles passing through a and b and tangent to $\mathbb{T}$, lies in X. One can observe that every horoycle is a D-convex set, so Theorem 3.3 is part of Heins's result. Barnard [**2**] generalized Heins's result to α-starlike functions. In addition, he proved that a subset of $\mathbb{D}$ is D-convex if and only if it is an arbitrary intersection of sets having oricycles as boundaries.

Using Theorem 3.3, we can characterize convex functions analytically through their property of mapping horocycles onto convex domains. Define $\mathcal{B}(0) = \{\omega \equiv 1\}$.

THEOREM 3.5. *If $f \in \mathcal{C}$, then there exists $\omega \in \mathcal{B}(\lambda)$, $\lambda \in [0, 1]$, such that*

$$(3.6) \qquad 2(1 - z) - (1 - z)^2 \frac{f''(z)}{f'(z)} = 4\frac{1 - \omega(z)}{1 + \omega(z)}, \qquad z \in \mathbb{D}.$$

PROOF. First, we show that

$$\operatorname{Re}\left\{2(1 - z) - (1 - z)^2 \frac{f''(z)}{f'(z)}\right\} > 0, \qquad z \in \mathbb{D}.$$

Fix $k > 0$, and use the parametrization (2.1) of O_k. By Theorem 3.3, $f(\mathbb{O}_k) \in \mathcal{K}$, which yields $(d/d\theta)\arg\{(f \circ o_k)'(\theta)\} \geq 0$ for $\theta \in (0, 2\pi)$. Using (2.1), we have

$$(1 - o_k(\theta))^2 = \frac{k^2}{(1+k)^2}\left(1 - e^{i\theta}\right)^2 = -\frac{4k\sin^2(\theta/2)}{(k+1)i}\left(\frac{k}{k+1}e^{i\theta}i\right)$$

$$= \frac{4k\sin^2(\theta/2)}{k+1}o'_k(\theta)i = 2\mathrm{Re}\{1 - o_k(\theta)\}o'_k(\theta)i, \qquad \theta \in (0, 2\pi).$$

Since the arc O_k is positively oriented, so is $f(O_k)$, as f is a conformal mapping. Hence we have

$$\frac{d}{d\theta}\arg\{o'_k(\theta)f'(o_k(\theta)))\} = \frac{d}{d\theta}\mathrm{Im}\,\log\{o'_k(\theta)f'(o_k(\theta))\}$$

$$= \mathrm{Im}\left\{\frac{o''_k(\theta)}{o'_k(\theta)} + o'_k(\theta)\frac{f''(o_k(\theta))}{f'(o_k(\theta))}\right\}$$

$$= \mathrm{Im}\left\{i + \frac{k+1}{4k\sin^2(\theta/2)i}(1 - o_k(\theta))^2\frac{f''(o_k(\theta))}{f'(o_k(\theta))}\right\}$$

$$= 1 - \frac{1}{2\mathrm{Re}\{1 - o_k(\theta)\}}\mathrm{Re}\left\{(1 - o_k(\theta))^2\frac{f''(o_k(\theta))}{f'(o_k(\theta))}\right\} \geq 0$$

for $\theta \in (0, 2\pi)$. Since any $z \in \mathbb{D}$ belongs to O_k for some k,

$$(3.7) \qquad \mathrm{Re}\left\{2(1-z) - (1-z)^2\frac{f''(z)}{f'(z)}\right\} \geq 0, \qquad z \in \mathbb{D}$$

for $z \in \mathbb{D}$.

Assume now that equality holds in (3.7) for some $z_0 \in \mathbb{D}$. By the maximum principle for harmonic functions, it holds in the whole disk $\mathbb{D}$, which implies that

$$(3.8) \qquad 2(1-z) - (1-z)^2\frac{f''(z)}{f'(z)} \equiv yi, \qquad z \in \mathbb{D},$$

for some $y \in \mathbb{R}$. For $y = 0$, the solution of (3.8) $f(z) = f_0(z) = az/(1-z) + b$, $a \in \mathbb{C} \setminus \{0\}$, $b \in \mathbb{C}$, belongs to $\mathcal{C}$. For $y \neq 0$, the solution of (3.8)

$$f(z) = f_1(z) = \frac{ai}{y}\left(\exp\left(\frac{yiz}{1-z}\right) - 1\right) + b, \qquad z \in \mathbb{D},$$

where $a \in \mathbb{C} \setminus \{0\}$ and $b \in \mathbb{C}$, is not univalent in $\mathbb{D}$, so $f_1 \notin \mathcal{C}$. For every $f \in \mathcal{C}$ different from f_0, strict inequality holds in (3.7) in $\mathbb{D}$.

Let $f \in \mathcal{C}$ and $p(z) = 2(1-z) - (1-z)^2 f''(z)/f'(z)$, and let

$$(3.9) \qquad \omega(z) = \frac{4 - p(z)}{4 + p(z)}, \qquad z \in \mathbb{D}.$$

If $f = f_0$, then f satisfies (3.6) with $\omega \equiv 1$, i.e., with $\omega \in \mathcal{B}(0)$.

Assume that $f \neq f_0$. Then $\omega(\mathbb{D}) \subset \mathbb{D}$. We now prove that $\omega \in \mathcal{B}(\lambda)$ for some $\lambda \in (0, 1]$.

Since f is a convex univalent function,

$$(3.10) \qquad \left|\frac{f''(z)}{f'(z)}\right| \leq \frac{4}{1 - |z|^2}, \qquad z \in \mathbb{D}$$

(see, e.g., [**19**]). Thus
$$|1-z|^2 \left|\frac{f''(z)}{f'(z)}\right| \le 4\frac{|1-z|^2}{1-|z|^2}.$$

Hence
$$\lim_{r\to 1^-} p(r) = \lim_{r\to 1^-}\left\{2(1-r) - (1-r)^2\frac{f''(r)}{f'(r)}\right\} = 0.$$

Now (3.9) yields $\lim_{r\to 1^-} \omega(r) = 1$ so the condition (2.2) of the Julia lemma is satisfied.

In view of (3.10), for every $r \in (0,1)$,
$$(1-r)\left|\frac{f''(r)}{f'(r)}\right| \le \frac{4}{1+r}.$$

Therefore, there exists $\lambda_0 \in [0,1]$ and a sequence (r_n) in $(0,1)$ with $\lim_{n\to\infty} r_n = 1$ such that
$$\lim_{n\to\infty}\left\{(1-r_n)\left|\frac{f''(r_n)}{f'(r_n)}\right|\right\} = 2\lambda_0.$$

Hence we have
$$\lim_{n\to\infty} \frac{|1-\omega(r_n)|}{1-r_n} = \lim_{n\to\infty}\left\{\frac{2}{|4+p(r_n)|}\left|\frac{p(r_n)}{1-r_n}\right|\right\}$$
$$= \lim_{n\to\infty}\left\{\frac{2}{|4+p(r_n)|}\left|2-(1-r_n)\frac{f''(r_n)}{f'(r_n)}\right|\right\} = 1 - \lambda_0 \in [0,1].$$

But $1 - |\omega(r_n)| \le |1 - \omega(r_n)|$, so we can find a subsequence (r_{n_k}) such that
$$\lim_{k\to\infty}\frac{1-|\omega(r_{n_k})|}{1-r_{n_k}} = \lambda \le 1 - \lambda_0.$$

By Lemma 2.1, $\lambda > 0$. Hence ω satisfies the assumptions of the Julia lemma, and by Remark 2.2 it belongs to $\mathcal{B}(\lambda)$ with $0 < \lambda \le \lambda_1$.

This ends the proof of the theorem. $\square$

REMARK 3.2. The above proof shows that $(d/d\theta)\arg\{o'_k(\theta)f'(o_k(\theta))\} > 0$ if and only if $\text{Re}\left\{2(1-z) - (1-z)^2 f''(z)/f'(z)\right\} > 0$ in $\mathbb{D}$.

3.1. If $f \in \mathcal{C}$, then there exists $\lambda \in [0,1]$ such that
$$\lim_{\Delta\ni z\to 1}\frac{(1-z)f''(z)}{f'(z)} = 2(1-\lambda)$$

for every Stolz angle Δ.

PROOF. Since, in view of Theorem 3.5,
$$\frac{(1-z)f''(z)}{f'(z)} = 2 - \frac{4}{1+\omega(z)}\frac{1-\omega(z)}{1-z}, \qquad z \in \mathbb{D},$$

for some $\omega \in \mathcal{B}(\lambda)$, $\lambda \in [0,1]$, and
$$\lim_{\Delta\ni z\to 1}\frac{1-\omega(z)}{1-z} = \omega'_\angle(1) = \Lambda \in [0,1]$$

exists for every Stolz angle Δ, the assertion holds with $\lambda = \Lambda$. $\square$

THEOREM 3.6. *Let $f \in \mathcal{A}$ have a finite unrestricted limit at 1. If there exists $\omega \in \mathcal{B}$ such that*

$$\lim_{z \to 1} \frac{1 - \omega(z)}{1 - z} = \lambda, \tag{3.11}$$

for some $\lambda \in [0, 1]$ and (3.6) holds, then $f \in \mathcal{C}$.

PROOF. The case $\lambda = 0$ is obvious. Let $\lambda \in (0, 1]$. Using Theorem 3.3, it is enough to show that f is univalent in $\mathbb{D}$ and $f(\mathbb{O}_k) \in \mathcal{K}$ for every $k > 0$.

Also, by Remark 3.2, $(d/d\theta)\arg\{o'_k(\theta)f'(o_k(\theta))\} > 0$, $\theta \in (0, 2\pi)$, where o_k is given by (2.1). Now we prove that $f(\mathbb{O}_k) \in \mathcal{K}$ for every $k > 0$. The method of proof comes from [**9**].

By monotonicity, the right-hand and left-hand limits

$$\varphi_k = \lim_{\theta \to 0^+} \arg\{o'_k(\theta)f'(o_k(\theta))\}, \qquad \psi_k = \lim_{\theta \to 2\pi^-} \arg\{o'_k(\theta)f(o_k(\theta))\}$$

(possibly infinite) both exist and $\psi_k - \varphi_k > 0$ (see [**14**], Theorem 1). We prove that $\psi_k - \varphi_k \leq 2\pi$ for every $k > 0$.

First we introduce some notation. For every $k > 0$ and $\varepsilon \in (0, 2k/(k+1))$, consider two arcs in $\mathbb{D}$:

$$O_{k,\varepsilon} = O_k \cap (\mathbb{D} \setminus \mathbb{D}(1, \varepsilon)), \qquad \sigma_{k,\varepsilon} = \mathbb{O}_k \cap \partial \mathbb{D}(1, \varepsilon).$$

Let z_1 and z_2 be the end points of the closed arc $O_{k,\varepsilon}$. The same points are the end points of the open arc $\sigma_{k,\varepsilon}$ so $O_{k,\varepsilon} \cup \sigma_{k,\varepsilon}$ is a closed curve in $\mathbb{D}$. We use the parametrization $\sigma_{k,\varepsilon} : z = \sigma(t) = 1 + \varepsilon e^{it}$, $t \in (t(k,\varepsilon), 2\pi - t(k,\varepsilon))$, where $t(k,\varepsilon) \in (\pi/2, \pi)$ is such that

$$\tan t(k,\varepsilon) = -\sqrt{\left(\frac{2k}{\varepsilon(1+k)}\right)^2 - 1}.$$

Hence $z_1 = \sigma(t(k,\varepsilon))$ and $z_2 = \sigma(2\pi - t(k,\varepsilon))$ The arc $O_{k,\varepsilon}$ is parametrized by (2.1), where $\theta \in (\theta(k,\varepsilon), 2\pi - \theta(k,\varepsilon))$, with $\theta(k,\varepsilon) \in (0, \pi/2)$ being such that

$$\cos \theta(k,\varepsilon) = 1 - \frac{1}{2}\left(\varepsilon \frac{1+k}{k}\right)^2.$$

Moreover, $z_1 = o_k(\theta(k,\varepsilon))$ and $z_2 = o_k(2\pi - \theta(k,\varepsilon))$.

In view of (3.11),

$$\lim_{z \to 1} \left(\frac{4}{1+\omega(z)} \frac{1-\omega(z)}{1-z}\right) = 2\lambda \in (0, 2]. \tag{3.12}$$

(a) Assume first that $\lambda \in (0, 1)$. Fix $k > 0$. Then by (3.12), there exists $\delta \in (0, 2k/(k+1))$ such that

$$\left|\frac{4}{1+\omega(z)} \frac{1-\omega(z)}{1-z}\right| \leq 2 \tag{3.13}$$

for $z \in \mathbb{D} \cap \mathbb{D}(1, \delta)$. Take $\varepsilon \in (0, \delta)$. Then $\mathbb{D}(1, \varepsilon) \subset \mathbb{D}(1, \delta)$. Since $O_{k,\varepsilon} \cup \sigma_{k,\varepsilon}$ is a closed curve lying in $\mathbb{D}$, from (3.13) we have

(3.14)
$$\arg\{(f \circ o_k)'(2\pi - \theta(k, \varepsilon))\} - \arg\{(f \circ o_k)'(\theta(k, \varepsilon))\}$$

$$= \int_{\theta(k,\varepsilon)}^{2\pi - \theta(k,\varepsilon)} \frac{d}{d\theta} \arg(f \circ o_k)'(\theta) d\theta$$

$$= \operatorname{Im} \int_{\theta(k,\varepsilon)}^{2\pi - \theta(k,\varepsilon)} \frac{d}{d\theta} \log\{o_k'(\theta) f'(o_k(\theta))\} d\theta$$

$$= 2(\pi - \theta(k, \varepsilon)) - \operatorname{Im} \int_{\theta(k,\varepsilon)}^{2\pi - \theta(k,\varepsilon)} o_k'(\theta) \frac{f''(o_k(\theta))}{f'(o_k(\theta))} d\theta$$

$$= 2(\pi - \theta(k, \varepsilon)) + \operatorname{Im} \int_{O_{k,\varepsilon}} \frac{f''(z)}{f'(z)} dz$$

$$= 2(\pi - \theta(k, \varepsilon)) + 2\operatorname{Im} \int_{O_{k,\varepsilon}} \frac{dz}{1-z} - \operatorname{Im} \int_{O_{k,\varepsilon}} \frac{4(1 - \omega(z))}{(1 + \omega(z))(1-z)^2} dz$$

$$= 2(\pi - \theta(k, \varepsilon)) - 4(\pi - 2t(k, \varepsilon)) + \int_{t(k,\varepsilon)}^{2\pi - t(k,\varepsilon)} \operatorname{Re}\left\{\frac{4(1 - \omega(\sigma(t)))}{(1 + \omega(\sigma(t)))(1 - \sigma(t))}\right\} dt$$

$$\leq 2(\pi - \theta(k, \varepsilon)) - 4(\pi - t(k, \varepsilon)) + 4(\pi - t(k, \varepsilon)) < 2\pi.$$

Since $\lim_{\varepsilon \to 0^+} t(k, \varepsilon) = \pi/2$ and $\lim_{\varepsilon \to 0^+} \theta(k, \varepsilon) = 0$, the above yields

(3.15)
$$0 < \psi_k - \varphi_k$$

$$= \lim_{\varepsilon \to 0^+} (\arg\{(f \circ o_k)'(2\pi - \theta(k, \varepsilon))\} - \arg\{(f \circ o_k)'(\theta(k, \varepsilon))\}) \leq 2\pi.$$

Since, as was observed, the function $(0, 2\pi) \ni \theta \mapsto \arg(f \circ o_k)'(\theta)$ is strictly increasing on O_k, from (3.15) we deduce that $f(\mathbb{O}_k) \in \mathcal{K}$. Also (3.15) implies that f is injective on O_k. But $f(1)$ exists and f is analytic in $\overline{\mathbb{O}}_k \setminus \{1\}$, so it has a continuous extension to $\overline{\mathbb{O}}_k$. Thus f is continuous and injective on $O_k \cup \{1\}$. Applying Corollary 9.5 of [15], we deduce that f is univalent in $\mathbb{O}_k$. Since this is true for every $k > 0$, f is univalent in $\mathbb{D}$.

(b) Let now $\lambda = 1$. Fix $k > 0$. For $n \in \mathbb{N}$, there exists $\delta(n) \in (0, 2k/(k+1))$ such that
$$\left|\frac{4}{1 + \omega(z)} \frac{1 - \omega(z)}{1 - z}\right| \leq 2 + \frac{1}{n}$$
for $z \in \mathbb{D} \cap \mathbb{D}(1, \delta(n))$. For each $n \in \mathbb{N}$, take $\varepsilon(n) \in (0, \delta(n))$ in such a way that $\lim_{n \to \infty} \varepsilon(n) = 0$. Clearly, $\mathbb{D}(1, \varepsilon(n)) \subset \mathbb{D}(1, \delta(n))$ for every $n \in \mathbb{N}$. Let $O_{k,\varepsilon(n)}$ and

$\sigma_{k,\varepsilon(n)}$ be defined as earlier. Since $O_{k,\varepsilon(n)} \cup \sigma_{k,\varepsilon(n)}$ is a closed curve lying in $\mathbb{D}$, we have, as in (3.19),

$$\arg\left\{(f \circ o_{k,\varepsilon(n)})'(2\pi - \theta(k,\varepsilon))\right\} - \arg\left\{(f \circ o_{k,\varepsilon(n)})'(\theta(k,\varepsilon))\right\}$$
$$\leq 2(\pi - \theta(k,\varepsilon(n))) - 4(\pi - t(k,\varepsilon(n))) + 2\left(2 + \frac{1}{n}\right)(\pi - t(k,\varepsilon(n)));$$

and we complete the proof of univalence as before. $\square$

REMARK 3.2. Since

$$\frac{1 - \omega(z)}{1 - z} = \frac{2}{4 + p(z)} \frac{p(z)}{1 - z}, \qquad z \in \mathbb{D},$$

the conditions $\omega(1) = 1$ and (3.11) can be translated as $p(1) = 0$ and

$$\lim_{z \to 1} \frac{p(z)}{1 - z} = 2\lambda \in [0, 2].$$

3. Now we present some examples.

EXAMPLES. 1. $f(z) = (1 + az)/(1 - az)$, $z \in \mathbb{D}$, $|a| \leq 1$.
Then $p(z) = 2(1-a)(1-z)/(1-az)$. Hence $\operatorname{Re} p(z) > 0$, $z \in \mathbb{D}$, and $p(1) = 0$. Since $\lim_{z \to 1} p(z)/(1-z) = 2$, $f \in \mathcal{C}$ by Theorem 3.6.

2. $f(z) = ((1+z)/(1-z))^\alpha$, $z \in \mathbb{D}$, $\alpha \in \mathbb{R}$.
Then $p(z) = 2(1-a)(1-z)/(1-az)$. Hence $\operatorname{Re} p(z) > 0$, $z \in \mathbb{D}$, and $p(1) = 0$. Since $\lim_{z \to 1} p(z)/(1-z) = 2$, $f \in \mathcal{C}$ by Theorem 3.6.

3. $f(z) = e^z$, $z \in \mathbb{D}$.
Then $p(z) = 1 - z^2$. So $\operatorname{Re} p(z) > 0$, $z \in \mathbb{D}$, $p(1) = 0$, and $\lim_{z \to 1} p(z)/(1-z) = 2$. Hence $f \in \mathcal{C}$ by Theorem 3.6.

4. $f(z) = \log\{(1+z)/(1-z)\}$, $z \in \mathbb{D}$.
Then $p(z) = 2(1-z^2)/(1+z^2)$. Hence $\operatorname{Re} p(z) > 0$, $z \in \mathbb{D}$, and $p(1) = 0$. Since $\lim_{z \to 1} p(z)/(1-z) = 2$, $f \in \mathcal{C}$ by Theorem 3.6.

5. $f(z) = 1 - z^n$, $z \in \mathbb{D}$, $n \in \mathbb{N}$.
Then $p(z) = (1-z)((n+1)z - (n-1))/z$. Hence we see that p is analytic in $\mathbb{D}$ only for $n = 1$ since then $p(z) = 2(1-z)$. Then $p(1) = 0$, $\operatorname{Re} p(z) > 0$, $z \in \mathbb{D}$, and $\lim_{z \to 1} p(z)/(1-z) = 2$. Consequently, $f \in \mathcal{C}$ by Theorem 3.4. For no $n \geq 2$ does the function f belong to $\mathcal{C}$.

References

[1] D. Aharonov, M. Elin, and D. Shoikhet, *Spiral-like functions with respect to a boundary point*, J. Math. Anal. Appl. **280** (2003), 17-29.
[2] R. W. Barnard, *A generalization of Study's theorem on convex maps*, Proc. Amer. Math. Soc. **72** (1978), 127-134.
[3] M. Elin, S. Reich, and D. Shoikhet, *Holomorphically accretive mappings and spiral-shaped functions of proper contractions*, Nonlinear Anal. Forum **5** (2000), 149-161.
[4] M. Elin, S. Reich, and D. Shoikhet, *Dynamics of inequalities in geometric function theory*, J. Inequal. Appl. **6** (2001), 651-664.
[5] M. Elin and D. Shoikhet, *Dynamic extension of the Julia-Wolff-Carathéodory Theorem*, Dynamic Systems Appl. **10** (2001), 421-438.

[6] J. B. Garnett, *Bounded Analytic Functions*, Academic Press, 1981.
[7] M. Heins, *On a theorem of Study concerning conformal maps with convex images*, Mathematical Essays Dedicated to A. J. Macintyre, Ohio Univ. Press, Athens, Ohio, 1970, pp. 171-176.
[8] G. Julia, *Extension nouvelle d'un lemme de Schwarz*, Acta Math. **42** (1918), 349-355.
[9] A. Lecko, *On the class of functions starlike with respect to a boundary point*, J. Math. Anal. Appl. **261** (2001), 649-664.
[10] _____, *On the class of functions convex in the negative direction of the imaginary axis*, J. Aust. Math. Soc. **72** (2002), 1-10.
[11] _____, *The class of functions convex in the negative direction of the imaginary axis of order* (α, β), Int. J. Math. Math. Sci. **29** (2002), 641-650.
[12] _____, *On the class of functions defined in a halfplane and starlike with respect to the boundary point*, Ann. Polon. Math. **79** (2002), 67-83.
[13] A. Lecko and A. Lyzzaik, *A note on the class of functions starlike with respect to a boundary point*, J. Math. Anal. Appl. **282** (2003), 846-851.
[14] S. Lojasiewicz, *Introduction to the Theory of Univalent Fuctions*, PWN, Warszawa, 1976. (Polish)
[15] Ch. Pommerenke, *Univalent Functions*, Vandenhoeck & Ruprecht, Göttingen, 1975.
[16] _____, *Boundary Behaviour of Conformal Maps*, Springer-Verlag, Berlin, Heidelberg, New York, 1992.
[17] M. S. Robertson, *On the theory of univalent functions*, Ann. of Math. (2) **37** (1936), 374-408.
[18] G. Sansone and J. Gerretsen, *Lectures on the Theory of Functions of a Complex Variable. II: Geometric Theory*, Wolters-Noordhoff, Groningen, 1969.
[19] S. Yamashita, *Norm estimates for function starlike or convex of order alpha*, Hokkaido Math. J. **28** (1998), 217-230.

DEPARTMENT OF MATHEMATICS, TECHNICAL UNIVERSITY OF RZESZÓW, UL. W. POLA 2, 35-959 RZESZÓW, POLAND

E-mail address: `alecko@prz.rzeszow.pl`

Regularization of a Solution to the Cauchy Problem for the System of Thermoelasticity

O. I. Makhmudov and I. E. Niyozov

Dedicated to Lawrence Zalcman's 60th birthday

We consider the problem of regularization of a solution of the system of thermo-elasticity equations in an unbounded three-dimensional domain, starting from the solution values and the values of the stress on a part of the boundary, i.e., the Cauchy problem.

The system of thermoelasticity equations is elliptic; the Cauchy problem for elliptic equations is unstable with respect to small variations of the data, i.e., it is an ill-posed problem. For ill-posed problems, one does not prove the existence theorem; existence is assumed a priori. Moreover, the solution is assumed to belong to some given subset of the function space, usually a compact one [3]. The uniqueness of the solution then follows from the general Holmgren theorem [7].

Let $x = (x_1, x_2, x_3)$, $y = (y_1, y_2, y_3)$ be points in three-dimensional Euclidean space E^3 and suppose that the domain $D \subset E^3$ lies in the strip $0 < y_3 < h$, $h = \frac{\pi}{\rho}$, $\rho > 0$, and the boundary of D consists of a piece Σ of the plane $y_3 = 0$ and a smooth surface S given by the equation $y_3 = F(y_1, y_2)$, which satisfies the conditions $0 < F(y_1, y_2) \leq h$, $|\operatorname{grad} F(y_1, y_2)| \leq \operatorname{const} < \infty$, $(y_1, y_2) \in E^2$, i.e., $\partial D = \Sigma \cup S$.

Consider the matrix differential operator

$$B(\partial_x, \omega) = \|B_{mj}(\partial_x, \omega)\|_{4\times 4},$$

$$B_{mj}(\partial_x, \omega) = (1 - \delta_{m4})(1 - \delta_{4j}) \left(\delta_{mj} \mu (\Delta(\partial_x) + \frac{\rho \omega^2}{\mu}) + (\lambda + \mu) \frac{\partial^2}{\partial x_m \partial x_j} \right)$$

$$- \delta_{4j}(1 - \delta_{m4}) \gamma \frac{\partial}{\partial x_m} + i\omega \eta \, \delta_{m4}(1 - \delta_{4j}) \frac{\partial}{\partial x_j} + \delta_{m4} \delta_{4j} \left(\Delta(\partial_x) + \frac{i\omega}{\kappa} \right),$$

$$m, j = 1, 2, 3, 4,$$

where $\Delta(\partial_x)$ is the Laplace operator, λ and μ are the Lamé constants, ρ is the density of the elastic medium, ω is the vibration frequency, δ_{mj} is the Kronecker delta, $\gamma = (3\lambda + 2\mu)/\alpha$, α is the coefficient of linear temperature expansion of the medium, κ is the temperature conductivity, i is the imaginary unit, K is the thermal conductivity $\eta = \gamma \theta_0/K$, and θ_0 is the temperature of the medium in

---

2000 *Mathematics Subject Classification.* Primary 74F05, 35Q72.

the nondeformed state. The equation of the thermoelastic vibrational state of the medium D in the mixture components acquires the form [2]

$$B(\partial_x, \omega) U = 0; \tag{1}$$

here $U = (u_1, u_2, u_3, u_4) = (u, u_4)$, where u_1, u_2, u_3 are the mixture components and $u_4 = \theta$ is the deviation of the temperature of the medium from θ_0. A solution U of system (1) in the domain D is said to be regular if $U \in C'(\bar{D}) \cap C(\bar{D})$.

In what follows, we use the matrix differential operator of thermal stress

$$R(\partial_x, n(y)) = \begin{pmatrix} & & & -\gamma n_1 \\ & T & & -\gamma n_2 \\ & & & -\gamma n_3 \\ 0 & 0 & 0 & \frac{\partial}{\partial n} \end{pmatrix},$$

where T is the stress operator

$$T = T(\partial_y, n(y)) = ||T_{mj}||_{3\times 3} = \left\| \lambda n_m \frac{\partial}{\partial y_j} + \mu n_j \frac{\partial}{\partial y_m} + \mu \delta_{mj} \frac{\partial}{\partial n} \right\|_{3\times 3},$$

and $n(y) = (n_1, n_2, n_3)$ is the unit outward normal vector on ∂D at the point y. We denote by $\tilde{R}(\partial_y, n(y))$ the matrix obtained from $R(\partial_y, n(y))$ if we replace γ by $i\omega\eta$ in the last column.

Statement of the problem. Find a regular solution U of system (1) in the domain D using its Cauchy data on the surface S:

$$U(y)|_S = f(y), \quad R(\partial_y, n(y)) U(y)|_S = g(y); \tag{2}$$

here $f = (f_1, f_2, f_3, f_4)$ and $g = (g_1, g_2, g_3, g_4)$ are given continuous vector functions. This problem is ill-posed. The ill-posed character is the same as for the Cauchy problem for the Helmholtz equation. For the case in which $w = 0$ and $\gamma = 0$ [9], problem (1), (2) was studied in [1], [4]-[6].

Our aim is to construct an approximate solution of problem (1), (2) using the Carleman function method [8]. By applying the results of Lavrent'ev and Yarmukhamedov [3, 8] on the Cauchy problem for the Laplace equation, we construct explicitly Carleman's matrix and the regularized solution to the Cauchy problem for systems of thermoelasticity equations.

Following [1], let us introduce the following definition.

DEFINITION. A (4×4) matrix $\Pi(y, x, \omega, \gamma, \sigma)$ is called a Carleman matrix for problem (1), (2) if it satisfies the following two conditions:

1) $\Pi(y, x, \omega, \gamma, \sigma) = \Gamma(y, x, \omega, \gamma) + G(y, x, \omega, \gamma, \sigma)$, where σ is a positive numerical parameter, the matrix $G(y, x, \omega, \gamma, \sigma)$ satisfies system (1) with respect to the variable y everywhere in the domain $\mathcal{D}$, and $\Gamma(y, x, \omega, \gamma)$ is a matrix of the fundamental solutions of equation (1);

2) $\int_\Sigma \left(|\Pi| + |(\tilde{R}\Pi^*)^*| \right) ds_y \leq \varepsilon(\sigma)$, where $\varepsilon(\sigma) \to 0$ as $\sigma \to \infty$.

Here and in what follows, Π^* is the adjoint of Π, and $|\Pi|$ is the Euclidean norm of the matrix $\Pi = ||\Pi_{mj}||$, i.e., $|\Pi| = \left(\sum_{m,i=1}^{4} \Pi_{mj}^2 \right)^{1/2}$.

In particular, for the vector $U = (u, u_4)$, we have $|U| = \left(\sum_{m=1}^{4} u_m^2\right)^{1/2}$. To construct an approximate solution of problem (1), (2), consider the matrix

$$\Pi(y, x, \omega, \gamma, \sigma) = ||\Pi_{mj}(y, x, \omega, \gamma, \sigma)||_{4 \times 4},$$

$$\Pi_{mj}(y, x, \omega, \gamma, \sigma) = \frac{(1 - \delta_{m4})(1 - \delta_{4j})}{2\pi} \left(\frac{\delta_{mj}}{\mu} \Phi_\sigma(y, x, i\lambda_3)\right.$$

$$\left. - \sum_{k=1}^{3} \alpha_k \frac{\partial^2}{\partial x_m \partial x_j} \Phi_\sigma(y, x, i\lambda_k)\right) + \frac{i\omega\eta}{2\pi} \delta_{m4}(1 - \delta_{j4}) \sum_{k=1}^{3} \beta_k \frac{\partial}{\partial x_j} \Phi_\sigma(y, x, i\lambda_k)$$

(3) $\quad -\frac{\gamma}{2\pi} \delta_{j4}(1 - \delta_{m4}) \sum_{k=1}^{3} \beta_k \frac{\partial}{\partial x_m} \Phi_\sigma(y, x, i\lambda_k) + \frac{\delta_{j4}\delta_{m4}}{2\pi} \sum_{k=1}^{3} \gamma_k \Phi_\sigma(y, x, i\lambda_k),$

where the constants α_k, β_k, γk and λ_k are explicitly expressed in terms of the thermoelasticity constants, that is, the coefficients of system (1) and

(4) $\quad\quad\quad \Phi_\sigma(y, x, \Lambda) = \frac{1}{2\pi} \int_0^\infty Im \frac{K(\sigma)(\omega - x_3))}{\omega - x_3} \frac{\cos \Lambda u \, du}{\sqrt{u^2 + \alpha^2}},$

$$K(\sigma\omega) = (\omega - x_3 + 2h)^{-1} exp(\sigma\omega + \omega^2),$$

$$\omega = i\sqrt{u^2 + \alpha^2} + y_3, \ \alpha^2 = (y_1 - x_1)^2 + (y_2 - x_2)^2, \ \alpha > 0, \ 0 < x_3 < h.$$

The results of [4] imply the following lemma.

LEMMA 1. *The function $\Phi_\sigma(y, x, \Lambda)$ is the Carleman function for the Helmholtz equation, i.e.,*

(5) $\quad\quad\quad\quad\quad \Phi_\sigma(y, x, \Lambda) = \frac{\exp \Lambda r}{4\pi z} + V(y, x, \Lambda, \sigma), z = |y - x|,$

where $V(y, x, \Lambda, \sigma)$ is a function that is defined for all y and x and satisfies Helmholtz' equation

$$\Delta(\partial y)V - \Lambda^2 V = 0, \ y \in D,$$

(6) $\quad\quad\quad\quad \int_\Sigma \left(|\Phi_\sigma| + \left|\frac{\partial \Phi_{(\sigma)}}{\partial n}\right|\right) ds_y \leq C(\Lambda, D)\sigma \exp(-\sigma x_3),$

where $C(\Lambda, D)$ is a constant.

By Lemma 1, we have

LEMMA 2. *The matrix $\Pi(y, x, \omega, \gamma, \sigma)$ given by (3) and (4) is the Carleman matrix for problem (1), (2).*

PROOF. By (3) and (5), we have

$$\Pi(y, x, w, \gamma, \sigma) = \Gamma(y, x, w, \gamma) + \left\|\frac{(1 - \delta_{mn})(1 - \delta_{jn})}{2\pi} \left(\frac{\delta_{mj}}{\mu} v(y, x, i\lambda_3, \sigma)\right.\right.$$

$$\left. - \sum_{k=1}^{3} \alpha_k \frac{\partial^2}{\partial x_m \partial x_j} v(y, x, i\lambda_k, \sigma)\right) + \frac{i\omega\eta}{2\pi} \delta_{mn}(1 - \delta_{mn}) \sum_{k=1}^{3} \beta_k \frac{\partial}{\partial x_j} v(y, x, i\lambda_k, \sigma)$$

$$\left. - \frac{\partial}{2\pi} \delta_{jn}(1 - \delta_{mn}) \sum_{k=1}^{3} \beta_k \frac{\partial}{\partial x_m} v(y, x, i\lambda_k, \sigma) - \frac{\delta_{jn}\delta_{mn}}{2\pi} \sum_{k=1}^{3} \gamma_k v(y, x, i\lambda_k, \sigma)\right\|$$

$$= \Gamma(y, x, w, \gamma) + ||G_{mj}(y, x, w, \gamma, \sigma)|| = \Gamma(y, x, w, \gamma) + G(y, x, w, \gamma, \sigma).$$

By a straightforward calculation, we can verify that the matrix $G(y,x,w,\gamma,\sigma)$ satisfies system (1) with respect to the variable y everywhere in D.

By using (3), (4) and (6) on each compact set of D, we obtain

(7) $$\int_\Sigma (|\Pi| + |\tilde{R}\Pi^*|)ds_y \leq C(\lambda,\mu,w,\gamma,D)\sigma^2 exp(-\sigma x_3),$$

where $C(\lambda,\mu,w,\gamma,D)$ is a constant. Here and in what follows, for brevity we omit the arguments of Π. The lemma is proved. □

Let us set

(8) $$2U_\sigma(x) = \int_S (\Pi R U(y) - (\tilde{R}\Pi^*)^* U(y))ds_y, \quad x \in D.$$

THEOREM 1. *Let $U(x)$ be a regular solution of the system (1) in D such that*

(9) $$|U(y)| + |R(\partial_y, n(y))U(y)| \leq M, \; y \in \Sigma,$$

Then for $\sigma \geq 1$, the following estimate is valid:

(10) $$|U(x)| - |U_\sigma(x)| \leq MC(\lambda,\mu,\omega,\gamma,D)\sigma^2 \exp(-\sigma x_3), \quad x \in D.$$

PROOF. Lemma 2 applied to a regular solution of system (1) in D implies the validity of the Green-Kupradze integral formula [**2**]:

$$2U(x) = \int_{\partial D} [(\Pi R U(y) - (\tilde{R}\Pi^*)^* U(y)]ds_y, \; x \in D.$$

We rewrite this relation in the form

$$2U(x) = \int_S [(\Pi R U(y) - (\tilde{R}\Pi^*)^* U(y)]ds_y + \int_\Sigma [(\Pi R U(y) - (\tilde{R}\Pi^*)^* U(y)]ds_y$$

(11) $$= 2U_\sigma(x) + \int_\Sigma [(\Pi R U(y) - (\tilde{R}\Pi^*)^* U(y)]ds_y.$$

From (11), we have

$$|v(x) - v_\sigma(x)| \leq \int_\Sigma [|\Pi R U| + |(\tilde{R}\Pi^*)^* U|]ds_y$$

(12) $$\leq \int_\Sigma [|\Pi| + |\tilde{R}\Pi^*|][|RU| + |U|]ds_y.$$

Inequality (10) follows from (7), (9) and (12).

Theorem 1 is proved. □

Now we state a result which allows us to calculate $U(x)$ approximately if, instead of $U(x)$ and $R(\partial_y, n(y))V(x)$, their continuous approximations $f_\delta(y)$ and $g_\delta(y)$ are given on the surface S:

(13) $$\max_S |f(y) - f_\delta(y)| + \max_S |g(y) - g_\delta(y)| \leq \delta, \; 0 < \delta < 1.$$

Suppose that S satisfies Lyapunov's conditions. We define a function $U_{\sigma\delta}$ by setting

(14) $$2U_{\sigma\delta}(x) = \int_S [(\Pi g_\delta - (\tilde{R}\Pi^*)^* f_\delta]ds_y, \; x \in D,$$

(15) $$\sigma = \frac{1}{x_3^o} \ln \frac{1}{\delta} \; x_3^o = \max_D x_3.$$

THEOREM 2. *Let $U(x)$ be a regular solution of system (1) in D satisfying condition (9). Then the following estimate is valid:*

$$|U(x)| - |U_{\sigma\delta}(x)| \leq C(\lambda, \mu, \omega, \gamma, D) \delta^{\frac{x_3}{x_3^o}} \left(\ln \frac{M}{\delta}\right)^3, \ x \in D' \subset D.$$

PROOF. For any $x \in D$, it follows from (7), (10) and (14) that

$$|U(x)| - |U_{\sigma\delta}(x)| \leq \int_S [(\Pi(RU - g_\delta(y)) - (\tilde{R}\Pi^*)^*(U(y) - f_\delta(y))] ds_y$$

$$+ \int_\Sigma [(\Pi RU(y) - (\tilde{R}\Pi^*)^* U(y)] ds_y, \ x \in D.$$

By the assumption of the theorem and inequalities (9) and (13), we have

$$|U(x)| - |U_{\sigma\delta}(x)| \leq C(\lambda, \mu, \omega, \gamma, D)\sigma^3 \exp(\sigma x_3^0 - \sigma x_3)$$

$$+ MC(\lambda, \mu, \omega, \gamma, D)\sigma^3 \exp(-\sigma x_3)$$

(16) $$\leq C_2(\lambda, \mu, \omega, \gamma, D)\sigma^3(M + \delta \exp(\sigma x_3^o)\exp(-\sigma x_3^0), \ x \in D.$$

For the choice (15) of the parameter σ, the assertion of the theorem follows from (16). $\square$

COROLLARY. *The limit relations*

$$\lim_{\sigma \to \infty} U_\sigma(x) = U(x), \lim_{\sigma \to 0} U_{\sigma\delta}(x) = U(x)$$

hold uniformly on each compact subset of D.

References

[1] T. I. Ishankulov and O. I. Makhmudov, *The Cauchy problem for the system of thermoelasticity equations in space*, Math. Notes **64** (1998), 181–185.

[2] V. D. Kupradze, T. G. Gegelia, M. O. Basheleishvili and T. V. Burchuladze, *Three-Dimensional Problems of the Mathematical Theory of Elasticity and Thermoelasticity*, Nauka, Moscow, 1976.

[3] M. M. Lavrent'ev, *Some Ill-posed Problems of Mathematical Physics*, Computer Center of the Sibirian Division of the Russian Academy of Sciences, Novosibirsk, 1962.

[4] O. I. Makhmudov and J. E. Niyozov, *Regularization of the solution of the Cauchy problem for a system of equations in the theory of elasticity in displacements*, Siberian Math. J. **39** (1998), 323–330.

[5] O. I. Makhmudov and I. E. Niyozov, *Regularization of solutions of the Cauchy problem for systems of elasticity theory in infinite domains*, Math. Notes **68** (2000), 471–475.

[6] O. I. Makhmudov and I. E. Niyozov, *On a Cauchy problem for a system of equations in elasticity theory*, Differ. Equ. **36** (2000), 749-754.

[7] I. G. Petrovskii, *Lectures on Partial Differential Equations*, Fizmatgiz, Moscow, 1961.

[8] Sh. Ya. Yarmukhamedov, *The Cauchy problem for Laplace's equation*, Dokl. Akad. Nauk SSSR **235** (1977), 281–283.

[9] Sh. Ya. Yarmukhamedov, T. I. Ishankulov and O. I. Makhmudov, *The Cauchy problem for a system of equations in the theory of elasticity in space*, Siberian Math. J. **33** (1992), 154-158.

DEPARTMENT OF MATHEMATICS, SAMARKAND STATE UNIVERSITY, SAMARKAND, UZBEKISTAN
E-mail address: makhmudovo@rambler.ru

DEPARTMENT OF MATHEMATICS, SAMARKAND STATE UNIVERSITY, SAMARKAND, UZBEKISTAN
E-mail address: iqboln@rambler.ru

Modules of Vector Measures on the Heisenberg Group

Irina Markina

Dedicated to Professor Lawrence Zalcman on his sixtieth birthday

ABSTRACT. We define the extremal length of horizontal vector measures on the Heisenberg group and study capacities associated with a linear sub-elliptic equation. The coincidence between the definition of the module of a horizontal vector measure system and two different definitions of the capacity is proved. We show the continuity property of the module of a curve family generated by a module of horizontal vector measures.

1. Introduction

The concept of the extremal length and the module of a family of curves goes back to H. Grötzsch, A. Beurling, L. V. Ahlfors [1, 16]. In 1957, B. Fuglede [14] introduced the module of a measure system. These notions play an important role and have numerous applications in analysis and potential theory. The interest in non-linear elliptic equations has inspired a more general notion of the module of a family of curves and the capacity associated with this type of equation [2, 19, 20, 21, 25].

Recently, analysis on homogeneous groups (the simplest example of which is the Heisenberg group) has been developed intensively. The fundamental role of such groups in analysis was pointed out by E. M. Stein [35] in his address to the International Congress of Mathematicians in 1970; see also his monograph [36]. Briefly, a homogeneous group is a simply connected nilpotent Lie group whose Lie algebra admits a grading. There is a natural family of dilations on the group, under which the metric behaves like the Euclidean metric under the Euclidean dilation [7, 13]. Analysis on homogeneous groups is a testing ground for the study of general sub-elliptic problems arising from vector fields $X_1, \ldots, X_k$ satisfying the Hörmander hypoellipticity condition [22]. An important motivation for the study of quasilinear sub-elliptic equations of the second order comes from the theory of quasiconformal and quasiregular mappings on stratified nilpotent groups [8, 15, 18, 31, 38]. Quasilinear sub-elliptic equations generate interest in a concept of capacity and extremal length associated with this type of equation. The foundation of the

---

2000 *Mathematics Subject Classification.* Primary 31B15; Secondary 22E30.
This work was supported by Projects Fondecyt (Chile) # 1040333, # 1030373.

theory of quasilinear sub-elliptic equations and non-linear potential theory can be found in the papers [3, 4, 5, 9, 12, 17, 28, 29] (see also the references therein).

In the present work, based on ideas of [2], we define a horizontal vector measure on the Heisenberg group. The non-Riemannian geometry of the group and properties of sub-elliptic equations force us to introduce some natural modifications. We prove the continuity property of the module of a family of curves related to the module of horizontal vector measures. The coincidence between the definitions of the module of a measure system and two definitions of the capacity associated with a linear sub-elliptic equation is established. We emphasize that the present article can be considered as model example for the studying relations between the module and the capacity on an arbitrary homogeneous group. In the next paragraph, the reader finds explicit definitions and detailed formulations of the main results.

2. Definitions and statements of main results

In our model of the Heisenberg group $\mathbb{H}^n$, we take $\mathbb{R}^{2n+1}$ as the underlying space and provide it with the non-commutative multiplication

$$pq = (x,t)(x',t') = \left(x + x', t + t' - 2\sum_{i=1}^{n}(x_i x'_{n+i} - x_{n+i} x'_i)\right),$$

where $x = (x_1, \ldots, x_{2n})$, $x' = (x'_1, \ldots, x'_{2n}) \in \mathbb{R}^{2n}$, $t, t' \in \mathbb{R}$; hence the left translation $L_p(q) = pq$ is defined. The left-invariant vector fields

$$X_i = \frac{\partial}{\partial x_i} + 2x_{n+i}\frac{\partial}{\partial t}, \quad X_{n+i} = \frac{\partial}{\partial x_{n+i}} - 2x_i\frac{\partial}{\partial t}, \quad i = 1, \ldots, n, \quad T = \frac{\partial}{\partial t},$$

form a basis of the Lie algebra of the Heisenberg group. There exist non-trivial relations $[X_i, X_{n+i}] = -4T$, $i = 1, \ldots, n$, and all other Poisson brackets vanish. Thus, the Heisenberg algebra $\mathcal{G}$ is of dimension $2n+1$ and splits into the direct sum $\mathcal{G} = V_1 \oplus V_2$. The vector space V_1 is generated by the vector fields X_i, $i = 1, \ldots, 2n$, and is called the *horizontal space*. The space V_2 is the one-dimensional center spanned by the vector field T. By definition, *the horizontal tangent space at* $q \in \mathbb{H}^n$ is the subspace HT_q of the tangent space T_q spanned by the vector fields $X_1(q), \ldots, X_{2n}(q)$ at q.

We use the Carnot-Carathéodory metric based on the length of horizontal curves. An absolutely continuous curve $\gamma : [0,b] \to \mathbb{H}^n$ is said to be *horizontal* if its tangent vector $\gamma'(t)$ lies in the horizontal tangent space $HT_{\gamma(t)}$, i.e., there exist functions $a_j(s)$, $s \in [0,b]$, such that

$$\sum_{j=1}^{2n} a_j^2(s) \leq 1 \quad \text{and} \quad \gamma'(s) = \sum_{j=1}^{2n} a_j(s) X_j(\gamma(s)).$$

The results in [6] imply that one can connect two arbitrary points $p, q \in \mathbb{H}^n$ by a horizontal curve. We fix a quadratic form $\langle \cdot, \cdot \rangle$ on HT with respect to which the vector fields $X_1(q), \ldots, X_{2n}(q)$ are orthonormal at every point $q \in \mathbb{H}^n$. Actually, we take the Euclidean quadratic form. Then the lengths of the curve $l(\gamma)$ is defined by the formula

$$l(\gamma) = \int_0^b \langle \gamma'(s), \gamma'(s) \rangle^{1/2} ds = \int_0^b \left(\sum_{j=1}^{2n} |a_j(s)|^2\right)^{1/2} ds.$$

The Carnot-Carathéodory distance $d_c(p,q)$ is the infimum of the lengths over all horizontal curves connecting p and $q \in \mathbb{H}^n$. Since the vector fields are left-invariant, the quadratic form is also left-invariant, and so is the Carnot-Carathéodory metric. We write the norm of a vector $\xi \in V_1$ as $|\xi| = \langle \xi, \xi \rangle^{1/2}$.

On the Heisenberg group, the natural dilation $\delta_\lambda q = \delta_\lambda(x,t) = (\lambda x, \lambda^2 t)$ is defined. One easily checks that the Jacobian determinant of the dilation δ_λ is λ^Q, where $Q = 2n + 2$ is the *homogeneous dimension* of the Heisenberg group. The metric space $(\mathbb{H}^n, d_c(\cdot, \cdot))$ has Hausdorff dimension $Q = 2n + 2$, which is greater than the topological dimension $N = 2n + 1$. We also use the homogeneous metric $d(p,q) = |p^{-1}q|$ generated by the homogeneous norm $|q| = \left(\left(\sum_{i=1}^{2n} x_i^2\right)^2 + t^2\right)^{\frac{1}{4}}$, $|\delta_\lambda q| = \lambda |q|$, $p, q \in \mathbb{H}^n$. The metrics $d_c(\cdot, \cdot)$ and $d(\cdot, \cdot)$ are equivalent [**23**]. The metric $d(\cdot, \cdot)$ is preferred, since it satisfies the triangle inequality: $|p^{-1}q| \le |p| + |q|$ [**24**], while the Carnot–Carathéodory metric $d_c(\cdot, \cdot)$ satisfies only the generalized triangle inequality: $d_c(p,q) \le K(d_c(p,w) + d_c(w,q))$ with a constant $K > 1$.

Lebesgue measure is the Haar measure on the Heisenberg group. We see that $m(B(r)) = r^Q m(B(1))$, where $m(\cdot)$ denotes Lebesgue measure and $B(r)$ is a ball of radius r. It is convenient to choose the normalization $m(B(1)) = 1$, so that balls of radius r have measure precisely r^Q.

A curve $\gamma : [0, l] \to \mathbb{H}^n$ is *rectifiable* if $\sup \left\{ \sum_{k=1}^{p} d_c\big(\gamma(s_k), \gamma(s_{k-1})\big) \right\}$ is finite, where the supremum ranges over all partitions $0 = s_0 \le s_1 \le \ldots \le s_p = l$ of the segment $[0, l]$. Note that the definition of a rectifiable curve is based on the Carnot-Carathéodory metric. For this reason, only horizontal curves are rectifiable (see [**23**]). Thus, henceforth we work only with horizontal curves.

We define an absolutely continuous function on curves of the horizontal fibration. For this, we consider a family of horizontal curves $\mathcal{X}$ that form a smooth fibration of an open set $U \subset \mathbb{H}^n$. Usually, one can think of a curve $\gamma \in \mathcal{X}$ as an orbit of a smooth horizontal vector field $X \in V_1$. If we denote by φ_s the flow associated with this vector field, then the fiber is of the form $\gamma(s) = \varphi_s(p)$. Here the point p belongs to the surface S which is transversal to the vector field X. The parameter s ranges over an open interval $J \in \mathbb{R}$. One can assume that there is a measure $d\gamma$ on the fibration $\mathcal{X}$ of the set $U \subset \mathbb{H}^n$. The measure $d\gamma$ on $\mathcal{X}$ is equal to the inner product of the vector field $X \in V_1$ and a biinvariant volume form dx (for more information, see, for instance, [**11, 24**]). The measure $d\gamma$ satisfies the inequality $k_0 R^{\frac{Q-1}{Q}} \le \int_{\gamma \in \mathcal{X}, \gamma \cap B(x,R) \ne \emptyset} d\gamma \le k_1 R^{\frac{Q-1}{Q}}$ for sufficiently small balls $B(x,R) \subset U$ with constants k_0, k_1 which do not depend on the ball $B(x,R)$ [**24, 37**].

We use the symbol Ω to denote a domain (open connected set) on the Heisenberg group.

DEFINITION 2.1. A function $u : \Omega \to \mathbb{R}$, $\Omega \subset \mathbb{H}^n$, is said to be *absolutely continuous on lines* ($u \in ACL(\Omega)$) if for any domain U, $\overline{U} \subset \Omega$, and for any fibration $\mathcal{X}$ defined by a left-invariant vector field X_j, $j = 1, \ldots, 2n$, the function u is absolutely continuous on $\gamma \cap U$ with respect to $\mathcal{H}^1$-Hausdorff measure for $d\gamma$-almost all curves $\gamma \in \mathcal{X}$.

The derivatives $X_j u$, $j = 1, \ldots, 2n$, exist almost everywhere in Ω for such functions u [**24**]. If they belong to $L_p(\Omega)$, $p \ge 1$, for all $X_j \in V_1$, then u is said to belong to $ACL_p(\Omega)$.

A function $u: \Omega \to \mathbb{R}$ is said to belong to the Sobolev space $L^1_p(\Omega)$ if its distributional derivatives $X_j u$ along horizontal vector fields X_j exist for all $j = 1, \ldots, 2n$, i.e., $\int_\Omega X_j u \varphi \, dx = \int_\Omega u X_j \varphi \, dx$ for all $\varphi \in C_0^\infty(\Omega)$, and have finite semi-norm $\|u\|_{L^1_p(\Omega)} = (\int_\Omega |\nabla_0 u|^p(x) \, dx)^{1/p}$. Here $\nabla_0 u = (X_1 u, \ldots, X_{2n} u)$ is *the horizontal gradient* of u and $|\nabla_0 u| = \left(\sum_{j=1}^{2n} |X_j u|^2 \right)^{1/2}$. If $u \in ACL_p(\Omega)$, then the derivatives $X_j u$, $j = 1, \ldots, 2n$ coincide with the distributional derivatives of u. Conversely, for any $u \in L^1_p(\Omega)$, there exists a function $v \in ACL_p(\Omega)$ such that $u = v$ almost everywhere in the domain Ω [**32, 39**].

Let $\mathcal{A}(x) = (a_{ij}(x))$, $x \in \Omega$, be a positive definite symmetric $(2n \times 2n)$-matrix with measurable components $a_{ij}(x)$ such that

(2.1) $$\alpha^{-1}|\xi| \leq |\mathcal{A}\xi| = \langle \mathcal{A}\xi, \mathcal{A}\xi \rangle^{1/2} \leq \alpha|\xi|$$

for any horizontal vector $\xi = (\xi_1, \ldots, \xi_{2n})$ and some constant $\alpha \geq 1$. We denote by $\mathcal{B}(x) = (b_{ij}(x))$ the inverse matrix to $\mathcal{A}(x)$. The matrix $\mathcal{B}(x)$ also satisfies inequality (2.1).

With $\mathcal{A}$ one can associate a second order sub-elliptic operator $\left(-\sum_{j=1}^{2n} X_j \mathcal{A}^2(x) \nabla_0 \right)$, where $\nabla_0 u = (X_1 u, \ldots, X_{2n} u)$ for any smooth function u. If $\mathcal{A}$ is the identity matrix then we obtain the sub-Laplacian operator on the Heisenberg group.

Recall the definition of the module of a system of measures [**14**]. Let $(X, \mathcal{N}, m)$ be a measure space with a non-negative measure m. Denote by $\mathcal{E}$ the system of measures μ in X whose domains contain the domain $\mathcal{N}$ of m. The quantity

$$M(\mathcal{E}) = \inf \left\{ \int f^2 \, dm : f \geq 0 \text{ an } m\text{-measurable function}, \int f \, d\mu \geq 1 \text{ for all } \mu \in \mathcal{E} \right\}$$

is called the *module* of the measure system $\mathcal{E}$.

In the present paper we deal with the module of a system of vector measures related to the stratified structure of the Lie algebra of the Heisenberg group. Let $\mu = (\mu_1, \ldots, \mu_{2n})$ be a vector measure whose components μ_i are signed measures defined for Borel sets in $\mathbb{H}^n$. The dimension of the vector measure coincides with the dimension of the horizontal space V_1, so we call this measure a *horizontal vector measure*. Define the total variation $|\mu|$ of μ by $|\mu|(E) = \sup \sum_j \left(\sum_{i=1}^{2n} \mu_i^2(E_j) \right)^{1/2}$, where the supremum is taken over all finite partitions of E into Borel sets E_j. The total variation $|\mu|$ is a non-negative measure.

Example 1. If Γ is a family of horizontal curves, then we have a family of horizontal vector measures $\{d\gamma = d\gamma_1, \ldots, d\gamma_{2n} : \gamma \in \Gamma\}$.

Example 2. The horizontal gradient of an ACL-function is another example of a horizontal vector measure $\{\nabla_0 u : u \in ACL(\Omega)\}$.

DEFINITION 2.2. Let $\mathcal{M}$ be a set of vector measures. Set $|\mathcal{M}| = \{|\mu| : \mu \in \mathcal{M}\}$. If $M(|\mathcal{M}|) = 0$, then we say that $\mathcal{M}$ is *exceptional*. If a statement with respect to vector measures fails only for an exceptional system, then we say that it holds *almost everywhere*.

Let Ω be a domain on $\mathbb{H}^n$ and K_0 and K_1 closed non-empty disjoint sets such that $K_0 \cap \overline{\Omega} \neq \emptyset$ and $K_1 \cap \overline{\Omega} \neq \emptyset$. We call the triple $(K_0, K_1; \Omega)$ a *condenser*.

Let $\lfloor a, b \rfloor$ be an interval of one of the following types: $[a, b]$, $[a, b)$, $(a, b]$, or (a, b). From now on, we suppose that a horizontal curve $\gamma : \lfloor a, b \rfloor \to \mathbb{H}^n$ is parametrized by its length element. We let

$$\Gamma = \Gamma(K_0, K_1 : \Omega) = \{\gamma : \overline{\gamma([a,b])} \cap K_i \neq \emptyset, \ i = 0, 1, \text{ and } \gamma(t) \in \Omega, \ t \in (a,b)\}$$

and call $\Gamma(K_0, K_1; \Omega)$ the family of curves that connect the compacts K_0 and K_1 in Ω.

Now we give two different definitions of the capacity of a condenser $(K_0, K_1; \Omega)$. In the first one, we use the notion of exceptional set with respect to the module of the family of horizontal curves that connect the compacts K_0 and K_1.

DEFINITION 2.3. Denote by $\mathcal{F}C(K_0, K_1; \Omega)$ the class of admissible functions $u \in ACL_2(\Omega)$ such that

$$u(x) \longrightarrow 0 \quad \text{as} \quad x \longrightarrow K_0 \cap \overline{\Omega}, \quad \text{along almost all curves from} \quad \Gamma(K_0, K_1; \Omega),$$
$$u(x) \longrightarrow 1 \quad \text{as} \quad x \longrightarrow K_1 \cap \overline{\Omega}, \quad \text{along almost all curves from} \quad \Gamma(K_0, K_1; \Omega).$$

The $\mathcal{A}$-capacity of the condenser $(K_0, K_1; \Omega)$ is

$$\operatorname{cap}_{\mathcal{A}}(K_0, K_1; \Omega) = \inf\Big\{\int_\Omega |\mathcal{A}\nabla_0 u|^2 \, dx \ : u \in \mathcal{F}C(K_0, K_1; \Omega)\Big\}.$$

DEFINITION 2.4. Let $\mathcal{F}C^\star(K_0, K_1; \Omega)$ be the class of functions $u \in ACL_2(\Omega)$ such that

$$u(x) = 0 \quad \text{on the intersection of} \quad \Omega \quad \text{with a neighborhood of} \quad K_0,$$
$$u(x) = 1 \quad \text{on the intersection of} \quad \Omega \quad \text{with a neighborhood of} \quad K_1.$$

The $\mathcal{A}^\star$-capacity of $(K_0, K_1; \Omega)$ is

$$\operatorname{cap}^\star_{\mathcal{A}}(K_0, K_1; \Omega) = \inf\Big\{\int_\Omega |\mathcal{A}\nabla_0 u|^2 \, dx \ : u \in \mathcal{F}C^\star(K_0, K_1; \Omega)\Big\}.$$

Results from [27, 33] imply that ACL_2-functions are absolutely continuous on almost all horizontal curves. So we may assume that admissible functions for both definitions are absolutely continuous on almost all horizontal curves with horizontal gradient $|\nabla_0 u|$ in the class L_2. The capacity associated with sub-elliptic equation is studied in [4, 5, 9, 10, 28, 29, 30].

We shall prove the equivalence of Definitions 2.3 and 2.4 in bounded domains of the Heisenberg group.

THEOREM 2.1. Let Ω be a bounded domain on the Heisenberg group $\mathbb{H}^n$ and $\mathcal{A}(x)$ a uniformly continuous matrix in Ω satisfying (2.1). Then

$$\operatorname{cap}_{\mathcal{A}}(K_0, K_1; \Omega) = \operatorname{cap}^\star_{\mathcal{A}}(K_0, K_1; \Omega).$$

Now we introduce the $\mathcal{A}$-module of a system of horizontal vector measures, associated with Definitions 2.3, 2.4. Let $\zeta = (\zeta_1, \ldots, \zeta_{2n})$ be a vector-valued function and $\mu = (\mu_1, \ldots, \mu_{2n})$ a signed horizontal vector measure. If $\int |\zeta_i| d|\mu_i| < \infty$ for all i, then we define $\int \zeta \, d\mu = \sum_{i=1}^{2n} \int \zeta_i \, d\mu_i$. We denote by $\mathcal{F}M(\mu)$ a class of vector-valued functions ζ such that $\int \zeta \, d\mu \geq 1$. If $\zeta \in \mathcal{F}M(\mu)$ for all $\mu \in \mathcal{M}$, then we write $\zeta \in \mathcal{F}M(\mathcal{M})$ and call ζ an admissible function for the measure system $\mathcal{M}$.

DEFINITION 2.5. Let $\xi = (\xi_1, \ldots, \xi_{2n})$ be a vector-valued function and $\mu \in \mathcal{M}$ a complete horizontal vector measure on $\Omega \subset \mathbb{H}^n$. We define the $\mathcal{A}$-module by

$$M_{\mathcal{A}}(\mathcal{M}) = \inf\left\{ \int_{\Omega} |\mathcal{A}\xi|^2 \, dx \ : \xi \in \mathcal{F}M(\mathcal{M}) \quad \text{almost everywhere}\right\}.$$

As was mentioned in Example 1, for a family of horizontal curves Γ, we have a system of horizontal vector measures $d\gamma = (d\gamma_1, \ldots, d\gamma_{2n})$, $\gamma \in \Gamma$, and non-negative measures $|d\gamma| = \langle d\gamma, d\gamma \rangle^{1/2}$. We write $d\Gamma = \{d\gamma : \gamma \in \Gamma\}$ and $|d\Gamma| = \{|d\gamma| : \gamma \in \Gamma\}$. More generally, for a positive definite $(2n \times 2n)$-matrix $Q(x) = (q_{ij}(x))$, we put $|Qd\gamma| = \langle Qd\gamma, Qd\gamma \rangle^{1/2}$ and $|Qd\Gamma| = \{|Qd\gamma| : \gamma \in \Gamma\}$.

We also prove that the $\mathcal{A}$-capacity of the condenser $(K_0, K_1; \Omega)$, the $\mathcal{A}$-module of the horizontal vector measure $d\Gamma(K_0, K_1; \Omega)$, and the module $|\mathcal{B}d\Gamma|$ all coincide.

THEOREM 2.2. *Let Ω be a domain on the Heisenberg group $\mathbb{H}^n$. Then*

$$\operatorname{cap}_{\mathcal{A}}(K_0, K_1; \Omega) = M_{\mathcal{A}}(d\Gamma) = M(|\mathcal{B}\, d\Gamma|) < \infty.$$

In the next theorem, we use the following notation. Let K_0 and K_1 be compact sets in $\overline{\Omega}$ and let K_0^j and K_1^j be sequences of compact sets such that $K_0^0 \cap K_1^0 = \emptyset$, $K_0^j \subset \operatorname{int} K_0^{j-1}$, $K_1^j \subset \operatorname{int} K_1^{j-1}$, $K_0 = \bigcap_{j=0}^{\infty} K_0^j$, and $K_1 = \bigcap_{j=0}^{\infty} K_1^j$.

THEOREM 2.3. *Suppose that $\mathcal{B}(x)$ is uniformly continuous in a bounded domain $\Omega \subset \mathbb{H}^n$. Then, $M(|\mathcal{B}\,d\Gamma|)$ possesses the continuity property. Namely, if $\Gamma_j = \Gamma(K_0^j, K_1^j; \Omega)$ and $\Gamma = \Gamma(K_0, K_1; \Omega)$, then*

$$\lim_{j \to \infty} M(|\mathcal{B}\, d\Gamma_j|) = M(|\mathcal{B}\, d\Gamma|).$$

From Theorems 2.1 and 2.2 follows

COROLLARY 2.1. *Let Ω be a bounded domain on the Heisenberg group $\mathbb{H}^n$ and $\mathcal{A} = \mathcal{B}$ be the identity matrix. Then*

$$\operatorname{cap}^{\star}(K_0, K_1; \Omega) = M(|d\Gamma|).$$

3. Proof of Theorem 2.2

We split the proof into two steps.

Step 1. We claim the inequalities

(3.1) $$M(|\mathcal{B}\,d\Gamma|) \leq M_{\mathcal{A}}(d\Gamma) \leq \operatorname{cap}_{\mathcal{A}}(K_0, K_1; \Omega) < \infty.$$

The set $\mathcal{F}C(K_0, K_1; \Omega)$ is not empty; hence $\operatorname{cap}_{\mathcal{A}}(K_0, K_1; \Omega) < \infty$. Let $u \in \mathcal{F}C(K_0, K_1; \Omega)$. Recall that curves connecting compacts K_0 and K_1 are parametrized by the length arc parameter $s \in I \subset \mathbb{R}$. Since γ is horizontal, its tangent vector $\dot{\gamma}$ has the form $(\dot{\gamma}_1, \ldots, \dot{\gamma}_{2n}, \dot{\gamma}_{2n+1})$ in the basis $\left(\frac{\partial}{\partial x_1}, \ldots, \frac{\partial}{\partial x_{2n}}, \frac{\partial}{\partial t}\right)$, where $\dot{\gamma}_{2n+1} = \sum_{i=1}^{n} 2\gamma_{n+i} \dot{\gamma}_i - 2\gamma_i \dot{\gamma}_{n+i}$. It follows that

$$\begin{aligned}
\frac{du(\gamma(s))}{ds} &= \sum_{i=1}^{2n} \frac{\partial u}{\partial x_i} \dot{\gamma}_i(s) + \frac{\partial u}{\partial t} \dot{\gamma}_{2n+1}(s) \\
&= \sum_{i=1}^{n} \left(\frac{\partial u}{\partial x_i} + 2\gamma_{n+i} \frac{\partial u}{\partial t}\right) \dot{\gamma}_i(s) + \sum_{i=1}^{n} \left(\frac{\partial u}{\partial x_{n+i}} - 2\gamma_i \frac{\partial u}{\partial t}\right) \dot{\gamma}_{n+i}(s) \\
&= \langle \nabla_0 u, \dot{\gamma} \rangle,
\end{aligned}$$

where $\dot\gamma = (\dot\gamma_1, \ldots, \dot\gamma_{2n}, 0)$ is the representation of the tangent vector of γ in the basis of vector fields $(X_1, \ldots, X_{2n}, T)$. Hence, we have

$$\int_\gamma \nabla_0 u \, d\gamma = \int_I \langle \nabla_0 u(\gamma(t)), \dot\gamma(t) \rangle \, dt = u(x_1) - u(x_0) = 1,$$

where $x_0 \in K_0$, $x_1 \in K_1$, and the equality holds except for some exceptional family of curves. Thus, $\nabla_0 u \in \mathcal{F}M(d\Gamma)$ for almost all curves of $d\Gamma$ and $M_\mathcal{A}(d\Gamma) \leq \int_\Omega |\mathcal{A}\nabla_0 u|^2 \, dx$. Taking the infimum with respect to u, we obtain the second inequality of (3.1).

Now let $\xi \in \mathcal{F}M(d\Gamma)$ almost everywhere. Then $1 \leq \int_\gamma \xi \, d\gamma = \int_\gamma \mathcal{A}\xi \mathcal{B} d\gamma \leq \int_\gamma |\mathcal{A}\xi||\mathcal{B}d\gamma|$, so $|\mathcal{A}\xi| \in \mathcal{F}M(|\mathcal{B}\,d\Gamma|)$ almost everywhere. Finally, we obtain

$$M_\mathcal{A}(d\Gamma) = \inf_\xi \int_\Omega |\mathcal{A}\xi|^2 \, dx \geq M(|\mathcal{B}\,d\Gamma|).$$

Step 2. Now we show that

(3.2) $$\operatorname{cap}_\mathcal{A}(K_0, K_1; \Omega) \leq M(|\mathcal{B}\,d\Gamma|).$$

Let $\rho \in \mathcal{F}M(|\mathcal{B}\,d\Gamma|)$. For each $x \in \Omega$, denote by Γ_0^x the family of curves starting at K_0 and terminating at x. Set

(3.3) $$u(x) = \inf_{\gamma \in \Gamma_0^x} \int_\gamma \rho |\mathcal{B}\,d\gamma|.$$

We construct an admissible function from $\mathcal{F}C(K_0, K_1; \Omega)$ making use of (3.3). First, we prove that u possesses the following properties:

   (i) $u \in ACL_2(\Omega)$;
   (ii) for almost all $x \in \Omega$,

(3.4) $$|\mathcal{A}\nabla_0 u(x)| \leq \rho(x);$$

   (iii) $\lim u(x) = 0$ as $x \to K_0$ along almost all curves $\gamma \in \Gamma(K_0, K_1; \Omega)$;
   (iv) $\liminf u(x) \geq 1$ as $x \to K_1$ along almost all curves $\gamma \in \Gamma(K_0, K_1; \Omega)$.

We have

(3.5) $$\int_\gamma \rho |\mathcal{B}\,d\gamma| \leq \alpha \int_\gamma \rho \, |d\gamma| < \infty,$$

from the property (2.1) for the matrix $\mathcal{B}$. The finiteness of the last integral for almost all curves follows from the properties of the measure system [14]. Then, by the definition of u, we get

(3.6) $$|u(x) - u(y)| \leq \int_\gamma \rho |\mathcal{B}\,d\gamma| \leq \alpha \int_\gamma \rho \, |d\gamma|$$

for arbitrary points $x, y \in \gamma$. We denote by β_i an orbit of $X_i \in V_1$, $i = 1, \ldots, 2n$. If we apply (3.6) to β_i, we obtain that u is absolutely continuous along almost all curves of the horizontal fibration. Thus the horizontal derivatives $X_i u$, $i = 1, \ldots, 2n$, exist for almost all points in Ω. We choose $x \in \Omega$, where $\nabla_0 u(x)$

exists, and a horizontal vector field $Y(x)$, $|Y(x)| = 1$. Then (3.6) implies

$$\left|\langle \nabla_0 u(x), Y(x)\rangle\right| = \left|\lim_{h\to 0} \frac{u(x \exp hY(x)) - u(x)}{h}\right|$$

(3.7)
$$\leq \lim_{h\to 0} \frac{1}{h} \int_0^h \rho(x \exp sY(x))|\mathcal{B}(x \exp sY(x))Y(x)|\,ds$$

$$= \rho(x)|\mathcal{B}(x)\,Y(x)|$$

for almost all points $x \in \Omega$. Applying (3.7) with $Y(x) = \frac{X_i(x)}{|X_i(x)|}$, we obtain $|X_i u| \leq \rho \mathcal{B} \frac{X_i}{|X_i|} \leq \alpha\rho$ by (2.1). The assumption $\rho \in L_2(\Omega)$ implies that $\nabla_0 u \in L_2(\Omega)$. We have shown property (i).

Now taking $Y(x) = \frac{\mathcal{A}^2 \nabla_0 u(x)}{|\mathcal{A}^2 \nabla_0 u(x)|}$, we get

$$\left|\langle \nabla_0 u(x), \frac{\mathcal{A}^2 \nabla_0 u(x)}{|\mathcal{A}^2 \nabla_0 u(x)|}\rangle\right| \leq \rho(x) \left|\mathcal{B}(x) \frac{\mathcal{A}^2 \nabla_0 u(x)}{|\mathcal{A}^2 \nabla_0 u(x)|}\right| = \rho(x) \left|\frac{\mathcal{A}\nabla_0 u(x)}{|\mathcal{A}^2 \nabla_0 u(x)|}\right|.$$

Since $|\langle \mathcal{A}^2 \nabla_0 u(x), \nabla_0 u(x)\rangle| = |\mathcal{A}\nabla_0 u(x)|^2$, we have property (ii).

Since the module of the family of nonrectifiable curves is equal to zero, we can assume that all curves under consideration are rectifiable. Making use of the arc length parameter s, we obtain

$$0 \leq u(\gamma(s)) \leq \int_\gamma \rho|\mathcal{B}\,d\gamma| \leq \alpha \int_0^s \rho\,ds \to 0, \quad \text{as} \quad s \to 0$$

from (3.3), (3.5). Thus, $\lim u(x) = 0$ as $x \to K_0$ along γ. This proves (iii).

We prove (iv) by contradiction. Suppose there exists a curve γ_1 such that $c = \liminf_{s \to l_{\gamma_1}} u(\gamma_1(s)) < 1$, where l_{γ_1} is the length of the curve γ_1. We fix $\varepsilon = 1-c > 0$. By definition, there is $\tilde{s} \in (0, l_{\gamma_1})$ such that

$$|u(\gamma_1(\tilde{s})) - c| < \frac{\varepsilon}{3}, \quad \text{and} \quad \int_{\tilde{s}}^{l_{\gamma_1}} \rho\,ds < \frac{\varepsilon}{3\alpha}.$$

We consider the family Γ_0^x, $x = \gamma_1(\tilde{s})$. The definition of the function $u(x)$ implies that we can find $\gamma_2 \in \Gamma_0^x$, with $\int_{\gamma_2} \rho|\mathcal{B}\,d\gamma| < u(x) + \frac{\varepsilon}{3}$. Let us denote by γ_3 the arc of the curve γ_1 between the points $\gamma_1(\tilde{s})$ and $\gamma_1(l_{\gamma_1})$. Then $\gamma_2 \cup \gamma_3 \in \Gamma(K_0, K_1; \Omega)$, and by (3.6), we get

$$\int_{\gamma_2 \cup \gamma_3} \rho|\mathcal{B}\,d\gamma| < u(x) + \frac{\varepsilon}{3} + \alpha \int_{\tilde{s}}^{l_{\gamma_1}} \rho\,ds < c + \frac{\varepsilon}{3} + \frac{\varepsilon}{3} + \frac{\varepsilon}{3} = 1,$$

which contradicts the assumption $\rho \in \mathcal{F}M(|\mathcal{B}\,d\Gamma|)$. Hence (iv) holds.

To complete the proof of Theorem 2.2, we set $\tilde{u}(x) = \min(u(x), 1)$. Then $\tilde{u} \in \mathcal{F}C(K_0, K_1; \Omega)$. By the definition of $\mathcal{A}$-capacity and property (ii), we have

$$\mathrm{cap}_\mathcal{A}(K_0, K_1; \Omega) \leq \int_\Omega |\mathcal{A}\nabla_0 \tilde{u}|^2\,dx \leq \int_\Omega |\mathcal{A}\nabla_0 u|^2\,dx \leq \int_\Omega \rho^2\,dx.$$

Taking the infimum with respect to ρ, we obtain (3.2).

Theorem 2.2 follows from (3.1) and (3.2). $\square$

4. Auxiliary results and proof of Theorem 2.3

In this section, we work under the assumption that K_0 and K_1 are disjoint non-empty compacts in the closure $\overline{\Omega}$ of a bounded domain $\Omega \subset \mathbb{H}^n$. Let K_0^j and K_1^j be sequences of closed sets such that $K_0^0 \cap K_1^0 = \emptyset$, $K_0^j \subset \operatorname{int} K_0^{j-1}$, $K_1^j \subset \operatorname{int} K_1^{j-1}$, $K_0 = \bigcap_{j=0}^\infty K_0^j$, and $K_1 = \bigcap_{j=0}^\infty K_1^j$. We need some auxiliary lemmas to prove Theorem 2.3.

In the case of $\mathbb{R}^n$, the next lemma goes back to the work [34] and has subsequently been revised by M. Ohtsuka and H. Aikawa (see [2]). We adopt its formulation for the Heisenberg group.

LEMMA 4.1. *Let $\rho \in L_p(\mathbb{H}^n)$ be a positive function which is continuous in $\Omega \setminus (K_0 \cup K_1)$, $p \in [1, \infty)$. For each $\varepsilon > 0$, we can construct a function ρ' on Ω, $\rho' \geq \rho$, with the following properties.*

(i) $\int_\Omega \rho'^p \, dx \leq \int_\Omega \rho^p \, dx + \varepsilon$.

(ii) *Suppose that for each j there is $\gamma_j \in \Gamma(K_0^j, K_1^j; \Omega)$, such that $\int_{\gamma_j} \rho' |\mathcal{B} \, d\gamma| \leq \alpha$. Then, there exists $\tilde{\gamma} \in \Gamma(K_0, K_1; \Omega)$, such that $\int_{\tilde{\gamma}} \rho |\mathcal{B} \, d\gamma| \leq \alpha + \varepsilon$.*

The proof of Lemma 4.1 on an arbitrary homogeneous group for the unit matrix $\mathcal{B}$ can be found in [27]. The case $\mathcal{B} \neq I$ is treated similarly.

LEMMA 4.2. *Suppose that D is a bounded domain in $\mathbb{H}^n$. Let $f \in L_p(D)$ and $\varepsilon > 0$. Then there exists a continuous function $\tilde{f}$ such that*

$$\|f - \tilde{f} \mid L_p(D)\| < \varepsilon.$$

PROOF. Let $D_n \subset \overline{D}_n \subset D_{n+1} \subset \overline{D}_{n+1} \subset \ldots \subset D$ be a sequence of open sets that exhaust the domain D. We assume that $D_{-1} = D_0 = \emptyset$. For each n, we find a positive function h_n such that $h_n \in C_0^\infty(D_{n+1} \setminus D_{n-2})$, $|\nabla h_n| \leq 1/6$, and $|h_n(x)| \leq 1/6 \min\{1, \operatorname{dist}(x, \partial D)\}$. Then the function $\eta(x) = \sum_{n=1}^\infty h_n(x)$ satisfies

(i) $|\nabla \eta(x)| \leq 1/2$,

(ii) $0 < \eta(x) \leq 1/2 \min\{1, \operatorname{dist}(x, \partial D)\}$.

Let $y \in \mathbb{H}^n$, $|y| \leq 1$, and $0 < t < \min\{1, C, \widehat{C}\}$, where the constants C, $\widetilde{C}$ are chosen later. Define a C^∞-map of the domain D onto itself by $T_{t,y}(x) = x \cdot \delta_{t\eta(x)} y$. We claim that $T_{t,y}$ is a homeomorphism. If $y = 0$, then $T_{t,0}$ is the identity map. Let $y \neq 0$. Since $0 < \eta(x) \leq 1/2 \operatorname{dist}(x, \partial D)$, the mapping $T_{t,y}$ transforms D to D. Let us show that $T_{t,y}$ is injective. Suppose there exist $x, x' \in D$ such that $T_{t,y}(x) = T_{t,y}(x')$. Applying the left translation and dilatation for the domain D, we can assume that $x' = 0$ and $|x| = 1$. In this case, we get $x \delta_{t\eta(x)} y = \delta_{t\eta(0)} y$ or $x = \delta_{t\eta(0)} y (\delta_{t\eta(x)} y)^{-1}$. The homogeneous norm $|\cdot|$ and the Euclidean norm $\|\cdot\|$ are connected by the inequality $C_1 \|x\| \leq |x| \leq C_2 \|x\|^{1/2}$, $x \in D$, where C_1, C_2 are positive constants (see, for instance, [17]). We deduce that

$$\begin{aligned}
(4.1) \quad |x| &= |\delta_{t\eta(0)} y (\delta_{t\eta(x)} y)^{-1}| \leq C_2 \|\delta_{t\eta(0)} y - \delta_{t\eta(x)} y\|^{1/2} \\
&\leq C_2 t^{1/2} |\eta(0) - \eta(x)|^{1/2} |P_1(\eta(0), \eta(x), y, t)|^{1/2}.
\end{aligned}$$

Here P_1 is a polynomial of the first order, which depends on $\eta(x)$, t, and the coordinates of the point y. Since $|y| \leq 1$, $0 < t \leq 1$, and $0 < \eta(x) \leq 1/2$, we have $|P_1| \leq C_3$. We estimate $|\eta(0) - \eta(x)| \leq |x|/2$ by (i). Taking into account these estimates, we conclude from (4.1) that $|x| \leq C_4 t^{1/2} |x|^{1/2}$. Since $|x| = 1$ for $t < C_0 = C_4^{-2}$, we obtain a contradiction.

Let us show that $T_{t,y}$ is surjective. Denote by $\omega(t)$ the curve $\delta_t y$. The intersection $\omega(t) \cap D$ is invariant under the map $T_{t,y}$ by (ii). This shows that the map is surjective.

The Jacobian matrix of $T_{t,y}(x)$ is equal to $I + t\hat{T}$, where I is the identity matrix and the elements of the matrix $\hat{T}$ depend on $t, x, y, \nabla \eta(x)$, and $\eta(x)$. Thus the Jacobian $J(T_{t,y})$ is of the form $1 + tH(t, x, y, \nabla \eta(x), \eta(x))$, where H is a polynomial. The properties of function η, the choice of y and t, and the boundedness of the domain D imply that $\max_{x \in D} |H| \leq C_5$, where the constant C_5 depends only on the diameter of D. If we choose $\widetilde{C} = 1/(2C_5)$, then we have $J(T_{t,y}) \geq 1 - tC_5 > 0$ for $t < \widetilde{C}$. This shows that the inverse map $T_{t,y}^{-1}$ is defined and smooth.

Let $\varphi(y)$ be a non-negative C^∞-function supported in the unit ball $|y| < 1$ such that $\int_{|y|<1} \varphi(y)\,dy = 1$. For $f \in L_p(D)$, we define

$$f_t(x) = \int_{|y|<1} f(x\delta_{t\eta(x)}y)\varphi(y)\,dy = \int_{\mathbb{H}^n} f(z)\varphi(\delta_{(t\eta(x))^{-1}}(x^{-1}z)) \frac{dz}{(t\eta(x))^{2n+2}}.$$

The function $f_t(x)$ is C^∞ in the domain D.

We show that $\|f_t - f \mid L_p(D)\| \to 0$ as $t \to 0$. Using the fact that continuous functions with compact support are dense in $L_p(D)$, we have

(4.2) $$\|f(xy) - f(x) \mid L_p(D)\| \to 0 \quad \text{as} \quad |y| \to 0.$$

Applying the Minkowski inequality to $f_t(x) - f(x) = \int_{|y|<1} (f(x\delta_{t\eta(x)}y) - f(x))\,\varphi(y)\,dy$, we obtain

$$\|f_t - f \mid L_p(D)\| \leq \int_{|y|<1} \|f(x\delta_{t\eta(x)}y) - f(x) \mid L_p(D)\| \varphi(y)\,dy.$$

From the property (4.2), the inequality

$$\|f(x\delta_{t\eta(x)}y) - f(x) \mid L_p(D)\| \leq 2\|f(x) \mid L_p(D)\|,$$

and the dominated convergence theorem, it follows that $\|f_t - f \mid L_p(D)\| \to 0$ as $t \to 0$. $\square$

THEOREM 4.1. *Let Ω be a bounded domain in $\mathbb{H}^n$. Let $\mathcal{B}(x)$ be uniformly continuous on $\Omega \setminus (K_0 \cup K_1)$ and $\mathcal{C} \subset \mathcal{F}M(|\mathcal{B}\,d\Gamma|)$ consist of continuous functions on $\Omega \setminus (K_0 \cup K_1)$. Then*

(4.3) $$M = \inf_{\hat{\rho} \in \mathcal{C}} \int_{\Omega \setminus (K_0 \cup K_1)} \hat{\rho}^2(x)\,dx = M(|\mathcal{B}\,d\Gamma|).$$

PROOF. Denote by D the domain $\Omega \setminus (K_0 \cup K_1)$ and let $\varepsilon \in (0, 1/2)$. Choose $\rho \in \mathcal{F}M(|\mathcal{B}\,d\Gamma|)$ with

(4.4) $$\int_D \rho^2(x)\,dx < \varepsilon + M(|\mathcal{B}\,d\Gamma|).$$

Then by Lemma 4.2, we can find a continuous function ρ_t in D such that

(4.5) $$\int_D \rho_t^2(x)\,dx < \varepsilon + \int_D \rho^2(x)\,dx.$$

We claim that for a sufficiently small t, the function $(1+\varepsilon)^2 \rho_t(x)$ is admissible for $M(|\mathcal{B}\,d\Gamma|)$.

The matrix $\mathcal{B}(x)$ is uniformly continuous. If $x, y \in \Omega \setminus (K_0 \cup K_1)$ and $d(x, y) < \varsigma(\varepsilon)$, then $|\mathcal{B}(x) - \mathcal{B}(y)| < \alpha^{-1}\varepsilon$. Hence, we obtain

$$(4.6) \quad |\mathcal{B}(y)\xi| \leq |\mathcal{B}(x)\xi| + |\mathcal{B}(x)\xi - \mathcal{B}(y)\xi| \leq |\mathcal{B}(x)\xi| + \alpha^{-1}\varepsilon|\xi| \leq (1+\varepsilon)|\mathcal{B}(x)\xi|$$

from the property (2.1) for the matrix $\mathcal{B}$.

We estimate

$$(4.7) \quad \int_\gamma \rho_t(x)\,|\mathcal{B}(x)\,d\gamma| = \int_\gamma \int_{|y|<1} \rho(x\delta_{t\eta(x)}y)\varphi(y)\,dy\,|\mathcal{B}(x)\,d\gamma|$$

$$= \int_{|y|<1} \varphi(y)\,dy \int_\gamma \rho(x\delta_{t\eta}y)\,|\mathcal{B}(x)\,d\gamma|.$$

Fix y for a moment and consider the integral $\int_\gamma \rho(x\delta_{t\eta}y)\,|\mathcal{B}(x)\,d\gamma|$. Denote by $\tilde{\gamma}$ the image of the curve γ under the map $T_{t,y}(x)|_\gamma$. Recall that the map $T_{t,y}$ has Jacobian matrix of the form $I + t\hat{T}$, where I is the identity matrix and the terms of the matrix $\hat{T}$ depend on $t, x, y, \nabla\eta(x)$, and $\eta(x)$. The properties of η, the choice of $|y|<1$ and $|t|<1$, and the boundedness of the domain D imply that the norm of $\hat{T}$ is bounded by a constant C that depends only on the diameter of D. It is obvious that the curve $\tilde{\gamma}$ connects the compacts K_0 and K_1. If the curve $\tilde{\gamma}$ is not horizontal, and is therefore not locally rectifiable,

$$\int_\gamma \rho(x\delta_{t\eta}y)\,|\mathcal{B}(x)\,d\gamma| \geq \alpha^{-1}\int_\gamma \rho(x\delta_{t\eta}y)\,|d\gamma| \geq \frac{1}{\alpha(1+tC)}\int_{\tilde{\gamma}} \rho(\tilde{\gamma})\,|d\tilde{\gamma}| = \infty.$$

Here α is the constant from (2.1). If the curve $\tilde{\gamma}$ is horizontal, then $\tilde{\gamma} \in \Gamma(K_0, K_1, \Omega)$. We choose t so small that $|x^{-1}x\delta_{t\eta(x)}y| = t\eta(x) \leq \varsigma(\varepsilon)$ and $t < \varepsilon/C$. Then

$$\int_\gamma \rho(x\delta_{t\eta}y)\,|\mathcal{B}(x)\,d\gamma| \geq \frac{1}{1+tC}\int_{\tilde{\gamma}} \rho(z)\,|\mathcal{B}(T_{t,y}^{-1}(z))\,d\tilde{\gamma}|$$

$$\geq \frac{1}{(1+\varepsilon)^2}\int_{\tilde{\gamma}} \rho(\tilde{\gamma})\,|\mathcal{B}((z))\,d\tilde{\gamma}| \geq \frac{1}{(1+\varepsilon)^2}.$$

Since $\int_{|y|<1}\varphi(y)\,dy = 1$, we conclude that $(1+\varepsilon)^2\rho_t(x) \in \mathcal{F}M(|\mathcal{B}d\Gamma|)$. It follows from (4.4) and (4.5) that

$$M = \inf_{\tilde{\rho}\in\mathcal{C}}\int_\Omega \tilde{\rho}^2(x)\,dx \leq (1+\varepsilon)^4\int_\Omega \rho_t^2(x)\,dx \leq (1+\varepsilon)^4\big(2\varepsilon + M(|\mathcal{B}.d\Gamma|)\big).$$

Since ε and $\rho \in \mathcal{F}M(|\mathcal{B}d\Gamma|)$ were arbitrary, we get $M \leq M(|\mathcal{B}d\Gamma|)$.

The reverse inequality is obvious, and we have (4.3) as desired. $\square$

PROOF OF THEOREM 2.3. Let $\varepsilon \in (0, 1/2)$. By definition, there is a non-negative function $\rho \in \mathcal{F}M(|\mathcal{B}\,d\Gamma|)$ with $\|\rho\|^2_{L_2(\Omega)} \leq M(|\mathcal{B}\,d\Gamma|) + \varepsilon$. We may assume that ρ is strictly positive on $\Omega \setminus K_0 \cup K_1$. If this were not so, we could consider the cut-off-function $\max(\rho, \frac{1}{m})$ instead of ρ. Moreover, we can suppose that ρ is continuous on $\Omega \setminus K_0 \cup K_1$ by Theorem 4.1.

Let ρ' be as in Lemma 4.1. Then $\int_\gamma \rho'\,|\mathcal{B}\,d\Gamma| > 1 - 2\varepsilon$ for any $\gamma \in \Gamma(K_0^j, K_1^j; \Omega)$ for sufficiently large j. In fact, supposing the contrary, we would have a sequence $\{j_k\}$ and curves $\gamma_{j_k} \in \Gamma(K_0^{j_k}, K_1^{j_k}; \Omega)$ such that $\int_{\gamma_{j_k}} \rho'\,|\mathcal{B}\,d\Gamma| \le 1 - 2\varepsilon$, so by Lemma 4.1, we could find $\tilde{\gamma} \in \Gamma(K_0, K_1; \Omega)$ with $\int_{\tilde{\gamma}} \rho\,|\mathcal{B}\,d\Gamma| \le 1 - 2\varepsilon + \varepsilon = 1 - \varepsilon$, which contradicts $\rho \in \mathcal{F}M(|\mathcal{B}\,d\Gamma|)$.

Now we can finish the proof. We have $(1 - 2\varepsilon)^{-1}\rho' \in \mathcal{F}M(|\mathcal{B}\,d\Gamma_j|)$, $\Gamma_j = \Gamma(K_0^j, K_1^j; \Omega)$ for sufficiently large j; therefore,

$$M(|\mathcal{B}\,d\Gamma_j|) \le \int_{\Omega \setminus K_0 \cup K_1} [(1 - 2\varepsilon)^{-1}\rho']^2\,dx \le (1 - 2\varepsilon)^{-2}(M(|\mathcal{B}\,d\Gamma|) + \varepsilon).$$

Hence, letting $j \to \infty$ and $\varepsilon \to 0$, we obtain $\limsup\limits_{j \to \infty} M(|\mathcal{B}\,d\Gamma_j|) \le M(|\mathcal{B}\,d\Gamma|)$. Since $M(|\mathcal{B}\,d\Gamma|) \le M(|\mathcal{B}\,d\Gamma_j|)$ for arbitrary j, we obtain the statement of Theorem 2.3. $\square$

5. Proof of Theorem 2.1

Let K_0^j and K_1^j be sequences of compacts which tend to K_0 and to K_1, respectively, and which satisfy the conditions stated at the beginning of Section 4. We take $u \in \mathcal{F}C(K_0^j, K_1^j; \Omega)$ and put

$$\overline{u} = \begin{cases} 0 & \text{on } K_0^j \cap \Omega, \\ 1 & \text{on } K_1^j \cap \Omega, \\ u & \text{on } \Omega \setminus (K_0^j \cup K_1^j). \end{cases}$$

By the definition of $\mathcal{F}C(K_0^j, K_1^j; \Omega)$, $u(x) \to 0$ along almost all curves as $x \to K_0^j$ and $u(x) \to 1$ on almost all curves as $x \to K_1^j$. Thus, $\overline{u}$ is absolutely continuous on almost all curves in Ω, and $\overline{u} \in \mathcal{F}C^\star(K_0, K_1; \Omega)$. Hence

$$\operatorname{cap}^\star_{\mathcal{A}}(K_0, K_1; \Omega) \le \int_\Omega |\mathcal{A}\nabla_0 \overline{u}|^2\,dx = \int_{\Omega \setminus (K_0^j \cup K_1^j)} |\mathcal{A}\nabla_0 u|^2\,dx.$$

Taking the infimum with respect to u, we obtain $\operatorname{cap}^\star_{\mathcal{A}}(K_0, K_1; \Omega) \le \operatorname{cap}_{\mathcal{A}}(K_0^j, K_1^j; \Omega)$. Theorem 2.2 implies that $\operatorname{cap}_{\mathcal{A}}(K_0^j, K_1^j; \Omega) = M(|\mathcal{B}\,d\Gamma_j|)$ and $\operatorname{cap}_{\mathcal{A}}(K_0, K_1; \Omega) = M(|\mathcal{B}\,d\Gamma|)$. The module $M(|\mathcal{B}\,d\Gamma_j|)$ tends to $M(|\mathcal{B}\,d\Gamma|)$ as $j \to \infty$ by Theorem 2.3. Hence

$$\operatorname{cap}^\star_{\mathcal{A}}(K_0, K_1; \Omega) \le \operatorname{cap}_{\mathcal{A}}(K_0, K_1; \Omega).$$

The reverse inequality holds by the inclusion $\mathcal{F}C^\star(K_0, K_1; \Omega) \subset \mathcal{F}C(K_0, K_1; \Omega)$. $\square$

References

[1] L. V. Ahlfors and A. Beurling, *Conformal invariants and function theoretic null sets,* Acta Math. **83** (1950), 101–129.

[2] H. Aikawa and M. Ohtsuka, *Extremal length of vector measures,* Ann. Acad. Sci. Fenn. Math. **24** (1999), 61–88.

[3] L. Capogna, D. Danielli and N. Garofalo, *Capacitary estimates and the local behavior of solutions of nonlinear subelliptic equations,* Amer. J. Math. **118** (1996), 1153–1196.

[4] V. M. Chernikov and S. K. Vodop'yanov, *Sobolev spaces and hypoelliptic equations. I,* Siberian Adv. Math. **6** (1996), no.3, 27–67.

[5] _____, *Sobolev spaces and hypoelliptic equations. II,* Siberian Adv. Math. **6** (1996), no.4, 64–96.

[6] W. L. Chow, *Über Systeme von linearen partiellen Differentialgleichungen erster Ordnung,* Math. Ann. **117** (1939), 98–105.

[7] L. Corwin and F. P. Greenleaf, *Representation of Nilpotent Lie Groups and their Applications, Part 1: Basic Theory and Examples,* Cambridge University Press, 1990.

[8] N. S. Dairbekov, *Mappings with bounded distortion on two-step Carnot groups,* Proceedings on Analysis and Geometry (Novosibirsk, Akademgorodok, 1999), Sobolev Institute Press, Novosibirsk, 2000, pp. 122–155.

[9] D. Danielli, *Regularity at the boundary for solutions of nonlinear subelliptic equations.* Indiana Univ. Math. J. **44** (1995), 269–286.

[10] D. Danielli and N. Garofalo, *Geometric properties of solutions to subelliptic equations in nilpotent Lie groups,* Reaction Diffusion Systems (Trieste, 1995), Dekker, New York, 1998, pp. 89–105.

[11] H. Federer, *Geometric Measure Theory,* Berlin, Springer-Verlag, 1969.

[12] G. B. Folland and E. M. Stein, *Estimates for the $\bar{\partial}_b$ complex and analysis on the Heisenberg group.* Comm. Pure Appl. Math. **27** (1974), 429–522.

[13] _____, *Hardy Spaces on Homogeneous Groups,* Princeton University Press, Princeton, NJ, 1982.

[14] B. Fuglede, *Extremal length and functional completion,* Acta. Math. **98** (1957), 171–219.

[15] A. V. Greshnov and S. K. Vodop'yanov, *Analytic properties of quasiconformal mappings on Carnot groups,* Sibirsk. Mat. Zh. **36** (1995), 1317–1327; translation in Siberian Math. J. **36** (1995), 1142–1151.

[16] H. Grötzsch, *Über einige Extremalprobleme der konformen Abbildungen. I–II,* Ber. Verh.-Sächs. Akad. Wiss. Leipzig, Math.–Phys. Kl. **80** (1928), 367–376.

[17] P. Hajłasz and P. Koskela, *Sobolev met Poincaré,* Mem. Amer. Math. Soc. **145** (2000), no. 688.

[18] J. Heinonen and I. Holopainen, *Quasiregular mappings on Carnot group,* J. Geom. Anal. **7** (1997), 109–148.

[19] J. Heinonen and T. Kilpeläinen, *Polar sets for supersolutions of degenerate elliptic equations,* Math. Scand. **63** (1988), 136–150.

[20] J. Heinonen, T. Kilpeläinen and O. Martio, *Nonlinear Potential Theory of Degenerate Elliptic Equations,* Oxford University Press, New York, 1993.

[21] I. Holopainen and S. Rickman, *Classification of Riemannian manifolds in nonlinear potential theory,* Potential Anal. **2** (1993), 37–66.

[22] L. Hörmander, *Hypoelliptic second order differential equations,* Acta Math. **119** (1967), 147–171.

[23] A. Korányi, *Geometric aspects of analysis on the Heisenberg group,* Topics in Modern Harmonic Analysis, Ist. Naz. Alta Mat. Francesco Severi, Roma, 1983, pp. 209-258.

[24] A. Korányi and H. M. Reimann, *Foundation for the theory of quasiconformal mappings on the Heisenberg group,* Adv. Math. **111** (1995), 1–87.

[25] P. Lindqvist and O. Martio, *Two theorems of N. Wiener for solutions of quasilinear elliptic equations,* Acta Math. **155** (1985), 153–171.

[26] I. Markina, *Extremal length for quasiregular mappings on Heisenberg groups,* J. Math. Anal. Appl. **284** (2003), 532–547.

[27] _____, *On coincidence of the p-module of a family of curves and the p-capacity on the Carnot group,* Rev. Mat. Iberoamericana **19** (2003), 143–160.

[28] I. G. Markina and S. K. Vodop'yanov, *Fundamentals of the nonlinear potential theory for subelliptic equations. I,* Siberian Adv. Math. **7** (1997), no.1, 32–62.

[29] _____, *Fundamentals of the nonlinear potential theory for subelliptic equations. II,* Siberian Adv. Math. **7** (1997), no.2, 18–63.

[30] _____, *Classification of sub-Riemannian manifolds,* Sibirsk. Mat. Zh. **39** (1998), 1271–1289; translation in Siberian Math. J. **39** (1998), 1096–1111.

[31] G. D. Mostow, *Quasi-conformal mappings in n-space and the rigidity of hyperbolic space forms,* Inst. Hautes Études Sci. Publ. Math. **34** (1968), 53–104.

[32] Yu. G. Reshetnyak, *Sobolev classes of functions with values in a metric space,* Sibirsk. Mat. Zh. **38** (1997), 657–675; translation in Siberian Math. J. **38** (1997), 567–583.

[33] N. Shanmugalingam, *Newtonian spaces: an extension of Sobolev spaces to metric measure spaces,* Rev. Mat. Iberoamericana **16** (2000), 243–279.

[34] V. A. Shlyk, *On the equality between p-capacity and p-modulus,* Sibirsk. Mat. Zh. **34** (1993), 216–221; translation in Siberian Math. J. **34** (1993), 1196–1200.

[35] E. M. Stein, *Some problems in harmonic analysis suggested by symmetric spaces and semi-simple groups,* Actes du Congrès International des Mathématiciens (Nice, 1970), Tome 1, Gauthier–Villars, Paris, 1971, pp. 173–189.

[36] E. M. Stein, *Harmonic Analysis: Real-Variable Methods, Orthogonality, and Oscillatory Integrals,* Princeton Univ. Press, Princeton, NJ, 1993.

[37] A. D. Ukhlov and S. K. Vodop'yanov, *Sobolev spaces and (P,Q)-quasiconformal mappings of Carnot groups,* Sibirsk. Mat. Zh. **39** (1998), 776–795, translation in Siberian Math. J. **39** (1998), 665–682.

[38] S. K. Vodop'yanov, *Monotone functions and quasiconformal mappings on Carnot groups,* Sibirsk. Mat. Zh. **37** (1996), 1269–1295; translation in Siberian Math. J. **37** (1996), 1113–1136.

[39] _____, $\mathcal{P}$-*differentiability of mappings of Sobolev classes on the Carnot group,* Mat. Sb. **194** (2003), 67–86.

DEPARTAMENTO DE MATEMÁTICA, UNIVERSIDAD TÉCNICA FEDERICO SANTA MARÍA, CASILLA 110-V, VALPARAÍSO, CHILE

E-mail address: `irina.markina@usm.cl`

An Analogue of the Fuglede Formula in Integral Geometry on Matrix Spaces

Elena Ournycheva and Boris Rubin

Dedicated to Professor Lawrence Zalcman on the occasion of his 60th birthday

ABSTRACT. The well-known formula of B. Fuglede expresses the mean value of the Radon k-plane transform on $\mathbb{R}^n$ as a Riesz potential. We generalize this formula for the space of $n \times m$ real matrices and show that the corresponding matrix k-plane transform $f \to \hat{f}$ is injective if and only if $n - k \geq m$. Various inversion formulas for this transform are obtained. We assume that $f \in L^p$ or f is a continuous function satisfying certain "minimal" conditions at infinity.

1. Introduction

In 1958, B. Fuglede [**Fu**] proved the remarkable formula

(1.1) $$c(\hat{f})^{\vee}(x) = (I^k f)(x), \qquad x \in \mathbb{R}^n,$$

where $\hat{f} \equiv \hat{f}(\tau)$ is the integral of $f(x)$ over a k-dimensional plane τ, $0 < k < n$; $(\hat{f})^{\vee}(x)$ is the mean value of $\hat{f}(\tau)$ over all k-planes through x, $(I^k f)(x)$ denotes the Riesz potential of order k, and $c = c(n, k)$ is a constant. For $k = n - 1$, when τ is a hyperplane, this formula was implicitly exhibited by J. Radon [**R**], who was indebted to W. Blaschke for this idea. A consequence of (1.1) is the inversion formula

(1.2) $$f = c(-\Delta)^{k/2}(\hat{f})^{\vee},$$

in which Δ denotes the Laplace operator.

Our aim is to extend (1.1) and (1.2) to the case where x is an $n \times m$ real matrix. We note that Riesz potentials of functions of matrix argument and their generalizations arise in different contexts in harmonic analysis, integral geometry, and PDE; see [**Far**], [**FK**], [**Ge**], [**Kh**], [**Ra**], [**St1**], [**Sh1**]-[**Sh3**]. A systematic study of Radon transforms on matrix spaces was initiated by E.E. Petrov [**Pe1**] and continued in [**Č**], [**Gr**], [**Pe2**]-[**Pe4**], [**Sh1**], [**Sh2**]. These publications traditionally

---

2000 *Mathematics Subject Classification.* Primary 44A12; Secondary 47G10.

Key words and phrases. Matrix Radon transform, Fourier transform, Riesz potential, inversion formulas.

The work was supported in part by the Edmund Landau Center for Research in Mathematical Analysis and Related Areas, sponsored by the Minerva Foundation (Germany).

deal with Radon transforms $f \to \hat{f}$ of C^∞ rapidly decreasing functions and employ decomposition in plane waves in order to recover f from $\hat{f}$.

We suggest another approach, which is based on a matrix analogue of (1.1) (see Theorem 5.4) and allows us to handle arbitrary continuous or locally integrable functions f subject to mild restrictions at infinity. These restrictions are minimal in a certain sense. We show that the inequality $n - k \geq m$ is necessary and sufficient for injectivity of the matrix k-plane transform and obtain two inversion formulas (in terms of the Fourier transform and in the form (1.2)).

Our key motivation is the following. Recasting integral-geometrical entities into their matrix counterparts has specific "higher rank" features and sheds new light on classical problems in multidimensional spaces (just note the space $\mathbb{R}^N$ with $N = nm$ can be treated as a collection of $n \times m$ matrices).

Acknowledgements. We are grateful to Prof. E.E. Petrov and Dr. S.P. Khekalo for the papers they sent us and useful discussions.

2. Preliminaries

We establish some notation and recall basic facts. Let $\mathfrak{M}_{n,m}$ be the space of real matrices $x = (x_{i,j})$ having n rows and m columns. We identify $\mathfrak{M}_{n,m}$ with the real Euclidean space $\mathbb{R}^{nm}$ and set $dx = \prod_{i=1}^{n} \prod_{j=1}^{m} dx_{i,j}$ for the Lebesgue measure on $\mathfrak{M}_{n,m}$. In the following, x' denotes the transpose of x, I_m is the identity $m \times m$ matrix, and 0 stands for zero entries. Given a square matrix a, we denote by $|a|$ the absolute value of the determinant of a, and by $\mathrm{tr}(a)$ the trace of a, respectively. $GL(m, \mathbb{R})$ is the group of real non-singular $m \times m$ matrices; $SO(n)$ is the group of orthogonal $n \times n$ real matrices of determinant one, endowed with the normalized invariant measure. We denote by $\mathcal{P}_m$ the cone of positive definite symmetric matrices $r = (r_{i,j})$ with the elementary volume $dr = \prod_{i \leq j} dr_{i,j}$. The Lebesgue space $L^p(\mathfrak{M}_{n,m})$ and the Schwartz space $\mathcal{S}(\mathfrak{M}_{n,m})$ are identified with respective spaces on $\mathbb{R}^{nm}$.

LEMMA 2.1. (see, e.g., [**Mu**, pp. 57–59]).

(i) If $x = ayb$, where $y \in \mathfrak{M}_{n,m}$, $a \in GL(n, \mathbb{R})$, $b \in GL(m, \mathbb{R})$, then $dx = |a|^m |b|^n dy$.

(ii) If $r = q'sq$, where $s \in \mathcal{P}_m$, $q \in GL(m, \mathbb{R})$, then $dr = |q|^{m+1} ds$.

(iii) If $r = s^{-1}$, where $s \in \mathcal{P}_m$, then $r \in \mathcal{P}_m$ and $dr = |s|^{-m-1} ds$.

The Siegel gamma function associated to the cone $\mathcal{P}_m$ is defined by

$$(2.1) \qquad \Gamma_m(\alpha) = \int_{\mathcal{P}_m} \exp(-\mathrm{tr}(r)) |r|^{\alpha - d} dr, \qquad d = (m+1)/2.$$

This integral converges absolutely if and only if $\mathrm{Re}\,\alpha > d - 1$, and represents a product of ordinary Γ-functions:

$$(2.2) \qquad \Gamma_m(\alpha) = \pi^{m(m-1)/4} \Gamma(\alpha) \Gamma\left(\alpha - \frac{1}{2}\right) \ldots \Gamma\left(\alpha - \frac{m-1}{2}\right).$$

If $1 \leq k < m$, $k \in \mathbb{N}$, then

$$(2.3) \qquad \Gamma_m(\alpha) = \pi^{k(m-k)/2} \Gamma_k(\alpha) \Gamma_{m-k}(\alpha - k/2),$$

$$\frac{\Gamma_m(\alpha)}{\Gamma_m(\alpha+k/2)} = \frac{\Gamma_k(\alpha+(k-m)/2)}{\Gamma_k(\alpha+k/2)}. \tag{2.4}$$

For $n \geq m$, let $V_{n,m} = \{v \in \mathfrak{M}_{n,m} : v'v = I_m\}$ be the Stiefel manifold of orthonormal m-frames in $\mathbb{R}^n$; $V_{n,n} = O(n)$ is the orthogonal group in $\mathbb{R}^n$. We fix the invariant measure dv on $V_{n,m}$ [**Mu**, p. 70] normalized by

$$\sigma_{n,m} \equiv \int_{V_{n,m}} dv = \frac{2^m \pi^{nm/2}}{\Gamma_m(n/2)}. \tag{2.5}$$

LEMMA 2.2. (polar decomposition; see, e.g., [**Mu**, pp. 66, 591], [**Ma**]). Let $x \in \mathfrak{M}_{n,m}$, $n \geq m$. If $\operatorname{rank}(x) = m$, then

$$x = vr^{1/2}, \qquad v \in V_{n,m}, \qquad r = x'x \in \mathcal{P}_m,$$

and $dx = 2^{-m}|r|^{(n-m-1)/2}drdv$.

3. Riesz potentials

The Riesz potential of order $\alpha \in \mathbb{C}$ on $\mathfrak{M}_{n,m}$, $n \geq m$, is defined by

$$(I^\alpha f)(x) = \frac{1}{\gamma_{n,m}(\alpha)} \int_{\mathfrak{M}_{n,m}} f(x-y)|y|_m^{\alpha-n} dy, \tag{3.1}$$

$$\gamma_{n,m}(\alpha) = \frac{2^{\alpha m} \pi^{nm/2} \Gamma_m(\alpha/2)}{\Gamma_m((n-\alpha)/2)}, \qquad |y|_m = \det(y'y)^{1/2},$$

provided this expression is finite. Note that for $m \geq 2$, the sets of poles of $\Gamma_m(\alpha/2)$ and $\Gamma_m((n-\alpha)/2)$ are $\{m-1, m-2, \ldots\}$, and $\{n-m+1, n-m+2, \ldots\}$ respectively. These sets overlap if and only if $2m \geq n+2$ (keep this inequality in mind!). For $m=1$, the set of poles of $\Gamma_m((n-\alpha)/2)$ is $\{n, n+2, n+4, \ldots\}$. Thus if

$$\alpha = \begin{cases} n-m+1, \, n-m+2, \ldots & \text{for } m \geq 2, \\ n, \, n+2, \ldots & \text{for } m = 1, \end{cases}$$

then the coefficient $1/\gamma_{n,m}(\alpha)$ is infinite. In the following, we exclude these values of α and focus on the case $m \geq 2$.

The Riesz distribution corresponding to (3.1) is defined by

$$(h_\alpha, f) = a.c. \frac{1}{\gamma_{n,m}(\alpha)} \int_{\mathfrak{M}_{n,m}} |x|_m^{\alpha-n} f(x)dx, \tag{3.2}$$

where "a.c." abbreviates analytic continuation in the α-variable and $f \in \mathcal{S}(\mathfrak{M}_{n,m})$. The following lemma summarizes the basic properties of h_α.

LEMMA 3.1. Let $m \geq 2$.
(i) The integral in (3.2) converges absolutely if and only if $\operatorname{Re}\alpha > m-1$.
(ii) The distribution h_α extends to all $\alpha \in \mathbb{C}$ as a meromorphic $\mathcal{S}'$-distribution whose only poles are at the points $\alpha = n-m+1, \, n-m+2, \ldots$. The order of these poles is the same as in $\Gamma_m((n-\alpha)/2)$.
(iii) If α belongs to the set

$$\mathcal{W} = \mathcal{W}_1 \cap \mathcal{W}_2, \tag{3.3}$$

$$\mathcal{W}_1 = \{0, 1, 2, \ldots, k_0\}, \qquad k_0 = \min(m-1, n-m),$$
$$\mathcal{W}_2 = \{\alpha : \alpha > m-1; \, \alpha \neq n-m+1, n-m+2, \ldots\},$$

then h_α is a positive measure.

(iv) If $\alpha \neq n-m+1, n-m+2, \ldots$, and

$$(3.4) \qquad (\mathcal{F}f)(y) = \int_{\mathfrak{M}_{n,m}} \exp(\mathrm{tr}(iy'x))f(x)dx$$

is the Fourier transform of $f \in \mathcal{S}(\mathfrak{M}_{n,m})$, then

$$(3.5) \qquad (h_\alpha, f) = (2\pi)^{-nm}(|y|_m^{-\alpha}, (\mathcal{F}f)(y))$$

in the sense of analytic continuation. In particular,

$$(3.6) \qquad (h_0, f) = f(0).$$

These statements can be found in [**Kh**], [**OR**], [**Ru3**]; see also [**FK**], [**Sh1**], [**Sh3**]. The set (3.3) is an analogue of the Wallach set in [**FK**, p. 137]. We adopt this name for our case too. The formula (3.5) is known as the functional equation for the zeta function $\alpha \to (h_\alpha, f)$. Different proofs of (3.5) can be found in [**Ge**, Chapter IV] and [**Ra**, Prop. II-9] for $n = m$. Concerning more general settings, see [**Far**], [**FK**, Theorem XVI.4.3], [**Ru3**], and references therein; see also [**Sh1**] and [**St1**].

Note that the number k_0 in (3.3) is chosen so that the discrete part $\{0, 1, 2, \ldots, k_0\}$ of $\mathcal{W}$ has no common points with either the continuous part $\{\alpha : \alpha > m - 1\}$ or the set of poles $\{n-m+1, n-m+2, \ldots\}$.

In the case $2m < n+2$, when poles of $\Gamma_m(\alpha/2)$ and $\Gamma_m((n-\alpha)/2)$ do not overlap, a simple proof of (3.5) can be given following a slight modification of the argument from [**St2**, Chapter III, Sec. 3.4]. Let $e_s(x) = \exp(\mathrm{tr}(-xsx')/4\pi)$, $s \in \mathcal{P}_m$. By the Plancherel formula,

$$(3.7) \qquad |s|^{-n/2} \int_{\mathfrak{M}_{n,m}} (\mathcal{F}f)(y) \exp(\mathrm{tr}(-\pi y s^{-1} y'))dy = \int_{\mathfrak{M}_{n,m}} f(x)e_s(x)dx.$$

Multiplying (3.7) by $|s|^{(n-\alpha)/2-d}$, $d = (m+1)/2$, and integrating in $s \in \mathcal{P}_m$, after changing the order of integration, we obtain

$$\int_{\mathfrak{M}_{n,m}} (\mathcal{F}f)(y) a(y) dy = \int_{\mathfrak{M}_{n,m}} f(x) b(x) dx,$$

where

$$\begin{aligned} a(y) &= \int_{\mathcal{P}_m} |s|^{-\alpha/2-d} \exp(\mathrm{tr}(-\pi y s^{-1} y'))ds \qquad (s = t^{-1}) \\ &= \int_{\mathcal{P}_m} |t|^{\alpha/2-d} \exp(\mathrm{tr}(-\pi t y' y))dt = \Gamma_m(\alpha/2)\pi^{-\alpha m/2}|y|_m^{-\alpha} \end{aligned}$$

if $\operatorname{Re}\alpha > m-1$ and

$$b(x) = \int_{\mathcal{P}_m} |s|^{(n-\alpha)/2-d} e_s(x) ds = \frac{(4\pi)^{m(n-\alpha)/2} \Gamma_m((n-\alpha)/2)}{|x|_m^{n-\alpha}}$$

if $\operatorname{Re}\alpha < n-m+1$. A simple computation of these integrals is performed using (2.1) and Lemma 2.1. Thus (3.5) follows if the set $m-1 < \operatorname{Re}\alpha < n-m+1$ is not vacuous, i.e., $2m < n + 2$. We note that if $2m \geq n + 2$, then the distributions on the two sides of (3.5) are not regular simultaneously.

Explicit representations for h_α in the discrete part of the Wallach set (3.3) play a vital role in our considerations.

LEMMA 3.2. *Let $f \in \mathcal{S}(\mathfrak{M}_{n,m})$, $k_0 = \min(m - 1, n - m)$, $m \geq 2$. Then for $k = 1, 2, \ldots, k_0$,*

$$(3.8) \qquad (h_k, f) = c \int_{\mathfrak{M}_{k,m}} du \int_{SO(n)} f\left(\gamma \begin{bmatrix} u \\ 0 \end{bmatrix}\right) d\gamma,$$

$$(3.9) \qquad c = 2^{-km} \pi^{-km/2} \Gamma_m\left(\frac{n-k}{2}\right) / \Gamma_m\left(\frac{n}{2}\right).$$

PROOF. We split $x \in \mathfrak{M}_{n,m}$ into two blocks $x = [y; b]$ where $y \in \mathfrak{M}_{n,k}$ and $b \in \mathfrak{M}_{n,m-k}$. Then for $\operatorname{Re}\alpha > m - 1$,

$$(h_\alpha, f) = \frac{1}{\gamma_{n,m}(\alpha)} \int_{\mathfrak{M}_{n,k}} dy \int_{\mathfrak{M}_{n,m-k}} f([y;b]) \left| \begin{matrix} y'y & y'b \\ b'y & b'b \end{matrix} \right|^{(\alpha-n)/2} db,$$

where $\left| \begin{matrix} * & * \\ * & * \end{matrix} \right|$ denotes the determinant of the respective matrix $\begin{bmatrix} * & * \\ * & * \end{bmatrix}$. By passing to polar coordinates (see Lemma 2.2) $y = vr^{1/2}$, $v \in V_{n,k}$, $r \in \mathcal{P}_k$, we have

$$(h_\alpha, f) = \frac{2^{-k}}{\gamma_{n,m}(\alpha)} \int_{V_{n,k}} dv \int_{\mathcal{P}_k} |r|^{(n-k-1)/2} dr$$

$$\times \int_{\mathfrak{M}_{n,m-k}} f([vr^{1/2}; b]) \left| \begin{matrix} r & r^{1/2}v'b \\ b'vr^{1/2} & b'b \end{matrix} \right|^{(\alpha-n)/2} db$$

$$= \frac{2^{-k}\sigma_{n,k}}{\gamma_{n,m}(\alpha)} \int_{SO(n)} d\gamma \int_{\mathcal{P}_k} |r|^{(n-k-1)/2} dr$$

$$\times \int_{\mathfrak{M}_{n,m-k}} f_\gamma([\lambda_0 r^{1/2}; b]) \left| \begin{matrix} r & r^{1/2}\lambda_0'b \\ b'\lambda_0 r^{1/2} & b'b \end{matrix} \right|^{(\alpha-n)/2} db.$$

Here

$$\lambda_0 = \begin{bmatrix} I_k \\ 0 \end{bmatrix} \in V_{n,k}, \qquad f_\gamma(x) = f(\gamma x).$$

We write

$$b = \begin{bmatrix} b_1 \\ b_2 \end{bmatrix}, \qquad b_1 \in \mathfrak{M}_{k,m-k}, \qquad b_2 \in \mathfrak{M}_{n-k,m-k}.$$

Since $\lambda_0'b = b_1$,

$$(h_\alpha, f) = \frac{2^{-k}\sigma_{n,k}}{\gamma_{n,m}(\alpha)} \int_{SO(n)} d\gamma \int_{\mathcal{P}_k} |r|^{(n-k-1)/2} dr \int_{\mathfrak{M}_{k,m-k}} db_1$$

$$\times \int_{\mathfrak{M}_{n-k,m-k}} f_\gamma\left(\begin{bmatrix} r^{1/2} & b_1 \\ 0 & b_2 \end{bmatrix}\right) \left| \begin{matrix} r & r^{1/2}b_1 \\ b_1'r^{1/2} & b_1'b_1 + b_2'b_2 \end{matrix} \right|^{(\alpha-n)/2} db_2.$$

Note that
$$\begin{bmatrix} r & r^{1/2}b_1 \\ b_1'r^{1/2} & b_1'b_1 + b_2'b_2 \end{bmatrix} = \begin{bmatrix} r & 0 \\ b_1'r^{1/2} & I_{m-k} \end{bmatrix} \begin{bmatrix} I_k & r^{-1/2}b_1 \\ 0 & b_2'b_2 \end{bmatrix}$$

and

$$\begin{vmatrix} r & r^{1/2}b_1 \\ b_1'r^{1/2} & b_1'b_1 + b_2'b_2 \end{vmatrix} = \det(r)\det(b_2'b_2);$$

see, e.g., [**Mu**, p. 577]. Therefore,

$$(3.10) \qquad (h_\alpha, f) = c_\alpha \int_{SO(n)} d\gamma \int_{\mathcal{P}_k} |r|^{(\alpha-k-1)/2} dr \int_{\mathfrak{M}_{k,m-k}} \psi_{\alpha-k}(\gamma, r, b_1) db_1,$$

where

$$c_\alpha = \frac{2^{-k} \sigma_{n,k} \gamma_{n-k,m-k}(\alpha-k)}{\gamma_{n,m}(\alpha)},$$

$$\psi_{\alpha-k}(\gamma, r, b_1) = \frac{1}{\gamma_{n-k,m-k}(\alpha-k)} \int_{\mathfrak{M}_{n-k,m-k}} f_\gamma\left(\begin{bmatrix} r^{1/2} & b_1 \\ 0 & b_2 \end{bmatrix}\right)$$
$$\times |b_2'b_2|^{\frac{(\alpha-k)-(n-k)}{2}} db_2.$$

The last expression is the Riesz distribution of order $\alpha - k$ in the b_2-variable. Owing to (3.6), analytic continuation of (3.10) at $\alpha = k$ reads

$$(h_\alpha, f) = c_k \int_{SO(n)} d\gamma \int_{\mathcal{P}_k} |r|^{-1/2} dr \int_{\mathfrak{M}_{k,m-k}} f_\gamma\left(\begin{bmatrix} r^{1/2} & b_1 \\ 0 & 0 \end{bmatrix}\right) db_1,$$

$c_k = \lim_{\alpha \to k} c_\alpha$. To transform this expression, we replace γ by $\gamma \begin{bmatrix} \beta & 0 \\ 0 & I_{n-k} \end{bmatrix}$, $\beta \in O(k)$, and integrate in β against the normalized measure $d\beta$ (so that $\int_{O(k)} d\beta = 1$). We have

$$(h_\alpha, f) = c_k \int_{\mathcal{P}_k} |r|^{-1/2} dr \int_{O(k)} d\beta \int_{\mathfrak{M}_{k,m-k}} db_1 \int_{SO(n)} f_\gamma\left(\begin{bmatrix} \beta r^{1/2} & \beta b_1 \\ 0 & 0 \end{bmatrix}\right) d\gamma$$

(set $\zeta = \beta b_1$, $\eta = b|r|^{1/2}$ and use Lemma 2.2)

$$= \frac{2^k c_k}{\sigma_{k,k}} \int_{\mathfrak{M}_{k,k}} d\eta \int_{\mathfrak{M}_{k,m-k}} d\zeta \int_{SO(n)} f_\gamma\left(\begin{bmatrix} \eta & \zeta \\ 0 & 0 \end{bmatrix}\right) d\gamma$$

$$= c \int_{\mathfrak{M}_{k,m}} du \int_{SO(n)} f\left(\gamma \begin{bmatrix} u \\ 0 \end{bmatrix}\right) d\gamma,$$

$$c = \frac{\sigma_{n,k}}{\sigma_{k,k}} \lim_{\alpha \to k} \frac{\gamma_{n-k,m-k}(\alpha-k)}{\gamma_{n,m}(\alpha)} = 2^{-km} \pi^{-km/2} \Gamma_m\left(\frac{n-k}{2}\right)/\Gamma_m\left(\frac{n}{2}\right).$$

(Here we have used formulae (2.3) and (2.4).) $\square$

According to Lemma 3.2 and (3.6), one can redefine the Riesz potential $I^\alpha f$ for any locally integrable function f as follows.

DEFINITION 3.3. Let $m \geq 2$, $k_0 = \min(m-1, n-m)$, $k = 1, \ldots, k_0$, and c be the constant (3.9). We set

$$(3.11) \quad (I^\alpha f)(x) = \begin{cases} \dfrac{1}{\gamma_{n,m}(\alpha)} \displaystyle\int\limits_{\mathfrak{M}_{n,m}} f(x-y)|y|_m^{\alpha-n} dy \\ \quad \text{if } \operatorname{Re}\alpha > m-1; \quad \alpha \neq n-m+1, n-m+2, \ldots, \\[1em] c \displaystyle\int\limits_{\mathfrak{M}_{k,m}} du \int\limits_{SO(n)} f\left(x - \gamma \begin{bmatrix} u \\ 0 \end{bmatrix}\right) d\gamma \text{ if } \alpha = k. \end{cases}$$

It is assumed that f is sufficiently regular that the corresponding integrals are absolutely convergent.

THEOREM 3.4. [**Ru3**] *For $f \in L^p(\mathfrak{M}_{n,m})$, the integrals in (3.11) converge absolutely if*

$$(3.12) \quad 1 \leq p < n/(\operatorname{Re}\alpha + m - 1).$$

In particular, if f is continuous and $f(x) = O(|I_m + x'x|^{-\lambda/2})$, then absolute convergence holds provided $\lambda > (\operatorname{Re}\alpha + m - 1)(n + m - 1)/n$.

REMARK 3.5. Another formula for h_k obtained in [**Sh1**] and [**Kh**] reads

$$(3.13) \quad (h_k, f) = c_1 \int\limits_{\mathfrak{M}_{n,k}} \frac{dy}{|y|_k^{n-m}} \int\limits_{\mathfrak{M}_{k,m-k}} f([y; yz]) dz, \quad k = 1, 2, \ldots, k_0.$$

$$(3.14) \quad c_1 = 2^{-km} \pi^{k(k-n-m)/2} \Gamma_k\left(\frac{n-m}{2}\right) / \Gamma_k\left(\frac{k}{2}\right).$$

This can be derived from (3.8). Indeed,

$$\begin{aligned}
(h_k, f) &= c \int\limits_{\mathfrak{M}_{k,m}} du \int\limits_{SO(n)} f\left(\gamma \begin{bmatrix} u \\ 0 \end{bmatrix}\right) d\gamma \\
&= c \int\limits_{\mathfrak{M}_{k,m}} du \int\limits_{SO(n)} f(\gamma \lambda_0 u) d\gamma \quad \left(\lambda_0 = \begin{bmatrix} I_k \\ 0 \end{bmatrix} \in V_{n,k}\right) \\
&= \frac{c}{\sigma_{n,k}} \int\limits_{\mathfrak{M}_{k,m}} du \int\limits_{V_{n,k}} f(vu) dv.
\end{aligned}$$

Now we represent u in the block form $u = [\eta; \zeta]$, $\eta \in \mathfrak{M}_{k,k}$, $\zeta \in \mathfrak{M}_{k,m-k}$, and change variable $\zeta = \eta z$. This gives

$$(h_k, f) = \frac{c}{\sigma_{n,k}} \int\limits_{\mathfrak{M}_{k,k}} |\eta|^{m-k} d\eta \int\limits_{\mathfrak{M}_{k,m-k}} dz \int\limits_{V_{n,k}} f(v[\eta; \eta z]) dv.$$

Using Lemma 2.2 repeatedly and changing variables, we obtain

$$(h_k, f) = \frac{c\,\sigma_{k,k}\,2^{-k}}{\sigma_{n,k}} \int_{\mathcal{P}_k} |r|^{(m-k-1)/2} dr \int_{\mathfrak{M}_{k,m-k}} dz \int_{V_{n,k}} f(v[r^{1/2}, r^{1/2}z]) dv$$

$$= c_1 \int_{\mathfrak{M}_{n,k}} \frac{dy}{|y|_k^{n-m}} \int_{\mathfrak{M}_{k,m-k}} f([y; yz]) dz,$$

where by (3.9), (2.5) and (2.4),

$$c_1 = \frac{c\,\sigma_{k,k}}{\sigma_{n,k}} = 2^{-km} \pi^{k(k-n-m)/2} \Gamma_k\left(\frac{n-m}{2}\right) / \Gamma_k\left(\frac{k}{2}\right).$$

4. Radon transforms

4.1. Matrix planes. Let k, n, and m be positive integers, $0 < k < n$, and $V_{n,n-k}$ be the Stiefel manifold of orthonormal $(n-k)$-frames in $\mathbb{R}^n$. For $\xi \in V_{n,n-k}$, $t \in \mathfrak{M}_{n-k,m}$, the linear manifold

(4.1) $$\tau = \tau(\xi, t) = \{x \in \mathfrak{M}_{n,m} : \xi' x = t\}$$

will be called *a matrix k-plane* in $\mathfrak{M}_{n,m}$. We denote by $G(n, k, m)$ the variety of all such planes. The parametrization $\tau = \tau(\xi, t)$ by the points (ξ, t) of the "matrix cylinder" $V_{n,n-k} \times \mathfrak{M}_{n-k,m}$ is not one-to-one because for any orthogonal transformation $\theta \in O(n-k)$, the pairs (ξ, t) and $(\xi\theta', \theta t)$ define the same plane. We identify functions $\varphi(\tau)$ on $G(n, k, m)$ with functions $\varphi(\xi, t)$ on $V_{n,n-k} \times \mathfrak{M}_{n-k,m}$ satisfying $\varphi(\xi\theta', \theta t) = \varphi(\xi, t)$ for all $\theta \in O(n-k)$, and supply $G(n, k, m)$ with the measure $d\tau$ such that

(4.2) $$\int_{G(n,k,m)} \varphi(\tau)\, d\tau = \int_{V_{n,n-k} \times \mathfrak{M}_{n-k,m}} \varphi(\xi, t)\, d\xi dt.$$

The matrix k-plane is, in fact, a usual km-dimensional plane in the Euclidean space $\mathbb{R}^{nm}$. To see this, we write $x = (x_{i,j}) \in \mathfrak{M}_{n,m}$ and $t = (t_{i,j}) \in \mathfrak{M}_{n-k,m}$ as column vectors

(4.3) $$\bar{x} = \begin{pmatrix} x_{1,1} \\ x_{1,2} \\ \ldots \\ x_{n,m} \end{pmatrix} \in \mathbb{R}^{nm}, \qquad \bar{t} = \begin{pmatrix} t_{1,1} \\ t_{1,2} \\ \ldots \\ t_{n-k,m} \end{pmatrix} \in \mathbb{R}^{(n-k)m},$$

and denote

(4.4) $$\bar{\xi} = \mathrm{diag}(\xi, \ldots, \xi) \in V_{nm,(n-k)m}.$$

Then (4.1) reads

(4.5) $$\tau = \tau(\bar{\xi}, \bar{t}) = \{\bar{x} \in \mathbb{R}^{nm} : \bar{\xi}'\bar{x} = \bar{t}\}.$$

The km-dimensional planes (4.5) form a subset of measure zero in the affine Grassmann manifold of *all* km-dimensional planes in $\mathbb{R}^{nm}$.

The manifold $G(n, k, m)$ can be regarded as a fibre bundle, the base of which is the ordinary Grassmann manifold $G_{n,k}$ of k-dimensional linear subspaces η of $\mathbb{R}^n$ and whose fibres are homeomorphic to $\mathfrak{M}_{n-k,m}$. Indeed, let $\pi : G(n, k, m) \to G_{n,k}$ be the canonical projection which assigns to each matrix k-plane $\tau(\xi, t)$ the subspace

(4.6) $$\eta = \eta(\xi) = \{y \in \mathbb{R}^n : \xi' y = 0\} \in G_{n,k}.$$

Let $\eta^\perp$ be the orthogonal complement of η in $\mathbb{R}^n$. The fiber $H_\eta = \pi^{-1}(\eta)$ is the set of all matrix planes (4.1), where t sweeps the space $\mathfrak{M}_{n-k,m}$.

Regarding $G(n,k,m)$ as a fibre bundle, one can utilize a parametrization which is alternative to (4.1) and one-to-one. Let

(4.7) $\qquad x = [x_1 \ldots x_m], \quad x_i \in \mathbb{R}^n, \quad t = [t_1 \ldots t_m], \quad t_i \in \mathbb{R}^{n-k}.$

For $\tau = \tau(\xi, t) \in G(n,k,m)$, we have
$$\tau = \{ x \in \mathfrak{M}_{n,m} : \xi' x_i = t_i, \quad i = 1, \ldots, m \}.$$
Each k-dimensional plane $\tau_i = \{ x_i \in \mathbb{R}^n : \xi' x_i = t_i \}$ can be parametrized by the pair (η, λ_i), where η is the subspace (4.6) and $\lambda_i \in \eta^\perp$, $i = 1, \ldots, m$, are columns of the matrix $\lambda = \xi t \in \mathfrak{M}_{n,m}$. The corresponding parametrization

(4.8) $\qquad \tau = \tau(\eta, \lambda), \qquad \eta \in G_{n,k}, \quad \lambda = [\lambda_1 \ldots \lambda_m], \quad \lambda_i \in \eta^\perp$

is one-to-one.

4.2. Definition of the Radon transform. The matrix k-plane Radon transform $\hat{f}$ of a function $f(x)$ on $\mathfrak{M}_{n,m}$ assigns to f a collection of integrals of f over all matrix planes $\tau \in G(n,k,m)$. Namely,
$$\hat{f}(\tau) = \int_{x \in \tau} f(x).$$
In order to give this integral precise meaning, we note that the matrix plane $\tau = \tau(\xi, t)$, $\xi \in V_{n,n-k}$, $t \in \mathfrak{M}_{n-k,m}$, consists of "points"
$$x = g_\xi \begin{bmatrix} u \\ t \end{bmatrix},$$
where $u \in \mathfrak{M}_{k,m}$ and $g_\xi \in SO(n)$ is a rotation satisfying

(4.9) $\qquad g_\xi \xi_0 = \xi, \qquad \xi_0 = \begin{bmatrix} 0 \\ I_{n-k} \end{bmatrix} \in V_{n,n-k}.$

This observation leads to the following

DEFINITION 4.1. The Radon transform of a function $f(x)$ on $\mathfrak{M}_{n,m}$ is defined as a function on the "matrix cylinder" $V_{n,n-k} \times \mathfrak{M}_{n-k,m}$ by the formula

(4.10) $\qquad \hat{f}(\tau) \equiv \hat{f}(\xi, t) = \int_{\mathfrak{M}_{k,m}} f\left(g_\xi \begin{bmatrix} u \\ t \end{bmatrix} \right) du.$

The reader is encouraged to check that (4.10) is independent of the choice of the rotation $g_\xi : \xi_0 \to \xi$. In terms of the one-to-one parametrization (4.8), where $\tau = \tau(\eta, \lambda)$, $\eta \in G_{n,k}$, $\lambda = [\lambda_1 \ldots \lambda_m] \in \mathfrak{M}_{n,m}$, and $\lambda_i \in \eta^\perp$, the Radon transform can be written as

(4.11) $\qquad \hat{f}(\tau) = \int_\eta dy_1 \ldots \int_\eta f([y_1 + \lambda_1 \ldots y_m + \lambda_m]) \, dy_m.$

If $m = 1$, then $\hat{f}(\xi, t)$ is the ordinary k-plane Radon transform which assigns to a function $f(x)$ on $\mathbb{R}^n$ a collection of integrals of f over all k-dimensional planes [**Hel**]. A different definition of the matrix Radon transform was given by E.E. Petrov [**Pe1**]–[**Pe3**] (the case $n - k = m$), and L.P. Shibasov [**Sh1**], [**Sh2**] (the general case).

The following properties can be easily checked.

LEMMA 4.2. *Suppose that the Radon transform*

$$f(x) \longrightarrow \hat{f}(\xi, t), \qquad x \in \mathfrak{M}_{n,m}, \quad (\xi, t) \in V_{n,n-k} \times \mathfrak{M}_{n-k,m},$$

exists (at least almost everywhere). Then

(4.12) $$\hat{f}(\xi\theta', \theta t) = \hat{f}(\xi, t), \qquad \forall \theta \in O(n-k).$$

If $g(x) = \gamma x \beta + y$, *where* $\gamma \in O(n)$, $\beta \in O(m)$, *and* $y \in \mathfrak{M}_{n,m}$, *then*

(4.13) $$(f \circ g)^\wedge(\xi, t) = \hat{f}(\gamma \xi, t\beta + \xi'\gamma' y).$$

In particular, if $f_y(x) = f(x + y)$, *then*

(4.14) $$\hat{f}_y(\xi, t) = \hat{f}(\xi, \xi' y + t).$$

Formula (4.12) is a matrix analogue of the "evenness property" of the classical Radon transform; cf. [**Hel**, p. 3].

LEMMA 4.3.
  (i) *If* $f \in L^1(\mathfrak{M}_{n,m})$, *then the Radon transform* $\hat{f}(\xi, t)$ *exists for all* $\xi \in V_{n,n-k}$ *and almost all* $t \in \mathfrak{M}_{n-k,m}$. *Furthermore,*

(4.15) $$\int_{\mathfrak{M}_{n-k,m}} \hat{f}(\xi, t) dt = \int_{\mathfrak{M}_{n,m}} f(x) \, dx, \qquad \forall \xi \in V_{n,n-k}.$$

  (ii) *Let* $\|x\| = (\mathrm{tr}(x'x))^{1/2} = (x_{1,1}^2 + \ldots + x_{n,m}^2)^{1/2}$. *If* f *is a continuous function satisfying*

(4.16) $$f(x) = O(\|x\|^{-a}), \qquad a > km,$$

then $\hat{f}(\xi, t)$ *exists for all* $\xi \in V_{n,n-k}$ *and all* $t \in \mathfrak{M}_{n-k,m}$.

PROOF. (i) is a consequence of Fubini's theorem:

$$\int_{\mathfrak{M}_{n-k,m}} \hat{f}(\xi, t) dt = \int_{\mathfrak{M}_{n-k,m}} dt \int_{\mathfrak{M}_{k,m}} f\left(g_\xi \begin{bmatrix} u \\ t \end{bmatrix}\right) du$$

$$= \int_{\mathfrak{M}_{n,m}} f(g_\xi x) \, dx = \int_{\mathfrak{M}_{n,m}} f(x) \, dx.$$

(ii) becomes obvious if we regard $\tau = \tau(\xi, t)$ as a km-dimensional plane (4.5) in $\mathbb{R}^{nm}$. □

A much deeper result is contained in the following.

THEOREM 4.4. *If* $f \in L^p(\mathfrak{M}_{n,m})$, *then the Radon transform* $\hat{f}(\xi, t)$ *is finite for almost all* $(\xi, t) \in V_{n,n-k} \times \mathfrak{M}_{n-k,m}$ *provided*

(4.17) $$1 \leq p < p_0 = \frac{n+m-1}{k+m-1}.$$

If f *is a continuous function satisfying* $f(x) = O(|I_m + x'x|^{-\lambda/2})$, $\lambda > k+m-1$, *then* $\hat{f}(\xi, t)$ *is finite for all* $(\xi, t) \in V_{n,n-k} \times \mathfrak{M}_{n-k,m}$.

The proof of this theorem was given in [**OR**] using an Abel type representation of the Radon transform of radial functions. The conditions for p and λ are sharp. For instance, one can show [**OR**] that

(4.18) $$f_0(x) = |2I_m + x'x|^{-(n+m-1)/2p}(\log|2I_m + x'x|)^{-1}$$

belongs to $L^p(\mathfrak{M}_{n,m})$, and $\hat{f}_0(\xi, t) \equiv \infty$ if $p \geq p_0$. For $m = 1$, the result of Theorem 4.4 is due to Solmon [**So**]; see also [**Ru2**] for another proof.

4.3. Connection with the Fourier transform. The Fourier transform of a function $f \in L^1(\mathfrak{M}_{n,m})$ is defined by (3.4). The following statement is a matrix generalization of the so-called Central Slice Theorem. It links together the Fourier transform (3.4) and the Radon transform (4.10). In the case $m = 1$, this theorem can be found in [**Na**, p. 11] (for $k = n - 1$) and [**Ke**, p. 283] (for any $0 < k < n$). For $y = [y_1 \ldots y_m] \in \mathfrak{M}_{n,m}$, let $\mathcal{L}(y) = \lin(y_1, \ldots, y_m)$ be the linear hull of the n-vectors $y_1, \ldots, y_m$, that is, the smallest linear subspace containing $y_1, \ldots, y_m$. Suppose that $\rank(y) = \ell$. Then $\dim \mathcal{L}(y) = \ell \leq m$.

THEOREM 4.5. *Let $f \in L^1(\mathfrak{M}_{n,m})$, $n - k \geq m$. If $y \in \mathfrak{M}_{n,m}$ and ζ is an $(n-k)$-dimensional plane in $\mathbb{R}^n$ containing $\mathcal{L}(y)$, then for any orthonormal frame $\xi \in V_{n,n-k}$ spanning ζ, there exists $b \in \mathfrak{M}_{n-k,m}$ such that $y = \xi b$. In this case,*

(4.19) $$(\mathcal{F}f)(y) = \int_{\mathfrak{M}_{n-k,m}} \exp(i\tr(b't)) \hat{f}(\xi, t)\, dt$$

or

(4.20) $$(\mathcal{F}f)(\xi b) = \mathcal{F}[\hat{f}(\xi, \cdot)](b), \quad \xi \in V_{n,n-k}, \quad b \in \mathfrak{M}_{n-k,m}.$$

PROOF. Since each vector y_j ($j = 1, \ldots, m$) lies in ζ, it decomposes as $y_j = \xi b_j$ for some $b_j \in \mathbb{R}^{n-k}$. Hence $y = \xi b$ where $b = [b_1 \ldots b_m] \in \mathfrak{M}_{n-k,m}$. Thus it remains to prove (4.20). By (4.10),

$$\mathcal{F}[\hat{f}(\xi, \cdot)](b) = \int_{\mathfrak{M}_{n-k,m}} \exp(i\tr(b't))\, dt \int_{\mathfrak{M}_{k,m}} f\left(g_\xi \begin{bmatrix} u \\ t \end{bmatrix}\right) du.$$

If $x = g_\xi \begin{bmatrix} u \\ t \end{bmatrix}$, then, by (4.9),

$$\xi'x = \xi_0' g_\xi' g_\xi \begin{bmatrix} u \\ t \end{bmatrix} = \xi_0' \begin{bmatrix} u \\ t \end{bmatrix} = t, \qquad \xi_0 = \begin{bmatrix} 0 \\ I_{n-k} \end{bmatrix} \in V_{n,n-k},$$

and Fubini's theorem yields

$$\mathcal{F}[\hat{f}(\xi, \cdot)](b) = \int_{\mathfrak{M}_{n,m}} \exp(i\tr(b'\xi'x))\, f(x)\, dx = (\mathcal{F}f)(\xi b).$$

$\square$

REMARK 4.6. It is clear that the matrices ξ and b in (4.19) are not uniquely defined. In the case $\rank(y) = m$, one can choose ξ and b as follows. By taking into account that $n - k \geq m$, we set

$$u_0 = \begin{bmatrix} 0 \\ I_m \end{bmatrix} \in V_{n-k,m}, \qquad v_0 = \begin{bmatrix} 0 \\ I_m \end{bmatrix} \in V_{n,m},$$

so that $\xi_0 u_0 = v_0$. Consider the polar decomposition

$$y = vr^{1/2}, \quad v \in V_{n,n-k}, \quad r = y'y \in \mathcal{P}_m,$$

and let g_v be a rotation with the property $g_v v_0 = v$. Then

$$y = vr^{1/2} = g_v v_0 r^{1/2} = g_v \xi_0 u_0 r^{1/2} = \xi b,$$

where

(4.21) $\quad \xi = g_v \xi_0 \in V_{n,n-k}, \quad b = u_0 r^{1/2} \in \mathfrak{M}_{n-k,m}.$

THEOREM 4.7.

(i) If $n-k \geq m$, then the Radon transform $f \to \hat{f}$ is injective on the Schwartz space $\mathcal{S}(\mathfrak{M}_{n,m})$, and f can be recovered by the formula

(4.22)
$$f(x) = \frac{2^{-m}}{(2\pi)^{nm}} \int_{\mathcal{P}_m} |r|^{\frac{n-m-1}{2}} dr$$
$$\times \int_{V_{n,m}} \exp(-i \operatorname{tr}(x' vr^{1/2}))(\mathcal{F}\hat{f}(g_v \xi_0, \cdot))(\xi_0' v_0 r^{1/2}) dv.$$

(ii) For $n - k < m$, the Radon transform is non-injective.

PROOF. By Theorem 4.5, given the Radon transform $\hat{f}$ of $f \in \mathcal{S}(\mathfrak{M}_{n,m})$, the Fourier transform $(\mathcal{F}f)(y)$ can be evaluated at each point $y \in \mathfrak{M}_{n,m}$ by the formula (4.19), so that if $\hat{f} \equiv 0$, then $\mathcal{F}f \equiv 0$. Since $\mathcal{F}$ is injective, $f \equiv 0$, and we are done. Remark 4.6 allows us to reconstruct f from $\hat{f}$, because (4.21) expresses ξ and b through $y \in \mathfrak{M}_{n,m}$ explicitly. This gives (4.22). To prove (ii), put

$$\mathfrak{L}_{n,m} = \{x \in \mathfrak{M}_{n,m} : \operatorname{rank}(x) = m\}.$$

This set is open in $\mathfrak{M}_{n,m}$. Let ψ be a Schwartz function with Fourier transform supported in $\mathfrak{L}_{n,m}$. By (4.20),

(4.23) $\quad \mathcal{F}[\hat{\psi}(\xi, \cdot)](b) = \hat{\psi}(\xi b) = 0, \quad \xi \in V_{n,n-k}, \ b \in \mathfrak{M}_{n-k,m},$

because $\xi b \notin \mathfrak{L}_{n,m}$ (note that since $n - k < m$, $\operatorname{rank}(\xi b) < m$). By the injectivity of the Fourier transform in (4.23), we obtain $\hat{\psi}(\xi, t) = 0$ for all ξ, t. Thus, for $n - k < m$, the injectivity of the Radon transform fails. □

REMARK 4.8. After the paper had been finished, we became aware of another account of the topic of this subsection in [Sh1],[Sh2], written in a different manner. Nevertheless, Remark 4.6, formula (4.22) and the statement (ii) of Theorem 4.7 seem to be new.

REMARK 4.9. For $n - k > m$, the dimension of the manifold $G(n, k, m)$ of all matrix k-planes in $\mathfrak{M}_{n,m}$ is greater than that of the ambient space $\mathfrak{M}_{n,m}$, and the inversion problem is overdetermined. In the case $n - k = m$ both dimensions coincide. The problem of reducing overdeterminedness by fixing a certain "invertible" mn-dimensional complex of matrix planes was studied in [Sh2].

5. The dual Radon transform and the Fuglede formula

DEFINITION 5.1. Let $\tau = \tau(\xi, t)$ be a matrix plane (4.1), $(\xi, t) \in V_{n,n-k} \times \mathfrak{M}_{n-k,m}$. The dual Radon transform $\check{\varphi}(x)$ assigns to a function $\varphi(\tau)$ on $G(n, k, m)$ its mean value over all matrix planes τ through x. Namely,

$$\check{\varphi}(x) = \int_{\tau \ni x} \varphi(\tau), \qquad x \in \mathfrak{M}_{n,m}.$$

This means that

$$(5.1) \qquad \check{\varphi}(x) = \frac{1}{\sigma_{n,n-k}} \int_{V_{n,n-k}} \varphi(\xi, \xi'x) d\xi$$

$$= \int_{SO(n)} \varphi(\gamma\xi_0, \xi_0'\gamma'x) d\gamma, \qquad \xi_0 = \begin{bmatrix} 0 \\ I_{n-k} \end{bmatrix} \in V_{n,n-k}.$$

The mean value $\check{\varphi}(x)$ apparently exists for all $x \in \mathfrak{M}_{n,m}$ if φ is a continuous function. Moreover, $\check{\varphi}(x)$ is finite a.e. on $\mathfrak{M}_{n,m}$ for any locally integrable function φ [OR].

REMARK 5.2. The dual Radon transform $\check{\varphi}(x)$ of a function $\varphi(\tau)$, $\tau \in G(n, k, m)$, is independent of the parametrization $\tau = \tau(\xi, t)$ in the sense that for any other parametrization $\tau = \tau(\xi\theta', \theta t)$, $\theta \in O(n-k)$, (see Sec. 4.1), (5.1) gives the same result:

$$\frac{1}{\sigma_{n,n-k}} \int_{V_{n,n-k}} \varphi(\xi\theta', \theta\xi'x) d\xi = \frac{1}{\sigma_{n,n-k}} \int_{V_{n,n-k}} \varphi(\xi_1, \xi_1'x) d\xi_1 = \check{\varphi}(x).$$

LEMMA 5.3. *The duality relation*

$$(5.2) \qquad \int_{\mathfrak{M}_{n,m}} f(x)\check{\varphi}(x) dx = \frac{1}{\sigma_{n,n-k}} \int_{V_{n,n-k}} d\xi \int_{\mathfrak{M}_{n-k,m}} \varphi(\xi, t)\hat{f}(\xi, t) dt$$

holds provided that either side of (5.2) is finite for f and φ replaced by $|f|$ and $|\varphi|$, respectively.

PROOF. By (4.10), the right hand side of (5.2) is

$$(5.3) \qquad \frac{1}{\sigma_{n,n-k}} \int_{V_{n,n-k}} d\xi \int_{\mathfrak{M}_{n-k,m}} \varphi(\xi, t) dt \int_{\mathfrak{M}_{k,m}} f\left(g_\xi \begin{bmatrix} u \\ t \end{bmatrix}\right) du.$$

Changing variables $x = g_\xi \begin{bmatrix} u \\ t \end{bmatrix}$, we have

$$\xi'x = (g_\xi \xi_0)' g_\xi \begin{bmatrix} u \\ t \end{bmatrix} = \xi_0' \begin{bmatrix} u \\ t \end{bmatrix} = t.$$

Hence, by Fubini's theorem, (5.3) reads

$$\frac{1}{\sigma_{n,n-k}} \int_{V_{n,n-k}} d\xi \int_{\mathfrak{M}_{n,m}} \varphi(\xi, \xi'x) f(x) dx = \int_{\mathfrak{M}_{n,m}} \check{\varphi}(x) f(x) dx.$$

$\square$

Now we state the main theorem.

THEOREM 5.4. *Let f be a locally integrable function on $\mathfrak{M}_{n,m}$ such that the Riesz potential $I^k f$ defined by (3.11) converges absolutely. Then*

$$(5.4) \qquad c(\hat{f})^{\vee}(x) = (I^k f)(x), \qquad c = 2^{-km} \pi^{-km/2} \Gamma_m\left(\frac{n-k}{2}\right) / \Gamma_m\left(\frac{n}{2}\right),$$

(the generalized Fuglede formula).

PROOF. Let $f_x(y) = f(x+y)$. By (5.1) and (4.14),

$$\begin{aligned}
(\hat{f})^{\vee}(x) &= \frac{1}{\sigma_{n,n-k}} \int_{V_{n,n-k}} \hat{f}(\xi, \xi'x) d\xi = \frac{1}{\sigma_{n,n-k}} \int_{V_{n,n-k}} \hat{f}_x(\xi, 0) d\xi \\
&= \int_{\mathfrak{M}_{k,m}} du \int_{SO(n)} f\left(x + \gamma \begin{bmatrix} u \\ 0 \end{bmatrix}\right) d\gamma.
\end{aligned}$$

This coincides with $c^{-1}(I^k f)(x)$ for $k < m$, see Definition 3.3. If $k \geq m$, $d = (m+1)/2$, we pass to polar coordinates and get

$$\begin{aligned}
(\hat{f})^{\vee}(x) &= 2^{-m} \int_{V_{k,m}} dv \int_{\mathcal{P}_m} |r|^{k/2-d} dr \int_{SO(n)} f_x\left(\gamma \begin{bmatrix} vr^{1/2} \\ 0 \end{bmatrix}\right) d\gamma \\
&= \frac{2^{-m} \sigma_{k,m}}{\sigma_{n,m}} \int_{\mathcal{P}_m} |r|^{k/2-d} dr \int_{V_{n,m}} f_x(wr^{1/2}) dw \\
&= \frac{\sigma_{k,m}}{\sigma_{n,m}} \int_{\mathfrak{M}_{n,m}} f(x+y) |y'y|^{(k-n)/2} dy \\
&= c^{-1}(I^k f)(x).
\end{aligned}$$

$\square$

REMARK 5.5. By Theorem 3.4, the formula (5.4) holds if $f \in L^p(\mathfrak{M}_{n,m})$, $1 \leq p < n/(k+m-1)$, in particular, for any continuous function f satisfying $f(x) = O(|I_m + x'x|^{-\lambda/2})$, $\lambda > (k+m-1)(n+m-1)/n$.

6. Inversion problem for the Radon transform

The Fuglede formula $c(\hat{f})^{\vee} = I^k f$ reduces the inversion problem for the Radon transform to Riesz potentials. This is exactly the same situation as in the rank-one case. Whereas for the ordinary k-plane transform and Riesz potentials, a variety of pointwise inversion formulas are available in a large scale of function spaces [**Ru1**], [**Ru2**], in the higher rank case, we cannot obtain pointwise inversion formulas other than for Schwartz functions via the Fourier transform (see (4.22)) or in terms of divergent integrals understood somehow in a regularized sense. This is still an open problem.

Below we show how the unknown "rough" function f can be recovered in the framework of the theory of distributions. First we specify the space of test functions. From the Fourier transform formula $(h_\alpha, f) = (2\pi)^{-nm}(|y|_m^{-\alpha}, (\mathcal{F}f)(y))$, it is evident that the Schwartz class $\mathcal{S} \equiv \mathcal{S}(\mathfrak{M}_{n,m})$ is not suitable because it is not invariant under multiplication by $|y|_m^{-\alpha}$. To get around this difficulty, we follow an idea of V.I. Semyanistyi [**Se**] for $m = 1$. Let $\Psi \equiv \Psi(\mathfrak{M}_{n,m})$ be the subspace of functions $\psi(y) \in \mathcal{S}$ vanishing on the set

(6.1) $\qquad \{y : y \in \mathfrak{M}_{n,m},\ \text{rank}(y) < m\} = \{y : y \in \mathfrak{M}_{n,m},\ |y'y| = 0\}$

with all derivatives. (The coincidence of both sets in (6.1) is clear because $\text{rank}(y) = \text{rank}(y'y)$; see, e.g., [**FZ**, p. 5].) The set Ψ is a closed linear subspace of $\mathcal{S}$. Therefore, it can be regarded as a linear topological space with the induced topology of $\mathcal{S}$. Let $\Phi \equiv \Phi(\mathfrak{M}_{n,m})$ be the Fourier image of Ψ. Since the Fourier transform $\mathcal{F}$ is an automorphism of $\mathcal{S}$ (i.e., a topological isomorphism of $\mathcal{S}$ onto itself), Φ is a closed linear subspace of $\mathcal{S}$. Having been equipped with the induced topology of $\mathcal{S}$, the space Φ becomes a linear topological space isomorphic to Ψ under the Fourier transform. We denote by $\Phi' \equiv \Phi'(\mathfrak{M}_{n,m})$ the space of all linear continuous functionals (generalized functions) on Φ. Since for any complex α, multiplication by $|y|_m^{-\alpha}$ is an automorphism of Ψ, according to the general theory [**GSh**], as a convolution with h_α, I^α is an automorphism of Φ, and we have

$$\mathcal{F}[I^\alpha f](y) = |y|_m^{-\alpha} \mathcal{F}[f](y)$$

for all Φ'-distributions f.

In the rank-one case, the spaces Φ, Ψ, their duals and generalizations were studied by P.I. Lizorkin, S.G. Samko and others in view of applications to the theory of function spaces and fractional calculus; see [**Sa**], [**SKM**], [**Ru1**] and references therein.

The Fuglede formula (5.4), Theorem 4.4, and Remark 5.5 imply the following

THEOREM 6.1. *Let $f \in L^p(\mathfrak{M}_{n,m})$, $1 \leq p < n/(k+m-1)$. Then the Radon transform $g = \hat{f}$ is well-defined, and f can be recovered from g in the sense of Φ'-distributions by the formula*

(6.2) $\qquad\qquad\qquad (f, \phi) = c(\check{g}, I^{-k}\phi), \qquad \phi \in \Phi,$

where

$$(I^{-k}\phi)(x) = (\mathcal{F}^{-1}|y|_m^k \mathcal{F}\phi)(x), \qquad c = 2^{-km}\pi^{-km/2}\Gamma_m\left(\frac{n-k}{2}\right)/\Gamma_m\left(\frac{n}{2}\right).$$

REMARK 6.2. For k even, the Riesz potential $I^k f$ can be inverted (in the sense of Φ'-distributions) by repeated application of the Cayley-Laplace operator $\Delta_m = \det(\partial'\partial)$, $\partial = (\partial/\partial x_{i,j})$ [**Kh**]. In Fourier terms, this operator agrees with multiplication by $(-1)^m |y|_m^2$, and therefore, $(-1)^m \Delta_m I^\alpha f = I^{\alpha-2} f$ in the Φ'-sense.

References

[Č]   V.G. Černov, *Homogeneous distributions and the Radon transform in the space of rectangular matrices over a continuous locally compact disconnected field*, Soviet Math. Dokl. **11** (1970), 415-418.

[FZ]   V.G. Fang and Y.-T. Zhang, *Generalized Multivariate Analysis*, Springer-Verlag, Berlin, 1990.

[Far]   J. Faraut, *Intégrales de Marcel Riesz sur un cône symétrique*, Actes du colloque Jean Braconnier (Lyon, 1986), 17–30, Publ. Dép. Math. Nouvelle Sér. B, 87-1, Univ. Claude-Bernard, Lyon, 1987.

[FK]   J. Faraut and A. Korányi, *Analysis on Symmetric Cones*, Clarendon Press, Oxford, 1994.

[Fu]   B. Fuglede, *An integral formula*, Math. Scand. **6** (1958), 207-212.

[Ge]   S.S. Gelbart, *Fourier Analysis on Matrix Space*, Memoirs Amer. Math. Soc., vol. 108, 1971.

[GSh] I.M. Gelfand and G.E. Shilov, *Generalized Functions, Vol. 2. Spaces of Fundamental and Generalized Functions*, Academic Press, New York, 1968.

[Gr] M. I. Graev, *Integral geometry on the space L^n, where L is a matrix ring*. Funktsional. Anal. i Prilozhen. **30** (1996), 71-74; translation in Funct. Anal. Appl. **30** (1996), 277–280.

[Hel] S. Helgason, *The Radon Transform*, second edition, Birkhäuser Boston, Boston, 1999.

[Ke] F. Keinert, *Inversion of k-plane transforms and applications in computer tomography*, SIAM Review **31** (1989), 273–289.

[Kh] S.P. Khekalo, *Riesz potentials in the space of rectangular matrices and iso-Huygens deformations of the Cayley-Laplace operator*, Doklady Mathematics **63** (2001), no. 1, 35-37.

[Ma] A.M. Mathai, *Jacobians of Matrix Transformations and Functions of Matrix Argument*, World Scientific, Singapore, 1997.

[Mu] R.J. Muirhead, *Aspects of Multivariate Statistical Theory*, Wiley, New York, 1982.

[Na] F. Natterer, *The Mathematics of Computerized Tomography*, Wiley, New York, 1986.

[OR] E. Ournycheva and B. Rubin, *Radon transform of functions of matrix argument*, Preprint, 2004.

[Pe1] E.E. Petrov, *The Radon transform in spaces of matrices and in Grassmann manifolds*, Dokl. Akad. Nauk SSSR **177** (1967), 782-785.

[Pe2] _____, *The Radon transform in spaces of matrices*, Trudy Sem. Vektor. Tenzor. Anal. **15** (1970), 279–315.

[Pe3] _____, *Paley-Wiener theorems for the matrix Radon transform*. Mat. Sb. 190 (1999), no. 8, 103–124; translation in Sb. Math. **190** (1999), 1173–1193.

[Pe4] _____, *Paley-Wiener theorems for the Radon complex*, Izv. Vysš. Učebn. Zaved. Matematika **1977**, no. 3 (178), 66–77.

[R] J. Radon, *Über die Bestimmung von Funktionen durch ihre Integralwerte längs gewisser Mannigfaltigkeiten*, Ber. Verh. Sächs. Akad. Wiss. Leipzig Math. - Nat. Kl. **69** (1917), 262–277.

[Ra] M. Raïs, *Distributions homogènes sur des espaces de matrices*, Bull. Soc. Math. France, Mem. **30** (1972), 3–109.

[Ru1] B. Rubin, *Fractional Integrals and Potentials*, Longman, Harlow, 1996.

[Ru2] _____, *Reconstruction of functions from their integrals over k-dimensional planes*, Israel J. Math. **141** (2004), 93-117.

[Ru3] _____, *Zeta integrals and integral geometry in the space of rectangular matrices*, http://arXiv.org/abs/math.FA/0406289.

[Sa] S.G. Samko, *Hypersingular Integrals and Their Applications*, Taylor & Francis, London, 2002.

[SKM] S.G. Samko, A.A. Kilbas and O.I. Marichev, *Fractional Integrals and Derivatives. Theory and Applications*, Gordon and Breach, New York, 1993.

[Se] V.I. Semyanistyi, *Homogeneous functions and some problems of integral geometry in spaces of constant cuvature*, Sov. Math. Dokl. **2** (1961), 59–61.

[Sh1] L.P. Shibasov, *Integral problems in a matrix space that are connected with the functional $X_{n,m}^\lambda$*. Izv. Vysš. Učebn. Zaved. Matematika **1973**, no. 8 (135), 101–112.

[Sh2] _____, *Integral geometry on planes of a matrix space. Harmonic analysis on groups*, Moskov. Gos. Zaočn. Ped. Inst. Sb. Naučn. Trudov Vyp. **39** (1974), 68–76.

[Sh3] _____, *Integral representation of a homogeneous spherically symmetric functional*, Functional Analysis, No. 19, 155–161, Ul'yanovsk. Gos. Ped. Inst., Ul'yanovsk, 1982.

[So] D. C. Solmon, *A note on k-plane integral transforms*, J. Math. Anal. Appl. **71** (1979), 351–358.

[St1] E. M. Stein, *Analysis in matrix spaces and some new representations of* $SL(N, C)$, Ann. of Math. (2) **86** (1967), 461–490.

[St2] _____, *Singular Integrals and Differentiability Properties of Functions*, Princeton Univ. Press, Princeton, NJ, 1970.

INSTITUTE OF MATHEMATICS, HEBREW UNIVERSITY, JERUSALEM 91904, ISRAEL
E-mail address: ournyce@math.huji.ac.il
E-mail address: boris@math.huji.ac.il

Characteristic Problems for the Spherical Mean Transform

V. P. Palamodov

This paper is dedicated to Larry Zalcman, in recognition of his contributions to integral geometry and complex analysis, on the occasion of his 60th birthday.

1. Introduction

The spherical transform of a function f in Euclidean space $\mathbf{E}^n$ of dimension $n > 1$ is the family of spherical integrals

$$M_{\mathbf{E}} f(S) = \int_S f \, \mathrm{d}S, \tag{1}$$

defined on the variety $\Sigma_{\mathbf{E}}$ of spheres in $\mathbf{E}^n$, where $\mathrm{d}S$ denotes the Euclidean surface area form. Replacing $\mathbf{E}^n$ by a Euclidean sphere $\mathbf{S}^n$, we obtain the spherical transform $M_{\mathbf{S}} f$ defined on the variety $\Sigma_{\mathbf{S}}$ of spheres in $\mathbf{S}^n$. The reconstruction problems for $\mathbf{E}^n$ and $\mathbf{S}^n$ are equivalent if all hyperplanes in $\mathbf{E}^n$ are included as spheres of infinite radius. In particular, Radon's reconstruction formula for the variety of straight lines in the plane is a version of Funk's formula for reconstruction from knowledge of integrals over great circles.

The reconstruction problem is of particular interest for non-redundant data of spherical integrals, for instance, if the spherical integrals are only known for an n-dimensional subvariety $\Sigma \subset \Sigma_{\mathbf{E}}$. In this paper, we focus on the case in which Σ is the family $\Sigma(Y)$ of spheres tangent to a smooth submanifold Y.

Gelfand and Graev [4] studied the horocycle integral transform in hyperbolic space. In the standard model of the hyperbolic space, horocycles are just all the spheres in a Euclidean ball tangent to the boundary. The inversion formulae look very much like those for the Radon transform in Euclidean space.

S. Gindikin [6],[3] and A. Goncharov [7] addressed the reconstruction problem for functions in a Euclidean domain D from data of integrals over spheres tangent to the boundary of D. Their methods are based on reduction to the Radon transform in a projective quadric. No explicit formula was known, except for the ball [3].

We prove here that for any odd n, a reconstruction can be done for *any* nonempty closed manifold Y. If Y is a point, the problem is reduced to the Radon transform by geometric inversion; see [1]. In the case of even n, we state an explicit

2000 *Mathematics Subject Classification.* Primary 53C65, 65R32.

reconstruction formula for a function f in an arbitrary bounded domain D with smooth boundary Y.

Our method is to consider a reconstruction as a solution of the Cauchy problem for the Darboux equation. This problem is characteristic, since the variety $\Sigma(Y)$ is always characteristic. Therefore, the reconstruction of a solution by means of the Green formula depends only on the data of Mf on $\Sigma(Y)$; one need not know the normal derivative[1]. Note that the class of "admissible" submanifolds in the manifold of spheres in the sense of [**3**] coincides with the class of characteristic hypersurfaces. For implementation of this approach, an explicit formula for the singular fundamental solution F_n of the adjoint Darboux equation is applied; cf. Section 6. For even n, the support of F_n is the half-space. The Cauchy problem for the Euler-Poisson-Darboux equation was studied in [**2**], which contains a survey of previous results.

2. Spheres tangent to a manifold

Normalizing the integral (1), yields the spherical mean

$$Nf(a,r) = s_n^{-1} r^{1-n} Mf(a,r) = \int_{|\omega|=1} f(a+r\omega)\,d\omega,$$

where $s_n = 2\pi^{n/2}/\Gamma(n/2)$ is the area of the unit sphere. For any continuous f, the function Nf is extended by continuity to spheres of zero radius; we have $Nf(x,0) = f(x)$. The spherical mean transform $g(x,r) \doteq Nf(x,r)$ satisfies the Darboux equation in $\Sigma_{\mathbf{E}} \cong \mathbf{E}^n \times \mathbb{R}_+$:

$$Ag \doteq \left(r\frac{\partial^2}{\partial r^2} + (n-1)\frac{\partial}{\partial r} - r\Delta_x\right)g = 0.$$

The reconstruction problem is as follows: to recover a function f from data of its spherical mean transform on the variety $\Sigma \subset \Sigma_{\mathbf{E}}$. If $\dim \Sigma = n$, the data of $Nf(S)$ for $S \in \Sigma$ are not redundant. For reconstruction of the function f, it is sufficient to find out the solution of the Darboux equation on the hyperplane $r = 0$. The principal part of the Darboux operator is equal to the wave operator $r\Box$, where $\Box = \partial_r^2 - \Delta_x$ is the wave operator with velocity 1. A hypersurface $\Sigma \subset \Sigma_{\mathbf{E}}$ is *characteristic* for the Darboux (and for the wave) operator at a point $(x,r) \in \Sigma_{\mathbf{E}}, r > 0$, if the principal symbol vanishes on the conormal vector ν to Σ at x, i.e., if $\nu_r^2 - \nu_x^2 = 0$, where $\nu_x \in \mathbf{E}^n$ and $\nu_r \in \mathbb{R}$ are the components of ν.

We now focus on the special case $\Sigma = \Sigma(Y)$; for a submanifold $Y \subset \mathbf{S}$, we denote by $\Sigma(Y)$ the submanifold of spheres in $\mathbf{S}$ tangent to Y.

PROPOSITION 1. *The variety $\Sigma(Y)$ is characteristic at each of its points.*

It follows that the reconstruction problem can be reduced to the characteristic Cauchy problem: to recover a solution g of the Darboux equation on the boundary $\{r=0\}$ of $\Sigma_{\mathbf{E}}$ from knowledge of its values on a characteristic hypersurface. We then have $f(x) = g(x,0)$.

Take an arbitrary bounded domain $D \subset \mathbf{E}^n$ with smooth boundary ∂D and define the function

$$\rho(y) \doteq \min_{z \in \partial D} |y-z|,\ y \in D.$$

---

[1] I proposed this idea for a similar problem in a discussion with I.M. Gelfand in the early sixties.

It is smooth almost everywhere in D.

THEOREM 2. *For arbitrary $n \geq 2$, any smooth function f in D can reconstructed from data of spherical integrals for spheres $S \subset D$ tangent to the boundary as follows:*

$$f(x) = \frac{1}{(2\pi \imath)^{n-1}} \int_D \delta^{(n-1)}(q) \, \Phi_n(x,y) \, Mf(y,r)|_{r=\rho} dy \tag{2}$$

for odd n where

$$\Phi_n(x,y) = 2 \langle y - x, \nabla \rho \rangle - 2\rho + q\left(\frac{1}{\rho} + \frac{\Delta \rho}{n-1}\right),$$

and

$$f(x) = \frac{(n-1)!}{(2\pi \imath)^n} \int_D [q^{-n}] \, \Phi_n(x,y) \, Mf(y,r)|_{r=\rho} dy \tag{3}$$

for even n, where $q = \left(r^2 - |y-x|^2\right)/2$ and the generalized function $[q^{-n}]$ is defined in Section 5.

REMARK. The integrals (2) and (3) have sense, at least for generic D and $x \in D$. Indeed, the wave front of $[q^{-n}]$ and of $\delta^{(n-1)}(q)$ coincides with the conormal bundle $N^*(Q_x)$ of the cone $Q_x \doteq \{q = 0\}$. The function $\rho(y)$ is smooth in D, except for the set C of points y where the sphere $S(y, \rho)$ has several points of contact with ∂D or y is a curvature centre of ∂D (multiple contact). The wave front $WF(G)$ of the function $G(y, r) \doteq r - \rho(y)$ is generically a conic Lagrange manifold in $T^*(\Sigma_{\mathbf{E}})$ (with singularities, because multiple contact points can appear). The two conic Lagrange manifolds $N^*(Q_x)$ and $WF(G)$ generically have no common point. In this case, the restriction of $[q^{-n}]$ and of $\delta^{(n-1)}(q)$ to the hypersurface $r = \rho(y)$ is well-defined. Moreover, the set of points x such that $N^*(Q_x) \cap WF(G) \neq \emptyset$ has zero measure in D.

THEOREM 3. *Let n be an odd integer, Y an arbitrary nonempty closed submanifold of $\mathbf{E}^n$, and $H(Y)$ the convex hull of Y. Any sufficiently smooth function f can be explicitly reconstructed in $\operatorname{int} H(Y)$ from data of spherical means on $\Sigma(Y)$ as follows:*

$$f(x) = K_{\mathbf{E}}(x)(Nf),$$

where for any $x \in \operatorname{int} H(Y)$, $K_{\mathbf{E}}(x)$ is a distribution in $\Sigma_{\mathbf{E}}$ with compact support in the manifold $\Sigma(Y) \cap \Sigma(x)$ of spheres tangent to Y and containing x.

The spherical mean transform satisfies the Darboux equation, and our key idea is to consider the reconstruction problem as a Cauchy problem with "initial" data on the characteristic variety $\Sigma(Y)$. A solution u is uniquely defined from knowledge of u on $\Sigma(Y)$; we do not need to know the normal derivative of u. The explicit reconstruction is done by application of the singular forward fundamental solution for the adjoint Darboux operator supported by the future cone.

The fundamental solution for the Darboux operator looks similar to the forward fundamental solution for the wave equation; however, the order of singularity is different. In particular, for odd n, the propagator is a derivative of the delta-function supported by the future cone, i.e., it possesses the strong Huygens property. In the case of even n, it is a homogeneous function, whose singularity on the cone is of integer degree, unlike the wave propagator, whose degree is half-integer.

THEOREM 4. *Let Y be an arbitrary nonempty closed submanifold of $\mathbf{S}^n$. An arbitrary sufficiently smooth function f in $\mathbf{S}^n$ can be reconstructed from data of spherical integrals $M_\mathbf{S} f(S)$, $S \in \Sigma(Y)$, as*

$$f(x) = K_\mathbf{S}(x)(Mf).$$

Here, for any $x \in \mathbf{S}$, $K_\mathbf{S}(x)$ is a distribution in $\Sigma_\mathbf{S}$ supported by $\Sigma(Y)$. For n odd, $K_\mathbf{S}(x)$ is supported by the set $\Sigma(Y) \cap \Sigma(x)$.

If Y is a point, the reconstruction problem is reduced to inversion of the Radon transform in $\mathbf{E}^n$ by application of the geometric inversion mapping J with centre at Y.

Let $\mathbf{S}^n$ be the unit sphere in $\mathbf{E}^{n+1}$; consider the inverse stereographic projection $\pi : \mathbf{E}^n \to \mathbf{S}^n$ with centre $(1,0)$, where

$$\pi(x) = (y_0, y), \qquad y_0 = \frac{|x|^2 - 1}{|x|^2 + 1}, \qquad y = \frac{2x}{|x|^2 + 1}.$$

The mapping π is conformal; and $\pi^*(\mathrm{d}\sigma) = (1 - y_0)\,\mathrm{d}s$ where $\mathrm{d}\sigma, \mathrm{d}s$ are line elements in $\mathbf{S}^n$ and $\mathbf{E}^n$, respectively. The image $\pi(S)$ of an arbitrary sphere S in $\mathbf{E}^n$ is a sphere in $\mathbf{S}^n$; the image of a hyperplane H is a sphere in $\mathbf{S}^n$ through the centre $(1,0)$ of the projection π. Therefore, $M_\mathbf{E} f(S) = M_\mathbf{S} g(\pi(S))$ for any sphere or hyperplane S, where g is a function on $\mathbf{S}^n$ and

$$f(x) = (1 - y_0)^{n-1} g(\pi(x)) = \left(\frac{2}{1 + |x|^2}\right)^{n-1} g(\pi(x))$$

is a function on $\mathbf{E}^n$. Therefore, inversion of the transform $M_\mathbf{E}$ is equivalent to inversion of $M_\mathbf{S}$, at least for functions f which satisfy the estimate $f(x) = O\left(|x|^{1-n-\varepsilon}\right)$ at infinity, for then g is integrable on any sphere.

Any sphere S divides $\mathbf{S}^n$ in two connected parts; we call each of them a *ball*.

PROOF OF THEOREM 4. We deduce this result from Theorem 2, which will be proved later. For a subset $G \subset \mathbf{S}^n$ and a point $p \in \mathbf{S}^n \setminus G$, we denote by $H_p(G)$ the convex hull of the set $\pi_p^{-1}(G)$ in the vector space $\mathbf{E}^n$, where $\pi_p : \mathbf{E}^n \to \mathbf{S}^n \setminus \{p\}$ is the stereographic projection with the centre p. Now we consider three cases.

Case 1: $\dim Y = n - 1$. The hypersurface Y divides $\mathbf{S}^n$ into several connected components. Take a centre z in one of the components. Then f can be reconstructed in any other component by means of Theorem 2.

Case 2: Y is a k-sphere for some k, $0 \leq k \leq n - 1$. In case $k = 0$, we apply inversion with centre at a point $y \in Y$ and reduce the problem to inversion of the Radon transform. The case $k \geq 1$ can be treated by simple arguments.

Case 3: Y is not a sphere and $\dim Y < n - 1$. Then for almost all points $p \in \mathbf{S}^n$, Y is contained in no $(n-1)$-sphere through p. Applying inversion with the centre p, we may assume that Y is contained in no hyperplane. We now require

LEMMA 5. *For each point $z \in \mathbf{S}^n \setminus Y$, there exists a point $p \in \mathbf{S}^n$ such that $z \in H_p(Y)$.*

By Theorem 2, the function f can be reconstructed in $H_p(Y)$ and hence at z. This completes the proof of Theorem 4. $\square$

PROOF OF LEMMA 5. First take a centre p such that the convex hull $K \doteq H_p(Y)$ is not contained in any hyperplane, i.e., K is an n-body. For a point $b \in \partial K$, we denote by C_b the intersection of all closed half spaces H such that $K \subset H$ and $b \in \partial H$. It is easy to see that if a point $q \in \mathbf{S}^n \backslash K$ approaches b, the set $H_q(Y)$ tends to C_b; hence the reconstruction is possible in int C_b. Then we note that for any convex n-body K, we have

$$\bigcup_{b \in \partial K} \text{int } C_b = \mathbf{E}^n.$$

This implies a reconstruction in $\mathbf{E}^n \simeq \mathbf{S}^n \backslash \{p\}$. Taking another centre p', we get a reconstruction at the point p. □

3. Characteristic Cauchy problem

To prove Theorem 2, we consider the Cauchy problem for the Darboux operator with initial data on the characteristic variety $\Sigma(Y)$. To clarify the idea, let us consider the general linear differential equation

(4) $$A(y, D) u = 0$$

with smooth coefficients in an open set U of a vector space $\mathbf{V}$ of dimension n with coordinates $y_1, ..., y_n$.

PROPOSITION 6. *Let K be a compact in U with Lipschitz boundary ∂K and Γ an open subset of ∂K. Suppose that*

(i) *F_x is a generalized function that satisfies the equation $A^* F_x = \delta_x$ for a point $x \in \text{int } K$,*
(ii) *supp $F_x \cap \partial K \Subset \Gamma$,*
(iii) *the trace $F_x|\Gamma$ is well-defined as a generalized function in Γ and*
(iv) *Γ is a characteristic hypersurface for A.*

Then an arbitrary sufficiently smooth solution u of (4) can be reconstructed at the point x from the data $u|\Gamma$.

PROOF. Take a smooth function ϕ in $\mathbf{V}$ such that $\phi \geq 0$ in K, $\phi < 0$ in $\mathbf{V} \backslash K$ and $\mathrm{d}\phi \neq 0$ on Γ. The composition $\Phi = \theta(\phi)$ is the indicator function of U, where $\theta(t) = 1$ for $t \geq 0$ and $\theta(t) = 0$ otherwise. By (i), we have

(5) $$u(x) = \delta_x (\Phi u) \, \mathrm{d}y = \int_U A^* (F_x) \, \Phi u \mathrm{d}y = \int_U A(\Phi u) F_x \mathrm{d}y,$$

where $\mathrm{d}y = \mathrm{d}y_1 \wedge ... \wedge \mathrm{d}y_n$. Write $A = a_2 + a_1 + a_0$, where a_j is a homogeneous differential operator in $\mathbf{V}$ of degree j. By the Leibniz formula,

$$A(y, D)(\Phi u) = \Phi A(y, D) u + B(y, \nabla \Phi, \nabla u) + u \left[a_2(y, D) + a_1(y, D) \right] \Phi,$$

where $2B(y, \xi, \eta) \doteq a_2(y, \xi + \eta) - a_2(y, \xi - \eta)$ is the symmetric bilinear form in ξ and η such that $B(y, \xi, \xi) = 2a_2(y, \xi)$. The first term vanishes in virtue of (4), while the second and the third are supported in ∂K. We have $\nabla \Phi = \nabla \phi \, \delta_0(\phi)$, where δ_0 is the Dirac function of one variable and $B(y, \nabla \Phi, \nabla u) = B(y, \nabla \phi, \nabla u) \delta_0(\phi)$. The operator $b \doteq B(y, \nabla \phi, \nabla \cdot)$ is a tangent field such that $b\phi = B(y, \nabla \phi, \nabla \phi) = 2a_2(y, \nabla \phi)$. The right side vanishes on Γ by (iv). This means that the field b is tangent to Γ. Any integral curve of b is a bicharacteristic of A. We have $a_1\Phi = a_1(y,D)(\phi)\,\delta_0(\phi)$ and $a_2\Phi = a_2(y,\nabla\phi)\,\delta'_0(\phi) + a_2(y,D)(\phi)\,\delta_0(\phi)$. By (iv), the function $a_2(y,\nabla\phi)$ vanishes on Γ; hence,

$$A(y,D)(\Phi u) = (b+a)\,u\,\delta_0(\phi), \quad a = a_1(\phi) + a_2(\phi).$$

According to (ii) and (iii), the product $F_x\delta_0(\phi)$ is well-defined as a distribution in U with support in Γ. This yields

$$A(\Phi u)\,F_x = (b+a)\,u\,\delta_0(\phi)\,F_x,$$

where $\alpha \doteq a_1(\phi) + a_2(\phi)$. By (5), we obtain

$$u(x) = \int_U (b+a)\,u\,\delta_0(\phi)\,F_x.$$

The derivatives of u can be eliminated by partial integration, taking in account that $b(\phi) = 0$:

(6) $\qquad u(x) = \int_U u\,(a - \operatorname{div} b - b)\,(F_x)\,\delta_0(\phi)\,y = \int_\Gamma u\,(a - \operatorname{div} b - b)\,F_x\,\dfrac{dy}{d\phi}.$

The right-hand side obviously does not depend on derivatives of u, and the second factor is a distribution on Γ. $\square$

PROOF OF THEOREM 2. Take the function $\phi(x,r) = \rho(x) - r$. The compact $K \doteq \{(x,r) : \phi(x,r) \geq 0\}$ is the set of all spheres $S \subset D$. Apply the formula (6) to the Darboux operator A, to the solution $u = Nf$ in the set K, to the function ϕ, and to the fundamental solution F_n as in Theorem 8 below. We have

$$b = -2r(\partial_r + \langle \nabla\rho, \nabla\rangle),\ a - \operatorname{div} b = 3 - n + r\Delta_y\rho,$$
$$-br_+^{n-2}(q+0\imath)^{1-n} = 2(n-2)\,r_+^{n-2}(q+0\imath)^{1-n}$$
$$+ 2(1-n)(r - \langle y - x, \nabla\rho\rangle)\,r_+^{n-1}(q+0\imath)^{-n}$$

and $dy\,dr/d\phi = dy$ in Γ. $\square$

PROOF OF THEOREM 3. Fix a point $x \in \operatorname{supp} f$ and choose a domain $U \subset \mathbf{E}^n \times \mathbb{R}_+$ which contains the point $(x,0)$ as follows. Take an arbitrary vector $v \in \mathbf{E}^n \setminus \{0\}$ and consider the family of spheres $S(x+rv, r),\ r \geq 0$. The set $\mathbf{E}_x(v) \doteq \cup_{r \geq 0} S(x+rv, r)$ is equal to $\mathbf{E}^n$ if $|v| < 1$ and contains the cone $K_v \doteq \{x : (x-y, v) > \varepsilon |x-y|\}$ in the case $|v| \geq 1$, where $\varepsilon^2 = |v|^2 - 1$. If ε is small, the cone K_v is close to an open half-space whose boundary contains x. By assumption, the point x is in the interior $H(Y)$, which implies that the cone K_v meets Y for any v in the ball $B \doteq \{|v|^2 \leq 1 + \varepsilon^2\}$ if ε is sufficiently small. For small r, the sphere $S(x+rv, r)$ does not meet Y; hence, for some r, it is tangent to Y, i.e., $S(x+rv, r) \in \Sigma(Y)$. Let $\rho(v) > 0$ be the minimal r that possesses this property. The function $v \mapsto \rho(v)$ is continuous in B; moreover, it is Lipschitz continuous. Consider the hypersurface $\Gamma(x) \doteq \{(y,r) : y = x + \rho(v)v, r = \rho(v),\ v \in B\}$ in $\Sigma_{\mathbf{E}}$. It is contained in $\Sigma(Y)$ and is Lipschitz continuous and smooth, except for a subset of zero measure area (spheres having multiple contact with Y and spheres of curvature of Y). Extend the function $\rho(v)$ to a positive continuous function in $\mathbf{E}^n$ that is smooth in $\mathbf{E}^n \setminus B$ and satisfies $\rho(v) = |v|^{-1}$ at infinity. Consider the domain

$$U(x) \doteq \operatorname{closure}\left\{(y,r) : 0 < r \leq \rho\left(\frac{y-x}{r}\right)\right\} \subset \Sigma_{\mathbf{E}}.$$

The intersection $U \cap \{r = 0\}$ contains a 1-neighborhood of x and $\Gamma(x) \subset \partial U(x)$. The generalized function $F_x(y,r) = F_n(y-x,r)$ is the singular fundamental solution for the Darboux operator with source at x. This kernel satisfies the condition (ii), since the support of F_x is contained in the future cone $V + x$. This cone is the union of all rays emanating from x. The hypersurface $\Gamma(x)$ is transversal to the cone $V + x$, and we see from the explicit formulae of the next sections that the trace of F_x on $\Gamma(x)$ is well-defined. This implies (iii). Thus all the conditions of Proposition 6 are fulfilled, and we can apply equation (6). □

4. Reconstruction in the ball

Let D be the unit ball in $\mathbf{E}^n$. Following the construction of Section 2, we obtain the function $\rho(y) = 1 - |y|$ and the cone $K = \{(y,r) : r + |y| \leq 1\}$. By direct calculation, we find

$$\Phi_n(x,y,r) = -\frac{1}{2} \frac{\left(1 - |x|^2 - 2\langle y,x\rangle + 2|y||x|^2\right)}{(1 - |y|)|y|}.$$

The integrals (3) and (2) have sense for all points $x \in D$, and we have from (3), for even n,

$$f(x) = -\frac{(n-1)!}{(2\pi\imath)^n} \int_D [q^{-n}] \frac{1 - |x|^2 - 2\langle y,x\rangle + 2|y||x|^2}{1 - |y|} \frac{Mf(y,\rho)\,dy}{|y|}.$$

where $q = (\rho^2 - |x-y|^2)/2$. From (2) for odd n, we have

$$f(x) = -\frac{1}{2(2\pi\imath)^{n-1}} \int_D \delta^{n-1}(q) \frac{1 - |x|^2 - 2\langle y,x\rangle + 2|y||x|^2}{1 - |y|} \frac{Mf(y,\rho)\,dy}{|y|}.$$

Similar formulae are given in [3].

5. Riesz kernels

Let q be a an arbitrary real nonsingular quadratic form in $\mathbf{V}$. Write $q(x) = 1/2 \sum q_{ij} x_i x_j$. The related Riesz kernels are powers of this quadratic form. We define complex powers of q as follows:

(7) $$(q \pm 0\imath)^\lambda(\rho) \doteq \lim_{\varepsilon \to +0} \int (q \pm \varepsilon\imath)^\lambda \rho.$$

Here ρ is an arbitrary test density in $\mathbf{V}$, i.e., a smooth density with compact support; and we take the branch of the function $\zeta^\lambda \doteq \exp(\lambda \ln \zeta)$, where $\ln \zeta$ is the branch of the logarithm in $\mathbb{C} \setminus (-\mathbb{R}_+)$ that is real for positive ζ.

Consider the quadratic form $q^*(\xi) = 1/2 \sum q^{ij} \xi_i \xi_j$ defined in the dual space $\mathbf{V}^*$, where the matrix $\{q^{ij}\}$ is inverse to $\{q_{ij}\}$. We call

$$q^*(D) = \frac{1}{2} \sum q^{ij} \frac{\partial^2}{\partial x_i \partial x_j}$$

the dual operator to the form q.

PROPOSITION 7. [8],[5] *The limits (7) exist for any ρ and define holomorphic families $\lambda \mapsto (q \pm 0\imath)^\lambda$, $\lambda \in \mathbb{C}$, of generalized functions in $\mathbf{V}$. They satisfy the equations*

(8) $$q^*(D)(q \pm 0\imath)^\lambda = \lambda(\lambda - 1 + n/2)(q \pm 0\imath)^{\lambda-1}.$$

PROOF. The Riesz identity states that

$$q^*(D) q^\lambda = \lambda (\lambda - 1 + n/2) q^{\lambda-1}. \tag{9}$$

It holds in the domain $\operatorname{Re} \lambda > 2$ if the power of q is properly defined; see [8, p.45]. Choose a positive quadratic form $p = p(x)$ in $\mathbf{V}$. Fix $\varepsilon > 0$ and consider the family of functions q_ε^λ, where $q_\varepsilon(x) = q(x) + \varepsilon p(x)$. They are locally integrable and holomorphic in λ for $\operatorname{Re} > 0$. Take the dual operator q_ε^*. By (9), we have

$$q_\varepsilon^{\lambda-1}(\rho) = \frac{1}{\lambda(\lambda-1+n/2)} q_\varepsilon^\lambda (q_\varepsilon^*(D) \rho). \tag{10}$$

The right-hand side has a meromorphic continuation to the half-plane $\{\lambda : \operatorname{Re} \lambda > -1\}$ with possible pole $\lambda = 0$ since the denominator vanishes there. Integrating by parts shows that the numerator also vanishes at $\lambda = 0$. Therefore, there is no pole at this point; and we get analytic extension of both sides to the half-plane $\{\operatorname{Re} \lambda > -1\}$. Using (10) again gives an analytic continuation to $\{\lambda : \operatorname{Re} \lambda > -2\}$, and so on. Another possible pole is $\lambda = 1 - n/2$. Integrating by parts again yields

$$\int q_\varepsilon^{1-n/2} (q_\varepsilon^*(D) \rho) = 0,$$

which means that there is no pole at all. Now let $\varepsilon \to 0$; the dual form q_ε^* tends to the dual form q^* since q is nonsingular. The right-hand side of (10) has a limit for $\operatorname{Re} \lambda > -1$, so the left side does also. Repeating the above arguments, we see that the family q_ε^λ has a limit as $\varepsilon \to 0$. It coincides with the function $(q + 0\imath)^\lambda$ for $\operatorname{Re} \lambda > 0$. This implies that the family $(q + 0\imath)^\lambda$ has analytic continuation to the whole plane. The same true for $(q - 0\imath)^\lambda$. □

Comparing various power functions, we get $(q \pm 0\imath)^\lambda = q_+^\lambda + \exp(\pm \pi \lambda \imath) q_-^\lambda$, where $q_-^\lambda = (-q)_+^\lambda$.

For an arbitrary natural integer k, we define the generalized function

$$[q^{-k}] = \frac{1}{2} \left[(q + 0\imath)^{-k} + (q - 0\imath)^{-k} \right].$$

Note the equation

$$\delta^{(k-1)}(q) = \frac{(n-1)!}{2\pi \imath} \left[(q - 0\imath)^{-k} - (q + 0\imath)^{-k} \right]. \tag{11}$$

6. Fundamental solution for the adjoint operator

The adjoint Darboux operator is

$$A^* = r \frac{\partial^2}{\partial r^2} + (3 - n) \frac{\partial}{\partial r} - r \Delta_x.$$

We call a generalized function F_n in $\mathbf{E}^n \times \mathbb{R}$ a *singular fundamental solution* for A^* if $A^* F_n = \delta_{0,0}$, where $\delta_{0,0} = \delta_{0,0}(x, r)$ is the Dirac function at the origin and $\operatorname{supp} F_n \subset \mathbf{E}^n \times \mathbb{R}_+$. We are going to find a singular fundamental solution for any n. We use the standard notation $r_+^\lambda = r^\lambda$, for $r > 0$, $\lambda \in \mathbb{C}$, and $r_+^\lambda = 0$ otherwise. Denote by dx the Euclidean volume form in $\mathbf{E}^n$.

THEOREM 8. *Set* $q = \left(r^2 - |x|^2 \right)/2$. *For $n \geq 3$ odd, the function*

$$F_n = \left(-\frac{1}{2\pi} \right)^{(n-1)/2} \frac{1}{2(n-2)!!} r_+^{n-2} \delta^{(n-2)}(q) \tag{12}$$

is a singular fundamental solution for the Darboux operator and is supported by the future cone. For $n \geq 2$ even, the function

$$F_n = \frac{(-1)^{n/2-1}(n-2)!}{2^{n-1}\pi^{n/2}\Gamma(n/2)} r_+^{n-2} \left[q^{1-n}\right] \tag{13}$$

is a singular fundamental solution for the Darboux operator.

LEMMA 9. *The function* $Q = r^{n-2}(q+0\imath)^{1-n}$ *satisfies the equation* $A^*Q = 0$ *in* $\mathbf{E}^n \times (\mathbb{R}\backslash\{0\})$.

PROOF. We have

$$A^*Q = r^{n-1}A(q+0\imath)^{1-n} = r^{n-1}\left(\partial_r^2 - \Delta_x + \frac{n-1}{r}\partial_r\right)(q+0\imath)^{1-n}.$$

Taking $\lambda = n-1$ in (8) yields $\left(\partial_r^2 - \Delta_x\right)(q+0\imath)^{1-n} = 2(n-1)^2(q+0\imath)^{-n}$. On the other hand,

$$(n-1)r^{-1}\partial_r(q+0\imath)^{1-n} = -2(n-1)^2(q+0\imath)^{-n},$$

which implies that the right-hand side vanishes. □

PROOF OF THEOREM 8. Take the function $Q_+ \doteq r_+^{n-2}(q+0\imath)^{1-n}$ and apply the operator

$$A^*Q_+ = A^*\left(r_+^0 Q\right) = 0 + 2r\delta_0(r)\partial_r Q + r\delta_0'(r)Q + (3-n)\delta_0(r)Q.$$

The second term vanishes, since the function $\partial_r Q$ is bounded. The third term is equal to $-2\delta_0(r)Q$ and $A^*Q = (1-n)\delta_0(r)Q$, where the product $\delta_0 Q$ can evaluated as the limit $\delta(r-s)Q$ as $s \to +0$. We have for an arbitrary test function ϕ in $\mathbf{E}^n$:

$$\delta(r-s)Q(\phi) = \int s^{n-2}\left(s^2 - |x|^2 + 0\imath\right)^{1-n}\phi(x)\,\mathrm{d}x$$

$$= \int \left(1 - |y|^2 + 0\imath\right)^{1-n}\phi(sy)\,\mathrm{d}y \to c_n\phi(0),$$

$$c_n \doteq \int_{\mathbb{R}^n} \left(1 - |y|^2 + 0\imath\right)^{1-n}\mathrm{d}y$$

as $s \to 0$, where we change variables $y = x/s$. To calculate the last integral, we use spherical coordinates $y = t\omega$ and again change variables $t = -\imath v$. This reduces the integral to the beta function:

$$c_n = \frac{2\pi^{n/2}(-\imath)^n}{\Gamma(n/2)} \int_0^\infty \frac{v^{n-1}\mathrm{d}v}{(1+v^2)^{n-1}} = -\frac{\pi^{n/2}(-\imath)^n \Gamma(n/2-1)}{\Gamma(n-1)}.$$

Therefore, $A^*Q_+ = c_n\delta_0(r)\delta_0(x) = c_n\delta_{0,0}(x,r)$; hence, $F_n \doteq \operatorname{Re} c_n^{-1}Q_+$ is a singular fundamental solution. The number c_n is imaginary for odd n and real for n even. Therefore, $F_n = c_n^{-1}\operatorname{Re} Q_+$ for even n and $F_n = -\imath c_n^{-1}\operatorname{Im} Q_+$ for n odd. We have

$$\operatorname{Re} Q_+ = r_+^{n-2}\left[q^{1-n}\right],\ \operatorname{Im} Q_+ = (-1)^{n-2}2\pi\left((n-2)!\right)^{-1}r_+^{n-2}\delta^{(n-2)}(q)$$

by Section 5. Elementary calculations yield (12) and (13). □

References

[1] A.M. Cormack and E.T. Quinto, *A Radon transform on spheres through the origin in $\mathbb{R}^n$ and applications to the Darboux equation*, Trans. Amer. Math. Soc. **260** (1980), 575-581.

[2] J.B. Diaz and H.F. Weinberger, *A solution of the singular initial value problem for Euler-Poisson-Darboux equation*, Proc. Amer. Math. Soc. **4** (1953), 703-715.

[3] I.M. Gelfand, S.G. Gindikin and M. I. Graev, *Selected Topics in Integral Geometry*, Amer. Math. Soc., Providence, RI, 2003.

[4] I.M. Gelfand, M.I. Graev, N.Ya. Vilenkin, *Integral Geometry and Representation Theory*, Academic Press, New York, 1966.

[5] I.M. Gelfand and G.E. Shilov, *Generalized Functions: Properties and Operators*, Academic Press, New York, 1964.

[6] S.G. Gindikin, *Integral geometry on real quadrics*, Lie Groups and Lie Algebras, Amer. Math. Soc. Transl. Ser. 2 **169**, 1995, pp. 23-31.

[7] A.B. Goncharov, *Differential equations and integral geometry*, Adv. Math. **131** (1997), 279-343.

[8] M. Riesz, *L'intégral de Riemann-Liouville et le problème de Cauchy*, Acta Math. **81** (1949), 1-223.

SCHOOL OF MATHEMATICAL SCIENCES, TEL-AVIV UNIVERSITY, 69978 RAMAT-AVIV, ISRAEL
E-mail address: `palamodo@post.tau.ac.il`

On the Essential Spectrum of Electromagnetic Schrödinger Operators

V. S. Rabinovich

Dedicated to Professor Lawrence Zalcman on the occasion of his sixtieth birthday

ABSTRACT. The aim of this paper is to present a new approach to the investigation of the essential spectrum of electromagnetic Schrödinger operators on $\mathbb{R}^N$. The well-known methods for the investigation of the location of the essential spectrum of Schrödinger operators are the Weyl criterion for the one particle problem and the Hunziker-van Winter-Zhislin Theorem for multiparticle problems, the modern proof of which is based on the Ruelle-Simon partition of unity. Our approach to the study of the location of the essential spectrum is different and is based on the theory of limit operators. For a Schrödinger operator $\mathcal{H}$, we introduce a family of limit operators and prove that the essential spectrum of $\mathcal{H}$ is the union of spectra of limit operators. Since the limit operators have a simpler structure than $\mathcal{H}$, we obtain an effective description of the location of the essential spectrum of $\mathcal{H}$.

1. Introduction

The aim of the paper is to present a new approach to the investigation of essential spectra of electromagnetic Schrödinger operators on $\mathbb{R}^N$ of the form

$$(1) \quad (\mathcal{H}u)(x) = \sum_{k=1}^{N}(i\partial_{x_k}+a_k(x))^2 u(x)+W(x)u(x) = (i\nabla+\mathbf{a}(x))^2 u(x)+W(x)u(x),$$

where $\mathbf{a} = (a_1, ..., a_N)$ is a vector potential of a magnetic field and W is a scalar potential of an electric field.

The well-known methods for the investigation of the location of essential spectra of Schrödinger operators are the Weyl criterion [6, Theorem XIII.14] for the one particle problem and the Hunziker-van Winter-Zhislin Theorem (HWZ Theorem) for multiparticle problems [1, Chap. 3], the modern proof of which is based on the Ruelle-Simon partition of unity [1, Chap.3] and references given there.

Our approach to the study of location of the essential spectrum is different and is based on the theory of limit operators. For the Schrödinger operator (1), we introduce a family of limit operators and prove that the essential spectrum of $\mathcal{H}$

2000 *Mathematics Subject Classification.* Primary 35J10, 35P05.

is the union of spectra of limit operators. Since the limit operators have a more simple structure then $\mathcal{H}$, we obtain an effective description of the location of the essential spectrum of $\mathcal{H}$.

Our consideration is based on the results [3] (see also [4], [5]) on the Fredholm theory of integral operators on $\mathbb{R}^N$ in the Wiener algebra $W(\mathbb{R}^N)$.

The structure of the paper is as follows. In Section 2, we introduce necessary definitions for the Schrödinger operator $\mathcal{H}$ and formulate a main result. Applying the limit operator method, in Section 3, we obtain a new and effective description of the location of the essential spectrum for Schrödinger operators with slowly oscillating and separately slowly oscillating potentials. We also consider the two-particle Schrödinger operators. Of course, not all applications of the limit operator method are exhausted by these examples. Indeed, the case of multi-particle systems is covered by our method. However, for simplicity and for lack of space, we restrict our attention to the case of two-particle systems. Section 4 is devoted to the proof of main theorem on the location of the essential spectrum of $\mathcal{H}$.

2. Schrödinger operators

We use the following notation.

- $C_b(\mathbb{R}^N)$ is the C^*-algebra of all bounded continuous functions on $\mathbb{R}^N$ with sup-norm,
- $UC_b(\mathbb{R}^N)$ is the closed subalgebra of $C_b(\mathbb{R}^N)$ of uniformly continuous functions,
- $C_b^{(1)}(\mathbb{R}^N)$ is the nonclosed subalgebra of $C_b(\mathbb{R}^N)$ of differentiable functions $a \in C_b(\mathbb{R}^N)$ with $\partial_{x_j} a \in C_b(\mathbb{R}^N)$, $j = 1, ..., N$.
- $SO(\mathbb{R}^N)$ is the closed subalgebra of $C_b(\mathbb{R}^N)$ of functions such that

$$\lim_{x \to \infty} \sup_{y \in K} |a(x+y) - a(x)| = 0$$

for arbitrary compact $K \subset \mathbb{R}^N$.

Note that $a \in SO(\mathbb{R}^N) \cap C_b^{(1)}(\mathbb{R}^N)$ if and only if

$$\lim_{x \to \infty} \partial_{x_j} a = 0, \; j = 1, ..., N.$$

- $Q(\mathbb{R}^N)$ is the subalgebra of $L^\infty(\mathbb{R}^N)$ of functions a satisfying the condition

$$\lim_{x \to \infty} \int_K |a(x+y)| \, dy = 0$$

for arbitrary compact $K \subset \mathbb{R}^N$.

One can see that $C_b^{(1)}(\mathbb{R}^N)$ and $SO(\mathbb{R}^N)$ are subalgebras of $UC_b(\mathbb{R}^N)$.

If a is a function on $\mathbb{R}^N$, we denote by aI the operator of multiplication by a. We denote by $\mathcal{B}(H)$ the space of all bounded linear operators acting on the Hilbert space H.

We consider a Schrödinger operator $\mathcal{H}$ satisfying the following assumptions.
- **A.** $a_j \in SO(\mathbb{R}^N) \cap C_b^1(\mathbb{R}^N)$, $j = 1, ..., N$.
- **B.** The electric potential W satisfies the following conditions:
  (i) the operator $W(1 - \Delta)^{-1}$ is bounded on $L^2(\mathbb{R}^N)$;
  (ii) every sequence $\mathbb{Z}^N \ni h_m \to \infty$ has a subsequence h_{m_k} for which there exists a measurable function W_h

$$\lim_{k \to \infty} \left\| (W(\cdot + h_k) - W_h(\cdot))(I - \Delta)^{-1} \psi_R I \right\|_{\mathcal{B}(L^2(\mathbb{R}^N))} = 0, \tag{2}$$

$$\lim_{k \to \infty} \left\| \psi_R (W(\cdot + h_k) - W_h(\cdot))(I - \Delta)^{-1} \right\|_{\mathcal{B}(L^2(\mathbb{R}^N))} = 0, \tag{3}$$

where ψ_R is a cut-off function, that is $\psi \in C_0^\infty(\mathbb{R}^N)$, $\psi(x) = 1$ if $|x| \le 1$, and $0 \le \psi(x) \le 1$, $\psi_R(x) = \psi(x/R)$.

The operator $\mathcal{H}$ can be considered an unbounded closed operator on the Hilbert space $L^2(\mathbb{R}^N)$ with domain $H^2(\mathbb{R}^N)$.

We recall that an unbounded closed operator on a Hilbert space H is a Fredholm operator if the spaces $\ker A$ and $\operatorname{coker} A = H/\operatorname{Range}(A)$ are finite dimensional. A point $\lambda \in \mathbb{C}$ is a point of the essential spectrum of A if the operator $A - \lambda I$ is not Fredholm. The essential spectrum of A is denoted by $\operatorname{sp}_{\mathrm{ess}} A$.

Applying the uniform ellipticity of $\mathcal{H}$, one can show that $\lambda \in \operatorname{sp}_{\mathrm{ess}} A$ if and only if
$$A - \lambda I : H^2(\mathbb{R}^N) \to L^2(\mathbb{R}^N)$$
is not a Fredholm operator in the usual sense.

Let $a \in UC_b(\mathbb{R}^N)$ and h_m be a sequence of points in $\mathbb{Z}^N$ tending to infinity. Then the sequence $a(\cdot + h_m)$ is uniformly bounded and equicontinuous on each compact $K \subset \mathbb{R}^N$. By the Arzela-Ascoli Theorem, there exists a subsequence $a(\cdot + h_{m_k})$ which converges uniformly on every compact $K \subset \mathbb{R}^N$ to a function a_h continuous on $\mathbb{R}^N$. One sees that $a_h \in C_b(\mathbb{R}^N)$.

Note that if $a \in SO(\mathbb{R}^N)$, then the limit functions a_h are constants.

We denote by $V_h, h \in \mathbb{R}^N$ the unitary shift operator on $L^2(\mathbb{R}^N)$
$$V_h u(x) = u(x - h).$$

Then
$$V_{-h} \mathcal{H} V_h = (i\nabla + \mathbf{a}(\cdot + h)I)^2 + W(\cdot + h)I. \tag{4}$$

Let $\mathcal{H}$ be a Schrödinger operator satisfying Conditions **A** and **B** and suppose $h_m \to \infty$. Then there exists a subsequence h_{m_k} such that
$$a_j(\cdot + h_{m_k}) \to a_j^h \in \mathbb{C}$$
uniformly on compacta in $\mathbb{R}^N$, and
$$W(\cdot + h_{m_k}) \to W_h(\cdot)$$
in the sense of the convergence defined by (2), (3).

The operator
$$\mathcal{H}_h = (i\nabla + \mathbf{a}^h I)^2 + W_h I,$$
where $\mathbf{a}^h = (a_1^h, ..., a_N^h)$, is called the limit operator of $\mathcal{H}$ defined by the sequence h_{m_k}. We denote by $\sigma(\mathcal{H})$ the set of all limit operators of $\mathcal{H}$.

The following propositions clarify the definition of limit operators.

PROPOSITION 1. *Let $\mathcal{H}_h$ be the limit operator defined by the sequence h_m. Then for every $R > 0$,*

$$\lim_{m \to \infty} \left\|(V_{-h_m}\mathcal{H}V_{h_m} - \mathcal{H}_h)(I - \Delta)^{-1}\psi_R I\right\|_{\mathcal{B}(L^2(\mathbb{R}^N))} = 0, \tag{5}$$

$$\lim_{m \to \infty} \left\|\psi_R(V_{-h_m}\mathcal{H}V_{h_m} - \mathcal{H}_h)(I - \Delta)^{-1}\right\|_{\mathcal{B}(L^2(\mathbb{R}^N))} = 0. \tag{6}$$

PROOF. It follows from the commutator formula for pseudodifferential operators and the Calderón-Vaillancourt Theorem (see, for instance, [7]) that

$$\lim_{R \to \infty} \left\|[\psi_R, (I - \Delta)^{-1}]\right\|_{\mathcal{B}(L^2(\mathbb{R}^N))} = 0. \tag{7}$$

Applying formulas (4) and (7), the uniform convergence $a_j(\cdot + h_{m_k}) \to a_j^h$ on compacta, and the condition

$$\lim_{x \to \infty} \partial_{x_k} a_j(x) = 0$$

which follows from $a_j \in SO(\mathbb{R}^N) \cap C_b^1(\mathbb{R}^N), j = 1, ..., N$, we obtain

$$\lim_{m \to \infty} \left\|(V_{-h_m}\mathcal{H}^0 V_{h_m} - \mathcal{H}_h^0)(I - \Delta)^{-1}\psi_R I\right\|_{\mathcal{B}(L^2(\mathbb{R}^N))} = 0,$$

where $\mathcal{H}^0 = (i\nabla + \mathbf{a})^2$ is the leading part of $\mathcal{H}$, that is, $\mathcal{H}_h^0 = (i\nabla + \mathbf{a}^h)^2$. □

PROPOSITION 2. *Let K be a compact operator on $L^2(\mathbb{R}^N)$. Then for every $R > 0$,*

$$\lim_{m \to \infty} \|V_{-h_m}KV_{h_m}\psi_R I\|_{\mathcal{B}(L^2(\mathbb{R}^N))} = \lim_{m \to \infty} \|\psi_R V_{-h_m}KV_{h_m}\|_{\mathcal{B}(L^2(\mathbb{R}^N))} = 0.$$

PROOF. Let $\chi \in C_0^\infty(\mathbb{R}^N)$, $\chi(x) = 1$ if $|x| \leq 1$, and $0 \leq \chi(x) \leq 1$; and put $\chi_R(x) = \chi(x/R), \chi_R' = 1 - \chi_R$. Note that

$$s - \lim_{R \to \infty} \chi_R' = 0.$$

Since K is a compact operator,

$$\lim_{R \to \infty} \|K\chi_R'\| = 0.$$

For fixed $\varepsilon > 0$, one can find R_0 such that

$$\|K\chi_{R_0}'\| < \varepsilon.$$

Since $h_m \to \infty$, one can find m_0 such that for all $m > m_0$,

$$\chi_{R_0} V_{h_m} \psi_R = 0.$$

Thus for $m > m_0$,

$$\begin{aligned}
\|V_{-h_m}KV_{h_m}\psi_R I\|_{\mathcal{B}(L^2(\mathbb{R}^N))} &= \|V_{-h_m}K(\chi_{R_0} + \chi_{R_0}')V_{h_m}\psi_R I\|_{\mathcal{B}(L^2(\mathbb{R}^N))} \\
&\leq \|K\chi_{R_0}' I\|_{\mathcal{B}(L^2(\mathbb{R}^N))} < \varepsilon.
\end{aligned}$$

Passing to the adjoint operator, we obtain

$$\lim_{m \to \infty} \|\psi_R V_{-h_m}KV_{h_m}\|_{\mathcal{B}(L^2(\mathbb{R}^N))} = 0.$$

□

COROLLARY 3. *Let $W(1-\Delta)^{-1}$ be a compact operator. Then all limit operators of $W_h I$ are zero operators.*

The main result of the paper is the following theorem.

THEOREM 4. *Let $\mathcal{H}$ be a Schrödinger operator with vector and scalar potentials satisfying Conditions* **A** *and* **B**. *Then*
$$\mathrm{sp}_{\mathrm{ess}} \mathcal{H} = \bigcup_{\mathcal{H}_h \in \sigma(\mathcal{H})} sp\, \mathcal{H}_h.$$

The proof of this theorem is given in Section 4.

3. Applications

3.1. Schrödinger operators with slowly oscillating potentials. We consider electromagnetic Schrödinger operators (1), with magnetic potential a, satisfying Condition **A** and electric potential W of the form
$$W = W^{(1)} + W^{(2)},$$
where $W^{(1)} \in SO(\mathbb{R}^N)$ and the operator $W^{(2)}(1-\Delta)^{-1}$ is a compact operator in $L^2(\mathbb{R}^N)$. In what follows, we suppose that a_j and W are real-valued functions.

For instance, if $W^{(2)} = W_1^{(2)} + W_2^{(2)}$, where $W_1^{(2)} \in Q(\mathbb{R}^N)$, $W_2^{(2)} \in L^q(\mathbb{R}^N)$, and $N/2 \leq q < \infty$ when $N \geq 4$ and $q = 2$ when $N \leq 4$, then $W^{(2)}(1-\Delta)^{-1}$ is a compact operator. (Note that in the case $N = 3$, the Colomb potential $W_3(x) = \frac{a}{|x|}$ belongs to this class).

Indeed, it is well-known (see, for instance, [2, p. 67]) that $W_2^{(2)}(1-\Delta)^{-1}$ is a compact operator on $L^2(\mathbb{R}^N)$. One can also show that the operator $W_1^{(2)}(1-\Delta)^{-1}$ is a compact operator on $L^2(\mathbb{R}^N)$. This implies by Corollary 3 that the limit operator $W_h^{(2)}$ for $W^{(2)}$ exists and equals 0 for an arbitrary sequence $h_m \to \infty$.

Let $h_m \to \infty$. Then there exists a subsequence g_m such that
$$a_k(x + g_m) \to a_k^g, \quad W^{(1)}(x + g_m) \to W_g^{(1)}$$
uniformly on every compact set in $\mathbb{R}^N$. Since $a_k, W^{(1)} \in SO(\mathbb{R}^N)$, $a_k^g, W_g^{(1)}$ are constants.

Thus the limit operators for $\mathcal{H}$ are the operators of the form
$$\mathcal{H}_g = (i\nabla + \mathbf{a}^g I)^2 + W_g^{(1)} I,$$
where $\mathbf{a}^g = (a_1^g, ..., a_N^g) \in \mathbb{R}^N, W_g^{(1)} \in \mathbb{R}$.

Let
$$T_{\mathbf{a}^g} f = e^{i\mathbf{a}^g \cdot x} f$$
be a unitary operator in $L^2(\mathbb{R}^N)$. Then
$$T_{\mathbf{a}^g} \mathbf{H}_g T_{\mathbf{a}^g}^{-1} = -\Delta + W_g^{(1)} I.$$
Thus
$$sp\, \mathcal{H}_g = \left\{ \lambda \in \mathbb{R} : \lambda = |\xi|^2 + W_g^{(1)}, \xi \in \mathbb{R}^N \right\} = [W_g^{(1)}, +\infty),$$
where $W_g^{(1)}$ is a partial limit of $W^{(1)}$, corresponding to the sequence $g_n \to \infty$.

Thus Theorem 4 implies the following result.

THEOREM 5. *Let the magnetic potential* **a** *satisfy Condition* **A** *and the electric potential W satisfy the conditions given above. Then*
(8) $$\mathrm{sp}_{\mathrm{ess}} \mathcal{H} = \bigcup_g [W_g^{(1)}, +\infty) = [\liminf_{x \to \infty} W^{(1)}(x), +\infty).$$

The following theorem locates the spectrum of $\mathcal{H}$ for a class of potentials $W \in SO(\mathbb{R}^N)$.

THEOREM 6. *Suppose the magnetic potential* **a** *satisfies Condition* **A**, *the electric potential* W *belongs to* $SO(\mathbb{R}^N)$, *and*

(9) $$\liminf_{x \to \infty} W(x) = \inf_{x \in \mathbb{R}^N} W(x).$$

Then

(10) $$\mathrm{sp}_{\mathrm{ess}} \mathcal{H} = \mathrm{sp}\, \mathcal{H} = [\inf_{x \in \mathbb{R}^N} W(x), \infty).$$

PROOF. Indeed, $\mathrm{sp}\mathcal{H} \subset [\inf_{x \in \mathbb{R}^N} W(x), \infty)$. Applying condition (9) and formula (8), we obtain $\mathrm{sp}\mathcal{H} \subset \mathrm{sp}_{\mathrm{ess}} \mathcal{H}$. But $\mathrm{sp}_{\mathrm{ess}} \mathcal{H} \subset \mathrm{sp}\mathcal{H}$ in all cases. Hence $\mathrm{sp}_{\mathrm{ess}} \mathcal{H} = \mathrm{sp}\mathcal{H}$. □

EXAMPLE 7. Let $W(x) = \sin|x|^\gamma, 0 < \gamma < 1$. Then

(11) $$\mathrm{sp}_{\mathrm{ess}} \mathcal{H} = \mathrm{sp}\, \mathcal{H} = [-1, +\infty).$$

Indeed, $\sin|x|^\gamma \in SO(R^N)$ and

$$\liminf_{x \to \infty} \sin|x|^\gamma = \inf_{x \in \mathbb{R}^N} \sin|x|^\gamma = -1.$$

Hence, (11) follows from (10).

3.2. Schrödinger operators with separately slowly oscillating potentials. For simplicity, we consider the case of 2-separately slowly oscillating potentials. Write

$$\mathbb{R}^N = \mathbb{R}^{N_1} \times \mathbb{R}^{N_2}.$$

We denote by $SO(\mathbb{R}^{N_1}) \otimes SO(\mathbb{R}^{N_2})$ the closure in $C_b(\mathbb{R}^N)$ of the set of functions of the form

$$a(x^1, x^2) = \sum_{j=1}^{M} a_j(x^1) b_j(x^2), \ x^1 \in \mathbb{R}^{N_1}, \ x^2 \in \mathbb{R}^{N_2},$$

where $a_j \in SO(\mathbb{R}^{N_1}), b_j \in SO(\mathbb{R}^{N_2})$.

We consider the Schrödinger operator

$$\mathcal{H} = (i\nabla + \mathbf{a}(x))^2 + WI,$$

where **a** satisfies Condition **A**, and the real-valued function W belongs to $SO(\mathbb{R}^{N_1}) \otimes SO(\mathbb{Z}^{N_2})$.

It is evident that $W(1 - \Delta_x)^{-1}$ is bounded in $L^2(\mathbb{R}^N)$. We show that each sequence $h_n \to \infty$ has a subsequence g_n defining a limit operator, and we study their structure.

Let $\mathbb{Z}^N \ni h_n = (h_n^1, h_n^2) \to \infty$. We consider the following cases.

1) $h_n^1 \to \infty, h_n^2$ is a bounded sequence. Then there exists a subsequence $g_n = (g_n^1, g_n^2)$, where $g_n^2 = g^2$ is a constant sequence, and a limit function W_g depending only on x^2, such that

$$\lim_{n \to \infty} \sup_{x \in K} \left| W(x^1 + g_n^1, x^2 + g_n^2) - W_g^2(x^2) \right| = 0$$

for arbitrary compacta $K \in \mathbb{R}^N$. We denote by J_1 the set of such sequences.

Let $g_n \in J_1$. Then the limit operator defined by this sequence is unitarily equivalent to the operator

$$\mathcal{H}_g^2 = -\Delta_{x^1} \otimes I_{L^2(\mathbb{R}^{N_2})} + I_{L^2(\mathbb{R}^{N_1})} \otimes \left(-\Delta_{x^2} + W_g^2 I_{L^2(\mathbb{R}^{N_2})} \right),$$

where W_g^2 depends only on x^2.

After applying the Fourier transform with respect to x^1, we obtain that $\mathcal{H}_g^2$ is unitarily equivalent to the operator

$$H_g^2 = \left|\xi^1\right|^2 \otimes I_{L^2(\mathbb{R}^{N_2})} + I_{L^2(\mathbb{R}^{N_1})} \otimes \left(-\Delta_{x^2} + W_g^2 I\right),$$

where $W_g^2 \in SO(\mathbb{R}^{N_2})$.

It follows from Theorem 5 that

$$\operatorname{sp}_{\operatorname{ess}}(-\Delta_{x^2} + W_g^2 I_{L^2(\mathbb{R}^{N_2})}) = [\liminf_{x^2 \to \infty} W_g^2(x^2), +\infty).$$

Outside of this set, the operator $-\Delta_{x^2} + W_g^2 I_{L^2(\mathbb{R}^{N_2})}$ can have a discrete spectrum. We set $\mu_g^2 = \lambda_{\min}(-\Delta_{x^2} + W_g^2 I_{L^2(\mathbb{R}^{N_2})})$, the minimal point of the discrete spectrum of $-\Delta_{x^2} + W_g^2 I_{L^2(\mathbb{R}^{N_2})}$ if this operator has a discrete spectrum, and $\mu_g^2 = \liminf_{x^2 \to \infty} W_g^2(x^2)$ otherwise. Then

$$\operatorname{sp} H_g^2 = \operatorname{sp} \mathcal{H}_g^2 = [\mu_g^2, +\infty).$$

2) h_n^1 is a bounded sequence, $h_n^2 \to \infty$. Then there exists a subsequence $g_n = (g_n^1, g_n^2)$ such that $g_n^1 = g^1$ is a constant sequence, and

$$\lim_{n \to \infty} \sup_{x \in K} \left|W(x^1 + g_n^1, x^2 + g_n^2) - W_g^1(x^1)\right| = 0$$

for arbitrary compact $K \in \mathbb{R}^N$. We denote by J_2 the set of such sequences.

Let $g_n \in J_2$. Then the limit operator defined by this sequence is unitarily equivalent to the operator

$$\mathcal{H}_g^1 = I_{L^2(\mathbb{R}^{N_1})} \otimes -\Delta_{x^2} + \left(-\Delta_{x^1} + W_g^1 I_{L^2(\mathbb{R}^{N_1})}\right) \otimes I_{L^2(\mathbb{R}^{N_2})},$$

where W_g^1 depends only on x^1. As above, we obtain that

$$\operatorname{sp} \mathcal{H}_g^1 = [\mu_g^1, +\infty),$$

where $\mu_g^1 = \lambda_{\min}(-\Delta_{x^1} + W_g^1 I_{L^2(\mathbb{R}^{N_1})})$ if the operator $-\Delta_{x^1} + W_g^1 I_{L^2(\mathbb{R}^{N_1})}$ has a discrete spectrum, and $\mu_g^1 = \liminf_{x^1 \to \infty} W_g^1(x^1)$ otherwise.

3) $h_n^1 \to \infty, h_n^2 \to \infty$. Then there exists a subsequence $g_n = (g_n^1, g_n^2)$ such that

$$\lim_{n \to \infty} \sup_{x \in K} \left|W(x^1 + g_n^1, x^2 + g_n^2) - W_g^{12}\right| = 0, \quad W_g^{12} \in \mathbb{R}$$

for every compact $K \subset \mathbb{R}^N$. We denote by J_{12} the set of such sequences.

Let $g_n - (g_n^1, g_n^2) \in J_{12}$. Then $\mathcal{H}_g = -\Delta + W_g^{12} I$ and

$$\operatorname{sp} \mathcal{H}_g = [W_g^{12}, +\infty).$$

By Theorem 4,

$$\operatorname{sp}_{\operatorname{ess}} \mathcal{H} = \bigcup_{g \in J_1 \cup J_2 \cup J_{12}} \operatorname{sp} \mathcal{H}_g.$$

Hence

$$\operatorname{sp}_{\operatorname{ess}} \mathcal{H} = \bigcup_{g \in J_1} [\mu_g^2, +\infty) \cup \bigcup_{g \in J_2} [\mu_g^1, +\infty) \cup \bigcup_{g \in J_{12}} [W_g^{12}, +\infty).$$

Let $g_n = (g_n^1, g_n^2) \in J_{12}$. We set $g_n' = (g_n^1, 0)$. Then

$$W_g^{12} \geq \liminf_{x^2 \to \infty} W_{g'}^2(x^2).$$

This implies that
$$\bigcup_{g \in J_{12}} [W_g^{12}, +\infty) \subset \bigcup_{g \in J_1} [\mu_g^2, +\infty),$$
and
$$(12) \quad \mathrm{sp}_{\mathrm{ess}} \mathcal{H} = \bigcup_{g \in J_1} [\mu_g^2, +\infty) \cup \bigcup_{g \in J_2} [\mu_g^1, +\infty) = [\inf_{g \in J_1} \mu_g^2, +\infty) \cup [\inf_{g \in J_2} \mu_g^1, +\infty).$$

Thus we have proved the following theorem.

THEOREM 8. *Suppose that the magnetic potential* **a** *satisfies Condition* **A** *and the electric potential* W *belongs to* $SO(\mathbb{R}^{N^1}) \otimes SO(\mathbb{R}^{N^2})$. *Then*
$$\mathrm{sp}_{\mathrm{ess}} \mathcal{H} = [\mu, +\infty),$$
where $\mu = \min \{\inf_{g \in J_1} \mu_g^2, \inf_{g \in J_2} \mu_g^1\}$.

EXAMPLE 9. Let $W(x^1, x^2) = \sin|x^1|^{\gamma_1} \sin|x^2|^{\gamma_2}, 0 < \gamma_1 < 1, 0 < \gamma_2 < 1$. It is evident that $W(x^1, x^2) \in SO(\mathbb{R}^{N_1}) \otimes SO(\mathbb{R}^{N_2})$; hence formula (12) applies. Further, by formula (11), we obtain $\inf_{g \in J_2} \mu_g^1 = \inf_{g \in J_1} \mu_g^2 = -1$. Thus
$$\mathrm{sp}_{\mathrm{ess}} \mathcal{H} = [-1, +\infty).$$

3.3. Two-particle Schrödinger operators.

We consider the 2–particle Schrödinger operator on $\mathbb{R}^{2l}$, where $x = (x^1, x^2), x^j \in \mathbb{R}^l, j = 1, 2$,
$$\mathcal{H}u(x) = -\Delta_{x^1} u(x) - \Delta_{x^2} u(x) + \sum_{1 \le j \le 2} W_j(x^j) u(x) + W_{12}(x^1 - x^2) u(x).$$

We suppose that W_1, W_2, W_{12} are measurable real-valued functions on $\mathbb{R}^l$ such that
$$W_1(1 - \Delta)^{-1}, W_2(1 - \Delta)^{-1}, W_{12}(1 - \Delta)^{-1}$$
are compact operators on $L^2(\mathbb{R}^l)$, where Δ is the Laplacian on $\mathbb{R}^l$.

We show that the potential
$$W(x) = W_1(x^1) + W_2(x^2) + W_{12}(x^1 - x^2)$$
satisfies Condition **B**.

Indeed, let $\Delta_x = \Delta_{x^1} + \Delta_{x^2}$. Then, since the operator $(1 - \Delta_{x^j})(1 - \Delta_x)^{-1}$ is bounded in $L^2(\mathbb{R}^{2l})$, applying the equality
$$W_j(1 - \Delta_x)^{-1} = W_j(1 - \Delta_{x^j})^{-1}(1 - \Delta_{x^j})(1 - \Delta_x)^{-1},$$
we see that the operator $W_1(1 - \Delta_x)^{-1}$ is bounded in $L^2(\mathbb{R}^{2l})$ also.

Let us consider the operator $W_{12}(x^1 - x^2)(1 - \Delta_x)^{-1}$. After the orthogonal linear transformation
$$y^1 = \frac{1}{\sqrt{2}}(x^1 - x^2), y^2 = \frac{1}{\sqrt{2}}(x^1 + x^2),$$
we obtain that $W_{12}(x^1 - x^2)(1 - \Delta_x)^{-1}$ is unitarily equivalent to the operator $W_{12}(y^1)(1 - \Delta_y)^{-1}$, which is bounded on $L^2(\mathbb{R}^{2l})$. Thus $W(x)(1 - \Delta_x)^{-1}$ is bounded on $L^2(\mathbb{R}^{2l})$.

Now we consider the limit operators for $\mathcal{H}$.

Let $\mathbb{Z}^{2l} \ni h_m = (h_m^1, h_m^2)$ be a sequence tending to infinity. Then the following cases are possible.

- $h_m^j \to \infty$, h_m^{3-j} is bounded, $j = 1, 2$. We can suppose that $h_m^{3-j} = h^{3-j}$ is a constant sequence; otherwise, we can pass to a subsequence.

  It follows from the proof of Proposition 1 that
  $$\lim_{m\to\infty} \left\| V_{-h_m^j} W_j(x^j) V_{h_m^j} (1 - \Delta_{x^j})^{-1} \psi_R(x^j) I \right\|$$
  $$= \lim_{m\to\infty} \left\| \psi_R(x^j) V_{-h_m^j} W_j V_{h_m^j} (1 - \Delta_{x^j})^{-1} \right\| = 0.$$

  Using the boundedness of $(1 - \Delta_{x^j})(1 - \Delta_x)^{-1}$ on $L^2(\mathbb{R}^{2l})$, we obtain
  $$\lim_{m\to\infty} \left\| V_{-h_m} W_j(x) V_{h_m} (1 - \Delta_x)^{-1} \psi_R(x) I \right\|$$
  $$= \lim_{m\to\infty} \left\| \psi_R V_{-h_m} W_j(x) V_{h_m} (1 - \Delta_x)^{-1} \right\| = 0.$$

  Thus the limit operator for W_j defined by a sequence h_m such that $h_m^j \to \infty$ is the 0−operator. For the case when h_m^{3-j} is a bounded sequence, the limit operators for W_{12} are also the 0-operator. Thus, in this case, the limit operators are
  $$\mathcal{H}_h = -\Delta_{x^1} - \Delta_{x^2} + W_j(x^j) I.$$

- Let $h_m^j \to \infty$, $h_m^{3-j} \to \infty$. Then we have to consider the following two cases.

  (i) $h_m^j - h_m^{3-j} \to \infty$. In this case, the limit operator is $-\Delta_{x^1} - \Delta_{x^2}$,

  (ii) $h_m^j - h_m^{3-j}$ is a bounded sequence, which is supposed to be a constant g. In this case, the limit operator is
  $$-\Delta_{x^1} - \Delta_{x^2} + W_{12}(x^1 - x^2 + g) I,$$

  which is unitarily equivalent to the operator
  $$-\Delta_{y^1} - \Delta_{y^2} + W_{12}(y^1) I.$$

Let
$$\mu_j = \inf \mathrm{sp}(-\Delta_y + W_j(y) I), j = 1, 2,$$
$$\mu_{1,2} = \inf \mathrm{sp}(-\Delta_y + W_{12}(y) I), y \in \mathbb{R}^l.$$

Hence the spectra of the operators $-\Delta_{x^1} - \Delta_{x^2} + W_j I$ are $[\mu_j, \infty)$, $j = 1, 2$, and the spectrum of $-\Delta_{y^1} - \Delta_{y^2} + W_{12}(y^1) I$ is $[\mu_{1,2}, \infty)$.

By the condition on the potentials W_j, W_{12}, the operators $-\Delta_y + W_j(y)$, $-\Delta_y + W_{12}(y)$ have essential spectrum $[0, \infty)$. Hence $\mu_j \leq 0, \mu_{1,2} \leq 0$.

Thus Theorem 4 implies the following result.

THEOREM 10. *Let the potential W satisfy the above given conditions. Then*
$$\mathrm{sp}_{\mathrm{ess}} \mathcal{H} = [\mu, +\infty), \tag{13}$$
where $\mu = \min\{\mu_1, \mu_2, \mu_{12}\} \leq 0$.

EXAMPLE 11. Let us consider the Hamiltonian
$$\mathcal{H} = -\Delta_{x^1} - \Delta_{x^2} - 2e^2 \left(\frac{1}{|x^1|} + \frac{1}{|x^2|} \right) + \frac{e^2}{|x^1 - x^2|}$$
acting in $L^2(\mathbb{R}^3_{x^1} \times \mathbb{R}^3_{x^2})$, where
$$x^1 = (x_1^1, x_2^1, x_3^1), x^2 = (x_1^2, x_2^2, x_3^2).$$

This operator describes the motion of two electrons of mass $1/2$ and charge $-e$ moving around a fixed nucleus of charge $+2e$.

Note that $|y|^{-1}(1-\Delta_y)^{-1}$ is a compact operator in $L^2(\mathbb{R}^3)$ [2, Corollary 3.6.6]. The operator
$$-\Delta_y - \frac{2e^2}{|y|}$$
acting in $L^2(\mathbb{R}_y^3)$ describes the hydrogen atom, for an electron of mass $1/2$ and charge $-e$, moving around fixed nucleus of charge $+e$. It is well-known [2, p. 164]) that
$$\mathrm{sp}\left(-\Delta_{x^j} - \frac{2e^2}{|x^j|}\right) = \left\{-\frac{e^4}{4n^2} : n \in \mathbb{N}\right\} \cup [0,+\infty).$$

The Colomb potential $\frac{e^2}{|y|}$ is positive, and $\frac{e^2}{|y|}(1-\Delta)^{-1}$ is a compact operator. Hence
$$\mathrm{sp}\left(-\Delta_y + \frac{e^2}{|y|}\right) = \mathrm{sp}_{\mathrm{ess}}\left(-\Delta_y + \frac{e^2}{|y|}\right) = [0,\infty).$$
Thus, applying formula (13), we obtain
$$\mathrm{sp}_{\mathrm{ess}}\mathcal{H} = [-\frac{e^4}{4},+\infty).$$
This result coincides with the result which follows from the HWZ Theorem.

4. Wiener algebra of operators on $\mathbb{R}^N$ and its application to Schrödinger operators

4.1. Fredholmness of operators in the Wiener algebra. In this subsection, we reformulate a result from [3] in a form convenient for us.

Denote by $C_0(\mathbb{R}^N)$ the C^*-algebra of functions a continuous on $\mathbb{R}^N$ such that
$$\lim_{x\to\infty} a(x) = 0.$$
with the sup-norm. To each function $a \in C_0(\mathbb{R}^N)$ corresponds the convolution operator
$$(14) \quad \mathcal{C}(a)u(x) = (2\pi)^{-N}\int_{\mathbb{R}^N} a(\xi)\hat{u}(\xi)e^{i(x,\xi)}d\xi, u \in C_0^\infty(\mathbb{R}^N),$$
where $\hat{u}(\xi) = \int_{\mathbb{R}^N} u(x)e^{-i(x,\xi)}d\xi$ is the Fourier transform of u. The operator $\mathcal{C}(a)$ is extended to a bounded operator on $L^2(\mathbb{R}^N)$ with norm
$$\|\mathcal{C}(a)\|_{B(L^2(\mathbb{R}^N))} = \sup_{\xi\in\mathbb{R}^N} |a(\xi)|.$$

We denote by $\mathfrak{C}_0(\mathbb{R}^N)$ the commutative C^*-algebra of operators (14), and by $\mathcal{A}(\mathbb{R}^N)$ the closure in $\mathcal{B}(L^2(\mathbb{R}^N))$ of the set of operators of the form
$$(15) \quad A = I + \sum_{j=1}^L \prod_{k=1}^M b_{jk}\mathcal{C}(a_{jk})c_{jk}I,$$
where $b_{jk}, c_{jk} \in L^\infty(\mathbb{R}^N)$, $\mathcal{C}(a_{jk}) \in \mathfrak{C}_0(\mathbb{R}^N)$.

Let χ_0 be the characteristic function of the set $\mathcal{I}_0 = [0,1)^N$. Denote by $\mathcal{W}(\mathbb{R}^N)$ the subclass of operators in $\mathcal{A}(\mathbb{R}^N)$ which satisfy the property
$$\|A\|_{\mathcal{W}(\mathbb{R}^N)} = \sum_{\gamma\in\mathbb{Z}^N}\sup_{\alpha\in\mathbb{Z}^N}\|\chi_0 V_{-\alpha}AV_{\alpha-\gamma}\chi_0 I\|_{\mathcal{B}(L^2(\mathcal{I}_0))}.$$

Then $\mathcal{W}(\mathbb{R}^N)$ is a Banach algebra, which is called the Wiener algebra. Note that $\mathcal{W}(\mathbb{R}^N) \subset \mathcal{B}(L^2(\mathbb{R}^N))$.

We say that the operator $A \in \mathcal{B}(L^2(\mathbb{R}^N))$ is a rich operator if every sequence $h_m \to \infty$ has a subsequence g_m for which there exists an operator $A_g \in B(L^2(\mathbb{R}^N))$ such that

$$\lim_{m\to\infty} \|(V_{-g_m} A V_{g_m} - A_g)\psi_R I\| = 0,$$
$$\lim_{m\to\infty} \|\psi_R (V_{-g_m} A V_{g_m} - A_g)\| = 0$$

for every $R > 0$. The operator A_g is called the limit operator of A defined by the sequence g_m. The set of all limit operators of a rich operator A is denoted by $\sigma(A)$.

The following theorem was proved in [**3**, Theorem 2.1].

THEOREM 12. *Let A be a rich operator in $\mathcal{W}(\mathbb{R}^N)$. Then $A: L^2(\mathbb{R}^N) \mapsto L^2(\mathbb{R}^N)$ is a Fredholm operator if and only if all limit operators $A_g \in \sigma(A)$ are invertible.*

4.2. Proof of Theorem 2.1. Since $\Lambda = (1-\Delta)^{-1}$ is a unitary operator from $L^2(\mathbb{R}^N)$ onto $H^2(\mathbb{R}^N)$, the operator $\mathcal{H} - \lambda I : H^2(\mathbb{R}^N) \to L^2(\mathbb{R}^N), \lambda \in \mathbb{C}$, is Fredholm if and only if

$$\mathcal{L}(\lambda) = (\mathcal{H} - \lambda)\Lambda = I + 2i(\mathbf{a}\cdot\nabla)\Lambda + (i\nabla\cdot\mathbf{a} + \mathbf{a}^2 + W - 1 - \lambda)\Lambda$$

is Fredholm on $L^2(\mathbb{R}^N)$.

The operators $\Lambda_j = i\nabla_j \Lambda, \Lambda$ are convolution operators, that is,

$$\Lambda_j u = k_j * u, \quad \Lambda u = k * u,$$

where $\hat{k}_j(\xi) = \xi_j(1-|\xi|^2)^{-1}, \hat{k}(\xi) = (1-|\xi|^2)^{-1} \in C_0(\mathbb{R}^N)$. Thus $\mathcal{L}(\lambda) \in \mathcal{A}(\mathbb{R}^N)$.

Moreover,

$$(16) \qquad \left|\partial_\xi^\alpha \hat{k}_j(\xi)\right| \le C_\alpha(1+|\xi|)^{-|\alpha|}, \left|\partial_\xi^\alpha \hat{k}(\xi)\right| \le C_\alpha(1+|\xi|)^{-|\alpha|}$$

with constants $C_\alpha > 0$. Integrating by parts in the Fourier integral and applying (16), we obtain

$$(17) \qquad |k_j(x)| \le C_{jm}|x|^{-m}, \quad |k(x)| \le C_m|x|^{-m}, \quad |x| > 0$$

for each nonnegative integer m.

The estimates (17) imply that $\mathcal{L}(\lambda) \in \mathcal{W}(\mathbb{R}^N)$, and by Conditions **A** and **B** $\mathcal{L}(\lambda)$ is a rich operator. Then by Theorem 12, $\mathcal{L}(\lambda)$ is Fredholm on $L^2(\mathbb{R}^N)$ if and only if all the limit operators

$$\mathcal{L}_h(\lambda) = I + (2i(\mathbf{a}^h\cdot\nabla) + (\mathbf{a}^h)^2 + W^h - 1 - \lambda)\Lambda$$

are invertible on $L^2(\mathbb{R}^N)$. Hence $\mathcal{H} - \lambda I : H^2(\mathbb{R}^N) \to L^2(\mathbb{R}^N)$ is a Fredholm operator if and only if all the limit operators

$$\mathcal{H}_h - \lambda I = (i\nabla - \mathbf{a}^h I)^2 + (W^h - \lambda) I : H^2(\mathbb{R}^N) \to L^2(\mathbb{R}^N)$$

are invertible. This implies that

$$\mathrm{sp}_{\mathrm{ess}}\mathcal{H} = \bigcup_{\mathcal{H}_h \in \sigma(\mathcal{H})} sp\mathcal{H}_h.$$

$\square$

References

[1] H.L. Cycon, R.G. Froese, W. Kirsch and B. Simon, *Schrödinger Operators with Applications to Quantum Mechanics and Global Geometry*, Springer-Verlag, 1987.
[2] E. B. Davies, *Spectral Theory and Differential Operators*, Cambridge University Press, Cambridge, 1995.
[3] V. S. Rabinovich, *Discrete operator convolutions and some of their applications*, Mat. Zametki **51** (1992), no. 5, 90-101; English translation in Math. Notes **51** (1992), 484-492.
[4] V. S. Rabinovich, S. Roch and B. Silbermann, *Fredholm theory and finite section method for band-dominated operators*, Integral Equations Operator Theory **30** (1998), 452-495.
[5] V. S. Rabinovich, S. Roch and B. Silbermann, *Band-dominated operators with operator-valued coefficients, their Fredholm properties and finite sections*, Integral Equations Operator Theory **40** (2001), 342-381.
[6] M. Reed and B. Simon, *Methods of Modern Mathematical Physics, Vol. IV. Analysis of Operators*, Academic Press, New York, 1978.
[7] M. E. Taylor, *Pseudodifferential Operators*, Princeton University Press, Princeton, New Jersey, 1981.

INSTITUTO POLITÉCNICO NACIONAL, ESIME-ZACATENCO, AV. IPN, EDIF.1, MÉXICO D.F., 07738, MÉXICO

E-mail address: vladimir_rabinovich@hotmail.com

A Critical Example for the Necessary and Sufficient Condition for Unique Quasiconformal Extremality

Edgar Reich

To my friend, Larry Zalcman

ABSTRACT. The necessary and sufficient condition for unique extremality of quasiconformal mappings with given boundary values is illustrated explicitly for the first time in a borderline case between uniqueness and non-uniqueness.

1. Introduction

Let Ω be a region of the complex plane. We denote by $L_a^1(\Omega)$ the class of functions $\varphi(z)$ analytic in Ω and belonging to $L^1(\Omega)$ and set

(1.1) $$\delta[\varphi] = \iint_\Omega [|\varphi(z)| - \operatorname{Re}\varphi(z)]\,dx\,dy,\quad \varphi \in L_a^1(\Omega).$$

We say that a sequence $\{\varphi_n(z)\}$, $\varphi_n \in L_a^1(\Omega)$, constitutes a *delta sequence* for Ω if

(1.2) $$\lim_{n\to\infty} \varphi_n(z) = 1,\ \textit{uniformly on every compact subset of } \Omega,$$

and

(1.3) $$\lim_{n\to\infty} \delta[\varphi_n] = 0.$$

When Ω has finite area, there is the delta sequence $\varphi_n(z) \equiv 1$; thus, if we want to know whether a delta sequence exists, only the case when Ω has infinite area is nontrivial. The existence of a delta sequence for a region Ω is related to quasiconformal mappings of Ω as follows. Let F denote the horizontal stretch of Ω onto $\Omega' = F(\Omega)$ by the factor $K > 1$; i.e.,

$$F(x+iy) = Kx + iy,\quad z = x+iy \in \Omega.$$

In line with standard terminology, F is called *extremal* if the maximal dilatation M of any arbitrary quasiconformal mapping ζ of Ω onto Ω' which agrees with F on $\partial\Omega$ satisfies $M \geq K$. The mapping $\zeta : \Omega \to \Omega'$ is called *uniquely extremal* if $M = K$ implies $\zeta(z) \equiv F(z)$. The following result ([1],[2]) holds.

2000 *Mathematics Subject Classification.* Primary 30A10.

THEOREM 1.1. *The horizontal stretch of Ω is uniquely extremal if and only if a delta sequence for Ω exists.*

Our purpose is to test Theorem 1.1 by attempting to verify the existence of a delta sequence explicitly, providing one exists, in the case of the family of "parabolic"-shaped regions, Ω_α, $1 < \alpha \leq \infty$, defined as

(1.4)
$$\Omega_\alpha = \{z = x + iy : x > |y|^\alpha\}, 1 < \alpha < \infty, \Omega_\infty = \{z = x + iy : x > 0, |y| < 1\}.$$

Ever since [5], the family (1.4) has been known to possess a precise transition point from non-unique extremality of the affine stretch to unique extremality, namely $\alpha = 3$. It is known ([5],[3]) that the horizontal stretch of Ω_α is uniquely extremal if and only if $3 \leq \alpha \leq \infty$. Therefore, the regions $\{\Omega_\alpha\}$, $3 \leq \alpha \leq \infty$, are particularly suitable for attempting to illustrate Theorem 1.1.

It is easy to check [4] that

(1.5)
$$\varphi_n(z) = e^{-z/n}$$

provides a delta sequence when $3 < \alpha \leq \infty$, but it fails to do so for the critical case, $\alpha = 3$. It is of course not surprising that a delicate situation arises in the critical case. This was already evidenced by the fact that for the proof ([5],[4]) of sufficiency for unique extremality some quite special techniques had to be used when $\alpha = 3$. The problem of existence of a delta sequence for Ω_3 was open for a long time. In *principle*, it was solved implicitly in [1]. Here we solve the problem for this critical case for the first time *explicitly*, namely by means of a sequence of completely elementary functions. The existence proof in [1] is very indirect and not sufficiently constructive to lead to an explicit sequence, and the present author gave up on that approach. Instead, a somewhat laborious "Edisonian" approach of trial and error turned out to work after the heuristic calculation of Section 2 yielded the key. What gives the problem additional special interest is the fact that not only is the existence of a delta sequence ruled out by Theorem 1.1 on the basis of non-unique extremality of the affine stretch when $\alpha < 3$, but that when $\alpha < 3$, for *any* sequence $\{\varphi_n(z)\}$, $\varphi_n \in L^1_a(\Omega)$, for which (1.2) holds, one actually obtains [4]

$$\lim_{n \to \infty} \delta[\varphi_n] = +\infty$$

instead of (1.3); that is, an ultra-abrupt change from (1.3) must occur. We claim

THEOREM 1.2. *The functions*

(1.6)
$$\varphi_n(z) = \operatorname{Exp}[-(2^{-n} z^{1/n})], \quad z \in \Omega_\alpha, \quad n = 1, 2, \ldots,$$

where $z^{1/n}$ denotes the branch in the right half-plane that is real on the positive x-axis, provide a delta sequence for Ω_α whenever $3 \leq \alpha \leq \infty$.

2. A heuristic computation

The material in this section serves as motivation for the construction of the delta sequence (1.6), which will be carried out in Section 3. The current section may be skipped by the reader who is solely interested in the actual proof of Theorem 1.2.

We note that the elements of the sequence (1.5) are of the form $G(tz)$, where $G(z)$ is analytic in Ω_α, $0 < t \leq 1$, $t = 1/n$, $n = 1, 2, \ldots$. Also, $G(z)$ is defined and continuous in $\Omega_\alpha \cup \partial\Omega_\alpha$, $G(0) = 1$, and $G(z) > 0$ on the positive real axis. Furthermore,

$$\iint_{\Omega_3} |G(z)|\, dx\, dy < \infty,$$

which, because Ω_3 is starlike with repect to $z = 0$, is enough to insure that $G(tz)$ belongs to $L_a^1(\Omega_3)$ as a function of z for any t, $0 < t \leq 1$.

In (1.5),

$$(2.1) \qquad G(z) = e^{-z}.$$

This is not good enough to give a delta sequence for Ω_3. In fact, setting

$$(2.2) \qquad \mathcal{I}_G(t) = \iint_{\Omega_3} [|G(tz)| - \operatorname{Re} G(tz)]\, dx\, dy,$$

the choice (2.1) gives [1]

$$(2.3) \qquad \lim_{t \searrow 0} \mathcal{I}_G(t) = \frac{1}{3}.$$

We now try to generalize the preceding to see if there might be better choices of G than (2.1). Since

$$(2.4) \qquad |w| - \operatorname{Re} w = \frac{(\operatorname{Im} w)^2}{|w| + \operatorname{Re} w}$$

we have

$$(2.5) \qquad \mathcal{I}_G(t) = \iint_{\Omega_3} \frac{[\operatorname{Im} G(tz)]^2}{|G(tz)| + \operatorname{Re} G(tz)}\, dx\, dy.$$

We use the sign $\approx$ to indicate some likely approximations when $t \searrow 0$, but we do not actually attempt to establish their validity rigorously.

Since $G(0)=1$, the value of the integrand in (2.2) can be made arbitrarily small when t is sufficiently small, given an upper bound on $x = \operatorname{Re} z$. So only "large" values of x matter. For large values of x, since $y < x^{1/3}$, the magnitude of $y = \operatorname{Im} z$ is small compared to the value of x. Thus,

$$|G(tz)| \approx \operatorname{Re} G(tz) \approx G(tx)$$

This will be used to approximate the denominator of the integrand of (2.5). In order to approximate the numerator we also include a term of order t in the approximation of $G(tz)$, that is,

$$G(tz) = G(tx + ity) \approx G(tx) + ityG'(tx).$$

Therefore,

$$\operatorname{Im} G(tz) \approx tyG'(tx).$$

---

[1] In [3], p. 296, the value of this limit is erroneously given as 1/6.

Hence,

$$\mathcal{I}_G(t) \approx \iint_{\Omega_3} \frac{t^2 y^2 [G'(tx)]^2}{2G(tx)} \, dx \, dy = t^2 \int_0^\infty \frac{[G'(tx)]^2}{G(tx)} dx \int_0^{x^{1/3}} y^2 \, dy$$

$$= \frac{t^2}{3} \int_0^\infty x \frac{[G'(tx)]^2}{G(tx)} dx = \frac{1}{3} \int_0^\infty \frac{u [G'(u)]^2}{G(u)} du$$

Note that the last expression no longer depends on t.

ROUGH CONCLUSION. *Suppose $G \in L_a^1(\Omega_3)$, G is continuous in $\Omega_3 \cup \partial\Omega_3$, $G(0) = 1$, $G(z) > 0$ on the positive real axis, and possibly further unspecified conditions are met. Then*

$$(2.6) \quad \lim_{t \searrow 0} \iint_{\Omega_3} [|G(tz)| - \operatorname{Re} G(tz)] \, dx \, dy = \frac{1}{3} \int_0^\infty \frac{u[G'(u)]^2}{G(u)} du \, .$$

EXAMPLE.

$$(2.7) \quad G_p(z) = \operatorname{Exp}[-(z^p)], \quad p > 0, \; z \in \Omega_3$$

Assuming that (2.6) actually holds, we may conjecture [2] that

$$(2.8) \quad \lim_{t \searrow 0} \iint_{\Omega_3} [|G_p(tz)| - \operatorname{Re} G_p(tz)] \, dx \, dy = \frac{p^2}{3} \int_0^\infty u^{2p-1} \operatorname{Exp}[-(u^p)] du = \frac{p}{3} \, .$$

Since the right side of (2.6) is non-zero, this is evidence indicating that there is *no* choice of G that can result in $\{G(z/n)\}$ being a delta sequence for Ω_3. On the other hand, in view of (2.8), a delta sequence might be produced by choosing a sequence of values of p that approach 0, together with a corresponding sequence of values of t that also go to 0. This is precisely what we do to prove Theorem 1.2 in Section 3.

3. Proof of Theorem 1.2

Let $G_p(z) = \operatorname{Exp}[-(z^p)]$, $p > 0$, $z \in \Omega_3$, as in (2.7), z^p being the branch analytic for $\operatorname{Re} z > 0$ and real on the positive real axis. We have $G_p(\bar{z}) = \overline{G_p(z)}$. It therefore suffices to consider the upper part of Ω_3. We focus mainly on the upper part of Ω_3 to the right of the line $\{\operatorname{Re} z = 1\}$,

$$\mathcal{R} = \{(x, y) : x > y^3, x > 1, y > 0\} \, .$$

Let

$$(3.1) \quad J_p(t) = \iint_{\mathcal{R}} [|G_p(tz)| - \operatorname{Re} G_p(tz)] \, dx \, dy \, .$$

We assume that $0 < t \leq 1$ throughout this section.

We start off by noting that

$$(3.2) \quad \operatorname{Re} G_p(tz) \geq 0 \; \text{for } z \in \mathcal{R}, \; \text{if } 0 < p \leq \frac{2}{3} \, .$$

Namely, with $\theta = \arg z$,

$$(3.3) \quad \operatorname{Re} G_p(tz) = \exp[-t^p |z|^p \cos(p\theta)] \cos[t^p |z|^p \sin(p\theta)] \, .$$

---

[2]Numerical experiments with Mathematica seem to indicate that the conjecture is correct.

In $\mathcal{R}$ we have $\theta < \pi/4$. Since
$$\frac{\sin\theta}{\theta} > \frac{2\sqrt{2}}{\pi}, \quad 0 < \theta < \frac{\pi}{4},$$
we have
$$\frac{\sin(p\theta)}{p\theta} < \frac{\pi}{2\sqrt{2}}\frac{\sin\theta}{\theta},$$
and, therefore,

(3.4) $$\sin(p\theta) \leq \frac{\pi}{2\sqrt{2}} p \sin\theta \leq \frac{10}{9} p \sin\theta, \quad z \in \mathcal{R}, \ p \geq 0.$$

Therefore, if $0 < p \leq 2/3$ and $z \in \mathcal{R}$,

$$\cos[t^p|z|^p \sin(p\theta)] \geq \cos\left(\frac{10}{9}pt^p|z|^{p-1}py\right) \geq \cos\left(\frac{10}{9}pt^p\frac{x^{1/3}}{x^{1-p}}\right)$$
$$\geq \cos\left(\frac{10}{9}\frac{p}{x^{\frac{2}{3}-p}}\right) > \cos(20/27) > 0.$$

In view of (3.3), (3.2) follows.

From now on, assume that $0 < p \leq 2/3$. Using (2.4), we have, in view of (3.2),

(3.5) $$J_p(t) \leq \iint_{\mathcal{R}} H_{pt}(z) \, dx \, dy,$$

where

(3.6) $$H_{pt}(z) = \frac{[\operatorname{Im} G_p(tz)]^2}{|G_p(tz)|} = \operatorname{Exp}[-t^p|z|^p \cos(p\theta)] \sin^2[t^p|z|^p \sin(p\theta)].$$

We now estimate the terms on the right side of (3.6) in $\mathcal{R}$. We have
$$\cos(p\theta) \geq \cos\left(\frac{2}{3}\cdot\frac{\pi}{4}\right) = \frac{\sqrt{3}}{2} = c.$$

Hence,
$$\operatorname{Exp}[-t^p|z|^p \cos(p\theta)] \leq \operatorname{Exp}[-ct^px^p].$$

Furthermore,
$$|z|^p \sin(p\theta) = |z|^{p-1}|z|\sin(p\theta) \leq \frac{\pi}{2\sqrt{2}} p|z|^{p-1}|z|\sin\theta$$
$$= \frac{\pi}{2\sqrt{2}}\frac{py}{|z|^{1-p}} \leq \frac{\pi}{2\sqrt{2}}\frac{py}{x^{1-p}}.$$

Thus,
$$\sin^2[t^p|z|^p \sin(p\theta)] \leq [t^p|z|^p \sin(p\theta)]^2 \leq \frac{\pi^2}{8}p^2t^{2p}x^{2p-2}y^2.$$

So, by (3.6),

(3.7) $$H_{pt}(z) \leq \frac{\pi^2}{8}p^2t^{2p}x^{2p-2}y^2 \operatorname{Exp}[-ct^px^p],$$

and therefore

(3.8) $$J_p(t) \leq \frac{\pi^2}{8}p^2t^{2p}\int_1^\infty x^{2p-2}\operatorname{Exp}[-ct^px^p]dx \int_0^{x^{1/3}} y^2 dy$$
$$< \frac{\pi^2}{24}p^2t^{2p}\int_0^\infty x^{2p-1}\operatorname{Exp}[-ct^px^p]dx = \frac{\pi^2 p}{24}\int_0^\infty u\,e^{-cu}du = \frac{\pi^2 p}{18}.$$

For convenience, let $S = \mathcal{A} \cup \Omega_3$, where $\mathcal{A} = \{(x,y) : 0 < x < 1, -1 < y < 1\}$. In $\mathcal{A}$, we make use of the fact that for any complex number w,

(3.9) $$|w| \leq \rho \Rightarrow |e^w| - \operatorname{Re}(e^w) \leq \frac{1}{2}\rho^2 e^\rho.$$

Evidently,
$$|t^p z^p| \leq (2t)^p \leq 1, \; if \; z \in \mathcal{A}, \; 0 < t \leq \frac{1}{2}.$$

Therefore, by (3.9),
$$|G_p(tz)| - \operatorname{Re} G_p(tz) < \frac{3}{2}(2t)^{2p}, \; z \in \mathcal{A}, \; 0 < t \leq \frac{1}{2},$$

and, hence,

(3.10) $$\iint_{\mathcal{A}} [|G_p(tz)| - \operatorname{Re} G_p(tz)] \leq 3(2t)^{2p}, \; 0 < t \leq \frac{1}{2}.$$

Combining (3.8) and (3.10), we have

(3.11) $$\iint_S [|G_p(tz)| - \operatorname{Re} G_p(tz)] \leq 3(2t)^{2p} + \frac{\pi^2 p}{9}, \; 0 < p \leq \frac{2}{3}, \; 0 < t \leq \frac{1}{2}.$$

Set $\varphi_n(z) = G_p(tz)$, where

(3.12) $$p = (1/n), \; t = 2^{-(n^2)}, \; (n = 1, 2, \ldots),$$

Since $\Omega_\alpha \subset S$ when $3 \leq \alpha \leq \infty$, inequality (3.11) now implies that

(3.13) $$\delta[\varphi_n] \leq 3 \cdot 2^{2/n} 2^{-2n} + \frac{\pi^2}{9n}, \; n = 2, 3, \ldots \; (3 \leq \alpha \leq \infty).$$

Since φ_n belongs to $L_a^1(\Omega_\alpha)$ and (1.2) is satisfied, the proof of Theorem 1.2 is complete.

References

[1] V. Božin, N. Lakic, V. Marković and M. Mateljević, *Unique extremality*, J. Anal. Math. **75** (1998), 299–338.

[2] Edgar Reich, *A criterion for unique extremality of Teichmüller mappings*, Indiana Univ. Math. J. **30** (1981), 441–447.

[3] Edgar Reich, *On criteria for unique extremality of Teichmüller mappings*, Ann. Acad. Sci. Fenn. Ser. A I Math. **6** (1981), 289–301.

[4] Edgar Reich, *On the L^1 approximation of $|f(z)|$ by $\operatorname{Re} f(z)$ for analytic functions*, Ann. Acad. Sci. Fenn. Math. **27** (2002), 373–380.

[5] Edgar Reich, Kurt Strebel, *On the extremality of certain Teichmüller mappings*, Comment. Math. Helv. **45** (1970), 353–362.

SCHOOL OF MATHEMATICS, UNIVERSITY OF MINNESOTA, MINNEAPOLIS, MN 55455, U.S.A.
E-mail address: reich@math.umn.edu

Generic Convergence of Iterates for a Class of Nonlinear Mappings in Hyperbolic Spaces

Simeon Reich and Alexander J. Zaslavski

Dedicated to Professor Lawrence Zalcman on the occasion of his sixtieth birthday

ABSTRACT. Let K be a bounded, closed and ρ-convex subset of a hyperbolic complete metric space (X, ρ). We show that the iterates of a typical (in the sense of Baire category) element of a class of continuous self-mappings of K converge uniformly on K to the unique fixed point of this typical element.

1. Introduction

In this paper, we are concerned with the behavior of certain discrete dynamical systems in a hyperbolic complete metric space, a concept which we recall in Section 2. In our study, we use the notion of σ-porosity, which we discuss in Section 3. It is a refinement of the notion of Baire's first category. Our main result, Theorem 4.1 below, is stated in Section 4. It extends the main result of [14] from Banach spaces to all hyperbolic complete metric spaces and is established in Section 5. Its proof is different from that in [14], because the method of proof we use there employs the vector space structure of Banach spaces and therefore is not applicable to metric spaces.

2. Hyperbolic spaces

Let (X, ρ) be a metric space and let R^1 denote the real line. We say that a mapping $c : R^1 \to X$ is a metric embedding of R^1 into X if $\rho(c(s), c(t)) = |s - t|$ for all real s and t. The image of R^1 under a metric embedding is called a metric line, and the image of a real interval $[a, b] = \{t \in R^1 : a \leq t \leq b\}$ under such a mapping is called a metric segment. Assume that (X, ρ) contains a family M of metric lines such that for each pair of distinct points x and y in X, there is a unique metric line

---

2000 *Mathematics Subject Classification.* Primary 47H09, 47H10, 54E50, 54E52.

Key words and phrases. Complete metric space, generic property, hyperbolic metric space, iteration, porous set.

The work of the first author was partially supported by the Israel Science Foundation founded by the Israel Academy of Sciences and Humanities (Grant 592/00), by the Fund for the Promotion of Research at the Technion, and by the Technion VPR Fund.

in M which passes through x and y. This metric line determines a unique metric segment joining x and y, which we denote by $[x, y]$. For each $0 \leq t \leq 1$, there is a unique point z in $[x, y]$ such that $\rho(x, z) = t\rho(x, y)$ and $\rho(z, y) = (1 - t)\rho(x, y)$. This point is denoted by $(1-t)x \oplus ty$. We say that X or, more precisely, (X, ρ, M), is a hyperbolic space if

$$\rho\left(\frac{1}{2}x \oplus \frac{1}{2}y, \frac{1}{2}x \oplus \frac{1}{2}z\right) \leq \frac{1}{2}\rho(y, z)$$

for all x, y and z in X. A set $K \subset X$ is called ρ-convex if $[x, y] \subset K$ for all x and y in K. It is clear that all normed linear spaces are hyperbolic. Discussions of more examples of hyperbolic spaces and, in particular, of the Hilbert ball (that is, the open unit ball of a complex Hilbert space, endowed with the hyperbolic metric), as well as of their connections to complex analysis and with the metric approach to the study of holomorphic maps, can be found, for example, in [**4-9**, **13**] and in references therein. We also note the useful fact (cf. [**4**], pp. 77, 104; and [**9**]) that if (X, ρ, M) is a hyperbolic space, then

$$\rho((1-t)x \oplus tz, (1-t)y \oplus tw) \leq (1-t)\rho(x, y) + t\rho(z, w)$$

for all x, y, z and w in X and $0 \leq t \leq 1$.

3. Porous sets

Let (Y, ρ) be a complete metric space. We denote by $B(y, r)$ the closed ball of center $y \in Y$ and radius $r > 0$. A subset $E \subset Y$ is called porous in (Y, ρ) if there exist $\alpha \in (0, 1)$ and $r_0 > 0$ such that for each $r \in (0, r_0]$ and each $y \in Y$, there exists $z \in Y$ for which

$$B(z, \alpha r) \subset B(y, r) \setminus E.$$

A subset of the space Y is called σ-porous in (Y, ρ) if it is a countable union of porous subsets in (Y, ρ).

REMARK. It is known that in the above definition of porosity, the point y can be assumed to belong to E.

Since porous sets are obviously nowhere dense, all σ-porous sets are of the first Baire category. If Y is a finite-dimensional Euclidean space, then σ-porous sets are also of Lebesgue measure 0.

To point out the difference between porous and nowhere dense sets, note that if $E \subset Y$ is nowhere dense, $y \in Y$ and $r > 0$, then there exist a point $z \in Y$ and a number $s > 0$ such that $B(z, s) \subset B(y, r) \setminus E$. If, however, E is also porous, then for small enough r, we can choose $s = \alpha r$, where $\alpha \in (0, 1)$ is a constant which depends only on E.

Recent applications of the concept of porosity to several diverse areas of analysis can be found, for instance, in [**1-3**], [**10-14**]. There are, in fact, quite a few types and variants of the notion of porosity. For a recent extensive survey of σ-porous sets in abstract spaces, see [**15**].

4. Main result

Let K be a nonempty, bounded, closed and ρ-convex subset of a hyperbolic complete metric space (X, ρ, M). Set

(4.1) $$\operatorname{diam}(K) = \sup\{\rho(x, y) : x, y \in K\}.$$

Denote by $\mathcal{A}$ the set of all continuous mappings $A : K \to K$ such that

(P1) for each $\epsilon > 0$, there exists $x_\epsilon \in K$ such that

(4.2) $$\rho(Ax, x_\epsilon) \leq \rho(x, x_\epsilon) + \epsilon \text{ for all } x \in K.$$

For $A, B \in \mathcal{A}$, set

(4.3) $$d(A, B) = \sup\{\rho(Ax, Bx) : x \in K\}.$$

Note that $\mathcal{A}$ certainly contains all those ρ-nonexpansive self-mappings of K which have a fixed point. Clearly, the metric space $(\mathcal{A}, d)$ is complete.

Our purpose in the present paper is to study the behavior of the iterates of the members of $\mathcal{A}$. Our main result is Theorem 4.1, which we now formulate. It shows, in particular, that the iterates of a typical (in the sense of Baire category) element in $\mathcal{A}$ converge uniformly on K to the unique fixed point of this typical element.

THEOREM 4.1. *There exists a set $\mathcal{F} \subset \mathcal{A}$ such that its complement $\mathcal{A} \setminus \mathcal{F}$ is σ-porous in $(\mathcal{A}, d)$ and each $A \in \mathcal{F}$ has the following properties:*
 (i) *the mapping A has a unique fixed point $x_A \in K$ such that*
$$A^n x \to x_A \text{ as } n \to \infty, \text{ uniformly for all } x \in K;$$
 (ii)
$$\rho(Ax, x_A) \leq \rho(x, x_A) \text{ for all } x \in K;$$
 (iii) *for each $\epsilon > 0$, there exist a natural number q and a number $\delta > 0$ such that for each integer $i \geq q$, each $x \in K$, and each $B \in \mathcal{A}$ satisfying $d(B, A) \leq \delta$,*
$$\rho(B^i x, x_A) \leq \epsilon.$$

For each $C \in \mathcal{A}$ and $x \in K$, set $C^0 x = x$. For each natural number n, denote by $\mathcal{F}_n$ the set of all $A \in \mathcal{A}$ with the property that

(P2) there exist $\bar{x} \in K$, a natural number q, and a positive number δ such that
$$\rho(Ax, \bar{x}) \leq \rho(x, \bar{x}) + 1/n \text{ for all } x \in K,$$

and such that for each $B \in \mathcal{A}$ satisfying $d(A, B) \leq \delta$ and each $x \in K$, the inequality $\rho(B^q x, \bar{x}) \leq 1/n$ is valid.

Define
$$\mathcal{F} = \bigcap_{n=1}^{\infty} \mathcal{F}_n.$$

In order to proceed, we need the following auxiliary result.

LEMMA 4.1. *Let $A \in \mathcal{F}$. Then there exists a unique fixed point $x_A \in K$ of A such that*
 (i) $A^n x \to x_A$ *as $n \to \infty$, uniformly on K;*
 (ii) $\rho(Ax, x_A) \leq \rho(x, x_A)$ *for all $x \in K$;*
 (iii) *for each $\epsilon > 0$, there exist a natural number q and a number $\delta > 0$ such that for each $B \in \mathcal{A}$ satisfying $d(B, A) \leq \delta$, each $x \in K$, and each integer $i \geq q$,*
$$\rho(B^i x, x_A) \leq \epsilon.$$

We omit the proof of Lemma 4.1, as it has already been proved in [14] in the setting of a Banach space; it is not difficult to check that it holds true for any hyperbolic complete metric space with essentially the same proof.

5. Proof of Theorem 4.1

In order to prove Theorem 4.1, it is sufficient, in view of Lemma 4.1, to show that for each natural number n, the set $\mathcal{A} \setminus \mathcal{F}_n$ is porous in $(\mathcal{A}, d)$.

To this end, let n be a natural number. Choose a positive number

(5.1) $$\alpha < (64)^{-2}(2n^2)^{-1}(\text{diam}(K)+1)^{-2}.$$

Let

(5.2) $$A \in \mathcal{A}, \ r \in (0,1].$$

Put

(5.3) $$\gamma_0 = 16^2 n^2 \alpha r (\text{diam}(K)+1).$$

Now choose a natural number q such that

(5.4) $$(4n\alpha r)^{-1} \geq q \geq (8n\alpha r)^{-1} \text{diam}(K)(\text{diam}(K)+1)^{-1}.$$

Set
$$\gamma = 2\gamma_0.$$

By (5.3) and (5.1),

(5.5) $$\gamma = 2\gamma_0 < 16^{-1} r (\text{diam}(K)+1)^{-1}.$$

Choose positive numbers δ_0 and ϵ_0 such that

(5.6) $$\epsilon_0 < \delta_0 < r/48,$$

$$\epsilon_0(1-\gamma) = \epsilon_0(1-2\gamma_0) < \gamma_0 \delta_0 = (\gamma - \gamma_0)\delta_0.$$

There exists $\bar{x} \in K$ such that

(5.7) $$\rho(Ax, \bar{x}) \leq \rho(x, \bar{x}) + \epsilon_0 \text{ for all } x \in K.$$

Define $A_\gamma : K \to K$ by

(5.8) $$A_\gamma x = \gamma \bar{x} \oplus (1-\gamma) Ax, \ x \in K.$$

Clearly, A_γ is a continuous mapping. By Urysohn's theorem, there exists a continuous function $\phi : K \to [0,1]$ such that

(5.9) $$\phi(x) = 1 \text{ for all } x \in K \text{ satisfying } \rho(x, \bar{x}) \leq \delta_0$$

and

(5.10) $$\phi(x) = 0 \text{ for all } x \in K \text{ satisfying } \rho(x, \bar{x}) \geq 2\delta_0.$$

Define

(5.11) $$Bx = \phi(x) \bar{x} \oplus (1 - \phi(x)) A_\gamma x, \ x \in K.$$

We claim that $B : K \to K$ is continuous. To see this, suppose

(5.12) $$\{x_i\}_{i=1}^\infty \subset K \text{ with } \lim_{i \to \infty} x_i = x.$$

By (5.11), for each integer $i \geq 1$,

$$\begin{aligned}
\rho(Bx, Bx_i) &= \rho(\phi(x)\bar{x} \oplus (1-\phi(x))A_\gamma x, \phi(x_i)\bar{x} \oplus (1-\phi(x_i))A_\gamma x_i) \\
&\leq \rho(\phi(x)\bar{x} \oplus (1-\phi(x))A_\gamma x, \phi(x_i)\bar{x} \oplus (1-\phi(x_i))A_\gamma x) \\
&\quad + \rho(\phi(x_i)\bar{x} \oplus (1-\phi(x_i))A_\gamma x, \phi(x_i)\bar{x} \oplus (1-\phi(x_i))A_\gamma x_i) \\
&\leq |\phi(x) - \phi(x_i)|\rho(\bar{x}, A_\gamma x) + \rho(A_\gamma x, A_\gamma x_i) \\
&\leq |\phi(x) - \phi(x_i)|\operatorname{diam}(K) + \rho(A_\gamma x, A_\gamma x_i) \to 0 \text{ as } i \to \infty.
\end{aligned}$$

Thus B is indeed continuous.

In view of (5.11) and (5.9),

(5.13) $\qquad Bx = \bar{x}$ for each $x \in K$ satisfying $\rho(x, \bar{x}) \leq \delta_0$.

Let $x \in K$ satisfy

(5.14) $\qquad\qquad\qquad \rho(x, \bar{x}) \geq \delta_0.$

It follows from (5.7), (5.8), (5.11) and (5.14) that

$$\begin{aligned}
\rho(\bar{x}, Bx) &= \rho(\bar{x}, \phi(x)\bar{x} \oplus (1-\phi(x))A_\gamma x) = (1-\phi(x))\rho(\bar{x}, A_\gamma x) \\
&= (1-\phi(x))\rho(\bar{x}, \gamma\bar{x} \oplus (1-\gamma)Ax) = (1-\phi(x))(1-\gamma)\rho(\bar{x}, Ax) \\
&\leq (1-\phi(x))(1-\gamma)[\rho(\bar{x}, x) + \epsilon_0] \leq (1-\gamma)\rho(\bar{x}, x) + (1-\gamma)\epsilon_0 \\
&= (1-\gamma)\rho(x, \bar{x}) + (\gamma - \gamma_0)\delta_0 \leq (1-\gamma)\rho(x, \bar{x}) + (\gamma - \gamma_0)\rho(x, \bar{x}) \\
&= (1-\gamma_0)\rho(x, \bar{x}).
\end{aligned}$$

Thus

(5.15) $\qquad\qquad\qquad \rho(\bar{x}, Bx) \leq (1-\gamma_0)\rho(x, \bar{x}).$

By (5.15) and (5.13),

(5.16) $\qquad\qquad \rho(\bar{x}, Bx) \leq (1-\gamma_0)\rho(x, \bar{x})$ for all $x \in K$,

$\qquad\qquad Bx = \bar{x}$ for all $x \in K$ satisfying $\rho(x, \bar{x}) \leq \delta_0$.

Clearly,

(5.17) $\qquad\qquad\qquad d(A, B) \leq d(A, A_\gamma) + d(A_\gamma, B).$

In view of (5.8), (4.1) and (5.5), we have

(5.18) $\quad d(A, A_\gamma) = \sup\{\rho(Ax, \gamma\bar{x} \oplus (1-\gamma)Ax) : x \in K\}$
$\qquad\qquad\qquad = \sup\{\gamma\rho(Ax, \bar{x}) : x \in K\} \leq \gamma \operatorname{diam}(K) \leq r/16.$

It now follows from (4.3), (5.11), (5.10), (5.8), (5.7) and (5.6) that

$$\begin{aligned}
d(A_\gamma, B) &= \sup\{\rho(A_\gamma x, Bx) : x \in K\} \\
&= \sup\{\rho(A_\gamma x, \phi(x)\bar{x} \oplus (1-\phi(x))A_\gamma x : x \in K\} \\
&\leq \sup\{\phi(x)\rho(A_\gamma x, \bar{x}) : x \in K\} \\
&= \sup\{\phi(x)\rho(A_\gamma x, \bar{x}) : x \in K, \rho(x, \bar{x}) \leq 2\delta_0\} \\
&= \sup\{\phi(x)\rho(\bar{x}, \gamma\bar{x} \oplus (1-\gamma)Ax) : x \in K, \rho(x, \bar{x}) \leq 2\delta_0\} \\
&\leq \sup\{\phi(x)(1-\gamma)\rho(\bar{x}, Ax) : x \in K, \rho(x, \bar{x}) \leq 2\delta_0\} \\
&\leq \sup\{\rho(\bar{x}, Ax) : x \in K, \rho(x, \bar{x}) \leq 2\delta_0\} \\
&\leq \sup\{\rho(\bar{x}, x) + \epsilon_0 : x \in K, \rho(x, \bar{x}) \leq 2\delta_0\} < 3\delta_0 < r/16.
\end{aligned}$$

When combined with (5.17) and (5.18), this implies that

(5.19) $$d(A, B) \leq r/8.$$

Assume now that

(5.20) $$C \in \mathcal{A}, \ d(C, B) \leq \alpha r.$$

By (5.20), (5.16) and (5.1), for each $x \in K$,

(5.21) $$\rho(Cx, \bar{x}) \leq \rho(Cx, Bx) + \rho(Bx, \bar{x}) \leq \alpha r + \rho(x, \bar{x}) \leq \rho(x, \bar{x}) + 1/n.$$

Relations (5.19), (5.20) and (5.1) imply that

(5.22) $$d(A, C) \leq d(A, B) + d(B, C) \leq r/8 + \alpha r \leq r/2.$$

Assume that $x \in K$. We show that there exists an integer $j \in [0, q-1]$ such that $\rho(C^j x, \bar{x}) \leq (8n)^{-1}$. Assume the contrary. Then

(5.23) $$\rho(C^j x, \bar{x}) > (8n)^{-1}, \ i = 0, \ldots, q-1.$$

Let an integer $i \in \{0, \ldots, q-1\}$. In view of (4.3), (5.20) and (5.16), we have

$$\rho(C^{i+1} x, \bar{x}) = \rho(C(C^i x), \bar{x}) \leq \rho(C(C^i x), B(C^i x))$$
$$+ \rho(B(C^i x), \bar{x}) \leq d(B, C) + \rho(B(C^i x), \bar{x})$$
$$\leq \alpha r + (1 - \gamma_0) \rho(C^i x, \bar{x}).$$

When combined with (5.23) and (5.3), this inequality implies that

$$\rho(C^i x, \bar{x}) - \rho(C^{i+1} x, \bar{x}) \geq \rho(C^i x, \bar{x}) - \alpha r - (1 - \gamma_0) \rho(C^i x, \bar{x})$$
$$= \gamma_0 \rho(C^i x, \bar{x}) - \alpha r \geq (8n)^{-1} \gamma_0 - \alpha r \geq (16n)^{-1} \gamma_0,$$

so that

$$\rho(C^i x, \bar{x}) - \rho(C^{i+1} x, \bar{x}) \geq (16n)^{-1} \gamma_0.$$

Together with (4.1), this last inequality implies that

$$\text{diam}(K) \geq \rho(x, \bar{x}) - \rho(C^q x, \bar{x}) \geq \sum_{i=0}^{q-1} (\rho(C^i x, \bar{x}) - \rho(C^{i+1} x, \bar{x})) \geq q(16n)^{-1} \gamma_0.$$

Thus, by (5.3),

$$q \leq \text{diam}(K) 16n/\gamma_0 = \text{diam}(K)(\text{diam}(K) + 1)^{-1}(16\alpha r n)^{-1},$$

a contradiction (see (5.4)).

The contradiction we have reached shows that indeed there exists an integer $j \in \{0, \ldots, q-1\}$ such that

(5.24) $$\rho(C^j x, \bar{x}) \leq (8n)^{-1}.$$

It follows from (5.20) and (5.16) that for each integer $i \in \{0, \ldots, q-1\}$,

$$\rho(C^{i+1} x, \bar{x}) \leq \rho(C(C^i x), \bar{x}) \leq \rho(C(C^i x), B(C^i x)) + \rho(B(C^i x), \bar{x}) \leq \alpha r + \rho(C^i x, \bar{x}).$$

This implies that for each integer s satisfying $j < s \leq q$,

$$\rho(C^s x, \bar{x}) \leq \rho(C^j x, \bar{x}) + \alpha r(s - j) \leq \rho(C^j x, \bar{x}) + \alpha r q.$$

Finally, it follows from this inequality, (5.24) and (5.4) that

$$\rho(C^q x, \bar{x}) \le (8n)^{-1} + \alpha r q \le (2n)^{-1}.$$

Thus we have shown that

$$\{C \in \mathcal{A}: \ d(C,B) \le \alpha r/2\} \subset \{C \in \mathcal{A}: \ d(A,C) \le r/2\} \cap \mathcal{F}_n.$$

This completes the proof of Theorem 4.1.

References

1. F. S. De Blasi and J. Myjak, *Sur la porosité de l'ensemble des contractions sans point fixe*, C. R. Acad. Sci. Paris Sér. I Math. **308** (1989), 51-54.
2. F. S. De Blasi and J. Myjak, *On a generalized best approximation problem*, J. Approx. Theory **94** (1998), 54-72.
3. F. S. De Blasi, J. Myjak and P. L. Papini, *Porous sets in best approximation theory*, J. London Math. Soc. (2) **44** (1991), 135-142.
4. K. Goebel and S. Reich, *Uniform Convexity, Hyperbolic Geometry, and Nonexpansive Mappings*, Marcel Dekker, New York and Basel, 1984.
5. W. A. Kirk, *Krasnosel'skii's iteration process in hyperbolic space*, Numer. Funct. Anal. Optim. **4** (1982), 371-381.
6. W. A. Kirk, *Geodesic geometry and fixed point theory*, Seminar of Mathematical Analysis, Univ. Sevilla, Seville, 2003, 195-225.
7. T. Kuczumow, S. Reich and D. Shoikhet, *Fixed points of holomorphic mappings: a metric approach*, Handbook of Metric Fixed Point Theory, Kluwer, Dordrecht, 2001, 437-515.
8. S. Reich, *The alternating algorithm of von Neumann in the Hilbert ball*, Dynam. Systems Appl. **2** (1993), 21-25.
9. S. Reich and I. Shafrir, *Nonexpansive iterations in hyperbolic spaces*, Nonlinear Anal. **15** (1990), 537-558.
10. S. Reich and A. J. Zaslavski, *The set of divergent descent methods in a Banach space is σ-porous*, SIAM J. Optim. **11** (2001), 1003-1018.
11. S. Reich and A. J. Zaslavski, *The set of noncontractive mappings is σ-porous in the space of all nonexpansive mappings*, C. R. Acad. Sci. Paris Sér. I Math. **333** (2001), 539-544.
12. S. Reich and A. J. Zaslavski, *Well-posedness and porosity in best approximation problems*, Topol. Methods Nonlinear Anal. **18** (2001), 395-408.
13. S. Reich and A. J. Zaslavski, *A porosity result for attracting mappings in hyperbolic spaces*, Complex Analysis and Dynamical Systems, Contemp. Math. **364** (2004), 237-242.
14. S. Reich and A. J. Zaslavski, *Generic convergence of iterates for a class of nonlinear mappings*, Fixed Point Theory Appl., in press.
15. L. Zajíček, *On σ-porous sets in abstract spaces*, Abstract Appl. Anal., to appear.

DEPARTMENT OF MATHEMATICAL AND COMPUTING SCIENCES, TOKYO INSTITUTE OF TECHNOLOGY, W8-49, 2-12-1 O-OKAYAMA, MEGURO-KU, TOKYO 152-8552, JAPAN

Current address: Department of Mathematics, The Technion - Israel Institute of Technology, 32000 Haifa, Israel

E-mail address: `sreich@tx.technion.ac.il`

DEPARTMENT OF MATHEMATICS, THE TECHNION - ISRAEL INSTITUTE OF TECHNOLOGY, 32000 HAIFA, ISRAEL

E-mail address: `ajzasl@tx.technion.ac.il`

The Beltrami Equation and FMO Functions

V. Ryazanov, U. Srebro, and E. Yakubov

To Larry Zalcman on his 60th birthday

ABSTRACT. We establish the existence of a homeomorphic ACL solution for the Beltrami equation when its dilatation has a majorant of finite mean oscillation.

1. Introduction

Let D be a domain in the complex plane $\mathbb{C}$, i.e., an open and connected subset of $\mathbb{C}$, and let $\mu : D \to \mathbb{C}$ be a measurable function with $|\mu(z)| < 1$ a.e. . The *Beltrami equation* can be written as

$$f_{\bar{z}} = \mu(z) \cdot f_z, \tag{1.1}$$

where $f_{\bar{z}} = \bar{\partial} f = (f_x + if_y)/2$, $f_z = \partial f = (f_x - if_y)/2$, $z = x + iy$, and f_x and f_y are the partial derivatives of f with respect to x and y. The function μ is the *complex coefficient* of the equation and

$$K_\mu(z) = \frac{1 + |\mu(z)|}{1 - |\mu(z)|} \tag{1.2}$$

is its *dilatation*. The Beltrami equation (1.1) is said to be *degenerate* if $||\mu||_\infty = 1$, or equivalently, $\operatorname{ess\,sup} K_\mu(z) = \infty$. In this paper, we prove the existence of homeomorphic ACL solutions in the degenerate case when a certain condition is imposed on K_μ.

Recall that a mapping $f : D \to \mathbb{C}$ is *absolutely continuous on lines*, briefly $f \in \operatorname{ACL}$, if for every closed rectangle R in D whose sides are parallel to the coordinate axes, $f|R$ is absolutely continuous on almost all line segments in R which are parallel to the sides of R. In particular, f is ACL if it belongs to the Sobolev class $W^{1,1}_{loc}$. Note that if $f \in \operatorname{ACL}$, then f has partial derivatives f_x and f_y a.e.

2000 *Mathematics Subject Classification.* Primary 30C65; Secondary 30C75.

The research of the first author was partially supported by grants from the University of Helsinki, from the Technion – Israel Institute of Technology and Holon Academic Institute of Technology and by Grant 01.07/00241 of Scientific Fund of Fundamental Investigations of Ukraine; the research of the second author was partially supported by the Israel Science Foundation (Grant 198/00-3) and by the Technion Fund for the Promotion of Research, and the third author was partially supported by the Israel Science Foundation (Grant 198/00-3).

and thus, by the well-known Gehring-Lehto theorem, every ACL homeomorphism $f : D \to \mathbb{C}$ is totally differentiable a.e.; see [**GL**] or [**LV**, p. 128]. For a sense-preserving ACL homeomorphism $f : D \to \mathbb{C}$, the Jacobian $J_f(z) = |f_z|^2 - |f_{\bar{z}}|^2$ is nonnegative a.e.; see [**LV**], p. 10. In this case, the *complex dilatation* of f is the ratio $\mu(z) = f_{\bar{z}}/f_z$ and $|\mu(z)| \leq 1$ a.e., and the *dilatation* of f is $K_\mu(z)$ and $K_\mu(z) \geq 1$ a.e. We set $\mu(z) = 0$ and $K_\mu(z) = 1$ if $f_z = 0$.

In [**RSY**$_2$], we showed that if μ is measurable in D and $K_\mu(z) \leq Q(z)$ a.e. for certain functions $Q(z)$ of bounded mean oscillation in D then (1.1) has a homeomorphic ACL solution. We now show that the same is true if $Q(z)$ is of finite mean oscillation in D. Recall that a real valued function $\varphi \in L^1_{loc}(D)$ is said to be of *bounded mean oscillation* in D, $\varphi \in \text{BMO}(D)$ or simply $\varphi \in \text{BMO}$, if

$$(1.3) \qquad \|\varphi\|_* = \sup_{B \subset D} \fint_B |\varphi(z) - \varphi_B|\, dxdy \; < \; \infty,$$

where the supremum is taken over all disks B in D and

$$(1.4) \qquad \varphi_B = \fint_B \varphi(z)\, dxdy = \frac{1}{|B|} \int_B \varphi(z)\, dxdy$$

is the mean value of the function φ over B. A function φ in BMO is of *vanishing mean oscillation*, $\varphi \in \text{VMO}$, if the supremum in (1.3) taken over all disks B in D with $|B| < \varepsilon$ converges to 0 as $\varepsilon \to 0$. It should be noted that the Sobolev class $W^{1,2}_{loc}$ is a subclass of VMO [**BN**], [**CP**], and that $\text{VMO}(D) \subset \text{BMO}(D) \subset L^p_{loc}(D)$ for all $1 \leq p < \infty$ [**RR**]. In contrast to BMO, the class FMO of functions of finite mean oscillation is not contained in L^p_{loc} for any $p > 1$. We will show that BMO is a proper subclass of FMO. Thus our existence theorem extends earlier results.

Functions of finite mean oscillation (FMO), were first introduced in [**IR**] and are defined below in Section 2. BMO was introduced by John and Nirenberg [**JN**] and soon became an important concept in harmonic analysis, complex analysis and partial differential equations. BMO functions are related in many ways to quasiconformal and quasiregular mappings ([**As**], [**AG**], [**Jo**], [**MRV**] and [**Rei**]), and to mappings with finite distortion ([**AIKM**] and [**IM**]). VMO was introduced by Sarason [**Sar**]. Existence, uniqueness and properties of solutions of various kinds of differential equations and, in particular, of elliptic type with coefficients of the class VMO have been studied by several authors [**CFL**], [**IS**], [**Pa**] and [**Ra**].

Conditions for the existence and uniqueness of ACL homeomorphic solutions for the Beltrami equation can be given in terms of suitable majorants for the dilatation $K_\mu(z)$ as here and in [**RSY**$_2$], or in terms of certain integral and measure constraints on the dilatation [**BJ**], [**Da**], [**GMSV**], [**IM**], [**Kr**], [**Le**], [**MM**], [**MS**], [**Pe**] and [**Tu**]. In the latter type of condition, one assumes either exponential integrability or at least high local integrability of the dilatation.

2. Finite mean oscillation

We say that a function $\varphi : D \to \mathbb{R}$ has *finite mean oscillation* at a point $z_0 \in D$ if

$$(2.1) \qquad d_\varphi(z_0) = \overline{\lim_{\varepsilon \to 0}} \fint_{D(z_0,\varepsilon)} |\varphi(z) - \overline{\varphi}_\varepsilon(z_0)|\, dxdy \; < \; \infty,$$

where

$$\overline{\varphi}_\varepsilon(z_0) = \fint_{D(z_0,\varepsilon)} \varphi(z) \, dxdy \tag{2.2}$$

is the mean value of the function $\varphi(z)$ over the disk

$$D(z_0, \varepsilon) = \{z \in \mathbb{C} : |z - z_0| < \varepsilon\}. \tag{2.3}$$

Condition (2.1) includes the assumption that φ is integrable in a neighborhood of the point z_0,; see [**IR**]. We call $d_\varphi(z_0)$ the *dispersion* of the function φ at the point z_0. We say that a function $\varphi : D \to \mathbb{R}$ is of *finite mean oscillation* in a domain D, $\varphi \in \mathrm{FMO}(D)$ or simply $\varphi \in \mathrm{FMO}$, if φ has finite dispersion at every point $z \in D$.

REMARK 2.1. Note that if a function $\varphi : D \to \mathbb{R}$ is integrable over $D(z_0, \varepsilon_0) \subset D$, then

$$\fint_{D(z_0,\varepsilon)} |\varphi(z) - \overline{\varphi}_\varepsilon(z_0)| \, dxdy \leq 2 \cdot \overline{\varphi}_\varepsilon(z_0); \tag{2.4}$$

and the right side in (2.4) is continuous in the parameter $\varepsilon \in (0, \varepsilon_0]$ by the absolute continuity of the indefinite integral. Thus, for every $\delta_0 \in (0, \varepsilon_0)$,

$$\sup_{\varepsilon \in [\delta_0, \varepsilon_0]} \fint_{D(z_0,\varepsilon)} |\varphi(z) - \overline{\varphi}_\varepsilon(z_0)| \, dxdy < \infty. \tag{2.5}$$

If (2.1) holds, then

$$\sup_{\varepsilon \in (0, \varepsilon_0]} \fint_{D(z_0,\varepsilon)} |\varphi(z) - \overline{\varphi}_\varepsilon(z_0)| \, dxdy < \infty. \tag{2.6}$$

The number in the left hand side of (2.6) is called the *maximal dispersion* of the function φ in the disk $D(z_0, \varepsilon_0)$.

PROPOSITION 2.1. *If for some set of numbers* $\varphi_\varepsilon \in \mathbb{R}$, $\varepsilon \in (0, \varepsilon_0]$,

$$\varlimsup_{\varepsilon \to 0} \fint_{D(z_0,\varepsilon)} |\varphi(z) - \varphi_\varepsilon| \, dxdy < \infty, \tag{2.7}$$

then φ is of finite mean oscillation at z_0.

PROOF. Indeed, by the triangle inequality,

$$\fint_{D(z_0,\varepsilon)} |\varphi(z) - \overline{\varphi}_\varepsilon(z_0)| \, dxdy \leq \fint_{D(z_0,\varepsilon)} |\varphi(z) - \varphi_\varepsilon| \, dxdy + |\varphi_\varepsilon - \overline{\varphi}_\varepsilon(z_0)|$$

$$\leq 2 \cdot \fint_{D(z_0,\varepsilon)} |\varphi(z) - \varphi_\varepsilon| \, dxdy.$$

$\square$

COROLLARY 2.2. *If*

$$\varlimsup_{\varepsilon \to 0} \fint_{D(z_0,\varepsilon)} |\varphi(z)| \, dxdy < \infty, \tag{2.8}$$

then φ has finite mean oscillation at z_0.

REMARK 2.2. The condition (2.7) is only sufficient but not necessary for a function φ to be of finite mean oscillation at z_0. Indeed, by John and Nirenberg, the function $\varphi(z) = \log \frac{1}{|z|}$ belongs to BMO in the unit disk Δ (cf. [**RR**, p. 5]) and hence also to FMO. However, $\overline{\varphi}_\varepsilon(0) \to \infty$ as $\varepsilon \to 0$, violating (2.7).

A point $z_0 \in D$ is called a *Lebesgue point* of a function $\varphi : D \to \mathbb{R}$ if φ is integrable in a neighborhood of z_0 and

$$\lim_{\varepsilon \to 0} \fint_{D(z_0,\varepsilon)} |\varphi(z) - \varphi(z_0)| \, dxdy = 0. \tag{2.9}$$

It is known that for every function $\varphi \in L^1(D)$, almost every point in D is a Lebesgue point.

COROLLARY 2.3. *Every function $\varphi : D \to \mathbb{R}$ which is locally integrable has finite mean oscillation at almost every point in D.*

We use the following notation: $D(r) = D(0,r) = \{z \in \mathbb{C} : |z| < r\}$ and

$$A(\varepsilon, \varepsilon_0) = \{z \in \mathbb{C} : \varepsilon < |z| < \varepsilon_0\}. \tag{2.10}$$

LEMMA 2.4. *Let D be a domain in $\mathbb{C}$ with $D(e^{-1}) \subset D$ and $\varphi : D \to \mathbb{R}$ a nonnegative function integrable in $D(e^{-1})$ with finite mean oscillation at 0. Then, for $\varepsilon \in (0, e^{-e})$,*

$$\int_{A(\varepsilon, e^{-1})} \frac{\varphi(z) \, dxdy}{\left(|z| \log \frac{1}{|z|}\right)^2} \leq C \cdot \log \log \frac{1}{\varepsilon}, \tag{2.11}$$

where

$$C = 2\pi \, (2\varphi_0 + 3e^2 d_0), \tag{2.12}$$

φ_0 *is the mean value of φ over the disk $D(e^{-1})$ and d_0 is the maximal dispersion of φ in the disk $D(e^{-1})$.*

Versions of this lemma were first established for BMO functions in [**RSY**$_1$] and [**RSY**$_2$] and later for FMO functions in [**IR**]. The proof of the present version goes along the same lines as in [**RSY**$_1$] and is omitted.

We close this section by constructing for every $p > 1$ a function $\varphi : \mathbb{C} \to \mathbb{R}$ which belongs to FMO but not to L^p_{loc} and hence not to BMO_{loc}. In the following example, $p = 1 + \delta$, where $\delta > 0$ is arbitrarily small.

EXAMPLE. The function

$$\varphi(z) = \begin{cases} e^{\frac{1}{|z|^2-1}} & \text{if } |z| < 1, \\ 0 & \text{if } |z| \geq 1 \end{cases} \tag{2.13}$$

belongs to C_0^∞ and hence $\varphi \in \text{BMO}$. For $\delta > 0$, let us consider the function

$$\varphi^*(z) = \begin{cases} \varphi_k(z) & \text{if } z \in D_k \\ 0 & \text{if } z \in \mathbb{C} \setminus \cup D_k, \end{cases} \tag{2.14}$$

where $D_k = D(z_k, r_k)$, $z_k = 2^{-k}$, $r_k = 2^{-(1+\delta)k^2}$, and

$$\varphi_k(z) = 2^{2k^2} \varphi\left(\frac{z - z_k}{r_k}\right), \quad z \in D_k, \ k = 2, 3, \ldots . \tag{2.15}$$

Then Q is smooth in $\mathbb{C} \setminus \{0\}$, so Q belongs to $\text{BMO}_{loc}(\mathbb{C} \setminus \{0\})$ and hence to $\text{FMO}(\mathbb{C} \setminus \{0\})$.

Now note that

$$\int_{D_k} \varphi_k(z)\, dx dy \;=\; 2^{-2\delta k^2} \int_{\mathbb{C}} \varphi(z)\, dx dy, \tag{2.16}$$

so by a straightforward computation,

$$\varlimsup_{\varepsilon \to 0} \fint_{D(\varepsilon)} \varphi^*(z)\, dx dy \;<\; \infty. \tag{2.17}$$

Thus, by Corollary 2.2, $\varphi \in$ FMO.

On the other hand,

$$\int_{D_k} \varphi_k^{1+\delta}(z)\, dx dy = \int_{\mathbb{C}} \varphi^{1+\delta}(z)\, dx dy; \tag{2.18}$$

hence $\varphi^* \notin L^{1+\delta}(U)$ for any neighborhood U of 0, as U contains infinitely many disks D_k.

3. Estimates of distortion

For points $z, \zeta \in \overline{\mathbb{C}}$, let $s(z, \zeta)$ denotes the spherical distance between z and ζ; and for a set $E \subset \overline{\mathbb{C}}$, let

$$\delta(E) = \sup_{z, \zeta \in E} s(z, \zeta) \tag{3.1}$$

denote the *spherical diameter* of E. Given a measurable function $Q : D \to [1, \infty]$ and $\Delta > 0$, let $\mathfrak{F}_Q^\Delta$ denote the class of all qc mappings $f : D \to \overline{\mathbb{C}}$ such that $\delta(\overline{\mathbb{C}} \setminus f(D)) \geq \Delta$ and such that

$$K_\mu(z) = \frac{1 + |\mu(z)|}{1 - |\mu(z)|} \leq Q(z) \quad \text{a.e.} \tag{3.2}$$

LEMMA 3.1. *Let $\Delta > 0$, D be a domain in $\mathbb{C}$ with $D(e^{-1}) \subset D$. If $Q : D \to [1, \infty]$ is integrable in $D(e^{-1})$ and $Q \in$ FMO at 0, then for every $f \in \mathfrak{F}_Q^\Delta$ and every point $z \in D(e^{-e})$,*

$$s(f(z), f(0)) \;\leq\; \alpha_0 \cdot \left(\log \frac{1}{|z|}\right)^{-\beta_0}, \tag{3.3}$$

where

$$\alpha_0 = \frac{32}{\Delta} \tag{3.4}$$

and

$$\beta_0 \;=\; (2q_0 + 3e^2 d_0)^{-1}; \tag{3.5}$$

here q_0 is the mean value of $Q(z)$ in $D(1/e)$ and d_0 is the maximal dispersion of $Q(z)$ in $D(1/e)$.

The proof follows the same lines as in [**RSY**$_2$] by standard modulus techniques, where here Lemma 2.4 is used.

4. Existence theorems

THEOREM 4.1. *Let D be a domain in $\mathbb{C}$ and let $\mu : D \to \mathbb{C}$ be a measurable function with $|\mu(z)| < 1$ a.e. such that*

(4.1) $$K_\mu(z) \leq Q(z)$$

for some FMO function $Q : D \to [1, \infty]$. Then the Beltrami equation (1.1) has a homeomorphic ACL solution $f_\mu : D \to \mathbb{C}$ with $f_\mu^{-1} \in W_{loc}^{1,2}$.

PROOF. Fix points z_1 and z_2 in D. For $n \in \mathbb{N}$, define $\mu_n : D \to \mathbb{C}$ by letting $\mu_n(z) = \mu(z)$ if $|\mu(z)| \leq 1 - 1/n$ and 0 otherwise. Then $\|\mu_n\| < 1$ and thus, by the classical existence theorem (see [**Ah**] or [**LV**]), (1.1) with μ_n instead of μ has a homeomorphic ACL solution $f_n : D \to \mathbb{C}$ which fixes the points z_1 and z_2. By the Arzela-Ascoli theorem and Lemma 3.1, the sequence f_n has a subsequence, denoted again by f_n, which converges locally uniformly to some nonconstant mapping f in D. Obviously $\mu_n \to \mu$ a.e. Then by the approximation and convergence theorems, Theorem 3.1 and Corollary 5.12 in [**RSY**$_1$], f is a homeomorphic ACL solution of (1.1).

Let $g_n = f_n^{-1}$, and let B be a relatively compact domain in D and B^* a relatively compact domain in $f(D)$ with $g(\bar{B}) \subset B^*$. Then for large n, one has by direct computation

$$\int_B |\partial g_n|^2 \, du dv = \int_{g_n(B)} \frac{dxdy}{1 - |\mu_n(z)|^2} \leq \int_{B^*} Q(z) \, dxdy < \infty$$

where the last relation implies that the sequence g_n is bounded in $W^{1,2}(B)$. Consequently, $f^{-1} \in W_{loc}^{1,2}(f(D))$. □

COROLLARY 4.2. *f_μ^{-1} is locally absolutely continuous and preserves nulls sets, and f_μ is regular, i.e., differentiable with $J_{f_\mu}(z) > 0$ a.e.*

Indeed, the assertion about f_μ^{-1} follows from the fact that $f_\mu^{-1} \in W_{loc}^{1,2}$; see [**LV**, pp. 131 and 150]. For the regularity of f_μ, let E denote the set of points of D where f_μ is differentiable and $J_{f_\mu}(z) = 0$, and suppose that $|E| > 0$. Then $|f_\mu(E)| > 0$, since $E = f_\mu^{-1}(f_\mu(E))$ and f_μ^{-1} preserves null sets. Clearly f_μ^{-1} is not differentiable at any point of $f_\mu(E)$, contradicting the fact that f_μ^{-1} is differentiable a.e.

Note that every ACL homeomorphic solution f of the Beltrami equation (1.1) with $K_\mu \in L_{loc}^1$ belongs to the class $W_{loc}^{1,1}$. This can be derived from the relation

$$|\bar\partial f| + |\partial f| = K_\mu^{1/2}(z) \cdot J_f^{1/2}(z) \quad \text{a.e.}$$

Thus we have also the following corollaries.

COROLLARY 4.3. *If $K_\mu(z)$ has a majorant $Q(z)$ for which every point $z \in D$ is a Lebesgue point, then the Beltrami equation has a homeomorphic $W_{loc}^{1,1}$ solution f_μ with $f_\mu^{-1} \in W_{loc}^{1,2}$.*

COROLLARY 4.4. *If, for every point $z \in D$,*

(4.2) $$\varlimsup_{\varepsilon \to 0} \fint_{D(z,\varepsilon)} \frac{1 + |\mu(z)|}{1 - |\mu(z)|} \, dxdy < \infty,$$

then the Beltrami equation has a homeomorphic $W_{loc}^{1,1}$ solution f_μ with $J_{f_\mu}(z) > 0$ a.e.

References

[Ah] L.V. Ahlfors, *Lectures on Quasiconformal Mappings*, Van Nostrand, Princeton, 1966.

[As] K. Astala, *A remark on quasiconformal mappings and BMO–functions*, Michigan Math. J. **30** (1983), 209–212.

[AG] K. Astala and F.W. Gehring, *Injectivity, the BMO norm and the universal Teichmüller space*, J. Anal. Math. **46** (1986), 16–57.

[AIKM] K. Astala, T. Iwaniec, P. Koskela and G. Martin, *Mappings of BMO-bounded distortion*, Math. Ann. **317** (2000), 703–726.

[BN] H. Brezis and L. Nirenberg, *Degree theory and BMO. I. Compact manifolds without boundaries*, Selecta Math. (N.S.) **1** (1995), 197-263.

[BJ] M.A. Brakalova and J.A. Jenkins, *On solutions of the Beltrami equation*, J. Anal. Math. **76** (1998), 67-92.

[CFL] F. Chiarenza, M. Frasca and P. Longo, $W^{2,p}$-*solvability of the Dirichlet problem for nondivergence elliptic equations with VMO coefficients*, Trans. Amer. Math. Soc. **336** (1993), 841–853.

[CP] A. Cianchi and L. Pick, *Sobolev embeddings into BMO, VMO, and L_∞*, Ark. Mat. **36** (1998), 317–340.

[Da] G. David, *Solutions de l'equation de Beltrami avec $\|\mu\|_\infty = 1$*, Ann. Acad. Sci. Fenn. Ser. A I Math. **13** (1988), 25-70.

[GL] F.W. Gehring and O. Lehto, *On the total differentiability of functions of a comlex variable*, Ann. Acad. Sci. Fenn. Ser. A I Math. **272** (1959).

[GMSV] V. Gutlyanskii, O. Martio, T. Sugawa and M. Vuorinen, *On the degenerate Beltrami equation*, Preprint of Department of Mathematics, University of Helsinki, Preprint 282 (2001).

[IM] T. Iwaniec and G. Martin, *Geometric Function Theory and Non–linear Analysis*, Clarendon Press, Oxford, 2001.

[IR] A. Ignat'ev and V. Ryazanov, *To the theory of removable singularities of space mappings*, Proc. Inst. Appl. Math. Mech. NASU **8** (2002), 1–14.

[IS] T. Iwaniec and C. Sbordone, *Riesz transforms and elliptic PDEs with VMO coefficients*, J. Anal. Math. **74** (1998), 183–212.

[Jo] P.M. Jones, *Extension theorems for BMO*, Indiana Univ. Math. J. **29** (1980), 41–66.

[JN] F. John and L. Nirenberg, *On functions of bounded mean oscillation*, Comm. Pure Appl. Math. **14** (1961), 415–426.

[Kr] V.I. Kruglikov, *The existence and uniqueness of mappings that are quasiconformal in the mean*, Metric Questions of the Theory of Functions and Mappings, No. IV, Naukova Dumka, Kiev, 1973, pp. 123–147.

[LV] O. Lehto and K. Virtanen, *Quasiconformal Mappings in the Plane*, Springer, New York, 1973.

[Le] O. Lehto, *Homeomorphisms with a given dilatation*, Proceedings of the Fifteenth Scandinavian Congress (Oslo, 1968), Springer Lecture Notes in Mathematics **118**, 1970, pp. 58-73.

[MM] O. Martio and V. Miklyukov, *On existence and uniqueness of the degenerate Beltrami equation*, Reports of Dept. Math., Univ. of Helsinki, Preprint 347 (2003).

[MRV] O. Martio, V. Ryazanov and M. Vuorinen, *BMO and injectivity of space quasiregular mappings*, Math. Nachr. **205** (1999), 149–161.

[MS] V.M. Miklyukov and G.D. Suvorov, *The existence and uniqueness of quasiconformal mappings with unbounded characteristics*, Studies in the Theory of Functions of a Complex Variable and its Applications, Kiev, 1972, pp. 45–53.

[Pa] D.K. Palagachev, *Quasilinear elliptic equations with VMO coefficients*, Trans. Amer. Math. Soc. **347** (1995), 2481–2493.

[Pe] I.N. Pesin, *Mappings quasiconformal in the mean*, Dokl. Akad. Nauk SSSR **187** (1969), 740–742.

[Ra] M.A. Ragusa, *Elliptic boundary value problem in vanishing mean oscillation hypothesis*, Comment. Math. Univ. Carolin. **40** (1999), 651–663.

[Rei] H.M. Reimann. *Functions of bounded mean oscillation and quasiconformal mappings*, Comment. Math. Helv. **49** (1974), 260–276.

[RR]   H.M. Reimann and T. Rychener, *Funktionen Beschränkter Mittlerer Oszillation*, Springer Lecture Notes in Mathematics **487**, 1975.
[Re]   Yu.G. Reshetnyak, *Space Mappings with Bounded Distortion*, Amer. Math. Soc., Providence, RI, 1989.
[RSY$_1$]   V. Ryazanov, U. Srebro and E. Yakubov, *Plane mappings with dilatation dominated by functions of bounded mean oscillation*, Siberian Adv. Math. **11** (2001), no. 2, 94–130.
[RSY$_2$]   V. Ryazanov, U. Srebro and E. Yakubov, *BMO-quasiconformal mappings*, J. Anal. Math. **83** (2001), 1–20.
[Sar]   D. Sarason, *Functions of vanishing mean oscillation*, Trans. Amer. Math. Soc. **207** (1975), 391–405.
[Tu]   P. Tukia, *Compactness properties of μ-homeomorphisms*, Ann. Acad. Sci. Fenn. Ser. AI Math. **16** (1991), 47-69.

INSTITUTE OF APPLIED MATHEMATICS AND MECHANICS, NAS OF UKRAINE, UL. ROZE LUXEMBURG 74, 83114, DONETSK, UKRAINE
E-mail address: `ryaz@iamm.ac.donetsk.ua`

TECHNION - ISRAEL, INSTITUTE OF TECHNOLOGY, , HAIFA 32000, ISRAEL
E-mail address: `srebro@math.technion.ac.il`

HOLON ACADEMIC INSTITUTE OF TECHNOLOGY, 52 GOLOMB ST., P.O.BOX 305, HOLON 58102, ISRAEL
E-mail address: `yakubov@hait.ac.il`

Pseudodifferential Operators with Operator-Valued Symbols

Bert-Wolfgang Schulze and Nikolai Tarkhanov

Dedicated to Professor Lawrence Zalcman on the occasion of his sixtieth birthday

ABSTRACT. Pseudodifferential calculus on manifolds with edges in the form developed by B.-W. Schulze is organised locally as Fourier calculus along the edges with symbols acting as operators in the transversal cone spaces. Thus, the index problem in the wedge algebra can be treated within a general framework of pseudodifferential operators with operator-valued symbols. In this paper, we introduce a concept of ellipticity for such operators and prove an index formula for elliptic elements.

1. Introduction

In the analysis on manifolds with singularities (conical points, edges, corners, etc.) there appear Mellin and Fourier pseudodifferential operators with operator-valued symbols.

A manifold with a conical singularity is locally of the form $\mathcal{C} = \mathbb{R}_+ \times X$ away from the conical point, where X is a smooth compact manifold. On $\mathcal{C}$ the pseudodifferential calculus of [15, 16] is based on the Mellin transform along the semiaxis $\mathbb{R}_+$. The symbols of cone operators take their values in the algebra of standard pseudodifferential operators on the cross-section X.

A manifold with edges is locally of the form $U \times \mathcal{C}$ close to an edge, where $\mathcal{C}$ is a stretched cone as above, and U is an open subset of $\mathbb{R}^q$. Rempel and Schulze [12] constructed pseudodifferential operators on $U \times \mathcal{C}$ in terms of the Fourier transform along the "edge" U with symbols taking their values in the algebra of cone operators on $\mathcal{C}$.

This gives rise to consideration of some aspects of the index problem on manifolds with singularities within a general framework of pseudodifferential operators with operator-valued symbols. In the present paper, we are interested in finding an index formula for edge operators with operator-valued symbols.

These operators have the form

$$(1.1) \qquad Au\,(y) = \frac{1}{(2\pi)^q} \iint e^{\imath \langle y-y', \eta \rangle}\, a(y,\eta)\, u(y')\, dy' d\eta,$$

2000 *Mathematics Subject Classification.* Primary 58J20; Secondary 58J05, 46L80.

where u is a C^∞ function of compact support on $\mathbb{R}^q$ with values in a Hilbert space H (throughout this paper, $C^\infty_{\text{comp}}(\mathbb{R}^q, H)$ stands for the set of such functions). The defining formula for a pseudodifferential operator thus makes sense when u is a Hilbert space-valued function and $a(y, \eta)$ is operator-valued. The theory is developed with the aim of discovering a natural class of "elliptic" operators, i.e., some generalisations of the usual elliptic operators, retaining the Fredholm property. We investigate those symbols which are themselves Fredholm operator-valued and invertible near infinity because it is clear that they already possess a topological index as defined in [1].

2. Edge symbols

The operator-valued Fourier symbols $a(y, \eta)$ are assumed to satisfy the following conditions:

1) $a(y, \eta) \in \mathcal{S}^m(T^*\mathbb{R}^q, \mathcal{L}(H, \tilde{H}))$, i.e., a is a C^∞ function of $(y, \eta) \in T^*\mathbb{R}^q$ taking its values in the space of continuous linear operator between Hilbert spaces H and $\tilde{H}$ and satisfying certain symbol estimates including group actions on H and $\tilde{H}$;
2) for $|y| \gg 1$, the symbol is independent of y, more precisely, $a(y, \eta) = a_\infty(\eta)$ for $|y| \geq R$.

As mentioned, the operators (1.1) are of great importance for the calculus of pseudodifferential operators on manifolds with edges.

EXAMPLE 2.1. Let
$$\begin{aligned} H &= H^{s_0,\gamma}(\mathcal{C}), \\ \tilde{H} &= H^{s_0-m,\gamma-m}(\mathcal{C}) \end{aligned}$$

be weighted Sobolev spaces on a stretched cone $\mathcal{C}$, both equipped with the group action

$$\kappa_\lambda u\,(r, x) = \lambda^{s_0-\gamma} u(\lambda r, x) \tag{2.1}$$

for $\lambda > 0$. Then by (1.1), we recover what are known as edge operators in the theory of [15, 16]. More precisely, given any weight data $w = (\gamma, \gamma - m)$, operators of the edge algebra $\Psi^m(\mathbb{R}^q \times \mathcal{C}, w)$ are described close to the "edge" $r = 0$ as those of the form (1.1), where

$$a(y, \eta) \in \bigcap_{s \in \mathbb{R}} \mathcal{S}^m(T^*\mathbb{R}^q, \mathcal{L}(H^{s,\gamma}(\mathcal{C}), H^{s-m, \gamma-m}(\mathcal{C}))).$$

It is well-known that, given any $s \in \mathbb{R}$, the operator (1.1) induces a continuous linear map $A : H^s_{\text{comp}}(\mathbb{R}^q, H) \to H^{s-m}_{\text{loc}}(\mathbb{R}^q, \tilde{H})$ between suitable Sobolev spaces on $\mathbb{R}^q$ whose definition involves relevant group actions. Moreover, under the above assumption on the symbol $a(y, \eta)$, we can even assign to (1.1) a continuous linear map of the global Hilbert spaces

$$A : H^s(\mathbb{R}^q, H) \to H^{s-m}(\mathbb{R}^q, \tilde{H}), \tag{2.2}$$

for any $s \in \mathbb{R}$.

3. Elliptic operators

Recall that the operator (2.2) is said to be Fredholm if both the null-space ker A of A and the cokernel coker A of A are finite dimensional. For a Fredholm operator A, the index is defined by

$$(3.1) \qquad \operatorname{ind} A = \dim \ker A - \dim \operatorname{coker} A.$$

The problem of evaluating the index of the operator A in terms of its symbol $a(y, \eta)$ has greatly influenced the construction of algebras of pseudodifferential operators in diverse contexts.

A Fredholm operator is invertible modulo compact operators, and the first step towards finding Fredholm operators is to find a relevant class of compact operators. We show that compact operator-valued symbols of order $-\infty$ give rise to compact operators and hence may take the place of smoothing operators in our theory. Then we carry out the familiar calculus for constructing parametrices.

DEFINITION 3.1. An operator A of form (1.1) is called elliptic if its symbol $a(y, \eta)$ satisfies the following conditions:
1) for any $(y, \eta) \in T^*\mathbb{R}^q$, the operator $a(y, \eta) : H \to \tilde{H}$ is Fredholm;
2) for any $(y, \eta) \in T^*\mathbb{R}^q$ with $|y|^2 + |\eta|^2 \gg 1$, the operator $a(y, \eta) : H \to \tilde{H}$ is invertible.

We prove that ellipticity implies the existence of a parametrix for A, i.e., an inverse up to smoothing operators. Hence it follows that A is Fredholm and the index of A does not depend on s.

In particular, in the special case $H = H^{s_0,\gamma}(\mathcal{C})$ and $\tilde{H} = H^{s_0-m,\gamma-m}(\mathcal{C})$, we obtain

$$\begin{aligned}
\operatorname{ind} A|_{H^s(\mathbb{R}^q, H^{s_0,\gamma}(\mathcal{C}))} &= \dim \ker A|_{H^s(\mathbb{R}^q, H^{s_0,\gamma}(\mathcal{C}))} - \dim \operatorname{coker} A|_{H^s(\mathbb{R}^q, H^{s_0,\gamma}(\mathcal{C}))} \\
&= \dim \ker A|_{H^{s_0}(\mathbb{R}^q, H^{s_0,\gamma}(\mathcal{C}))} - \dim \operatorname{coker} A|_{H^{s_0}(\mathbb{R}^q, H^{s_0,\gamma}(\mathcal{C}))} \\
&= \operatorname{ind} A|_{H^{s_0,\gamma}(\mathbb{R}^q \times \mathcal{C})},
\end{aligned}$$

where $H^{s_0,\gamma}(\mathbb{R}^q \times \mathcal{C}) := H^{s_0}(\mathbb{R}^q, H^{s_0,\gamma}(\mathcal{C}))$. It follows that one may keep "frozen" the exponent s in $H^{s,\gamma}(\mathcal{C})$ when studying the index of edge operators in the spaces $H^s(\mathbb{R}^q, H^{s,\gamma}(\mathcal{C}))$.

4. The index of a family

We obtain an index formula for elliptic operators A. A basic observation is that the ellipticity condition implies that the Fredholm family $a(y, \eta)$ parametrised by $(y, \eta) \in T^*\mathbb{R}^q$ is trivial outside a compact subset of $T^*\mathbb{R}^q$. Thus it defines an index bundle $\operatorname{ind} a \in K_{\text{comp}}(T^*\mathbb{R}^q)$, where K_{comp} means the K-functor with compact support. Its Chern character is represented by a closed differential form of compact support, and we prove that

$$(4.1) \qquad \operatorname{ind} A = \int_{T^*\mathbb{R}^q} \operatorname{ch}(\operatorname{ind} a) \, \mathcal{T}(T^*\mathbb{R}^q),$$

where $\mathcal{T}(T^*\mathbb{R}^q)$ is the Todd class of the manifold $\mathbb{R}^q$ (which is trivial).

If $a(y, \eta) \in \Psi^m(M; \mathbb{R}^q)$ is a C^∞ family of parameter-dependent pseudodifferential operators of order m on a smooth compact manifold M, then

$$a(y, \eta) \in \bigcap_{s \in \mathbb{R}} S^m(T^*\mathbb{R}^q, \mathcal{L}(H^s(M), H^{s-m}(M)))$$

relative to the identity group action on $H^s(M)$ and $H^{s-m}(M)$. If moreover $a(y,\eta)$ is parameter-dependent elliptic, then the operator A given by (1.1) is elliptic. By the Atiyah-Singer theorem on the index of a family of elliptic operators, cf. [3], we have

$$\text{ch}(\text{ind}\, a) \sim \int_{T^*M} \text{ch}(d(a_m))\, \mathcal{T}(T^*(M)), \tag{4.2}$$

where $d(a_m) \in K_{\text{comp}}(T^*(\mathbb{R}^q \times M))$ is the difference element determined by the principal symbol of $a(y,\eta)$, and $\sim$ means the coincidence of cohomology classes with compact support. Now, formula (4.1) follows from the Atiyah-Singer Index Theorem of [1, 2], for

$$\begin{aligned}\mathcal{T}(T^*\mathbb{R}^q)\,\mathcal{T}(T^*M) &= \mathcal{T}(T^*\mathbb{R}^q \times T^*M) \\ &= \mathcal{T}(T^*(\mathbb{R}^q \times M)),\end{aligned}$$

which is due to the multiplicativity of the Todd class.

Another interesting case is when $a(y,\eta) \in \Psi^m(\mathcal{C},w;\mathbb{R}^q)$ is a C^∞ family of parameter-dependent operators in the algebra of Schulze [15, 16] on a stretched cone $\mathcal{C}$. If $a(y,\eta)$ is parameter-dependent elliptic in the sense of the cone algebra, then the operator A given by (1.1) is elliptic. It follows that formula (4.1) holds, while the Atiyah-Singer index theorem for families is not applicable.

5. Operators of finite rank

Let $A : L^2(\mathbb{R}^q, H) \to L^2(\mathbb{R}^q, \tilde{H})$ be a linear operator of finite rank. This means that

$$Au(y) = \sum_j (u, b_j)_{L^2(\mathbb{R}^q, H)} e_j(y)$$

for some

$$\begin{aligned}(b_j)_{j=1,\ldots,N} &\subset L^2(\mathbb{R}^q, H), \\ (e_j)_{j=1,\ldots,N} &\subset L^2(\mathbb{R}^q, \tilde{H}).\end{aligned}$$

If $u \in L^2(\mathbb{R}^q, H)$, we get

$$Au(y) = \sum_j \frac{1}{(2\pi)^q} (\mathcal{F}_{y' \mapsto \eta} u, \mathcal{F}_{y' \mapsto \eta} b_j)_{L^2(\mathbb{R}^q, H)} e_j(y)$$

by Parseval's formula. Hence, if the $b_j \in \mathcal{S}(\mathbb{R}^q, H)$ are functions of rapid descent and the $e_j \in C^\infty(\mathbb{R}^q, \tilde{H})$ are smooth functions, then $A = \text{op}(a)$ is a pseudodifferential operator with symbol

$$a(y,\eta)h = e^{-\imath\langle y,\eta\rangle} \sum_j e_j(y)\,(h, \mathcal{F}_{y' \mapsto \eta} b_j)_H.$$

It is easy to verify that the symbol $a(y,\eta)$ is actually of order $-\infty$ with values in $\mathcal{K}(H, \tilde{H})$.

6. The continuity property

In this section, we prove that the operator (1.1) is well-defined on the global Sobolev spaces, as in (2.2).

To this end, fix a function $\omega \in C^\infty_{\text{comp}}(\mathbb{R}^q)$ such that $\omega(y) = 1$ for $|y| < R$. If $u \in C^\infty_{\text{comp}}(\mathbb{R}^q, H)$, then

$$
\begin{aligned}
Au &= \omega \operatorname{op}(a) u + (1-\omega) \operatorname{op}(a) u \\
&= \operatorname{op}(\omega a) u + (1-\omega) \operatorname{op}(a_\infty) u,
\end{aligned}
\tag{6.1}
$$

because $(1-\omega(y)) a(y,\eta) = (1-\omega(y)) a_\infty(y)$.

THEOREM 6.1. *Let $s \in \mathbb{R}$. The operator A defined by (6.1) extends to a continuous linear map of*

$$A_s : H^s(\mathbb{R}^q, H) \to H^{s-m}(\mathbb{R}^q, \tilde{H}).$$

PROOF. Pick $u \in H^s(\mathbb{R}^q, H)$. By Proposition 2.1.12 of Schulze and Tarkhanov [17], there exists a sequence (u_ν) in $C^\infty_{\text{comp}}(\mathbb{R}^q, H)$ such that $u_\nu \to u$ in $H^s(\mathbb{R}^q, H)$.

As the symbol $\omega(y) a(y,\eta)$ is compactly supported in y, a familiar argument of tensor products (cf. Schulze [15, 3.2.1]) shows that $\operatorname{op}(\omega a)$ induces a continuous map $H^s(\mathbb{R}^q, H) \to H^{s-m}(\mathbb{R}^q, \tilde{H})$, for each $s \in \mathbb{R}$. Hence the sequence $(\operatorname{op}(\omega a) u_\nu)$ converges in $H^{s-m}(\mathbb{R}^q, \tilde{H})$ as $\nu \to \infty$.

Similarly, the sequence $(\operatorname{op}(a_\infty) u_\nu)$ converges to $\operatorname{op}(a_\infty) u$ in $H^{s-m}(\mathbb{R}^q, \tilde{H})$, since the symbol a_∞ does not depend on y. Moreover, $(1-\omega) \operatorname{op}(a_\infty) u$ lies again in $H^{s-m}(\mathbb{R}^q, \tilde{H})$ because the multiplication operator $f \mapsto \omega f$ acts continuously on this space.

Hence it follows that the sequence (Au_ν) converges in $H^{s-m}(\mathbb{R}^q, \tilde{H})$. Setting $Au = \lim_{\nu \to \infty} Au_\nu$, we get a continuous map $H^s(\mathbb{R}^q, H) \to H^{s-m}(\mathbb{R}^q, \tilde{H})$, as desired. □

7. A parametrix construction and corrections to the ellipticity

In this section, we indicate a natural candidate for a parametrix of the operator A and discuss its properties.

It will cause no confusion if we use the same notation (κ_λ) to designate group actions in both H and $\tilde{H}$. As usual, $\eta \mapsto \langle \eta \rangle$ stands for a smoothed norm function on $\mathbb{R}^q$.

In the sequel we invoke a general framework of pseudodifferential calculus on non-compact manifolds, taking into account the behaviour of symbols as $y \to \infty$. Nowdays this calculus is standard. For more details and bibliography, we refer the reader for instance to the book [16].

LEMMA 7.1. *Under the above assumptions, the symbol $a(y,\eta)$ belongs to the symbol class $\mathcal{S}^{m,0}(T^*\mathbb{R}^q, \mathcal{L}(H, \tilde{H}))$.*

PROOF. We have to verify that for any multi-indices $\alpha, \beta \in \mathbb{Z}_+^q$, there is a constant $c_{\alpha,\beta}$ such that

$$\|\kappa^{-1}_{\langle \eta \rangle} D_y^\alpha D_\eta^\beta a(y,\eta) \kappa_{\langle \eta \rangle}\|_{\mathcal{L}(H,\tilde{H})} \leq c_{\alpha,\beta} \langle y \rangle^{-|\alpha|} \langle \eta \rangle^{m-|\beta|} \tag{7.1}$$

whenever $(y,\eta) \in T^*\mathbb{R}^q$.

Indeed, if $\alpha = 0$, then (7.1) is satisfied because the symbol $a(y,\eta)$ belongs to $\mathcal{S}^m(T^*\mathbb{R}^q, \mathcal{L}(H, \tilde{H}))$ and does not depend on y for y large enough.

On the other hand, if $\alpha \neq 0$, then the derivative $D_y^\alpha D_\eta^\beta a(y,\eta)$ is compactly supported in y. Estimate (7.1) again follows from the fact that $a(y,\eta)$ lies in $\mathcal{S}^m(T^*\mathbb{R}^q, \mathcal{L}(H, \tilde{H}))$. □

Having disposed of this preliminary step, we can now return to constructing a parametrix for A on the symbolic level.

Over any point $(y, \eta) \in T^*\mathbb{R}^q$, the symbol $a(y, \eta)$ has a regulariser $p(y, \eta)$, i.e., an operator $\tilde{H} \to H$ such that

(7.2) $$\begin{aligned} p(y,\eta)\, a(y,\eta) &= 1 - s_0(y,\eta) \quad \text{on} \quad H, \\ a(y,\eta)\, p(y,\eta) &= 1 - s_1(y,\eta) \quad \text{on} \quad \tilde{H}, \end{aligned}$$

where $s_0(y, \eta)$ and $s_1(y, \eta)$ are compact operators. Moreover, there exists a regulariser $p(y, \eta)$ such that both $s_0(y, \eta)$ and $s_1(y, \eta)$ are of finite rank.

A familiar argument shows that we can choose $p(y, \eta)$ to depend smoothly on $(y, \eta) \in T^*\mathbb{R}^q$. Thus, we obtain a family $p(y, \eta) \in C^\infty_{\text{loc}}(T^*\mathbb{R}^q, \mathcal{L}(H, \tilde{H}))$ inverting $a(y, \eta)$ on all of $T^*\mathbb{R}^q$ up to operators

$$\begin{aligned} s_0(y,\eta) &\in C^\infty_{\text{loc}}(T^*\mathbb{R}^q, \mathcal{K}(H)), \\ s_1(y,\eta) &\in C^\infty_{\text{loc}}(T^*\mathbb{R}^q, \mathcal{K}(\tilde{H})). \end{aligned}$$

By the above, we can even ensure that both $s_0(y, \eta)$ and $s_1(y, \eta)$ take their values in the respective spaces $\mathcal{L}_1(H)$ and $\mathcal{L}_1(\tilde{H})$ of trace class operators.

On the other hand, there exists a closed ball $\bar{B}$ in the cotangent space $T^*\mathbb{R}^q$ such that the symbol $a(y, \eta)$ is invertible outside of $\bar{B}$. Again, the inverse $a^{-1}(y, \eta)$ is a smooth family away from the ball with values in $\mathcal{L}(\tilde{H}, H)$.

Now fix a function $\phi(y, \eta) \in C^\infty_{\text{comp}}(T^*\mathbb{R}^q)$ such that $\phi \equiv 1$ in a neighbourhood of $\bar{B}$. Given any $(y, \eta) \in T^*\mathbb{R}^q$, set

(7.3) $$p^{(0)}(y, \eta) = \phi(y, \eta)\, p(y, \eta) + (1 - \phi(y, \eta))\, a^{-1}(y, \eta).$$

LEMMA 7.2. *Assume that*

(7.4) $$\|\kappa_{\langle\eta\rangle}^{-1}\, a^{-1}(y, \eta)\, \kappa_{\langle\eta\rangle}\|_{\mathcal{L}(\tilde{H}, H)} \leq c\, \langle\eta\rangle^{-m}$$

for all $|\eta| \gg 1$. Then $p^{(0)}(y, \eta) \in S^{-m,0}(T^*\mathbb{R}^q, \mathcal{L}(\tilde{H}, H))$.

PROOF. First of all, observe that $p^{(0)}(y, \eta) \in C^\infty_{\text{loc}}(T^*\mathbb{R}^q, \mathcal{L}(\tilde{H}, H))$. Since $p^{(0)}(y, \eta)$ is independent of y, for y large enough, the lemma is proved once we show that $p^{(0)}(y, \eta) \in S^{-m}(T^*\mathbb{R}^q, \mathcal{L}(\tilde{H}, H))$. We treat the summands of $p^{(0)}(y, \eta)$ separately.

As $\phi(y, \eta)\, p(y, \eta)$ is compactly supported, this summand actually belongs to $S^{-\infty}(T^*\mathbb{R}^q, \mathcal{L}(\tilde{H}, H))$.

On the other hand, the second summand $(1 - \phi(y, \eta))\, a^{-1}(y, \eta)$ coincides with $a^{-1}(y, \eta)$ away from the support of ϕ. Hence we are reduced to proving an estimate

$$\|\kappa_{\langle\eta\rangle}^{-1}\, D_y^\alpha D_\eta^\beta\, a^{-1}(y, \eta)\, \kappa_{\langle\eta\rangle}\|_{\mathcal{L}(\tilde{H}, H)} \leq c_{\alpha,\beta}\, \langle\eta\rangle^{-m-|\beta|}$$

for all $\eta \in \mathbb{R}^q$ with $|\eta| \gg 1$.

We give the proof only for the case $\alpha = 0$ and $|\beta| = 1$; the other cases follow by induction. From $a^{-1} a = 1$ we deduce that $(D^\beta a^{-1})\, a + a^{-1}\, (D^\beta a) = 0$, whence $D^\beta a^{-1} = a^{-1}\, (D^\beta a)\, a^{-1}$. Applying (7.1) and (7.4) thus yields

$$\begin{aligned} &\|\kappa_{\langle\eta\rangle}^{-1}\, D_\eta^\beta a^{-1}(y, \eta)\, \kappa_{\langle\eta\rangle}\|_{\mathcal{L}(\tilde{H}, H)} \\ &= \|(\kappa_{\langle\eta\rangle}^{-1}\, a^{-1}(y, \eta)\, \kappa_{\langle\eta\rangle})(\kappa_{\langle\eta\rangle}^{-1}\, D_\eta^\beta a(y, \eta)\, \kappa_{\langle\eta\rangle})(\kappa_{\langle\eta\rangle}^{-1}\, a^{-1}(y, \eta)\, \kappa_{\langle\eta\rangle})\|_{\mathcal{L}(\tilde{H}, H)} \\ &\leq \|\kappa_{\langle\eta\rangle}^{-1}\, a^{-1}(y, \eta)\, \kappa_{\langle\eta\rangle}\|^2_{\mathcal{L}(\tilde{H}, H)}\, \|\kappa_{\langle\eta\rangle}^{-1}\, D_\eta^\beta a(y, \eta)\, \kappa_{\langle\eta\rangle}\|_{\mathcal{L}(H, \tilde{H})} \\ &\leq c^2\, c_{0,\beta}\, \langle\eta\rangle^{-m-1} \end{aligned}$$

for $|\eta| \gg 1$, as desired. □

Henceforth, we assume that the symbol $a(y, \eta)$ satisfies condition (7.4). Then the family $p^{(0)}(y, \eta)$ is a natural candidate for the symbol of a "soft" parametrix of the operator A.

LEMMA 7.3. *The symbol $p^{(0)}(y, \eta)$ defined by (7.3) satisfies*

$$p^{(0)}(y, \eta) \, a(y, \eta) = 1 \mod \mathcal{S}^{-\infty, -\infty}(T^*\mathbb{R}^q, \mathcal{L}_1(H)),$$
$$a(y, \eta) \, p^{(0)}(y, \eta) = 1 \mod \mathcal{S}^{-\infty, -\infty}(T^*\mathbb{R}^q, \mathcal{L}_1(\tilde{H})).$$

PROOF. By (7.2), an easy computation shows that

(7.5)
$$\begin{aligned} p^{(0)}(y, \eta) \, a(y, \eta) &= 1 - \phi(y, \eta) \, s_0(y, \eta), \\ a(y, \eta) \, p^{(0)}(y, \eta) &= 1 - \phi(y, \eta) \, s_1(y, \eta). \end{aligned}$$

To complete the proof, it suffices to observe that the remainders on the right side are compactly supported. □

8. From symbols to operators

As in the classical theory, it is to be expected that the pseudodifferential $P^{(0)}$ with complete symbol $p^{(0)}$ gives us a "soft" parametrix for A. Thus, we consider

(8.1)
$$\begin{aligned} P^{(0)} f(y) &= \operatorname{op}(p^{(0)}) f(y) \\ &= \frac{1}{(2\pi)^q} \int e^{\imath \langle y, \eta \rangle} \, p^{(0)}(y, \eta) \, \hat{f}(\eta) \, d\eta, \end{aligned}$$

where

$$\hat{f}(\eta) = \int e^{-\imath \langle \eta, y' \rangle} f(y') \, dy'$$

is the Fourier transform of $f \in C^\infty_{\text{comp}}(\mathbb{R}^q, \tilde{H})$.

Combining Lemma 7.2 and Theorem 6.1, we can assert that, given any $s \in \mathbb{R}^q$, the operator $P^{(0)}$ extends to a continuous linear map of global Sobolev spaces

$$P^{(0)} : H^{s-m}(\mathbb{R}^q, \tilde{H}) \to H^s(\mathbb{R}^q, H).$$

Hence it follows that both compositions $P^{(0)} A$ and $A P^{(0)}$ are well-defined as operators in the spaces $H^s(\mathbb{R}^q, H)$ and $H^{s-m}(\mathbb{R}^q, \tilde{H})$, respectively.

Now, we may use the Neumann series argument for improving properties of the "soft" parametrix $P^{(0)}$. Namely, given any $N = 1, 2, \ldots$, we set

(8.2)
$$P^{(N)} = \left(\sum_{j=0}^{N} (1 - P^{(0)} A)^j \right) P^{(0)}.$$

LEMMA 8.1. *The operator $P^{(N)}$ satisfies*

(8.3)
$$\begin{aligned} P^{(N)} A &= 1 - (1 - P^{(0)} A)^{N+1}, \\ A P^{(N)} &= 1 - (1 - A P^{(0)})^{N+1}. \end{aligned}$$

PROOF. Indeed,
$$\begin{aligned} P^{(N)} A &= \sum_{j=0}^{N} (1 - P^{(0)} A)^j \left(1 - 1 + P^{(0)} A\right) \\ &= \sum_{j=0}^{N} (1 - P^{(0)} A)^j - \sum_{j=0}^{N} (1 - P^{(0)} A)^{j+1} \\ &= 1 - \left(1 - P^{(0)} A\right)^{N+1}. \end{aligned}$$

Moreover, it is easy to see that
$$P^{(N)} = P^{(0)} \Big(\sum_{j=0}^{N} (1 - A P^{(0)})^j \Big),$$

whence the second equality of (8.3) follows as before. □

Recall that if
$$\begin{aligned} a_1(y,\eta) &\in \mathcal{S}^{m_1,0}(T^*\mathbb{R}^q, \mathcal{L}(H_1, H_2)), \\ a_2(y,\eta) &\in \mathcal{S}^{m_2,0}(T^*\mathbb{R}^q, \mathcal{L}(H_2, H_3)), \end{aligned}$$

then the composition $\operatorname{op}(a_2) \operatorname{op}(a_1)$ is again a pseudodifferential operator with symbol in $\mathcal{S}^{m_1+m_2,0}(T^*\mathbb{R}^q, \mathcal{L}(H_1, H_3))$, given by
(8.4)
$$e^{i\langle D_{y'}, D_{\eta'}\rangle} a_2(y, \eta') a_1(y', \eta) \Big|_{\substack{y'=y \\ \eta'=\eta}} = \sum_{|\alpha|\le J} \frac{1}{\alpha!} \partial_\eta^\alpha a_2(y,\eta) \, D_y^\alpha a_1(y,\eta) + r^{(J+1)}(y,\eta),$$

where $r^{(J+1)}(y,\eta) \in \mathcal{S}^{m_1+m_2-(J+1),0}(T^*\mathbb{R}^q, \mathcal{L}(H_1, H_3))$.

We often write formula (8.4) in the symbolic form
(8.5)
$$a_2 \circ a_1 \sim e^{i\langle D_{y'}, D_{\eta'}\rangle} a_2(y,\eta') a_1(y',\eta)\big|_{\substack{y'=y \\ \eta'=\eta}},$$

denoting by ∘ the composition of symbols (in this way, we obtain what is known as the Leibniz product). The right-hand side means a formal series obtained by formal Taylor expansion of the exponent. This series breaks off if either $a_1(y,\eta)$ is polynomial in y or $a_2(y,\eta)$ is polynomial in η. For general symbols from our symbol classes, it is an asymptotic series in the sense that (8.4) holds for each $J = 0, 1, \ldots$. Thus, this series can be summed modulo symbols of order $-\infty$.

Hence it follows that $1 - P^{(0)} A$ and $1 - A P^{(0)}$ are pseudodifferential operators with symbols in $\mathcal{S}^{-1,0}(T^*\mathbb{R}^q, \mathcal{L}(H))$ and $\mathcal{S}^{-1,0}(T^*\mathbb{R}^q, \mathcal{L}(\tilde{H}))$, respectively, given by

(8.6)
$$\operatorname{symbol}\left(1 - P^{(0)} A\right) = \phi s_0 - \sum_{\substack{|\alpha|\le J \\ \alpha \ne 0}} \frac{1}{\alpha!} \partial_\eta^\alpha p^{(0)} D_y^\alpha a - r_0^{(J+1)},$$

$$\operatorname{symbol}\left(1 - A P^{(0)}\right) = \phi s_1 - \sum_{\substack{|\alpha|\le J \\ \alpha \ne 0}} \frac{1}{\alpha!} \partial_\eta^\alpha a \, D_y^\alpha p^{(0)} - r_1^{(J+1)},$$

where
$$\begin{aligned} r_0^{(J+1)}(y,\eta) &\in \mathcal{S}^{-(J+1),0}(T^*\mathbb{R}^q, \mathcal{L}(H)), \\ r_1^{(J+1)}(y,\eta) &\in \mathcal{S}^{-(J+1),0}(T^*\mathbb{R}^q, \mathcal{L}(\tilde{H})). \end{aligned}$$

We continue in this fashion, concluding that $(1 - P^{(0)} A)^{N+1}$ and $(1 - A P^{(0)})^{N+1}$ are operators with symbols in $\mathcal{S}^{-(N+1),0}(T^*\mathbb{R}^q, \mathcal{L}(H))$ and $\mathcal{S}^{-(N+1),0}(T^*\mathbb{R}^q, \mathcal{L}(\tilde{H}))$, respectively. In other words, the operator $P^{(N)}$ is an inverse for A up to pseudo-differential operators of order $-N-1$.

Our next objective is to show that the operators $P^{(N)}$ have a limit in a suitable asymptotic sense as $N \to \infty$.

THEOREM 8.1. *There exists a symbol $p^{(\infty)}(y,\eta) \in \mathcal{S}^{-m,0}(T^*\mathbb{R}^q, \mathcal{L}(\tilde{H}, H))$ such that*

(8.7) $$\begin{aligned} p^{(\infty)} \circ a &= 1 \mod \mathcal{S}^{-\infty,-\infty}(T^*\mathbb{R}^q, \mathcal{L}(H)), \\ a \circ p^{(\infty)} &= 1 \mod \mathcal{S}^{-\infty,-\infty}(T^*\mathbb{R}^q, \mathcal{L}(\tilde{H})). \end{aligned}$$

PROOF. We define the symbol $p^{(\infty)}(y,\eta)$ by the asymptotic series

(8.8) $$p^{(\infty)}(y,\eta) \sim \Big(\sum_{j=0}^{\infty} (1 - p^{(0)} \circ a)^{\circ j}\Big) \circ p^{(0)};$$

we now give a precise meaning to this sum.

Since $a(y,\eta)$ is independent of y for $|y| \geq R$, the symbol

$$1 - p^{(0)} \circ a \sim \phi s_0 - \sum_{\alpha \neq 0} \partial_\eta^\alpha p^{(0)} \, D_y^\alpha a$$

vanishes for y large enough. We can certainly assume, by increasing R if necessary, that $1 - p^{(0)} \circ a$ is supported on $|y| \leq R$. Hence it follows that the iteration $(1 - p^{(0)} \circ a)^{\circ j}$ is supported on $|y| \leq R$ for every $j = 1, 2, \ldots$. Thus, each symbol $(1 - p^{(0)} \circ a)^{\circ j}$ with $j \geq 1$ actually belongs to $\mathcal{S}^{-j,-\infty}(T^*\mathbb{R}^q, \mathcal{L}(H))$.

We now invoke a standard result on asymptotic sums of symbols (cf., for instance, [**10**, 18.1.3]) to conclude that there is a symbol $\sigma \in \mathcal{S}^{-1}(T^*\mathbb{R}^q, \mathcal{L}(H))$ supported on $|y| \geq R$ such that

$$\sigma - \sum_{j=1}^{J} (1 - p^{(0)} \circ a)^{\circ j} \in \mathcal{S}^{-(J+1)}(T^*\mathbb{R}^q, \mathcal{L}(H))$$

for each $J = 1, 2, \ldots$. We write this as $\sigma \sim \sum_{j=1}^{\infty}(1 - p^{(0)} \circ a)^{\circ j}$. Note that we have actually proved that both σ and the asymptotic sum are in $\mathcal{S}^{-1,-\infty}(T^*\mathbb{R}^q, \mathcal{L}(H))$.

Set $p^{(\infty)} = (1 + \sigma) \circ p^{(0)}$. Then $p^{(\infty)}(y,\eta) \in \mathcal{S}^{-m,0}(T^*\mathbb{R}^q, \mathcal{L}(\tilde{H}, H))$, and we readily obtain

$$\begin{aligned} p^{(\infty)} \circ a &= (1+\sigma) \circ (p^{(0)} \circ a) \\ &= (1+\sigma) - (1+\sigma) \circ (1 - p^{(0)} \circ a) \\ &= 1 \end{aligned}$$

modulo $\mathcal{S}^{-\infty,-\infty}(T^*\mathbb{R}^q, \mathcal{L}(H))$, since

$$(1+\sigma) \circ (1 - p^{(0)} \circ a) = \sigma$$

modulo $\mathcal{S}^{-\infty,-\infty}(T^*\mathbb{R}^q, \mathcal{L}(H))$. This establishes the first equality in (8.7).

On the other hand, an easy computation shows that

$$p^{(\infty)}(y,\eta) \sim p^{(0)} \circ \Big(\sum_{j=0}^{\infty}(1 - a \circ p^{(0)})^{\circ j}\Big),$$

whence the second equality in (8.7) follows by the same method. $\square$

9. Regularity properties of elliptic operators

Set
$$H^{-\infty}(\mathbb{R}^q, H) = \cup_s H^s(\mathbb{R}^q, H),$$
$$H^{\infty}(\mathbb{R}^q, H) = \cap_s H^s(\mathbb{R}^q, H).$$

As defined by (6.1), the operator A induces continuous linear maps
$$A_{\pm\infty} : H^{\pm\infty}(\mathbb{R}^q, H) \to H^{\pm\infty}(\mathbb{R}^q, \tilde{H}),$$
by Theorem 6.1.

THEOREM 9.1. *If $A = \mathrm{op}(a)$ is an elliptic operator whose symbol satisfies (7.4), then the canonical maps*
$$\ker A_\infty \hookrightarrow \ker A_{-\infty},$$
$$\operatorname{coker} A_\infty \hookrightarrow \operatorname{coker} A_{-\infty}$$
are topological isomorphisms.

It is worth pointing out that the first assertion of this theorem is known in the classical setting as Weyl's Lemma.

PROOF. To prove the first part of the theorem, it suffices to have a left parametrix $P^{(N)}$ for A similar to that given by Lemma 8.1. Indeed, if $u \in H^s(\mathbb{R}^q, H)$ satisfies $Au = 0$ then it follows from Lemma 8.1 that
$$u = \left(1 - P^{(0)} A\right)^{N+1} u.$$

By the above, $\left(1 - P^{(0)} A\right)^{N+1}$ is a pseudodifferential operator with symbol in $\mathcal{S}^{-(N+1),0}(T^*\mathbb{R}^q, \mathcal{L}(H))$. Applying Theorem 6.1, we can thus assert that u lies in $H^{s+(N+1)}(\mathbb{R}^q, H)$. Since N is arbitrary, we get $u \in H^\infty(\mathbb{R}^q, H)$, as required.

In order to prove the second part of the theorem, we invoke Theorem 8.1. Set $P^{(\infty)} = \mathrm{op}(p^{(\infty)})$. Given any $f \in H^s(\mathbb{R}^q, \tilde{H})$, write
$$f = \left(1 - A P^{(\infty)}\right) f + A P^{(\infty)} f.$$

Combining Theorems 8.1 and 6.1, we deduce that
$$\left(1 - A P^{(\infty)}\right) f \in H^\infty(\mathbb{R}^q, \tilde{H}),$$
$$P^{(\infty)} f \in H^{s+m}(\mathbb{R}^q, H).$$

This completes the proof. □

Theorem 9.1 at once yields information about the null-space and the cokernel of the operator $A_s : H^s(\mathbb{R}^q, H) \to H^{s-m}(\mathbb{R}^q, \tilde{H})$. Namely, they are independent of the exponent s and equal to those of the operator A_∞. Thus, the index of A is well-defined independently of indicating the edge Sobolev spaces in which A has to be considered.

REMARK 9.1. That $\mathrm{ind}\, A_{s'} = \mathrm{ind}\, A_{s''}$ for finite s' and s'' follows from Lemma 8.1.

10. Does ellipticity imply the Fredholm property?

Our aim in studying pseudodifferential operators on Hilbert space-valued functions was to find an interesting class of Fredholm operators. The following simple example due to Luke [11] illuminates some of the problems.

EXAMPLE 10.1. Let $a(y,\eta) = (1 + \sigma(y,\eta)) \otimes 1_H$, where σ is a C^∞ function of compact support on $T^*\mathbb{R}^q$ and 1_H stands for the identity operator on H. Then a belongs to the symbol space $\mathcal{S}^0(T^*\mathbb{R}^q, \mathcal{L}(H))$ independently of any group action on H. We consider the identity action on H. Since the scalar operator $\mathrm{op}(1 + \sigma)$ is obviously elliptic, it is Fredholm in the Sobolev spaces $H^s(\mathbb{R}^q)$. It follows that

$$\begin{aligned} A &= \mathrm{op}(a) \\ &= \mathrm{op}(1+\sigma) \hat{\otimes} 1_H \end{aligned}$$

is well-defined on $H^s(\mathbb{R}^q, H) = H^s(\mathbb{R}^q) \hat{\otimes} H$. The null-space is now

$$\ker A = (\ker \mathrm{op}(1+\sigma)) \, \hat{\otimes} H.$$

If H is of finite dimension, then the operator A is elliptic and Fredholm. If, however, H fails to be finite dimensional, then the operator A is elliptic if and only if the function $1 + \sigma(y, \eta)$ vanishes nowhere on $T^*\mathbb{R}^q$. (As for condition (7.4), it is satisfied independently of any group action on H.) On the other hand, the null-space of A is either zero or infinite dimensional. Hence the operator A on $H^s(\mathbb{R}^q, H)$ is only Fredholm if it is invertible. When σ is chosen in such a way that $1 + \sigma(y, \eta) \neq 0$ for $(y, \eta) \in T^*\mathbb{R}^q$ and $\ker \mathrm{op}(1 + \sigma)$ is non-trivial, the operator A is elliptic but non-Fredholm.

This example shows that if we are to find a class of Fredholm operators, then our calculus needs to be more accurate than merely modulo symbols of order $-\infty$.

11. Symbols of compact fibre variation

Symbols of negative order taking values in the compact operators are known to give rise to compact operators (cf. Luke [11], Schulze [16, Theorem 1.3.61]). Thus, we may use them for the error terms in our calculus. This restriction on error terms, in turn, restricts the class of symbols with which we can work. As soon as this restriction is imposed, the calculus formulas all go through, and we are able to find our Fredholm operators and construct their parametrices.

DEFINITION 11.1. A pseudodifferential operator A is said to be smoothing if its symbol belongs to $\mathcal{S}^{-\infty,-\infty}(T^*\mathbb{R}^q, \mathcal{K}(H, \tilde{H}))$.

With this definition, we need some restrictions on our operators in order that the asymptotic expansions given by the calculus converge modulo smoothing operators. The following class of symbols seems to have first been considered by Luke [11].

DEFINITION 11.2. A symbol $a(y, \eta) \in \mathcal{S}^m(T^*\mathbb{R}^q, \mathcal{L}(H, \tilde{H}))$ has compact fibre variation (class CV) if, for each $y \in \mathbb{R}^q$, $a(y,\eta) - a(y,0) \in \mathcal{K}(H, \tilde{H})$ for all $\eta \in T_y^*\mathbb{R}^q$.

The space of such symbols is denoted by $\mathcal{S}_{\mathrm{CV}}^m(T^*\mathbb{R}^q, \mathcal{L}(H, \tilde{H}))$. Differentiations in η within this class lead to symbols taking values in the compact operators, since

$$\begin{aligned}\frac{\partial}{\partial \eta_j} a(y, \eta) &= \lim_{\Delta \eta_j \to 0} \frac{a(y, \eta + \Delta \eta_j) - a(y, \eta)}{\Delta \eta_j} \\ &= \lim_{\Delta \eta_j \to 0} \frac{(a(y, \eta + \Delta \eta_j) - a(y, 0)) - (a(y, \eta) - a(y, 0))}{\Delta \eta_j}\end{aligned}$$

and the limit of compact operators in the operator norm is compact. The converse statement is also quite obvious.

LEMMA 11.1. *A symbol $a(y, \eta) \in \mathcal{S}^m(T^*\mathbb{R}^q, \mathcal{L}(H, \tilde{H}))$ is of class CV if for each multi-index β of length 1, $D_\eta^\beta a(y, \eta) \in \mathcal{S}^{m-1}(T^*\mathbb{R}^q, \mathcal{K}(H, \tilde{H}))$.*

PROOF. The short Taylor expansion

$$a(y, \eta) - a(y, 0) = \sum_{j=1}^q \eta_j \int_0^1 \frac{\partial a}{\partial \eta_j}(x, \theta \eta) \, d\theta$$

displays $a(y, \eta)$ in the required form. □

We can perform the usual calculus modulo smoothing operators within the class of pseudodifferential operators of class CV. The class is closed under transposition and change of coordinates.

Moreover, it is closed under composition and the asymptotic formula (8.5) for the symbol of a composition still holds. Indeed, all the terms of this expansion, other than the first, are compact operator-valued; and hence the first term determines whether the composition is of class CV. As

$$\begin{aligned}&a_2(y, \eta) a_1(y, \eta) - a_2(y, 0) a_1(y, 0) \\ &= (a_2(y, \eta) - a_2(y, 0)) a_1(y, \eta) + a_2(y, 0) (a_1(y, \eta) - a_1(y, 0)),\end{aligned}$$

this is certainly the case.

We may extend the definition of pseudodifferential operators of class CV to operators defined on manifolds and Hilbert bundles. Given Hilbert bundles H and $\tilde{H}$ over a compact manifold Y, an operator $A \in \Psi_{\mathrm{CV}}^m(H, \tilde{H})$ is elliptic if it is defined locally in terms of Fredholm operator-valued symbols which are invertible for large $|\eta|$ with "bounded inverses."

The calculus allows us to construct parametrices (inverses modulo smoothing operators) for the elliptic operators.

THEOREM 11.1. *Let $a(y, \eta) \in \mathcal{S}_{\mathrm{CV}}^m(T^*\mathbb{R}^q, \mathcal{L}(H, \tilde{H}))$ be independent of y for $|y| \gg 1$. Then A is elliptic if and only if there is a symbol $p \in \mathcal{S}_{\mathrm{CV}}^{-m,0}(T^*\mathbb{R}^q, \mathcal{L}(\tilde{H}, H))$ such that*

(11.1) $$\begin{aligned} p \circ a &= 1 \mod \mathcal{S}^{-\infty, -\infty}(T^*\mathbb{R}^q, \mathcal{K}(H)), \\ a \circ p &= 1 \mod \mathcal{S}^{-\infty, -\infty}(T^*\mathbb{R}^q, \mathcal{K}(\tilde{H})).\end{aligned}$$

PROOF. The usual construction of a parametrix of Theorem 8.1 carries through with very little modification. The error terms $p \circ a - 1$ and $a \circ p - 1$ are in fact compactly supported in y.

Assume that p exists. Then

(11.2)
$$p(y,\eta)\,a(y,\eta) + \sum_{\alpha \neq 0} \frac{1}{\alpha!}\, \partial_\eta^\alpha p(y,\eta)\, D_y^\alpha a(y,\eta) \;\sim\; 1,$$
$$a(y,\eta)\,p(y,\eta) + \sum_{\alpha \neq 0} \frac{1}{\alpha!}\, \partial_\eta^\alpha a(y,\eta)\, D_y^\alpha p(y,\eta) \;\sim\; 1,$$

and hence $p\,a - 1$ and $a\,p - 1$ are both compact operator valued. It follows that a and p must be Fredholm operator valued symbols.

Only the leading terms in the expansions above are of non-negative order. This implies that, for large $|\eta|$,

$$\kappa_{\langle \eta \rangle}^{-1}\, p(y,\eta)\, a(y,\eta)\, \kappa_{\langle \eta \rangle},$$
$$\kappa_{\langle \eta \rangle}^{-1}\, a(y,\eta)\, p(y,\eta)\, \kappa_{\langle \eta \rangle}$$

are close to the identity and hence $a(y,\eta)$ is invertible.

Also, $a^{-1}(y,\eta)$ approximates $p(y,\eta)$ and hence is bounded. Therefore, $a(y,\eta)$ is an elliptic symbol. $\square$

Let $P = \mathrm{op}(p)$ be the pseudodifferential operator with symbol p. Then (11.1) just amounts to saying that
$$PA = 1,$$
$$AP = 1$$
modulo smoothing operators.

COROLLARY 11.1. *Suppose $a(y,\eta) \in \mathcal{S}_{\mathrm{CV}}^m(T^*\mathbb{R}^q, \mathcal{L}(H, \tilde{H}))$ is an elliptic symbol independent of y for $|y| \gg 1$. Then the operator $A_s : H^s(\mathbb{R}^q, H) \to H^{s-m}(\mathbb{R}^q, \tilde{H})$ is Fredholm for each $s \in \mathbb{R}$.*

PROOF. Letting $P_{s-m} : H^{s-m}(\mathbb{R}^q, \tilde{H}) \to H^s(\mathbb{R}^q, H)$ denote the extension of P, we deduce from the above that the operators
$$P_{s-m}\,A_s - 1 \;:\; H^s(\mathbb{R}^q, H) \to H^s(\mathbb{R}^q, H),$$
$$A_s\,P_{s-m} - 1 \;:\; H^{s-m}(\mathbb{R}^q, \tilde{H}) \to H^{s-m}(\mathbb{R}^q, \tilde{H})$$
are compact. This means that P_{s-m} is the required parametrix for A_s, which proves the corollary. $\square$

In the sequel, we tacitly assume that all symbols under consideration are of compact fibre variation.

12. Parameter-dependent ellipticity

The quality of symbols $a(y,\eta) \in \mathcal{S}_{\mathrm{CV}}^m(T^*\mathbb{R}^q, \mathcal{L}(H, \tilde{H}))$ is improved under differentiation in η. Not only the order of $D_\eta^\beta a(y,\eta)$ is $|\beta|$ less than the order of $a(y,\eta)$, but also the values of $D_\eta^\beta a(y,\eta)$ are improved, becoming compact operators. In organising a refined calculus of pseudodifferential operators over $\mathbb{R}^q$, it is necessary to control qualitatively the values of the symbols. To this end, one can invoke the scale of Schatten ideals of compact operators in Hilbert spaces; cf. [5], [14], and [13].

Let $A : H \to \tilde{H}$ be a compact operator. Denote by $s_j = s_j(A)$ the eigenvalues of the operator $\sqrt{A^*A}$. These are called the singular numbers (or s-numbers) of the operator A.

DEFINITION 12.1. Let $0 < p < \infty$. The pth Schatten norm of A is the l^p-norm of the sequence of its singular numbers, i.e.,

$$\|A\|_p := \Big(\sum_{j=1}^{\infty} s_j^p\Big)^{1/p}.$$

It follows that $\|A\|_p^p = \mathrm{tr}(A^*A)^{p/2}$. The Schatten class $\mathfrak{S}^p(H,\tilde{H})$ consists of all operators $A \in \mathcal{K}(H,\tilde{H})$ with finite norm $\|A\|_p$. It is convenient to set $\mathfrak{S}^\infty(H,\tilde{H}) = \mathcal{L}(H,\tilde{H})$ with $\|A\|_\infty = \|A\|$.

THEOREM 12.1.

(12.1) $$\|A\|_p^p = \sup \sum_{j=1}^{\infty} \|Ae_j\|_{\tilde{H}}^p,$$

the supremum being over all orthonormal bases $\{e_1, e_2, \ldots\}$ of the space H.

PROOF. If $\{e_1, e_2, \ldots\}$ is a proper basis of the operator A^*A, i.e.,

$$A^*Ae_j = (s_j(A))^2 e_j,$$

then

$$\begin{aligned}\|Ae_j\|_{\tilde{H}}^2 &= (Ae_j, Ae_j)_{\tilde{H}} \\ &= s_j^2.\end{aligned}$$

In this case,

$$\begin{aligned}\sum_{j=1}^{\infty} \|Ae_j\|_{\tilde{H}}^p &= \sum_{j=1}^{\infty} s_j^p \\ &= \|A\|_p^p.\end{aligned}$$

Let us prove that for any other orthonormal basis $\{b_1, b_2, \ldots\}$ of H, we have

$$\sum_{j=1}^{\infty} \|Ab_j\|_{\tilde{H}}^p \leq \sum_{j=1}^{\infty} \|Ae_j\|_{\tilde{H}}^p.$$

Let $b_j = \sum_k c_{jk} e_k$ where (c_{jk}) is a unitary matrix. Then

$$\begin{aligned}(Ab_j, Ab_j)_{\tilde{H}} &= \sum_k \sum_l c_{jk}\overline{c_{jl}}(Ae_k, Ae_l)_{\tilde{H}} \\ &= \sum_k |c_{jk}|^2 s_k^2,\end{aligned}$$

whence

$$\sum_{j=1}^{\infty} \|Ab_j\|_{\tilde{H}}^p = \sum_{j=1}^{\infty} \Big(\sum_{k=1}^{\infty} |c_{jk}|^2 s_k^2\Big)^{p/2}.$$

We think of $S = \{s_1^2, s_2^2, \ldots\}$ as a vector in the space $l^{p/2}$, and of $C = (|c_{jk}|^2)$ as the matrix of a linear transformation in $l^{p/2}$. We aim at evaluating the $l^{p/2}$-norm of the vector CS. Since (c_{jk}) is a unitary matrix, the matrix of squares of absolute values of its entries is such that its elements are non-negative and the sum of the entries of any row or any column is equal to 1. Such matrices are called doubly stochastic. The Kirchhoff Theorem states that each doubly stochastic matrix is actually a convex linear combination of matrices of permutations, i.e.,

$$C = \sum_\pi k_\pi T_\pi$$

where T_π is the matrix of a permutation π, and $k_\pi \geq 0$ satisfy $\sum_\pi k_\pi = 1$. Hence

$$\begin{aligned}
\|CS\|_{l^{p/2}} &= \|\sum_\pi k_\pi T_\pi S\|_{l^{p/2}} \\
&\leq \sum_\pi k_\pi \|T_\pi S\|_{l^{p/2}} \\
&= \sum_\pi k_\pi \|S\|_{l^{p/2}} \\
&= \|S\|_{l^{p/2}},
\end{aligned}$$

as desired. We have used the fact that permutations of coordinates do not change the $l^{p/2}$-norm. □

We next mention several important properties of Schatten classes which are well-known.

LEMMA 12.1. *If $A \in \mathfrak{S}^p(H, \tilde{H})$, then $A^* \in \mathfrak{S}^p(\tilde{H}, H)$ and the Schatten norm of A^* is equal to that of A.*

PROOF. This is obvious, for the non-zero eigenvalues of AA^* and A^*A are easily seen to coincide. □

LEMMA 12.2. *If $A, B \in \mathfrak{S}^p(H, \tilde{H})$, then $A + B \in \mathfrak{S}^p(H, \tilde{H})$.*

PROOF. We prove the lemma only for $0 < p \leq 1$; the proof for $p > 1$ is simpler. By Theorem 12.1, we have

$$\begin{aligned}
\|A+B\|_p^p &= \sup \| \{\|(A+B)e_j\|_{\tilde{H}}\} \|_{l^p}^p \\
&\leq \sup \| \{\|Ae_j\|_{\tilde{H}}\} + \{\|Be_j\|_{\tilde{H}}\} \|_{l^p}^p \\
&\leq \|A\|_p^p + \|B\|_p^p,
\end{aligned}$$

where the supremum is taken over all orthonormal bases $\{e_1, e_2, \ldots\}$ in H. □

LEMMA 12.3. *If $A \in \mathfrak{S}^p(H_1, H_2)$ and $B \in \mathcal{L}(H_2, H_3)$ then $BA \in \mathfrak{S}^p(H_1, H_3)$ and $\|BA\|_p \leq \|B\| \|A\|_p$.*

PROOF. We have

$$\begin{aligned}
\|BA\|_p &= \sup \| \{\|(BA)e_j\|_{H_3}\} \|_{l^p} \\
&\leq \sup \| \{\|B\| \|Ae_j\|_{H_2}\} \|_{l^p} \\
&= \|B\| \|A\|_p,
\end{aligned}$$

which implies the desired statement. □

This property just amounts to saying that $\mathfrak{S}^p(H)$ is a right ideal in $\mathcal{L}(H)$. Furthermore, since

$$\begin{aligned}
\|AB\|_p &= \|(AB)^*\|_p \\
&= \|B^*A^*\|_p \\
&\leq \|B^*\| \|A^*\|_p \\
&= \|B\| \|A\|_p,
\end{aligned}$$

$\mathfrak{S}^p(H)$ is also a left ideal in $\mathcal{L}(H)$.

LEMMA 12.4. *If $A \in \mathfrak{S}^p(H_1, H_2)$ and $B \in \mathfrak{S}^q(H_2, H_3)$ then $BA \in \mathfrak{S}^r(H_1, H_3)$, where $1/r = 1/p + 1/q$, and*
$$\|BA\|_r \leq \|B\|_q \|A\|_p.$$

PROOF. The operators $(BA)^*BA$ and AA^*B^*B have the same non-zero eigenvalues; hence it suffices to estimate
$$\sum_j (s_j(AA^*B^*B))^{r/2} = \sup \sum_j (AA^*B^*Be_j, e_j)_{H_2}^{r/2},$$
the supremum being over all orthonormal bases $\{e_1, e_2, \ldots\}$ in H_2. Since
$$\begin{aligned}|(AA^*B^*Be_j, e_j)_{H_2}| &= |(B^*Be_j, AA^*e_j)_{H_2}| \\ &\leq \|B^*Be_j\| \|AA^*e_j\|\end{aligned}$$
we get
$$\begin{aligned}\|BA\|_r^r &\leq \sum_j \|B^*Be_j\|^{\frac{r}{2}} \|AA^*e_j\|^{\frac{r}{2}} \\ &\leq \left(\sum_j \|B^*Be_j\|^{\frac{r}{2}\frac{p+q}{p}}\right)^{\frac{p}{p+q}} \left(\sum_j \|AA^*e_j\|^{\frac{r}{2}\frac{p+q}{q}}\right)^{\frac{q}{p+q}}\end{aligned}$$
by Hölder's inequality. Substituting
$$r = \frac{pq}{p+q}$$
in this estimate establishes the lemma. □

We may now specify the fibre values of symbols $a(y, \eta) \in \mathcal{S}^m(T^*\mathbb{R}^q, \mathcal{L}(H, \tilde{H}))$ by requiring

(12.2) $$D_y^\alpha D_\eta^\beta a(y, \eta) \in \mathcal{S}^{m-|\beta|}(T^*\mathbb{R}^q, \mathfrak{S}^{\frac{\delta}{-m+|\beta|}}(H, \tilde{H}))$$

for all $\alpha \in \mathbb{Z}_+^q$ and $m \leq |\beta| \leq N$, where $\delta > 0$ is a fixed constant and N is large enough. Under this condition, the parametrix construction of Theorem 8.1 leads to parametrices of elliptic operators up to trace class remainders; cf. [**9**] and [**13**].

If $a(y, \eta)$ takes its values in classical pseudodifferential operators of order $m \leq 0$ on an n-dimensional manifold X which depends on the parameter η, then the estimate (12.2) with $H = \tilde{H} = L^2(X)$ and $\delta = n$ holds for all N. More involved examples arise in the calculus of pseudodifferential operators on manifolds with singularities; cf. [**9**].

13. Analytical index

Let Y be a compact Riemannian manifold, T^*Y the cotangent bundle of Y and let H, $\tilde{H}$ be separable Hilbert spaces equipped with group actions. Recall that a Hilbert bundle G with an infinite-dimensional fibre H and the structure group $\text{GL}(H)$ over a compact manifold Y is trivial; cf. [**4**].

DEFINITION 13.1. *A smooth function $a : T^*Y \to \mathcal{L}(H, \tilde{H})$ with the property that for each coordinate patch $U \subset Y$, the restriction $a|_{T^*U}$ defines an elliptic operator of order m on U is called an elliptic symbol of order m on Y.*

Let us denote by $\mathcal{S}^m_{\mathrm{CV},E}(T^*Y, \mathcal{L}(H,\tilde{H}))$ the space of all elliptic symbols of order m on T^*Y.

Given any symbol $a \in \mathcal{S}^m_{\mathrm{CV},E}(T^*Y, \mathcal{L}(H,\tilde{H}))$, we may define the analytical index of a as follows.

Select a partition of unity $(\varphi_\nu)_{\nu=1,\ldots,N}$ subordinate to some coordinate cover $(U_\nu)_{\nu=1,\ldots,N}$ of the manifold Y. Then select functions $(\psi_\nu)_{\nu=1,\ldots,N}$ in $C^\infty_{\mathrm{comp}}(X)$ such that the support of ψ_ν is contained in U_ν and ψ_ν restricted to the support of φ_ν is equal to 1. Choose trivialisations for T^*U_ν and let $A_\nu = \mathrm{op}\,(a\,|_{T^*U_\nu})$. Now put

$$A = \sum_{\nu=1}^N \varphi_\nu\, A_\nu\, \psi_\nu.$$

The map $a \mapsto A$ induces a well-defined map

(13.1) $$\mathcal{S}^m_{\mathrm{CV},E}(T^*Y, \mathcal{L}(H,\tilde{H})) \to \frac{\Psi^m_{\mathrm{CV},E}(Y\times H, Y\times \tilde{H})}{\Psi^{m-1}_{\mathcal{K}}(Y\times H, Y\times \tilde{H})},$$

the subscript $\mathcal{K}$ indicating that the operators originate from compact operator-valued symbols.

Notice that we may use a similar patching procedure to define a map from $\Psi^m_{\mathrm{CV},E}(Y\times H, Y\times \tilde{H})$ to $\mathcal{S}^m_{\mathrm{CV},E}(T^*Y, \mathcal{L}(H,\tilde{H}))$. The various choices involved give symbols which differ only by compact operator-valued symbols of order $m-1$.

Our next claim is that this correspondence, modulo compact operator-valued symbols of order $m-1$, between elliptic operators and symbols implies that the analytical index is well-defined for both operators and symbols. To see this, it suffices to use the following result.

LEMMA 13.1. *Let $a(y,\eta) \in \mathcal{S}^{m-1}(T^*\mathbb{R}^q, \mathcal{K}(H,\tilde{H}))$ have compact support in y. Then*

$$A_s : H^s(\mathbb{R}^q, H) \to H^{s-m}(\mathbb{R}^q, \tilde{H})$$

is compact for each $s \in \mathbb{R}$.

PROOF. Consider the operator-valued symbol $\langle \eta \rangle^m 1_H$ on $T^*\mathbb{R}^q$, where 1_H stands for the identity operator in H. It is easy to see that this symbol belongs to the symbol space $\mathcal{S}^m(T^*\mathbb{R}^q, \mathcal{L}(H))$, independent of any group action on H.

As it does not depend on y, the corresponding operator $\mathrm{op}\,(\langle \eta \rangle^m 1_H)$ is a topological isomorphism of $H^s(\mathbb{R}^q, H) \overset{\cong}{\to} H^{s-m}(\mathbb{R}^q, H)$. Moreover, the inverse is given by the operator with symbol $\langle \eta \rangle^{-m} 1_H$. Hence it follows that the operator A_s acts through the diagram

(13.2)
$$H^s(\mathbb{R}^q, H) \xrightarrow{\mathrm{op}(\langle\eta\rangle^m 1_H)} H^{s-m}(\mathbb{R}^q, H) \xrightarrow{\mathrm{op}(\langle\eta\rangle^{-m} 1_H)} H^s(\mathbb{R}^q, H) \xrightarrow{\mathrm{op}(a)} H^{s-m}(\mathbb{R}^q, \tilde{H}).$$

Since the composition of the last two operators is the operator with symbol $\langle \eta \rangle^{-m} a(y,\eta)$ in $\mathcal{S}^{-1}(T^*\mathbb{R}^q, \mathcal{K}(H,\tilde{H}))$, Theorem 1.3.61 of [16] shows that this composition is a compact operator $H^{s-m}(\mathbb{R}^q, H) \to H^{s-m}(\mathbb{R}^q, \tilde{H})$. It follows that A_s is compact as an operator $H^s(\mathbb{R}^q, H) \to H^{s-m}(\mathbb{R}^q, \tilde{H})$, which completes the proof. □

14. Topological index

Following Atiyah [4], we have a map

(14.1) $$\text{ind} : \mathcal{S}^m_{\text{CV},E}(T^*Y, \mathcal{L}(H, \tilde{H})) \to K_{\text{comp}}(T^*Y).$$

It may be defined as follows.

Given any $a \in \mathcal{S}^m_{\text{CV},E}(T^*Y, \mathcal{L}(H, \tilde{H}))$, there is a subspace C of H, closed and of finite codimension, such that

1) for each $(y, \eta) \in T^*Y$, the restriction $a(y, \eta)|_C$ is injective; and
2) the image of C under $a(y, \eta)$ is of finite codimension in $\tilde{H}$, and the quotient

$$\frac{\tilde{H}}{a(y, \eta)\, C}$$

is a locally trivial vector bundle over T^*Y.

Then the element $\text{ind}\, a \in K_{\text{comp}}(T^*Y)$ is given by the virtual bundle

$$\left(\frac{H}{C}, \frac{\tilde{H}}{a\, C}\right)$$

and depends only on the homotopy class of the map $a : T^*Y \to \mathcal{F}(H, \tilde{H})$. Here, $\mathcal{F}(H, \tilde{H})$ denotes the space of all Fredholm operators from H to $\tilde{H}$. In this way, we obtain what is known as the index bundle of the Fredholm family $a(y, \eta)$; cf. [8, 4.1].

In particular, if $\ker a(y, \eta)$ and $\operatorname{coker} a(y, \eta)$ are locally trivial vector bundles themselves, then the index bundle $\text{ind}\, a$ actually coincides with the virtual bundle $(\ker a, \operatorname{coker} a)$.

DEFINITION 14.1. The topological index of $a \in \mathcal{S}^m_{\text{CV},E}(T^*Y, \mathcal{L}(H, \tilde{H}))$ is the topological index of $\text{ind}\, a$, i.e.,

$$\int_{T^*(Y)} \text{ch}(\text{ind}\, a)\, \mathcal{T}(T^*Y);$$

cf. [1].

The correspondence (13.1) implies that the topological index is also well-defined for elliptic operators.

15. The index theorem

The next step should be to prove that the analytical and topological indices are equal.

THEOREM 15.1 (Conjectured Theorem). *Let $a \in \mathcal{S}^m_{\text{CV},E}(T^*Y, \mathcal{L}(H, \tilde{H}))$. Then the topological and analytical indices of a coincide.*

IDEAS TOWARDS A PROOF. The proof of [11] for $m = 0$ and trivial group actions in both H and $\tilde{H}$ consists of performing certain deformations of a that preserve both the topological and analytical indices. It deforms a to an ordinary elliptic symbol $\tilde{a} \in \mathcal{S}^m(T^*Y, \text{Hom}(V, \tilde{V}))$, where V and $\tilde{V}$ are vector bundles of finite rank over Y such that the analytical and topological indices of a and $\tilde{a}$ are equal. It would be of great interest to perform such a deformation in the general case, but we have not been able to do this. The non-trivial group actions included make the problem much more difficult. In [9], we prove the index theorem for the edge algebra presented in Example 2.1. The proof is based on an analytical approach suggested in [6] and algebraic machinery developed in [7]. The analytical part of the paper consists in the theorem on the regularised trace of a product. However, we have not succeeded in deriving the theorem on the regularised trace of the product in the general context of (12.2). Hence Theorem 15.1 remains a conjecture in the general case, which we hope to prove in future. □

References

1. M. F. Atiyah and I. M. Singer, *The index of elliptic operators. I*, Ann. of Math. (2) **87** (1968), 484–530.
2. M. F. Atiyah and I. M. Singer, *The index of elliptic operators. III*, Ann. of Math. (2) **88** (1968), 127–182.
3. M. F. Atiyah and I. M. Singer, *The index of elliptic operators. IV*, Ann. of Math. (2) **93** (1971), 119–138.
4. M. F. Atiyah, *K-Theory*, Benjamin, New York, 1967.
5. J. W. Calkin, *Two-sided ideals and congruences in the ring of bounded operators in Hilbert space*, Ann. of Math. (2) **42** (1941), 839–873.
6. B. V. Fedosov, *Analytic formulas for the index of elliptic operators*, Trans Moscow Math. Soc. **30** (1974), 159–241.
7. B. V. Fedosov, *A periodicity theorem in the algebra of symbols*, Mat. Sb. **105** (1978), 622–637.
8. B. V. Fedosov, *Deformation Quantization and Index Theory*, Akademie-Verlag, Berlin, 1995.
9. B. Fedosov, B.-W. Schulze, and N. Tarkhanov, *On the index of elliptic operators on a wedge*, J. Funct. Anal. **157** (1998), 164–209.
10. L. Hörmander, *The Analysis of Linear Partial Differential Operators. Vol. 3: Pseudodifferential Operators*, Springer-Verlag, Berlin et al., 1985.
11. G. Luke, *Pseudodifferential operators on Hilbert bundles*, J. Differential Equations **12** (1972), 566–589.
12. St. Rempel and B.-W. Schulze, *Asymptotics for Elliptic Mixed Boundary Problems*, Math. Research 50, Akademie-Verlag, Berlin, 1989.
13. G. Rozenblum, *On Some Analytical Index Formulas Related to Operator-Valued Symbols*, Preprint 2000/16, Univ. of Potsdam, Potsdam, 2000.
14. R. Schatten, *Norm Ideals of Completely Continuous Operators*, Springer-Verlag, Berlin et al., 1960.
15. B.-W. Schulze, *Pseudo-Differential Operators on Manifolds with Singularities*, North-Holland, Amsterdam, 1991.
16. B.-W. Schulze, *Boundary Value Problems and Singular Pseudo-Differential Operators*, J. Wiley, Chichester, 1998.
17. B.-W. Schulze and N. Tarkhanov, *Wedge Sobolev Spaces*, Preprint MPI/95-122, Max-Planck-Inst. für Math., Bonn, 1995, 67 pp.

UNIVERSITÄT POTSDAM, INSTITUT FÜR MATHEMATIK, POSTFACH 60 15 53, 14415 POTSDAM, GERMANY
E-mail address: schulze@math.uni-potsdam.de

UNIVERSITÄT POTSDAM, INSTITUT FÜR MATHEMATIK, POSTFACH 60 15 53, 14415 POTSDAM, GERMANY
E-mail address: tarkhanov@math.uni-potsdam.de

Pluripolar Sets and Pseudocontinuation

Józef Siciak

Dedicated to Larry Zalcman on the occasion of his 60th birthday

ABSTRACT. In their recent book [6], the authors study various types of *generalized analytic continuations* (GAC) of meromorphic functions of a complex variable in situations where the classical theory says there is a *natural boundary*. An important representative of GAC is *pseudocontinuation*. In this paper, we study a new type of GAC (not discussed in [6]), expressed in terms of a pluripolar hull of the graph of a meromorphic function. In particular, we give an example of a function $f \in A^\infty(\mathbb{D})$ such that f does not have a pseudocontinuation across any subset E of $\partial \mathbb{D}$ of positive measure, while there is a meromorphic function F in $\mathbb{D}_e := \{|z| > 1\}$ such that for every function U plurisubharmonic on $\mathbb{C}^2$: if $U(z, f(z)) = -\infty$ on $\mathbb{D}$ then $U(z, f(z)) = -\infty$ on the unit circle $\partial \mathbb{D}$, and $U(z, F(z)) = -\infty$ for all $z \in \mathbb{D}_e$ with $F(z) \neq \infty$.

1. Introduction

Let E be a pluripolar set in $\mathbb{C}^N$. The *pluripolar hull* E^* of E is defined by the formula

(1) $$E^* := \cap \{z \in \mathbb{C}^N : U(z) = -\infty\},$$

the intersection being taken over all plurisubharmonic functions U on $\mathbb{C}^N$ with $U(z) = -\infty$ on E.

A subset E of $\mathbb{C}^N$ is called *pluripolar* (respectively, *complete pluripolar*) if there exists a function U plurisubharmonic on $\mathbb{C}^N$ such that $U \not\equiv -\infty$ and $U(z) = -\infty$ on E (respectively, $E = \{U(z) = -\infty\}$).

By Zeriahi [10] a pluripolar set E which is simultaneously of type F_σ and G_δ is complete pluripolar if and only if $E^* = E$.

Put $\mathbb{D} := \{z \in \mathbb{C}; |z| < 1\}$, $\mathbb{D}_e := \{z \in \hat{\mathbb{C}}; |z| > 1\}$. Let E be a subset of the unit circle $\partial \mathbb{D}$ of positive linear Lebesgue measure. Let $\Omega_1 \subset \mathbb{D}$, $\Omega_2 \subset \mathbb{D}_e$ be domains such that there exists a nonempty open set I_o of the unit circle such that $E \subset I_o$ and $\Omega := \Omega_1 \cup I_o \cup \Omega_2$ is a domain.

2000 *Mathematics Subject Classification.* Primary 30B40, 32D15, 32F05.

Key words and phrases. plurisubharmonic function, pluripolar set, pluripolar hull, pseudocontinuation, generalized analytic continuation.

Supported by KBN Grant No. 2 PO3A 047 22.

We say that a function $F \in \mathcal{M}(\Omega_2)$ meromorphic in Ω_2 is a *pseudocontinuation* of a function $f \in \mathcal{M}(\Omega_1)$ across E, if the angular limits of f and F exist and are equal for all points of E.

The main result of this paper is given by the following two theorems.

THEOREM 1. *Let a function F meromorphic in a domain $\Omega_2 \subset \mathbb{D}_e$ be a pseudocontinuation of a function f meromorphic in a domain $\Omega_1 \subset \mathbb{D}$ across a subset E of $\partial \mathbb{D}$ of positive measure.*

Then either the graph $G_f(E) := \{(z, f(z)) : z \in E\}$ of f over E is not pluripolar[1], or

$$G_f^*(\Omega_1 \cup E) = G_F^*(\Omega_2 \cup E),$$

i.e., $G_F(\Omega_2 \cup E) := \{(z, F(z)) : z \in \Omega_2 \cup E, F(z) \neq \infty\} \subset G_f^*(\Omega_1 \cup E)$ and $G_f(\Omega_1 \cup E) \subset G_F^*(\Omega_2 \cup E)$.

THEOREM 2. *Let $A := \{a_n\}$ be a countable subset of $\mathbb{D}_e$ such that A has no limit point in $\mathbb{D}_e$ and each point of $\partial \mathbb{D}$ is a limit point of A. One can choose non-zero complex constants c_n so that the series $\sum_1^\infty |c_n|$ is convergent, and the function*

(*) $$f(z) := \sum_1^\infty c_n \frac{1 - \bar{a}_n z}{z - a_n}, \quad z \in \hat{\mathbb{C}} \setminus A,$$

respectively,

(**) $$f(z) := \sum_1^\infty \frac{c_n}{z - a_n}, \quad z \in \hat{\mathbb{C}} \setminus A,$$

has the following properties.

   (a) *f is holomorphic in $\hat{\mathbb{C}} \setminus \bar{A}$, meromorphic on $\mathbb{D}_e$, and $f_{|\bar{\mathbb{D}}} \in \mathcal{C}^\infty(\bar{\mathbb{D}})$.*
   (b) *$G_f(\mathbb{C} \setminus A) \subset G_f^*(\mathbb{D}) \cap G_f^*(\mathbb{D}_e \setminus A) \cap G_f^*(E)$ (i.e., $G_f^*(\mathbb{D}) = G_f^*(\mathbb{D}_e \setminus A) = G_f^*(E)$) for every subset E of $\partial \mathbb{D}$ of positive measure.*
   (c) *If $\sum_{n=1}^\infty (|a_n| - 1) < \infty$, then there exists a subset E of $\partial \mathbb{D}$ such that $m(E) = 2\pi$, and $f_{|\mathbb{D}}$ and $f_{|\mathbb{D}_e}$ are pseudocontinuations of each other across E but $f|\mathbb{D}$ has no analytic continuation across any boundary point of $\mathbb{D}$.*
   (d) *If each point ζ of an open set $I_o \subset \partial \mathbb{D}$ is an angular accumulation point of A, then the function f does not have a pseudocontinuation across any subset E of I_o of positive measure.*

REMARKS. 1°. Given $c > 2$, put $r_n = 1 + \frac{c}{n}$, $A_n = \{z \in \mathbb{C} : z^n = r_n^n\}$, $A = \cup_1^\infty A_n$. Then there exists $\theta \in (0, \pi)$ such that each point $\zeta \in \partial \mathbb{D}$ is an accumulation point of the set $A \cap S(\zeta, \theta)$, where $S(\zeta, \theta) := \{z \in \mathbb{C}; |Arg(z - \zeta)| < \theta\}$, i.e., each point ζ of $\partial \mathbb{D}$ is a non-tangential accumulation point of A.

2°. Let f be a holomorphic function on $\mathbb{D}$ which does not have analytic continuation across any boundary point of the unit disc. The authors of [4] asked whether $G_f(\mathbb{D})$ is complete pluripolar. The authors of [2] found a function f given by (**) with suitably chosen A and $\{c_n\}$ such that f satisfies the conditions (a), f does not have analytic continuation across any point of the unit circle but $G_f(\mathbb{D})$ is not complete pluripolar. So they have proved that the answer to the question is no.

---

[1]Here $f(z)$ denotes the angular limit value of f at $z \in E$.

By (a), (b), (d) of Theorem 2, there is a function $f \in A^\infty(\mathbb{D})$ that has no pseudocontinuation across any set $E \subset \partial \mathbb{D}$ with $m(E) > 0$ (in particular, f has no analytic continuation across any point of the unit circle), while the pluripolar hull of the graph of f over $\mathbb{D}$ contains $G_F(\mathbb{D}_e \setminus F^{-1}(\infty))$, where F is a meromorphic function in the exterior unit disc.

3°. Both the claim and proof of (d) are essentially contained in [**1**]. To make our article self-contained, we give a proof of (d) which is an adaptation of the proof of Theorem 6.2 in [**1**].

4° The authors of [**4**] have proved that if $f(z) := \sum_1^\infty c_{n_k} z^{n_k}$, $|z| \leq 1$, where $\lim_{k \to \infty} n_k/n_{k+1} = 0$, and the coefficients satisfy suitable conditions, then the graph of f over $\bar{\mathbb{D}}$ is complete pluripolar. By Theorem 1, such a function has no pseudocontinuation across any subset of $\partial \mathbb{D}$ of positive measure. For instance, the function $f(z) := \sum_1^\infty 2^{\frac{-2k^2}{k}} z^{2^{k^2}}$ has the required property.

5°. Let $A = \{a_n\}$ be a countable subset of $\mathbb{D} \setminus \{0\}$ such that $\sum_1^\infty (1 - |a_n|) < \infty$, and let c_n be complex numbers such that $\sum_1^\infty |c_n| < \infty$. Consider the *Borel series*

$$f(z) := \sum_n c_n \frac{1}{z - a_n}, \quad z \in \mathbb{D} \cup \mathbb{D}_e \setminus A$$

and the *Blaschke product*

$$b(z) := \prod_n \frac{|a_n|}{a_n} \frac{z - a_n}{\bar{a}_n z - 1}, \quad z \in \mathbb{D} \cup \mathbb{D}_e \setminus \{1/\bar{a}_1, 1/\bar{a}_2, \cdots\}.$$

By Corollary 4.2.18 and Proposition 4.2.19 in [**6**], there exists a subset E of the unit circle such that $m(E) = 2\pi$ ($m(E)$ being the Lebesgue measure of E on the unit circle) and $f_{|\mathbb{D}_e}$ (resp., $b_{|\mathbb{D}_e}$) is a pseudocontinuation across E of $f_{|\mathbb{D}}$ (resp., $b_{|\mathbb{D}}$). If there exists a subset E_o of E of positive measure such that $G_f(E_o)$ (resp., $G_b(E_o)$) is pluripolar, then by Theorem 1

$$G_f\left((\mathbb{D} \setminus A) \cup E_0 \cup (\mathbb{D}_e \setminus \{\infty\})\right) \subset G_f^*((\mathbb{D} \setminus A) \cup E_0) = G_f^*(\mathbb{D}_e \setminus \{\infty\}) = G_f^*(E_0),$$

respectively,

$$G_b\left(\mathbb{D} \cup E_0 \cup (\mathbb{D}_e \setminus \{\infty, 1/\bar{a}_1, 1/\bar{a}_2, \cdots\})\right) \subset G_b^*(\mathbb{D} \cup E_0)$$
$$= G_b^*((\mathbb{D}_e \setminus \{\infty, 1/\bar{a}_1, 1/\bar{a}_2, \cdots\}) \cup E_0) = G_b^*(E_0).$$

6°. When discussing pluripolar continuations of graphs of meromorphic functions, it is worth keeping in mind the following recent result due to Shcherbina [**9**]:

Let f be a complex-valued continuous function on a domain Ω in $\mathbb{C}^N$. Then f is holomorphic if and only if its graph $G_f(\Omega)$ is a pluripolar subset of $\mathbb{C}^{N+1}$.

2. Lusin-Privalov Lemma

Given α with $0 < \alpha < \pi/4$, let $T(\alpha)$ be the closed isosceles "triangle" with the arc base $J := \{e^{it} : -\alpha \leq t \leq \alpha\}$, and linear sides of equal length $[e^{-i\alpha}, R]$, $[e^{i\alpha}, R]$, where $R = R(\alpha) > 1$ is the unique real number such that $Arg[e^{i\alpha}(R - e^{-i\alpha})] = \pi/4$, $Arg[e^{-i\alpha}(R - e^{i\alpha})] = -\pi/4$. One can check that there exists a positive constant C such that

$$|R - e^{i\alpha}| = |R - e^{-i\alpha}| \leq C\alpha, \quad 0 < \alpha < \pi/4.$$

Given a real number $r > 1$, the set
$$\Delta(r) := \left\{ z : 1 < |z| < r,\ -\frac{\pi}{4} < Arg(z-1) < \frac{\pi}{4} \right\}$$
is an open isosceles "triangle" with arc base $J := \{z : |z| = r,\ Arg(z-1) < \frac{\pi}{4}\}$, and the two linear sides of equal length which meet at a right angle at the vertex 1.

Given $r > 1$ and an open non-empty arc I of the unit circle, put
$$W(I, r) := \{t\zeta : \zeta \in I,\ 1 < t < r\} \equiv \{z : 1 < |z| < r,\ z/|z| \in I\}.$$

LUSIN-PRIVALOV LEMMA. *Let F be a function meromorphic in a domain $W = W(I, r)$ such that the finite angular limit of F exists at each point of a compact subset E of I of positive measure.*

Then there exist a number $r_o > 0$ and a relatively compact subset E_o of E such that

$1°$ $1 < r_o < r$ *and E_o has positive measure;*

$2°$ $E_o = J \cap E$ *where $J := (e^{i\theta_1}, e^{i\theta_2})$ is an open sub-arc of I with the endpoints belonging to E;*

$3°$ *the set*
$$G := W(J, r_0) \cap \left(\bigcup_{\zeta \in E_o} \zeta \Delta(r_o) \right)$$
is a bounded simply connected domain whose boundary is a Jordan rectifiable curve containing the set E_o;

$4°$ *F is bounded on G and has finite angular limit value (denoted $F(\zeta)$) at each point $\zeta \in E_o$.*

PROOF. Choose r' with $1 < r' < r$ so close to 1 that for every $\zeta \in E$, one has $\zeta\Delta(r') \subset W$. Observe that the function
$$\varphi_n(\zeta) := \sup\left\{ |F(\zeta z)| : z \in \Delta\left(1 + \frac{r'-1}{n}\right) \right\},\quad \zeta \in E,$$
is lower semicontinuous (since for every $z \in \Delta(1 + \frac{r'-1}{n})$, the function $E \ni \zeta \mapsto |F(\zeta z)| \in [0, +\infty]$ is continuous). Moreover, $|F(\zeta)| = \lim_{n\to\infty} \varphi_n(\zeta)$ for all $\zeta \in E$, where $F(\zeta)$ is the angular limit value of F at ζ. The set
$$E_n := \{\zeta \in E : \varphi_n(\zeta) \leq n\}$$
is measurable, $E_n \subset E_{n+1}$, and $\cup E_n = E$. Hence there exists $k \geq 1$ so large that the measure of E_k is positive, and
$$|F(z)| \leq k \quad \text{for all} \quad z \in \bigcup_{\zeta \in E_k} \zeta\Delta\left(1 + \frac{r'-1}{k}\right).$$

Without loss of generality, we may assume that E_k is compact. Choose two points $e^{i\theta_1}$, $e^{i\theta_2}$ in E_k such that the triangles $e^{i\theta_1}\Delta(r_o)$, $e^{i\theta_2}\Delta(r_o)$ with $r_o := 1 + \frac{r'-1}{k}$ have non-empty intersection, and the set
$$E_0 := J \cap E_k$$

with $J =: \{e^{it} : \theta_1 < t < \theta_2\}$ has positive measure. Write $J \setminus E_0$ as the disjoint union of open arcs $J_n = \{e^{it} : \alpha_n < t < \beta_n\}$. Put

$$G := W(J, r_o) \setminus \bigcup_n T_n \equiv W(J, r_o) \cap \left(\bigcup_{\zeta \in E_o} \zeta \Delta(r_o) \right),$$

where

$$T_n := e^{i\frac{\alpha_n + \beta_n}{2}} T\left(\frac{\beta_n - \alpha_n}{2} \right).$$

The set G is a simply connected bounded domain whose boundary is a rectifiable Jordan curve Γ. The curve Γ is a union of the arc $\{r_o e^{it} : \theta_1 \leq t \leq \theta_2\}$, two intervals $[e^{i\theta_j}, r_o e^{i\theta_j}]$ $(j = 1, 2)$, the set E_o, and the linear sides of the triangles T_n $(n \geq 1)$. Therefore, the length of Γ is estimated by

$$r_o(\theta_2 - \theta_1) + 2(r_o - 1) + m(E_o) + C(\theta_2 - \theta_1) < \infty.$$

It is clear that $G \subset \bigcup_{\zeta \in E_k} \zeta \Delta(r_o)$. Therefore, $|F(z)| \leq k$ for all $z \in G$. □

REMARK. The construction of the domain G is due to Lusin and Privalov (see the proof of their identity theorem in [5]).

3. Proof of Theorem 1

Let I_0 be an open subset of $\partial \mathbb{D}$ such that $E \subset I_0$, $\Omega_1 \cup I_0 \cup \Omega_2$ is connected, and $f(\zeta) = F(\zeta)$ for all $\zeta \in E$, where $f(\zeta)$ and $F(\zeta)$ are angular limit values at ζ of f and F, respectively.

Suppose that $G_f(E)$ is pluripolar. Let U be a plurisubharmonic function on $\mathbb{C}^2$ such that

$$U(z, f(z)) = -\infty, \quad \text{for all} \quad z \in \Omega_1 \cup E.$$

We need to show that $U(z, F(z)) = -\infty$ for all $z \in \Omega_2 \cup E$.

Choose $r > 0$ and a non-empty open arc $I \subset I_0$ such that $m(E \cap I) > 0$ and $W(I, r) \subset \Omega_2$. Without loss of generality, we may assume that $E \cap I$ is compact. For $F|_W$ and $E \cap I$, let $G \subset W$ and $E_0 \subset E \cap I$ be the sets given by the Lusin-Privalov Lemma. Then the function $v(z) := U(z, F(z))$ is subharmonic and bounded in G. Moreover, the angular limit of v exists and is equal $-\infty$ at all points of E_o. Therefore, by the two constant theorem, $v(z) = -\infty$ on $G \cup E_o$, and consequently on $\Omega \cup E$ (observe that U is continuous at each point $(\zeta, f(\zeta))$ and $f(\zeta) = F(\zeta)$ for all $\zeta \in E$). We have proved that $G_F(\Omega_2 \cup E) \subset G_f^*(\Omega_1 \cup E)$.

The inclusion $G_f(\Omega_1 \cup E) \subset G_F^*(\Omega_2 \cup E)$ can be proved by using the conformal mapping $z \to \frac{1}{z}$.

We already know that $G_F^*(E) = G_f^*(E) \subset G_f^*(\Omega_1 \cup E) = G_F^*(\Omega_2 \cup E)$. It remains to show that $G_F^*(\Omega_2 \cup E) \subset G_F^*(E)$. Fix $(a, b) \in G_F^*(\Omega_2 \cup E)$. Given $U \in PSH(\mathbb{C}^2)$ with $U = -\infty$ on $G_F(E)$, it follows from the Lusin-Privalov Lemma that $U(z, F(z)) = -\infty$ on Ω_2. Hence $U(a, b) = -\infty$, i.e., $(a, b) \in G_F^*(E)$.

4. Proof of Theorem 2

Following the proof of Theorem 6.2 in [1], for $a \in \mathbb{D}_e$ and $0 < r < 1$, let

(1) $$H(a, r) := \left\{ z \in \mathbb{D}_e : \left| \frac{z - a}{1 - \bar{a}z} \right| < r \right\}$$

be a pseudohyperbolic disk around a. Observe that

$$h(z) \equiv h(z; a, r) := \frac{\log\left|\frac{1-\bar{a}z}{z-a}\right|}{\log\frac{1}{r}} \tag{2}$$

is the harmonic measure of $\partial H(a,r)$ with respect to the domain $\mathbb{D}_e \setminus \overline{H(a,r)}$, i.e., $h(z) = 0$ on $\partial \mathbb{D}_e$ and 1 on $\partial H(a,r)$.

CLAIM 1. *Let J be a union of two closed Jordan arcs J_1, J_2 such that $J_1 \subset \mathbb{D}$ and $J_2 \subset \{1 < |z| < R\} \setminus A$, where R is so large that $|a_n| < R$, $n \geq 1$. We claim that there exists a decreasing sequence of positive numbers $\{\rho_n\}$ such that the following conditions hold.*

    $1°$ *The closures of the discs $H(a_n, \rho_n)$ are disjoint, all of them are contained in $\{1 < |z| < R\} \setminus J_2$, and $H(a_n, \rho_n) \subset \{|z - a_n| < \frac{1}{3}(|a_n| - 1)\}$, $n \geq 1$.*

    $2°$ *For every n, the open set $G_n := \mathbb{D}(0, R) \setminus \bigcup\limits_{j}^{n} \overline{H(a_j, \rho_j)}$ is connected.*

    $3°$ *There exists a positive constant c such that for every $n \geq 1$,*

$$\omega(z, J_\ell, G_n) \geq c, \quad z \in (J \setminus J_\ell) \cup \partial \mathbb{D}, \quad \ell = 1, 2,$$

where $\omega(\cdot) = \omega(\cdot, J_\ell, G_n)$ denotes the harmonic measure of J_ℓ in G_n (i.e., ω is the solution of the Dirichlet problem for $G_n \setminus J_\ell$ with boundary values 0 on ∂G_n and 1 on J_ℓ).

PROOF OF CLAIM 1. Put

$$2c := \min\{\omega(z, J_\ell, \mathbb{D}(0, R)) : z \in (J \setminus J_\ell) \cup \partial \mathbb{D}, \quad \ell = 1, 2\}.$$

It is clear that $c > 0$. Now take $\rho_1 > 0$ so small that $1°$ (resp., $2°$) is satisfied for $n = 1$ and

$$\omega(z, J_\ell, G_1) > 2c - 2^{-1}c, \quad z \in (J \setminus J_\ell) \cup \partial \mathbb{D}, \quad \ell = 1, 2.$$

Assuming that the numbers $\rho_1, \ldots, \rho_m$ ($m \geq 1$) have been already chosen in such a way that

    (i)   the condition $1°$ (resp., $2°$) is satisfied for $n = 1, \ldots, m$;
    (ii) $G_n \supset A \setminus \{a_1, \ldots, a_n\}$ for $n = 1, \ldots, m$;
    (iii) $\omega(z, J_\ell, G_n) > 2c - c(2^{-1} + 2^{-2} + \cdots + 2^{-n})$, $z \in (J \setminus J_\ell) \cup \partial \mathbb{D}$, $n = 1, \cdots, m$, $\ell = 1, 2$,

we can find $\rho_{m+1} > 0$ such that (i), (ii) and (iii) are satisfied with m replaced by $m + 1$. The induction argument implies that the claim is true. $\square$

For each $n \geq 1$, choose real numbers ρ_n' and r_n such that $0 < r_n \leq \rho_n' < \rho_n$, $\mathbb{D}(a_n, \rho_n') \subset H(a_n, \rho_n)$, and

$$\sum_{1}^{\infty} \frac{\log \frac{1}{\rho_n}}{\log \frac{1}{r_n}} < \infty. \tag{3}$$

Let c_n be non-zero complex numbers such that

$$|c_n| \left(\frac{1}{r_n} + \|\varphi_n\|_\mathbb{D} + \|\varphi_n'\|_\mathbb{D} + \cdots + \|\varphi_n^{(n)}\|_\mathbb{D}\right) \leq \frac{1}{n^2}, \quad n \geq 1, \tag{***}$$

where $\varphi_n(z) := \frac{1-\bar{a}_n z}{z-a_n}$ (resp., $\varphi_n(z) = \frac{1}{z-a_n}$).

We claim that the function f given by (*) (resp., (**)) with c_n satisfying (***) has the required properties (a), (b), (c), (d).

PROOF OF (a). By (***), the series $\sum_n c_n$ is absolutely convergent. Therefore $f \in \mathcal{O}(\hat{\mathbb{C}} \setminus \bar{A}) \cap \mathcal{M}(\mathbb{D}_e)$.

By (***),
$$|c_n| \|\varphi_n^{(k)}\|_{\mathbb{D}} \leq \frac{1}{n^2}, \quad n \geq k.$$

Therefore, $f_{\bar{\mathbb{D}}} \in \mathcal{C}^\infty(\bar{\mathbb{D}})$. □

PROOF OF (b). First we show that if U is a plurisubharmonic function on $\mathbb{C}^2$ such that $U(z, f(z)) = -\infty$ on J_1 then $U(z, f(z)) = -\infty$ for all $z \in \mathbb{C} \setminus A$. To this end, fix a point $a \in \partial \mathbb{D} \cup J_2$ and put $f_n(z) := \sum_1^n c_j \frac{1-\bar{a}_j z}{z - a_j} + \sum_{n+1}^\infty c_j \frac{1-\bar{a}_j a}{a - a_j}$ (resp., $f_n(z) := \sum_1^n \frac{1}{z-a_j} + \sum_{n+1}^\infty \frac{1}{a-a_j}$). The function f_n is holomorphic on $\hat{\mathbb{C}} \setminus \{a_1, \cdots, a_n\}$ and $f_n(a) = f(a)$. Moreover, by (***)
$$\sup\{|f_n(z)| : z \in \partial G_n\} \leq \sum_1^\infty \frac{|c_j|}{\rho'_j} =: C_1 < \infty,$$
where C_1 does not depend on n. Put
$$C_2 := \sup\{U(z, w) : z \in \partial G_n, \ |w| \leq C_1\}.$$
Then
$$\sup\{U(z, f_n(z)) : z \in \partial G_n\} \leq C_2, \quad n \geq 1,$$
where C_2 is a positive constant. Put
$$M_n := \sup\{U(z, f_n(z)) : z \in J_1\}, \quad n \geq 1.$$
It is clear that $\lim_{n \to \infty} M_n = -\infty$. The function $U(z, f_n(z))$ is subharmonic on $\mathbb{C} \setminus \{a_1, \cdots, a_n\}$. By the two constant theorem and by 3° of Claim 1,
$$U(z, f_n(z)) \leq C_2 + (M_n - C_2)\omega(z, J_1, G_n) \leq C_2 + (M_n - C_2)c$$
for all sufficiently large n (say $n \geq n_o$, so that $M_n - C_2 < 0$ for $n > n_o$) and for all $z \in J_2 \cup \partial \mathbb{D}$. In particular,
$$U(a, f(a)) = U(a, f_n(a)) \leq C_2 + (M_n - C_2)c, \quad n > n_o.$$
It follows that $U(a, f(a)) = -\infty$ which (since a arbitrary) implies that $U(z, f(z)) = -\infty$ on $J_2 \cup \partial \mathbb{D}$. The function $U(z, f(z))$ is subharmonic on the domain $\mathbb{D}_e \setminus A$ and takes the value $-\infty$ at each point of the non-polar subset J_2. It is also clear that the function $U(z, f(z))$ is subharmonic on $\mathbb{D}$ and takes the value $-\infty$ at each point of the non-polar subset J_1. Hence $U(z, f(z)) = -\infty$ on $\mathbb{C} \setminus A$.

Analogously, we can show that if U is a plurisubharmonic function on $\mathbb{C}^2$ such that $U(z, f(z)) = -\infty$ on J_2, then $U(z, f(z)) = -\infty$ on $\mathbb{C} \setminus A$.

Now let the subset E of the unit circle have positive measure, and let U be a plurisubharmonic function on $\mathbb{C}^2$ such that $U(z, f(z)) = -\infty$, $z \in E$. Without loss of generality, we may assume that E is compact. Fix a point $a \in \mathbb{D}$. Then $U(z, f_n(z))$ is subharmonic on a neighborhood of $\bar{\mathbb{D}}$ and $\sup\{U(z, f_n(z)); |z| \leq 1\} =: C_2 < \infty, n \geq 1$. It is clear that $M_n := \sup\{U(z, f_n(z)); z \in E\} \to -\infty$ as $n \to \infty$. Hence
$$U(a, f(a)) = U(a, f_n(a)) \leq \frac{1}{2\pi} \frac{1+|a|}{1-|a|}(C_2 + M_n m(E)), \quad n \geq 1.$$

Therefore, $U(a, f(a)) = -\infty$ for all $a \in \mathbb{D}$; as we already know, this implies that $U(z, f(z)) = -\infty$ on $\mathbb{C} \setminus A$. □

PROOF OF (c). This condition is a direct consequence of Corollary 4.2.18 and Proposition 4.2.19 of [6] (see also Remark 5°). □

PROOF OF (d). Put

$$(4) \quad h(z) := \sum_n \frac{\log\left|\frac{1-\bar{a}_n z}{z-a_n}\right|}{\log \frac{1}{r_n}}, \quad z \in \hat{\mathbb{C}} \setminus (\{a_1, a_2, \cdots\} \cup \{1/\bar{a}_1, 1/\bar{a}_2, \cdots\}).$$

CLAIM 2. *The function h has the following properties:*

1° h *is harmonic on*

$$(\mathbb{D} \setminus \{1/\bar{a}_1, 1/\bar{a}_2, \cdots\}) \cup (\mathbb{D}_e \setminus \{a_1, a_2, \cdots\});$$

2° $h(z) > 0$ *on* $\mathbb{D}_e \setminus A$, $h(z) = 0$ *on* $\{|z| = 1\}$, $h(z) = -h(\frac{1}{\bar{z}})$ *for* $z \in \mathbb{D}$;

3° h *is continuous on the compact subset* $\bar{H}$ *of* $\hat{\mathbb{C}} \setminus \mathbb{D}$, *where*

$$H := \left\{ z \in \mathbb{D}_e : \left|\frac{z - a_n}{1 - \bar{a}_n z}\right| > \rho_n, \quad n \geq 1 \right\}$$

is a connected open subset of $\mathbb{D}_e$ *such that* $\partial \mathbb{D} \subset \partial H$.

By (3), the series (4) is uniformly convergent on $\bar{H}$ and $h(z) = 0$ on $\partial \mathbb{D}$. Therefore h is continuous on $\bar{H}$, and by Harnack's theorem h is a positive harmonic function on $\mathbb{D}_e \setminus A$. It is clear that $h(z) = -h(\frac{1}{\bar{z}})$ for $z \in \mathbb{D} \setminus \{1/\bar{a}_1, 1/\bar{a}_2, \cdots\}$. Hence, h is harmonic and negative on $\mathbb{D} \setminus \{1/\bar{a}_1, 1/\bar{a}_2, \cdots\}$.

Let us now assume that for every point $\zeta \in I_o$, there is a subsequence of A which converges to ζ non-tangentially. Suppose, contrary to (d), that there exists a function F meromorphic in a domain $\Omega \subset \mathbb{D}_e$ such that F is a pseudocontinuation of f across a subset E of I_0 of positive measure.

Without loss of generality, we may assume that E is a compact subset of an arc $I := \{e^{it} : \theta_1 < t < \theta_2\} \subset I_o$ such that the angular limits of f and F exist and are equal at each point ζ of E. We can also assume that

$$W \equiv W(I, r) := \{t\zeta : 1 < t < r, \zeta \in I\} \subset \Omega,$$

where r is sufficiently close to 1.

For the meromorphic function $(F - f)_{|W}$ and E, let $G \subset W$ and $E_0 \subset E$ denote a pair of sets given by the Lusin-Privalov Lemma. Put

$$\mathcal{R} := G \setminus \bigcup_1^\infty \overline{H(a_j, r_j)}.$$

One can show that $\mathcal{R}$ is an open connected set such that $E_o = \partial \mathbb{D} \cap \partial \mathcal{R}$. By (***), the function f given by (*) (resp., by (**)) is continuous on $\bar{\mathcal{R}}$.

Put

$$X := \{n : \partial G \cap H(a_n, r_n) \neq \emptyset\}, \quad Y := \{n : H(a_n, r_n) \subset G\}.$$

One can check that the boundary of $\mathcal{R}$ is a union of the Jordan curve

$$\Gamma := (\partial G) \setminus \left(\bigcup_{n \in X} \partial G \cap H(a_n, r_n)\right) \cup \bigcup_{n \in X} (\bar{G} \cap \partial H(a_n, r_n))$$

and the set

$$\bigcup_{n \in Y} \partial H(a_n, r_n).$$

Given $n \geq 1$, the set $\mathcal{R}_n := G \setminus \bigcup_1^n \overline{H(a_j, r_j)}$ is an open connected set regular with respect to the Dirichlet problem such that $E_o \subset \partial \mathcal{R}_n$, $\mathcal{R}_{n+1} \subset \mathcal{R}_n$ and $\mathcal{R} = \cap_1^\infty G_n$.

Observe that
$$g_n(z) := \omega(z, E_o, \mathcal{R}_n), \quad z \in \bar{\mathcal{R}}, \quad n \geq 1$$
is a decreasing sequence of positive harmonic functions in $\mathcal{R}$. Its limit g is a harmonic function in $\mathcal{R}$ with $0 \leq g(z) \leq 1$ on $\bar{\mathcal{R}}$. Moreover, g is continuous on $\bar{\mathcal{R}} \cap \mathbb{D}_e$ and $g(z) = 0$ for all $z \in \mathbb{D}_e \cap \partial \mathcal{R}$.

We claim that $g(z) > 0$ for all $z \in \mathcal{R}$. Indeed, by the maximum principle,
$$\omega(z, E_0, G) \leq g_n(z) + h_n(z), \quad z \in \mathcal{R}_n, \quad n \geq 1,$$
where
$$h_n(z) := \sum_1^n \frac{\log|\frac{1-\bar{a}_j z}{z - a_j}|}{\log \frac{1}{r_j}}.$$
Letting n tend to ∞, we get
$$\omega(z, E_o, G) - h(z) \leq g(z), \quad z \in \mathcal{R}.$$

The function $\omega(z, E_o, G)$ has angular limit equal 1 at almost all points of E_o. Therefore, by 2° and 3° of Claim 2, $g(z) > 0$ for all $z \in \mathcal{R}$ near E_o, and consequently for all $z \in \mathcal{R}$.

Let $\varphi : \mathbb{D} \mapsto \mathcal{R}$ be an analytic covering of $\mathcal{R}$. We claim that the set
$$\varphi^{-1}(E_o) = \{\zeta \in \partial \mathbb{D} : \varphi(\zeta) \in E_o\}$$
(where $\varphi(\zeta)$ denotes the angular limit of φ at ζ) has positive measure. Indeed, the harmonic function $g \circ \varphi$ has a finite angular limit at almost all points of $\partial \mathbb{D}$. If $|\varphi(\zeta)| > 1$, then $g(\varphi(\zeta)) = 0$. Therefore, $g(\varphi(\zeta)) > 0$ on a subset of $\varphi^{-1}(E_o)$ having positive measure. Otherwise, we would have $g(\varphi(z)) \equiv 0$ in $\mathbb{D}$, which implies that $g \equiv 0$ in $\mathcal{R}$, which gives a contradiction.

By the Lusin-Privalov Lemma, the function F is bounded on G, and it has the angular limit value $F(\zeta)$ at almost every point ζ of E_0. By (***), the function f is continuous on the set $\mathbb{C} \setminus \bigcup_1^\infty H(a_n, r_n)$. Therefore $F(\zeta) = f(\zeta)$ for all $\zeta \in E_0$, because $f(\zeta)$ is equal to the angular limit value of $f_{|\mathbb{D}}$ at ζ. Thus the function $F \circ \varphi - f \circ \varphi$, being bounded on $\mathbb{D}$, has zero angular limit value at almost each point ζ of $\varphi^{-1}(E_0)$. Hence $F \circ \varphi = f \circ \varphi$ on $\mathbb{D}$, and consequently $F = f$ on $\mathcal{R}$. Since $\mathcal{R}$ is a non-empty open subset of the domain Ω, $F = f$ in Ω. But since f has poles $\{a_n : n \geq 1\}$ which accumulate non-tangentially to every point of the unit circle, F cannot have angular limit for almost all points of E_o, a contradiction. $\square$

References

1. A. Aleman, S. Richter and W.T.Ross, *Pseudocontinuations and the backward shift*, Indiana Univ. Math. J. **47** (1998), 223–276.
2. A. Edigarian and J. Wiegerinck, *Graphs that are not complete pluripolar*, Proc. Amer. Math. Soc. **131** (2003), 2459–2465.
3. M. Klimek, *Pluripotential Theory*, Clarendon Press, Oxford, 1991.
4. N. Levenberg, G. Martin and E.A. Poletsky, *Analytic discs and pluripolar sets*, Indiana Univ. Math. J. **41** (1992), 515–532.
5. I.I. Privalov, *Boundary Properties of Analytic Functions*, Moscow-Leningrad, 1950 (Russian).
6. W. T. Ross and H.S. Shapiro, *Generalized Analytic Continuation*, University Lecture Series **25**, Amer. Math. Soc., Providence, RI, 2002.

7. A. Sadullaev, *P-regularity of sets in $\mathbb{C}^n$*, Analytic Functions, Kozubnik 1979 (Proc. Seventh Conf. Kozubnik, 1979), Springer Lecture Notes in Mathematics **798**, 1980, pp. 402–408.
8. J. Siciak, *Functions with complete pluripolar graphs*, preprint (2002).
9. N.V. Shcherbina, *Pluripolar graphs are holomorphic*, http://arXiv.org/abs/math.CV/0301181.
10. A. Zériahi, *Ensembles pluripolaires exceptionnels pour la croissance partielle des fonctions holomorphes*, Ann. Polon. Math. **50** (1989), 81–91.

INSTITUTE OF MATHEMATICS, REYMONTA 4, 30-059 KRAKÓW, POLAND
E-mail address: Siciak@im.uj.edu.pl

Convolution Inverses

Herb Silverman and Evelyn M. Silvia

Dedicated to Larry Zalcman on his 60th

ABSTRACT. Let $\mathcal{A}$ denote the class of functions f analytic in $\Delta = \{z : |z| < 1\}$ and normalized by $f(0) = f'(0) - 1 = 0$. The subclasses of $\mathcal{A}$ consisting of functions univalent in Δ, starlike with respect to the origin, and convex are denoted by S, S^* and K, respectively. In this paper, we investigate conditions under which $f \in S^*$ has a starlike inverse, i.e., there exists $g \in S^*$ for which the convolution $f * g = \frac{z}{1-z}$. We also determine conditions under which a fixed $h \in K$ can be expressed as $h = f * g$, where f and g are in S^* (or S).

Let $\mathcal{A}$ denote the class of functions f analytic in $\Delta = \{z : |z| < 1\}$ and normalized by $f(0) = f'(0) - 1 = 0$. The subclasses of $\mathcal{A}$ consisting of functions univalent in Δ, starlike with respect to the origin, and convex are denoted by S, S^* and K, respectively. For $h(z) = z + \sum_{n=2}^{\infty} a_n z^n \in \mathcal{A}$ and $\delta > 0$, a δ-neighborhood of h is defined by

$$N_\delta(h) = \left\{ z + \sum_{n=2}^{\infty} b_n z^n \in \mathcal{A} : \sum_{n=2}^{\infty} n |a_n - b_n| \leq \delta \right\}.$$

In [13], St. Ruscheweyh introduced the notion of δ-neighborhoods and proved the following result.

THEOREM A. *If $f \in K$, then $N_{1/4}(f) \subset S^*$.*

Theorem A shows that the well-known result $N_1(z) \subset S^*$ [6] can be extended to the existence of neighborhoods of arbitrary convex functions that consist of starlike functions. For extensions and generalizations of the work that was initiated by Ruscheweyh, see [2], [4], [5], and [14].

For $f(z) = z + \sum_{n=2}^{\infty} b_n z^n \in \mathcal{A}$ and $g(z) = z + \sum_{n=2}^{\infty} c_n z^n \in \mathcal{A}$, the convolution or Hadamard product of f and g is $(f * g)(z) = z + \sum_{n=2}^{\infty} b_n c_n z^n$. One motivation for looking at convolutions over various subclasses of $\mathcal{A}$ was the Pólya-Schoenberg conjecture [9] that the convolution of two convex functions is convex. In addition to proving the conjecture, Ruscheweyh and Sheil-Small [10] showed that convolution with convex functions also preserves the classes of starlike and close-to-convex functions. Such results enable us to determine geometric properties associated with

---

2000 *Mathematics Subject Classification.* Primary 30C45; Secondary 30C50.
Key words and phrases. Convex, Starlike, Neighborhoods.

various operators which can be realized as convolutions with specific convex functions. For example, $g_\gamma(z) = \sum_{n=1}^\infty \frac{1+\gamma}{n+\gamma} z^n \in K$ whenever $\operatorname{Re}\gamma \geq 0$ [11]. Hence, setting $J(f) = f * g_\gamma$ for $\operatorname{Re}\gamma \geq 0$, we see that $J(f) = \frac{1+\gamma}{z^\gamma} \int_0^z t^{\gamma-1} f(t)\, dt \in S^*$ whenever $f \in S^*$. Other operator applications can be found in [1] and [16].

In another direction, one can specify a function g and define a class $\mathcal{F}$ consisting of all $f \in \mathcal{F}$ for which $f * g$ satisfies a particular property. The best known example of this is with the class of prestarlike functions of order α, denoted $\mathcal{R}_\alpha$, introduced by Ruscheweyh [12]. A function $f \in \mathcal{A}$ is prestarlike of order α for $0 \leq \alpha < 1$ if $f * \left(z/(1-z)^{2(1-\alpha)}\right)$ is starlike of order α. It is known [15] that $\mathcal{R}_\alpha \subset S$ if and only if $\alpha \leq \frac{1}{2}$.

Note that $I(z) = \frac{z}{1-z}$ is the identity function under convolution. In this paper, we investigate conditions under which $f \in S^*$ has a starlike inverse; i.e., there exists $g \in S^*$ for which $f * g = I$. We also determine conditions under which a fixed $h \in K$ can be expressed nontrivially as $h = f * g$, where f and g are in S^* (or S).

1. Some Preliminaries

In this section, we identify some subclasses $\mathcal{F}$ and $\mathcal{G}$ of S^* for which corresponding to each $f \in \mathcal{F}$, there exists a $g \in \mathcal{G}$ such that $f * g = I$.

LEMMA 1. *Let $F_B(z) = \frac{z+Bz^2}{1-z}$ for $B \in \mathbb{C}$. Then $F_B \in S^*$ if and only if $|B| \leq \frac{1}{3}$ or $B = -1$.*

PROOF. If $|B| \leq \frac{1}{3}$ and $|z| < 1$, then $\operatorname{Re}\left\{\frac{zF_B'(z)}{F_B(z)}\right\} = \operatorname{Re}\left\{1 + \frac{1}{1-z} - \frac{1}{1+Bz}\right\} > \frac{3}{2} - \frac{1}{1-|B|} \geq 0$, while $F_{-1}(z) = z$.

To prove the necessity, we consider two cases. If $|B| > 1$, then F_B is not univalent in Δ because $F_B(0) = F_B\left(-\frac{1}{B}\right) = 0$ and $-\frac{1}{B} \in \Delta$. Now, suppose $B = \left(\frac{1}{3} + \varepsilon\right) e^{i\beta}$, $\beta \neq \pi$, and set $z = re^{i(\pi-\beta)}$. If $0 < \varepsilon < \frac{2}{3}$ and $r = 1$, then

$$\operatorname{Re}\left\{\frac{zF_B'(z)}{F_B(z)}\right\} = \operatorname{Re}\left\{\frac{1}{1+e^{-i\beta}} - \frac{\left(\frac{1}{3}+\varepsilon\right)}{1-\left(\frac{1}{3}+\varepsilon\right)}\right\} = \frac{1}{2} - \frac{1+3\varepsilon}{2-3\varepsilon} = \frac{-9\varepsilon}{2(2-3\varepsilon)} < 0;$$

when $\varepsilon = \frac{2}{3}$,

$$\operatorname{Re}\left\{\frac{zF_B'(z)}{F_B(z)}\right\} = \operatorname{Re}\left\{\frac{1}{1+re^{-i\beta}} - \frac{r}{1-r}\right\} = \frac{1+r\cos\beta}{1+r^2+2r\cos\beta} - \frac{r}{1-r} \longrightarrow -\infty$$

as $r \to 1^-$.

Finally, suppose that $\beta = \pi$; i.e., $-1 < B < -\frac{1}{3}$. Then $(1-3|B|)(1-|B|) < 0$, so that $\frac{1+3B^2}{4|B|} < 1$; for $z = e^{i\theta}$, we have that

$$\operatorname{Re}\left\{\frac{zF_B'(z)}{F_B(z)}\right\} = \frac{3}{2} - \frac{1-|B|\cos\theta}{1+B^2-2|B|\cos\theta} < 0$$

whenever $\frac{1+3B^2}{4|B|} < \cos\theta < 1$. It follows that, for B satisfying $B \neq -1$ and $|B| > \frac{1}{3}$, $F_B \notin S^*$. This completes the proof. □

Lemma 1 immediately yields a subclass of starlike functions whose inverses with respect to convolution are also starlike.

THEOREM 1. *Let* $\Omega = \{\zeta \in \mathbb{C} : |\zeta - 1| \leq \frac{1}{3}\} \cap \{\zeta \in \mathbb{C} : |\zeta - \frac{9}{8}| \leq \frac{3}{8}\}$. *Then*

$$z + A\sum_{n=2}^{\infty} z^n \in S^* \quad \text{and} \quad z + \frac{1}{A}\sum_{n=2}^{\infty} z^n \in S^*$$

if and only if $A \in \Omega$.

PROOF. Since $z + A\sum_{n=2}^{\infty} z^n = z + \frac{Az^2}{1-z} = F_{(A-1)}$, taking $B = A - 1$ in Lemma 1 yields that $z + A\sum_{n=2}^{\infty} z^n \in S^*$ if and only if A satisfies $|A - 1| \leq \frac{1}{3}$ or $A = 0$. Similarly, taking $B = \frac{1}{A} - 1$ yields that $z + \frac{1}{A}\sum_{n=2}^{\infty} z^n = F_{(A^{-1}-1)}$ is starlike if and only if A satisfies $|\frac{1}{A} - 1| \leq \frac{1}{3}$ which is equivalent to A satisfying $|A - \frac{9}{8}| \leq \frac{3}{8}$. Combining the conditions leads to the desired conclusion. □

REMARK 1. The circles that form the boundary of Ω intersect at $\frac{17 \pm i\sqrt{35}}{18}$ which are on $\partial\Delta$.

COROLLARY 1. *For* $|\varepsilon| \leq \frac{1}{4}$, $f_\varepsilon(z) = z + (1+\varepsilon)\sum_{n=2}^{\infty} z^n \in S^*$ *and has the starlike inverse* $g_\varepsilon = z + \frac{1}{1+\varepsilon}\sum_{n=2}^{\infty} z^n \in S^*$. *The result is sharp.*

PROOF. From Theorem 1, $f_\varepsilon \in S^*$ and $g_\varepsilon \in S^*$ if and only if

$$\varepsilon \in \Omega = \left\{\zeta \in \mathbb{C} : |\zeta| \leq \frac{1}{3}\right\} \cap \left\{\zeta \in \mathbb{C} : \left|\zeta - \frac{1}{8}\right| \leq \frac{3}{8}\right\}.$$

The largest disk centered at the origin that is contained in Ω has radius $\frac{1}{4}$. □

The following lemma will be used to show sharpness in several results.

LEMMA 2. *For* $A_k = \frac{(-1)^{k+1}}{k}\alpha$, $\alpha > \frac{1}{4}$, *the function* $g_\alpha(z) = \frac{z}{1-z} - A_k z^k \notin S$.

PROOF. Note that $g'_\alpha(z) = \frac{1}{(1-z)^2} - (-1)^{k+1}\alpha z^{k-1}$ is real for z real. Since $g'_\alpha(0) = 1$ and $g'_\alpha(-1) = \frac{1}{4} - (-1)^{2k}\alpha = \frac{1}{4} - \alpha < 0$, the function g_α isn't locally univalent in Δ. □

REMARK 2. From Theorem A, we see that $g(z) = \frac{z}{1-z} - \frac{\alpha}{k}z^k$ is in S^* for all α, $|\alpha| \leq \frac{1}{4}$.

THEOREM 2. *For n a fixed integer, $n \geq 2$, let*

$$\Omega_n = \left\{\zeta \in \mathbb{C} : |\zeta| \leq \frac{1}{4n}\right\} \cap \left\{\zeta \in \mathbb{C} : \left|\zeta - \frac{1}{16n^2 - 1}\right| \leq \frac{4n}{16n^2 - 1}\right\}.$$

If $\varepsilon \in \Omega_n$, *then* $F_\varepsilon(z) = \frac{z}{1-z} + \varepsilon z^n \in S^*$ *and its inverse* $G_\varepsilon(z) = \frac{z}{1-z} - \frac{\varepsilon}{1+\varepsilon}z^n$ *is also in* S^*.

PROOF. Since $n|\varepsilon| \leq \frac{1}{4}$, $F_\varepsilon \in N_{1/4}\left(\frac{z}{1-z}\right)$ and, by Theorem A, $F_\varepsilon \in S^*$. Now $G_\varepsilon = F_{-\varepsilon/(1+\varepsilon)} \in N_{1/4}\left(\frac{z}{1-z}\right)$ whenever $\left|\frac{\varepsilon}{1+\varepsilon}\right| \leq \frac{1}{4n}$ which is equivalent to $\left|\varepsilon - \frac{1}{16n^2 - 1}\right| \leq \frac{4n}{16n^2 - 1}$. This completes the proof. □

REMARK 3. To see that the result of Theorem 2 is sharp, note from Lemma 2 that $F_\varepsilon \notin S$ for $\varepsilon = \frac{(-1)^n}{n}\alpha, \alpha > \frac{1}{4}$.

REMARK 4. If $|\varepsilon| \leq \frac{1}{4n+1}$, then $\varepsilon \in \Omega_n$.

The last example given in this section makes use of the following result due to Lewis [7].

THEOREM B. *The function $f_\lambda(z) = \sum_{n=1}^{\infty} n^{-\lambda} z^n \in K$ when $\lambda \geq 0$.*

In view of Theorem B, if $0 \leq \delta \leq 1$, $\phi_\delta(z) = z + \sum_{n=2}^{\infty} n^\delta z^n \in S^*$ because $\int_0^z \zeta^{-1} \phi_\delta(\zeta) d\zeta = z + \sum_{n=2}^{\infty} \frac{z^n}{n^{1-\delta}} \in K$. Hence, if $0 \leq \delta \leq 1$, $\phi_\delta \in S^*$ and $\phi_\delta * \phi_{-\delta} = \frac{z}{1-z}$ with $\phi_{-\delta} \in K$.

REMARK 5. Note that $\phi_1(z) = z + \sum_{n=2}^{\infty} n z^n = \frac{z}{(1-z)^2}$, the well-known Koebe function, has the convex function $\phi_{-1}(z) = z + \sum_{n=2}^{\infty} \frac{1}{n} z^n = -\log(1-z)$ as its inverse.

2. Convex Functions

Next we illustrate the important role played by the identity function under convolutions when determining if neighborhoods must contain starlike functions.

THEOREM 3. *If $f \in N_{2/9}(I)$, then f has an inverse g and $g \in S^*$.*

PROOF. If $f \in N_{2/9}(I)$, we may set $f(z) = \frac{z}{1-z} + \sum_{n=2}^{\infty} \varepsilon_n z^n$ where $\sum_{n=2}^{\infty} n |\varepsilon_n| \leq \frac{2}{9}$. Then $g(z) = \frac{z}{1-z} - \sum_{n=2}^{\infty} \frac{\varepsilon_n}{1+\varepsilon_n} z^n$. Since $|\varepsilon_n| \leq \frac{1}{9}$ for each n, it follows that

$$\sum_{n=2}^{\infty} n \left| \frac{\varepsilon_n}{1+\varepsilon_n} \right| \leq \frac{9}{8} \sum_{n=2}^{\infty} n |\varepsilon_n| \leq \frac{1}{4};$$

i.e., $g \in N_{1/4}(I)$ and, by Theorem A, $g \in S^*$.

To see that this is best possible, set $f(z) = \frac{z}{1-z} - \left(\frac{1}{9} + \beta\right) z^2$ for some β, $0 < \beta < \frac{1}{72}$. Then $f \in N_{1/4}(I)$ while its inverse $g(z) = \frac{z}{1-z} + \frac{\left(\frac{1}{9}+\beta\right)}{\left(\frac{8}{9}-\beta\right)} z^2 \notin S$ by Lemma 2. □

REMARK 6. For $I_x(z) = \frac{z}{1-xz}$, $|x| = 1$, a rotation of $I(z)$, a similar argument shows that any $f \in N_{2/9}(I_x)$ also has a starlike inverse.

For functions in K other than rotations of $I(z) = \frac{z}{1-z}$, the question arises as to whether the result of Theorem 3 remains valid for perhaps a smaller than 2/9-neighborhood. We show that a theorem of Eenigenburg and Keogh (see Theorem 4 in [3]) answers this question in the negative.

THEOREM C. *If $f(z) = z + \sum_{n=2}^{\infty} a_n z^n \in K$ is not a rotation of $I(z)$, then $a_n \longrightarrow 0$ as $n \to \infty$.*

THEOREM 4. *Suppose that $F(z) = z + \sum_{n=2}^{\infty} a_n z^n \in K$ is not a rotation of $I(z)$. Then for every $\delta > 0$, there exists $f \in N_\delta(F)$ and $g \in \mathcal{A} - S$ for which $f * g = F$.*

PROOF. By Theorem C, $a_n \longrightarrow 0$ as $n \to \infty$. Choose k so large that $|a_k| \leq \delta$. Now $f(z) = F(z) + \varepsilon_k z^k \in N_\delta(F)$ as long as $|\varepsilon_k| \leq \frac{\delta}{k}$. If $a_k = 0$, set $\varepsilon_k = 0$ and $g(z) = \frac{z}{1-z} - (-1)^{k+1} z^k$. For $a_k \neq 0$, $g(z) = \frac{z}{1-z} - \frac{\varepsilon_k}{a_k + \varepsilon_k} z^k$ satisfies $f * g = F$. If $0 < |a_k| \leq \frac{\delta}{k}$, set $\varepsilon_k = \frac{(-1)^{k+1} a_k}{2}$. For $\frac{\delta}{k} < |a_k| \leq \delta$, let $\varepsilon_k = (-1)^{k+1} \frac{\delta}{k} \exp(i \arg a_k)$. From Lemma 2, in all cases, $g \notin S$ as needed. □

In Theorem 4, the only use made of the convexity of F was that we could choose $|a_k| \leq \delta$. This leads to the following more general result.

COROLLARY 2. *Suppose $F(z) = z + \sum_{n=2}^{\infty} a_n z^n \in \mathcal{A}$ is such that $\inf_n |a_n| = 0$. Then for every $\delta > 0$, there exists $f \in N_\delta(F)$ and $g \in \mathcal{A} - S$ for which $f * g = F$.*

3. Some Open Questions

1. Characterize $f, g \in S^*$ for which $f * g = I(z) = \frac{z}{1-z}$.

   We have shown that such functions may be found when $f_A(z) = z + A \sum_{n=2}^{\infty} z^n$ and $f_\varepsilon(z) = \frac{z}{1-z} - \varepsilon z^n$, with appropriate restrictions on A and ε. We also know that all functions in the $\frac{2}{9}$-neighborhood of the identity function have this property and that $f_\lambda(z) = \sum_{n=1}^{\infty} n^{-\lambda} z^n$ has a starlike inverse for $0 \leq \lambda \leq 1$.

2. Characterize the subclass $\mathcal{B}$ of starlike functions where $h \in \mathcal{B}$ admits a δ-neighborhood for which $f \in N_\delta(h)$ implies that $f * g = h$ $(f, g \in S^*)$.

   Note that if $h = I$, we may take $\delta = \frac{2}{9}$. If $h(z) = z$, then $\delta = 1$ with $g(z) = z$. Theorems 1 and 2 enable us to find neighborhoods under special conditions.

3. Are there "primes" under convolution?

   For every $h \in S$, $h * \frac{z}{1-z} = h$. Is this ever unique? That is, can we find an $h \in S$ for which the univalent convolution $f * g = h$ occurs only if f or g is the identity $I(z) = \frac{z}{1-z}$? Although the Koebe function is extremal for many problems in univalent functions, it is not prime. It may be expressed as the convolution of two starlike functions

$$z + \sum_{n=2}^{\infty} nz^n = \left(z + \sum_{n=2}^{\infty} n^\lambda z^n\right) * \left(z + \sum_{n=2}^{\infty} n^{1-\lambda} z^n\right)$$

for $0 < \lambda < 1$. Our last result gives other cases where functions are not prime.

THEOREM 5. *Let* $h(z) = z + \sum_{n=2}^{\infty} a_n z^n \in S$ *and suppose (i)* $a_k = 0$ *for some* $k \geq 2$ *or (ii)* $N_\delta(h) \subset S$ *for some* $\delta > 0$. *Then there exist functions* f *and* g *in* $S - \left\{\frac{z}{1-z}\right\}$ *such that* $f * g = h$.

PROOF. (i) Set $f(z) = \frac{z}{1-z} + \frac{1}{4k} z^k$, $g = h$, and note, from Theorem A, that $f \in N_{1/4}\left(\frac{z}{1-z}\right) \subset S^*$.

(ii) From (i), we may assume $a_2 \neq 0$. Set $\varepsilon = \min\left\{\frac{\delta}{2}, \frac{|a_2|}{9}\right\}$ and note that $h = f * g$ for $f = h + \varepsilon z^2$ and $g = \left(\frac{z}{1-z} - \frac{\varepsilon}{a_2+\varepsilon} z^k\right)$. Now $f \in S$ by hypothesis and, since $\left|\frac{\varepsilon}{a_2+\varepsilon}\right| \leq \left(\frac{|a_2|}{9}\right)\left(|a_2| - \frac{|a_2|}{9}\right)^{-1} = \frac{1}{8}$, we see from Theorem A that $g \in S^*$. □

REMARK 7. Applying Theorem A to Theorem 5, we see that no function in K can be prime. In fact, for $f \in K$, we can find explicit nontrivial factors. Setting $\phi(z) = -\log(1-z) \in K$, we may express f as $zf'(z) * \phi(z)$. This is a nontrivial solution as long as $f(z) \neq \phi(z)$. But in view of Theorem B, we may express ϕ as the convolution of two nontrivial convex functions, namely,

$$\phi(z) = \left(\sum_{n=1}^{\infty} \frac{z^n}{n^\lambda}\right) * \left(\sum_{n=1}^{\infty} \frac{z^n}{n^{1-\lambda}}\right), \quad 0 < \lambda < 1.$$

In [8], Pohl showed that no convex function that maps onto a convex polygon may be realized as the convolution of two nontrivial functions in K.

It is known [13] that, for $h \in S^*\left(\frac{1}{2}\right)$, the class of functions that are starlike of order $1/2$, $N_{1/4}(h)$ consists of close-to-convex functions. This leads to the following

COROLLARY 3. If $h \in S^*\left(\frac{1}{2}\right)$, then there exist functions f and g in $S - \left\{\frac{z}{1-z}\right\}$ such that $f * g = h$.

References

[1] R.W. Barnard and C. Kellogg, *Applications of convolution operators to problems in univalent function theory*, Michigan Math. J. **27** (1980), 81-94.
[2] J.E. Brown, *Some sharp neighborhoods of univalent functions*, Trans. Amer. Math. Soc. **287** (1985), 475-482.
[3] P.J. Eenigenburg and F.R. Keogh, *The Hardy class of some univalent functions and their derivatives*, Michigan Math. J. **17** (1970), 335-346.
[4] R. Fournier, *A note on neighborhoods of univalent functions*, Proc. Amer. Math. Soc. **87** (1983), 117-120.
[5] R. Fournier, *On neighborhoods of univalent convex functions*, Rocky Mountain J. Math. **16** (1986), 579-589.
[6] A.W. Goodman, *Univalent functions and nonanalytic curves*, Proc. Amer. Math. Soc. **8** (1957), 598-601.
[7] John L. Lewis, *Convexity of a certain series*, J. London Math. Soc. (2) **27** (1983), 435-446.
[8] M. Pohl, *On the inverse of the Pólya-Schoenberg conjecture*, Complex Variables Theory Appl. **15** (1990), 293-306.
[9] G. Pólya and I. J. Schoenberg, *Remarks on the de la Vallée Poussin means and convex conformal mapping of the circle*, Pacific J. Math. **8** (1958), 295-334.
[10] St. Ruscheweyh and T. Sheil-Small, *Hadamard products of schlicht functions and the Pólya-Schoenberg conjecture*, Comment. Math. Helv. **48** (1973), 119-135.
[11] St. Ruscheweyh, *New criteria for univalent functions*, Proc. Amer. Math. Soc. **49** (1975), 109-115.
[12] St. Ruscheweyh, *Linear operators between classes of prestarlike functions*, Comment. Math. Helv. **52** (1977), 497-509.
[13] St. Ruscheweyh, *Neighborhoods of univalent functions*, Proc. Amer. Math. Soc. **81** (1981), 521-527.
[14] T. Sheil-Small and E.M. Silvia, *Neighborhoods of analytic functions*, J. Analyse Math. **52** (1989), 210-240.
[15] H. Silverman and E.M. Silvia, *The influence of the second coefficient on prestarlike functions*, Rocky Mountain J. Math. **10** (1980), 469-474.
[16] H. Silverman and E.M. Silvia, *Subclasses of starlike functions subordinate to convex functions*, Canad. J. Math. **37** (1985), 48-61.

DEPARTMENT OF MATHEMATICS, COLLEGE OF CHARLESTON, CHARLESTON, SC 29424-0001, U.S.A.
E-mail address: `silvermanh@cofc.edu`

DEPARTMENT OF MATHEMATICS, UNIVERSITY OF CALIFORNIA, DAVIS, CA 95616-8633, U.S.A.
E-mail address: `emsilvia@math.ucdavis.edu`

Composition Operators on Sobolev Spaces

S. K. Vodopyanov

Dedicated to Lawrence Zalcman on his 60th birthday

ABSTRACT. We obtain necessary and sufficient conditions for the composition operator defined by a mapping to be bounded (or an isomorphism) on Sobolev spaces of functions having generalized first derivatives.

Consider an arbitrary open set D in the Euclidean space $\mathbb{R}^n$, $n \geq 1$. The Lebesgue space $L_p(D)$, $1 \leq p \leq \infty$, consists of functions measurable on D having finite norm $\|f \mid L_p(D)\| = \left(\int_D |f(x)|^p \, dx\right)^{1/p}$, where dx is Lebesgue measure. The Sobolev space $W_p^1(D)$ ($L_p^1(D)$), $1 \leq p \leq \infty$, consists of functions locally summable on D having generalized first derivatives and finite norm (semi-norm)

$$\|f \mid W_p^1(D)\| = \|f \mid L_p(D)\| + \|\nabla f \mid L_p(D)\|$$
$$(\|f \mid L_p^1(D)\| = \|\nabla f \mid L_p(D)\|),$$

where $\nabla f = \left(\frac{\partial f}{\partial x_1}, \dots, \frac{\partial f}{\partial x_n}\right)$ is the generalized gradient of f.

Let D and D' be open sets in $\mathbb{R}^n$, $n \geq 1$, and $\varphi : D \to D'$ be a measurable mapping.

DEFINITION. *A mapping φ induces a bounded operator $\varphi^* : W_p^1(D') \to W_p^1(D)$ by composition* if the following properties hold.

(a) If two quasicontinuous functions $f_1, f_2 \in W_p^1(D')$ differ on a set of p-capacity zero, then the functions $f_1 \circ \varphi$, $f_2 \circ \varphi$ differ on a set of measure zero.

(b) If $f \in W_p^1(D')$ is a quasicontinuous representative, then $f \circ \varphi \in W_p^1(D)$ but $f \circ \varphi$ is not required to be quasicontinuous.

(c) The mapping $\varphi^* : f \mapsto \tilde{f} \circ \varphi$, where $\tilde{f}$ is a quasicontinuous representative of f, is a bounded operator $W_p^1(D') \to W_p^1(D)$.

2000 *Mathematics Subject Classification*. Primary 46E35, 3065.

Key words and phrases. Sobolev space, embedding theorem, quasiconformal mapping.

This research was carried out with the partial support of the Russian Foundation for Basic Research (Grant no. 03–01–00899) and the Program of Support of Leading Scientific Schools of the Russian Federation (Grant no. 311.2003.1).

If the operator φ^* is surjective, then we say that φ induces an isomorphic operator $\varphi^* : W_p^1(D') \to W_p^1(D)$ by composition.

For the definition and properties p-capacity, see [**CV, GR, M3, R1**].

The main goal of the paper is to obtain necessary and sufficient conditions for the mapping φ to induce a bounded operator (or an isomorphism) $\varphi^* : W_p^1(D') \to W_p^1(D)$ by composition. In particular, we give a positive solution of the problem in the following cases:

1) $\varphi^* : W_p^1(D') \to W_p^1(D)$ is a bounded operator, $1 \leq p < n$, in Theorem 1;
2) $\varphi^* : W_p^1(\mathbb{R}^n) \to W_p^1(\mathbb{R}^n)$ is a bounded operator, $n < p < \infty$, in Theorem 2;
3) $\varphi^* : W_n^1(\mathbb{R}^n) \to W_n^1(\mathbb{R}^n)$ is a bounded operator in Theorem 3;
4) $\varphi^* : W_p^1(D') \to W_p^1(D)$ is an isomorphism, $p \neq n$, in Theorem 4

Note that an analytic description of mappings $\varphi : D \to D'$, inducing a bounded operator $\varphi^* : L_p^1(D') \to L_p^1(D)$ of homogeneous Sobolev spaces, is obtained in [**V1, VU1, VU2**], where one can also find a detailed bibliography on this question. The main difference between papers [**V1, VU1-VU3**] and those on this subject written earlier is that we do not assume that φ is a diffeomorphism or a Lipschitz homeomorphism, as was done in [**M1, M2**], nor even a homeomorphism [**V2–V4**].

1. To formulate the main results we need the following concepts. A function $f : D \to \mathbb{R}$ is said to be *absolutely continuous on a straight line l* having non-empty intersection with D if it is absolutely continuous on an arbitrary closed segment of this line which is contained in D. The function $f : D \to \mathbb{R}$ belongs to the class ACL(D) (*absolutely continuous on almost all lines*) if it is absolutely continuous on almost all straight lines parallel to the jth coordinate axis, $j = 1, \ldots, n$. Recall that f belongs to the Sobolev space $L_1^1(D)$ if and only if f is locally summable and can be modified on a set of measure zero to belong to the class ACL(D) and its partial derivatives $\frac{\partial f}{\partial x_i}(x)$, $i = 1, \ldots, n$, existing almost everywhere, are just the first generalized derivatives of f in D.

A mapping $\varphi : D \to D'$ belongs to the class ACL(D) if its coordinate functions φ_j belong to ACL(D), $j = 1, \ldots, n$. Then the formal Jacobian matrix $D\varphi(x) = \left(\frac{\partial \varphi_i}{\partial x_j}(x)\right)$, $i, j = 1, \ldots, n$, and its Jacobian determinant are defined at almost all points of D. The norm $|D\varphi(x)|$ of the matrix $D\varphi(x)$ is the norm of the linear operator determined by the matrix in the Euclidean space $\mathbb{R}^n$.

A mapping $\varphi : D \to D'$ has *Luzin's condition $\mathcal{N}$* if the image of any set of measure zero is a set of measure zero.

We formulate below a change-of-variable formula in the Lebesgue integral in the form which is used in proofs of the basic results of the paper.

PROPOSITION 1. ([**V5**], Corollary 5.1) *Let A be a measurable set. Assume that a mapping $\varphi : A \to \mathbb{R}^n$ has approximate partial derivatives in A. Then there exists a set $\Sigma_\varphi \subset A$ of measure zero such that for any non-negative measurable function $u : A \to \mathbb{R}$, one has the change-of-variable formula*

$$(1) \qquad \int_A u(x)|J(x,\varphi)|\,dx = \int_{\mathbb{R}^n} \left(\sum_{x \in \varphi^{-1}(y) \cap (A \setminus \Sigma_\varphi)} u(x) \right) dy.$$

If φ has Luzin's condition $\mathcal{N}$, then $\Sigma_\varphi = \emptyset$.

Note that any mapping whose coordinate functions belong to $W^1_{q,\text{loc}}(D)$, $q > n$, (are monotone and belong to $W^1_{n,\text{loc}}(D)$) has a continuous representative possessing Luzin's condition $\mathcal{N}$ [**MM, V5–V7**].

We say that a mapping $\varphi : D \to D'$ of class $\text{ACL}(D)$ has the *finite distortion* if $D\varphi(x) = 0$ almost everywhere in $Z = \{x \in D : J(x, \varphi) = 0\}$.

For a mapping $\varphi : D \to D'$ of the class $\text{ACL}(D)$, we define the *distortion function*

$$(2) \qquad D' \ni y \mapsto H_p(y) = \begin{cases} \left(\displaystyle\sum_{x \in \varphi^{-1}(y) \setminus \Sigma_\varphi, \, J(x,\varphi) \neq 0} \frac{|D\varphi|^p(x)}{|J(x,\varphi)|} \right)^{\frac{1}{p}}, \\ 0, \text{ if } \{x \in \varphi^{-1}(y) \setminus \Sigma_\varphi : J(x, \varphi) \neq 0\} = \emptyset \end{cases}$$

(henceforth $\Sigma_\varphi \subset D$ is the set from Proposition 1). The distortion function was introduced in [**V1**].

Two functions $f : D' \to \mathbb{R}$ and $g : D' \to \mathbb{R}$ are said to be equivalent ($f \sim g$) if $\alpha f(x) \leq g(x) \leq \beta f(x)$ for almost all $x \in D'$, where $0 < \alpha \leq \beta < \infty$ are constants independent of the choice of $x \in D'$.

To prove the main statements of the paper we need the following result, separate fragments of which are contained in [**VU1, VU2**]. As usual, we denote by $B(a,r)$ the ball $\{y \in \mathbb{R}^n : |y - a| < r\}$ in $\mathbb{R}^n$.

LEMMA 1. *Let D and D' be open sets in $\mathbb{R}^n$. If $\varphi : D \to D'$ induces a bounded embedding operator $\varphi^* : W^1_p(D') \to W^1_p(D)$ (or $\varphi^* : L^1_p(D') \to L^1_p(D)$), $1 \leq p < \infty$, then φ can be modified on a set of measure zero to belong to $\text{ACL}(D)$, has finite distortion and $\|H_p(\cdot) \mid L_\infty(D')\| \leq T\|\varphi^*\|$, where T is some constant.*

PROOF. For the operator $\varphi^* : L^1_p(D') \to L^1_p(D)$, the lemma is proved in [**VU2**]. For the reader's convenience, we provide the details of the proof in the case of non-homogeneous Sobolev spaces. For proving $\varphi \in \text{ACL}(D)$, we consider finite functions $\xi_N \in C^\infty_0(\mathbb{R}^n)$, $N \in \mathbb{N}$, equal to 1 on the compact set $\overline{B(0,N)}$. For any coordinate function $y_j : D' \to \mathbb{R}$, $j = 1, \ldots, n$, $\xi_N(y) \cdot y_j \in W^1_p(D')$. Then the functions $\varphi^*(\xi_N y_j)(x) = \xi_N(\varphi(x))\varphi_j(x)$ can be modified on a set of measure zero to belong to the class $W^1_q(D) \cap \text{ACL}(D)$.

Fix arbitrary coordinate axes and consider the family L_N consisting of all straight lines l which are parallel to the chosen coordinate axes and such that on every line $l \in L_N$, the modified function $\varphi^*(\xi_N y_j)$, $N \in \mathbb{N}$, is absolutely continuous. As the set $\{\varphi^*(\xi_N y_j)\}_{N \in \mathbb{N}}$ is countable, the projection of the set $L = \bigcap_{N=1}^{\infty} L_N$ along the chosen axes to the subspace $\mathbb{R}^{n-1}$ has full Lebesgue $(n-1)$-measure.

The modification of the function $\varphi^*(\xi_N y_j)(x) = \xi_N(\varphi(x))\varphi_j(x)$, $N \in \mathbb{N}$, implies also a modification of the mapping φ. Now we show that the modified mapping $\widetilde{\varphi}$ is absolutely continuous on any line $l \in L$. First and foremost, we notice that the set $\widetilde{\varphi}^{-1}(B(0,N)) \cap l$ is open on a line l since $\xi_N(y) \cdot y_j = y_j$ for $y \in \overline{B(0,N)}$, and the modified functions $\varphi^*(\xi_N y_j)(x)$, $N \in \mathbb{N}$, are continuous on the line l. We choose two arbitrary points $x_1, x_2 \subset D$ on the line l in such a way that the segment $[x_1, x_2] \subset D$. Because the set $\widetilde{\varphi}^{-1}(B(0,N)) \cap l$ is open on the line l and $[x_1, x_2] \subset \bigcup_{N=1}^{\infty} \widetilde{\varphi}^{-1}(B(0,N))$, there exists a number N_1 such that

$[x_1, x_2] \subset \widetilde{\varphi}^{-1}(B(0, N_1))$. Consequently, $\widetilde{\varphi}$ is absolutely continuous on $[x_1, x_2]$ since

$$\widetilde{\varphi}|_{[x_1,x_2]} = (\varphi^*(\xi_{N_1} y_1)|_{[x_1,x_2]}, \ldots, \varphi^*(\xi_{N_1} y_n)|_{[x_1,x_2]}) \quad \text{a. e.}$$

and the modified functions $\varphi^*(\xi_{N_1} y_j) \in \mathrm{ACL}([x_1, x_2])$, $j = 1, \ldots, n$. Thus we have proved $\widetilde{\varphi} \in \mathrm{ACL}(D)$.

Fix a cut-off function $\eta \in C_0^\infty(\mathbb{R}^n)$ equal to 1 on $B(0,1)$ and 0 outside the ball $B(0, 2)$. By substituting the functions $h_j(z) = (z_j - y_j)\eta(\frac{z-y}{r})$, $j = 1, \ldots, n$, where $z_j - y_j$ is the j-th coordinate of the vector $z - y \in \mathbb{R}^n$, $B(y, 2r) \subset D'$, into the inequality $\|\nabla \varphi^* f \mid L_p(D)\| \le \|\varphi^*\| \|f \mid W_p^1(D')\|$, we arrive at the inequality

$$(3) \qquad \left(\int_{\varphi^{-1}(B(y,r))} |D\varphi|^p(x)\, dx \right)^{1/p} \le C \|\varphi^*\| |B(y, 2r)|^{1/p}, \quad r \in (0, 1),$$

since $\|h_j \mid L_p(D')\| \le C|B(y, 2r)|^{1/p}$ for all the balls $B(y, 2r) \subset D'$, $r \in (0, 1)$ (here C is a constant depending only on the dimension n and the index p).

Let $Z = \{x \in D : J(x, \varphi) = 0\}$. We want to show that

$$(4) \qquad \int_Z |D\varphi|^p(x)\, dx = 0.$$

By formula (1), $|\varphi(Z \setminus \Sigma_\varphi)| = 0$. Fix $\varepsilon > 0$ and an open set $U \subset D'$ such that $U \supset \varphi(Z \setminus \Sigma_\varphi)$ and $|U| < \varepsilon$. We can choose a covering $\{B(y_i, r_i)\}$, $r_i < \frac{1}{2} \mathrm{dist}(y_i, \partial U)$, of U of finite multiplicity such that $B(y_i, 2r_i) \subset U$, $i \in \mathbb{N}$, and this collection of balls also constitutes a covering of U of finite multiplicity. Moreover, the multiplicity M of the covering $\{B(y_i, 2r_i)\}$ is independent of U, and $\sum_i |B(y_i, 2r_i)| < M\varepsilon$. Then, applying the inequality (3), we derive

$$\int_Z |D_h \varphi|^p(x)\, d = \int_{Z \setminus \Sigma_\varphi} |D_h \varphi|^p(x)\, dx \le \sum_{i=1}^\infty \int_{\varphi^{-1}(B(y_i, r_i))} |D_h \varphi|^q(x)\, dx$$

$$\le C^p \|\varphi^*\|^p \sum_{i=1}^\infty |B(y_i, 2r_i)| \le C_1^p \|\varphi^*\|^p |U|.$$

As $\varepsilon > 0$ is arbitrary, (4) is proved. Hence $|D_h \varphi| = 0$ almost everywhere on Z.

Changing variables on the left hand side of (3) by formula (1), we obtain

$$(5) \qquad \int_{B(y,r)} \sum_{\substack{x \in \varphi^{-1}(y) \setminus \Sigma_\varphi, \\ |J(x,\varphi)| \ne 0}} \frac{|D\varphi|^p(x)}{|J(x,\varphi)|}\, dy \le C'^p \|\varphi^*\|^p |B(y,r)|.$$

Applying the Lebesgue differentiability theorem, we infer $\operatorname*{ess\,sup}_{y \in D'} H_p(y) \le C' \|\varphi^*\|$.
The lemma is proved. $\square$

Lemma 1, Theorems 1 and 3 and Proposition 4 of [**VU2**] imply the following result.

THEOREM 1. *Let D and D' be open sets in $\mathbb{R}^n$. A mapping $\varphi : D \to D'$ induces a bounded embedding operator $\varphi^* : W_p^1(D') \to W_p^1(D)$, $1 \leq p < n$, if and only if one of the following conditions holds*:
  1) φ *can be modified on a set of measure zero to belong to the class* ACL(D), *has finite distortion and* $H_p(\cdot) \in L_\infty(D')$;
  2) φ *induces a bounded embedding operator* $\varphi^* : L_p^1(D') \to L_p^1(D)$.

The norm of the operator $\varphi^* : W_p^1(D') \to W_p^1(D)$, $1 \leq p < n$, is equivalent to the value

$$\operatorname*{ess\,sup}_{y \in D'} \left(\sum_{x \in \varphi^{-1}(y) \setminus \Sigma_\varphi} \frac{1}{|J(x,\varphi)|} \right)^{\frac{1}{p}} + \|H_p(\cdot) \mid L_\infty(D')\|.$$

PROOF. By Lemma 1, the boundedness of the operator $\varphi^* : W_p^1(D') \to W_p^1(D)$ (or $\varphi^* : L_p^1(D') \to L_p^1(D)$), $1 \leq p < n$, implies that φ can be modified on a set of measure zero to belong to ACL(D), φ has finite distortion and $H_p(\cdot) \in L_\infty(D')$. Moreover, if the operator $\varphi^* : W_p^1(D') \to W_p^1(D)$ is bounded then the modified mapping $\widetilde{\varphi}$ induces a bounded operator $\varphi^* : L_p^1(D') \to L_p^1(D)$, $1 \leq p < n$, by Theorem 1 of [**VU2**].

Now suppose that the operator $\varphi^* : L_p^1(D') \to L_p^1(D)$, $1 \leq p < n$, is bounded. The essential part of the converse assertion is a verification that the boundedness of the quantity $\|H_p(\cdot) \mid L_\infty(D')\|$ implies the boundedness of the operator $\varphi^* : W_p^1(D') \to W_p^1(D)$. Indeed, by the pointwise relation $|J(x,\varphi)| \leq |D\varphi|^n(x)$ and Jensen's inequality, we have

$$H_p(y) = \left(\sum_{x \in \varphi^{-1}(y) \setminus \Sigma_\varphi} \frac{|D\varphi|^p(x)}{|J(x,\varphi)|} \right)^{\frac{1}{p}}$$

$$= \left(\sum_{x \in \varphi^{-1}(y) \setminus \Sigma_\varphi} \left(\frac{|D\varphi|^n(x)}{|J(x,\varphi)|} \right)^{\frac{p}{n}} \cdot \frac{1}{|J(x,\varphi)|^{1-\frac{p}{n}}} \right)^{\frac{1}{p}}$$

$$\geq \left(\sum_{x \in \varphi^{-1}(y) \setminus \Sigma_\varphi} \frac{1}{|J(x,\varphi)|^{1-\frac{p}{n}}} \right)^{\frac{1}{p}} \geq \left(\sum_{x \in \varphi^{-1}(y) \setminus \Sigma_\varphi} \frac{1}{|J(x,\varphi)|} \right)^{\frac{1}{p} - \frac{1}{n}}$$

for almost all $y \in D'$. It remains to observe ([**VU2**], Proposition 4) that the essential boundedness of $J_{\varphi^{-1}}(y) = \sum_{x \in \varphi^{-1}(y) \setminus \Sigma_\varphi} \frac{1}{|J(x,\varphi)|}$ is a necessary and sufficient condition for the operator $\varphi^* : L_p(D') \to L_p(D)$ to be bounded with norm equivalent to $\|J_{\varphi^{-1}}(\cdot)^{\frac{1}{p}} \mid L_\infty(D')\|$. If φ has the property that $\operatorname{cap}_p(A) = 0$, $A \subset D'$, implies $|\varphi^{-1}(A)| = 0$, then $\varphi^* : L_p(D') \cap L_p^1(D') \to L_p(D) \cap L_p^1(D)$ is a bounded operator $\varphi^* : W_p^1(D') \to W_p^1(D)$ with the above-mentioned estimate of its norm. □

REMARK. In Theorem 1 and other statements, the sufficient part of an assertion is proved by the following scheme. We consider a mapping $\varphi : D \to D'$ having the ACL-property. For any smooth function $f : D' \to \mathbb{R}$, the composition $f \circ \varphi : D \to \mathbb{R}$ possesses the ACL-property and has first partial derivatives $\frac{\partial}{\partial x_i}(f \circ \varphi)$ calculated by the classical chain rule. Applying formula (1) we see that the operator $\varphi^* : W_p^1(D') \cap C^1(D') \to W_p^1(D)$ is bounded. It follows that the preimage of a set of capacity zero is a set of measure zero. It follows that the extension of the operator $\varphi^* : W_p^1(D') \cap C^1(D') \to W_p^1(D)$ by continuity coincides with the composition operator.

Let $E \subset \mathbb{R}^n$ be a measurable set. A mapping $\varphi : E \to \mathbb{R}^n$ is called *Lipschitz* if $|\varphi(y) - \varphi(x)| \leq L|y - x|$ for some constant L and all points $x, y \in E$.

PROPOSITION 2. *Suppose that the mapping $\varphi : D \to D'$ induces a bounded operator $\varphi^* : L_p^1(D') \to L_p^1(D)$, $n < p < \infty$. Then*
1) $|J(x, \varphi)|^{\frac{1}{n}} \leq |D\varphi|(x) \leq \|H_p(\cdot) \mid L_\infty(D')\|^{\frac{p}{p-n}}$ *for almost all $x \in D$;*
2) $\varphi \in W_{\infty,\text{loc}}^1(D)$;
3) *the mapping φ can be modified on a set of measure zero to be a continuous mapping $\Phi : D \to \mathbb{R}^n$ which is Lipschitz on every ball $B \subset D$;*
4) *the mapping Φ has Luzin's property $\mathcal{N}$.*

PROOF. Lemma 1 implies that φ can be modified on a set of measure zero to belong to ACL(D), to have the finite distortion and $H_p(\cdot) \in L_\infty(D')$. In the case $J(x, \varphi) = 0$ for almost all $x \in D$, the statement of the proposition is evident because the ACL-property and finite distortion imply that φ differs from a mapping which is constant on every connected component of the open set D on a set of measure zero. Therefore, we assume that $J(x, \varphi) \neq 0$ on a set of positive measure.

By the pointwise relation $|J(x, \varphi)| \leq |D\varphi|^n(x)$ and Jensen's inequality, we have

$$(6) \quad H_p(y) = \left(\sum_{\substack{z \in (\varphi^{-1}(y) \cap V) \setminus \Sigma_\varphi, \\ J(z,\varphi) \neq 0}} \frac{|D\varphi|^p(z)}{|J(z, \varphi)|} \right)^{\frac{1}{p}}$$

$$= \left(\sum_{\substack{z \in (\varphi^{-1}(y) \cap V) \setminus \Sigma_\varphi, \\ J(z,\varphi) \neq 0}} \left(\frac{|D\varphi|^n(z)}{|J(z, \varphi)|} \right)^{\frac{p}{n}} \cdot \frac{1}{|J(z, \varphi)|^{1-\frac{p}{n}}} \right)^{\frac{1}{p}} \geq |J(x, \varphi)|^{\frac{1}{n} - \frac{1}{p}}$$

for almost all $y = \varphi(x) \in D'$. It follows that

$$|J(x, \varphi)|^{\frac{1}{n} - \frac{1}{p}} \leq \|H_p(\cdot) \mid L_\infty(D')\|$$

for almost all $x \in D$. Indeed, if

$$E = \{y \in D' : H_p(y) < |J(x, \varphi)|^{\frac{1}{n} - \frac{1}{p}}, \; \varphi(x) = y\}$$

then $|E| = 0$. Hence, by the change-of-variable formula, $J(x, \varphi) = 0$ almost everywhere on the set $\varphi^{-1}(E)$.

Further,

$$H_p(y) = \left(\sum_{\substack{z \in (\varphi^{-1}(y) \cap V) \setminus \Sigma_\varphi, \\ J(z,\varphi) \neq 0}} \frac{|D\varphi|^p(z)}{|J(z, \varphi)|} \right)^{\frac{1}{p}}$$

$$\geq \left(\frac{1}{\|H_p(\cdot) \mid L_\infty(D')\|} \right)^{\frac{n}{p-n}} \left(\sum_{\substack{z \in (\varphi^{-1}(y) \cap V) \setminus \Sigma_\varphi, \\ J(z,\varphi) \neq 0}} |D\varphi|^p(z) \right)^{\frac{1}{p}}$$

$$\geq \left(\frac{1}{\|H_p(\cdot) \mid L_\infty(D')\|} \right)^{\frac{n}{p-n}} |D\varphi|(x).$$

Evaluating the quantity $H_p(y)$ by means of $\|H_p(\cdot) \mid L_\infty(D')\|$, we obtain assertion 1.

The boundedness of the norm of the gradient and the ACL-property of the mapping φ imply assertion 2.

Modifying the mapping on a set of measure zero, we obtain assertion 3. Assertion 4 is a consequence of the fact that Φ is locally Lipschitz. $\square$

THEOREM 2. *The mapping* $\varphi : \mathbb{R}^n \to \mathbb{R}^n$ *induces a bounded embedding operator* $\varphi^* : W_p^1(\mathbb{R}^n) \to W_p^1(\mathbb{R}^n)$, $n < p < \infty$, *if and only if φ has the following properties:*
1) φ *can be modified on a set of measure zero to be Lipschitz on every ball* $B \subset D$, φ *has the finite distortion and* $H_p(\cdot) \in L_\infty(\mathbb{R}^n)$;
2) $V = \sup\limits_{z \in \mathbb{R}^n} \int_{B(z,1)} \sum\limits_{x \in \varphi^{-1}(y),\, J(x,\varphi) \neq 0} \frac{1}{|J(x,\varphi)|}\, dy \leq c\|\varphi^*\|^p$, *for some constant c.*

The norm of the operator $\varphi^* : W_p^1(\mathbb{R}^n) \to W_p^1(\mathbb{R}^n)$, $n < p < \infty$, *is equivalent to* $V^{\frac{1}{p}} + \|H_p(\cdot) \mid L_\infty(\mathbb{R}^n)\|$.

PROOF. Let the embedding operator $\varphi^* : W_p^1(\mathbb{R}^n) \to W_p^1(\mathbb{R}^n)$, $n < p < \infty$, be bounded. We proved in Lemma 1 that the mapping φ can be modified on a set of measure zero to belong to the class $\mathrm{ACL}(D)$, that it has finite distortion and that $\|H_p(\cdot) \mid L_\infty(D')\| \leq c\|\varphi^*\|$.

Proposition 2 implies that φ can be modified on a set of measure zero to be Lipschitz on every ball $B \subset D$. We obtain condition 2 and the estimate $V \leq c\|\varphi^*\|^p$, where c is some constant, if we substitute the function $h(\cdot) = \eta(\cdot - z)$ instead of f in the inequality $\|\varphi^* f \mid W_p^1(\mathbb{R}^n)\|^p \leq \|\varphi^*\|^p \|f \mid W_p^1(\mathbb{R}^n)\|^p$, where $\eta \in C_0^\infty(B(0,2))$ is a test function such that $\eta \equiv 1$ on $B(0,1)$ and $z \in \mathbb{R}^n$ is arbitrary.

To prove the sufficiency of conditions it remains only to estimate $\|\varphi^*(f) \mid L_p(\mathbb{R}^n)\|$ since by Theorem 1 of [**VU2**], the operator $\varphi^* : L_p^1(\mathbb{R}^n) \to L_p^1(\mathbb{R}^n)$ is bounded with norm $\|\varphi^*\|$ equivalent to $\|H_p(\cdot) \mid L_\infty(\mathbb{R}^n)\|$. To do this, we fix a covering $\{B_j = B(z_j, 1/2)\}$ of the space $\mathbb{R}^n$ of a finite multiplicity such that the covering $\{2B_j = B(z_j, 1)\}$ also has finite multiplicity. Let $h_j(\cdot) = \xi(\cdot - z_j)$, where $\xi \in C_0^\infty(B(0,1))$ is a test function such that $\xi \equiv 1$ on $B(0, 1/2)$ and $0 \leq \xi \leq 1$. Assuming φ and $f \in W_p^1(\mathbb{R}^n)$ to be continuous, we have

$$\|\varphi^*(f) \mid L_p(\mathbb{R}^n)\|^p \leq \sum_j \|\varphi^*(fh_j) \mid L_p(\mathbb{R}^n)\|^p$$

$$\leq \sum_j |\varphi^{-1}(2B_j)| \|fh_j \mid C(\varphi^{-1}(2B_j))\|^p \leq \sum_j |\varphi^{-1}(2B_j)| \|f \mid C(2B_j)\|^p$$

$$\leq \sup_j |\varphi^{-1}(2B_j)| \sum_j \|f \mid W_p^1(2B_j)\|^p \leq VC \|f \mid W_p^1(\mathbb{R}^n)\|^p,$$

since by (1),

$$|\varphi^{-1}(2B_j)| = \int_{B(z_j, 1)} \sum_{x \in \varphi^{-1}(y), J(x,\varphi) \neq 0} \frac{1}{|J(x,\varphi)|}\, dy \leq V.$$

Here C depends only on the multiplicity of the covering and the norm of the embedding operator $i : W_p^1(\mathbb{R}^n) \to C(\mathbb{R}^n)$. $\square$

For a mapping $\varphi : D \to D'$ of the class $\mathrm{ACL}(D)$, we introduce the characteristic

$$K_p(x) = \inf\{k : |D\varphi|(x) \leq k|J(x,\varphi)|^{1/p}\}, \quad x \in \Omega.$$

Recall that the *capacity* $\operatorname{cap}(e; W_p^1(\Omega))$ of the compact $e \subset \Omega$ in the space $W_n^1(\Omega)$ is the quantity

$$\operatorname{cap}(e; W_p^1(\Omega)) = \inf \|g \mid W_p^1(\Omega)\|^p,$$

where the infimum is taken over all continuous functions $g \in W_p^1(\Omega)$ such that $g \geq 1$ on e. For properties and applications of capacity see, for instance, [**CV, M3, MS, R1**].

THEOREM 3. *A mapping $\varphi : \mathbb{R}^n \to \mathbb{R}^n$ induces a bounded embedding operator $\varphi^* : W_n^1(\mathbb{R}^n) \to W_n^1(\mathbb{R}^n)$ if and only if*
  1) *φ can be modified on a set of measure zero to belongs to* $\operatorname{ACL}(\mathbb{R}^n)$ *and has the finite distortion;*
  2) $K_n(\cdot) \in L_\infty(\mathbb{R}^n)$;
  3) *the Banach indicatrix* $M(y, \varphi) = \#\{x \in \varphi^{-1}(y) \setminus \Sigma_\varphi\}$ *belongs to* $L_\infty(\mathbb{R}^n)$;
  4) *the least constant in the inequality*

$$\int_e \sum_{x \in \varphi^{-1}(y) \setminus \Sigma_\varphi,\, J(x,\varphi) \neq 0} \frac{1}{|J(x,\varphi)|}\, dy \leq C \operatorname{cap}(e; W_n^1(\mathbb{R}^n))$$

  is bounded where e is an arbitrary compact, the diameter of which does not exceed 1.

  The norm of the operator $\varphi^* : W_n^1(\mathbb{R}^n) \to W_n^1(\mathbb{R}^n)$ does not exceed $C^{\frac{1}{n}} + \|M(\cdot, \varphi) \mid L_\infty(\mathbb{R}^n)\|^{\frac{1}{n}} \cdot \|K_n(\cdot) \mid L_\infty(\mathbb{R}^n)\|$.

  If the Jacobian $J(x,\varphi)$ has a single sign, then φ can be modified on a set of measure zero to be a mapping with bounded distortion of finite multiplicity [**R2**]. *If, additionally, $\|M(\cdot, \varphi) \mid L_\infty(\mathbb{R}^n)\| = 1$, then φ is a quasiconformal mapping.*

PROOF. By Lemma 1, the boundedness of the operator $\varphi^* : W_n^1(D') \to W_n^1(D)$ implies that φ can be modified on a set of measure zero to belong to $\operatorname{ACL}(D)$, φ has the finite distortion and $H_n(\cdot) \in L_\infty(D')$. By Theorem 1 of [**VU2**], the operator $\varphi^* : L_n^1(D') \to L_n^1(D))$ is bounded and its norm is equivalent to the value $\|H_n(\cdot) \mid L_\infty(D')\|$. From Theorem 5 of [**VU2**], it follows that conditions 1–3 are necessary and sufficient for the operator $\varphi^* : L_n^1(\mathbb{R}^n) \to L_n^1(\mathbb{R}^n)$ to be bounded. Moreover, its norm does not exceed

$$\|M(\cdot, \varphi) \mid L_\infty(\mathbb{R}^n)\|^{\frac{1}{n}} \cdot \|K_n(\cdot) \mid L_\infty(\mathbb{R}^n)\|.$$

Now let $f \in W_n^1(\mathbb{R}^n)$. As is known [**MS**],

$$\|f \circ \varphi \mid L_n(\mathbb{R}^n)\|^n = \int_{\mathbb{R}^n} |f(y)|^n \sum_{x \in \varphi^{-1}(y) \setminus \Sigma_\varphi,\, J(x,\varphi) \neq 0} \frac{1}{|J(x,\varphi)|}\, dy$$
$$\leq \|i\| \cdot \|f \mid W_n^1(\mathbb{R}^n)\|^n$$

if and only if condition 4 of the theorem holds. The least constant $\|i\|$ in this inequality is equivalent to the least constant C of condition 4. $\square$

To complete our exposition, we formulate the following result.

COROLLARY 1. ([**V8, V9**]) *A homeomorphism $\varphi : \mathbb{R}^n \to \mathbb{R}^n$ induces a bounded embedding operator $\varphi^* : W_n^1(\mathbb{R}^n) \to W_n^1(\mathbb{R}^n)$ if and only if*
  1) $\varphi \in W_{n,\operatorname{loc}}^1(\mathbb{R}^n)$;
  2) *the mapping φ is quasiconformal and $K_n \leq c_1 \|\varphi^*\|$;*

3) $R = \sup\limits_{x \in \mathbb{R}^n} \sup\limits_{y \in B(x,1)} |\varphi^{-1}(x) - \varphi^{-1}(y)| \leq c_2 \|\varphi^*\|$ where c_2 depends on the coefficient of distortion K_n.

The norm of the operator $\varphi^* : W_n^1(\mathbb{R}^n) \to W_n^1(\mathbb{R}^n)$ does not exceed the quantity $(c_3 R + 1)K_n$. (Here c_1, c_3 are constants.)

PROOF. If a homeomorphism $\varphi : \mathbb{R}^n \to \mathbb{R}^n$ induces a bounded embedding operator $\varphi^* : W_n^1(\mathbb{R}^n) \to W_n^1(\mathbb{R}^n)$, then it is evident that $\varphi \in W_{n,\mathrm{loc}}^1(\mathbb{R}^n)$.

Suppose that the embedding operator $\varphi^* : W_n^1(\mathbb{R}^n) \to W_n^1(\mathbb{R}^n)$ is bounded. Then Lemma 1 and ([**VU2**], Theorem 1) imply the boundedness of the operator $\varphi^* : L_n^1(\mathbb{R}^n) \to L_n^1(\mathbb{R}^n)$ and the estimate on its norm. This condition is equivalent to the quasiconformality of the mapping φ; see, for instance, [**Mo**, **Vä**]. Moreover, the mapping $\psi = \varphi^{-1}$ is also quasiconformal; therefore,

$$\sup_{z \in \mathbb{R}^n} \frac{\max\limits_{y \in S(z,1)} |\psi(z) - \psi(y)|}{\min\limits_{y \in S(z,1)} |\psi(z) - \psi(y)|} = D < \infty,$$

where D depends on the quasiconformality coefficient K_n [**Mo**, **Vä**]. (We may express this by saying that the image of a ball is a quasiball.) We now obtain condition 3 and the estimate

$$R = \sup_{x \in \mathbb{R}^n} \sup_{y \in B(x,1)} |\varphi^{-1}(x) - \varphi^{-1}(y)| \leq c_4 \|\varphi^*\|,$$

where c_4 is a constant, by substituting the function $h(\cdot) = \eta(\cdot - z)$ in place of f in the inequality

$$\|\varphi^* f \mid W_n^1(\mathbb{R}^n)\| \leq \|\varphi^*\| \|f \mid W_n^1(\mathbb{R}^n)\|,$$

where $\eta \in C_0^\infty(B(0,2))$ is a test function such that $\eta \equiv 1$ on $B(0,1)$ and $z \in \mathbb{R}^n$ is arbitrary.

To verify the sufficiency of the conditions, it remains only to estimate $\|\varphi^*(f) \mid L_n(\mathbb{R}^n)\|$ since in view of [**V8**, **V9**] (see also [**VU2**], Theorem 1), the operator $\varphi^* : L_n^1(\mathbb{R}^n) \to L_n^1(\mathbb{R}^n)$ is bounded and has norm K_n. To do this, we fix a covering $\{B_j = B(z_j, 1/2)\}$ of the space $\mathbb{R}^n$ of finite multiplicity such that the covering $\{2B_j = B(z_j, 1)\}$ is also of finite multiplicity. Let $h_j(\cdot) = \xi(\cdot - z_j)$, where $\xi \in C_0^\infty(B(0,1))$ is a test function such that $\xi \equiv 1$ on $B(0, 1/2)$ and $0 \leq \xi \leq 1$. Applying Poincaré's inequality, we obtain

$$\|\varphi^*(f) \mid L_n(\mathbb{R}^n)\|^n \leq C_1 \sum_j \|\varphi^*(fh_j) \mid L_n(\mathbb{R}^n)\|^n$$

$$\leq C_1 C_2 R^n \sum_j \|\nabla(\varphi^*(fh_j)) \mid L_n(\varphi^{-1}(2B_j))\|^n$$

$$\leq C_1 C_2 R^n K_n^n \sum_j \|\nabla(fh_j) \mid L_n(2B_j)\|^n$$

$$\leq C_2 C_3 R^n K_n^n \|f \mid W_n^1(\mathbb{R}^n)\|^n,$$

where C_1, C_2, C_3 are constants depending on the constant in Poincaré's inequality and the multiplicity of the covering. The boundedness of the operator $\varphi^* : W_n^1(\mathbb{R}^n) \to W_n^1(\mathbb{R}^n)$ and the estimate on its norm are proved. $\square$

PROPOSITION 3. *Let f be a function of Sobolev class, as in Lemma 1, Theorems 1–3, Proposition 2 and Corollary 1. The partial derivatives of the composition $f \circ \varphi$ are calculated by the formula*

$$\frac{\partial(f \circ \varphi)}{\partial x_i}(x) = \begin{cases} \sum_{j=1}^{n} \frac{\partial f}{\partial y_j}(\varphi(x)) \frac{\partial \varphi_j}{\partial x_i}(x), & \text{if } J(x, \varphi) \neq 0, \\ 0 & \text{otherwise} \end{cases}$$

for almost all $x \in D$.

PROOF. We can assume that φ has the ACL-property. In the case of a smooth function f, the chain rule is derived by taking into consideration the ACL-property of the mapping φ and its bounded distortion. The general case is obtained as a result of approximation of an arbitrary function of Sobolev class by a sequence of smooth functions. Indeed, if, for example, $f_n \to f$ in $L_p^1(D')$ (or $W_p^1(D')$), then we can suppose that $\frac{\partial f_n}{\partial y_j}(y) \to \frac{\partial f}{\partial y_j}(y)$ everywhere in $D' \setminus \Sigma$ where $\Sigma \subset D'$ is a set of measure zero. Then by (1), the Jacobian $J(x, \varphi)$ vanishes on $\varphi^{-1}(\Sigma)$ a.e. Thus we have $f_n \circ \varphi \to f \circ \varphi$ in $L_p^1(D)$ (or $W_p^1(D)$) and

$$\frac{\partial(f_n \circ \varphi)}{\partial x_i}(x) = \sum_{j=1}^{n} \frac{\partial f_n}{\partial y_j}(\varphi(x)) \frac{\partial \varphi_j}{\partial x_i}(x) \to \sum_{j=1}^{n} \frac{\partial f}{\partial y_j}(\varphi(x)) \frac{\partial \varphi_j}{\partial x_i}(x) \quad \text{a.e.}$$

on $D \setminus (\varphi^{-1}(\Sigma) \cup \{x \in D : J(x, \varphi) = 0\})$ and $0 = \frac{\partial(f_n \circ \varphi)}{\partial x_i}(x) = \frac{\partial(f \circ \varphi)}{\partial x_i}(x)$ a.e. on $\varphi^{-1}(\Sigma) \cup \{x \in D : J(x, \varphi) = 0\}$, since φ has finite distortion. $\square$

2. Let $E \subset \mathbb{R}^n$ be a measurable set. A mapping $\varphi : E \to \mathbb{R}^n$ is called *bi-Lipschitz* if it is Lipschitz and has an inverse mapping $\varphi^{-1} : \varphi(E) \to E$ which is also Lipschitz.

A homeomorphism $\varphi : D \to D'$ of two open sets is called *quasi-isometric* if for some constant M the relations

$$\varlimsup_{y \to x} \frac{|\varphi(y) - \varphi(x)|}{|y - x|} \leq M, \quad \varlimsup_{y \to z} \frac{|\varphi^{-1}(y) - \varphi^{-1}(z)|}{|y - z|} \leq M$$

hold at all points $x \in D$ and $z \in D'$.

It is well-known (see, for instance, [**VGR**]) that any quasi-isometric mapping $\varphi : D \to D'$ induces an isomorphism $\varphi^* : W_p^1(D') \to W_p^1(D)$, $1 \leq p < \infty$, of Sobolev spaces. Conversely, results of [**G**] may be interpreted to say that if a homeomorphism $\varphi : D \to D'$ induces an isomorphism $\varphi^* : L_p^1(D') \to L_p^1(D)$, $1 \leq p < \infty$, $p \neq n$, of homogeneous Sobolev spaces, then $\varphi : D \to \mathbb{R}^n$ is quasi-isometric.

Recall that two open sets Ω_1 and Ω_2 are said to be $(1, p)$-*equivalent* [**VG1**] if, for any function $f \in W_p^1(\Omega_1)$ ($f \in W_p^1(\Omega_2)$), the restriction $f|_{\Omega_1 \cap \Omega_2}$ can be extended in a unique way to be a function $\tilde{f} \in W_p^1(\Omega_2)$ ($\tilde{f} \in W_p^1(\Omega_1)$). It follows that the operator $W_p^1(\Omega_1) \ni f \mapsto \tilde{f} \in W_p^1(\Omega_2)$ is an isomorphism of vector spaces. This definition implies that not only open sets Ω_1 and Ω_2 are $(1, p)$-equivalent but Ω_2 and Ω_1 are also $(1, p)$-equivalent. Moreover, if open sets Ω_1 and Ω_2 are $(1, p)$-equivalent and open sets Ω_2 and Ω_3 are $(1, p)$-equivalent, then the open sets Ω_1 and Ω_3 are $(1, p)$-equivalent.

In [**VG1**], a stronger result is obtained comparing with [**G**]: if a measurable mapping $\varphi : D \to D'$ induces an isomorphism $\varphi^* : W_p^1(D') \to W_p^1(D)$, $p > n$, of

Sobolev spaces, then $\varphi : D \to \mathbb{R}^n$ can be modified on a set of measure zero in such a way that the modified mapping $\Phi : D \to \mathbb{R}^n$ becomes quasi-isometric. Moreover, the open sets D' and $\Phi(D)$ are $(1,p)$-equivalent. In this paper, we extend this result to the whole scale of Sobolev spaces.

THEOREM 4. *A mapping $\varphi : D \to D'$ induces an isomorphism $\varphi^* : W_p^1(D') \to W_p^1(D)$, $1 \leq p < \infty$, $p \neq n$, if and only if $\varphi : D \to D'$ coincides almost everywhere with a quasi-isometric homeomorphism $\Phi : D \to \mathbb{R}^n$. Moreover, the open sets D' and $\Phi(D)$ are $(1,p)$-equivalent.*

PROOF. The sufficiency of the condition is verified in the standard way; for details, see [**V9, VG1**].

The necessity of the theorem in the case $1 \leq p < n$ and $D \neq \mathbb{R}^n$ or $D' \neq \mathbb{R}^n$ is a new result. The proof given below is also new. Other proofs of the necessity for $p > n$ ($1 < p < n$ and $D = D' = \mathbb{R}^n$), known earlier, can be found in [**VG1, GR**] ([**R**]).

In the proof given below, we reduce the cases $p > n$ and $p < n$ to the same situation.

Common part of arguments. By Lemma 1, we can suppose that the mapping φ belongs to the class ACL(D), has finite distortion and $H_p(\cdot) \in L_\infty(D')$, $1 \leq p < \infty$, $p \neq n$. Let
$$Z = \{x \in D : J(x, \varphi) = 0\}.$$
There exists a family $\{E_j\}$, $j = 0, 1, \ldots$, of pairwise disjoint bounded sets $E_j \subset D \setminus Z$ such that $|E_0| = 0$, $|E_j| > 0$ for all $j \in \mathbb{N}$, $\varphi|_{E_j}$ is a bi-Lipschitz mapping for all $j \geq 1$, and $D = Z \cup \bigcup_{j=0}^{\infty} E_j$ (see, for instance, [**V5, VU4**]). We assume that the set Σ_φ from Proposition 1 is included in E_0 and $Z \cap E_0 = \emptyset$.

Observe that the assumption $|\varphi(E_i) \cap \varphi(E_j)| > 0$ for some $i, j \geq 1$, $i \neq j$, leads immediately to a contradiction with the isomorphism of the operator φ^*. Indeed, by this assumption, for any compact set $A \subset \varphi(E_i) \cap \varphi(E_j)$ of positive measure, the pre-images $\varphi^{-1}(A) \cap E_i$ and $\varphi^{-1}(A) \cap E_j$ are disjoint, and each of them is compact and has positive measure. Then a finite smooth function ψ, which equals 1 (0) on $\varphi^{-1}(A) \cap E_i$ ($\varphi^{-1}(A) \cap E_j$) belongs to the Sobolev space $W_p^1(D)$ and cannot be a composition of any function $f \in W_p^1(D')$ with the mapping φ.

It follows that outside of some set $F \subset D \setminus (Z \cup E_0)$ of measure zero the mapping φ is bijective on $D \setminus (Z \cup E_0 \cup F)$. Hence, on $\varphi(D) \setminus \varphi(E_0 \cup Z \cup F)$, there exists an inverse mapping ψ. From this, it follows that $\varphi^{*-1}(g)(y) = g \circ \psi(y)$ for all $y \in \varphi(D) \setminus \varphi(E_0 \cup Z \cup F)$.

Since the mapping ψ is Lipschitz on every set $\varphi(E_i \setminus F)$, it has an approximate differential (see, for instance, [**F**]) almost everywhere on its range of definition.

Fix a cut-off function $\eta \in C_0^\infty(\mathbb{R}^n)$ which equals 1 on $B(0,1)$ and 0 outside of $B(0,2)$. Let x be a density point of $D \setminus (Z \cup E_0 \cup F)$. Substituting the function $h_j(z) = (z - x)_j \eta(\frac{z-x}{r})$, $j = 1, \ldots, n$, in place of f in the inequality $\|\varphi^{*-1} f \mid L_p^1(D')\| \leq \|\varphi^{*-1}\| \|f \mid L_p^1(D)\|$, where $(z - x)_j$ denotes the j-th coordinate of the vector $z - x$ and $B(x, 2r) \subset D$, we have

$$(7) \qquad \int_{\varphi(B(x,r) \setminus (E_0 \cup Z \cup F))} |D\psi|^p(y)\, dy \leq C \|\varphi^{*-1}\|^p |B(x, 2r)|,$$

in which C is a constant depending only on the dimension n and the index p. Applying formula (1) to the left-hand side of (7), we obtain

$$\int_{\varphi(B(x,r)\setminus(E_0\cup Z\cup F))}|D\psi|^p(y)\,dy = \int_{B(x,r)\setminus(E_0\cup Z\cup F)}\frac{|D\psi|^p(\varphi(z))}{|J(\varphi(z),\psi)|}\,dz$$

$$\leq C\|\varphi^{*-1}\|^p|B(x,2r)|.$$

By the Lebesgue differentiability theorem, it follows that

$$\frac{|D\psi|^p(y)}{|J(y,\psi)|} \leq C\|\varphi^{*-1}\|^p$$

for almost all $y \in \varphi(D\setminus(E_0\cup Z\cup F))$, since the mapping φ is injective and possesses Luzin's property $\mathcal{N}$ on $D\setminus(E_0\cup Z\cup F)$. Hence we have the pointwise estimate

(8) $$|D\psi|^p(y) \leq C\|\varphi^{*-1}\|^p|J(y,\psi)|$$

between the norm of the formal differential $D\psi(y)$ and its Jacobian for almost all $y \in \varphi(D\setminus(E_0\cup Z\cup F))$.

Case $p > n$. In this case, the mapping φ has the following additional properties.

1) By Proposition 2, φ can be modified on a set of measure zero to a mapping $\Phi : D \to \mathbb{R}^n$ which is Lipschitz on every ball $B \subset D$, has finite distortion and whose Jacobian satisfies $|J(y,\Phi)| \leq \gamma < \infty$ almost everywhere in D for some constant γ. Moreover, Φ possesses Luzin's property $\mathcal{N}$ and is differentiable almost everywhere in D. Therefore, $|\Phi(E_0 \cup F)| = 0$. Since $\Phi(Z)$ has measure zero, by formula (1), $|\Phi(E_0 \cup F \cup Z)| = 0$. Thus, the mapping ψ is defined almost everywhere on $\varphi(D)$.

2) Applying arguments used in proving of (6), we obtain from (8)

$$C\|\varphi^{*-1}\|^p \geq \frac{D\psi|^p(y)}{|J(y,\psi)|}$$
$$= \left(\frac{|D\psi|^n(y)}{|J(y,\psi)|}\right)^{\frac{p}{n}} \cdot \frac{1}{|J(y,\psi)|^{1-\frac{p}{n}}} \geq |J(y,\psi)|^{\frac{p}{n}-1}$$

for almost all $y = \varphi(x) \in \varphi(D)$. Therefore, the Jacobian $|J(y,\psi)|$ is bounded from above: $|J(y,\psi)| \leq \alpha < \infty$ almost everywhere in $\varphi(D)$.

3) $|J(x,\Phi)| = |J^{-1}(y,\psi)|$, $y = \Phi(x)$, for almost all points $x \in D\setminus Z$. It follows that $\alpha^{-1} \leq |J(x,\Phi)| \leq \gamma$ for almost all points $x \in D\setminus Z$.

4) From the previous property, it follows that the locally Lipschitz mapping $\Phi : D \to \mathbb{R}^n$ is invertible on $D\setminus Z$ outside a set of measure zero and satisfies not only the pointwise estimate $|D\Phi|^p(x) \leq K_p|J(x,\Phi)|$ for almost all $x \in D$, where K_p is some constant, but also the pointwise estimate $|D\Phi|^n(x) \leq K_n|J(x,\Phi)|$ for almost all $x \in D$, where K_n is a constant depending on n and $\|\varphi^*\|$. By [**VU2**], Theorem 5, the mapping φ induces a bounded operator $\varphi^* : L_n^1(D') \to L_n^1(D)$. By [**VU2**], Corollary 4, we have $|Z| = 0$, where $Z = \{x \in D : J(x,\varphi) = 0\}$.

Case $1 \leq p < n$. In this case, both the set $Z = \{x \in D : J(x,\varphi) = 0\}$ of zeros of the Jacobian and its image $\varphi(Z)$ have measure zero by [**VU2**], Theorem 4, and Proposition 1. In addition, the mapping φ has the following additional properties.

5) Since φ is almost bijective, it follows from Proposition 1 (see also [**VU2**], Proposition 4) that $|J^{-1}(x,\varphi)| = |J(y,\psi)|$, $y = \varphi(x)$, for almost all points $x \in D$.

6) By Theorem 3 of [**VU2**], the mapping $\psi : \varphi(D \setminus (E_0 \cup F)) \to D \setminus (E_0 \cup F)$ has Luzin's property $\mathcal{N}$ and its Jacobian is bounded from above almost everywhere: $|J(y, \psi)| \leq \alpha < \infty$. It follows that $|J(x, \varphi)| = |J(y, \psi)|^{-1} \geq \alpha^{-1} > 0$, $y = \varphi(x)$, for almost all points $x \in D \setminus (E_0 \cup F)$.

7) $|D\psi|^p(y) \leq C\|\varphi^{*-1}\|^p |J(y, \psi)| \leq \alpha C\|\varphi^{*-1}\|^p$ and $|J(y, \psi)| \geq \gamma^{-1} > 0$ for almost all $y \in \varphi(D \setminus \varphi(E_0 \cup F))$. Here the constant C is independent of the choice of domains and the mapping φ.

We use estimate (8) to prove the last property. Indeed,

$$C\|\varphi^{*-1}\|^p \geq \frac{|D\psi|^p(y)}{|J(y, \psi)|} = \left(\frac{|D\psi|^n(y)}{|J(y, \psi)|}\right)^{\frac{p}{n}} \frac{1}{|J(y, \psi)|^{1-\frac{p}{n}}} \geq \frac{1}{|J(y, \psi)|^{1-\frac{p}{n}}};$$

therefore, $|J(y, \psi)| \geq \gamma^{-1} > 0$ for almost all $y \in \varphi(D \setminus (E_0 \cup F))$.

8) $|D\varphi|(x) \leq L$ for almost all $x \in D$.

This property follows from the inequalities $|D\varphi|^p(x) \leq C_1 |J(x, \varphi)|$ and $|J(x, \varphi)| \leq \gamma$, which are valid for almost all $x \in D$ (the first property follows from Lemma 1 and the second one is a consequence of properties 5 and 7).

9) On every ball $B \subset D$, the mapping φ coincides almost everywhere with a Lipschitz mapping Φ.

10) The image $\Phi(E_0 \cup F)$ has measure zero.

11) The mapping $\Phi : D \to \mathbb{R}^n$ is Lipschitz on every ball $B \subset D$ and invertible outside of a set of measure zero; and $|D\Phi|^n(x) \leq K_n |J(x, \Phi)|$ for almost all $x \in D$, where K_n is some constant depending on n and $\|\Phi^*\|$. By Theorem 5 of [**VU2**], the operator $\Phi^* : L_n^1(D') \to L_n^1(D)$ is bounded.

Note that the findings of 4) and 11) coincide. Further, we have *a common part of arguments*.

12) The mapping Φ is topologically non-degenerate, i.e., if a point $\Phi(z)$ does not belong to the image $\Phi(S(z, r))$ of the sphere $S(z, r)$, where $B(z, r) \Subset D$, then the topological degree $\mu(\Phi(z), \Phi, B(z, r))$ does not vanish.

To prove this, we show that the assumption $\mu(\Phi(z), \Phi, B(x, r)) = 0$ contradicts the condition of injectivity almost everywhere. Indeed, in this case, we substitute the characteristic function χ_V of those connected component V of the complement $\mathbb{R}^n \setminus \Phi(B(z, r))$ to which the point $\Phi(z)$ belongs, in place of f in the change-of-variable formula with topological degree (see, for instance, [**VG2**, **V5**])

$$(9) \qquad \int_{B(z,r)} f(\Phi(x)) J(x, \Phi) \, dx = \int_{\mathbb{R}^n} f(y) \mu(y, \Phi, B(z, r)) \, dy$$

to obtain

$$\int_{B(z,r) \cap \Phi^{-1}(V)} J(x, \Phi) \, dx = 0.$$

Since, by the continuity of Φ, we have $|B(z, r) \cap \Phi^{-1}(V)| > 0$, formula (1) implies $|\Phi(B(z, r)) \cap V| > 0$ (as $J(x, \Phi) \neq 0$ almost everywhere). On the other hand, the sets $Z_+ = \{x \in B(z, r) \cap \Phi^{-1}(V) : J(x, \Phi) > 0\}$ and $Z_- = \{x \in B(z, r) \cap \Phi^{-1}(V) : J(x, \Phi) < 0\}$ do not intersect, and each of them has positive measure (if one of them had measure zero, then from (9) we would obtain $\mu(\Phi(z), \Phi, B(z, r)) \neq 0$). It follows that each of the images $\Phi(Z_+)$ and $\Phi(Z_-)$ has positive measure and they intersect on a set of positive measure (as we have

$$0 = \mu(y, \Phi, B(z,r)) = \sum_{x \in \Phi^{-1}(y) \cap B(z,r)} \operatorname{sign} J(x, \Phi) \text{ for almost all } y \in V; \text{ see, for}$$
instance, [**V5**]).

13) For any point $y \in D'$, the one-dimensional Hausdorff measure of the pre-image $\Phi^{-1}(y)$ equals zero.

Indeed, it was mentioned above that the operator $\Phi^* : L_n^1(D') \to W_n^1(D)$ is bounded and therefore the pre-image of a set of n-capacity zero in D' is a set of n-capacity zero in D (see, for instance, [**V10**]). In particular, the pre-image of a point has n-capacity zero. Hence, it has one-dimensional Hausdorff measure zero [**CV, R1, R2**].

Observe that the last property means that the mapping Φ is light: the pre-image $\Phi^{-1}(y)$ is totally disconnected for any point $y \in D'$.

14) The mapping Φ is a homeomorphism.

We first show that Φ is an open mapping. Let $x \in D$. Then, by the previous property, there exists a ball $B(x,r)$ such that $\Phi(x) \notin \Phi(S(x,r))$. By property 12, we have $\mu(\Phi(x), \Phi, B(x,r)) \neq 0$. Hence, by properties of the topological degree, $\Phi(x)$ is an interior point of the image $\Phi(B(x,r))$.

It is evident that a mapping which is continuous, open and injective almost everywhere is injective. Hence, Φ is a homeomorphism.

15) The mapping Φ is a quasi-isometric homeomorphism.

Indeed, Φ is locally Lipschitz and the inequality $|D\Phi|^n(x) \leq K_n J(x, \Phi)$ holds a.e. if Φ is a sense-preserving homeomorphism (in other words, if its topological degree is positive); otherwise, $|D\Phi|^n(x) \leq -K_n J(x, \Phi)$ a.e. By properties of quasi-conformal mappings [**Mo**], the inverse mapping $\Phi^{-1} = \Psi$ is also quasiconformal. As $\alpha^{-1} \leq |J(x, \Phi)| \leq \gamma$ a.e. in D, the mapping Φ is a quasi-isometric homeomorphism.

16) The domains D' and $\Phi(D)$ are $(1,p)$-equivalent.

Indeed, on the one hand, by assumption, the operator $\varphi^* : W_p^1(D') \to W_p^1(D)$ is an isomorphism. On the other hand, the operator $\Phi^* : W_p^1(\Phi(D)) \to W_p^1(D)$, $\Phi^*(f) = f \circ \Phi$, is also an isomorphism. Therefore, $\Phi^{*-1} \circ \varphi^* : W_p^1(D') \to W_p^1(\Phi(D))$ is an isomorphism such that $g|_{D' \cap \Phi(D)} = (\Phi^{*-1} \circ \varphi^*)(g)|_{D' \cap \Phi(D)}$. □

Properties of $(1,p)$-equivalent domains can be found in [**VG3**].

Generalizations of some results of the paper for mappings defined on Carnot groups can be found in [**VU5**].

References

[CV] V. M. Chernikov and S. K. Vodopyanov, *Sobolev Spaces and Hypoelliptic Equations.* II, Siberian Adv. Math. **6** (1996), no. 4, 64–96.

[F] H. Federer, *Geometric Measure Theory*, Springer-Verlag, Berlin, 1969.

[G] F. W. Gehring, *Lipschitz mappings and the p-capacity of rings in n-space*, Advances in the Theory of Riemann Surfaces, Ann. Math. Studies, vol. 66, 1971, pp. 175–193.

[GR] V. M. Gol'dshteĭn and Yu. G. Reshetnyak, *Quasiconformal Mappings and Sobolev Spaces*, Kluwer Academic Publishers, Dordrecht, 1990.

[MM] O. Martio and J. Malý, *Luzin's condition (N) and mappings of the class $W^{1,n}$*, J. Reine Angew. Math. **485** (1995), 19–36.

[M1] V. G. Maz'ya, *Classes of Sets and Embedding Theorems of Function Spaces. Some Problems of the Theory of Elliptic Operators*, Ph. D. Thesis, Leningrad Univ., Leningrad, 1961.

[M2] V. G. Maz'ya, *On weak solutions of Dirichlet and Neumann problems*, Trudy Moskovskogo Matematicheskogo Obschestva. T. 20, Moscow Univ., Moscow, 1969, pp. 137–172.

[M3] V. G. Maz'ya, *Sobolev Spaces*, Leningrad Univ., Leningrad, 1985.

[MS] V. G. Maz'ya and T. O. Shaposhnikova, *Multipliers in Spaces of Differentiable Functions*, Leningrad Univ., Leningrad, 1986.

[Mo] G. D. Mostow, *Quasi-conformal mappings in n-space and the rigidity of hyperbolic space forms*, Inst. Hautes Etudes Sci. Publ. Math. **34** (1968), 53–104.

[R1] Yu. G. Reshetnyak, *The concept of capacity in the theory of functions with generalized derivatives*, Sibirsk. Mat. Ž. **10** (1969), 1109–1138.

[R2] Yu. G. Reshetnyak, *Space Mappings with Bounded Distortion*, American Mathematical Society, Providence, RI, 1989.

[R] A. S. Romanov, *On the change of variable in spaces of Bessel and Riesz potentials*, Functional Analysis and Mathematical Physics, Academy of Science of USSR, Siberian branch, Institute of Mathematics, Novosibirsk, 1985, pp. 117–133.

[Vä] J. Väisälä, *Lectures on n-dimensional Quasiconformal Mappings*, Springer-Verlag, Berlin-New York, 1971.

[V1] S. K. Vodopyanov, *Composition operators of Sobolev spaces*, Modern Problems of Function Theory and its Applications, Saratov, Jan. 2002, Saratov, 2002, pp. 42–43.

[V2] S. K. Vodopyanov, *Taylor Formula and Function Spaces*, Novosibirsk State University, Novosibirsk, 1988.

[V3] S. K. Vodopyanov, *Weighted Sobolev spaces and the theory of mappings*, Tezisy dokladov Vsesouznoi matematicheskoi shkoly "Potential Theory". Katciveli, June 26 — July 3, 1991, Kiev: Institute of Mathematics AN Ukrainy, 1991, p. 7.

[V4] S. K. Vodopyanov, *Geometric Properties of Function Spaces with a Generalized Smoothness. Dr. Sc. Thesis*, Sobolev Institute of Mathematics, Novosibirsk, 1992.

[V5] S. K. Vodopyanov, *$\mathcal{P}$-differentiability on Carnot groups in different topologies and related topics*, Proceedings on Analysis and Geometry (S. K. Vodopyanov, ed.), Sobolev Institute Press, Novosibirsk, 2000, pp. 603–670.

[V6] S. K. Vodopyanov, *Monotone functions and quasiconformal mappings on Carnot groups*, Siberian Math. J. **37** (1996), 1113–1136.

[V7] S. K. Vodopyanov, *Differentiability of maps of Carnot group of Sobolev spaces*, Sb. Math. **194** (2003), 857–877.

[V8] S. K. Vodopyanov, *L_p-theory of potential and quasiconformal mappings on homogeneous groups*, Sovrem. Probl. Geom. Analiz, (Trudy Inst, Mat.), vol. 14, Nauka, Novosibirsk, 1989, pp. 45–89.

[V9] S. K. Vodopyanov, *Mappings of homogeneous groups and embeddings of function spaces*, Siberian Math. J. **30** (1989), 685–698.

[V10] S. K. Vodopyanov, *Topological and geometric properties of mappings with an integrable Jacobian in Sobolev classes. I*, Siberian Math. J. **41** (2000), 19–39.

[VG1] S. K. Vodopyanov and V. M. Gol'dšteĭn, *Functional characterizations of quasi-isometric mappings*, Sibirsk. Mat. Ž. **17** (1976), 768–773.

[VG2] S. K. Vodopyanov and V. M. Gol'dšteĭn, *Quasiconformal mappings and spaces of functions with the first generalized derivatives*, Sibirsk. Mat. Ž. **17** (1976), 515–531.

[VG3] S. K. Vodopyanov and V. M. Gol'dšteĭn, *A test of the removability of sets for L_p^1 spaces of quasiconformal and quasi-isomorphic mappings*, Sibirsk. Mat. Ž. **18** (1977), 48–68.

[VGR] S. K. Vodopyanov, V. M. Gol'dshtein, and Yu. G. Reshetnyak, *The geometric properties of functions with generalized first derivatives*, Uspekhi Mat. Nauk **34** (1979), 17–65.

[VU1] S. K. Vodopyanov and A. D. Ukhlov, *Substitution operators in Sobolev spaces*, Dokl. Akad. Nauk **386** (2002), 730–734.

[VU2] S. K. Vodopyanov and A. D. Ukhlov, *Superposition operators in Sobolev spaces*, Russian Math. (Iz. VUZ) **46** (2002), no. 10, 9–31.

[VU3] S. K. Vodopyanov and A. D. Ukhlov, *Sobolev spaces and (P,Q)-quasiconformal mappings of Carnot groups*, Siberian Math. J. **39** (1998), 665–682.

[VU4] S. K. Vodopyanov and A. D. Ukhlov, *Approximately differentiable transformations and the change of variables on nilpotent groups*, Siberian Math. J. **37** (1996), 62–78.

[VU5] S. K. Vodopyanov and A. D. Ukhlov, *Set functions and their applications in the theory of Lebesgue and Sobolev spaces. I*, Siberian Adv. Math. **14** (2004), no. 4, 1–48.

SOBOLEV INSTITUTE OF MATHEMATICS, ACADEMICIAN KOPTUG PR., 4, 630090 NOVOSIBIRSK, RUSSIA

E-mail address: vodopis@math.nsc.ru

New Results in Integral Geometry

V. V. Volchkov and Vit. V. Volchkov

Dedicated to Professor Lawrence Zalcman on the occasion of his sixtieth birthday

ABSTRACT. Among the results of this paper, we point out the definitive version of the local two-radii theorem on symmetric spaces and the reduced Heisenberg group, examples of Pompeiu sets with non-Lipschitz (and even fractal) boundary, the extreme variants of the Pompeiu problem on a sphere, and various definitive results related to the spherical Radon transform.

1. Introduction

Integral geometry deals with the problem of determining functions by their integrals over given families of sets. Most of the question arising here relate, in one way or another, to convolution equations. Some of the well-known publications in this field include works by J. Radon, F. John, J. Delsarte, L. Zalcman, C.A. Berenstein, M.L. Agranovsky and the monographs by L. Hörmander and S. Helgason. Until recently, research in this area was carried out mostly using the technique of the Fourier transform and corresponding methods of complex analysis.

In recent years, the first author has worked out an essentially different methodology, which has enabled him to establish best possible results for several well-known problems. Among these results, we can point out the definitive version of the two-radii theorem, the complete solution of Zalcman's problem on the holomorphy of a function with vanishing integrals over a conformally invariant family of circles, the solution of the support problem for mean values on the ball, the extreme variant of the Pompeiu problem, a description of injectivity sets of the Pompeiu transform for a broad class of distributions, definitive versions of uniqueness theorems for multiple trigonometric series with gaps, and many others.

The recent book [27] contains all these results and many others, with a number of applications to various branches of mathematics.

Here we present some new results in this field not included in [27]. The proof of these results is based on ideas proposed in [27] and their development.

The results in Sections 2.1, 3.1, 5.1, 5.2 are due to V.V. Volchkov. The results in Sections 2.2, 3.2, 3.3, 4.1, 4.2, 4.3, 5.3, are due to Vit.V. Volchkov.

2000 *Mathematics Subject Classification.* Primary 26B15, 43A85, 53C65; Secondary 53C35.

2. Uniqueness sets for the spherical Radon transform

2.1. Zalcman conjecture. Let $\mathbb{R}^n$ be a real Euclidean space of dimension $n \geq 2$ with the Euclidean norm $|\cdot|$, let $\mathbb{S}^{n-1} = \{x \in \mathbb{R}^n : |x| = 1\}$, and let $L_{\text{loc}}(\mathbb{R}^n)$ be the class of functions which are locally summable in $\mathbb{R}^n$. For every $f \in L_{\text{loc}}(\mathbb{R}^n)$, for any $x \in \mathbb{R}^n$, and almost all $r \in (0, +\infty)$ (with respect to Lebesgue measure), the spherical Radon transform of f is defined by

$$\mathcal{R}f(x,r) = \int_{\mathbb{S}^{n-1}} f(x+r\sigma)\, d\omega(\sigma),$$

where $d\omega$ is area measure on $\mathbb{S}^{n-1}$. Let $\mathfrak{X}$ be some class of locally integrable functions in $\mathbb{R}^n$. The kernel of $\mathcal{R}$ in $\mathfrak{X}$ with respect to $E \subset \mathbb{R}^n$ is defined to be the set of functions $f \in \mathfrak{X}$ such that $\mathcal{R}f(x,r) = 0$ for all $x \in E$ and almost all $r \in (0, +\infty)$. A set $E \subset \mathbb{R}^n$ is called a uniqueness set for $\mathcal{R}$ in $\mathfrak{X}$ if the kernel of $\mathcal{R}$ in $\mathfrak{X}$ with respect to E contains only the zero function.

The problem of describing of the uniqueness sets is in general fairly difficult. The only complete result was obtained by M.L. Agranovsky and E.T. Quinto for the class of finitary functions in $\mathbb{R}^2$ (see [3]). For other results on uniqueness sets, see the book [27], which contains an extensive bibliography.

The following method of constructing of sets of non-injectivity for $\mathcal{R}$ was suggested by L. Zalcman. Let f be a solution of the Helmholtz equation

(2.1) $$\Delta f + \lambda f = 0.$$

Then $\mathcal{R}f(x,r) = 0$ for all $r > 0$ as soon as $f(x) = 0$.

Given an $\mathfrak{X} \subset L_{\text{loc}}(\mathbb{R}^n)$, the following problem arises.

PROBLEM 2.1. Assume that the kernel of $\mathcal{R}$ in $\mathfrak{X}$ with respect to E contains a nontrivial function. Does this imply that the class $\mathfrak{X}$ contains an eigenfunction of the Laplace operator vanishing on E?

L. Zalcman conjectured that the answer is "yes" in the cases $\mathfrak{X} = L_{\text{loc}}(\mathbb{R}^n)$ and $\mathfrak{X} = L^\infty(\mathbb{R}^n)$ (see [5]). It is known that for such $\mathfrak{X}$ the answer is in the positive if E is a cone or E has spherical symmetry (see [27, Chapter 5.1], [5]).

For $1 \leq p \leq \infty$, $\alpha \geq 0$, let

$$L^{p,\alpha}(\mathbb{R}^n) = \{f \in L_{\text{loc}}(\mathbb{R}^n) : (1+|x|)^{-\alpha}|f(x)| \in L^p(\mathbb{R}^n)\}.$$

The main result of this section is the following.

THEOREM 2.1. *Let $1 \leq p \leq \infty$, $\alpha \geq 0$, and let $\mathfrak{X} = L^{p,\alpha}(\mathbb{R}^n)$. Also let E be a subset in $\mathbb{R}^n$ which contains the boundary of some bounded domain in $\mathbb{R}^n$. Assume that the kernel of $\mathcal{R}$ in $\mathfrak{X}$ with respect to E contains a nontrivial function. Then the class $\mathfrak{X}$ contains a solution of equation (2.1) for some $\lambda > 0$. Moreover, any function from the kernel of $\mathcal{R}$ in $\mathfrak{X}$ is the limit of a sequence of linear combinations of such eigenfunctions in the space of tempered distributions in $\mathbb{R}^n$.*

Our method of the proof allows us to obtain similar results for other $\mathfrak{X}$.

For $R > 0$ and $1 \leq p \leq 2n/(n-1)$, we set $\eta_p(R) = R^{n-(n-1)p/2}$ for $1 \leq p < 2n/(n-1)$, $\eta_p(R) = \ln R$ for $p = 2n/(n-1)$.

THEOREM 2.2. *Let $1 \leq p \leq 2n/(n-1)$, and suppose that $E \subset \mathbb{R}^n$ is the boundary of some bounded domain in $\mathbb{R}^n$. Assume that f is in the kernel of $\mathcal{R}$ in*

$\mathfrak{X} = L^{p,\alpha}(\mathbb{R}^n)$ for some $\alpha \geq 0$. Then the condition

(2.2) $$\liminf_{R \to +\infty} \frac{1}{\eta_p(R)} \int_{|x| \leq R} |f(x)|^p \, dx = 0$$

implies $f = 0$. If (2.2) is replaced by

$$\int_{|x| \leq R} |f(x)|^p \, dx = O(\eta_p(R)) \quad \text{as} \quad R \to +\infty,$$

then this statement becomes false.

Theorem 2.2 is a refinement of the result by M.L. Agranovsky, C. Berenstein, and P. Kuchment [2]. For other results in this direction, see [27, Chapter 5.1].

2.2. Spherical Radon transform on spheres. Throughout this section, $d(\cdot, \cdot)$ is the inner metric on $\mathbb{S}^n$, $n \geq 2$, and $d\xi$ is the area element of the sphere $\mathbb{S}^n$. The Cartesian coordinates of a point $\xi \in \mathbb{S}^n$ are denoted by $(\xi_1, \ldots, \xi_{n+1})$, and we use the standard system of spherical coordinates $(\theta_1, \ldots, \theta_n)$ on $\mathbb{S}^n$ in which $\xi_1 = \sin\theta_n \sin\theta_{n-1} \ldots \sin\theta_2 \sin\theta_1$, $\xi_2 = \sin\theta_n \sin\theta_{n-1} \ldots \sin\theta_2 \cos\theta_1$, ..., $\xi_n = \sin\theta_n \cos\theta_{n-1}$, $\xi_{n+1} = \cos\theta_n$, ($0 \leq \theta_1 \leq 2\pi$ and $0 \leq \theta_k \leq \pi$ for $k \neq 1$). For $R > 0$, we set $B_R = \{\xi \in \mathbb{S}^n : d(o, \xi) < R\}$, where $o = (0, \ldots, 0, 1) \in \mathbb{S}^n$. Let χ_R be the characteristic function (indicator) of the ball B_R. Denote by $L_{\mathrm{loc}}(B_R)$ the class of functions locally integrable (with respect to $d\xi$) in B_R. Let $\{Y_l^{(k)}\}$, $1 \leq l \leq a_k$, be a fixed orthonormal basis in the space of spherical harmonics of degree k on $\mathbb{S}^{n-1}$, which is regarded as a subspace of $L^2(\mathbb{S}^{n-1})$.

For $\xi \in \mathbb{S}^n \setminus \{o\}$, we set $\xi' = (\xi_1, \ldots, \xi_n)$, $\sigma = \xi'/|\xi'| \in \mathbb{S}^{n-1}$. To any function $f(\xi) = f(\sigma \sin\theta_n, \cos\theta_n) \in L_{\mathrm{loc}}(B_R)$, there corresponds the Fourier series

$$f(\xi) \sim \sum_{k=0}^{\infty} \sum_{l=1}^{a_k} f_{k,l}(\theta_n) Y_l^{(k)}(\sigma), \quad \theta_n \in (0, R),$$

where

$$f_{k,l}(\theta_n) = \int_{\mathbb{S}^{n-1}} f(\sigma \sin\theta_n, \cos\theta_n) \overline{Y_l^{(k)}(\sigma)} \, d\omega(\sigma).$$

Denote by $\varphi_1 \times \varphi_2$ the convolution of distributions φ_1, φ_2 on $\mathbb{S}^n$. For $0 < r < R \leq \pi$, we set $\mathcal{V}_r(B_R) = \{f \in L_{\mathrm{loc}}(B_R) : (f \times \chi_t)(\xi) = 0, \xi \in S_r, 0 < t < R - r\}$, where $S_r = \{\xi \in \mathbb{S}^n : d(o, \xi) = r\}$.

We now present a number properties of the class $\mathcal{V}_r(B_R)$.

THEOREM 2.3. *Assume that $f \in \mathcal{V}_r(B_R)$, and let $f = 0$ in B_r. Then $f = 0$ in B_R.*

As a consequence of Theorem 2.3, we have the following statement.

COROLLARY 2.1. *Let $f \in \mathcal{V}_r(B_R)$ and let $R < 2r$. Then*
1) *if $f(\xi) = 0$ in the domain $2r - R < d(o, \xi) < r$, then $f(\xi) = 0$ for $2r - R < d(o, \xi) < R$;*
2) *if $f(\xi) = 0$ in the domain $r < d(o, \xi) < R$, then $f(\xi) = 0$ for $2r - R < d(o, \xi) < R$.*

For $\nu \in \mathbb{C}$, $\theta \in (0, \pi)$, we set
$$\psi_{\nu,k}(\theta) = (\sin \theta)^{1-(n/2)} P_{\nu+(n/2)-1}^{-(n/2)-k+1}(\cos \theta),$$
where $P_{\nu+(n/2)-1}^{-(n/2)-k+1}$ is the Legendre function of the first kind on $(-1,1)$. For $r \in (0, \pi)$, $k \in \mathbb{Z}_+$, let $N_k(r) = \{\nu \in (k, +\infty) : \psi_{\nu,k}(r) = 0\}$. This set can be represented as an infinite increasing sequence.

THEOREM 2.4. *Let $f \in L_{\text{loc}}(B_R)$. Then $f \in \mathcal{V}_r(B_R)$ if and only if for all $k \in \mathbb{Z}_+$, $1 \leq l \leq a_k$,*
$$f_{k,l}(\theta_n) Y_l^{(k)}(\sigma) = \sum_{\nu \in N_k(r)} c_{\nu,k,l} \psi_{\nu,k}(\theta_n) Y_l^{(k)}(\sigma), \tag{2.3}$$
where $c_{\nu,k,l} \in \mathbb{C}$ and the series (2.3) is converges in the space of distributions $\mathcal{D}'(B_R)$.

Let the numbers $r_1, r_2 \in (0, \pi)$ be fixed. For $R \in (\max(r_1, r_2), \pi]$, we set $\mathcal{V}_{r_1,r_2}(B_R) = \mathcal{V}_{r_1}(B_R) \cap \mathcal{V}_{r_2}(B_R)$. We also set $N_k(r_1, r_2) = N_k(r_1) \cap N_k(r_2)$, $N(r_1, r_2) = \bigcup_{k=0}^{\infty} N_k(r_1, r_2)$. Theorem 2.4 enables us to obtain the following local two-radii theorem for the spherical Radon transform on spheres.

THEOREM 2.5. *Let $0 < r_1 < r_2 < R \leq \pi$. Then*
1) *if $r_1 + r_2 \leq R$, $N(r_1, r_2) = \emptyset$ and $f \in \mathcal{V}_{r_1,r_2}(B_R)$, then $f = 0$;*
2) *if $r_1 + r_2 > R$ or $N(r_1, r_2) \neq \emptyset$, then there exists a non-trivial function $f \in \mathcal{V}_{r_1,r_2}(B_R) \cap C^\infty(B_R)$.*

The analogues of Theorems 2.3–2.5 for $\mathbb{R}^n$ and non-compact rank-one symmetric spaces were obtained by V.V. Volchkov (see [**27**, Chapter 5.1], [**22**], [**26**]).

Consider some applications of Theorems 2.3–2.5 to the spherical Radon transform in $\mathbb{R}^{n+1}$.

For $\alpha \in (0, \pi]$, let
$$K_\alpha = \left\{ x = (x_1, \ldots, x_{n+1}) \in \mathbb{R}^{n+1} : x_{n+1} > (\cot \alpha) \sqrt{x_1^2 + \cdots + x_n^2} \right\}$$
if $\alpha \in (0, \pi)$ and
$$K_\alpha = \mathbb{R}^{n+1} \setminus \left\{ x \in \mathbb{R}^{n+1} : x_1 = \cdots = x_n = 0, \ x_{n+1} \leq 0 \right\}$$
if $\alpha = \pi$. We write $\overline{K}_\alpha$ for the closure of K_α. Denote by ∂K_α the boundary of K_α. We set $\text{dist}(x, \partial K_\alpha) = \inf\{|x - y| : y \in \partial K_\alpha\}$.

THEOREM 2.6. *Let $0 < r < R \leq \pi$. Assume that $f \in L_{\text{loc}}(K_R)$. Suppose that $f = 0$ in K_r and $\mathcal{R}f(x, t) = 0$ for any $x \in \partial K_r$ and almost all $t \in (0, \text{dist}(x, \partial K_R))$. Then $f = 0$ in K_R.*

THEOREM 2.7. *Let $0 < r_1 < r_2 < R \leq \pi$. Then the following assertions hold.*
1) *Assume that $f \in L_{\text{loc}}(K_R)$ and*
$$\mathcal{R}f(x,t) = 0 \tag{2.4}$$
   for any $x \in \partial K_{r_1} \cup \partial K_{r_2}$ and almost all $t \in (0, \text{dist}(x, \partial K_R))$. If $r_1 + r_2 \leq R$ and $N(r_1, r_2) = \emptyset$, then $f = 0$.
2) *If $r_1 + r_2 > R$ or $N(r_1, r_2) \neq \emptyset$, then there exists a non-trivial function $f \in C^\infty(K_R)$ satisfying the condition (2.4).*

Theorems 2.6 and 2.7 allow us to obtain new applications to partial differential equations (see [**27**, Chapter 5.6], [**3**]). For other results relating to conical uniqueness sets for the spherical Radon transform, see [**5**].

3. Pompeiu problem

3.1. Examples of Pompeiu sets with non-Lipschitz boundary.
Throughout this section, we assume that A is a bounded set in $\mathbb{R}^n$, $n \geq 2$, of positive Lebesgue measure. Denote by $\mathbf{M}(n)$ the group of all rigid motion of $\mathbb{R}^n$. For each $\lambda \in \mathbf{M}(n)$, we set $\lambda A = \{x \in \mathbb{R}^n : \lambda^{-1}x \in A\}$. As usual, $\overline{A}$ is the closure of A in $\mathbb{R}^n$.

We say that A has the Pompeiu property if every locally integrable function $f : \mathbb{R}^n \to \mathbb{C}$ satisfying the condition

$$(3.1) \qquad \int_{\lambda A} f(x)\,dx = 0 \quad \text{for all} \quad \lambda \in \mathbf{M}(n)$$

vanishes almost everywhere in $\mathbb{R}^n$. One says also that such A is a Pompeiu set. The Pompeiu problem, in its pristine form, asks: under what conditions does the set A have the Pompeiu property? A large amount of research has gone into this problem (see the survey papers [**37**], [**10**] and the book [**27**], which contain an extensive bibliography) but it remains open.

It is easily seen that each ball in $\mathbb{R}^n$ fails to possess the Pompeiu property (see, for instance, [**37**]). Several examples of multiply-connected domains lacking the Pompeiu property can be found in [**35**, p. 186].

The most general sufficient condition for a set to have the Pompeiu property have been obtained by S.A. Williams (see [**35**], [**36**]). He has shown that if a bounded domain $A \subset \mathbb{R}^n$ with Lipschitz boundary ∂A fails to have the Pompeiu property and the complement of $\overline{A}$ is connected, then ∂A must be a real-analytic submanifold in $\mathbb{R}^n$. It follows that if the boundary of A is Lipschitz but not real-analytic, then A is a Pompeiu set. Sets with real-analytic boundary which have the Pompeiu property exist, for instance, any ellipsoid which is not a ball (see [**37**]). Other Pompeiu sets with real-analytic boundary can be found in [**13**], [**12**].

Assume that $n = 2$. In what follows, each point $(x,y) \in \mathbb{R}^2$ is identified with the complex number $z = x + iy$. As usual, we set $|z| = \sqrt{x^2 + y^2}$. For $z \neq 0$, let $\arg z$ be a principal value of the argument, that is, $-\pi < \arg z \leq \pi$. For $\varphi_1, \varphi_2 \in \mathbb{R}^1$ and $\delta > 0$, we set $U(\varphi_1, \varphi_2) = \{z \in \mathbb{C} \setminus \{0\} : \varphi_1 < \arg z < \varphi_2\}$, $V(\delta) = \{z \in \mathbb{C} : 0 < |z| < \delta\}$. The central result of this section is the following.

THEOREM 3.1. *Let A be a bounded subset in $\mathbb{R}^2$. Assume that there exist $\alpha, \beta, \gamma, \delta \in \mathbb{R}^1$ such that $0 \leq \alpha < \beta \leq \pi/2$, $\beta < \gamma < 2\beta - \alpha$, $\delta > 0$, and $U(\alpha, \beta) \cap V(\delta) \subset A \subset U(0, \gamma)$. Then A has the Pompeiu property.*

The result stated above deals with sets without any regularity conditions for boundary points. Thus Theorem 3.1 enables us to construct Pompeiu sets with non-Lipschitz (and even fractal) boundary. A specific example is the famous von Koch snowflake domain, obtained by starting with an open regular triangle in $\mathbb{R}^2$ and repeatedly replacing each line segment of the boundary by four equal segments. The limiting set is the von Koch snowflake domain.

COROLLARY 3.1. *The von Koch snowflake domain is a Pompeiu set.*

Corollary 3.1 provides the first example of a domain with fractal boundary which has the Pompeiu property.

To conclude this section, we mention that the method of the proof of Theorem 3.1 enables us to obtain some analogues of this result for $n \geq 3$.

3.2. The local Pompeiu property on a sphere. Of particular interest are local versions of the Pompeiu problem, in which a function f is defined on a bounded domain $\mathcal{U} \subset \mathbb{R}^n$ and (3.1) is required to hold only when $\lambda A \subset \mathcal{U}$. In this case, the object is to determine conditions on the set A under which (3.1) implies that $f = 0$ on $\mathcal{U}$. Many definitive results in this direction can be found in [27, Part 4]. Here we consider similar problems on a sphere.

As in Section 1.2, we set $B_r = \{\xi \in \mathbb{S}^n : d(o, \xi) < r\}$, where $o = (0, \ldots, 0, 1) \in \mathbb{S}^n$. We note that $B_r = \mathbb{S}^n$ for each $r > \pi$. Let A be a compact subset of $\mathbb{S}^n$. Denote by $r^*(A)$ the radius of the smallest closed ball on $\mathbb{S}^n$ containing the set A. Assume that $0 < r^*(A) < \pi$, and let $r > r^*(A)$. We denote by $\mathfrak{P}(A, B_r)$ the set of functions $f \in L_{\mathrm{loc}}(B_r)$ satisfying the condition

$$\int_{\tau A} f(\xi) \, d\xi = 0$$

for every rotation τ of $\mathbb{S}^n$ such that $\tau A \subset B_r$. We say that the set A has the local Pompeiu property with respect to the ball B_r if every function $f \in \mathfrak{P}(A, B_r)$ vanishes almost everywhere in B_r. Such a set A is also called a Pompeiu set in B_r.

Denote by $\mathrm{Pomp}\,(B_r)$ the collection of all Pompeiu sets in the ball B_r. The following problem arises.

PROBLEM 3.1. ([27]). Suppose that $A \in \mathrm{Pomp}\,(B_r)$ for some $r > r^*(A)$. Find

$$\mathcal{R}(A) = \inf\{r > r^*(A) : A \in \mathrm{Pomp}\,(B_r)\}$$

and investigate when the value $\mathcal{R}(A)$ is attainable, that is, $A \in \mathrm{Pomp}\,(B_r)$ for $r = \mathcal{R}(A)$.

So far, the exact value of $\mathcal{R}(A)$ is not known for any A. (However, we point out that analogues of Problem 3.1 in $\mathbb{R}^n$ have been completely solved by V.V. Volchkov for a broad class of sets; see [27, Part 4]). We now present some results in this direction.

For $\alpha \in (0, 2\pi)$, set $A_\alpha = \{\xi \in \mathbb{S}^2 : 0 \leq \theta_1 \leq \alpha, \ 0 \leq \theta_2 \leq \pi\}$. Note that $A_\pi \notin \mathrm{Pomp}\,(\mathbb{S}^2)$.

THEOREM 3.2. *The following relation holds:*

$$\mathcal{R}(A_\alpha) = \begin{cases} \pi/2 & \text{if } 0 < \alpha \leq \pi/2, \\ \alpha & \text{if } \pi/2 < \alpha < \pi, \\ \pi & \text{if } \pi < \alpha < 2\pi. \end{cases}$$

This result is a considerable refinement of Laquer's theorem ([19]).

THEOREM 3.3. *Assume that $\alpha \in (0, \pi)$, and let*

$$A(\alpha) = \{\xi \in \mathbb{S}^2 : 0 \leq \theta_1 \leq \pi, \ 0 \leq \theta_2 \leq \alpha\}.$$

Then

$$\mathcal{R}(A(\alpha)) = \begin{cases} \arccos(\cos\alpha \cdot \cos(\alpha/2)) & \text{if } 0 < \alpha \leq \pi/2 \\ \alpha & \text{if } \pi/2 < \alpha < \pi. \end{cases}$$

For other results relating to the Pompeiu problem on a sphere, see [**27**], [**37**], [**6**], [**11**].

3.3. Functions with zero integrals over spherical caps.
Here we retain the notation of Section 2.2. For $0 \leq a < b \leq \pi$, we set
$$B_{a,b} = \{\xi \in \mathbb{S}^n : a < d(o,\xi) < b\}.$$
Let $0 < r < (b-a)/2$. Denote by $V_r(B_{a,b})$ the set of functions $f \in L_{\mathrm{loc}}(B_{a,b})$ such that $f \times \chi_r = 0$ in $B_{a+r,b-r}$. Analogously, for $0 < r < R \leq \pi$, we define the class $V_r(B_R)$ as follows: $V_r(B_R) = \{f \in L_{\mathrm{loc}}(B_R) : (f \times \chi_r)(\xi) = 0, \xi \in B_{R-r}\}$. For $\theta \in (0,\pi)$, we set
$$\Psi_{\nu,k}(\theta) = \begin{cases} (\sin\theta)^{1-(n/2)} Q_{\nu+(n/2)-1}^{(n/2)+k-1}(\cos\theta) & \text{if } n \text{ is even and } 2-n-k-\nu \notin \mathbb{N}, \\ (\sin\theta)^{1-(n/2)} P_{\nu+(n/2)-1}^{(n/2)+k-1}(\cos\theta) & \text{if } n \text{ is odd and } \nu \in \mathbb{C}, \end{cases}$$
where $Q_{\nu+(n/2)-1}^{(n/2)+k-1}$ is the Legendre function of the second kind on $(-1,1)$.

THEOREM 3.4. *Let $f \in C^\infty(B_{a,b})$. In order that f belong to $V_r(B_{a,b})$, it is necessary and sufficient that for all $k \in \mathbb{Z}_+$ and $1 \leq l \leq a_k$,*
$$f_{k,l}(\theta_n) = \sum_{\nu \in N_1(r)} \alpha_{\nu,k,l} \psi_{\nu,k}(\theta_n) + \beta_{\nu,k,l} \Psi_{\nu,k}(\theta_n),$$
where $\alpha_{\nu,k,l}, \beta_{\nu,k,l} \in \mathbb{C}$ and $|\alpha_{\nu,k,l}| + |\beta_{\nu,k,l}| = O(\nu^{-c})$ as $\nu \to +\infty$ for any fixed $c > 0$.

A similar result for the class $V_r(B_R)$ is contained in [**27**, Chapter 2.3].

Let σ_r be the δ–function distributed over the sphere S_r. For $0 < r < R \leq \pi$, set $U_r(B_R) = \{f \in L_{\mathrm{loc}}(B_R) : (f \times \sigma_r)(\xi) = 0, \xi \in B_{R-r}\}$, $W_r(B_R) = V_r(B_R) \cap U_r(B_R)$.

The following result is an analogue of the local two-radii theorem on $\mathbb{S}^n$ [**27**, Chapter 2.3].

THEOREM 3.5. *Let $f \in L_{\mathrm{loc}}(B_R)$.*

1) *If $R > 2r$ and $f \in W_r(B_R)$, then $f = 0$ in B_R.*
2) *If $R \leq 2r$, then $f \in W_r(B_R)$ if and only if*
$$\int_{B_{2r-R}} f(\xi)\, d\xi = 0$$
and
$$f_{k,l}(\theta) = \sum_{m=0}^{k-2} c_{m,k,l}(\cos\theta)^m (\sin\theta)^{2-n-k}$$
for all $k \in \mathbb{Z}_+$, $1 \leq l \leq a_k$, and almost all $\theta \in (2r-R, R)$, where the sum is set equal to zero for $k = 0, 1$.

We now present some consequences of Theorems 3.4 and 3.5. As a consequence of Theorem 3.4, we have the following result on the removal of singularities for radial functions in the class $V_r(B_R)$.

THEOREM 3.6. 1) *Assume that $R \geq 2r$ and suppose $f \in V_r(B_R)$ and f has the form $f(\xi) = f_0(d(o,\xi))$. Then there exists a function $g \in C^\infty(B_R)$ equal to f in $B_{0,R}$.*

2) Let $R < 2r$. Then there exists a radial function $f \in V_r(B_R) \cap C^\infty(B_{0,R})$ which is unbounded in a neighbourhood of the point $o \in \mathbb{S}^n$.

Consider some consequences of Theorem 3.5. Let $s \in \mathbb{Z}_+$ or $s = \infty$. We say that $f \in C^s(\{o\})$ if $f \in C^s(B_\varepsilon)$ for some $\varepsilon \in (0, \pi)$ depending of f.

THEOREM 3.7. *Let $R = 2r$. Then*
1) *if $f \in W_r(B_R) \cap C^\infty(\{o\})$, then $f = 0$ in B_R;*
2) *for each $s \in \mathbb{Z}_+$, there exists a non-trivial function $f \in W_r(B_R) \cap C^s(\{o\})$.*

THEOREM 3.8. *Let $R = 2r$. Then*
1) *if $f \in W_r(B_R)$, and f is continuous in a neighbourhood of the point o, and $f(\xi) = o\left((d(o,\xi))^\alpha\right)$ as $\xi \to o$ for each fixed $\alpha > 0$, then $f = 0$ in B_R;*
2) *for each $\alpha > 0$, there exists a non-trivial function $f \in W_r(B_R) \cap C(B_R)$ such that $f(\xi) = o\left((d(o,\xi))^\alpha\right)$ as $\xi \to o$.*

THEOREM 3.9. *Let $R = 2r$ and let $f \in W_r(B_R)$. Assume that there exists a set $E \subset (0, R)$ of positive measure such that $f(\xi) = 0$ for $d(o,\xi) \in E$. Then $f = 0$ in B_R.*

The analogues of Theorems 3.4–3.9 for $\mathbb{R}^n$ and non-compact rank-one symmetric spaces were obtained by V.V. Volchkov in [**27**, Chapter 2.1], [**26**].

4. Spherical means on the Heisenberg group

4.1. Uniqueness theorem. Suppose that $\mathbb{C}^n$ is the complex Euclidean space of dimension $n \geq 1$ endowed with the Hermitian inner product $\langle \cdot, \cdot \rangle$. Consider the Heisenberg group H^n as the set $\{(z,t): z \in \mathbb{C}^n, t \in \mathbb{R}\}$ with the group operation $(z,t)(w,s) = \left(z+w, t+s+\frac{1}{2}\mathrm{Im}\langle z,w\rangle\right)$. Let Γ be the discrete subgroup $\{(0, 2\pi k): k \in \mathbb{Z}\}$ of H^n. The quotient group H^n/Γ is called the reduced Heisenberg group.

For $R > 0$, we set $B_R = \{z \in \mathbb{C}^n : |z| = \sqrt{\langle z,z \rangle} < R\}$,
$$C_R = \{\{(z, t + 2\pi k): k \in \mathbb{Z}\} \in H^n/\Gamma : z \in B_R,\ 0 \leq t < 2\pi\}.$$
In what follows, functions on C_R will be identified with functions on $B_R \times \mathbb{R}$ which are 2π-periodic in the t variable. We write $f \in L_{\mathrm{loc}}(C_R)$ if
$$\int_0^{2\pi} \int_{B_r} |f(z,t)|\, dm(z)\, dt < \infty$$
for any $r \in (0, R)$, where dm is Lebesgue measure on $\mathbb{C}^n$. For $f \in L_{\mathrm{loc}}(C_R)$ and $k \in \mathbb{Z}$, we put
$$f_k(z) = \frac{1}{2\pi} \int_0^{2\pi} f(z,t) e^{-ikt}\, dt, \quad z \in B_R.$$

Let ρ, σ be polar coordinates in $\mathbb{C}^n$ (for any $z \in \mathbb{C}^n$, we assume that $\rho = |z|$ and if $z \neq 0$, then $\sigma = z/|z|$). Let $\{S_{p,q}^l\}_{l=1}^{d(n,p,q)}$ be a fixed orthonormal basis in the space of spherical harmonics of bidegree (p,q), which is regarded as a subspace of $L^2(\mathbb{S}^{2n-1})$. To any function $f \in L_{\mathrm{loc}}(B_R)$, there corresponds the Fourier series
$$f(z) \sim \sum_{p,q=0}^{\infty} \sum_{l=1}^{d(n,p,q)} f_{p,q}^l(\rho) S_{p,q}^l(\sigma),$$

where $f_{p,q}^l(\rho) = \int_{\mathbb{S}^{2n-1}} f(\rho\sigma)\overline{S_{p,q}^l(\sigma)}\,d\sigma$.

Denote by μ_r the normalized surface measure on $\{(z,0)\colon |z|=r\}$. Let $U_r(C_R)$ be the set of functions $f \in L_{\text{loc}}(C_R)$ such that the integral

$$\int_{|w|=r} f\bigl(z-w, t - \tfrac{1}{2}\operatorname{Im}\langle z,w\rangle\bigr)\,d\mu_r(w)$$

is zero for almost all $(z,t) \in B_{R-r} \times \mathbb{R}$.

For a non-negative integer s and for $s = \infty$, we set $U_r^s(C_R) = U_r(C_R) \cap C^s(C_R)$. Now we can state the following uniqueness theorem for the class $U_r(C_R)$.

THEOREM 4.1. *The following assertions are valid.*
1) *Let $f \in U_r^s(C_R)$, and suppose that $f = 0$ in C_r. Then $(f_k)_{p,q}^l(\rho) = 0$ in C_R for all $k \in \mathbb{Z}$, $0 \le p+q \le s$, $1 \le l \le d(n,p,q)$.*
2) *Let $f \in U_r(C_R)$ and $f = 0$ in $C_{r+\varepsilon}$ for some $\varepsilon \in (0, R-r)$. Then $f = 0$ in C_R. If $f \in C^\infty(C_R)$, then this assertion is true also for $\varepsilon = 0$.*
3) *For each integer $s \ge 0$, there exists a non-trivial function $f \in U_r^s(C_R)$ such that $f = 0$ in C_r.*
4) *For each $\varepsilon \in (0, r)$, there exists a non-trivial function $f \in U_r^\infty(C_R)$ such that $f = 0$ in $C_{r-\varepsilon}$.*

The first uniqueness theorems for functions with zero spherical means were studied by F. John (see, for example, [27, Chapter 2.4]).

The proof of Theorem 4.1 is based on a new method of investigation of uniqueness sets with spherical symmetry of solutions of convolution equations.

4.2. Description of functions in the class $U_r(C_R)$. For $r > 0$, let $N(r) = \{\lambda \in \mathbb{C}\colon {}_1F_1(-\lambda; n; r^2/2) = 0\}$, where ${}_1F_1$ is the degenerate hypergeometric function. This set can be represented as an increasing sequence of positive numbers. Let Λ be the sequence of all positive zeros of the Bessel function J_{n-1}, indexed in increasing order. For $f \in L_{\text{loc}}(C_R)$, we set

$$_k f = \begin{cases} f_k & \text{if } k \le 0 \\ \overline{f_k} & \text{if } k > 0. \end{cases}$$

Theorem 4.1 enables us to obtain the following description of smooth functions in the class $U_r(C_R)$.

THEOREM 4.2. *Let $f \in C^\infty(C_R)$. Then, in order for f to belong to $U_r(C_R)$, it is necessary and sufficient that for all $k \in \mathbb{Z}$, $p, q \ge 0$ and $1 \le l \le d(n,p,q)$, we have*

$$\frac{(_k f)_{p,q}^l(\rho)}{\rho^{p+q}} e^{\rho^2 |k|/4} = \sum_{\lambda \in N(\sqrt{|k|}r)} c_{\lambda,k,p,q,l}\, {}_1F_1(p-\lambda; n+p+q; \rho^2|k|/2), \quad k \ne 0$$

$$(_0 f)_{p,q}^l(\rho) = \rho^{1-n} \sum_{\lambda \in \Lambda} c_{\lambda,0,p,q,l} J_{n+p+q-1}(\lambda \rho/r), \quad k = 0,$$

where $c_{\lambda,k,p,q,l} \in \mathbb{C}$ and $c_{\lambda,k,p,q,l} = O(\lambda^{-c})$ as $\lambda \to +\infty$ for any fixed $c > 0$.

This result plays a crucial role for the solution of many problems related to spherical means on H^n/Γ.

4.3. A local two-radii theorem on H^n/Γ. Let E_n be the set of numbers of the form α/β, where $\alpha, \beta \in \Lambda$. We say that a number $\tau > 0$ is well approximated by elements of E_n if for each $c > 0$ there exists numbers $\alpha, \beta \in \Lambda$ such that $|\tau - (\alpha/\beta)| < (1+\beta)^{-c}$. We set $\mathcal{N}_0 = \{(r_1, r_2) : r_1/r_2 \notin E_n\}$, $\Omega_0 = \{(r_1, r_2) : r_1/r_2 \in \mathcal{A}\}$, where $\mathcal{A}$ is the set of points well approximated by elements of E_n.

Let $k \in \mathbb{N}$. We put $\mathcal{N}_k = \{(r_1, r_2) : N(\sqrt{k}r_1) \cap N(\sqrt{k}r_2) = \emptyset\}$. Denote by Ω_k the set of pairs (r_1, r_2) with the property that for any $c > 0$, there exists $\lambda \in N(\sqrt{k}r_1)$ such that $\left|{}_1F_1(-\lambda; n; kr_2^2/2)\right| < (1+\lambda)^{-c}$. For the structure of these sets, see [27, Proof of Lemma 2.2.18]. We now introduce the sets $\mathcal{N} = \bigcap_{k=0}^{\infty} \mathcal{N}_k$ and $\Omega = \bigcap_{k=0}^{\infty} \Omega_k$. Let $r_1, r_2 \in (0, R)$. For $s \in \mathbb{Z}_+$ or $s = \infty$, we set $U_{r_1,r_2}^s(C_R) = U_{r_1,r_2}(C_R) \cap C^s(C_R)$, where $U_{r_1,r_2}(C_R) = U_{r_1}(C_R) \cap U_{r_2}(C_R)$.

The main result of this section is as follows.

THEOREM 4.3. *Let $0 < r_1 < r_2 < R$. Then the following assertions are true:*
1) *if $r_1 + r_2 < R$, $(r_1, r_2) \in \mathcal{N}$ and $f \in U_{r_1,r_2}(C_R)$, then $f = 0$;*
2) *if $r_1 + r_2 = R$, $(r_1, r_2) \in \mathcal{N}$ and $f \in U_{r_1,r_2}^\infty(C_R)$, then $f = 0$;*
3) *if $r_1 + r_2 = R$, $(r_1, r_2) \in \mathcal{N} \cap \Omega$ and $f \in U_{r_1,r_2}(C_R)$, then $f = 0$;*
4) *if $r_1 + r_2 = R$ and $(r_1, r_2) \notin \Omega$, then for each integer $s \geq 0$ there exists a non-trivial function $f \in U_{r_1,r_2}^s(C_R)$;*
5) *if $r_1 + r_2 > R$, then there exists a non-trivial function $f \in U_{r_1,r_2}^\infty(C_R)$;*
6) *if $(r_1, r_2) \notin \mathcal{N}$, then there exists a non-trivial real-analytic function $f \in U_{r_1,r_2}(H^n/\Gamma)$.*

The proof of Theorem 4.3 is based on the description of the class $U_r(B_R)$.

The two-radii theorem on the whole group H^n/Γ for tempered continuous functions was obtained by S. Thangavelu [20] in a different way.

We mention also that the method of the proof of Theorem 4.3 enables us to obtain analogues of other results from [27, Parts 2,3,5] for the group H^n/Γ.

5. Related problems on symmetric spaces

5.1. Uniqueness theorem for the convolution equation on a non-compact symmetric space. For basic notations and facts from the theory of symmetric spaces, see [16].

Throughout, G is a non-compact connected semisimple Lie group of real rank 1 with finite centre, $K \subset G$ is a maximal compact subgroup, and $X = G/K$ is the corresponding rank-one symmetric space of non-compact type. As usual, $\mathfrak{g} = \mathfrak{k} + \mathfrak{p}$ is the Cartan decomposition of the Lie algebra $\mathfrak{g}$ of G into the direct sum of the Lie algebra $\mathfrak{k}$ of K and its orthogonal complement $\mathfrak{p}$ with respect to the Killing form in $\mathfrak{g}$. Let $\mathfrak{a} \subset \mathfrak{p}$ be a maximal Abelian subspace (one-dimensional in our case) and let $\mathfrak{a}_\mathbb{C}^*$ be the set of complex-valued linear functionals on $\mathfrak{a}$. The Killing form in $\mathfrak{g}$ induces a G-invariant Riemannian structure on X, with the corresponding distance function $d(\cdot, \cdot)$.

Let $o = \pi(e)$, where $\pi : G \to X$ is the canonical projection and e is the identity element of G. For $0 \leq R \leq +\infty$, denote by $B_R = \{x \in X : d(x, o) < R\}$ the open geodesic ball of radius R with centre at o. Let $\mathcal{D}'(B_R)$ and $\mathcal{E}'(B_R)$ be the spaces of distributions and compactly supported distributions on B_R, respectively. Denote by $\mathcal{E}'_K(B_R)$ the set of all K-invariant distributions in $\mathcal{E}'(B_R)$.

Let $\mathfrak{e}$ be a fixed nonzero vector in $\mathfrak{a}$. For $T \in \mathcal{E}'_K(B_R)$, denote by $\widetilde{T}(\lambda)$, $\lambda \in \mathfrak{a}^*_{\mathbb{C}}$, the spherical transform of T (see [**16**, p. 281]). For $z \in \mathbb{C}$, we put $F_T(z) = \widetilde{T}(\lambda)$, where $\lambda \in \mathfrak{a}^*_{\mathbb{C}}$ such that $\lambda(\mathfrak{e}) = z$. We set $r(T) = \inf\{r > 0 \colon \operatorname{supp} T \subset B_r\}$, $Z_T = \{z \in \mathbb{C} \colon F_T(z) = 0\} \setminus \{z \in \mathbb{C} \setminus \{0\} \colon 0 \leq \arg(-z) < \pi\}$. For $\zeta \in Z_T$, denote by n_ζ the multiplicity of the zero ζ of the entire function F_T. Let $\mathfrak{N}(B_R)$ be the set of distributions $T \in \mathcal{E}'_K(B_R)$ such that for any $\zeta \in Z_T$ we have the estimates

$$|\operatorname{Im} \zeta| \leq \alpha \ln(2 + |\zeta|), \quad \left|\left(\frac{d}{dz}\right)^{n_\zeta} F_T(z)\bigg|_{z=\zeta}\right| \geq (2+|\zeta|)^{n_\zeta - \beta}$$

with constants $\alpha, \beta > 0$ independent of ζ. We point out that the class $\mathfrak{N}(B_R)$ is fairly extensive (see [**27**, Chapter 3.2]). For example, the characteristic function of each geodesic ball with centre at o is in the class $\mathfrak{N}$.

Denote by $f_1 \times f_2$ the convolution of distributions f_1, f_2 on X. For $T \in \mathcal{E}'_K(B_R)$, we set $\mathcal{D}'_T(B_R) = \{f \in \mathcal{D}'(B_R) \colon f \times T = 0 \text{ in } B_{R-r(T)}\}$. We have the following uniqueness theorem for the class $\mathcal{D}'_T(B_R)$.

THEOREM 5.1.   1) *Let $T \in \mathcal{E}'_K(B_R)$, $f \in \mathcal{D}'_T(B_R)$, and suppose that $f = 0$ in $B_{r(T)+\varepsilon}$ for some $\varepsilon \in (0, R - r(T))$. Then $f = 0$ in B_R.*
2) *Let $T \in \mathcal{E}'_K(B_R)$, $r(T) > 0$, $f \in \mathcal{D}'_T(B_R) \cap C^\infty(B_R)$, and assume that $f = 0$ in $B_{r(T)}$. Then $f = 0$ in B_R.*
3) *Let $T \in \mathcal{E}'_K(B_R) \cap C^\infty(B_R)$, $r(T) > 0$, $f \in \mathcal{D}'_T(B_R)$, and suppose that $f = 0$ in $B_{r(T)}$. Then $f = 0$ in B_R.*
4) *If $T \in \mathfrak{N}(B_R)$, then for each integer $s \geq 0$ there exists a non-trivial function $f \in C^s(B_R) \cap \mathcal{D}'_T(B_R)$ such that $f = 0$ in $B_{r(T)}$.*
5) *If $T \in \mathfrak{N}(B_R)$ and $r(T) > 0$, then for each $\varepsilon \in (0, r(T))$, there exists a non-trivial function $f \in C^\infty(B_R) \cap \mathcal{D}'_T(B_R)$ such that $f = 0$ in $B_{r(T)-\varepsilon}$.*

The Euclidean analogue of Theorem 5.1 was obtained in [**27**, Chapter 3.2]. Some special cases of Theorem 5.1 are contained in [**27**, Part 2], [**26**], [**17**]–[**34**].

5.2. A definitive version of the local two-radii theorem.

One well-known result in integral geometry is Zalcman's two-radii theorem: each function with vanishing integrals over all balls in $\mathbb{R}^n$ with radii in a fixed set $\{r_1, r_2\}$ vanishes if r_1/r_2 is not a ratio of two zeros of the Bessel function $J_{n/2}$. Examples show that this condition on r_1/r_2 is also necessary (see [**37**]).

The two-radii theorem was further developed and improved by many authors (see [**27**], [**26**], [**10**], [**11**], [**24**], [**8**]–[**15**] and the references therein). Here we present the definitive version of the local two-radii theorem on a non-compact rank-one symmetric space.

Denote by $\mathfrak{M}_1$ the set of all pairs (T_1, T_2) of distributions $T_1, T_2 \in \mathcal{E}'_K(X)$ such that for any $\zeta \in \mathcal{Z}_{T_1}$, we have the estimate

$$|F_{T_2}(\zeta)| \geq (2+|\zeta|)^{-\gamma}$$

with constant $\gamma > 0$ independent of ζ. Also let $\mathfrak{M}_2$ be the set of all pairs (T_1, T_2) of distributions $T_1, T_2 \in \mathcal{E}'_K(X)$ for which there exists a constant $c > 0$ with the property that for each $\varepsilon > 0$, there exists a nonzero entire function w of exponential type at most ε such that $w(\zeta) = 0$ for each $\zeta \in \{z \in \mathcal{Z}_{T_1} \colon |F_{T_2}(z)| \leq (1+|z|)^{-c}\}$. For example, $(T_1, T_2) \in \mathfrak{M}_2$ if T_1, T_2 are characteristic functions of geodesic balls with different radii (see [**27**, Proof of Theorem 2.1.6]). Some sufficient conditions for a pair (T_1, T_2) to be in $\mathfrak{M}_2$ can be found in [**27**, Chapter 5.3].

We set $\mathcal{D}'_{T_1,T_2}(B_R) = \mathcal{D}'_{T_1}(B_R) \cap \mathcal{D}'_{T_2}(B_R)$, $\mathcal{Z}_{T_1,T_2} = \mathcal{Z}_{T_1} \cap \mathcal{Z}_{T_2}$.

THEOREM 5.2. *Let $T_1 \in \mathfrak{N}(B_R)$, $T_2 \in \mathcal{E}'_K(B_R)$. Then the following assertions hold.*

1) *If $r(T_1) + r(T_2) < R$, $\mathcal{Z}_{T_1,T_2} = \emptyset$ and $f \in \mathcal{D}'_{T_1,T_2}(B_R)$, then $f = 0$.*
2) *If $r(T_1) + r(T_2) = R$, $\mathcal{Z}_{T_1,T_2} = \emptyset$ and $f \in \mathcal{D}'_{T_1,T_2}(B_R) \cap C^\infty(B_R)$, then $f = 0$.*
3) *If $r(T_1) + r(T_2) = R$, $\mathcal{Z}_{T_1,T_2} - \emptyset$, $(T_1, T_2) \notin \mathfrak{M}_1$ and $f \in \mathcal{D}'_{T_1,T_2}(B_R)$, then $f = 0$.*
4) *If $r(T_1) + r(T_2) = R$ and $(T_1, T_2) \in \mathfrak{M}_1$, then for each integer $s \geq 0$ there exists a non-trivial function $f \in C^s(B_R) \cap \mathcal{D}'_{T_1,T_2}(B_R)$.*
5) *If $r(T_1) + r(T_2) > R$ and $(T_1, T_2) \in \mathfrak{M}_2$, then there exists a non-trivial function $f \in C^\infty(B_R) \cap \mathcal{D}'_{T_1,T_2}(B_R)$. Moreover, if $(T_1, T_2) \notin \mathfrak{M}$, then this assertion is not true, generally speaking.*
6) *If $\mathcal{Z}_{T_1,T_2} \neq \emptyset$, then there exists a non-trivial real-analytic function $f \in \mathcal{D}'_{T_1,T_2}(B_R)$.*

The Euclidean analogue of Theorem 5.2 was obtained by V.V. Volchkov [**27**, Chapter 3.4]. Some special cases of Theorem 5.2 are contained in [**27**], [**26**], [**10**], [**11**], [**24**], [**8**]–[**15**]. Various other definitive results related to the convolution equations can be found in [**27**], [**10**], [**24**], [**21**], [**32**], [**34**], [**29**], [**30**].

5.3. Theorems on ball mean values in compact symmetric spaces.

Suppose now that $X = \mathcal{U}/K$ is a classical compact symmetric space of rank one, i.e., that X is a sphere $\mathbb{S}^n$ or a projective space $\mathbb{P}^n(\mathbb{K})$, where $\mathbb{K} = \mathbb{R}, \mathbb{C}, \mathbb{Q}$ (here $\mathbb{Q}$ is the quaternion algebra). In [**27**], analogues of Theorems 5.1, 5.2 for spherical means on $\mathbb{S}^n$ and $\mathbb{P}^n(\mathbb{R})$ are contained. We state similar results for $\mathbb{P}^n(\mathbb{C})$ and $\mathbb{P}^n(\mathbb{Q})$.

In what follows, $\mathbb{K} = \mathbb{C}$ or $\mathbb{K} = \mathbb{Q}$, $\mathbb{P}^n(\mathbb{K}) = \mathbb{K}^n \cup \mathbb{P}^{n-1}(\mathbb{K})$ is the corresponding projective space with the Fubini-Study metric normalized by the condition $\operatorname{diam} \mathbb{P}^n(\mathbb{K}) = \pi/2$, and $o = (0, \ldots, 0) \in \mathbb{K}^n$, see [**18**, Part 9, Chapter 6]. (We regard $\mathbb{Q}^n$ as the left quaternionic Euclidean space.) Let $d(\cdot, \cdot)$ be the distance between points of $\mathbb{P}^n(\mathbb{K})$ in the metric indicated above and for $R \in (0, \pi/2]$, let $B_R = \{x \in \mathbb{P}^n(\mathbb{K}) : d(o, x) < R\}$. Note that B_R coincides with Euclidean ball of radius $\tan R$ centered at $o \in \mathbb{K}^n$. For $0 < r < R \leq \pi/2$, denote by $V_r(B_R)$ the class of functions $f \in L_{\text{loc}}(B_R)$ satisfying the convolution equation $(f \times \chi_r)(x) = 0$, $x \in B_{R-r}$, where χ_r is the characteristic function of B_r. As above, for $s \in \mathbb{Z}_+$ or $s = \infty$, we set $V_r^s(B_R) = V_r(B_R) \cap C^s(B_R)$.

Let $\mathfrak{G} = \{a\tau : a \in \mathbb{Q}, |a| = 1, \tau \in \operatorname{Sp}(n)\}$, where $\operatorname{Sp}(n)$ is the isometry group of $\mathbb{Q}^n$, and let $T(g)$ be the quasi-regular representation of the group $\mathfrak{G}$ on $\mathbb{S}^{4n-1}$ (see [**31**]). In the constructions below, an important role is played by a suitable realization of the irreducible components of $T(g)$. We identify $\mathbb{Q}^n$ with $\mathbb{C}^{2n}$ using the correspondence $(a_1, \ldots, a_n) \to (\alpha_1, \ldots, \alpha_{2n})$, where $a_k = \alpha_k + \alpha_{n+k} j$, $\alpha_k, \alpha_{n+k} \in \mathbb{C}$, $1 \leq k \leq n$, $j \in \mathbb{Q}$. We realize the quaternion hyperbolic space $\mathbb{H}^n(\mathbb{Q})$ as the open unit ball of the space $\mathbb{C}^{2n}$ (see [**31**]). Let L be the Laplace-Beltrami operator on $\mathbb{H}^n(\mathbb{Q})$. Denote by $[t]$ the integer part of $t \in \mathbb{R}^1$. For integers $p \geq 0$,

$0 \leq m \leq [p/2]$, we set
$$\mathcal{H}^{p,m} = \left\{ f \in \mathcal{H}_p \colon (Lf)(z) = 4(m-1)(m-p)(1-|z|^2)f(z) \right\},$$
where $\mathcal{H}_p$ is the space of homogeneous harmonic polynomials of degree p in $\mathbb{C}^{2n}$. Denote by $T^{p,m}(g)$ the restriction of $T(g)$ to $\mathcal{H}^{p,m}|_{\mathbb{S}^{4n-1}}$.

THEOREM 5.3. *The quasi-regular representation $T(g)$ is the orthogonal direct sum of the irreducible unitary representations $T^{p,m}(g)$, $p \geq 0$, $0 \leq m \leq [p/2]$. The representations $T^{p_i,m_i}(g)$, $i=1,2$, are equivalent if and only if $p_1 = p_2$, $m_1 = m_2$.*

A similar realization of the irreducible components of the quasi-regular representation of the group $\operatorname{Sp}(n)$ on $\mathbb{S}^{4n-1}$ is contained in [31].

Let $\{S_l^{p,m}\}_{l=1}^{b(n,p,m)}$ be a fixed orthonormal basis in $\mathcal{H}^{p,m}|_{\mathbb{S}^{4n-1}}$. (We regard $\mathcal{H}^{p,m}|_{\mathbb{S}^{4n-1}}$ as a subspace of $L^2(\mathbb{S}^{4n-1})$.) For $f \in L_{\mathrm{loc}}(B_R)$, we set
$$f_l^{p,m}(\rho) = \int_{\mathbb{S}^{4n-1}} f(\rho\sigma)\overline{S_l^{p,m}(\sigma)}\, d\sigma.$$

In the next theorem, we use the notation of Section 4.1.

THEOREM 5.4. *Let $0 < r < R \leq \pi/2$. Then*
1) *if $\mathbb{K} = \mathbb{C}$, $f \in V_r^s(B_R)$ and $f = 0$ in B_r, then $f_{p,q}^l(\rho) = 0$ in B_R for all $0 \leq p+q \leq s$, $1 \leq l \leq d(n,p,q)$;*
2) *if $\mathbb{K} = \mathbb{Q}$, $f \in V_r^s(B_R)$ and $f = 0$ in B_r, then $f_l^{p,m}(\rho) = 0$ in B_R for all $0 \leq p \leq s$, $0 \leq m \leq [p/2]$, $1 \leq l \leq b(n,p,m)$;*
3) *if $f \in V_r^\infty(B_R)$ and $f = 0$ in B_r, then $f = 0$ in B_R;*
4) *for each integer $s \geq 0$, there exists a non-trivial function $f \in V_r^s(B_R)$ such that $f = 0$ in B_r;*
5) *for each $\varepsilon \in (0,r)$, there exists a non-trivial function $f \in V_r^\infty(B_R)$ such that $f = 0$ in $B_{r-\varepsilon}$.*

For $\lambda \in \mathbb{C}$, $\rho \geq 0$, we set
$$P_{\lambda,p,q}(\rho) = \frac{\rho^{p+q}}{(1+\rho^2)^q} F\left(q - \lambda, n+q+\lambda; n+p+q; \frac{\rho^2}{1+\rho^2}\right),$$
$$P_\lambda^{p,m}(\rho) = \frac{\rho^p}{(1+\rho^2)^{p-m}} F\left(p - m - \lambda, 2n+1+p-m+\lambda; 2n+p; \frac{\rho^2}{1+\rho^2}\right),$$
where F is the hypergeometric function.

Also let $M(r) = \{\lambda > 1 : P_\lambda(\tan r) = 0\}$, where $P_\lambda = P_{\lambda,0,1}$ if $\mathbb{K} = \mathbb{C}$ and $P_\lambda = P_\lambda^{1,0}$ if $\mathbb{K} = \mathbb{Q}$.

THEOREM 5.5. *Let $0 < r < R \leq \pi/2$ and assume that $f \in C^\infty(B_R)$. For f to belong to $V_r(B_R)$, it is necessary and sufficient that*
$$f(\rho\sigma) = \sum_{p,q=0}^{\infty} \sum_{l=1}^{d(n,p,q)} \sum_{\lambda \in M(r)} c_{\lambda,p,q,l} P_{\lambda,p,q}(\rho) S_{p,q}^l(\sigma) \quad \text{if} \quad \mathbb{K} = \mathbb{C}$$
$$f(\rho\sigma) = \sum_{k=0}^{\infty} \sum_{m=1}^{[k/2]} \sum_{i=1}^{b(n,k,m)} \sum_{\lambda \in M(r)} d_{\lambda,k,m,i} P_\lambda^{k,m}(\rho) S_i^{k,m}(\sigma) \quad \text{if} \quad \mathbb{K} = \mathbb{Q},$$
where $|c_{\lambda,p,q,l}| + |d_{\lambda,k,m,i}| = O(\lambda^{-c})$ as $\lambda \to +\infty$ for any fixed $c > 0$.

For $r_1, r_2 \in (0, \pi/2)$, let $M(r_1, r_2) = M(r_1) \cap M(r_2)$ and let Π be the set of pairs (r_1, r_2) with the property that for any $c > 0$ there exists $\lambda \in M(r_1)$ such that $|P_\lambda(\tan r_2)| < (1 + \lambda)^{-c}$. For $R \in (\max(r_1, r_2), \pi/2]$, $s \in \mathbb{Z}_+ \cup \{\infty\}$, we set $V^s_{r_1, r_2}(B_R) = V_{r_1, r_2}(B_R) \cap C^s(B_R)$, where $V_{r_1, r_2}(B_R) = V_{r_1}(B_R) \cap V_{r_2}(B_R)$.

The following result is a local version of the two-radii theorem on $\mathbb{P}^n(\mathbb{K})$ (see [11]).

THEOREM 5.6. *Let $0 < r_1 < r_2 < R \le \pi/2$. Then*
1) *if $r_1 + r_2 < R$, $M(r_1, r_2) = \emptyset$ and $f \in V_{r_1, r_2}(B_R)$, then $f = 0$;*
2) *if $r_1 + r_2 = R$, $M(r_1, r_2) = \emptyset$ and $f \in V^\infty_{r_1, r_2}(B_R)$, then $f = 0$;*
3) *if $r_1 + r_2 = R$, $M(r_1, r_2) = \emptyset$, $(r_1, r_2) \in \Pi$ and $f \in V_{r_1, r_2}(B_R)$, then $f = 0$;*
4) *if $r_1 + r_2 = R$ and $(r_1, r_2) \notin \Pi$, then for each $s \in \mathbb{Z}_+$ there exists a non-trivial function $f \in V^s_{r_1, r_2}(B_R)$;*
5) *if $r_1 + r_2 > R$, then there exists a non-trivial function $f \in V^\infty_{r_1, r_2}(B_R)$;*
6) *if $M(r_1, r_2) \ne \emptyset$, then there exists a non-trivial real-analytic function $f \in V_{r_1, r_2}(B_{\pi/2})$.*

To conclude this section, we point out that the method of proof of Theorems 5.4–5.6 allows us to obtain similar results for convolution equations on $\mathcal{U}/K$ of the form $f \times T = 0$, where T is K-invariant distribution in $\mathcal{E}'(B_R)$ satisfying certain additional requirements (see Sections 5.1, 5.2). Moreover, analogous results hold for K-invariant solutions f of the equation indicated above on the Cayley projective plane $\mathbb{P}^{16}(\mathbb{C}ay)$. For other aspects of the Pompeiu problem on symmetric spaces, see [26], [10], [11], [24], [23], [25]–[30], [9], [4], and the references there.

6. Acknowledgement

The first author is deeply grateful to L. Zalcman, who invited him to come and work at his Seminar in 1993, 1996, and 2001 and to the Department of Mathematics and Computer Science of Bar-Ilan University (Israel) for its hospitality and library facilities during the stay. Thanks are also due to M. Agranovsky for the literature placed at our disposal.

References

1. M.L. Agranovsky, C. Berenstein, and D.C. Chang, *Morera theorem for holomorphic H^p spaces in the Heisenberg group*, J. Reine Angew. Math. **443** (1993), 49–89.
2. M.L. Agranovsky, C. Berenstein, and P. Kuchment, *Approximation by spherical waves in L^p spaces*, J. Geom. Anal. **6** (1996), 365–383.
3. M.L. Agranovsky and E.T. Quinto, *Injectivity sets for the Radon transform over circles and complete systems of radial functions*, J. Funct. Anal. **139** (1996), 383–414.
4. M.L. Agranovsky and A.M. Semenov, *Deformations of balls in Schiffer's conjecture for Riemannian symmetric spaces*, Israel. J. Math. **95** (1996), 43–59.
5. M.L. Agranovsky, V.V. Volchkov and L.A. Zalcman, *Conical injectivity sets for the spherical Radon transform*, Bull. London Math. Soc. **31** (1999), 231–236.
6. D.H. Armitage, *The Pompeiu problem for spherical polygons*, Proc. Royal Irish Acad. **96A** (1996), 25–32.
7. C.A. Berenstein, D.C. Chang, D. Pascuas, and L. Zalcman, *Variations on the theorem of Morera*, The Madison Symposium on Complex Analysis, Contemp. Math. **137** (1992), 63–78.
8. C.A. Berenstein and R. Gay, *A local version of the two-circles theorem*, Israel J. Math. **55** (1986), 267–288.
9. C.A. Berenstein and M. Shahshahani, *Harmonic analysis and the Pompeiu problem*, Amer. J. Math. **105** (1983), 1217–1229.

10. C.A. Berenstein and D.C. Struppa, *Complex analysis and convolution equations*, Several Complex Variables, V (G.M. Henkin, ed.), Springer-Verlag, 1993, pp. 1-108.
11. C.A. Berenstein and L. Zalcman, *Pompeiu's problem on symmetric spaces*, Comment Math. Helv. **55** (1980), 593–621.
12. R. Dalmasso, *A new result on the Pompeiu problem*, Trans. Amer. Math. Soc. **352** (2000), 2723–2736.
13. P. Ebenfelt, *Propagation of singularities from singular and infinite points in certain complex-analytic Cauchy problems and an application to the Pompeiu problem*, Duke Math. J. **73** (1994), 561–582.
14. N. Garofalo and F. Segala, *Univalent functions and the Pompeiu problem*, Trans. Amer. Math. Soc. **346** (1994), 137–146.
15. M. El Harchaoui, *Inversion de la transformation de Pompéiu locale dans les espaces hyperboliques réel et complexe (Cas de deux boules)*, J. Analyse Math. **67** (1995), 1–37.
16. S. Helgason, *Geometric Analysis on Symmetric Spaces*, Amer. Math. Soc., Providence, RI, 1994.
17. F. John, *Plane Waves and Spherical Means Applied to Partial Differential Equations*, Interscience, New York-London, 1955.
18. S. Kobayashi and K. Nomizu, *Foundations of Differential Geometry*, Vol. II, Interscience, 1969.
19. H.T. Laquer, *The Pompeiu problem*, Amer. Math. Monthly **100** (1993), 461–467.
20. S. Thangavelu, *Spherical means and CR functions on the Heisenberg group*, J. Analyse Math. **63** (1994), 255–286.
21. V.V. Volchkov, *Problems of Pompeiu type on manifolds*, Dokl. Akad. Nauk Ukraïni **1993**, no. 11, 9–13.
22. V.V. Volchkov, *Two-radii theorems on spaces of constant curvature*, Dokl. Akad. Nauk. **347** (1996), 300–302. English transl.: Russian Acad. Sci. Dokl. Math. **53** (1996), 199–201.
23. V.V. Volchkov, *Theorems on ball means values in symmetric spaces*, Mat. Sb. **192** (2001), 17–38. English transl.: Sb. Math. **192** (2001), 1275–1296.
24. V.V. Volchkov, *Convolution equation on symmetric spaces*, Vestnik Dneprop. Univ. Ser. Mat. **6** (2001), 50–57.
25. V.V. Volchkov, *A local two-radii theorem on symmetric spaces*, Dokl. Akad. Nauk **381** (2001), 727–731.
26. V.V. Volchkov, *Ball means values on symmetric spaces*, Dopov. Nats. Akad. Nauk Ukr. Mat. Prirodozn. Tekh. Nauki **2002**, no. 3, 15–19.
27. V.V. Volchkov, *Integral Geometry and Convolution Equations*, Kluwer Academic Publishers, Dordrecht, 2003.
28. Vit.V. Volchkov, *Theorems on spherical means in complex hyperbolic spaces*, Dopov. Nats. Akad. Nauk Ukr. Mat. Prirodozn. Tekh. Nauki **2000**, no. 4, 7–10.
29. Vit.V. Volchkov, *Convolution equations on complex hyperbolic spaces*, Dopov. Nats. Akad. Nauk Ukr. Mat. Prirodozn. Tekh. Nauki **2001**, no. 2, 11–14.
30. Vit.V. Volchkov, *Convolution equation on quaternionic hyperbolic space*, Dopov. Nats. Akad. Nauk Ukr. Mat. Prirodozn. Tekh. Nauki **2002**, no. 12, 12–14.
31. Vit.V. Volchkov, *Functions with zero ball means on quaternionic hyperbolic space*, Izv. Ross. Akad. Nauk Ser. Mat. **66** (2002), no. 5, 3–32.
32. Vit.V. Volchkov, *Uniqueness theorems for mean periodic functions on complex hyperbolic spaces*, Anal. Math. **28** (2002), 61–76.
33. Vit.V. Volchkov, *A definitive version of the local two-radii theorem on quaternionic hyperbolic space*, Dokl. Akad. Nauk **384** (2002), 449–451.
34. Vit.V. Volchkov. *Uniqueness theorems for periodic (in mean) functions on quaternionic hyperbolic space*, Mat. Zametki **74** (2003), no. 1, 32–40; English transl. in Math. Notes **74** (2003), 30–37.
35. S.A. Williams, *A partial solution of the Pompeiu problem*, Math. Ann. **223** (1976), 183–190.
36. S.A. Williams, *Analyticity of the boundary for Lipschitz domains without the Pompeiu property*, Indiana Univ. Math. J. **30** (1981), 357–369.
37. L. Zalcman, *A bibliographic survey of the Pompeiu problem*, Approximation by Solutions of Partial Differential Equations, (B. Fuglede et al., eds.), Kluwer Academic Publishers, Dordrecht, 1992, pp. 185–194.

38. L. Zalcman, *Supplementary bibliography to 'A bibliographic survey of the Pompeiu problem'*, Radon Transforms and Tomography, Contemp. Math. **278** (2001), 69–74.

DEPARTMENT OF MATHEMATICS, DONETSK NATIONAL UNIVERSITY, UNIVERSITETSKAYA, 24, DONETSK 83055, UKRAINE
E-mail address: volchkov@univ.donetsk.ua

DEPARTMENT OF MATHEMATICS, DONETSK NATIONAL UNIVERSITY, UNIVERSITETSKAYA, 24, DONETSK 83055, UKRAINE
E-mail address: volchkov@univ.donetsk.ua

Titles in This Series

382 Mark Agranovsky, Lavi Karp, and David Shoikhet, Editors, Complex analysis and dynamical systems II, 2005

381 David Evans, Jeffrey J. Holt, Chris Jones, Karen Klintworth, Brian Parshall, Olivier Pfister, and Harold N. Ward, Editors, Coding theory and quantum computing, 2005

380 Andreas Blass and Yi Zhang, Editors, Logic and its applications, 2005

379 Dominic P. Clemence and Guoqing Tang, Editors, Mathematical studies in nonlinear wave propagation, 2005

378 Alexandre V. Borovik, Editor, Groups, languages, algorithms, 2005

377 G. L. Litvinov and V. P. Maslov, Editors, Idempotent mathematics and mathematical physics, 2005

376 José A. de la Peña, Ernesto Vallejo, and Natig Atakishiyev, Editors, Algebraic structures and their representations, 2005

375 Joseph Lipman, Suresh Nayak, and Pramathanath Sastry, Variance and duality for cousin complexes on formal schemes, 2005

374 Alexander Barvinok, Matthias Beck, Christian Haase, Bruce Reznick, and Volkmar Welker, Editors, Integer points in polyhedra—geometry, number theory, algebra, optimization, 2005

373 O. Costin, M. D. Kruskal, and A. Macintyre, Editors, Analyzable functions and applications, 2005

372 José Burillo, Sean Cleary, Murray Elder, Jennifer Taback, and Enric Ventura, Editors, Geometric methods in group theory, 2005

371 Gui-Qiang Chen, George Gasper, and Joseph Jerome, Editors, Nonlinear partial differential equations and related analysis, 2005

370 Pietro Poggi-Corradini, Editor, The p-harmonic equation and recent advances in analysis, 2005

369 Jaime Gutierrez, Vladimir Shpilrain, and Jie-Tai Yu, Editors, Affine algebraic geometry, 2005

368 Sagun Chanillo, Paulo D. Cordaro, Nicholas Hanges, Jorge Hounie, and Abdelhamid Meziani, Editors, Geometric analysis of PDE and several complex variables, 2005

367 Shu-Cheng Chang, Bennett Chow, Sun-Chin Chu, and Chang-Shou Lin, Editors, Geometric evolution equations, 2005

366 Bernhelm Booß-Bavnbek, Gerd Grubb, and Krzysztof P. Wojciechowski, Editors, Spectral geometry of manifolds with boundary and decompositon of manifolds, 2005

365 Robert S. Doran and Richard V. Kadison, Editors, Operator algebras, quantization, and non-commutative geometry, 2004

364 Mark Agranovsky, Lavi Karp, David Shoikhet, and Lawrence Zalcman, Editors, Complex analysis and dynamical systems, 2004

363 Anthony To-Ming Lau and Volker Runde, Editors, Banach algebras and their applications, 2004

362 Carlos Concha, Raul Manasevich, Gunther Uhlmann, and Michael S. Vogelius, Editors, Partial differential equations and inverse problems, 2004

361 Ali Enayat and Roman Kossak, Editors, Nonstandard models of arithmetic and set theory, 2004

360 Alexei G. Myasnikov and Vladimir Shpilrain, Editors, Group theory, statistics, and cryptography, 2004

359 S. Dostoglou and P. Ehrlich, Editors, Advances in differential geometry and general relativity, 2004

TITLES IN THIS SERIES

358 **David Burns, Christian Popescu, Jonathan Sands, and David Solomon, Editors,** Stark's Conjectures: Recent work and new directions, 2004
357 **John Neuberger, Editor,** Variational methods: open problems, recent progress, and numerical algorithms, 2004
356 **Idris Assani, Editor,** Chapel Hill ergodic theory workshops, 2004
355 **William Abikoff and Andrew Haas, Editors,** In the tradition of Ahlfors and Bers, III, 2004
354 **Terence Gaffney and Maria Aparecida Soares Ruas, Editors,** Real and complex singularities, 2004
353 **M. C. Carvalho and J. F. Rodrigues, Editors,** Recent advances in the theory and applications of mass transport, 2004
352 **Marek Kubale, Editor,** Graph colorings, 2004
351 **George Yin and Qing Zhang, Editors,** Mathematics of finance, 2004
350 **Abbas Bahri, Sergiu Klainerman, and Michael Vogelius, Editors,** Noncompact problems at the intersection of geometry, analysis, and topology, 2004
349 **Alexandre V. Borovik and Alexei G. Myasnikov, Editors,** Computational and experimental group theory, 2004
348 **Hiroshi Isozaki, Editor,** Inverse problems and spectral theory, 2004
347 **Motoko Kotani, Tomoyuki Shirai, and Toshikazu Sunada, Editors,** Discrete geometric analysis, 2004
346 **Paul Goerss and Stewart Priddy, Editors,** Homotopy theory: Relations with algebraic geometry, group cohomology, and algebraic K-theory, 2004
345 **Christopher Heil, Palle E. T. Jorgensen, and David R. Larson, Editors,** Wavelets, frames and operator theory, 2004
344 **Ricardo Baeza, John S. Hsia, Bill Jacob, and Alexander Prestel, Editors,** Algebraic and arithmetic theory of quadratic forms, 2004
343 **N. Sthanumoorthy and Kailash C. Misra, Editors,** Kac-Moody Lie algebras and related topics, 2004
342 **János Pach, Editor,** Towards a theory of geometric graphs, 2004
341 **Hugo Arizmendi, Carlos Bosch, and Lourdes Palacios, Editors,** Topological algebras and their applications, 2004
340 **Rafael del Río and Carlos Villegas-Blas, Editors,** Spectral theory of Schrödinger operators, 2004
339 **Peter Kuchment, Editor,** Waves in periodic and random media, 2003
338 **Pascal Auscher, Thierry Coulhon, and Alexander Grigor'yan, Editors,** Heat kernels and analysis on manifolds, graphs, and metric spaces, 2003
337 **Krishan L. Duggal and Ramesh Sharma, Editors,** Recent advances in Riemannian and Lorentzian geometries, 2003
336 **José González-Barrios, Jorge A. León, and Ana Meda, Editors,** Stochastic models, 2003
335 **Geoffrey L. Price, B. Mitchell Baker, Palle E.T. Jorgensen, and Paul S. Muhly, Editors,** Advances in quantum dynamics, 2003
334 **Ron Goldman and Rimvydas Krasauskas, Editors,** Topics in algebraic geometry and geometric modeling, 2003

For a complete list of titles in this series, visit the
AMS Bookstore at **www.ams.org/bookstore/**.